LOW-DIMENSIONAL HALIDE PEROVSKITES

LOW-DIMENSIONAL HALIDE PEROVSKITES

Structure, Synthesis, and Applications

Edited by

YIQIANG ZHAN
MOHAMMAD KHALID
PAOLA VIVO
NUMAN ARSHID

ELSEVIER

Elsevier
Radarweg 29, PO Box 211, 1000 AE Amsterdam, Netherlands
The Boulevard, Langford Lane, Kidlington, Oxford OX5 1GB, United Kingdom
50 Hampshire Street, 5th Floor, Cambridge, MA 02139, United States

ISBN: 978-0-323-88522-5

For information on all Elsevier publications
visit our website at https://www.elsevier.com/books-and-journals

Publisher: Matthew Deans
Acquisitions Editor: Stephen Jones
Editorial Project Manager: Tom Means
Production Project Manager: Anitha Sivaraj
Cover Designer: Vicky Esser Pearson

Typeset by STRAIVE, India

Contents

Contributors

Yaser Abdi Department of Physics, Nanophysics Research Laboratory, University of Tehran, Tehran, Iran

Irfan Ahmed State Key Laboratory of ASIC and System, Centre of Micro-Nano System, School of Information Science and Technology, Fudan University, Shanghai, China; Department of Physics, Government Postgraduate College, Higher Education Department-HED, Mansehra, Khyber Pakhtunkhwa, Pakistan

Jamilur R. Ansari University School of Basic and Applied Sciences, Guru Gobind Singh Indraprastha University, Dwarka, Delhi; Department of Applied Science & Humanities, Dronacharya College of Engineering, Gurugram, Haryana, India

A. Arulraj Instituto de Investigaciones Científicas y Tecnológicas (IDICTEC), Universidad de Atacama, Copiapó; Faculty of Engineering & Sciences, Universidad Adolfo Ibáñez, Peñalolén, Santiago, Chile

Fatemeh Behrouznejad State Key Laboratory of ASIC and System, Centre of Micro-Nano System, School of Information Science and Technology, Fudan University, Shanghai, China; Nanoparticles and Coatings Lab, Department of Physics, Sharif University of Technology, Tehran, Iran

Cristina Besleaga National Institute of Materials Physics, Magurele, Ilfov, Romania

Rachel Elizabeth Brophy Reykjavik University, Icelandic School of Energy, Reykjavik, Iceland

Paulina Carmona-Monroy Instituto de Investigaciones en Materiales, Universidad Nacional Autónoma de México, Mexico

Amreesh Chandra Department of Physics, Indian Institute of Technology Kharagpur, Kharagpur, 721302, India

Yonghua Chen Key Laboratory of Flexible Electronics (KLoFE) & Institute of Advanced Materials (IAM), Nanjing Tech University (Nanjing Tech), Nanjing, Jiangsu, China

Faisal Islam Chowdhury Department of Chemistry, University of Chittagong, Chittagong, Bangladesh

Towhid H. Chowdhury Organic Printed Electronics Laboratory (OPEL), Department of Chemical Engineering, Pohang University of Science & Technology, Pohang, Republic of Korea

Alejandro Cortés-Villena Institute of Molecular Science, University of Valencia, Valencia, Spain

Marina Cuzminschi Horia Hulubei National Institute for Physics and Nuclear Engineering; Faculty of Physics, MDEO Research Center, Research Institute of the University of Bucharest (ICUB), University of Bucharest, Magurele, Ilfov, Romania

Sarah Derbali National Institute of Materials Physics, Magurele, Ilfov, Romania

Nicolae Filipoiu Horia Hulubei National Institute for Physics and Nuclear Engineering; Faculty of Physics, MDEO Research Center, Research Institute of the University of Bucharest (ICUB), University of Bucharest, Magurele, Ilfov, Romania

Mihaela Florea National Institute of Materials Physics, Magurele, Ilfov, Romania

Raquel E. Galian Institute of Molecular Science, University of Valencia, Valencia, Spain

Somayeh Gholipour METU GUNAM Center, Middle East Technical University, Ankara, Turkey

Subrata Ghosh Functional Materials and Device Laboratory, School of Energy Science and Engineering, Indian Institute of Technology Kharagpur, Kharagpur, 721302, India

G. Krishnamurthy Grandhi Hybrid Solar Cells, Faculty of Engineering and Natural Sciences, Tampere University, Tampere, Finland

Meenakshi Gussain State Key Laboratory of ASIC and System, Centre of Micro-Nano System, School of Information Science and Technology, Fudan University, Shanghai, China

Mahboubeh Hadadian Department of Mechanical and Materials Engineering, Faculty of Technology, University of Turku, Turku, Finland

Waseem Haider School of Engineering and Technology, Central Michigan University, Mt. Pleasant, MI, United States

Mohammad Hatamvand Key Laboratory of Flexible Electronics (KLoFE) & Institute of Advanced Materials (IAM), Nanjing Tech University (Nanjing Tech), Nanjing, Jiangsu, China

Wei Huang Key Laboratory of Flexible Electronics (KLoFE) & Institute of Advanced Materials (IAM), Nanjing Tech University (Nanjing Tech), Nanjing, Jiangsu, China

Neha Jain Department of Physics, Dr. H. S. Gour Central University, Sagar, Madhya Pradesh, India

Sakshi Kansal Functional Materials and Device Laboratory, School of Energy Science and Engineering, Indian Institute of Technology Kharagpur, Kharagpur, 721302, India

Movaffaq Kateb Reykjavik University, Department of Engineering, Reykjavik, Iceland

Mohammad Khalid Graphene & Advanced 2D Materials Research Group (GAMRG), School of Engineering and Technology, Sunway University, No. 5, Jalan Universiti, Petaling Jaya, Selangor, Malaysia

Dinesh Kumar Functional Materials and Device Laboratory, School of Energy Science and Engineering, Indian Institute of Technology Kharagpur, Kharagpur, 721302, India

Maning Liu Hybrid Solar Cells, Faculty of Engineering and Natural Sciences, Tampere University, Tampere, Finland

Chung-Hsin Lu Department of Chemical Engineering, National Taiwan University (NTU); Department of Chemical Engineering, National Taiwan University Science and Technology (NTUST); Advanced Research Centre of Green Materials Science & Technology, Taipei, Taiwan

R.V. Mangalaraja Faculty of Engineering & Sciences, Universidad Adolfo Ibáñez, Peñalolén, Santiago, Chile

Andrei Manolescu Reykjavik University, Department of Engineering, Reykjavik, Iceland

Pandiyarajan Mariyappan Department of Electrical and Electronics Engineering, Sri Venkateswara College of Engineering (SVCE), Chennai, Tamil Nadu, India

Mohan Lal Meena Department of Chemical Engineering, National Taiwan University Science and Technology (NTUST), Taipei, Taiwan

Tudor Luca Mitran Horia Hulubei National Institute for Physics and Nuclear Engineering, Magurele, Ilfov, Romania

Md. Farhan Naseh University School of Basic and Applied Sciences, Guru Gobind Singh Indraprastha University, Dwarka, Delhi, India

George Alexandru Nemnes Horia Hulubei National Institute for Physics and Nuclear Engineering; Faculty of Physics, MDEO Research Center, Research Institute of the University of Bucharest (ICUB), University of Bucharest, Magurele, Ilfov, Romania

Julia Pérez-Prieto Institute of Molecular Science, University of Valencia, Valencia, Spain

Ioana Pintilie National Institute of Materials Physics, Magurele, Ilfov, Romania

Shivam Porwal Functional Materials and Device Laboratory, School of Energy Science and Engineering, Indian Institute of Technology Kharagpur, Kharagpur, 721302, India

Surbhi Priya Functional Materials and Device Laboratory, School of Energy Science and Engineering, Indian Institute of Technology Kharagpur, Kharagpur, 721302, India

M. Atikur Rahman Department of Electrical & Electronic Engineering, University of Chittagong, Chittagong, Bangladesh

C. Rajkumar Center for High Energy Systems and Sciences (CHESS), DRDO, Hyderabad, India

Bart Roose Department of Physics, Cavendish Laboratory, University of Cambridge, Cambridge, United Kingdom

Ignacio Rosa-Pardo Institute of Molecular Science, University of Valencia, Valencia, Spain

Mohammad Sajedi Alvar Max Planck Institute for Polymer Research, Mainz, Germany

T.S. Shyju Centre for Nanoscience and Nanotechnology, Sathyabama Institute of Science and Technology (Deemed to be University), Chennai, India

Anupriya Singh Research Center for Applied Science, Academia Sinica, Taipei, Taiwan

Rajan Kumar Singh Department of Chemical Engineering, National Taiwan University (NTU), Taipei, Taiwan

Trilok Singh Functional Materials and Device Laboratory, School of Energy Science and Engineering, Indian Institute of Technology Kharagpur, Kharagpur, 721302, India

Diego Solis-Ibarra Instituto de Investigaciones en Materiales, Universidad Nacional Autónoma de México, Mexico

Sudipta Som Department of Chemical Engineering, National Taiwan University (NTU), Taipei, Taiwan

Viorica Stancu National Institute of Materials Physics, Magurele, Ilfov, Romania

Choudhary Arjun Sunilbhai Faculty of Physical Sciences, PDM University, Bahadurgarh, Haryana, India

Radha Tamrakar Department of Physics, Govt. K. N. M. M., Damoh, Madhya Pradesh, India

Andrei G. Tomulescu National Institute of Materials Physics, Magurele, Ilfov, Romania

Brenda Vargas Instituto de Investigaciones en Materiales, Universidad Nacional Autónoma de México, Mexico

P. Vengatesh Centre for Nanoscience and Nanotechnology, Sathyabama Institute of Science and Technology (Deemed to be University), Chennai, India

Paola Vivo Hybrid Solar Cells, Faculty of Engineering and Natural Sciences, Tampere University, Tampere, Finland

Mozhgan Yavari Department of Electrical and Electronic Engineering, Advanced Technology Institute, University of Surrey, Guildford, Surrey, United Kingdom

Yiqiang Zhan State Key Laboratory of ASIC and System, Centre of Micro-Nano System, School of Information Science and Technology, Fudan University, Shanghai, China

CHAPTER

1

Introduction

Maning Liu, G. Krishnamurthy Grandhi, and Paola Vivo

Hybrid Solar Cells, Faculty of Engineering and Natural Sciences, Tampere University, Tampere, Finland

1. Introduction

Metal halide perovskites (MHPs) have been the focus of intense research worldwide over the last decade, due to their enormous potential as semiconductors for highly performing optoelectronics, and in particular photovoltaics. Perovskite solar cells (PSCs) have already marked a certified power conversion efficiency (PCE) record above 25% [1]. Such an unprecedented pace of development of PSCs promises to revolutionize the prospects of third-generation photovoltaics. The outstanding optoelectronic properties of MHPs include high absorption coefficient, wide tunable bandgap (400–800nm), long charge carrier diffusion length ($\sim 1\,\mu m$), and high charge carrier conductivity ($1–10\,cm^2\,V^{-1}\,s^{-1}$) [2–5]. The exceptional interest of the community in MHPs is also motivated by their processability in solutions starting from low-cost and easily available precursors. The rich chemistry of MHPs enables a plethora of structural, optical, and electronic properties as a result of the complex physicochemical processes underlying the apparently simple solution-based syntheses of these materials [6–8]. Finally, the tunability of the optoelectronic properties of MHPs has opened the possibility for other applications beyond photovoltaics, such as light-emitting diodes (LEDs), photodetectors, and lasers, just to mention a few of them [9–11].

In this chapter, we introduce the reader to the topic of MHPs, starting from some highlights on the historical development of these materials and their intense recent investigation. We also clarify which materials can be appropriately classified as halide perovskites, given the frequent misuse of this term in the literature. We then focus on an overview of low-dimensional MHPs and their key features and optoelectronic properties.

https://doi.org/10.1016/B978-0-323-88522-5.00010-7

Finally, we provide our view on the key open challenges for the future advancement of this class of materials and their development of yet unexplored optoelectronics.

2. History of MHPs

The term "perovskite" was coined when the calcium titanate ($CaTiO_3$) mineral was discovered by Gustav Rose in 1839 and later named in honor of the Russian mineralogist L.A. Perovski. Since then, the concept of perovskite has been expanded to all the compounds that possess an identical or similar crystal structure to that of $CaTiO_3$. MHPs have a typical chemical formula of ABX_3 derived from the original $CaTiO_3$ structure (Fig. 1), where A represents either organic or inorganic cations with different sizes such as $CH_3NH_3^+$ (methylammonium, MA), $CH(NH_2)_2^+$ (FA), or Cs^+, B stands for metal cations (e.g., Pb^{2+} and Sn^{2+}), and X represents heavy halide anions (Cl^-, Br^-, I^-). The first MHPs were presented by Moller in 1958 and had a fully inorganic structure, i.e., $CsPbX_3$ [12]. Later in 1978, organometal hybrid halide perovskites were discovered by Weber and coworkers with MA as A-site cation [13]. Despite some early applications for field-effect transistors (FETs) and LEDs in the 1990s, MHPs did not initially catch much attention of the researchers owing to their unsatisfied device performance compared to that of other organic or inorganic semiconductors. In 2009, Miyasaka and coworkers first employed $MAPbBr_3$ or $MAPbI_3$ as the light absorber in dye-sensitized solar cells, achieving an encouraging PCE of 3.8%. This groundbreaking work quickly promoted an enormous research interest in MHPs, both from academia and industry [14]. The skyrocketing increase in the PCE over the years has led to values now highly competitive with those of conventional silicon-based photovoltaics. Since 2014, a few years after the pioneering solar cells work by Miyasaka et al., MHPs have been also employed for the development of

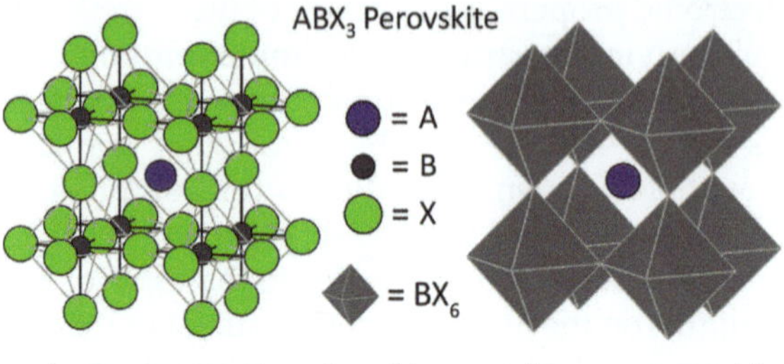

FIG. 1 The standard representation of a cubic perovskite structure with an ABX_3 crystal structure. *Reprinted with permission from Q.A. Akkerman, L. Manna, What Defines a Halide Perovskite? ACS Energy Lett. (2020) 604–610. https://doi.org/10.1021/acsenergylett.0c00039. Copyright (2020) American Chemical Society.*

electroluminescence devices at room temperature, stemming from the promising study by Tan and coworkers [15] Currently, MHP-based LEDs display external quantum efficiency values of >20% [16–18].

3. What are MHPs?

MHPs differ from their conventional chalcogenide counterparts as the latter contain large bivalent cations (e.g., Ca^{2+}, Sr^{2+}) at the A-site, small tetravalent cations (e.g., Ti^{4+}, Zr^{4+}) at the B-site, and oxygen anions at the X-site. In contrast, MHP structures include monovalent cations at the A-site, divalent metal cations at the B-site, and halogen anions at the X-site. A so-called three-dimensional (3D) network forms into the ABX_3 perovskite structure, in which the cavity between four adjacent corner-sharing BX_6 octahedra is occupied by A-site cations. By calculating the Goldschmidt tolerance factor (t) and the octahedral factor (μ), the probability of forming a perovskite structure can be evaluated [19] Typically, t is determined by the ionic radii (r) of the A, B, and X and is expressed as Eq. (1):

$$t = (r_A + r_X)/\sqrt{2} \times (r_B + r_X) \quad (1)$$

The octahedral factor is defined as $\mu = r_B/r_X$, where r_B and r_X are the radii of B-site and X-site ions, respectively. Generally, 3D networks of stable perovskite structures are formed when $0.76 < t < 1.13$. If t is outside this range, other perovskite-related structures (not 3D) can be also stably formed [20]. Thus, only the incorporation of small cations (e.g., Cs^+, MA^+, FA^+) at the A-site can result in the formation of stable 3D halide perovskite structures. Other potential alternatives at the A-site are either too small (e.g., K^+, Rb^+) or too large (e.g., ethylammonium, guanidinium). Some MHPs with a boundary on the tolerance factor range, i.e., $CsPbI_3$ ($t = 0.8$) and $FAPbI_3$ ($t = 1$), readily induce a phase transition even at room temperature (RT) toward more stable orthorhombic and hexagonal phases (Fig. 2A–D), respectively [21]. Accordingly, the transited phases are often defined as "yellow phases" [22]. The stability of the BX_6 octahedra is typically assessed by the value of μ, which is related to the radius comparison of the B-site and X-site ions. It is empirically found that, when μ falls in the range of 0.442–0.895, stable cubic perovskite structures can be formed [23]. Thus, both t and μ factors are commonly used to predict the stability of newly designed halide perovskite compositions (Fig. 2E). Besides the above-discussed ABX_3 stoichiometry defining a 3D network of corner-sharing BX_6 octahedra of the typical MHP structure, recently, more and more deviations from the ABX_3 stoichiometry have been found, which still belong to the family of MHPs. However, some general misunderstanding and misuse of the term "halide perovskites" are acknowledged in the

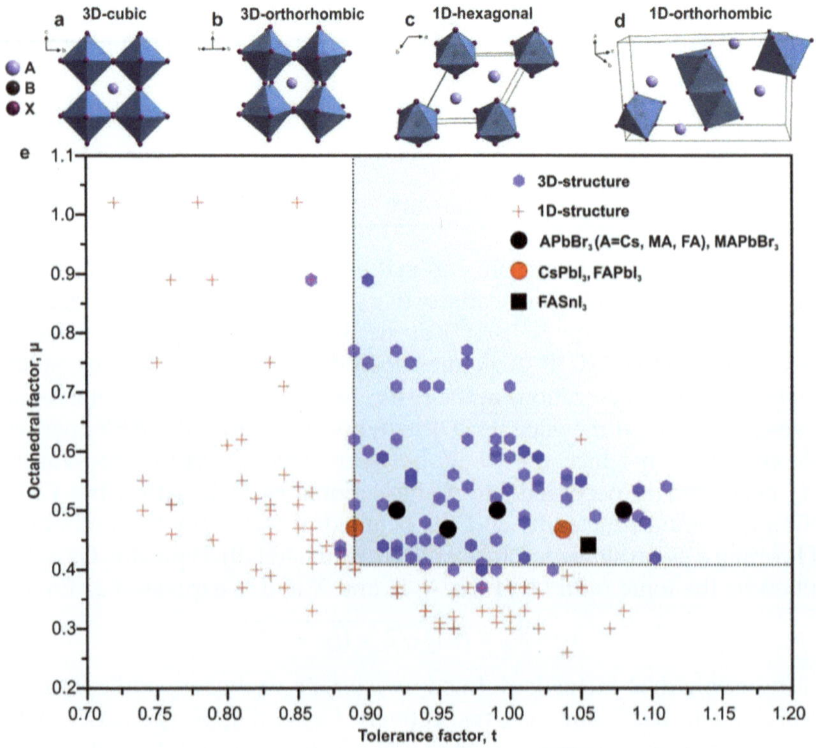

FIG. 2 Schematic representatives of (A) 3D cubic structure, observed in α-$FAPbI_3$; (B) orthorhombically distorted 3D structure, reported for $CsPbBr_3$; (C) one-dimensional (1D) hexagonal lattice, observed in the yellow phase of $FAPbI_3$; (D) 1D orthorhombic structure, found in the yellow phase of $CsPbI_3$; (E) 3D and 1D structures reported for various fully inorganic and hybrid organic-inorganic ABX_3 MHPs. The light blue square area highlights the region where stable MHPs are located. *Reprinted with permission from J. Shamsi, A.S. Urban, M. Imran, L. De Trizio, L. Manna, Metal halide perovskite nanocrystals: synthesis, post-synthesis modifications, and their optical properties. Chem. Rev. 119 (5) (2019) 3296–3348. https://doi.org/10.1021/acs.chemrev.8b00644. Copyright (2019) American Chemical Society.*

field. Thus, it is necessary to clearly specify which class of materials can be termed metal halide perovskites according to current mainstream viewpoints. Fig 3 shows an overview of the different MHPs that have been reported in the literature [24].

Compared to the standard ABX_3 cubic structure (Fig. 3A), the so-called halide antiperovskites can form when the A- and B-sites are occupied with anions (halide and chalcogenide for A and B, respectively) and the X-site with a monovalent cation (Fig. 3B), e.g., Li_3OBr and Ag_3SI. Although these antiperovskites are still classified as halide perovskites, this book will not focus on them being the halide not effectively part of the BX network in

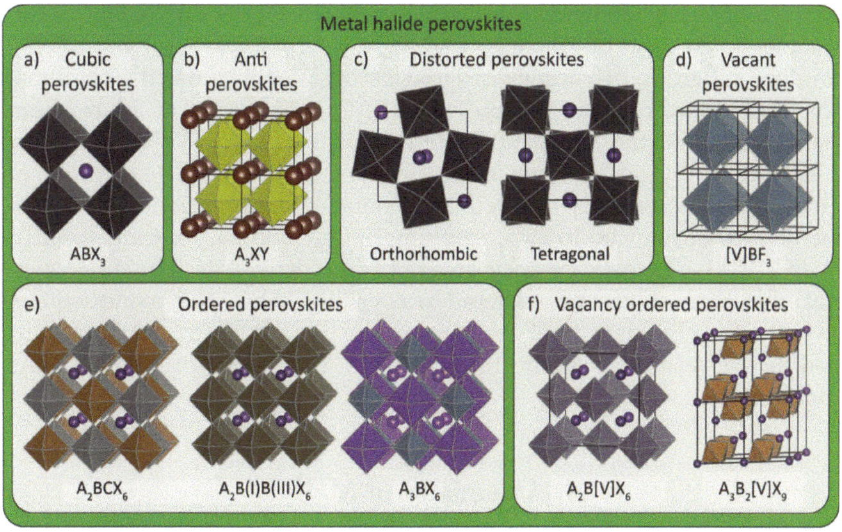

FIG. 3 Overview of various metal halide perovskites. (A) Standard ABX_3 cubic halide perovskites, (B) Antiperovskites, (C) Common orthorhombic and tetragonal disordered perovskites, (D) Vacant BX_3 perovskites, (E) Ordered perovskites, (F) Vacancy-ordered perovskites. *Reprinted with permission from Q.A. Akkerman, L. Manna, What Defines a Halide Perovskite? ACS Energy Lett. (2020) 604–610. https://doi.org/10.1021/acsenergylett.0c00039. Copyright (2020) American Chemical Society.*

that case. The size difference between the A-, B-, and X-site ions results in a distorted lattice. These distortions can induce the single octahedra to tilt, leading to either orthorhombic or tetragonal crystal structures (Fig. 3C). When ABX_3 stoichiometry is partially or fully replaced or even removed for A- and/or B-site cations, some more halide perovskite derivatives can be obtained. Fig. 3D shows a so-called vacant BX_3 crystal structure with a completely vacant A-site cation, which is only applicable in the case of fluorides (e.g., AlF_3 and MnF_3) [25]. In addition, another derivative of the ABX_3 perovskite, namely ordered perovskite, can be attained by replacing the B-site cation with a combination of two cations that are positioned at particular crystallographic sites (Fig. 3E) [26]. This structure is often termed "A_2BCX_6" or "$A_2BB'X_6$," and it is commonly called "double perovskite" due to the featured double occupancy at the B-site cation. Recently, this type of ordered perovskites has been intensively investigated as a lead-free candidate, e.g., $Cs_2AgBiBr_6$, $Cs_2AgInCl_6$, and MA_2AgSbI_6 [26]. Fig. 3F shows the last type of halide perovskites, namely vacancy-ordered perovskites, which are similar to ordered perovskites, but a vacancy is involved to partially replace the B-site cation. The most widely investigated subgroup of the vacancy-ordered perovskites has an A_2BX_6 stoichiometry (e.g., Cs_2BX_6). In this case, the B-site cations are half occupied by tetravalent metals (M(IV)) and another half occupied by vacancies (occupied ratio of 1:1), e.g., Cs_2SnI_6, Cs_2PdBr_6, and Cs_2TiBr_6 [27,28]. Another subgroup of vacancy-ordered perovskites includes the combination of trivalent metals (e.g., Sb^{3+} and Bi^{3+}) with a ratio of 2:1 between the vacancy and the occupied B-site cations, e.g., $Cs_3Sb_2[V]I_9$ [29]. More interestingly, by adding one more divalent metal cation (e.g., Cd^{2+} or Cu^{2+}) into these vacancy-ordered perovskites, the proportion of vacancies can be further reduced to 25%, resulting in $Cs_4CdSb_2Cl_{12}$ or $Cs_4CuSb_2Cl_{12}$ ($A_4BC_2[V]X_{12}$), namely vacancy-ordered triple perovskites [30,31]. In general, this type of vacancy-ordered perovskites show poor conductivity, owing to the high density of vacancies in the perovskite structure and thus leading to restricted applications, i.e., in the fields of photovoltaics and LEDs.

4. Chemistry of MHPs

Most MHP structures are held together by ionic bonding, which is one of the main causes for the easy formation of highly crystalline structures even at room temperature. These soft ionic crystals are prone to defects and susceptible to structural disorder at the surface. However, the high defect tolerance of MHP is the key responsible for their capability to behave as high-performing semiconductor materials. This can enable them to maintain the electronic structure in their pure states, despite the existence of a high defect density. The first-principles density-functional theory (DFT) calculations are a common and reliable tool to simulate the defect tolerance, in particular for the formation energy of point defects and their associated effect on the electronic structure [32]. The unique nature of the electronic structure of MHPs results in one of the most attractive properties, i.e., their optical tunability. As an example, the formation of energetic bands in a lead iodide-based perovskite is shown in Fig. 4. The hybrid antibonding orbitals of the Pb 6*p* orbitals and the outer *p* orbitals of the halide (3*p* for Cl, 4*p* for Br, and 5*p* for I) form the conduction band of the perovskite, exhibiting a more *p*-like structure owing to the major contribution from the lead side with a high density of states (DOS) [33]. Similarly, the hybrid antibonding states of the Pb 6*s* and the identical halide *p* orbitals together form the valence band. This featured electronic structure of MHPs is different from that of conventional semiconductor materials, i.e., gallium arsenide (GaAs), with their bandgaps forming in between bonding and antibonding orbitals [25]. Consequently, the valence band edge can shift at various energy levels as the halide component varies (Cl, Br, and I). However, their conduction band edge shifts restrictedly. In contrast to the roles of B-site and X-site in tuning electronic structure, A-site cation plays a minor role in both changing the

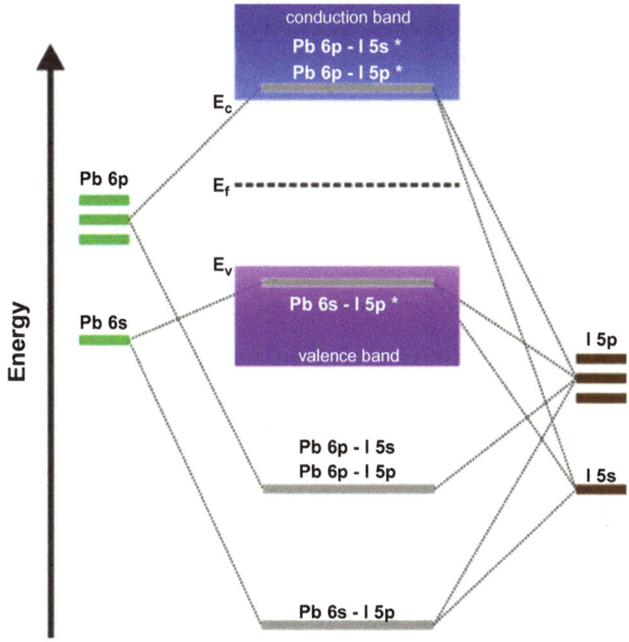

FIG. 4 Formation of energetic bands in a lead iodide perovskite through the hybridization of lead and iodide orbitals. *Reprinted with permission from J. Shamsi, A.S. Urban, M. Imran, L. De Trizio, L. Manna, Metal halide perovskite nanocrystals: synthesis, post-synthesis modifications, and their optical properties. Chem. Rev. 119 (5) (2019) 3296–3348. https://doi.org/10.1021/acs.chemrev.8b00644. Copyright (2019) American Chemical Society.*

conduction and valence bands (according to its corresponding DOS) and tailoring the bandgap of the perovskite [34].

5. Dimensionality of MHPs

MHPs can be classified into 3D, quasi-2D, 2D, 1D, and 0D based on the metal halide octahedra connectivity in their crystal structures, as shown in Fig. 5 [35]. Perovskites with dimensionality lower than 3D consist of corner-, edge-, or face-sharing BX_6 octahedra and are derived from the reference 3D ABX_3 structure. It is worth noting that the dimensionality here discussed has nothing to do with the nanostructured morphology with spatial confinement in different directions (e.g., quantum dots, nanoplatelets/nanorods, nanowires, and nanocubes).

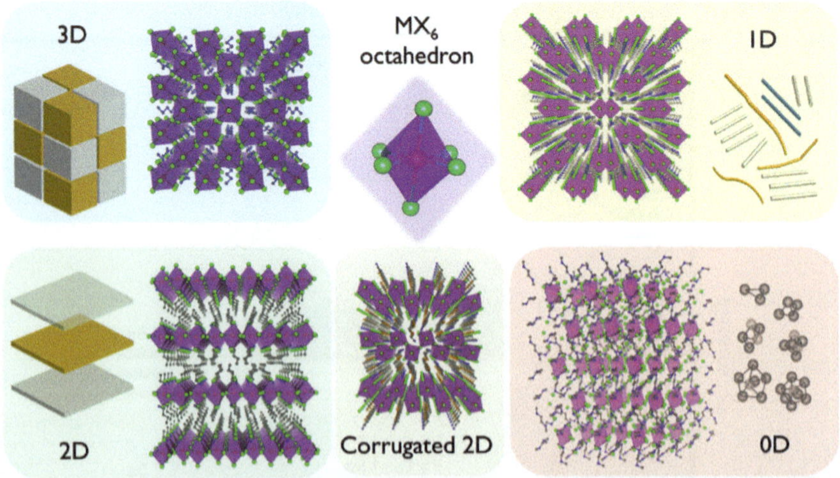

FIG. 5 Classification of metal halide perovskites based on the metal halide octahedra connectivity. The 2D, 1D, and 0D perovskites can be assumed as bulk assemblies of 2D quantum wells, 1D quantum wires, and 0D clusters, respectively. *Reproduced with permission from H. Lin, C. Zhou, Y. Tian, T. Siegrist, B. Ma, Low-dimensional organometal halide perovskites. ACS Energy Lett. 3 (1) (2018) 54–62. https://doi.org/10.1021/acsenergylett.7b00926*

The 2D layered perovskites can be assumed as one-layer thick BX_6 octahedral sheets cut from 3D ABX_3 perovskite along <100>, <110>, and <111> crystallographic directions, which are linked with each other via bulky organic cations. Their general formula is $(RNH_3)_2A_{n-1}B_nX_{3+1}$, where R is the organic linker and n represents the number of BX_6 layers. For instance, atomically thin 2D $(C_4H_9NH_3)_2PbX_4$ crystals with $n = 1$ and $R{=}C_4H_9$ were obtained by following a solution-based protocol [36]. The thickness of these monolayer perovskite sheets was found to be 1.6 ± 0.2 nm, where ~0.6 nm corresponds to the length of the single BX_6 layer and the organic linkers present on both sides of the inorganic layer contribute to the remaining thickness. The thickness of 2D layered perovskites increases by increasing the number of BX_6 layers in the inorganic framework. Also, the organic constituent comprises a monolayer or bilayer of organic cations. For instance, in the case of monoammonium ion (monolayer), NH_3^+ functional group binds to the halogens in one inorganic layer, and the organic (hydrophobic) group extends into the space between the inorganic layers. The van der Waals interaction between the two organic moieties keeps the BX_6 layers together. When n value approaches ∞, the formation of cubic 3D halide perovskites, such as $MAPbI_3$ or $CsPbI_3$, takes place. Intermediate n values ($n = 10, 20,\ldots$) produce quasi-2D perovskites, such as Ruddlesden-Popper (RP) structures. In this case, the inorganic framework consists of a few 3D ABX_3 layers, which are separated by the alkylammonium cations (van der Waals gap), which, in turn, form a 2D+3D (quasi-2D) layered perovskite structure (A_2BX_4-type). On the other hand, organic linkers with two alkylammonium groups at both ends form Dian-Jacobson (DJ) ABX_4 structure [37]. For example, while (PDA)$(MA)_3Pb_4I_{13}$ adopts the DJ structure, $(PA)_2(MA)_3Pb_4I_{13}$ crystallizes in RP structure [37]. Other well-known examples of RP and DJ phases are $(BA)_2(MA)_2Pb_3I_{10}$ and (3AMP)$(MA)_2Pb_3I_{10}$, respectively [38]. In RP structures, the van der Waals gaps with monoammonium cations promote weak interactions between the ABX_3 layers, destabilizing the layered perovskite structure when compared to the DJ structures. DJ perovskites not only display better stability but also lead to higher solar cell efficiencies than the corresponding RP structures [37].

The 2D perovskites can be further scissored in the direction perpendicular to BX_6 sheets to ensure that the metal halide octahedra remain in contact with each other only along one axis. This produces 1D analogs. Only a few 1D perovskites have been reported so far due to their limited optoelectronic efficiencies. The first known family of 1D halide perovskites, $[NH_2C(I){=}NH_2]_2(CH_3NH_3)_mSn_mI_{3m+2}$, was synthesized in the 1990s by Mitzi and coworkers [39]. These 2D structures break down into a 1D structure, with corner sharing of tin iodide octahedra when $m=1$ [40]. Pb-based 1D perovskites, such as $C_4N_2H_{14}PbBr_4$, are also known [41]. If the 1D perovskites are further sliced to obtain metal halide octahedra fully isolated from each other, the resulting structures are termed "0D metal halides." This class of structures is well known for their near-unity photoluminescence efficiencies and robustness, which led to the discovery of a plethora of organic-inorganic and all-inorganic 0D metal halides [42,43]. The type of cation also influences the geometry of the metal halide. For instance, Cu^+ ions form 0D halide with copper halide tetrahedra, while Pb^{2+}, Sn^{2+}, and Mn^{2+} mostly form structures with metal halide octahedra [44]. With an appropriate choice of the organic linker and metal halide, low-dimensional perovskite with the desired dimensionality can be designed. The existence of a large library of organic molecules that can fit into the van der Waals space of low-dimensional organic-inorganic perovskites may offer the opportunity to further engineer the structural and optoelectronic properties of these materials.

To synthesize high-quality low-dimensional halide perovskite films, the most common protocol includes a conventional one-step spin coating; metal and organic halides dissolved in a solvent are deposited on a substrate under spinning [45]. In a modified hot-casting method, the substrates are preheated (70-150 °C) just before the spin-coating process [46]. In some cases, antisolvent (solvent that dissolves the halide precursors but not the perovskite film) is dripped onto the spinning substrate right after applying the precursor solution to accelerate the perovskite crystallization by inducing a rapid supersaturation state in the thin film.

Another widely employed approach is the subsequential deposition of metal halide and organic halide (two-step method) [47]. Low-dimensional perovskite single crystals have been grown using various techniques, and one of them is the temperature-cooling method. For example, to prepare $(CH_3(CH_2)_3NH_3)_2(CH_3NH_3)_{n-1}Pb_nI_{3n+1}$ perovskite crystals, butylamine is added into a boiling hydroiodic (HI) acid solution containing the remaining precursors to form a transparent yellow-colored solution. The layered perovskite crystals are precipitated out of the solution upon cooling it down to room temperature [48].

5.1 Stability of low-dimensional perovskites

One of the driving points for exploiting low-dimensional organic-inorganic perovskites for optoelectronic applications, particularly 2D and quasi-2D structures, is their strong resistance to moisture degradation. One of the early reports by Karunadasa and coworkers demonstrates the enhanced moisture stability of 2D layered perovskite when used as the photo-absorbing layer in solar cells [45]. They found that films of $(PEA)_2(MA)_2[Pb_3I_{10}]$ (PEA = $C_6H_5(CH_2)_2NH_3{}^+$, MA = $CH_3NH_3{}^+$) are more moisture-resistant than $MAPbI_3$ films, and, thus, solar cells using 2D films can be fabricated even under high humid environment. The following report by Liao et al. compared the stability of 3D $MAPbI_3$ and $CsPbI_3$ with that of a low-dimensional $BA_2CsPb_2I_7$ (BA = butyl ammonium; $n=2$) perovskite (Fig. 6) [49]. In the XRD pattern of the $CsPbI_3$ film, the degradation peaks started appearing after 30 min of air storage. Similarly, $MAPbI_3$ partially decomposes to PbI_2 after 48 h. On the contrary, the $BA_2CsPb_2I_7$ film displayed superior stability in the air for a month under 30% relative humidity. The $BA_2CsPb_2I_7$ film exhibited thermal stability also. Moreover, $BA_2CsPb_2I_7$-based solar cells retained 92% of their initial efficiencies under the same storage conditions. The hydrophobic nature of long organic cations (linker molecules) of the low-dimensional perovskites provides a great repulsion from the moisture.

Also, the geometry of the spacer cations affects the stability of low-dimensional perovskites. For instance, Chen et al. compared the stability and solar cell efficiency when the spacer cation changed from linear-BA to iso-BA [44]. The 2D perovskite ($n = 4$) with the branched spacer ($(iso\text{-}BA)_2(MA)_3Pb_4I_{13}$) displays better stability over the one with a linear spacer cation ($(BA)_2(MA)_3Pb_4I_{13}$) by preserving its initial optical absorption after storage for over a month at 20°C with a relative humidity of 60%. Here, the reduced separation between the inorganic (ABX_3) layers with shorter iso-BA units might be the reason for the improved stability of $(iso\text{-}BA)_2(MA)_3Pb_4I_{13}$. A better charge transport from reduced ABX_3 layer separation, out-of-plane crystal orientation, and better crystallinity

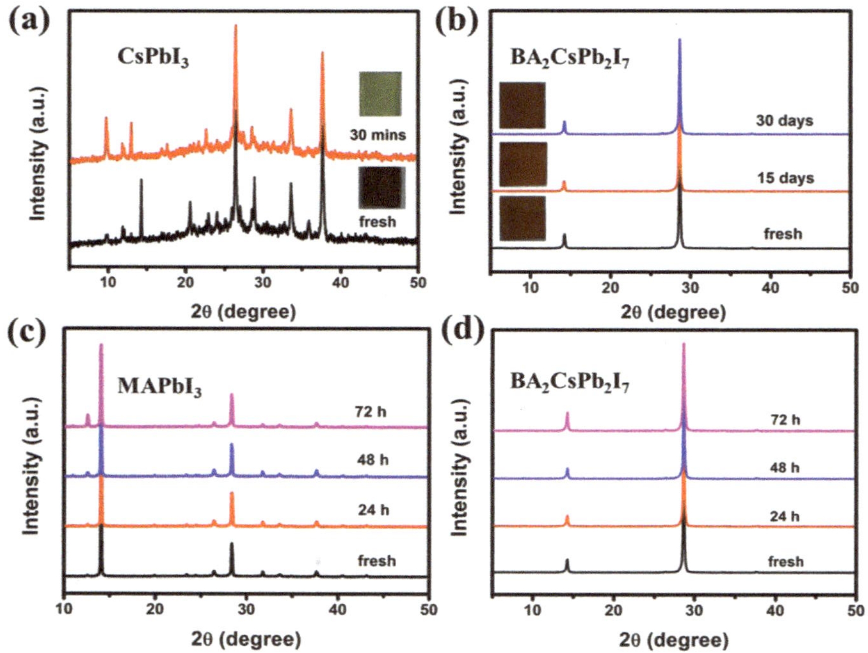

FIG. 6 XRD patterns of thin films of (A) $CsPbI_3$, (B) $BA_2CsPb_2I_7$ stored under 30% RH, (C) $MAPbI_3$, and (D) $BA_2CsPb_2I_7$ stored at 85°C in an inert environment for 72h. *Reproduced with permission from J.-F. Liao, H.-S. Rao, B.-X. Chen, D.-B. Kuang, C.-Y. Su, Dimension engineering on cesium lead iodide for efficient and stable perovskite solar cells, J. Mater. Chem. A. 5 (2017) 2066–2072. https://doi.org/10.1039/c6ta09582h.*

of $(iso\text{-}BA)_2(MA)_3Pb_4I_{13}$ ensured a maximum PCE of 10.63% in solar cells. Note that, though low n values produce low-dimensional perovskites with superior stability, however, their PCE values are still lower than those of 3D perovskites.

The stability of 3D perovskites can be enhanced by making a 2D/3D interface. Grancini et al. mixed a long organic cation, $(HOOC(CH_2)_4NH_3)_2$ (AVA), with 3D $MAPbI_3$ to form a highly stable AVAI/$MAPbI_3$ perovskite junction (quasi-2D with large n) [50]. This 2D/3D perovskite solar cell maintained nearly 60% of initial PCE after 300 h of continuous AM 1.5G illumination under argon atmosphere, which proves its higher stability than devices employing the standard 3D $MAPbI_3$. The 10 x 10 cm^2 2D/3D solar modules obtained via a fully printable industrial-scale process delivered 11.2% efficiency, and they were stable for more than 10,000 h with no loss in the efficiency measured under controlled (standard) conditions, showing the stability enhancement due to the 2D component of the 2D/3D perovskite. Similarly, $CA_2PbI_4/MAPbI_xCl_{3-x}$ (CA = cyclopropylammonium) 2D/3D film exhibited superior stability: no degradation was observed for 40 days when stored at a relative humidity of 63%. On the other hand, 3D $MAPbI_xCl_{3-x}$ fully degraded after 8 days under the same storage conditions [51]. In another report, BA cations were introduced into a double-cation lead mixed-halide $FA_{0.83}Cs_{0.17}Pb(I_yBr_{1-y})_3$ 3D perovskite. As a result, layered 2D crystallites were embedded between 3D perovskite grains and produced a quasi-2D (2D/3D) structure [52]. An optimal BA concentration ensured $17.5 \pm 1.3\%$ solar cell efficiency for a 1.61 eV bandgap perovskite. The solar cells lost only 20% of their PCE after 1000 h under 1 sun illumination in the air. Since the exciton binding energy values of quasi-2D (2D/3D) perovskites are close to those of 3D perovskites, they have a higher potential to generate efficient and stable solar cells than their 2D counterparts.

The addition of 2D perovskites has been known to improve the phase stability of 3D perovskites as well. For instance, α-$CsPbI_3$ (or γ-$CsPbI_3$) films with a suitable bandgap (1.73 eV) for tandem solar cells are known to degrade instantaneously into undesirable δ-$CsPbI_3$ phase under ambient environment. Zhang et al. showed that the addition of a small amount of 2D $EDAPbI_4$ (EDA = ethylenediamine cation) perovskite into 3D $CsPbI_3$ prevents its transformation into the nonperovskite δ-$CsPbI_3$ phase [53]. The enhanced phase stability of 3D $CsPbI_3$ was ascribed to the reduced crystallite size of $CsPbI_3 \cdot 0.025EDAPbI_4$ (2D+3D) perovskite and the unique (110) layered structure of the 2D $EDAPbI_4$. Very recently, intact 2D/3D halide junction perovskite solar cells with a record and certified solar cell efficiency of 24.35% were reported [54]. A thick and highly crystalline 2D $(C_4H_9NH_3)_2PbI_4$ film was grown on top of 3D $(FAPbI_3)_{0.95}(MAPbBr_3)_{0.05}$. These intact 2D/3D halide junction solar cells retained 94% and 98% of their initial efficiencies after ~1000 h at 85°C with a relative humidity of 85% and under 1 sun illumination, respectively. Such outstanding thermal and moisture stabilities of the device were achieved via encapsulation with 2D $(C_4H_9NH_3)_2PbI_4$. Both 2D and quasi-2D perovskites exhibit better moisture stability due to the presence of excess hydrophobicity, coming from the long organic spacer cations. However, their thermal stability is reduced. Therefore, the incorporation of inorganic cations, e.g., Cs, in 2D or 2D+3D perovskites, improves their thermal stability. All-inorganic 0D metal halides are well known for their robust thermal stabilities [55].

5.2 Optical properties of low-dimensional perovskites

An important parameter of low-dimensional perovskites is their high exciton binding energy, which is in the order of a few 100 meVs, hence typically higher than that of 3D perovskites. For instance, the exciton binding energy values of 2D $(PEA)_2(MA)_{n-1}[Pb_nI_{3n+1}]$ for $n = 1$ and $n = 2$ are 220 and 170 meV, respectively, which are much higher than that of the 3D $MAPbI_3$ perovskite (40 meV) [45]. The exciton binding energy of low-dimensional perovskites depends on the difference between the dielectric constants of the spacer cation (barrier) and the inorganic framework (quantum well), and this is known as the dielectric confinement effect. The inorganic framework has a higher dielectric constant (~6.1) compared to the cation spacers (dielectric constant ~2.1). Organic spacer cations with a different dielectric constant can modify the binding energy value.

The exciton binding energy values increase when decreasing the perovskite dimension; 0D Cs_4PbBr_6 has an exciton binding energy value of ~375 meV [55]. This shows that the excitons in low-dimensional perovskites are strongly bound; i.e., they are Frenkel excitons. On the other hand, the binding energies of quasi-2D perovskites (for large n; $n = 40, 50, 60, \ldots$) are smaller than those of other low-dimensional perovskites and thus can be suitable for achieving high solar cell efficiencies similarly to those of 3D perovskites [56]. While 2D and quasi-2D perovskites are suitable for photovoltaic applications, strongly confined 2D (low n values), 1D, and 0D perovskites have the potential to perform efficiently in LEDs.

Since the spacer cations used in low-dimensional perovskites are optically inert when compared to the inorganic sheets, the electronic (optical) transition energies, and particularly the bandgap of these materials, are determined by the number of the BX_6 layers in the inorganic framework. The optical bandgap usually increases when decreasing the number of layers or dimensionality of the perovskite material, due to increased quantum confinement. For instance, Stoumpos et al. synthesized a series of 2D $(CH_3(CH_2)_3NH_3)_2(CH_3NH_3)_{n-1}Pb_nI_{3n+1}$ perovskites with varying thickness (n) by adjusting the ratio between the spacer and small organic cations [48]. The bandgaps change gradually from 2.43 eV for the $n = 1$ member to 1.50 eV for the $n = \infty$ (3D) member. The intermediate members, $n = 2$, 3, and 4, exhibit bandgaps of 2.17 eV, 2.03 eV, and 1.91 eV, respectively. Accordingly, the appearance of these materials varies from orange ($n = 1$) to red ($n = 2$) to dark red ($n = 3$) up to black ($n = 4,5, \ldots\infty$). Also, they exhibit intense PL at 2.35 eV ($n = 1$), 2.12 eV ($n = 2$), 2.01 eV ($n = 3$), and 1.90 eV ($n = 4$).

2D organic-inorganic lead bromides display white-light emission spectra comprising free exciton (FE) emission and a broad emission, known as self-trapped exciton (STE) emission. A very high electron-phonon coupling in (001) and (110) 2D lead bromides creates an STE state as additional excited state [57]. No potential barrier exists for the exciton self-trapping from the FE state in 0D materials, where distortion of isolated and flexible polyhedra requires low energy [58]. Negligible exciton migration and large Stokes shifts are essential for achieving bright STE emission. For instance, 0D Sn bromides (Stokes shift ~1.5 eV) exhibit bright STE PL at room temperature, while the 0D Pb Cs_4PbBr_6 (Stokes shift 0.69 eV) is almost nonemissive [43].

6. Conclusions and prospects

Low-dimensional MHPs have attracted extensive attention during the last several years in different research directions, i.e., dimensional tunability, novel compositional design, understanding of fundamental mechanisms, and various applications such as LEDs, solar cells, and other optoelectronics. Their tunable structures and related diverse properties reveal that there is still a big room for the advancement of materials engineering and yet unexplored applications. Existing low-dimensional MHPs with promising properties (e.g., 0D cluster and 1D nanotube) have already demonstrated the universal opportunities of this type of nanomaterials [59,60]. However, we foresee the following key challenges to be further addressed to promote their application in optoelectronics: (i) Design and synthesis of new low-dimensional MPs. The precise control of dimensionality, nano-to-micro structure, and crystallinity is still questionable most of the time. Thus, it would be beneficial for chemists to develop some guidelines for this purpose. Instead of current common octahedrons, other metal halide polyhedrons (e.g., tetrahedrons and hexahedrons) could be considered as building blocks to construct nanostructures in low dimensions (0D–2D). The size control of metal halide clusters is crucial to realize an effective conversion from their molecular forms to the crystal phases. Besides the known solution process, some other synthetic routes (e.g., mechanical grind and vapor method) could be further exploited. The computational approaches should be utilized more frequently to anticipate the self-assembly process of the organic and inorganic compounds before the actual synthesis. (ii) Understanding of the photophysical and electronic properties of low-dimensional MHPs. Although several mechanisms and theoretical studies have already been established, there is still a large space to comprehensively understand the dimension-dependent charge carrier kinetics, especially in their excited states. The interaction between the organic and inorganic constituents and the low-dimensional network and their influence on the steady- and dynamic-state properties must be further investigated. (iii) Stability enhancement related to a wide scope of applications. In contrast to conventional inorganic or even some of the organic semiconductors, the relatively low stability of low-dimensional MHPs is still the major bottleneck for their massive applications. A deep understanding of the degradation mechanisms, which are linked to, e.g., moisture, oxygen, heat, pressure, and illumination, is urgently required. In addition, other types of applications based on the uncovered characteristics of various low-dimensional MHPs, e.g., solar concentrators, light guiding, nonlinear optics, memory devices, and scintillators, could be continuously explored as promising research orientations in the near future.

References

[1] NREL, Best research-cell efficiency chart, Photovoltaic Research, https://www.nrel.gov/pv/cell-efficiency.html (Accessed 30, Best Research-Cell Efficiency Chart, Photovoltaic Research, NREL, April 2021. https://www.nrel.gov/pv/cell-efficiency.html (accessed April 30, 2021).

[2] N.K. Noel, S.D. Stranks, A. Abate, C. Wehrenfennig, S. Guarnera, A.-A. Haghighirad, A. Sadhanala, G.E. Eperon, S.K. Pathak, M.B. Johnston, A. Petrozza, L.M. Herz, H.J. Snaith, Lead-free organic–inorganic tin halide perovskites for photovoltaic applications, Energ. Environ. Sci. 7 (2014) 3061–3068.

[3] M. Liu, H. Zhang, D. Gedamu, P. Fourmont, H. Rekola, A. Hiltunen, S.G. Cloutier, R. Nechache, A. Priimagi, P. Vivo, Halide perovskite nanocrystals for next-generation optoelectronics, Small 15 (2019) 1900801.

[4] J.H. Noh, S.H. Im, J.H. Heo, T.N. Mandal, S., Il Seok, Chemical management for colorful, efficient, and stable inorganic–organic hybrid nanostructured solar cells, Nano Lett. 13 (2013) 1764–1769.

[5] K.W. Tan, D.T. Moore, M. Saliba, H. Sai, L.A. Estroff, T. Hanrath, H.J. Snaith, U. Wiesner, Thermally induced structural evolution and performance of mesoporous block copolymer-directed alumina perovskite solar cells, ACS Nano 8 (2014) 4730–4739.

[6] E. Radicchi, E. Mosconi, F. Elisei, F. Nunzi, F. De Angelis, Understanding the solution chemistry of lead halide perovskites precursors, ACS Appl. Energy Mater. 2 (2019) 3400–3409.

[7] M. Yang, T. Zhang, P. Schulz, Z. Li, G. Li, D.H. Kim, N. Guo, J.J. Berry, K. Zhu, Y. Zhao, Facile fabrication of large-grain $CH_3NH_3PbI_{3-x}Br_x$ films for high-efficiency solar cells via CH_3NH_3Br-selective Ostwald ripening, Nat. Commun. 7 (2016) 1–9.

[8] J. Seo, S. Park, Y. Chan Kim, N.J. Jeon, J.H. Noh, S.C. Yoon, S., Il Seok, Benefits of very thin PCBM and LiF layers for solution-processed p-i-n perovskite solar cells, Energ. Environ. Sci. 7 (2014) 2642–2646.

[9] M. Liu, M.B. Johnston, H.J. Snaith, Efficient planar heterojunction perovskite solar cells by vapour deposition, Nature 501 (2013) 395–398.

[10] K. Lin, J. Xing, L.N. Quan, F.P.G. de Arquer, X. Gong, J. Lu, L. Xie, W. Zhao, D. Zhang, C. Yan, W. Li, X. Liu, Y. Lu, J. Kirman, E.H. Sargent, Q. Xiong, Z. Wei, Perovskite light-emitting diodes with external quantum efficiency exceeding 20 per cent, Nature 562 (2018) 245–248.

[11] X. Qi, Y. Zhang, Q. Ou, S.T. Ha, C.-W. Qiu, H. Zhang, Y.-B. Cheng, Q. Xiong, Q. Bao, Photonics and optoelectronics of 2D metal-halide perovskites, Small 14 (2018) 1800682.

[12] C.K. Møller, Crystal structure and photoconductivity of caesium plumbohalides, Nature 182 (1958) 1436.

[13] D. Weber, $CH_3NH_3SnBr_xI_{3-x}$ (x = 0-3), ein Sn (II)-System mit kubischer Perowskitstruktur, Zeitschrift Fur Naturforsch. - Sect. B, J. Chem. Sci. 33 (1978) 862–865.

[14] A. Kojima, K. Teshima, Y. Shirai, T. Miyasaka, Organometal halide perovskites as visible-light sensitizers for photovoltaic cells, J. Am. Chem. Soc. 131 (2009) 6050–6051.

[15] Z.-K. Tan, R.S. Moghaddam, M.L. Lai, P. Docampo, R. Higler, F. Deschler, M. Price, A. Sadhanala, L.M. Pazos, D. Credgington, F. Hanusch, T. Bein, H.J. Snaith, R.H. Friend, Bright light-emitting diodes based on organometal halide perovskite, Nat. Nanotechnol. 9 (2014) 687–692.

[16] Y. Cao, N. Wang, H. Tian, J. Guo, Y. Wei, H. Chen, Y. Miao, W. Zou, K. Pan, Y. He, H. Cao, Y. Ke, M. Xu, Y. Wang, M. Yang, K. Du, Z. Fu, D. Kong, D. Dai, Y. Jin, G. Li, H. Li, Q. Peng, J. Wang, W. Huang, Perovskite light-emitting diodes based on spontaneously formed submicrometre-scale structures, Nature 562 (2018) 249–253.

[17] T. Fang, T. Wang, X. Li, Y. Dong, S. Bai, J. Song, Perovskite QLED with an external quantum efficiency of over 21% by modulating electronic transport, Sci. Bull. 3 (2020) 1–8.

[18] Z. Wei, J. Xing, The rise of perovskite light-emitting diodes, J. Phys. Chem. Lett. 10 (2019) 3035–3042.

[19] M.A. Green, A. Ho-Baillie, H.J. Snaith, The emergence of perovskite solar cells, Nat. Photonics. 8 (2014) 506–514.

[20] J. Zhang, X. Yang, H. Deng, K. Qiao, U. Farooq, M. Ishaq, F. Yi, H. Liu, J. Tang, H. Song, Low-dimensional halide perovskites and their advanced optoelectronic applications, Nano-Micro Lett. 9 (2017) 36.

[21] W. Travis, E.N.K. Glover, H. Bronstein, D.O. Scanlon, R.G. Palgrave, On the application of the tolerance factor to inorganic and hybrid halide perovskites: a revised system, Chem. Sci. 7 (2016) 4548–4556.

[22] J.P. Correa-Baena, M. Saliba, T. Buonassisi, M. Grätzel, A. Abate, W. Tress, A. Hagfeldt, Promises and challenges of perovskite solar cells, Science (80-.) 358 (2017) 739–744.

[23] C. Zhou, H. Lin, Q. He, L. Xu, M. Worku, M. Chaaban, S. Lee, X. Shi, M.H. Du, B. Ma, Low dimensional metal halide perovskites and hybrids, Mater. Sci. Eng. R Reports 137 (2019) 38–65.

[24] Q.A. Akkerman, L. Manna, What defines a halide perovskite? ACS Energy Lett. (2020) 604–610.

[25] Y. Zhang, J.D. Lin, V. Vijayaragavan, K.K. Bhakoo, T., Thatt Yang Tan, Tuning sub-10 nm single-phase $NaMnF_3$ nanocrystals as ultrasensitive hosts for pure intense fluorescence and excellent T1 magnetic resonance imaging, Chem. Commun. 48 (2012) 10322–10324.

[26] G. Volonakis, M.R. Filip, A.A. Haghighirad, N. Sakai, B. Wenger, H.J. Snaith, F. Giustino, Lead-free halide double Perovskites via heterovalent substitution of noble metals, J. Phys. Chem. Lett. 7 (2016) 1254–1259.

[27] M. Chen, M.-G. Ju, A.D. Carl, Y. Zong, R.L. Grimm, J. Gu, X.C. Zeng, Y. Zhou, N.P. Padture, Cesium titanium (IV) bromide thin films based stable lead-free perovskite solar cells, Joule 2 (2018) 558–570.

[28] A.E. Maughan, A.M. Ganose, D.O. Scanlon, J.R. Neilson, Perspectives and design principles of vacancy-ordered double perovskite halide semiconductors, Chem. Mater. 31 (2019) 1184–1195.

[29] B. Saparov, F. Hong, J.-P. Sun, H.-S. Duan, W. Meng, S. Cameron, I.G. Hill, Y. Yan, D.B. Mitzi, Thin-film preparation and characterization of $Cs_3Sb_2I_9$: a lead-free layered perovskite semiconductor, Chem. Mater. 27 (2015) 5622–5632.

[30] Y.P. Lin, S. Hu, B. Xia, K.Q. Fan, L.K. Gong, J.T. Kong, X.Y. Huang, Z. Xiao, K.Z. Du, Material design and optoelectronic properties of three-dimensional quadruple perovskite halides, J. Phys. Chem. Lett. 10 (2019) 5219–5225.

[31] T. Cai, W. Shi, S. Hwang, K. Kobbekaduwa, Y. Nagaoka, H. Yang, K. Hills-Kimball, H. Zhu, J. Wang, Z. Wang, Y. Liu, D. Su, J. Gao, O. Chen, Lead-free $Cs_4CuSb_2Cl_{12}$ layered double perovskite nanocrystals, J. Am. Chem. Soc. 142 (2020) 11927–11936.

[32] W.J. Yin, T. Shi, Y. Yan, Unusual defect physics in $CH_3NH_3PbI_3$ perovskite solar cell absorber, Appl. Phys. Lett. 104 (2014), 063903.

[33] T. Umebayashi, K. Asai, T. Kondo, A. Nakao, Electronic structures of lead iodide based low-dimensional crystals, Phys. Rev. B 67 (2003) 155405.

[34] G.E. Eperon, S.D. Stranks, C. Menelaou, M.B. Johnston, L.M. Herz, H.J. Snaith, Formamidinium lead trihalide: a broadly tunable perovskite for efficient planar heterojunction solar cells, Energ. Environ. Sci. 7 (2014) 982.

[35] H. Lin, C. Zhou, Y. Tian, T. Siegrist, B. Ma, Low-dimensional organometal halide perovskites, ACS Energy Lett. 3 (2018) 54–62.

[36] L. Dou, A.B. Wong, Y. Yu, M. Lai, N. Kornienko, S.W. Eaton, A. Fu, C.G. Bischak, J. Ma, T. Ding, Atomically thin two-dimensional organic-inorganic hybrid perovskites, Science 349 (2015) 1518–1521.

[37] S. Ahmad, P. Fu, S. Yu, Q. Yang, X. Liu, X. Wang, X. Wang, X. Guo, C. Li, Dion-Jacobson phase 2D layered perovskites for solar cells with ultrahigh stability, Joule 3 (2019) 794–806.

[38] L. Mao, W. Ke, L. Pedesseau, Y. Wu, C. Katan, J. Even, M.R. Wasielewski, C.C. Stoumpos, M.G. Kanatzidis, Hybrid Dion–Jacobson 2D lead iodide perovskites, J. Am. Chem. Soc. 140 (2018) 3775–3783.

[39] D.B. Mitzi, S. Wang, C.A. Feild, C.A. Chess, A.M. Guloy, Conducting layered organic-inorganic halides containing <110>-oriented perovskite sheets, Science 267 (1995) 1473–1476.

[40] S. Wang, D.B. Mitzi, C.A. Feild, A. Guloy, Synthesis and characterization of $[NH_2C(I){:}NH_2]_3MI_5$ (M = Sn, Pb): stereochemical activity in divalent tin and lead halides containing single <110> perovskite sheets, J. Am. Chem. Soc. 117 (1995) 5297–5302.

[41] Z. Yuan, C. Zhou, Y. Tian, Y. Shu, J. Messier, J.C. Wang, L.J. Van De Burgt, K. Kountouriotis, Y. Xin, E. Holt, One-dimensional organic lead halide perovskites with efficient bluish white-light emission, Nat. Commun. 8 (2017) 1–7.

[42] M. Li, Z. Xia, Recent progress of zero-dimensional luminescent metal halides, Chem. Soc. Rev. 50 (2021) 2626–2662.

[43] D. Han, H. Shi, W. Ming, C. Zhou, B. Ma, B. Saparov, Y.-Z. Ma, S. Chen, M.-H. Du, Unraveling luminescence mechanisms in zero-dimensional halide perovskites, J. Mater. Chem. C 6 (2018) 6398–6405.

[44] G. Zhou, B. Su, J. Huang, Q. Zhang, Z. Xia, Broad-band emission in metal halide perovskites: mechanism, materials, and applications, Mater. Sci. Eng. R Reports 141 (2020) 100548.

[45] I.C. Smith, E.T. Hoke, D. Solis-Ibarra, M.D. McGehee, H.I. Karunadasa, A layered hybrid perovskite solar-cell absorber with enhanced moisture stability, Angew. Chemie 126 (2014) 11414–11417.

[46] Y. Chen, Y. Sun, J. Peng, W. Zhang, X. Su, K. Zheng, T. Pullerits, Z. Liang, Tailoring organic cation of 2D air-stable organometal halide perovskites for highly efficient planar solar cells, Adv. Energy Mater. 7 (2017) 1700162.

[47] T.M. Koh, V. Shanmugam, J. Schlipf, L. Oesinghaus, P. Müller-Buschbaum, N. Ramakrishnan, V. Swamy, N. Mathews, P.P. Boix, S.G. Mhaisalkar, Nanostructuring mixed-dimensional perovskites: a route toward tunable, efficient photovoltaics, Adv. Mater. 28 (2016) 3653–3661.

[48] C.C. Stoumpos, D.H. Cao, D.J. Clark, J. Young, J.M. Rondinelli, J.I. Jang, J.T. Hupp, M.G. Kanatzidis, Ruddlesden–Popper hybrid lead iodide perovskite 2D homologous semiconductors, Chem. Mater. 28 (2016) 2852–2867.

[49] J.-F. Liao, H.-S. Rao, B.-X. Chen, D.-B. Kuang, C.-Y. Su, Dimension engineering on cesium lead iodide for efficient and stable perovskite solar cells, J. Mater. Chem. A 5 (2017) 2066–2072.

[50] G. Grancini, C. Roldán-Carmona, I. Zimmermann, E. Mosconi, X. Lee, D. Martineau, S. Narbey, F. Oswald, F. De Angelis, M. Graetzel, One-year stable perovskite solar cells by 2D/3D interface engineering, Nat. Commun. 8 (2017) 1–8.

[51] C. Ma, C. Leng, Y. Ji, X. Wei, K. Sun, L. Tang, J. Yang, W. Luo, C. Li, Y. Deng, 2D/3D perovskite hybrids as moisture-tolerant and efficient light absorbers for solar cells, Nanoscale 8 (2016) 18309–18314.

[52] Z. Wang, Q. Lin, F.P. Chmiel, N. Sakai, L.M. Herz, H.J. Snaith, Efficient ambient-air-stable solar cells with 2D–3D heterostructured butylammonium-caesium-formamidinium lead halide perovskites, Nat. Energy 2 (2017) 17135.

[53] T. Zhang, M.I. Dar, G. Li, F. Xu, N. Guo, M. Grätzel, Y. Zhao, Bication lead iodide 2D perovskite component to stabilize inorganic α-$CsPbI_3$ perovskite phase for high-efficiency solar cells, Sci. Adv. 3 (2017) 1700841.

[54] Y.-W. Jang, S. Lee, K.M. Yeom, K. Jeong, K. Choi, M. Choi, J.H. Noh, Intact 2D/3D halide junction perovskite solar cells via solid-phase in-plane growth, Nat. Energy 6 (2021) 63–71.

[55] D. Chen, Z. Wan, X. Chen, Y. Yuan, J. Zhong, Large-scale room-temperature synthesis and optical properties of perovskite-related Cs_4PbBr_6 fluorophores, J. Mater. Chem. C 4 (2016) 10646–10653.

[56] T. Ishihara, X. Hong, J. Ding, A.V. Nurmikko, Dielectric confinement effect for exciton and biexciton states in PbI4-based two-dimensional semiconductor structures, Surf. Sci. 267 (1992) 323–326.

[57] E.R. Dohner, A. Jaffe, L.R. Bradshaw, H.I. Karunadasa, Intrinsic white-light emission from layered hybrid perovskites, J. Am. Chem. Soc. 136 (2014) 13154–13157.

[58] C. Zhou, H. Lin, J. Neu, Y. Zhou, M. Chaaban, S. Lee, M. Worku, B. Chen, R. Clark, W. Cheng, Green emitting single-crystalline bulk assembly of metal halide clusters with near-unity photoluminescence quantum efficiency, ACS Energy Lett. 4 (2019) 1579–1583.

[59] W. Liu, K. Zhu, S.J. Teat, G. Dey, Z. Shen, L. Wang, D.M. O'Carroll, J. Li, All-in-one: achieving robust, strongly luminescent and highly dispersible hybrid materials by combining ionic and coordinate bonds in molecular crystals, J. Am. Chem. Soc. 139 (2017) 9281–9290.

[60] H. Lin, C. Zhou, Y. Tian, T. Besara, J. Neu, T. Siegrist, Y. Zhou, J. Bullock, K.S. Schanze, W. Ming, M.H. Du, B. Ma, Bulk assembly of organic metal halide nanotubes, Chem. Sci. 8 (2017) 8400–8404.

CHAPTER

2

Fundamentals and classification of halide perovskites

Sarah Derbali[a], Viorica Stancu[a], Andrei G. Tomulescu[a], Cristina Besleaga[a], George Alexandru Nemnes[b,c], Ioana Pintilie[a], and Mihaela Florea[a]

[a]National Institute of Materials Physics, Magurele, Ilfov, Romania, [b]Horia Hulubei National Institute for Physics and Nuclear Engineering, Magurele, Ilfov, Romania, [c]Faculty of Physics, MDEO Research Center, Research Institute of the University of Bucharest (ICUB), University of Bucharest, Magurele, Ilfov, Romania

1. Discovery and fundamental structure of halide perovskites

The material called perovskite was pioneered by the mineralogist Gustav Rose for the first time in 1839 and has been attributed to calcium titanate $CaTiO_3$ [1,2]. The term perovskite was given to honor the Russian mineralogist Lev Alexeïevich Perovski. Since then, this term has been used for all compounds with a similar crystallographic structure to calcium titanate.

A large number of metal-oxides including lead titanate ($PbTiO_3$), strontium titanate ($SrTiO_3$), bismuth ferrite ($BiFeO_3$), etc., are reported to have a perovskite structure, making perovskite materials more generally known as metal oxides.

The perovskite has a general formula of ABX_3, where A represents a monovalent organic and/or inorganic cation, then B stands for a bivalent cation, and X is a monovalent anion. Most of the ABX_3 perovskite materials have an ideal cubic crystal structure with the space group $Pm3m\text{-}O_h$ (see Fig. 1) [3].

https://doi.org/10.1016/B978-0-323-88522-5.00005-3

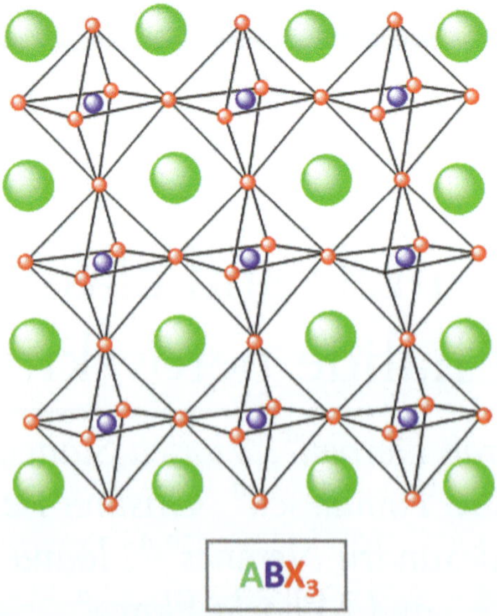

FIG. 1 Schematic representation of 3D halide perovskite structure. *No permission required.*

Perovskites present an atomic arrangement as a three-dimensional (3D) lattice of octahedra with shared corners. In this crystalline structure, ionic bonds are created among the B atoms and the X ions, resulting in BX_6 octahedra that are linked via their corners to form the 3D structure. B cations are placed in the middle of the corner-shared X anion octahedra, while A cations are positioned in the 3D corner-sharing frame of the BX_6 octahedra. Hundreds of perovskites are known, most of them being metal oxides but also metal halide perovskites that possess important properties to be used in different applications such as catalysis, optoelectronics, microelectronic, multiferroics, photovoltaics, batteries, fuel cells, resistive switches, thermoelectrics, and sensing materials (see Fig. 2) [4–6].

The distinctive adaptability of the perovskite crystal structure derives from its ability to accommodate an incredible variety of elemental combinations. In fact, the perovskite structure may accommodate almost all periodic table elements, and compounds formed are classifiable as depicted in Fig. 3. This wide flexibility in composition induces valuable properties and therefore the perovskites were thoroughly studied from an electrical, magnetic, optical, and chemical point of view [5]. This aspect will be discussed in the next sections.

Perovskite stability can be predicted by calculating two factors: (i) Goldschmidt's tolerance factor, t, [7] and (ii) octahedral factor, μ [8]. The tolerance factor (t) is an empirical index that can be calculated, for

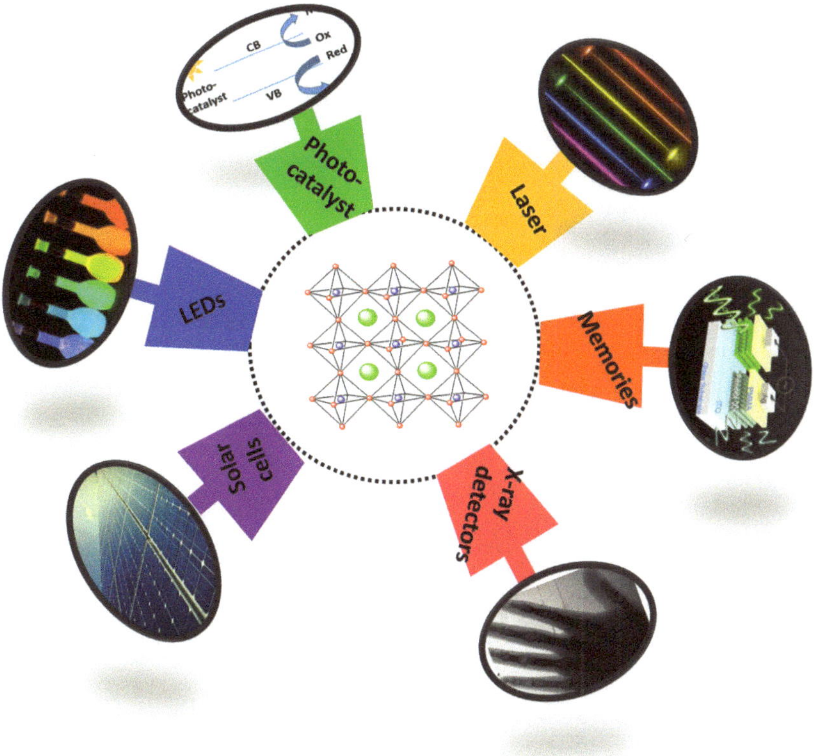

FIG. 2 Applications based on halide perovskite materials. *No permission required.*

a sphere model, from the ratio of the unit cell axis (a) expressions using the following equation (Eq. 1):

$$t = (r_A + r_X)/\left(\sqrt{2}(r_B + r_X)\right) \quad (1)$$

with r_A, r_B, and r_X being the ionic radii of ions A, B, and X, respectively. If the t is within the interval of $0.85 < t < 1.11$, the cubic perovskite structure is stable [9]. A smaller ($t < 0.85$) and larger ($t > 1.11$) tolerance factor value leads to orthorhombic and hexagonal noncubic perovskite structures, respectively [10].

The octahedral factor is considered as a second feature that can be used to assess the perovskite stability and enables the determination of the coordination numbers of the B cation as well as X anion. The octahedral factor can be written as Eq. (2).

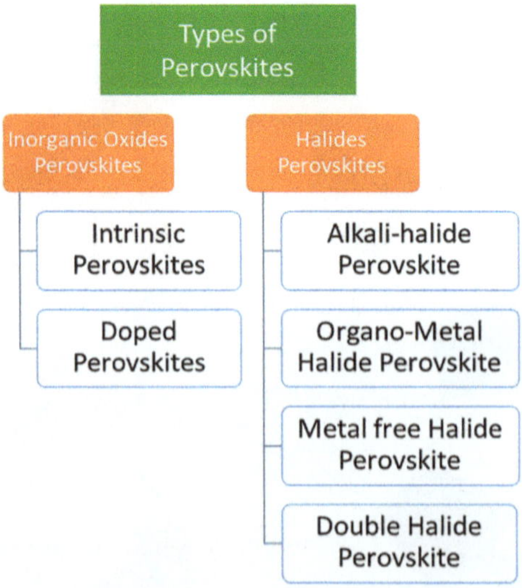

FIG. 3 Classification of perovskites. *No permission required.*

$$\mu = r_B/r_X \tag{2}$$

The class of halide perovskites was discovered in the 1890s. They have a structure identical to that of oxide perovskites, with a monovalent halide ion instead of a divalent oxygen ion. Therefore, to satisfy charge neutrality, halide perovskites can accommodate only metal cations with a valence state of 2+, limiting the diversity of halide perovskite in terms of composition as compared with oxide perovskites, which can also allow trivalent and tetravalent B cations. The most known halide perovskites are the ones containing Pb^{2+} or Sn^{2+}. However, the important feature of the halide perovskite is it can include in its composition organic cations, thus a new family of materials, named hybrid halide perovskites, was developed. Moreover, taking into account that the organic cations can have different sizes will provide an easy way to convert a three-dimensional (3D) perovskite to a two-dimensional (2D) perovskite. This feature of the halide perovskite will be presented in detail further. The first paper about halide perovskite based on Cs^+, Pb^{2+}, and halides was published by Welles in 1892 [11]. Later in 1978, Weber includes an organic molecule within the perovskite structure and synthesized for the first time the hybrid halide perovskites (ABX_3, A=methylammonium: $CH_3NH_3^+$ (MA^+), B=Pb^{2+} or Sn^{2+} and X=Cl^-, I^-, Br^-) [12,13]. During the late 20th century, the synthesis of halide-based perovskites gained more attention and a large variety were synthesized using both small and big organic cations [14,15]. These types of perovskites exhibit semiconducting properties that are well suited for different applications. In halide perovskites, the ionic bond linking the monovalent halide anion and the divalent cation is weak; therefore, they have soft, polarizable, and dynamically unorganized lattices, and as a consequence, point defects are easily formed. These defects play an important role in the different applications. For instance, it can result in considerable luminescence quenching in perovskite-based light-emitting diodes (PeLEDs) [16]. Theoretical and experimental researches on their properties (semiconductors, carrier transport mechanisms, etc.) and their photovoltaic applications have been widely studied in the last few years [17].

There is another kind of halide perovskite which was reported by Laidler et al. in 1968 [18]. This type of perovskite is called halide double perovskite having the following chemical formula: $A_2B'B''X_6$ (see Fig. 4), where A is a monovalent cation as MA^+ or Cs^+, B′ is a trivalent metal cation such Bi^{3+} or Sb^{3+}, B″ is a monovalent metal cation such as Ag^+, Na^+, or Cu^+, and X is a halide anion. The most studied double halide perovskites crystallize in a face-centered cubic structure corresponding to the *Fm*-3*m* space group [19]. Due to their adjustable properties and superior stability, double halide perovskites have high potential to realize stable and environmentally friendly optoelectronic devices for various applications such as photo-detectors, solar cells, photo-catalyst, LEDs, and X-ray detectors [20]. Among the reported double perovskite in the literature, we can mention Cs_2NaMCl_6 (M=Am, Bk, Bi, Lu, etc.), $Cs_2AgBiCl_6$, $(MA)_2AgBiBr_6$, $(MA)_2SnI_6$, and other compositions.

Later in 2002, Bremner et al. discovered a new molecular perovskite hydrate (perazinium ammonium chloride hydrate $(C_4H_{12}N_2)$ $(NH_4Cl_3)\cdot$

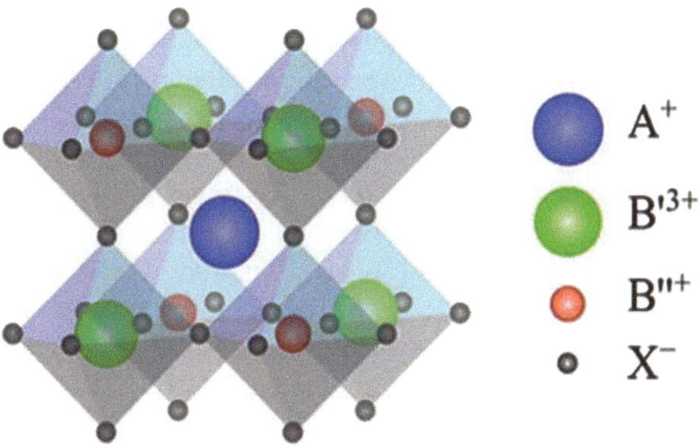

FIG. 4 Structure of double $A_2B'B''X_6$ perovskite. *O.A. Lozhkina, A.A. Murashkina, M.S. Elizarov, V.V. Shilovskikh, A.A. Zolotarev, Y.V. Kapitonov, R. Kevorkyants, A.V. Emeline, T. Miyasaka, Microstructural analysis and optical properties of the halide double perovskite Cs2BiAgBr6 single crystals, Chem. Phys. Lett. 694 (2018) 18–22. https://doi.org/10.1016/j.cplett.2018.01.031.*

H_2O) which consists of cubic perovskite like an array of vertex-sharing $(NH_4)Cl_6$ octahedra extending in three orthogonal directions [21]. This discovery marked the beginning of metal-free halide perovskite. The formula of this kind of perovskite is also ABX_3, with A representing a divalent organic cation such as R-3ammoniopyrrolidinium, rac-3-ammonioquinuclidinium, and other cyclic molecules; B is an ammonium cation (NH_4); and X is a halide anion [22]. This class of 3D materials possesses outstanding ferroelectric properties making them attractive for flexible devices, biomedical devices, soft robotics, and others applications [22].

2. Dimensions of halide perovskites

Recently, organometallic halide perovskites were identified as a significantly promising category of functional materials to be applied in various ranges of processes. The remarkable chemical tunability affords these materials to have 3D, 2D, 1D, and 0D dimensional structures (see Fig. 5). However, until now, research has mainly focused on 3D and 2D structures for solar cell and optoelectronic applications [23].

In three-dimensional ABX_3 halide perovskite, the A-cation size drives the expansion or contraction of the perovskite lattice. There are various perovskites (halides, oxides, or nitrides) with several compositions published; however, perovskites based on halide has been proven to offer the most valuable optoelectronic properties.

In halide perovskite, the A-site could be occupied by MA^+, formamidinium: $HC(NH_2)_2^+$ (FA^+), Cs^+, K^+, or Rb^+, the B-site by a bivalent cation such as Pb^{2+}, Sn^{2+}, Mg^{2+}, Ca^{2+}, etc., and the X-site by monovalent anion such as halides (I^-, Cl^-, and Br^-). Moreover, some other anions such as formate ($HCOO^-$), thiocyanate (SCN^-), and tetrafluoroborate ($BF4^-$) were successfully incorporated as molecular anions [24]. The A, B, and X sites can each be occupied by one or more ions, forming the so-called mixed halide perovskite such as $MAPbI_{3-x}(BF_4)_x$ [25], $K_xCs_{0.05}(FA_{0.85}MA_{0.15})_{0.95}Pb(I_{0.85}Br_{0.15})_3$ [26], $MAPb_{1-x}Sb_{2x/3}I_3$ [27]. Through changing the composition of perovskite by replacement of ions at the ABX_3 sites, various properties can be explored and tuned.

When large cations occupy the A-site (e.g., long-chain alkyl amines), it will not be able to insert into gaps in the octahedral units, resulting in a phase transition to a layered Ruddlesden-Popper structure called "2D perovskite," whose general formula is $(R{-}NH_3)_2A_{n-1}B_nX_{3n+1}$ [28,29].

The general chemical formula of 2D perovskite is the following: $(R{-}NH_3)_2A_{n-1}B_nX_{3n+1}$ [16], where $(R{-}NH_3)^+$ represents long organic cation (typically an aryl or aliphatic ammonium cation) as butylammonium (BA), ammoniumvaleric acid (AVA), phenethylammonium (PEA), or iodoethylammonium (IEA). The n represents the number of metal

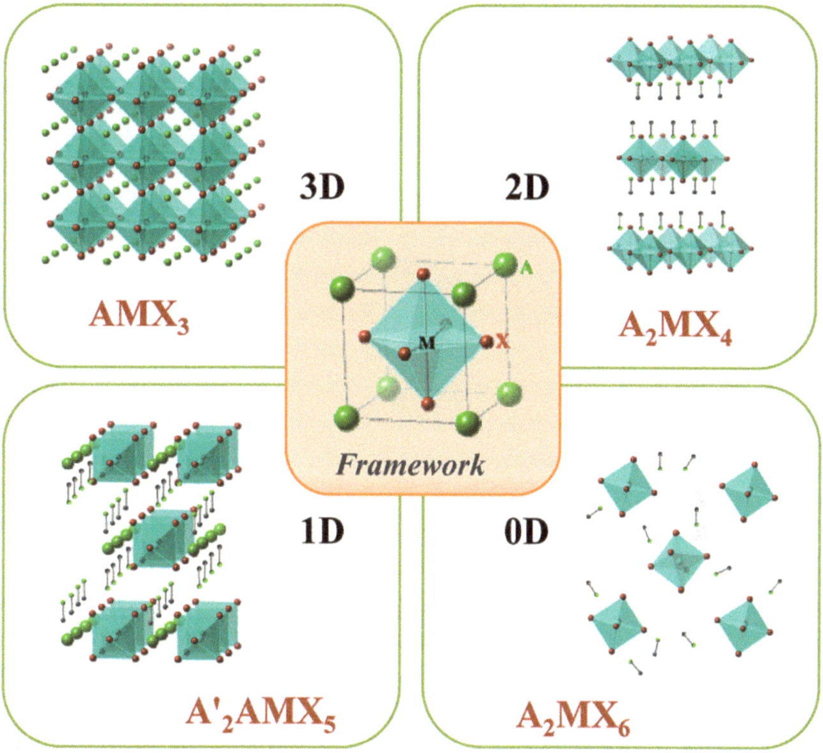

FIG. 5 A representation of the perovskite frameworks in various dimensionalities: 0D, 1D, 2D, and 3D. *S. Gonzalez-Carrero, R.E. Galian, J. Pérez-Prieto, Organic-inorganic and all-inorganic lead halide nanoparticles (Invited), Opt. Express 24 (2016) A285. https://doi.org/10.1364/oe.24.00a285.*

halide sheets; in the case where $n = 1$ (see Fig. 6), it represents 2D perovskites, and when $n = \infty$, it describes 3D perovskites. Perovskite 2D family has a two-dimensional structure which is an alternation of an inorganic layer consisting of BX_6^{4-} octahedral and an organic layer. The cohesion of the structure is provided by different kinds of interactions; the R—NH_3 parts are assembled by Van Der Waals type bonds if the R is an alkyl chain or π-π type in the case where R contains an aromatic nucleus. In 2D perovskite material, the halide anions interact with the organic part via hydrogen-ionic bonding [15,30].

2D halide perovskites possess unique optical properties and charge transport properties, which have not been detected on 3D halide perovskites.

Moreover, the perovskites materials can also self-organize in zero-dimensional (0D) or one-dimensional (1D) structure. For the 1D structure, the octahedra are connected in a chain by faces or edges. Selected

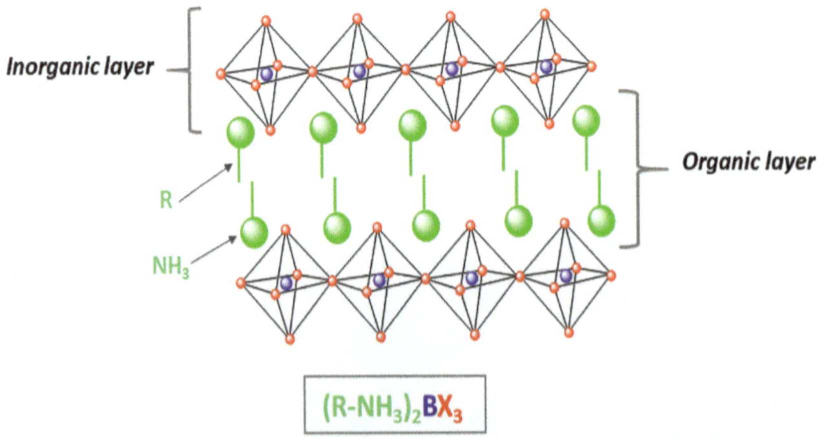

FIG. 6 Typical structure of 2D hybrid perovskite for $n = 1$. *No permission required.*

examples of perovskites with 1D structure: $((CH_3)_2C{=}NHCH_2CH_2CH_3)$ $[PbI_3]$ and $(C_4H_{13}N_3)[PbI_4]$ [31,32]. Regarding 0D perovskites, it is characterized by a structure formed by isolated octahedra or bi-octahedra meaning that they do not exhibit any connection between them, as in the case of materials: $[(CH_3)_2NH_2]_3[BiI_6]$ and $(C_6H_{11}NH_3)_3[Bi_2I_9]_{40}$ [33,34].

2.1 Applications of 3D and 2D perovskite materials

Because of interesting optoelectronic properties, 3D and 2D perovskite materials are nowadays of great interest for manufacturing optoelectronic systems including LEDs, photovoltaic devices, laser, and more devices.

2.1.1 Solar cells

3D and 2D perovskites materials have remarkably different physical and chemical properties. The 3D perovskite has a large absorption coefficient [35], tunable and narrow bandgap (~1.5 eV) [36], small exciton binding energy (~50 meV) [37], large carrier mobilities for holes (12.5–66 cm^2/Vs) and electrons (7.5 cm^2/Vs) [38], long carrier diffusion lengths (100–1000 nm) [39], and facile fabrication process, which explains their excellent performance in photovoltaic field. Despite their outstanding properties and the incredible increase in performance, the 3D perovskite-based solar cells (PSCs) still suffer from poor stability under real operational conditions.

At the same time, 2D perovskite possesses appealing properties, which are beneficial for photovoltaic applications. Indeed, the 2D perovskites materials are thermally resilient; big cations hinder internal ionic motion and can provide the necessary hydroscopic character to the absorber by suitable organic moieties, which improves the chemical stability compared to the 3D perovskites counterparts [40]. This outstanding property promotes the 2D perovskite a good candidate for solar cells applications.

The potential of hybrid perovskite was revealed in 2009 when they were first used as light absorber materials in photovoltaic devices. The integration of 3D perovskite material ($MAPbX_3$) into dye cells has been initiated by Miyasaka's group as a light absorber as well as a possible material for solar cells application. A power conversion efficiency (PCE) of 2.2% was achieved with $MAPbBr_3$-based solar cells [41]. Regarding 2D halide perovskite, they were integrated for the first time in 2014 into a planar device using $PEA_2MA_2Pb_3I_{10}$ as an absorber layer. The resulting solar device exhibits a high open circuit voltage of 1.18 V and PCE of 4.73% [42]. More importantly, this material remained quite stable for about 46 days in the air at 52% relative humidity. Since then, the PCE of these solar cells has skyrocketed, with laboratory records of 25.2% for 3D-PSCs and 18.20% for 2D-PSCs [43,44]. However, PSCs based on 2D perovskite materials display low PCE, attributed to their large bandgap, large exciton binding energy, and poor charge transport. As for PSCs based on 3D halide perovskite, they suffer from low long-term stability against extrinsic factors (heat, humidity, and light) and other intrinsic factors (linear and punctual defects). Several approaches have been used to overcome these problems, including encapsulation, compositional, and interfacial engineering [45–47]. In recent research on PSCs, attention has been focused on a new concept of using mixed 2D and 3D perovskites in order to combine their advantages to provide efficient and stable PSCs. Indeed, Yao et al. described a simple two-step method of forming 3D/2D perovskite films: $(MAPbI_3)_{1-x}[(PEI)_2PbI_4]_x$. Solar cells based on this composition achieved a champion PCE of 15.2% with high reproducibility and excellent stability allowing to maintain >84% of the initial PCE after 14 days [48]. Later, Nazeeruddin et al. reported a PCE of 11.2% by using 2D/3D perovskite film: $(AVA_2PbI_4)_{0.03}(MAPbI_3)_{0.97}$. The device exhibited 0% of degradation in performance after >10,000 h at 55°C under continuous illumination [49]. In another work, Snaith et al. incorporated n-butyl ammonium cations into $FA_{0.83}Cs_{0.17}Pb(I_yBr_{1-y})_3$ 3D perovskite. They observed the formation of 2D perovskite flakes intercalated between high-oriented 3D perovskite grains. The resulted solar cells exhibited an average PCE of 17.5 ± 1.3%. These devices retain 80% of their efficiency after 1000 h in air and nearly 4000 h when encapsulated [29].

2D perovskite films could be applied also on top of a 3D perovskite absorber as an interfacial layer. This approach has been found to be a very successful route to achieve PSCs with enhanced performance and stability.

Ma et al. have succeeded in fabricating ($CA_2PbI_4/MAPbI_xCl_{3-x}$) by covering 3D perovskite with a 2D perovskite film produced in situ. The resulting photovoltaic devices keep 54% of the initial efficiency after 220 h [50]. Since this paper, more research efforts have been deployed by integrating 2D perovskite as an interfacial layer on 3D perovskite. These studies yielded excellent results related to stability and efficiency [29,51–56].

In conclusion, the elaboration of multidimensional 2D/3D perovskites is an effective strategy toward efficient and stable PSCs, offering the opportunity for their commercialization soon.

2.1.2 *Light-emitting diodes*

The controllable photoluminescence property of halide perovskites, in particular the tunable band gap, provides a great interest for LEDs fabrication. The ambipolar carrier migration and low defect density in such materials allow an efficient charges injection and transport at the interfaces, thus enhancing device performance. Light-emitting devices based on perovskites materials have been proposed since the 1990s [57–60]. The outcomes of these studies were encouraging to further improve the efficiency of this type of device later. Recently, perovskite-based light-emitting diodes have demonstrated great progress in performance and achieved external quantum efficiencies (EQEs) of 20.3% and 21.6% [61,62].

2.1.3 *Lasers*

The high photoluminescence quantum yields (PLQYs), tunable wavelengths, and low defect density of halide perovskite make them more promising as laser materials. Deschler et al. were the first to demonstrate the emission of stimulated light from a cavity laser that enclosed a solution-processed $MAPbI_{3-x}Cl_x$ film between a highly reflective gold mirror and dielectric mirror [63]. Afterward, investigations of lasing and amplified spontaneous emission (ASE) were extended to FA^+- and Cs^+-based perovskites.

Although research has been strongly focused on photovoltaics, LEDs, and laser, further electronic applications involving memories, X-ray detectors, gas sensors, and transistors are also explored using such types of materials.

3. Compositional engineering of perovskites

Year by year, as a consequence of the growth of the population of the planet, the global energy demand is increasing and is predicted to double by 2050 [64,65]. The continuous development of renewable sources of energy (e.g., hydro, wind, and solar) becomes an imminent requirement. Converting the sunlight directly into electricity is a viable option but it is a major challenge. In this way, it is necessary to develop reasonable cheap, environmentally friendly materials for devices and integrated photovoltaic systems.

Hybrid perovskites presented notable semiconducting and photoelectric properties in multiple architectures of solar cells, such as mesoporous [66,67] and planar [36,68].

Compositional engineering of the halide perovskite active layer has become an easy-to-use approach for optimizing these photovoltaic devices, as it allows direct modification of other intrinsic properties of the material, resulting in high-quality perovskite with increased photovoltaic performances. By using compositional engineering, it will be possible to improve the stability, for instance, by preventing the degradation by UV radiation or hindering anion migration. Strain develops in the structure in the case of a dimensional mismatch between the A cation size and the halogen cage in the lead halide, and the extra energy in the system is alleviated by the appearance of point defects like vacancies or ions in interstitial positions.

During the years, the efficiency of organic-inorganic perovskite solar cells (PSCs) was the starting point for optimizing the perovskite structure. The remarkable progress in organic-inorganic solar cells was achieved mainly by a trial-and-error approach. In the organic light emitting diodes (OLEDs) case, to obtain an efficient and stable device with a low power usage [69,70] which is an important key in practical applications, a hole injection layer (HIL) is deposited before the hole transport layer (HTL) to lower the anode/HTL energy barrier, which enhances hole injection to the HTL from the anode and helps equilibration of the charge balance in the OLED [71]. Fabrication of uniform large grain stable perovskite layer has been done by partial substitution of both cations and halide ions leading to high efficiency and durability in the solar cell case. For example, single cations like perovskites methyl ammonium halide ($MAPbX_3$), formamidinium lead halide ($FAPbX_3$), cesium lead halide ($CsPbX_3$, X = I or Cl or Br) rapidly degrade when exposed to ambient conditions. Comparatively, mixed perovskite (with either at least two organic cations or multiple halogen ions) presents significant resistance to moisture and heat [72,73]. There are many studies worth to be mentioned: $(FAPbI_3)_{1-x}$ $(MAPbBr_3)_x$ have been investigated by Seok and coworkers [74] and a high efficiency beyond 20% was obtained by Graetzel and coworkers [75] for FA/MA mix cations. The influence of dopants was studied by Park and coworkers [76] who investigated the effect of Cs^+ addition to FA^+ in planar heterojunction. They concluded that Cs grants higher stability than the pure $FAPbI_3$ against humidity and UV light exposure. The addition of Cs^+ into FA^+/MA^+ was described by Saliba et al. providing good results in the growth and morphology of perovskite crystals, which has a profound effect on charge collection and stability of the PSC [77]. Tan et al. [78] using triple cation-based perovskite and chlorine-capped TiO_2 (Cl-TiO_2) colloidal nanocrystal as a compact layer created highly efficient planar devices.

3.1 A-site cations mixture

Increasing the perovskite composition is driven by the need for better stability through the addition of more inorganic elements and increase the entropy of the system, which can stabilize ordinarily unstable phases (e.g., the "yellow" phase of $FAPbI_3$ which does not present a suitable bandgap for solar cell applications) [76,79]. $MAPbI_3$ with a bandgap of about 1.5–1.6eV presents very good light absorption properties in the visible spectrum and, therefore, is one of the most studied inorganic-organic lead halide perovskite for application in solar cells [74,80]. Gao et al. presented one diagram with all factors which are responsible for the instability of solar cell and how the stability problem can be solved [81].

Excellent properties of halide perovskite have been investigated by Ma et al. [82] and Feng et al. [71] when they explored the OLED configuration with PeLED.

Surprising new properties were encountered when three cations Cs^+, FA^+, and MA^+ were used in a perovskite simultaneously [77,79,83]. For example, employing a Cs/FA mixture leads to a secondary phase inhibition of lead halide formation, opening the possibility of perovskite/silicon tandem solutions [84]. Comparing $MAPbBr_3$ with $FAPbBr_3$ [85], an increase of the organic cation radius of 3.3% is observed (from 2.70 Å in the case of MA^+ to 2.79 Å for FA^+). The consequence of this difference is the modification of the Pb-Br-Pb bond angle by 15 degrees leading to better stability of the perovskite structure by a better bounding of the lead in the compound. Many studies have been performed and it should be noted that Graetzel published results pertaining to a triple cation perovskite (Cs/FA/MA) of high-performance PSCs (PCE = 18.02%) that presented improved thermal stability [77,79]. They did not stop here and continued to introduce inorganic cation with even smaller radius (Rb^+ cation) to form new mixed perovskite of Rb/Cs/MA/FA, obtaining a new record stabilized efficiency of 21.6% [77].

The positive role of cesium as A-site in the perovskite structure was also studied by Ling et al. [86]. Comparing LEDs based on $MAPbBr_3$ and $CsPbBr_3$, they observed that the one based on $MAPbBr_3$ degraded fast in ambient conditions, while the device based on $CsPbBr_3$ shows stable luminance when bias voltage scanned multiple times. Hou et al. [87] were interested in perovskite nanocrystals. They reported that the addition of Mn ions in the perovskite leads to greater stability of $CsPbBr_3$ nanocrystals in green LEDs.

This path of the research field was continued; organic cations other than FA^+ have been tested. Yang et al. reported that the inclusion of guanidinium (GA^+) cation in the perovskite film will decrease the defect density. Instead of directly replacing MA^+ cation, the GA^+ cation tends to form coordination bonds with $MAPbI_3$ lattices [88] owing to its relatively larger size (278 pm); it passivates the crystal defects [89]. Troshin et al. studied other cations, such as hydrazinium (HI) cation, to form a mixed HI/MA cation and revealed the beneficial role in improving the crystallinity of the perovskite and its performances [90].

Many authors suggest other alternatives to stabilize the perovskite structure, such as the use of larger cations like the ones from the ionic liquids [91]. Larger cations are introduced as additives in the perovskite precursor solution and used as interface modifiers at both perovskite interfaces with the charge selective layers [92,93]. There versatility in the molecular structure of the mixed compound induces an increase in the ionic conductivity and mechanical stability.

Imidazolium iodide ($C_3N_2H_5I$, IMI) was used to improve the PCE in dye-sensitized solar cells by increasing the conductivity of the electrolyte. Thus, the charge transfers within the electrolyte as well as between the electrolyte and the other layers at interfaces were improved and the dye-sensitized long-term stability was enhanced. The beneficial effect of IMI is to prevent dye aggregation and regenerating the oxidized dyes [94,95] and/or to passivate volume defects because it acts as a Lewis base, counter-balancing the excess of the positive charge of Pb atoms in the crystal and forming a coordinated covalent bond [96,97]. Tomulescu et al. [98] reported that the substitution of MA^+ by IM^+ constitutes a favorable route to reduce recombination processes and a great potential to achieve high stability (see Fig. 7). It was observed that the pristine $MAPbI_{2.6}Cl_{0.4}$-based-solar devices degrade faster than their $MA_{0.94}IM_{0.06}PbI_{2.6}Cl_{0.4}$-doped counterparts. After 2000 h, $MA_{0.94}IM_{0.06}PbI_{2.6}Cl_{0.4}$-based devices retain 70% of their initial efficiency, while the pristine $MAPbI_{2.6}Cl_{0.4}$-based cells maintain only 40% of their initial PCE.

Metal halide perovskites are contenders with encouraging properties for the next generation of commercial solar cells. Despite the high PCEs obtained, the mechanism generating this enhancement is not, at this point, completely clear.

3.2 X-site anions mixture

Varying the X-site anions is another way to achieve new compositions for perovskite solar cells. The type of halogen in the structure of the perovskite crystal determines the optical absorption range of the material, while the organic cation is responsible for its stability. Stoumpos et al. [99] demonstrated that an ion with a small radius in X-site favors the formation of a cubic structure. By combining ions in the X-site, the properties of the perovskite materials can be tailored by meticulous optimization of stoichiometry leading to achieving solar cells with high efficiency. The bandgap of $MAPbI_3$ can be modified by using Br instead of I from 1.59 to 2.31 eV.

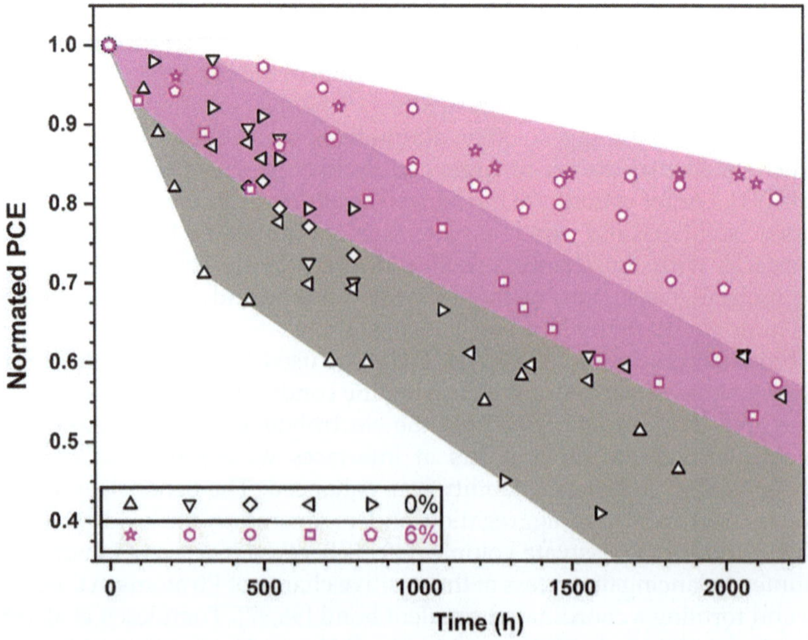

FIG. 7 Solar cells efficiency over 2000h of shelf storage with $MAPbI_{2.6}Cl_{0.4}$ and $MA_{0.94}IM_{0.06}PbI_{2.6}Cl_{0.4}$ perovskites. *A. Gabriel Tomulescu, L. Nicoleta Leonat, F. Neațu, V. Stancu, V. Toma, S. Derbali, Ș. Neațu, A. Mihai Rostas, C. Beșleagă, R. Pătru, I. Pintilie, M. Florea, Enhancing stability of hybrid perovskite solar cells by imidazolium incorporation, Sol. Energy Mater. Sol. Cells. 227 (2021) 111096. https://doi.org/10.1016/j.solmat.2021.111096.*

Hao et al. optimized the bandgap by chemical substitution of iodide with bromide and reported a PCE of 5.73% under AM 1.5G light [100]. Furthermore, Zhao et al. successfully fabricate 10% efficient perovskite solar cells from nanosheets of $MAPbI_2Br$ in a planar structure with a bandgap of 1.8 eV [101].

According to literature, the bandgap of perovskite can be effortlessly modified by altering the ratio of halides [102]. Sutherland and Sargent show the coverage of the hole visible of emission spectra of different $MAPbX_3$ perovskites by varying the ratio of Cl/Br and Br/I [103].

The tunability in the bandgap (see Fig. 8) allows the design of multicolor LED of interest and for any special device application requiring a specific emission bandgap [104]. $CsPbBr_3$ nanocrystals (NCs) with a bandgap of 2.43 eV can be modified, the halide-substituted perovskite NCs present progressively increased bandgap to 3.03 eV with Cl full substitution and decreased bandgap to 1.88 eV with the replacement of Br with I [102].

Another aspect is ferroelectric properties which is possible to be present in perovskite with I and Cl. In literature, it was speculated the presence of

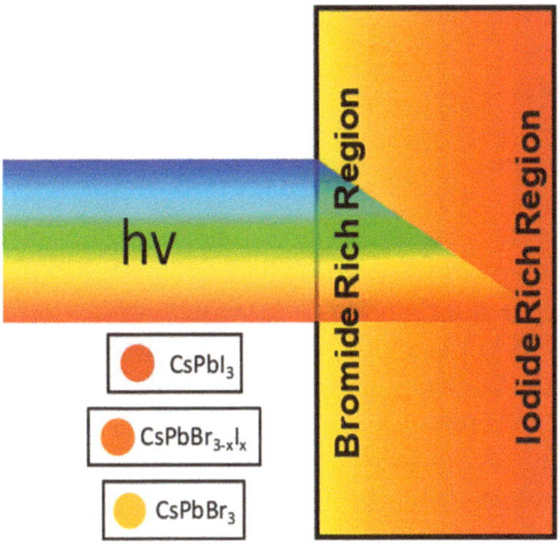

FIG. 8 The tunability in bandgap for $CsPbX_3$ materials with X=Cl, I, Br. *R.A. Scheidt, P.V Kamat, Temperature-driven anion migration in gradient halide perovskites, J. Chem. Phys. 151 (2019) 134703. https://doi.org/10.1063/1.5120270.*

the polar order (ferroelectricity) in the halide perovskite concurring to the high values of PCE in perovskite solar cells [105,106]. Pintilie et al. [107] illustrated this fact that, from the experimental results, the positive polarization due to the ferroelectric layer attracts the electrons photogenerated in $MAPbX_3$, with X=Cl, I have to be located near the collecting FTO electrode. The results confirm the theoretical calculations conducted for ferroelectric-$MAPbX_3$, with X=Cl, I interface [108].

The combination of both chloride and bromide in perovskite structure promotes carrier generation and extends carrier diffusion while reducing the charge recombination in perovskite absorber film [109]. Solar cell devices with chlorine or bromine have superior photovoltaic properties and lower hysteresis effects compared to $MAPbI_3$-based solar cells containing only iodine [110,111]. An explanation for this behavior is that the Br^- presents higher activation energy of the ionic migration; therefore, its movement within the perovskite film is slower while Cl^- ions improve the diffusion length of the free carriers [112]. Mohebpour et al. [113] and Kawano et al. [114] explained this behavior from absorption spectra: "the blue shift of the E_g from Cl to Br is due to the decrease in the covalent character of the halogen bonding with Pb, which means a decrease in the electron negativity of the halogen" [113,114].

Another aspect evidenced by Colella et al. [115] was that halogen doping improves cells performances and Cl doping increases the charge transport within the perovskite layer. An increase of the PCE by adding $PbCl_2$ to the precursor solution was also reported [116,117]. Furthermore, $PbCl_2$

promotes the decrease of the residual PbI_2 in the perovskite solar cell [118]. Chen et al. [119] reached a PCE of 17.91% and found that chlorine plays a significant role as it enhances carrier transport through the heterojunction interfaces, and has a comparatively small effect on the charge transport in the perovskite crystals.

Moreover, Wang et al. [120] reported that the performance of the PeLEDs depends also on the type of halogen. They observed also that the quantum efficiency of their devices is lower than those of the infrared devices, linked to the inadequate compactness of the perovskite film. The coverage of the $MAPbI_{3-x}Cl_x$ films (≈95%) is higher than that of the $MAPbBr_3$ films (≈90%). One explanation is that the incomplete surface coverage caused pinholes in the perovskite films.

3.3 Concurrent mixing of A-cations and halides-anions

The advantages of using cations-halides mixtures have been highlighted as follows [121]: (i) increased performance, (ii) better stability, (iii) heightened carriers charge transport, (iv) bandgap tunability, and (v) reduced hysteresis. A more in-depth understanding of the fundamental aspects of Pb-based mixed perovskites can lead to alternative Pb-free perovskites through deliberate designs. In this regard, the most efficient and stable perovskite solar cells reported so far show efficiency above 22% [74,77]. In the next paragraphs, the most interesting combinations used in photovoltaic applications are presented.

Gkini et al. [122] presented a comparative study of $Cs_{0.05}(MA_{0.17}FA_{0.83})_{0.95}Pb(I_{0.83}Cl_{0.17})_3$ with $Cs_{0.05}(MA_{0.17}FA_{0.83})_{0.95}Pb(I_{0.83}Br_{0.17})_3$. They found that photovoltaic performance is improved when using perovskites containing I—Br. The resulting solar cells reached a PCE of ~17% and ~13.5%, for I—Br and I—Cl, respectively, with short-circuit photocurrents of 23.15 mA cm^{-2} for I-Br-based solar cells. However, the PSCs based on I—Cl show greater in-depth stability in normal atmospheric conditions. In 2018, Liu et al. [123] reported that the introduction of a small amount of additive pyridine in $Cs_{0.05}(MA_{0.17}FA_{0.83})_{0.95}Pb(I_{0.83}Br_{0.17})_3$ precursor mixed solution leads to large crystal size, dense, and pinhole-free perovskite films. Adding pyridine retards the crystal growth and controls rapid nucleation. In addition, it also passivates the surface defects owing to the rapid formation of ion coordinate bonds between Cs^+ and the perovskite absorbers. The maximum PCE obtained for the resulting solar devices was 19.03% [123]. $Cs_{0.05}(MA_{0.17}FA_{0.83})_{0.95}Pb(I_{0.83}Br_{0.17})_3$ was also investigated by Yavari et al. [124] using an environmentally greener friendly antisolvent, anisole (ANS). The corresponding solar cells achieve a PCE of ~16%.

Huyen et al. [125] present the research results performed on the $Rb_{0.03}Cs_{0.07}FA_{0.765}MA_{0.135}PbI_{2.55}Br_{0.45}$ hybrid perovskite. The impact of inorganic cations Rb^+ and Cs^+ on the morphology, crystal structure, and chemical composition of perovskite materials was reported. It was noted that Rb^+ cation exists separately in the perovskite film only as a segregated phase, while the Cs^+ cation was integrated into the perovskite structure but was also found segregated within an Rb- and Cs-rich phase. Thereby, it was beneficial for the stability of the film, since Rb- and Cs-rich phase tends to form at the grain boundary of the perovskite. Hu et al. [126] also studied the effect of Rb^+ and Cs^+ on the performance enhancement of solar cell. For a clear understanding, they illustrated the effect of RbI and CsI in mixed halide perovskite (see Fig. 9).

It was concluded [126] that the incorporation of Cs^+ in the mixed-halide compound $(FA_{0.83}MA_{0.17})Pb(I_{0.83}Br_{0.17})_3$ leads to a smaller lattice volume and a more stable structure of the perovskite phase. RbI addition into the perovskite presents several problems. First, during film crystallization, a secondary phase of $RbPb(I_{0.84}Br_{0.16})_3$ appears, which does not present photovoltaic properties. Second, under humidity exposure, a nonperovskite compound ($RbPb_2I_4Br$) can be formed. The origin of the secondary phase occurrence is the larger affinity between Rb and Br caused by their more compatible ionic radii, with the appearance of more stable compounds than an Rb containing perovskite. They demonstrate that the instability of Rb-containing mixed-halide perovskites causes degradation by rapid phase segregation, which can affect the device performance before any notable chemical decomposition of the perovskite occurs.

In 2021, Derbali et al. [127] explore the incorporation of the alkali metals in $FA_{0.80}MA_{0.20}PbI_{2.8}Cl_{0.2}$ perovskite. The role of alkali metals was to

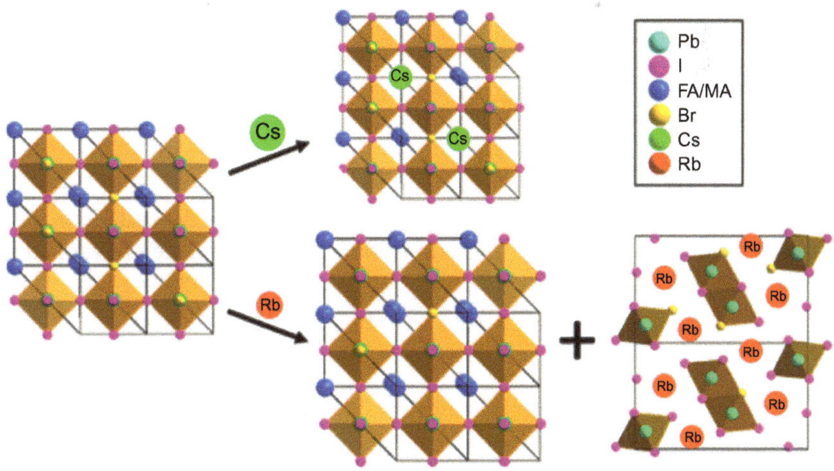

FIG. 9 Schematic representation of the incorporation of RbI and CsI in the perovskite crystal. *Y. Hu, M.F. Aygüler, M.L. Petrus, T. Bein, P. Docampo, Impact of rubidium and cesium cations on the moisture stability of multiple-cation mixed-halide perovskites, ACS Energy Lett. 2 (2017) 2212–2218.*

stabilize the perovskite in order to improve the photovoltaic performance of PSCs and reduce the *J-V* hysteresis. The partial substitution of FA^+ by K^+ in $K_xFA_{0.80-x}MA_{0.20}PbI_{2.8}Cl_{0.2}$ improves the morphology, prevents the chemical decomposition of perovskite film, and stabilizes the α-phase of $FAPbI_3$. The obtained results proved that the crystallization of the nonperovskite phase was significantly inhibited by the addition of potassium which also prevented the appearance of defects. The perovskite film containing K demonstrated a better crystallinity and reduced bandgap that induce a significant improvement of the PCE, from 10.99% to 12.51% with low hysteresis. These devices retain more than 70% of their initial efficiency after more than a month in ambient air, without encapsulation.

The doping of the "A" and "B" sites was the subject of the Stancu et al. [128]. The authors of this paper present findings for complex perovskite compounds obtained by using simultaneously two different dopants in B-site (Eu^{3+}/Sb^{3+}) and in the A-site (IMI^+). The performance of the resulted compounds $[MA]_{1-x}[C_3N_2H_5]_xPb_{1-y}Eu_yI_{2.6}Cl_{0.4}$ and $[MA]_{1-x}[C_3N_2H_5]_xPb_{1-y}Sb_yI_{2.6}Cl_{0.4}$ with y varied from 0 to 0.05 was studied relative to the pristine $MAPbI_{2.6}Cl_{0.4}$ perovskite. The effect of inorganic Eu^{3+} and Sb^{3+} substitution of Pb^{2+} in the perovskite on the photovoltaic characteristics has been investigated by current density-voltage (*J-V*) measurements performed in standard conditions. The *J-V* curves revealed that the cells with high Eu^{3+} concentration show a more pronounced hysteresis effect compared to $MAPbI_{2.6}Cl_{0.4}$ based cells, indicating higher defect concentration. Almost the same effect has been observed when Sb^{3+} was incorporated in the active layer. In addition, the decrease in PCE is not substantial, 3% Sb results in only a 0.48% decrease in PCE (or 6% of the standard cell efficiency) caused by a decrease in the open circuit voltage (V_{oc}) of 0.011 V and short-circuit current density (J_{sc}) of 1.3 mA cm^{-2}. At the same time, the solar cells show similar series resistance (R_s) and fill factor (FF), while the shunt resistance (R_{sh}) decreases by 9%. One possible explanation for the small reduction in performance of Sb-doped perovskite is the presence of several defects in the active layer [128–130]. Increasing the percentage of Sb leads to a decrease in efficiency. According to literature, solar cells with 100% Sb (where Pb^{2+} is completely replaced by Sb^{3+}) do not exceed 1% PCE.

4. Transport phenomena in light absorber materials: Impact on the measurement protocols

There is no definitive answer, yet, about how the halide perovskite solar cells work. A series of theories aiming to elucidate the peculiar characteristics of this type of device has been advanced to date. The prominent ones are delineated hereunder.

In 2013, Edri et al. [131] stated that halide perovskite solar cells work as a p-i-n device, where *i* is a near-intrinsic (low-doped) semiconductor acting as light absorber (the halide perovskite layer). The high-power conversion efficiency of the material is determined by the large diffusion lengths for both electrons and holes of 1.9 ± 0.1 and $1.5 \pm 0.2\,\mu m$, respectively. The p-i-n halide perovskite devices differ from the classical ones as the quasi-Fermi level splits only within the i layer and not in the adjacent n- and p-layers acting as the selective contacts. In the same year, Stranks et al. [132] indicated that the comparable values for diffusion coefficients and lengths can be explained either by the similar mobility of electrons and holes or by the fact that the predominant diffusion species are weakly bound excitons.

In 2014, Juarez-Perez et al. [133] showed that the halide perovskite material exhibits an apparent carrier-induced giant dielectric constant (GDC). The GDC in halide perovskite material is induced by the large freedom of rotation of the polar MA^+ cations. The dipolar domains tend to align to the external field, and when carriers are injected, the polarizability is boosted by the free charges from the conduction and the valence bands. In the same year, Frost et al. [134] revealed that the halide perovskite material contains ferroelectric domains at the nanometer scale which, due to their built-in electric fields, promote the separation of the light-generated carriers. Due to the small dimensions, the probability of carrier recombination within the ferroelectric domains is highly reduced with respect to "classical" p-n heterojunctions, hence the large diffusion lengths of the carriers in these types of materials. Baumann et al. [135] demonstrated that V_{oc} decay has two very different regimes: a fast one, within 10–100 μs, corresponding to carriers with a short lifetime and a second one, within 1–100 ms, revealing the very long lifetime of the remaining carriers. The intriguing dependence of the V_{oc} decay with the light irradiance intensity, in the short timescales, could be explained by the light-induced polarization showed by Juarez-Perez et al. [133] and the presence of ferroelectric domains predicted by Frost et al. [134]. The second regime (long lifetimes), linked to the recombination of the slowly de-trapped carriers, is characterized by a V_{oc} decay independent of illumination.

Hoke et al. [136] showed that the chemical instability of halide perovskite can be photo-induced. As a result of light-soaking, iodine-enriched domains are formed which are acting as carrier traps, facilitating the radiative electron-hole recombination. The formation of halide segregation is reversed once the traps are filled with holes, and thus, the lattice strain makes the segregated phase to relax into initial crystalline structure.

A "dynamical" bandgap of the halide perovskite is proposed first in 2015, by Motta et al., and later, in 2016, by Wang et al. [137,138], meaning that under certain conditions of molecular alignment, the bandgap of halide perovskite can alternate between a direct and an indirect one. This property, induced by the molecular motion, suppresses the carrier recombination while enhancing the exciton separation. Such phenomenology could explain the extended lifetimes of the photo-generated carriers (large diffusion lengths) and the high conversion efficiency observed for $MAPbI_3$.

In 2017, Jacobs et al. [139] indicated that the ion migration and accumulation toward the halide perovskite interfaces with the transport layers disturb the charge distribution in the solar cell and, thus, affects the photovoltaic current. The accumulation of ions at the interfaces depends on the cell's bias history and it can be either "negative" or "positive." Negative accumulation designates the case when the negatively charged ions gather at the electron transporter layer side while the holes accumulate at the hole transporter side. The ions in negative accumulation screen the build-in electric field responsible for charge extraction while also affecting the heterojunction interfaces between halide perovskite and selective blocking layers leading to a reduced short-circuit current and open-circuit voltage, and therefore, decreasing the PCE. On the contrary, the positive accumulation appears if the samples are kept some time under voltages higher than V_{oc} and it manifests in instantaneously higher J_{sc}, V_{oc}, and FF during *J-V* reverse scans.

Tsai et al. [140] advanced the theory that the interaction with light can cause lattice expansion in halide perovskites. This effect is possible because the photo-generated electrons and holes can populate the bonding and antibonding states resulting in weak covalent bonds, leading thus to either less-distorted Pb—I—Pb bonds or elongated Pb—I bonds. The lattice expansion is independent of ions migration and has a direct positive impact over the V_{oc}, FF, and therefore, PCE. In 2019, nanoscale ferroelastic domains were empirically evidenced in halide perovskite thin films by piezoresponse force microscopy [141]. The piezoelectricity in halide perovskites is enhanced under illumination with white LED light and is diminishing under dark or high temperature (it almost vanishes at 72° C). The disappearing of ferroelastic domains under dark or thermal annealing is a reversible phenomenon. The switching of the domains when different polarities of bias voltages are applied cannot be assigned with ferroelectricity but with ion migration through the halide perovskite, which leads to the generation of elastic stress in the structure, affecting thus the ferroelastic domains. It was experimentally demonstrated that the movement of organic ions causes larger strain than the halide anions.

The general consensus is that the halide perovskite solar cells are much different than the ideal p-n junction photovoltaic cells. The nonideality of the halide perovskite solar cell is related to the transport phenomena, which governs the halide perovskite material (according to the theories mentioned above: ion migration, charge trapping/detrapping, light-induced instability, dipolar domains, ferroelectricity, ferroelasticity, lattice expansion). Ultimately, the complexity of halide perovskite physics is reflected in the *J-V* characteristics which are consistently affected by the dynamic hysteresis phenomena [142–144]. For field-effect transistor structures where the halide perovskite plays the role of semiconductor channel, the hysteresis observed in *J-V* curves is explained by ion migration theory [145,146]. The hysteretic effects pose a challenge for an accurate determination of the PCE because the *J-V* characteristics can be altered by factors like bias prepolling, bias scan rate, and light exposure prior to measurement [147]. For example, the PCE determined from reverse *J-V* scans will be increasingly overestimated, as the initial voltage applied in the *J-V* scan overpasses the V_{oc}. At the same time, the hysteresis is maximized for intermediate bias scan rates and vanishes for very fast or very slow scan rates. Light exposure prior measurement can also under- or overestimate the PCE by either degrading halide perovskite material (causing halogen segregation) or by enhancing the charge separation and making unfavorable the carrier recombination.

In order to explain the peculiar *J-V* characteristics for halide perovskite cells, Nemnes et al. developed a dynamic electrical model (DEM) [148] which, in addition to the static one, introduce a nonlinear parallel capacitance in accordance with a non-linear polarization (P_{nl}) resulting from the response of the charge to the light-induced effects and electric field within halide perovskite or at its interfaces with the electrodes. The equivalent circuit of DEM is given in Fig. 10.

The measured current I is given by Eq. (3):

$$I = I_{ph} - I_{d} - I_{sh} - I_{C} \tag{3}$$

where $I_c = C_0 \times dU_c(t)/t + A \times dP_{nl}(t)/dt$; $C_0 = \varepsilon \times A/d$ is the geometrical capacitance; $U_c = V + I \times R_s$ is the bias on the capacitor; A is the capacitor area; $P_{nl}(t)$ is the nonlinear polarization accounting for the dynamic accumulation of charges (ions, electrons holes) inside the device and at the interfaces. As shown by the authors [148], the contribution of $P_{nl}(t)$ to the measured current has a crucial role in producing the observed *J-V* hysteresis.

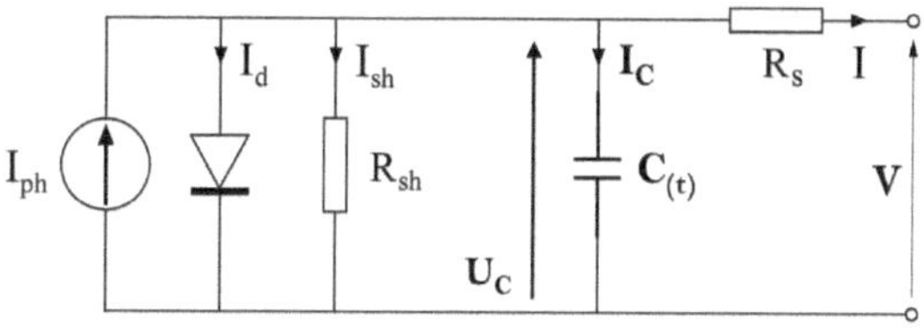

FIG. 10 Schematics of the equivalent circuit for a perovskite solar cell, with the standard elements: current source, a diode element, a series, and a shunt resistance. In addition, a time-dependent (i.e., voltage-dependent) capacitance is introduced to model the hysteretic effects. *From G.A. Nemnes, C. Besleaga, A.G. Tomulescu, I. Pintilie, L. Pintilie, K. Torfason, A. Manolescu, Dynamic electrical behavior of halide perovskite based solar cells, Solar Energy Mater. Solar Cells 159 (2017) 197–203. https://doi.org/10.1016/j.solmat.2016.09.012.*

The dynamic model describes in detail the hysteresis of the *J-V* curve, either normal (NH), inverted (IH), or mixt (MH) on the same PSCs samples [149]. Examples are given in Figs. 11 and 12 where the electrical response of the halide perovskite cell under different bias scan rates (from 5mV/s. to 100V/s) and upon several prepoling conditions (from −5V to 2V), respectively, are very well described by the DEM fits [148–150]. For $V_{pol} > V_{oc}$, one obtains NH, while for $V_{pol} < 0$, IH is found. When the prepoling voltage lies between 0V and V_{oc}, a mixed hysteresis typically appears. Both types of hysteretic effects are reproduced by the DEM

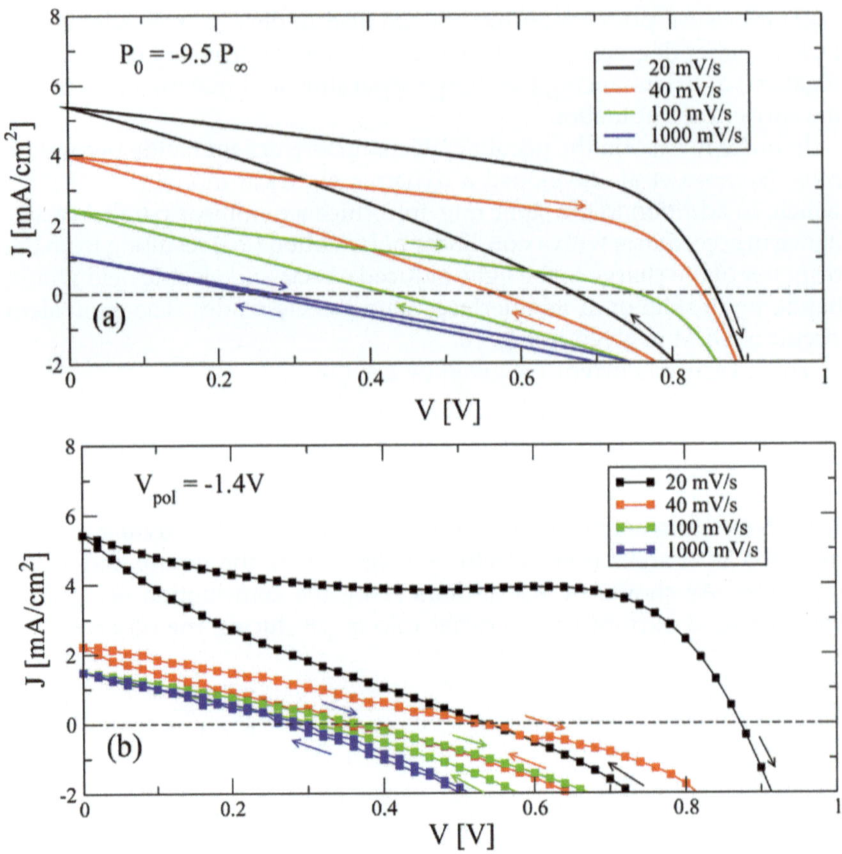

FIG. 11 *J-V* characteristics showing inverted hysteresis: (A) numerical simulations using a negative initial polarization, where P_∞ corresponds to steady state polarization; (B) *J-V* measurements, using $V_{pol} = -1.4$V. The same sequence of scan rates was considered: 20, 40, 100, 1000mV/s. A systematic decrease in J_{sc} is observed. *From G.A. Nemnes, C. Besleaga, V. Stancu, D.E. Dogaru, L.N. Leonat, L. Pintilie, K. Torfason, M. Ilkov, A. Manolescu, I. Pintilie, Normal and inverted hysteresis in perovskite solar cells, J. Phys. Chem. C 121 (21) (2017) 11207–11214. https://doi.org/10.1021/acs.jpcc.7b04248.*

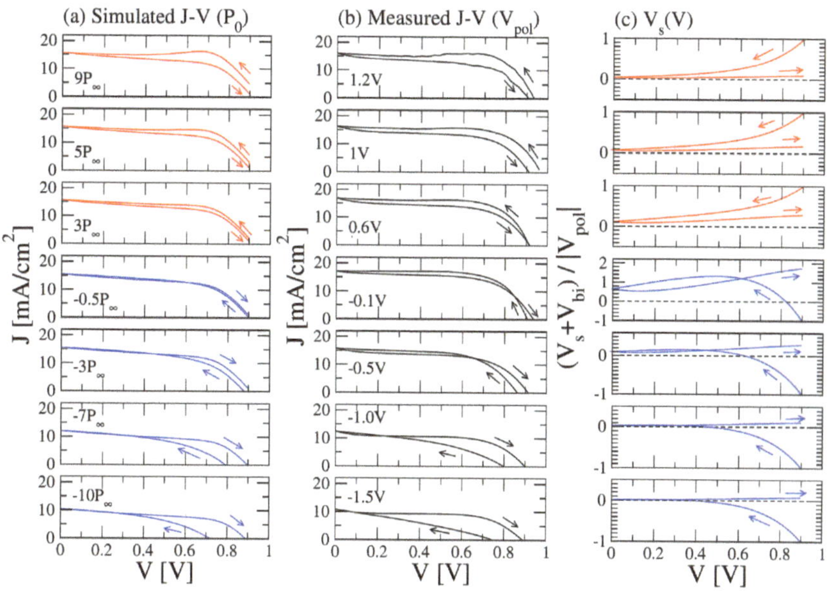

FIG. 12 Simulated and measured *J-V* characteristics, indicating the possibility to tune the hysteresis from normal to inverted hysteresis, based on the initial prepoling: (A) simulations (changing P_0); (B) measurements (changing V_{pol}); (C) the variation of the surface polarization voltage V_s, where $V_{pol}=\chi_s V_{oc}|P_0|/P_\infty$, V_{bi} is the built-in voltage and χ_s is an interface-dependent parameter. *From G.A. Nemnes, C. Besleaga, V. Stancu, D.E. Dogaru, L.N. Leonat, L. Pintilie, K. Torfason, M. Ilkov, A. Manolescu, I. Pintilie, Normal and inverted hysteresis in perovskite solar cells, J. Phys. Chem. C 121 (21) (2017) 11207–11214. https://doi.org/10.1021/acs.jpcc.7b04248.*

model, by tuning a single parameter (initial polarization—P_0), which corresponds to the prepoling voltage (V_{pol}) applied in a typical measurement. The DEM reproduces the different hysteretic behaviors by changing only the initial polarization parameter, related to experimental V_{pol}.

It is worth discussing the appearance of both NH and IH in the same sample due to different poling conditions. Nemnes et al. [150] showed that the inverted hysteresis can be found applying negative poling voltages. The band diagrams during the solar cell operation are depicted in Fig. 13 for both NH and IH.

During poling stage, the ionic and free carrier charges accumulate at the interfaces (opposite behavior for NH and IH cases) leading to strong electric fields at the two interfaces, while the electric field inside the perovskite is rather small. Subsequently, as the voltage is set to V_{oc} and the reverse scan begins, an electric field is established within the perovskite, as the ions do not redistribute immediately. The orientation of the electric field is different for NH and IH. Besides the discharge of the two capacitors associated with the interfaces, one should note that in the case of NH,

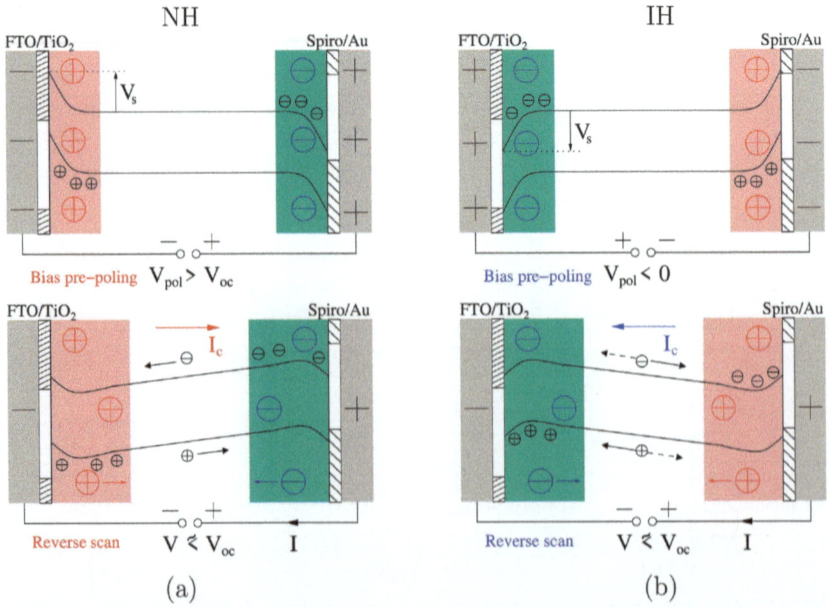

FIG. 13 Band-bending under operation conditions for normal and inverted hysteresis. The hysteresis type is set by the sign of V_{pol}: (A) $V_{pol} > V_{oc}$ (NH) and (B) $V_{pol} < 0$ (IH). The electrons/holes and positive/negative ions first accumulate at the respective interfaces in accordance with V_{pol}. Then, as the reverse scan progresses (starting with $V = V_{oc}$), the ion charges are slowly drifting back to the bulk of the perovskite. The established electric field in the perovskite changes during the scan and it can enhance or diminish the charge collection. *From G.A. Nemnes, C. Besleaga, V. Stancu, D.E. Dogaru, L.N. Leonat, L. Pintilie, K. Torfason, M. Ilkov, A. Manolescu, I. Pintilie, Normal and inverted hysteresis in perovskite solar cells, J. Phys. Chem. C 121 (21) (2017) 11207–11214. https://doi.org/10.1021/acs.jpcc.7b04248.*

the electric field enhances the electron collection into the TiO_2 layer, while in the case of IH, the electric field initially opposes the charge collection. This leads to the appearance of the bump in the current (low recombination) in the NH case, while the current is lower (even opposite sign) in the reverse scan for the IH poling conditions (see Figs. 11 and 12). As the scan progresses toward low voltages, the ions redistribute and the normal operation is established for the IH case.

To conclude, the initial poling of PSCs creates conditions for charge accumulations (free carriers and ionic) which influence the dynamic *J-V* characteristics by tuning the recombination in the bulk of the perovskite and/or at the interfaces (see Fig. 14).

Fig. 15 shows how PCE is actually influenced by the prepoling of the samples, even when the samples are measured at a low bias scan rate

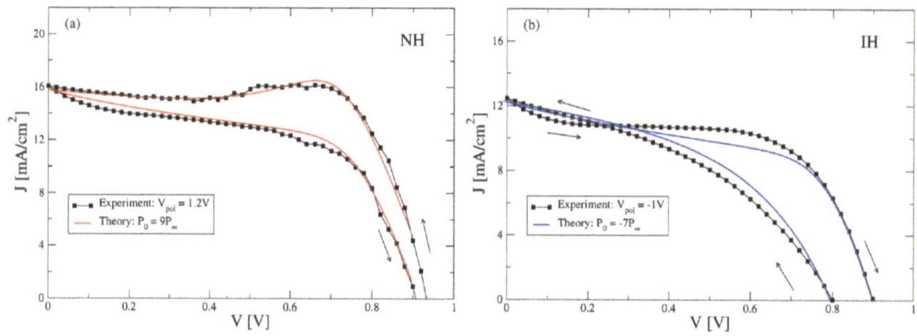

FIG. 14 Simulated and measured *J-V* characteristics, subsequent to prepoling phase, presenting NH and IH: (A) $V_{pol}=1.2\,V$, corresponding to $P_0=9P_\infty$; (B) $V_{pol}=-1\,V$, corresponding to $P_0=-7P_\infty$. *From G.A. Nemnes, C. Besleaga, V. Stancu, D.E. Dogaru, L.N. Leonat, L. Pintilie, K. Torfason, M. Ilkov, A. Manolescu, I. Pintilie, Normal and inverted hysteresis in perovskite solar cells, J. Phys. Chem. C 121 (21) (2017) 11207–11214. https://doi.org/10.1021/acs.jpcc.7b04248.*

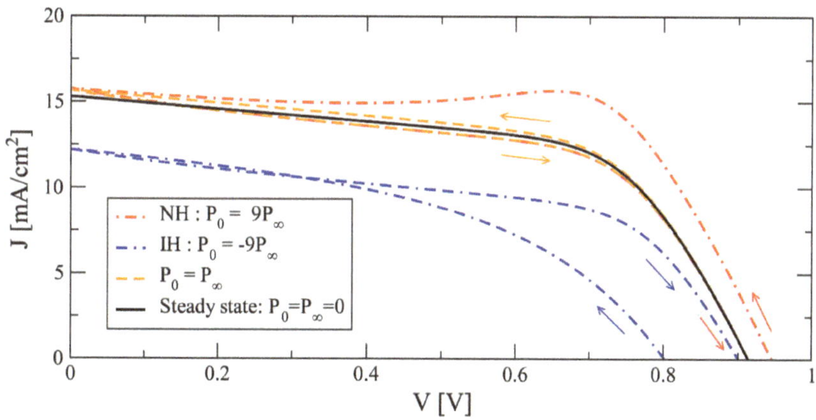

FIG. 15 The PCE depends on the prepoling conditions, showing in the reverse scan an increased value for NH (10.75%) and a decreased value for IH (4.47%), compared to the steady-state scan (8.44%). *From G.A. Nemnes, C. Besleaga, V. Stancu, D.E. Dogaru, L.N. Leonat, L. Pintilie, K. Torfason, M. Ilkov, A. Manolescu, I. Pintilie, Normal and inverted hysteresis in perovskite solar cells, J. Phys. Chem. C 121 (21) (2017) 11207–11214. https://doi.org/10.1021/acs.jpcc.7b04248.*

in the bias range $[0, V_{oc}]$. Four cases are represented in the figure, corresponding to different prepoling conditions: $P_0=9P_\infty$ (NH), $P_0=-9P_\infty$ (IH), $P_0=P_\infty$ (minimal polarization case), all for a 20 mV/s bias scan rate, and $P_\infty=0$ (the steady-state case). The PCE varies substantially in forward and reverse scans leading to important overestimations for NH and underestimations for IH. For the minimal polarization case, however, only a small hysteresis remains, and this happens because $P_\infty \neq 0$.

The study of hysteresis behavior in halide perovskite solar cells led to the following set of guidelines for a proper characterization of these types of devices:

- The bias scan rate should be as low as possible (down to a value at which the halide perovskite degradation in time does not take place).
- The applied bias should not exceed the V_{oc} or be lower than 0 V because it can introduce strong positive or negative polarization, leading then to an over- or underestimation of the PCE.
- Consecutive forward (from 0 V to V_{oc})–reverse (from V_{oc} to 0 V) and reverse-forward bias sweeps are recommended. Generally, the current measured in the "reverse" direction overestimates the PCE (NH) and the one measured in the "forward" direction underestimates it. However, if negative biases are applied, the opposite can happen (IH).
- When performing a "reverse-forward" sweep, the starting voltage should be the stabilized V_{oc} (following V_{oc} decay and start when its value is constant); similarly, for the "forward-reverse" sweep, the I_{sc} current should be stabilized before starting the *I-V* measurement.

5. Concluding remarks

This chapter covers the available knowledge on halide perovskite classifications, dimensions, applications, composition engineering, and transport phenomena in perovskites for solar cells application. As the literature is extremely rich, we have limited ourselves to mention only the selected findings.

Perovskite's materials have seen rapid advances over the past few years in many different fields. These materials can be classified according to different criteria, namely, the nature of the atoms forming the inorganic part (oxide/halide perovskite), the nature of the A cation (inorganic/organic), and their dimensionality (0D, 1D, 2D, or 3D). Due to their unique properties, perovskite materials have received much attention from scientists and industries, which has driven their use mainly in the photovoltaic field but also in many other applications, as X-ray detectors, lasers, LEDs, transistors, memories, and gas sensors. To explore more the potential of low-dimensional perovskite for these applications, further research is required targeting the material design, namely, the composition or dimensionality as well as the fundamental chemical and physical properties of halide perovskite. In addition, this will enable the application and integration of perovskite materials in new advanced technologies.

Regarding the application of perovskite materials in solar devices, numerous scientific challenges and issues must be overcome to develop industrial perovskite solar cells. Currently, the adoption of 3D/2D multidimensional perovskites materials is considered to be the most successful strategy. The investigation of new combinations of 2D and 3D halide perovskite materials and the study of their structural, electronic, and optical properties by using density functional theory (DFT) are desirable to achieve more stable and efficient PSCs.

Acknowledgments

This work was supported by the EEA Grants 2014–2021, under Project contract no. 36/2021.

References

[1] N.K. Elumalai, M.A. Mahmud, D. Wang, A. Uddin, Perovskite solar cells: progress and advancements, Energies 9 (11) (2016), https://doi.org/10.3390/en9110861.

[2] G. Rose, Beschreibung einiger neuen Mineralien des Urals, Ann. Phys. 124 (12) (1839) 551–573, https://doi.org/10.1002/andp.18391241205.

[3] Catalytic Chemistry of Solid State Inorganics. (1976). New York Academy of Sciences.

[4] M. Florea, M. Alifanti, V. Kuncser, D. Macovei, N. Apostol, P. Granger, V.I. Parvulescu, Evidence of A-B site cooperation in the $EuFeO_3$ perovskite from ^{151}Eu and ^{57}Fe Mössbauer spectroscopy, EXAFS, and toluene catalytic oxidation, J. Catal. 316 (2014), https://doi.org/10.1016/j.jcat.2014.04.016.

[5] M.A. Peña, J.L.G. Fierro, Chemical structures and performance of perovskite oxides, Chem. Rev. 101 (7) (2001) 1981–2018, https://doi.org/10.1021/cr980129f.

[6] W.-J. Yin, B. Weng, J. Ge, Q. Sun, Z. Li, Y. Yan, Oxide perovskites, double perovskites and derivatives for electrocatalysis, photocatalysis, and photovoltaics, Energ. Environ. Sci. 12 (2) (2019) 442–462, https://doi.org/10.1039/C8EE01574K.

[7] V.M. Goldschmidt, Die Gesetze der Krystallochemie, Naturwissenschaften 14 (21) (1926) 477–485, https://doi.org/10.1007/BF01507527.

[8] C.J. Bartel, C. Sutton, B.R. Goldsmith, R. Ouyang, C.B. Musgrave, L.M. Ghiringhelli, M. Scheffler, New tolerance factor to predict the stability of perovskite oxides and halides, Sci. Adv. 5 (2) (2019), https://doi.org/10.1126/sciadv.aav0693.

[9] C. Li, X. Lu, W. Ding, L. Feng, Y. Gao, Z. Guo, Formability of ABX 3 (X = F, Cl, Br, I) halide perovskites, Acta Crystallogr. Sect. B: Struct. Sci. 64 (6) (2008) 702–707, https://doi.org/10.1107/S0108768108032734.

[10] H. Kim, J.S. Han, J. Choi, S.Y. Kim, H.W. Jang, Halide perovskites for applications beyond photovoltaics, Small Methods 2 (3) (2018) 1700310, https://doi.org/10.1002/smtd.201700310.

[11] H.L. Wells, Über die Cäsium- und Kalium-Bleihalogenide, Z. Anorg. Chem. 3 (1893) 195–210, https://doi.org/10.1002/zaac.18930030124.

[12] W. Dieter, CH3NH3PbX3, a Pb(II)-system with cubic perovskite structure, Z. Naturforsch., B 33 (12) (1978) 1443–1445.

[13] D. Weber, C. Perovskite, CH3NH3SnBrI(x = 0–3), ein Sn (II) -System mit kubischer Perowskitstruktur, Z. Naturforsch., B 33 (Ii) (1978) 862–865.

[14] D.B. Mitzi, Synthesis, structure, and properties of organic-inorganic perovskites and related materials, in: Progress in Inorganic Chemistry, John Wiley & Sons, Ltd., 1999, pp. 1–121, https://doi.org/10.1002/9780470166499.ch1.

[15] D.B. Mitzi, Templating and structural engineering in organic-inorganic perovskites, J. Chem. Soc. Dalton Trans. 1 (2001) 1–12, https://doi.org/10.1039/b007070j.

[16] S.I. Seok, T.-F. Guo, Halide perovskite materials and devices, MRS Bull. 45 (6) (2020) 427–430, https://doi.org/10.1557/mrs.2020.140.

[17] A.K. Jena, A. Kulkarni, T. Miyasaka, Halide perovskite photovoltaics: background, status, and future prospects, Chem. Rev. 119 (5) (2019) 3036–3103, https://doi.org/10.1021/acs.chemrev.8b00539.
[18] K.W. Bagnall, J.B. Laidler, M.A.A. Stewart, Americium chloro-complexes, J. Chem. Soc. A 0 (1968) 133–136, https://doi.org/10.1039/J19680000133.
[19] E. Meyer, D. Mutukwa, N. Zingwe, R. Taziwa, Lead-free halide double perovskites: a review of the structural, optical, and stability properties as well as their viability to replace lead halide perovskites, Metals 8 (9) (2018), https://doi.org/10.3390/met8090667.
[20] L. Chu, W. Ahmad, W. Liu, J. Yang, R. Zhang, Y. Sun, J. Yang, X. Li, Lead-free halide double perovskite materials: a new superstar toward green and stable optoelectronic applications, Nano-Micro Lett. 11 (1) (2019) 1–18, https://doi.org/10.1007/s40820-019-0244-6.
[21] C.A. Bremner, M. Simpson, W.T.A. Harrison, New moleculara perovskites: cubic C(4)N(2)H(12).NH(4)Cl(3).H(2)O and 2-H hexagonal C(6)N(2)H(14).NH(4)Cl(3), J. Am. Chem. Soc. (2002) 10960, https://doi.org/10.1021/ja027484e.
[22] H.Y. Ye, Y.Y. Tang, P.F. Li, W.Q. Liao, J.X. Gao, X.N. Hua, H. Cai, P.P. Shi, Y.M. You, R.G. Xiong, Metal-free three-dimensional perovskite ferroelectrics, Science 361 (6398) (2018) 151–155, https://doi.org/10.1126/science.aas9330.
[23] J. Yang, P. Zhang, J. Wang, S.H. Wei, Theoretical investigation of halide perovskites for solar cell and optoelectronic applications, Chin. Phys. B 29 (10) (2020), https://doi.org/10.1088/1674-1056/abb3f6.
[24] G. Kieslich, S. Sun, A.K. Cheetham, An extended tolerance factor approach for organic-inorganic perovskites, Chem. Sci. 6 (6) (2015) 3430–3433, https://doi.org/10.1039/c5sc00961h.
[25] S. Nagane, U. Bansode, O. Game, S. Chhatre, S. Ogale, CH3NH3PbI(3−x)(BF4)x: molecular ion substituted hybrid perovskite, Chem. Commun. 50 (68) (2014) 9741–9744, https://doi.org/10.1039/c4cc04537h.
[26] T. Bu, X. Liu, Y. Zhou, J. Yi, X. Huang, L. Luo, J. Xiao, Z. Ku, Y. Peng, F. Huang, Y.B. Cheng, J. Zhong, A novel quadruple-cation absorber for universal hysteresis elimination for high efficiency and stable perovskite solar cells, Energ. Environ. Sci. 10 (12) (2017) 2509–2515, https://doi.org/10.1039/c7ee02634j.
[27] J. Zhang, M.H. Shang, P. Wang, X. Huang, J. Xu, Z. Hu, Y. Zhu, L. Han, N-type doping and energy states tuning in CH3NH3Pb1-xSb2x/3I3 Perovskite solar cells, ACS Energy Lett. 1 (3) (2016) 535–541, https://doi.org/10.1021/acsenergylett.6b00241.
[28] C.C. Stoumpos, C.M.M. Soe, H. Tsai, W. Nie, J.C. Blancon, D.H. Cao, F. Liu, B. Traoré, C. Katan, J. Even, A.D. Mohite, M.G. Kanatzidis, High members of the 2D ruddlesden-popper halide perovskites: synthesis, optical properties, and solar cells of (CH3(CH2)3NH3)2(CH3NH3)4Pb5I16, Chem 2 (3) (2017) 427–440, https://doi.org/10.1016/j.chempr.2017.02.004.
[29] Z.Z. Wang, Q. Lin, F.P. Chmiel, N. Sakai, L.M. Herz, H.J. Snaith, Efficient ambient-air-stable solar cells with 2D-3D heterostructured butylammonium-caesium-formamidinium lead halide perovskites, Nat. Energy 2 (9) (2017) 1–10, https://doi.org/10.1038/nenergy.2017.135.
[30] Y. Tabuchi, K. Asai, M. Rikukawa, K. Sanui, K. Ishigure, Preparation and characterization of natural lower dimensional layered perovskite-type compounds, J. Phys. Chem. Solid 61 (6) (2000) 837–845, https://doi.org/10.1016/S0022-3697(99)00402-3.
[31] S. Elleuch, H. Boughzala, A. Driss, Y. Abid, A one-dimensional organic-inorganic hybrid compound [(CH3) 2C=NHCH2CH2CH3][PbI3], Acta Crystallogr. Sect. E: Struct. Rep. Online 63 (2) (2007) 306–308, https://doi.org/10.1107/S160053680605392X.

[32] E. Samet Kallel, H. Boughzala, A. Driss, Y. Abid, The one-dimensional organic inorganic hybrid compound poly[(diethyl-ene- triamine)tetra - Iodido-dilead(II)], Acta Crystallogr., Sect. E: Struct. Rep. Online 64 (7) (2008), https://doi.org/10.1107/S1600536808013913.

[33] H. Dammak, A. Yangui, S. Triki, Y. Abid, H. Feki, Structural characterization, vibrational, optical properties and DFT investigation of a new luminescent organic-inorganic material: (C6H14N)3Bi2I9, JOL 161 (2015) 214–220, https://doi.org/10.1016/j.jlumin.2015.01.010.

[34] A. Samet, A.B. Ahmed, A. Mlayah, H. Boughzala, E.K. Hlil, Y. Abid, Optical properties and ab initio study on the hybrid organic-inorganic material [(CH3)2NH2]3[BiI 6], J. Mol. Struct. 977 (1–3) (2010) 72–77, https://doi.org/10.1016/j.molstruc.2010.05.016.

[35] W. Ke, G. Fang, J. Wang, P. Qin, H. Tao, H. Lei, Q. Liu, X. Dai, X. Zhao, Perovskite solar cell with an efficient TiO2 compact film, ACS Appl. Mater. Interfaces 6 (18) (2014) 15959–15965, https://doi.org/10.1021/am503728d.

[36] G.E. Eperon, S.D. Stranks, C. Menelaou, M.B. Johnston, L.M. Herz, H.J. Snaith, Formamidinium lead trihalide: a broadly tunable perovskite for efficient planar heterojunction solar cells, Energ. Environ. Sci. 7 (3) (2014) 982–988, https://doi.org/10.1039/c3ee43822h.

[37] L.J. Phillips, A.M. Rashed, R.E. Treharne, J. Kay, P. Yates, I.Z. Mitrovic, A. Weerakkody, S. Hall, K. Durose, Maximizing the optical performance of planar CH3NH3PbI3 hybrid perovskite heterojunction stacks, Sol. Energy Mater. Sol. Cells 147 (2016) 327–333, https://doi.org/10.1016/j.solmat.2015.10.007.

[38] L.K. Ono, S.R. Raga, S. Wang, Y. Kato, Y. Qi, Temperature-dependent hysteresis effects in perovskite-based solar cells, J. Mater. Chem. A 3 (17) (2015) 9074–9080, https://doi.org/10.1039/c4ta04969a.

[39] L. Zheng, D. Zhang, Y. Ma, Z. Lu, Z. Chen, S. Wang, L. Xiao, Q. Gong, Morphology control of the perovskite films for efficient solar cells, Dalton Trans. 44 (23) (2015) 10582–10593, https://doi.org/10.1039/c4dt03869j.

[40] A. Krishna, S. Gottis, M.K. Nazeeruddin, F. Sauvage, Mixed dimensional 2D/3D hybrid perovskite absorbers: the future of perovskite solar cells? Adv. Funct. Mater. 29 (8) (2019) 1–20, https://doi.org/10.1002/adfm.201806482.

[41] A. Mahapatra, D. Prochowicz, M.M. Tavakoli, S. Trivedi, P. Kumar, P. Yadav, A review of aspects of additive engineering in perovskite solar cells, J. Mater. Chem. A 8 (1) (2020) 27–54, https://doi.org/10.1039/c9ta07657c.

[42] I.C. Smith, E.T. Hoke, D. Solis-Ibarra, M.D. McGehee, H.I. Karunadasa, A layered hybrid perovskite solar-cell absorber with enhanced moisture stability, Angew. Chem. 126 (42) (2014) 11414–11417, https://doi.org/10.1002/ange.201406466.

[43] C. Ortiz-Cervantes, P. Carmona-Monroy, D. Solis-Ibarra, Two-dimensional halide perovskites in solar cells: 2D or not 2D? ChemSusChem 12 (8) (2019) 1560–1575, https://doi.org/10.1002/cssc.201802992.

[44] K.J. Prince, M. Nardone, S.P. Dunfield, G. Teeter, M. Mirzokarimov, E.L. Warren, D.T. Moore, J.J. Berry, C.A. Wolden, L.M. Wheeler, Complementary interface formation toward high-efficiency all-back-contact perovskite solar cells, Cell Rep. Phys. Sci. (2021) 100363, https://doi.org/10.1016/j.xcrp.2021.100363.

[45] W. Chi, S.K. Banerjee, Stability improvement of perovskite solar cells by compositional and interfacial engineering, Chem. Mater. 33 (5) (2021) 1540–1570, https://doi.org/10.1021/acs.chemmater.0c04931.

[46] J. Li, R. Xia, W. Qi, X. Zhou, J. Cheng, Y. Chen, G. Hou, Y. Ding, Y. Li, Y. Zhao, X. Zhang, Encapsulation of perovskite solar cells for enhanced stability: structures, materials and characterization, J. Power Sources 485 (December 2020) (2021) 229313, https://doi.org/10.1016/j.jpowsour.2020.229313.

[47] A. Urbina, The balance between efficiency, stability and environmental impacts in perovskite solar cells: a review, J. Phys. Energy 2 (2) (2020), https://doi.org/10.1088/2515-7655/ab5eee.
[48] K. Yao, X. Wang, F. Li, L. Zhou, Mixed perovskite based on methyl-ammonium and polymeric-ammonium for stable and reproducible solar cells, Chem. Commun. 51 (84) (2015) 15430–15433, https://doi.org/10.1039/c5cc05879a.
[49] G. Grancini, C. Roldán-Carmona, I. Zimmermann, E. Mosconi, X. Lee, D. Martineau, S. Narbey, F. Oswald, F. De Angelis, M. Graetzel, M.K. Nazeeruddin, One-year stable perovskite solar cells by 2D/3D interface engineering, Nat. Commun. 8 (2017) 1–8, https://doi.org/10.1038/ncomms15684.
[50] C. Ma, C. Leng, Y. Ji, X. Wei, K. Sun, L. Tang, J. Yang, W. Luo, C. Li, Y. Deng, S. Feng, J. Shen, S. Lu, C. Du, H. Shi, 2D/3D perovskite hybrids as moisture-tolerant and efficient light absorbers for solar cells, Nanoscale 8 (43) (2016) 18309–18314, https://doi.org/10.1039/c6nr04741f.
[51] P. Chen, Y. Bai, S. Wang, M. Lyu, J.H. Yun, L. Wang, In situ growth of 2D perovskite capping layer for stable and efficient perovskite solar cells, Adv. Funct. Mater. 28 (17) (2018) 1–10, https://doi.org/10.1002/adfm.201706923.
[52] K.T. Cho, G. Grancini, Y. Lee, E. Oveisi, J. Ryu, O. Almora, M. Tschumi, P.A. Schouwink, G. Seo, S. Heo, J. Park, J. Jang, S. Paek, G. Garcia-Belmonte, M.K. Nazeeruddin, Selective growth of layered perovskites for stable and efficient photovoltaics, Energ. Environ. Sci. 11 (4) (2018) 952–959, https://doi.org/10.1039/c7ee03513f.
[53] H. Min, M. Kim, S.U. Lee, H. Kim, G. Kim, K. Choi, J.H. Lee, S.I. Seok, Efficient, stable solar cells by using inherent bandgap of a-phase formamidinium lead iodide, Science 366 (6466) (2019) 749–753, https://doi.org/10.1126/science.aay7044.
[54] T. Niu, J. Lu, X. Jia, Z. Xu, M.C. Tang, D. Barrit, N. Yuan, J. Ding, X. Zhang, Y. Fan, T. Luo, Y. Zhang, D.M. Smilgies, Z. Liu, A. Amassian, S. Jin, K. Zhao, S. Liu, Interfacial engineering at the 2D/3D heterojunction for high-performance perovskite solar cells, Nano Lett. 19 (10) (2019) 7181–7190, https://doi.org/10.1021/acs.nanolett.9b02781.
[55] J.J. Yoo, S. Wieghold, M.C. Sponseller, M.R. Chua, S.N. Bertram, N.T.P. Hartono, J.S. Tresback, E.C. Hansen, J.P. Correa-Baena, V. Bulović, T. Buonassisi, S.S. Shin, M.G. Bawendi, An interface stabilized perovskite solar cell with high stabilized efficiency and low voltage loss, Energ. Environ. Sci. 12 (7) (2019) 2192–2199, https://doi.org/10.1039/c9ee00751b.
[56] Q. Zhou, L. Liang, J. Hu, B. Cao, L. Yang, T. Wu, X. Li, B. Zhang, P. Gao, High-performance perovskite solar cells with enhanced environmental stability based on a (p-FC 6H4C2H4NH3)2 [PbI4] capping layer, Adv. Energy Mater. 9 (12) (2019) 1–11, https://doi.org/10.1002/aenm.201802595.
[57] K. Chondroudis, D.B. Mitzi, Electroluminescence from an organic-inorganic perovskite incorporating a quaterthiophene dye within lead halide perovskite layers, Chem. Mater. 11 (11) (1999) 3028–3030, https://doi.org/10.1021/cm990561t.
[58] M. Era, S. Morimoto, T. Tsutsui, S. Saito, Organic-inorganic heterostructure electroluminescent device using a layered perovskite semiconductor (C6H5C2H 4NH3)2PbI4, Appl. Phys. Lett. 65 (6) (1994) 676–678, https://doi.org/10.1063/1.112265.
[59] T. Hattori, T. Taira, M. Era, T. Tsutsui, S. Saito, Highly efficient electroluminescence from a heterostructure device combined with emissive layered-perovskite and an electron-transporting organic compound, Chem. Phys. Lett. 254 (1–2) (1996) 103–108, https://doi.org/10.1016/0009-2614(96)00310-7.
[60] X. Hong, T. Ishihara, A.V. Nurmikko, Photoconductivity and electroluminescence in lead iodide based natural quantum well structures, Solid State Commun. 84 (6) (1992) 657–661, https://doi.org/10.1016/0038-1098(92)90210-Z.
[61] K. Lin, J. Xing, L.N. Quan, F.P.G. de Arquer, X. Gong, J. Lu, L. Xie, W. Zhao, D. Zhang, C. Yan, W. Li, X. Liu, Y. Lu, J. Kirman, E.H. Sargent, Q. Xiong, Z. Wei, Perovskite

light-emitting diodes with external quantum efficiency exceeding 20 per cent, Nature 562 (7726) (2018) 245–248, https://doi.org/10.1038/s41586-018-0575-3.

[62] W. Xu, Q. Hu, S. Bai, C. Bao, Y. Miao, Z. Yuan, T. Borzda, A.J. Barker, E. Tyukalova, Z. Hu, M. Kawecki, H. Wang, Z. Yan, X. Liu, X. Shi, K. Uvdal, M. Fahlman, W. Zhang, M. Duchamp, F. Gao, Rational molecular passivation for high-performance perovskite light-emitting diodes, Nat. Photonics 13 (6) (2019) 418–424, https://doi.org/10.1038/s41566-019-0390-x.

[63] F. Deschler, M. Price, S. Pathak, L.E. Klintberg, D.D. Jarausch, R. Higler, S. Hüttner, T. Leijtens, S.D. Stranks, H.J. Snaith, M. Atatüre, R.T. Phillips, R.H. Friend, High photoluminescence efficiency and optically pumped lasing in solution-processed mixed halide perovskite semiconductors, J. Phys. Chem. Lett. 5 (8) (2014) 1421–1426, https://doi.org/10.1021/jz5005285.

[64] World Energy Consumption by Energy Source 2050 | Statista. (n.d.).

[65] EIA Projects 50% Increase of World Energy Consumption by 2050 Led by Growth in Asia—Energy Global News. (n.d.).

[66] N.J. Jeon, J.H. Noh, Y.C. Kim, W.S. Yang, S. Ryu, S.I. Seok, Solvent engineering for high-performance inorganic–organic hybrid perovskite solar cells, Nat. Mater. 13 (9) (2014) 897–903, https://doi.org/10.1038/nmat4014.

[67] M.M. Lee, J. Teuscher, T. Miyasaka, T.N. Murakami, H.J. Snaith, Efficient hybrid solar cells based on meso-superstructured organometal halide perovskites, Science 338 (6107) (2012) 643–647, https://doi.org/10.1126/science.1228604.

[68] M. Kaltenbrunner, G. Adam, E.D. Głowacki, M. Drack, R. Schwödiauer, L. Leonat, D.H. Apaydin, H. Groiss, M.C. Scharber, M.S. White, N.S. Sariciftci, S. Bauer, Flexible high power-per-weight perovskite solar cells with chromium oxide–metal contacts for improved stability in air, Nat. Mater. 14 (10) (2015) 1032–1039, https://doi.org/10.1038/nmat4388.

[69] D. Dong, D. Fang, H. Li, C. Zhu, X. Zhao, J. Li, L. Jin, L. Xie, L. Chen, J. Zhao, H. Zhang, W. Huang, C – H direct arylated 6H-indolo[2,3-b]quinoxaline derivative as a thickness-dependent hole-injection layer, Chem. Asian J. 12 (8) (2017) 920–926, https://doi.org/10.1002/asia.201700112.

[70] M. Fathollahi, M. Ameri, E. Mohajerani, E. Mehrparvar, M. Babaei, Organic/organic heterointerface engineering to boost carrier injection in OLEDs, Sci. Rep. 7 (1) (2017) 42787, https://doi.org/10.1038/srep42787.

[71] Y. Feng, J. Xu, H. Shan, L. Dong, X. Sun, Q. Hu, Y. Wang, V.A.L. Roy, Z.-X. Xu, Organometal halide perovskite as hole injection enhancer in organic light-emitting diode, Org. Electron. 51 (2017) 257–263, https://doi.org/10.1016/j.orgel.2017.09.027.

[72] Q.A. Akkerman, V. D'Innocenzo, S. Accornero, A. Scarpellini, A. Petrozza, M. Prato, L. Manna, Tuning the optical properties of cesium lead halide perovskite nanocrystals by anion exchange reactions, J. Am. Chem. Soc. 137 (32) (2015) 10276–10281, https://doi.org/10.1021/jacs.5b05602.

[73] R.J. Sutton, G.E. Eperon, L. Miranda, E.S. Parrott, B.A. Kamino, J.B. Patel, M.T. Hörantner, M.B. Johnston, A.A. Haghighirad, D.T. Moore, H.J. Snaith, Bandgap-tunable cesium lead halide perovskites with high thermal stability for efficient solar cells, Adv. Energy Mater. 6 (8) (2016) 1502458, https://doi.org/10.1002/aenm.201502458.

[74] N.J. Jeon, J.H. Noh, W.S. Yang, Y.C. Kim, S. Ryu, J. Seo, S.I. Seok, Compositional engineering of perovskite materials for high-performance solar cells, Nature 517 (7535) (2015) 476–480, https://doi.org/10.1038/nature14133.

[75] M. Saliba, S. Orlandi, T. Matsui, S. Aghazada, M. Cavazzini, J.-P. Correa-Baena, P. Gao, R. Scopelliti, E. Mosconi, K.-H. Dahmen, F. De Angelis, A. Abate, A. Hagfeldt, G. Pozzi, M. Graetzel, M.K. Nazeeruddin, A molecularly engineered hole-transporting material for efficient perovskite solar cells, Nat. Energy 1 (2) (2016) 15017, https://doi.org/10.1038/nenergy.2015.17.

[76] J.-W. Lee, D.-H. Kim, H.-S. Kim, S.-W. Seo, S.M. Cho, N.-G. Park, Formamidinium and cesium hybridization for photo- and moisture-stable perovskite solar cell, Adv. Energy Mater. 5 (20) (2015) 1501310, https://doi.org/10.1002/aenm.201501310.
[77] M. Saliba, T. Matsui, J.-Y. Seo, K. Domanski, J.-P. Correa-Baena, M.K. Nazeeruddin, S.M. Zakeeruddin, W. Tress, A. Abate, A. Hagfeldt, M. Grätzel, Cesium-containing triple cation perovskite solar cells: improved stability, reproducibility and high efficiency, Energ. Environ. Sci. 9 (6) (2016) 1989–1997, https://doi.org/10.1039/C5EE03874J.
[78] H. Tan, A. Jain, O. Voznyy, X. Lan, F.P. García de Arquer, J.Z. Fan, R. Quintero-Bermudez, M. Yuan, B. Zhang, Y. Zhao, F. Fan, P. Li, L.N. Quan, Y. Zhao, Z.-H. Lu, Z. Yang, S. Hoogland, E.H. Sargent, Efficient and stable solution-processed planar perovskite solar cells via contact passivation, Science 355 (6326) (2017) 722–726, https://doi.org/10.1126/science.aai9081.
[79] C. Yi, J. Luo, S. Meloni, A. Boziki, N. Ashari-Astani, C. Grätzel, S.M. Zakeeruddin, U. Röthlisberger, M. Grätzel, Entropic stabilization of mixed A-cation ABX3 metal halide perovskites for high performance perovskite solar cells, Energ. Environ. Sci. 9 (2) (2016) 656–662, https://doi.org/10.1039/C5EE03255E.
[80] S.J. Adjogri, E.L. Meyer, A review on lead-free hybrid halide perovskites as light absorbers for photovoltaic applications based on their structural, optical, and morphological properties, Molecules 25 (21) (2020), https://doi.org/10.3390/molecules25215039.
[81] L. Gao, G. Yang, Organic-inorganic halide perovskites: from crystallization of polycrystalline films to solar cell applications, Solar RRL 4 (2) (2020) 1900200, https://doi.org/10.1002/solr.201900200.
[82] Y. Tian, Y. Ling, Y. Shu, C. Zhou, T. Besara, T. Siegrist, H. Gao, B. Ma, A solution-processed organometal halide perovskite hole transport layer for highly efficient organic light-emitting diodes, Adv. Electron. Mater. 2 (7) (2016) 1600165, https://doi.org/10.1002/aelm.201600165.
[83] H. Choi, J. Jeong, H.-B. Kim, S. Kim, B. Walker, G.-H. Kim, J.Y. Kim, Cesium-doped methylammonium lead iodide perovskite light absorber for hybrid solar cells, Nano Energy 7 (2014) 80–85, https://doi.org/10.1016/j.nanoen.2014.04.017.
[84] D.P. McMeekin, G. Sadoughi, W. Rehman, G.E. Eperon, M. Saliba, M.T. Hörantner, A. Haghighirad, N. Sakai, L. Korte, B. Rech, M.B. Johnston, L.M. Herz, H.J. Snaith, A mixed-cation lead mixed-halide perovskite absorber for tandem solar cells, Science 351 (6269) (2016) 151–155, https://doi.org/10.1126/science.aad5845.
[85] A. Amat, E. Mosconi, E. Ronca, C. Quarti, P. Umari, M.K. Nazeeruddin, M. Grätzel, F. De Angelis, Cation-induced band-gap tuning in organohalide perovskites: interplay of spin–orbit coupling and octahedra tilting, Nano Lett. 14 (6) (2014) 3608–3616, https://doi.org/10.1021/nl5012992.
[86] Y. Ling, Y. Tian, X. Wang, J.C. Wang, J.M. Knox, F. Perez-Orive, Y. Du, L. Tan, K. Hanson, B. Ma, H. Gao, Enhanced optical and electrical properties of polymer-assisted all-inorganic perovskites for light-emitting diodes, Adv. Mater. 28 (40) (2016) 8983–8989, https://doi.org/10.1002/adma.201602513.
[87] S. Hou, M.K. Gangishetty, Q. Quan, D.N. Congreve, Efficient blue and white perovskite light-emitting diodes via manganese doping, Joule 2 (11) (2018) 2421–2433, https://doi.org/10.1016/j.joule.2018.08.005.
[88] N. De Marco, H. Zhou, Q. Chen, P. Sun, Z. Liu, L. Meng, E.-P. Yao, Y. Liu, A. Schiffer, Y. Yang, Guanidinium: a route to enhanced carrier lifetime and open-circuit voltage in hybrid perovskite solar cells, Nano Lett. 16 (2) (2016) 1009–1016, https://doi.org/10.1021/acs.nanolett.5b04060.
[89] B. Saparov, D.B. Mitzi, Organic–inorganic perovskites: structural versatility for functional materials design, Chem. Rev. 116 (7) (2016) 4558–4596, https://doi.org/10.1021/acs.chemrev.5b00715.

[90] A.F. Akbulatov, L.A. Frolova, D.V. Anokhin, K.L. Gerasimov, N.N. Dremova, P.A. Troshin, Hydrazinium-loaded perovskite solar cells with enhanced performance and stability, J. Mater. Chem. A 4 (47) (2016) 18378–18382, https://doi.org/10.1039/C6TA08215G.
[91] Y. Zhang, Z. Fei, P. Gao, Y. Lee, F.F. Tirani, R. Scopelliti, Y. Feng, P.J. Dyson, M.K. Nazeeruddin, A strategy to produce high efficiency, high stability perovskite solar cells using functionalized ionic liquid-dopants, Adv. Mater. 29 (36) (2017) 1702157, https://doi.org/10.1002/adma.201702157.
[92] E.A. Alharbi, A.Y. Alyamani, D.J. Kubicki, A.R. Uhl, B.J. Walder, A.Q. Alanazi, J. Luo, A. Burgos-Caminal, A. Albadri, H. Albrithen, M.H. Alotaibi, J.-E. Moser, S.M. Zakeeruddin, F. Giordano, L. Emsley, M. Grätzel, Atomic-level passivation mechanism of ammonium salts enabling highly efficient perovskite solar cells, Nat. Commun. 10 (1) (2019) 3008, https://doi.org/10.1038/s41467-019-10985-5.
[93] S. Ghosh, T. Singh, Role of ionic liquids in organic-inorganic metal halide perovskite solar cells efficiency and stability, Nano Energy 63 (2019) 103828, https://doi.org/10.1016/j.nanoen.2019.06.024.
[94] C.-T. Li, New Class of Ionic Liquids for Dye-Sensitized Solar Cells (p. Ch. 24-Ch. 24), IntechOpen, 2015, https://doi.org/10.5772/59057.
[95] A. Syairah, M.H. Khanmirzaei, N.M. Saidi, N.K. Farhana, S. Ramesh, K. Ramesh, S. Ramesh, Effect of different imidazolium-based ionic liquids on gel polymer electrolytes for dye-sensitized solar cells, Ionics 25 (5) (2019) 2427–2435, https://doi.org/10.1007/s11581-018-2603-6.
[96] N.K. Noel, A. Abate, S.D. Stranks, E.S. Parrott, V.M. Burlakov, A. Goriely, H.J. Snaith, Enhanced photoluminescence and solar cell performance via lewis base passivation of organic–inorganic lead halide perovskites, ACS Nano 8 (10) (2014) 9815–9821, https://doi.org/10.1021/nn5036476.
[97] T. Zhao, C.-C. Chueh, Q. Chen, A. Rajagopal, A.K.-Y. Jen, Defect passivation of organic–inorganic hybrid perovskites by diammonium iodide toward high-performance photovoltaic devices, ACS Energy Lett. 1 (4) (2016) 757–763, https://doi.org/10.1021/acsenergylett.6b00327.
[98] A. Gabriel Tomulescu, L. Nicoleta Leonat, F. Neațu, V. Stancu, V. Toma, S. Derbali, Ș. Neațu, A. Mihai Rostas, C. Beşleagă, R. Pătru, I. Pintilie, M. Florea, Enhancing stability of hybrid perovskite solar cells by imidazolium incorporation, Sol. Energy Mater. Sol. Cells 227 (March) (2021) 111096, https://doi.org/10.1016/j.solmat.2021.111096.
[99] C.C. Stoumpos, C.D. Malliakas, M.G. Kanatzidis, Semiconducting tin and lead iodide perovskites with organic cations: phase transitions, high mobilities, and near-infrared photoluminescent properties, Inorg. Chem. 52 (15) (2013) 9019–9038, https://doi.org/10.1021/ic401215x.
[100] F. Hao, C.C. Stoumpos, D.H. Cao, R.P.H. Chang, M.G. Kanatzidis, Lead-free solid-state organic–inorganic halide perovskite solar cells, Nat. Photonics 8 (6) (2014) 489–494, https://doi.org/10.1038/nphoton.2014.82.
[101] Y. Zhao, K. Zhu, Efficient planar perovskite solar cells based on 1.8 eV band gap CH3NH3PbI2Br nanosheets via thermal decomposition, J. Am. Chem. Soc. 136 (35) (2014) 12241–12244, https://doi.org/10.1021/ja5071398.
[102] Q. Ou, X. Bao, Y. Zhang, H. Shao, G. Xing, X. Li, L. Shao, Q. Bao, Band structure engineering in metal halide perovskite nanostructures for optoelectronic applications, Nano Mater. Sci. 1 (4) (2019) 268–287, https://doi.org/10.1016/j.nanoms.2019.10.004.
[103] B.R. Sutherland, E.H. Sargent, Perovskite photonic sources, Nat. Photonics 10 (5) (2016) 295–302, https://doi.org/10.1038/nphoton.2016.62.
[104] R.A. Scheidt, P.V. Kamat, Temperature-driven anion migration in gradient halide perovskites, J. Chem. Phys. 151 (13) (2019) 134703, https://doi.org/10.1063/1.5120270.

[105] M. Coll, A. Gomez, E. Mas-Marza, O. Almora, G. Garcia-Belmonte, M. Campoy-Quiles, J. Bisquert, Polarization switching and light-enhanced piezoelectricity in lead halide perovskites, J. Phys. Chem. Lett. 6 (8) (2015) 1408–1413, https://doi.org/10.1021/acs.jpclett.5b00502.

[106] G.A. Sewvandi, K. Kodera, H. Ma, S. Nakanishi, Q. Feng, Antiferroelectric nature of CH3NH3PbI3 – xClx perovskite and its implication for charge separation in perovskite solar cells, Sci. Rep. 6 (1) (2016) 30680, https://doi.org/10.1038/srep30680.

[107] I. Pintilie, V. Stancu, A. Tomulescu, R. Radu, C. Besleaga Stan, L. Trinca, L. Pintilie, Properties of perovskite ferroelectrics deposited on F doped SnO2 electrodes and the prospect of their integration into perovskite solar cells, Mater. Des. 135 (2017) 112–121, https://doi.org/10.1016/j.matdes.2017.09.013.

[108] N. Plugaru, G.A. Nemnes, L. Filip, I. Pintilie, L. Pintilie, K.T. Butler, A. Manolescu, Atomistic simulations of methylammonium lead halide layers on PbTiO3 (001) surfaces, J. Phys. Chem. C 121 (17) (2017) 9096–9109, https://doi.org/10.1021/acs.jpcc.7b00399.

[109] Y. Umemoto, A. Suzuki, T. Oku, Effects of halogen doping on the photovoltaic properties of HC(NH2)2PbI3 perovskite solar cells, AIP Conf. Proc. 1807 (1) (2017) 20011, https://doi.org/10.1063/1.4974793.

[110] K. Cao, H. Li, S. Liu, J. Cui, Y. Shen, M. Wang, MAPbI3 – xBrx mixed halide perovskites for fully printable mesoscopic solar cells with enhanced efficiency and less hysteresis, Nanoscale 8 (16) (2016) 8839–8846, https://doi.org/10.1039/C6NR01043A.

[111] R. Sedighi, F. Tajabadi, S. Shahbazi, S. Gholipour, N. Taghavinia, Mixed-halide CH3NH3PbI3 – xXx (X = Cl, Br, I) perovskites: vapor-assisted solution deposition and application as solar cell absorbers, ChemPhysChem 17 (15) (2016) 2382–2388, https://doi.org/10.1002/cphc.201600230.

[112] E. Mosconi, A. Amat, M.K. Nazeeruddin, M. Grätzel, F. De Angelis, First-principles modeling of mixed halide organometal perovskites for photovoltaic applications, J. Phys. Chem. C 117 (27) (2013) 13902–13913, https://doi.org/10.1021/jp4048659.

[113] M.A. Mohebpour, M. Saffari, H.R. Soleimani, M.B. Tagani, High performance of mixed halide perovskite solar cells: role of halogen atom and plasmonic nanoparticles on the ideal current density of cell, Physica E 97 (2018) 282–289, https://doi.org/10.1016/j.physe.2017.11.022.

[114] Y. Kawano, J. Chantana, T. Nishimura, T. Minemoto, Influence of halogen content in mixed halide perovskite solar cells on cell performances through device simulation, Sol. Energy Mater. Sol. Cells 205 (2020) 110252, https://doi.org/10.1016/j.solmat.2019.110252.

[115] S. Colella, E. Mosconi, P. Fedeli, A. Listorti, F. Gazza, F. Orlandi, P. Ferro, T. Besagni, A. Rizzo, G. Calestani, G. Gigli, F. Angelis, R. Mosca, MAPbI3-xClx mixed halide perovskite for hybrid solar cells: the role of chloride as dopant on the transport and structural properties, Chem. Mater. 25 (2013) 4613–4618, https://doi.org/10.1021/cm402919x.

[116] M.I. Dar, N. Arora, P. Gao, S. Ahmad, M. Grätzel, M.K. Nazeeruddin, Investigation regarding the role of chloride in organic–inorganic halide perovskites obtained from chloride containing precursors, Nano Lett. 14 (12) (2014) 6991–6996, https://doi.org/10.1021/nl503279x.

[117] E. Edri, S. Kirmayer, M. Kulbak, G. Hodes, D. Cahen, Chloride inclusion and hole transport material doping to improve methyl ammonium lead bromide perovskite-based high open-circuit voltage solar cells, J. Phys. Chem. Lett. 5 (3) (2014) 429–433, https://doi.org/10.1021/jz402706q.

[118] Y. Pei, X. Zou, Y. Guan, G. Teng, Effect of codoping cl anion and 5-AVA cation on performance of large-area perovskite solar cells with double-mesoporous layers, Int. J. Photoenergy 2016 (2016) 2953592, https://doi.org/10.1155/2016/2953592.

[119] Q. Chen, H. Zhou, Y. Fang, A.Z. Stieg, T.-B. Song, H.-H. Wang, X. Xu, Y. Liu, S. Lu, J. You, P. Sun, J. McKay, M.S. Goorsky, Y. Yang, The optoelectronic role of chlorine in CH3NH3PbI3 (Cl)-based perovskite solar cells, Nat. Commun. 6 (1) (2015) 7269, https://doi.org/10.1038/ncomms8269.

[120] J. Wang, N. Wang, Y. Jin, J. Si, Z.-K. Tan, H. Du, L. Cheng, X. Dai, S. Bai, H. He, Z. Ye, M.L. Lai, R.H. Friend, W. Huang, Interfacial control toward efficient and low-voltage perovskite light-emitting diodes, Adv. Mater. 27 (14) (2015) 2311–2316, https://doi.org/10.1002/adma.201405217.

[121] L.K. Ono, E.J. Juarez-Perez, Y. Qi, Progress on perovskite materials and solar cells with mixed cations and halide anions, ACS Appl. Mater. Interfaces 9 (2017) 30197.

[122] K.E. Gkini, M. Antoniadou, N. Balis, A. Kaltzoglou, A.G. Kontos, P. Falaras, Mixing cations and halide anions in perovskite solar cells, Mater. Today: Proc. 19 (2019) 73–78, https://doi.org/10.1016/j.matpr.2019.07.660.

[123] X. Liu, J. Wu, Y. Yang, T. Wu, Q. Guo, Pyridine solvent engineering for high quality anion-cation-mixed hybrid and high performance of perovskite solar cells, J. Power Sources 399 (2018) 144–150, https://doi.org/10.1016/j.jpowsour.2018.07.093.

[124] M. Yavari, M. Mazloum-Ardakani, S. Gholipour, M.M. Tavakoli, S.-H. Turren-Cruz, N. Taghavinia, M. Grätzel, A. Hagfeldt, M. Saliba, Greener, nonhalogenated solvent systems for highly efficient perovskite solar cells, Adv. Energy Mater. 8 (21) (2018) 1800177, https://doi.org/10.1002/aenm.201800177.

[125] H.T. Pham, T. Duong, W.D.A. Rickard, F. Kremer, K.J. Weber, J. Wong-Leung, Understanding the chemical and structural properties of multiple-cation mixed halide perovskite, J. Phys. Chem. C 123 (43) (2019) 26718–26726, https://doi.org/10.1021/acs.jpcc.9b09035.

[126] Y. Hu, M.F. Aygüler, M.L. Petrus, T. Bein, P. Docampo, Impact of rubidium and cesium cations on the moisture stability of multiple-cation mixed-halide perovskites, ACS Energy Lett. 2 (10) (2017) 2212–2218, https://doi.org/10.1021/acsenergylett.7b00731.

[127] S. Derbali, K. Nouneh, M. Florea, L.N. Leonat, V. Stancu, A.G. Tomulescu, A.C. Galca, M. Secu, L. Pintilie, M.E. Touhami, Potassium-containing triple-cation mixed-halide perovskite materials: toward efficient and stable solar cells, J. Alloys Compd. 858 (2021), https://doi.org/10.1016/j.jallcom.2020.158335.

[128] V. Stancu, L.N. Leonat, A.G. Tomulescu, S. Derbali, L. Pintilie, C. Besleaga, A.C. Galca, F. Neațu, Ș. Neațu, M. Florea, I. Pintilie, Influence of doping the inorganic cation with Eu or Sb on the properties of perovskite films, Phys. Scr. 95 (7) (2020) 75707, https://doi.org/10.1088/1402-4896/ab90be.

[129] J.-C. Hebig, I. Kühn, J. Flohre, T. Kirchartz, Optoelectronic properties of (CH3NH3) 3Sb2I9 thin films for photovoltaic applications, ACS Energy Lett. 1 (1) (2016) 309–314, https://doi.org/10.1021/acsenergylett.6b00170.

[130] F. Jiang, D. Yang, Y. Jiang, T. Liu, X. Zhao, Y. Ming, B. Luo, F. Qin, J. Fan, H. Han, L. Zhang, Y. Zhou, Chlorine-incorporation-induced formation of the layered phase for antimony-based lead-free perovskite solar cells, J. Am. Chem. Soc. 140 (3) (2018) 1019–1027, https://doi.org/10.1021/jacs.7b10739.

[131] E. Edri, S. Kirmayer, S. Mukhopadhyay, K. Gartsman, G. Hodes, D. Cahen, Elucidating the charge carrier separation and working mechanism of CH3NH3PbI3 − xClx perovskite solar cells, Nat. Commun. 5 (1) (2014) 3461, https://doi.org/10.1038/ncomms4461.

[132] S.D. Stranks, G.E. Eperon, G. Grancini, C. Menelaou, M.J.P. Alcocer, T. Leijtens, L.M. Herz, A. Petrozza, H.J. Snaith, Electron-hole diffusion lengths exceeding 1 micrometer in an organometal trihalide perovskite absorber, Science 342 (6156) (2013) 341–344, https://doi.org/10.1126/science.1243982.

[133] E.J. Juarez-Perez, R.S. Sanchez, L. Badia, G. Garcia-Belmonte, Y.S. Kang, I. Mora-Sero, J. Bisquert, Photoinduced giant dielectric constant in lead halide perovskite solar cells, J. Phys. Chem. Lett. 5 (13) (2014) 2390–2394, https://doi.org/10.1021/jz5011169.

[134] J.M. Frost, K.T. Butler, F. Brivio, C.H. Hendon, M. van Schilfgaarde, A. Walsh, Atomistic origins of high-performance in hybrid halide perovskite solar cells, Nano Lett. 14 (5) (2014) 2584–2590, https://doi.org/10.1021/nl500390f.
[135] A. Baumann, K. Tvingstedt, M.C. Heiber, S. Väth, C. Momblona, H.J. Bolink, V. Dyakonov, Persistent photovoltage in methylammonium lead iodide perovskite solar cells, APL Mater. 2 (8) (2014) 81501, https://doi.org/10.1063/1.4885255.
[136] E.T. Hoke, D.J. Slotcavage, E.R. Dohner, A.R. Bowring, H.I. Karunadasa, M.D. McGehee, Reversible photo-induced trap formation in mixed-halide hybrid perovskites for photovoltaics, Chem. Sci. 6 (1) (2015) 613–617, https://doi.org/10.1039/C4SC03141E.
[137] C. Motta, F. El-Mellouhi, S. Kais, N. Tabet, F. Alharbi, S. Sanvito, Revealing the role of organic cations in hybrid halide perovskite CH3NH3PbI3, Nat. Commun. 6 (1) (2015) 7026, https://doi.org/10.1038/ncomms8026.
[138] T. T. Wang, B. Daiber, J.M. Frost, S.A. Mann, E.C. Garnett, A. Walsh, B. Ehrler, Indirect to direct bandgap transition in methylammonium lead halide perovskite, Energ. Environ. Sci. 10 (2) (2017) 509–515, https://doi.org/10.1039/C6EE03474H.
[139] D.A. Jacobs, Y. Wu, H. Shen, C. Barugkin, F.J. Beck, T.P. White, K. Weber, K.R. Catchpole, Hysteresis phenomena in perovskite solar cells: the many and varied effects of ionic accumulation, Phys. Chem. Chem. Phys. 19 (4) (2017) 3094–3103, https://doi.org/10.1039/C6CP06989D.
[140] H. Tsai, R. Asadpour, J.-C. Blancon, C.C. Stoumpos, O. Durand, J.W. Strzalka, B. Chen, R. Verduzco, P.M. Ajayan, S. Tretiak, J. Even, M.A. Alam, M.G. Kanatzidis, W. Nie, A.D. Mohite, Light-induced lattice expansion leads to high-efficiency perovskite solar cells, Science 360 (6384) (2018) 67–70, https://doi.org/10.1126/science.aap8671.
[141] D. Kim, J.S. Yun, P. Sharma, D.S. Lee, J. Kim, A.M. Soufiani, S. Huang, M.A. Green, A.W.Y. Ho-Baillie, J. Seidel, Light- and bias-induced structural variations in metal halide perovskites, Nat. Commun. 10 (1) (2019) 444, https://doi.org/10.1038/s41467-019-08364-1.
[142] N.K. Elumalai, A. Uddin, Hysteresis in organic-inorganic hybrid perovskite solar cells, Sol. Energy Mater. Sol. Cells 157 (2016) 476–509, https://doi.org/10.1016/j.solmat.2016.06.025.
[143] H.J. Snaith, A. Abate, J.M. Ball, G.E. Eperon, T. Leijtens, N.K. Noel, S.D. Stranks, J.T.-W. Wang, K. Wojciechowski, W. Zhang, Anomalous hysteresis in perovskite solar cells, J. Phys. Chem. Lett. 5 (9) (2014) 1511–1515, https://doi.org/10.1021/jz500113x.
[144] W. Tress, N. Marinova, T. Moehl, S.M. Zakeeruddin, M.K. Nazeeruddin, M. Grätzel, Understanding the rate-dependent J–V hysteresis, slow time component, and aging in CH3NH3PbI3 perovskite solar cells: the role of a compensated electric field, Energ. Environ. Sci. 8 (3) (2015) 995–1004, https://doi.org/10.1039/C4EE03664F.
[145] F. Haque, M. Mativenga, Halide perovskite memtransistor enabled by ion migration, Jpn. J. Appl. Phys. 59 (8) (2020) 81002, https://doi.org/10.35848/1347-4065/aba5e1.
[146] Z. Hong, J. Zhao, K. Huang, B. Cheng, Y. Xiao, S. Lei, Controllable switching properties in an individual CH3NH3PbI3 micro/nanowire-based transistor for gate voltage and illumination dual-driving non-volatile memory, J. Mater. Chem. C 7 (14) (2019) 4259–4266, https://doi.org/10.1039/C9TC00382G.
[147] G.A. Nemnes, C. Besleaga, A.G. Tomulescu, A. Palici, L. Pintilie, A. Manolescu, I. Pintilie, How measurement protocols influence the dynamic J-V characteristics of perovskite solar cells: theory and experiment, Sol. Energy 173 (2018) 976–983, https://doi.org/10.1016/j.solener.2018.08.033.
[148] G.A. Nemnes, C. Besleaga, A.G. Tomulescu, I. Pintilie, L. Pintilie, K. Torfason, A. Manolescu, Dynamic electrical behavior of halide perovskite based solar cells, Sol. Energy Mater. Sol. Cells 159 (2017) 197–203, https://doi.org/10.1016/j.solmat.2016.09.012.

[149] G.A. Nemnes, C. Besleaga, A.G. Tomulescu, L.N. Leonat, V. Stancu, M. Florea, A. Manolescu, I. Pintilie, The hysteresis-free behavior of perovskite solar cells from the perspective of the measurement conditions, J. Mater. Chem. C 7 (18) (2019), https://doi.org/10.1039/c8tc05999c.

[150] G.A. Nemnes, C. Besleaga, V. Stancu, D.E. Dogaru, L.N. Leonat, L. Pintilie, K. Torfason, M. Ilkov, A. Manolescu, I. Pintilie, Normal and inverted hysteresis in perovskite solar cells, J. Phys. Chem. C 121 (21) (2017) 11207–11214, https://doi.org/10.1021/acs.jpcc.7b04248.

CHAPTER

3

Structural effects on halide perovskite properties

Irfan Ahmed[a,b], Meenakshi Gussain[a], Fatemeh Behrouznejad[a,c], Waseem Haider[d], and Yiqiang Zhan[a]

[a]State Key Laboratory of ASIC and System, Centre of Micro-Nano System, School of Information Science and Technology, Fudan University, Shanghai, China, [b]Department of Physics, Government Postgraduate College, Higher Education Department-HED, Mansehra, Khyber Pakhtunkhwa, Pakistan, [c]Nanoparticles and Coatings Lab, Department of Physics, Sharif University of Technology, Tehran, Iran, [d]School of Engineering and Technology, Central Michigan University, Mt. Pleasant, MI, United States

1. Introduction

Over the past few decades, organic-inorganic lead halide perovskite has shown outstanding optoelectronic properties that capitulated their fate toward highly researched topics for scientists all around the globe. Owing to their distinctive and extraordinary potential, perovskite material paved its way beyond photovoltaics where it has achieved the highest PCE of around 26% [1], to other optoelectronic applications such as light-emitting diodes [2], photodetectors [3], and lasers [4,5]. Furthermore, organic-inorganic perovskite materials attain excellent optoelectronic properties such as (i) high absorption coefficient ($4.3 \times 10^5\,cm^{-1}$), which reduces the required thickness of this light-absorbing material, and eventually, material's final cost is decreased, (ii) excellent carrier mobility (hole $\sim$12.5–66 $cm^2\,V^{-1}s^{-1}$, electron $\sim$7.5 $cm^2\,V^{-1}s^{-1}$), and long-range charge diffusion length (0.1–1 μm), (iii) low defect state density of single crystal prepared by solution method is of the order of $10^{10}\,cm^{-3}$ comparable to the value of pure silicon, and (iv) wide-range tunability in color and color purity with high luminous efficiency [6,7].

Low-Dimensional Halide Perovskites
https://doi.org/10.1016/B978-0-323-88522-5.00008-9

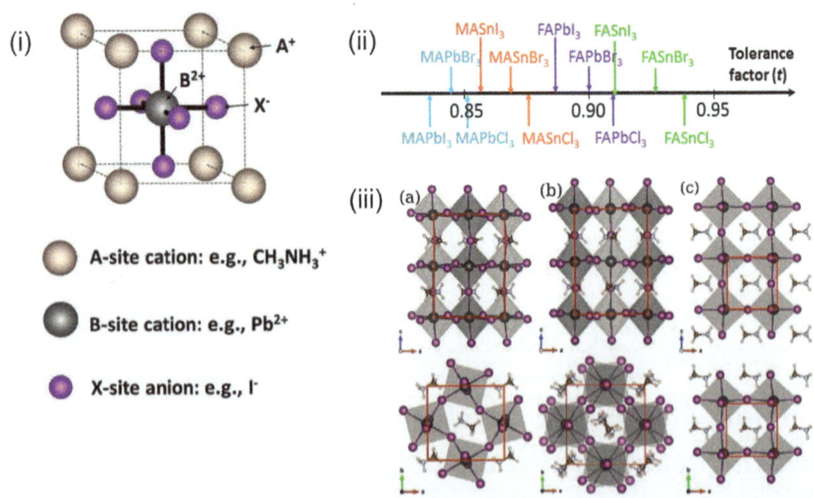

FIG. 1 The schematics representing (i) the ABX_3 structure of halide perovskite crystal, (ii) the tolerance factors (t) of halide perovskite, and (iii) the comparison of three types of perovskite ($MAPbI_3$) phases as (a) orthorhombic, (b) tetragonal, and (c) cubic (the top row and bottom row represent ac-plane and ab-plane, respectively). *(i) From K. Wang, D. Yang, C. Wu, M. Sanghadasa, S. Priya, Recent progress in fundamental understanding of halide perovskite semiconductors, Prog. Mater. Sci. 106 (2019) 100580, https://doi.org/10.1016/j.pmatsci.2019.100580, with permission, (ii) From Z. Fan, K. Sun, J. Wang, Perovskites for photovoltaics: a combined review of organic-inorganic halide perovskites and ferroelectric oxide perovskites, J. Mater. Chem. A* **3** *(2015) 18809–18828. https://doi.org/10.1039/c5ta04235f and (iii) From K. Korshunova, L. Winterfeld, W.J.D. Beenken, E. Runge, Thermodynamic stability of mixed Pb:Sn methyl-ammonium halide perovskites, Phys. Status Solidi B 253(10) (2016) 1907–1915, https://doi.org/10.1002/pssb.201600136, with permission,*

The hybrid crystal structure of halide perovskite materials is represented by the formulation of ABX_3, where A and B represent the monovalent and bivalent cations, respectively ($A = CH_3NH_3^+$ or known as MA^+), and the monovalent halide anion is represented by X ($X^- = Cl$, I, Br) (Fig. 1(i)) [8]. Based on their crystal structure, perovskite materials can be cataloged as two-dimensional (2D) and three-dimensional (3D) materials. The 2D materials are termed as the layered structures that include pure-2D and quasi-2D and are represented by $(RH_3)_2A_{n-1}B_nX_{3n+1}$ (if $n = 1$, pure-2D; and if n is definite integer > 1, quasi-2D) [9,10]. The 2D class has different crystallographic orientations such as <100> oriented 2D, which can be further divided into Dion-Jacobson (DJ), Ruddlesden-Popper (RP), and the mixture of the two (DJ-RP) layered structures [10,11]. There are two main strategies mainly employed to tailor the optoelectronic properties of perovskite materials such as (i) compositional engineering and (ii) dimensional tailoring; however, compositional tuning also affects the dimensionality as well. For instance, if the cation (A) is smaller, the 3D framework of ABX_3 is retained, while larger cation (A) promotes the formation of 2D layered nanosheets, 1D chains, or 0D isolated clusters, which are represented by formulae as A_2BX_4, A_3BX_5, or A_4BX_6, respectively [10,12–15]. In most of the remarkable optoelectronic devices such as light-emitting diodes [16], photodetectors [17], lasers [18], photosensors, and chemosensors [19], halide perovskite serves as charge generation and transport medium. However, depending on the structural factor at various length scales that include crystal structure, microstructure, hierarchical architecture, and morphology, the suitability of this material varies accordingly, which effects its optoelectronic properties [20–23] and device efficiency, significantly [24]. Therefore, the investigation of the influence of structural effect on the properties of halide perovskite has a wide implication in both fundamental and applied research. The precise control of the structure of perovskite and an accurate well-defined structure are the required strategies for outstanding performance in unprecedented applications. The current chapter details the structural effect of halide perovskite materials on various optoelectronic properties as well as identifies their future challenges.

2. Composition and crystal structure of halide perovskite

As mentioned earlier, the structure of halide perovskite materials is represented by the formulation of ABX_3, where the bivalent cation "B" localizes at the body center of the octahedron that consists of six monovalent halide X ions $[(PbX_{-6})^{4-}]$ corner-shared to form a 3D crystalline network with a stoichiometry of Pb X_3^- (Fig. 1(i)). In perovskite, the 3D structure forming possibility is predicted by Goldschmidt's tolerance factor (t) and octahedral factor (μ) where $t = \frac{R_A + R_X}{\sqrt{2(R_B + R_X)}}$ and $\mu = \frac{R_B}{R_X}$, respectively. The R_A, R_B, and R_X are the effective radii of the corresponding ions, and the majority of the 3D perovskite structure forms in the approximate range $0.8 \leq t \leq 1.0$ [25,26] and μ range $0.44 \leq t \leq 0.90$ [27]. The cubic phase appears when t lies in a narrow range of 0.89–0.1; however, with less value of t, the less symmetric orthorhombic and tetragonal phases are observed. It seems that t increases by replacing Pb with Sn but at the expanse of decrease in stability (Fig. 1(i)). The geometry of the three phases is illustrated in Fig. 1(ii), and it can be seen that PbI_3 (octahedra) are corner-connected in all the phases. Korshunova et al. investigated the mixed $MA(Pb{:}Sn)X_3$ perovskite structures using density functional theory (DFT) and found that halogen substitution has a much stronger influence on both structures and cohesive energies of cubic, orthorhombic, and tetragonal phases as compared to the Sn-based perovskite [28]. Despite these limitations, the transitions between the crystallographic structures take place on heating as well. The comparison of various crystalline phases of $MAPbI_3$ is illustrated in Fig. 1(iii). The cubic ($a=b=c$) and tetragonal ($a=b\neq c$) phases are mostly observed at high temperature in experiments; however, at low temperature, the orthorhombic Bravais lattice, i.e., ($a\neq b\neq c$) but orthogonal, is a preferred structure [28]. At high temperature, the volume of the unit cell (cubic phase) is observed and is larger as compared to an orthorhombic phase, which is found at low temperature. The volume decreases with an increase in electronegativity of the halogen substitution. Thus, the geometric properties of perovskite structure are influenced by chemical composition, temperature, and temperature-driven phase transitions. The unit cell volume is increased by increase in temperature, which is further linked with the mobility of organic cation. Increase in electronegativity of halogen results in deformation of octahedra (BX_6) and is stronger for Sn as compared to Pb. For a device under working condition (temperature), it is possible to synthesize a highly stable mixed compound with various Sn: Pb ratios for three halogen groups.

3. Optoelectronic properties

The optical properties of perovskite materials (2D or 3D structures) depend on organic or inorganic components' characteristics. In 3D halide perovskite materials, the exciton binding energy (E_b) for room-temperature tetragonal structures and low-temperature orthorhombic structures has been reported <50 and 37–50 meV [29,30], respectively, by considering the dielectric constant value of 6.5 in the calculations. This low value for E_b leads to the dissociation of excitons spontaneously (2 ps) after photoexcitation [31,32]. In excitonic solar cells such as organic solar cells, excitons dissociate at the interface of heterojunctions, but in perovskite solar cells that are predominantly nonexcitonic solar cells [33], the functionality of electron transfer layers and hole transfer layers is just transferring charge carriers. Nevertheless, at low temperature, based on the solar cell structure, the fingerprint of Wannier-Mott excitons[a] is observable. Fig. 2 shows optical density spectra for the perovskite layer deposited on the mesoporous Al_2O_3 layer (with and without capping layer) and on the flat substrate at 4.2 K. The signature of Wannier-Mott excitons is observable only for the perovskite layer on a flat substrate. The 2D structured organic-inorganic perovskite materials have the conventional chemical structure of (H_3N-R-NH_3) MX_4 (R is an organic group, M is a metal such as Pb, and X is a halogen (I, Br, or Cl)). The dielectric constant of the organic layer and MX_6^{-2} layers differs significantly, and E_b is as large as a few hundred meV [34], which can limit their photovoltaic applications. Utilizing an electron acceptor layer, such as perylene diimide organic chromophore on the top of these 2D structures, leads to an efficient generation of charge carriers [35].

[a]Wannier-Mott excitons are related to materials with relatively high dielectric value. In these materials, the Coulomb interaction between electrons and holes will be reduced via electric field screening, and the exciton radius (electron-hole pair distance) for them is more than the lattice constant.

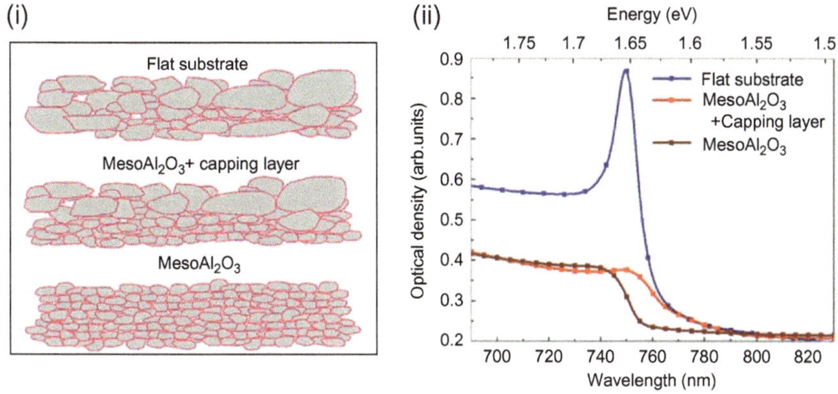

FIG. 2 The schematic demonstration of the average crystalline perovskite material deposited on the flat substrate infiltrated into mesoporous Al_2O_3 under layer with upper capping layer, and infiltrated into mesoporous Al_2O_3 under layer (i) and the related optical density spectra (ii). *No permission required.*

In the case of solar cells, certain optoelectronic properties such as absorption range and the upper theoretical limit of open-circuit voltage (V_{oc}) can be predicted by the bandgap of active material. Semiconductors possess two types of bandgaps: direct and indirect. Direct bandgap refers to the excitation where the momentum energy of the electrons is unaffected by excitation from the valence band to the conduction band; however, in the case of indirect bandgap, some energy of phonon is consumed to alter the momentum of electrons. In 1972, Tauc stated a mathematical relation between the absorption tail and bandgap of material, written based on the absorption coefficient [36].

$$\alpha_{direct} = C/h\nu\left[\left(h\nu - E_g\right)\right]^{1/2}$$

$$\alpha_{indirect} = C/h\nu\left[\left(hv - E_g\right)\right]^{1}$$

C is the constant, $h\nu$ is the photon energy, and E_g is the bandgap energy of active material. The distinction between direct and indirect bandgap material can be obtained by plotting the material's light absorption energy values against the momentum of the crystal, and minimum conduction and maximum valence band values are observed. In the direct bandgap materials, the majority of photons are harvested with a minimum loss in incident energy, and an equal momentum value is obtained for both maximum and minimum band values. However, in the direct bandgap materials, light may need to interact with phonon before getting absorbed. In fact, for better absorption of light, in direct bandgap materials a thin layer (<1 μm) is used as compared to the indirect bandgap material which is relatively thick (100 μm). The suitable E_g value for PV devices is from 1.1 to 2 eV. Nevertheless, the ideal value for single-junction solar cell applications is 1.3 eV [37]. In the case of the top cell in tandem structures (e.g., Si/perovskite structure), the ideal value of E_g is 1.67–1.75 eV [38]. Therefore, different approaches have already been utilized to tune the bandgap for perovskite materials. As represented in Fig. 3, the E_g can be affected by changing stoichiometry (in the case of 3D perovskite selecting A, B, and X in ABX_3 chemical formula, crystal symmetry, doping, and reducing dimensions via synthesizing nanostructures or depositing material on a nano-porous scaffold [39,40]. It is noteworthy that due to defect tolerance of perovskite materials, defects do not form states in bandgap that can lead to increasing nonradiative recombination. This means that the predominant recombination in perovskite-based devices is radiation recombination. Room-temperature and low-temperature quantum efficiencies up to 70% [41] and 100% are reported, respectively, for perovskite materials. For an efficient optoelectronic solar cell or light-emitting diode (LED), the main layer whether as a light-absorbent layer or a light-emitting layer must have high photoluminescence intensity; in other words, it must approach the Shockley-Queisser limit in which recombination is radiative.

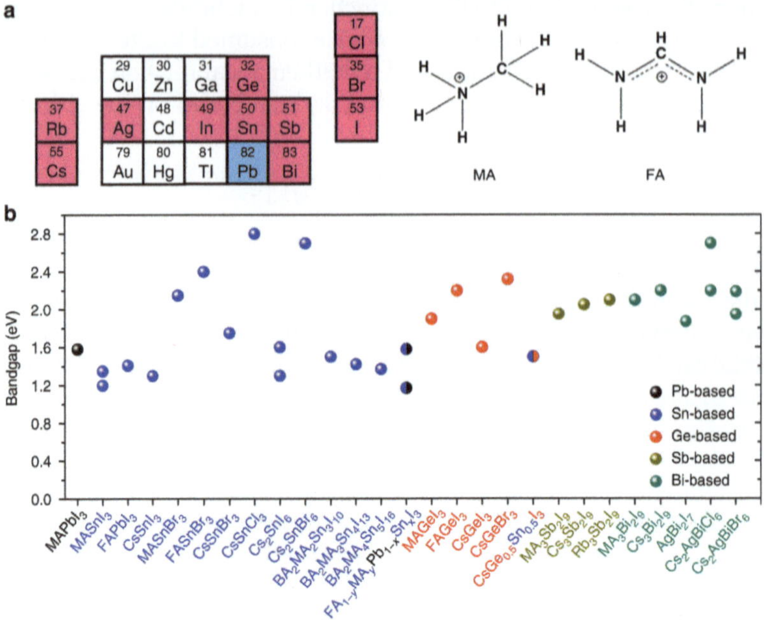

FIG. 3 Bandgap (E_g) of various perovskite multidimensional materials. *No permission required.*

3.1 Effects of changing stoichiometry (ABX_3) on bandgap energy

3.1.1 *Changing halide atoms (X)*

Xia et al. [42] have reported E_g values of 1.53 eV, 2.24 eV, and 2.97 eV for single-crystal $MAPbI_3$, $MAPbCl_3$, and $MAPbBr_3$, respectively, concluding that E_g increases by increasing the size of halide. Suarez et al. [43] showed that in the case of perovskite material with $MAPb(Br_xI_{(1-x)})_{(3-y)}Cl_y$ ($x<=1$), E_g has a direct relation with the "x" value, and the color of perovskite changes from yellow ($E_g=2.27$ eV) to brown (1.57 eV) by varying x from "0" to "1." They have mentioned that this significant change in E_g is observable only if perovskite materials contain chlorine. Nevertheless, Seok et al. have shown that bandgap value increases in $MAPbI_{(3-x)}Br_x$ structure (generally $ABX_{(3-x)}Br_x$ structure) by increasing the "x" value [44]. Kanatzidis et al. have indicated that in $MASnI_{(3-x)}Br_x$, increasing E_g with increasing x value is related to increasing conduction band minimum rather than decreasing valence band maximum. In the case of the chemical structure of $FAPbI_yBr_{(3-y)}$, it has been shown that varying y from 3 to 0 leads to changing E_g from 1.48 eV to 2.23 eV, respectively (Fig. 4(i)) [45].

3.1.2 *Changing metal atoms (B)*

Silver bismuth iodide materials (Ag-Bi-I, $AgBi_2I_7$ as an example) have been introduced as one of the low-toxic lead-free light absorbents for photovoltaic applications. Johansson et al. have shown that partially exchanging Bi by Sn in the structure results in increasing both direct and indirect E_g values for this material [46]. Kanatzidis et al. indicated that by replacing Pb with Sn in $CH_3NH_3SnI_{(3-x)}Br_x$ structure (Fig. 4(ii)), the value of E_g can be tuned from 1.3 eV (for $x=1$) to 2.15 eV ($x=3$) [47]. Ogomi et al. have demonstrated that by varying x value in the $CH_3NH_3Pb_{(1-x)}Sn_xI_3$ from 0.3 to 1, conduction band energy shifts from −3.81 to −3.63 eV, and valence band energy shifts from −5.12 to −4.73 eV[40].

3.1.3 *Changing cations (A)*

Seok et al. have indicated that by increasing the size of the cation, the crystalline structure will be more cubic and E_g will be reduced [44]. Nevertheless, Mosconi et al. have shown that E_g of $FASnBr_3$ with cubic structure is larger than E_g value of $FASnI_3$ with tetragonal structure, even though its constant lattice is smaller [48]. Giustino et al. have investigated $MAPbI_3$ structure and have shown that without changing metal halide network and only by changing cation, the Pb-I-Pb bond angle and therefore E_g change [49].

Rothlisberger et al. utilized density functional theory (DFT) to investigate how changing chemical composition (ABX_3, A=Na, Li, B=Sn, Pb, and X=I, Br, Cl) affects E_g [50]. Fig. 5 demonstrates the cubic structures of $CsPbI_3$ (A) and its tetragonal (B) and orthorhombic (C) structures after BX_6 tilting. Replacing Cs as a cation in the cubic structure of $CsPbI_3$ with

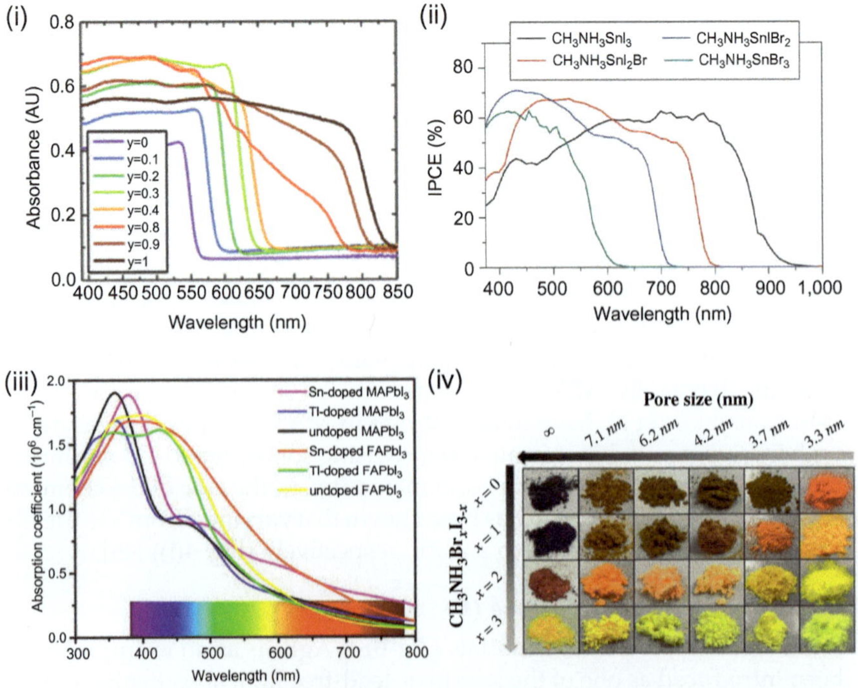

FIG. 4 Absorbance spectra of $FAPbI_yBr_{3-y}$ with varying y (i), IPCE spectra of $MASnI_xBr_{3-x}$ with varying x (ii), the absorption coefficient of Sn- and Ti-doped $MAPbI_3$ and $FAPbI_3$ in comparison with the undoped ones (iii), and color of $MABr_xI_{3-x}$ with varying x deposited on the nanoporous scaffold with a different pore size (iv). *(i) From G.E. Eperon, S.D. Stranks, C. Menelaou, M.B. Johnston, L.M. Herz, H.J. Snaith, Formamidinium lead trihalide: a broadly tunable perovskite for efficient planar heterojunction solar cells, Energ. Environ. Sci. 7(3) (2014) 982–988, https://doi.org/10.1039/c3ee43822h, with permission, (ii) From F. Hao, C.C. Stoumpos, D.H. Cao, R.P.H. Chang, M.-G. Kanatzidis, Lead-free solid-state organic-inorganic halide perovskite solar cells, Nat. Photonics 8(6) (2014) 489–494, https://doi.org/10.1038/nphoton.2014.82, with permission, (iii) From Z.K. Tang, Z.F. Xu, D.Y. Zhang, S.X. Hu, W.M. Lau, L.M. Liu, Enhanced optical absorption via cation doping hybrid lead iodine perovskites. Sci. Rep. 7(1) (2017), https://doi.org/10.1038/s41598-017-08215-3, with permission, and (iv) From V. Malgras, S. Tominaka, J.W. Ryan, J. Henzie, T. Takei, K. Ohara, Y. Yamauchi, Observation of quantum confinement in monodisperse methylammonium lead halide perovskite nanocrystals embedded in mesoporous silica, J. Am. Chem. Soc. 138(42) (2016) 13874–13881, https://doi.org/10.1021/jacs.6b05608, with permission.*

NH_4, PH_4, and Li leads to cubic structures of NH_4PbI_3, PH_4PbI_3, and $LiPbI_3$ respectively. In the cubic, tetragonal, and orthorhombic structures, if the "A" cation is replaced with a larger cation, the lattice constant increases, and if replaced with a smaller one, the lattice constant decreases. In the tetragonal and orthorhombic structures, by replacing "A" cation with a larger cation, the tilting angle of BX_6 decreases and increases for smaller cations. If Cs^+ is replaced with NH_4^+, due to strong hydrogen bonds in the structure, BX_6 octahedra deform, but in the case of PH_4^+, due to weak hydrogen bonds, the structure does not deform. They have

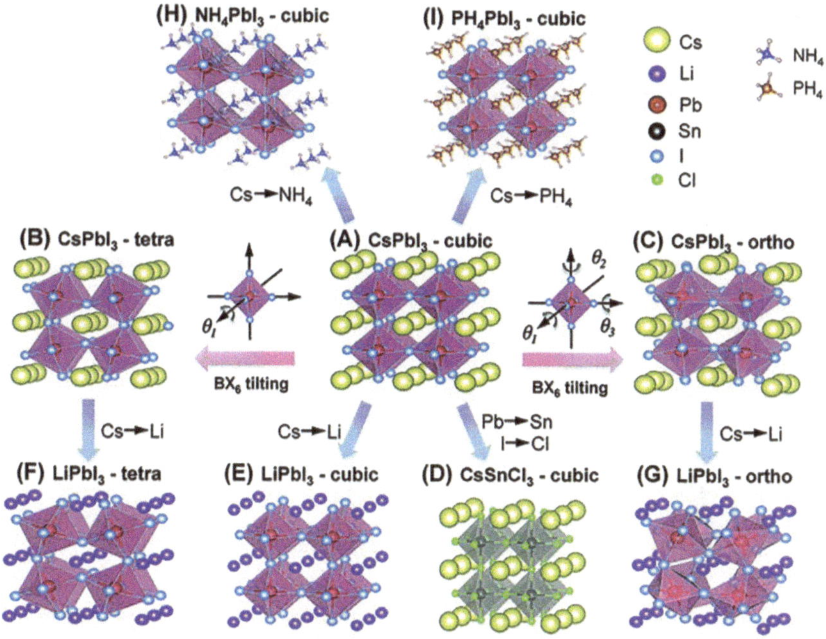

FIG. 5 The configuration of equilibrium structure of different perovskite materials replacing Cs (A) with Li (E), NH_4 (H), or PH_4 (I), or Pb with Sn and I with Cl (D) to obtain a cubic structure with another stoichiometry or tilting octahedra BX_6 structure to obtain a tetragonal (B and F) or orthogonal (C and G) structures. *From S. Meloni, G. Palermo, N. Ashari-Astani, M. Grätzel, U. Rothlisberger, Valence and conduction band tuning in halide perovskites for solar cell applications, J. Mater. Chem. A, 4(41) (2016) 15997–16002. https://doi.org/10.1039/c6ta04949d, with permission.*

mentioned that valence band maximum is formed by an antibonding combination of Pb-6p or Sn-4s, and I-5p, Br-4p, or Cl-3p. The size of the cation affects B-ns/X-mp orbitals. In both cubic and orthorhombic structures, by reducing lattice constant, overlap increases. Nevertheless, in orthorhombic structures, also by reducing BX_6 tilting angle, overlap decreases and the effect of tilting angle is dominant. Therefore, by reducing the size of the cation, overlap decreases. They have shown that by increasing negative overlap between orbitals of B and X ions, E_g decreases.

As another effect of cation on the value of E_g, Mc Gehee et al. have investigated $Cs_xFA_{(1-x)}Pb(Br_yI_{(1-y)})_3$ structure and showed that by increasing either x value or y value, bandgap almost increases linearly, but in the case of tuning bandgap by Cs content, perovskite material has higher photostability [51].

3.1.4 *Doping*

Doping is a well-known strategy to tune the optical, absorption, and electric structure of perovskite materials by introducing new energy states in the bandgap. One of the most important applications of doping is tuning photovoltaic characteristics in the case of utilizing perovskite materials as a light absorbent. Light-absorbent materials with a higher bandgap lead to a higher open-circuit voltage (V_{oc}) but lower current density because in comparison with a light-absorbent material with a lower bandgap materials. Utilizing a light-absorbent material with low bandgap results in higher current density but lower Voc. Thus, doping is the best possible solution in this case. Song et al. have observed a near-infrared photoluminescence peak by introducing "Bi" in the $CH_3NH_3PbI_3$ structure [52]. Fig. 4(iii) shows doping in $MAPbI_3$, and $FAPbI_3$ by "Sn" results in increasing light absorption in the near-infrared part of the spectrum significantly [53]. In the case of light-emitting materials, doping by Mn_2^+ ions is well-known due to Mn d-d orange emission. This doping has already been reported in several articles for doping nanocrystalline $CsPbCl_3$ or $CsPbBr_3$ and other structures of inorganic perovskite materials that have a wide bandgap [53]. The significant redshift of the photoluminescence peak is observable by increasing the content of Mn in the structure [54].

3.1.5 *Reducing the size (quantum confinement effect)*

For various optoelectronic applications, the layered perovskites have attracted much research attention due to their moisture resistance and moisture stability [55–57]. By the incorporation of larger cation (A^+) in 3D structure, additional alternatives such as low-dimensional layered structures (2D, 1D, and 2D-3D mixed) are observed. This cationic incorporation leads to the reconstruction of a 3D framework into a 2D-3D mixed perovskite phase, which is known as a "quasi-2D" structure [58]. The Ruddlesden-Popper (RP) and Dion-Jacobson (DJ) with $n=2$ layered phases are the two types of quasi-2D structures whose formation depends on the choice of A^+. As represented in Fig. 6(i and ii) by inserting a pair of monovalent organic cations such as benzyl ammonium (BZA), octylammonium (OA), and phenylethyl ammonium (PEA), in between the inorganic octahedral layers, results in the formation of RP perovskite structure. However, by inserting a single layer of a divalent organic cation such as 3-(aminomethyl) piperidinium (3-AMP) or 4-(aminomethyl) piperidinium (4-AMP) or diammonium cation such as 1,4-benzenediamethanamonium), the D-J phase of perovskite is formed.

The quasi-2D structure is stable as compared to its counterpart 3D structure. The geometry of the quasi-2D structure can be conceptualized by cutting the 3D structure in-plane along <100> crystallographic direction (Fig. 7 (i)). Typically, the large organic amine groups incorporated during the crystal growth cannot enter into the octahedron, thus inhibiting the growth of octahedron along out-of-plane direction [59–61]. The chemical formula of quasi-2D perovskite is represented by $A'_2A_{n-1}B_nX_{3n+1}$ ($1 \leq n \leq \infty$), where the large cation is denoted by A' or A'' ($^+H_3N\text{-}R\text{-}NH_3^+$) with R representing the aromatic ligand of alkyl chain (Fig. 7(ii)). Similarly, the

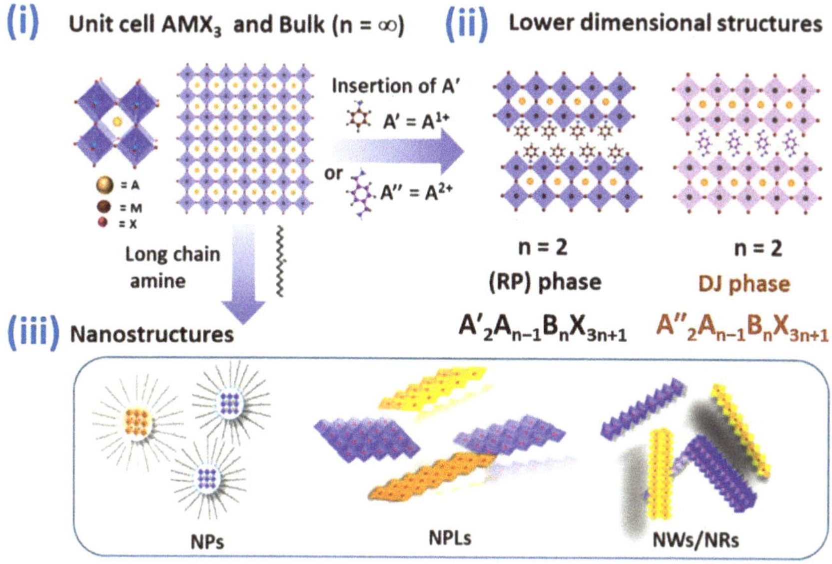

FIG. 6 The schematics of low-dimensional layered perovskite. The schematic illustration of (i) AMX_3 unit cell and 3D bulk form ($n = \infty$), (ii) low-dimensional perovskite phases with a layered structure, e.g., the Ruddlesden-Popper (RP) and Dion-Jacobson (DJ) with $n = 2$ layered phases, and (iii) various morphologies such as nanoparticles (NPs), nanoplatelets (NPLs), and nanowires/nanorods (NWs/NRs) can be synthesized by judicially adjusting the parameters. *From S.A. Kulkarni, N. Yantara, K.S. Tan, et al., Perovskite nanostructures: leveraging quantum effects to challenge optoelectronic limits, Mater. Today 33 (2020) 122–140.*

monovalent small cations such as methylammonium (MA^+), formamidine (FA^+), and cesium (Cs^+); B stands for divalent cation, e.g., lead (Pb^+) or tin (Sn^+); X refers to halide ions such as chloride, bromide, and iodide. Here, n stands for the octahedral units. Typically, A′ or A″ serves as an insulating layer that isolates the inorganic metal halide layers (octahedral units ($[BX_6]^{4-}$)) connected together by corner-sharing halide ions, and the voids between the framework by A cations [62,63]. By judicially controlling the various parameters, the nanostructures with different morphologies and specific dimensionalities such as quantum dots (QDs), nanoplatelets (NPLs), nanowires (NWs), or nanorods (NRs) can be synthesized (Fig. 6 (iii)) [64–66]. The typical quantum confinement effect arises due to a reduction in layer thickness up to several unit cells. Thus, both layered perovskite and nanostructures leverage the quantum confinement effect and show extraordinary photophysical properties [67] to yield slow hot carrier cooling, high-performance LEDs, and lower threshold polarized lasing. In layered structure, the series of layers of organic and inorganic sheets are arranged alternatively where the inorganic octahedral layers are sandwiched in between two layers of organic spacer (conventional; $C_6H_5(CH_2)_2NH_3^+$ and $C_4H_9NH_3^+$), which has a relatively low dielectric constant [68]. These layered

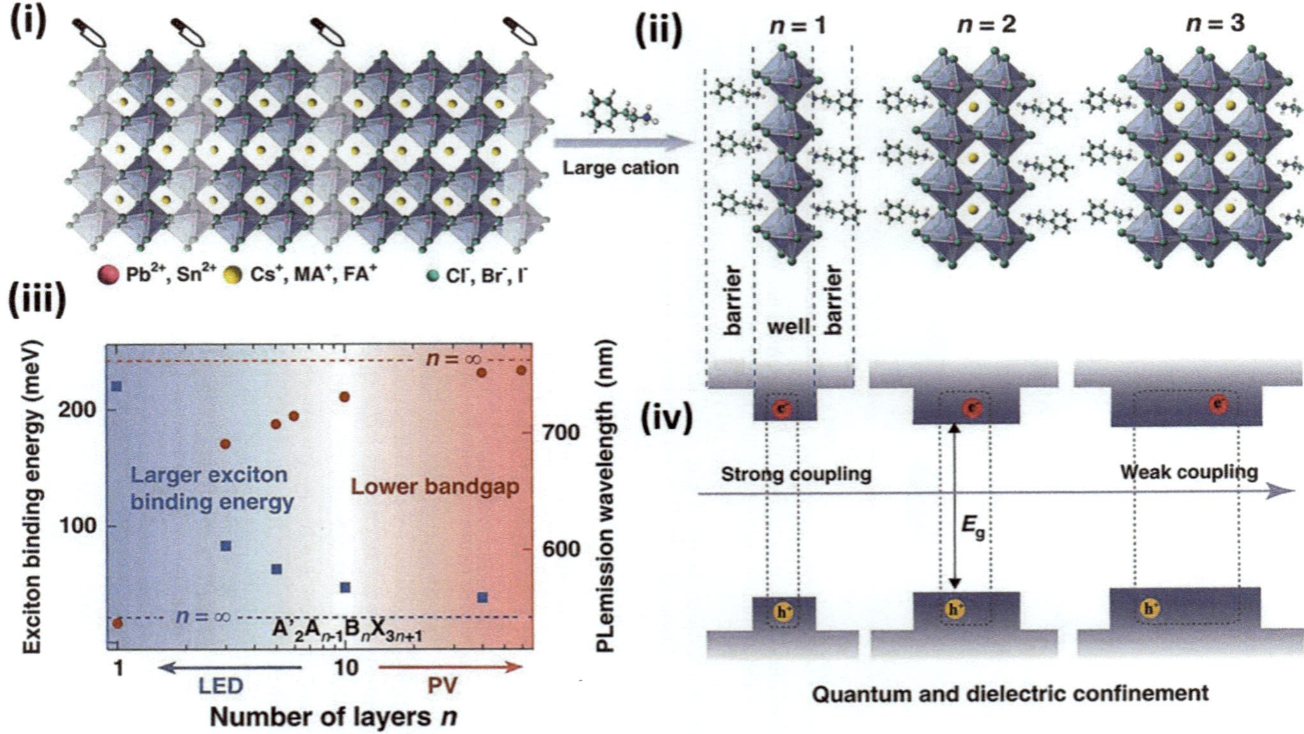

FIG. 7 (i) The schematics show the slicing of 3D perovskite along with the specific crystallographic direction (<100>) to obtain quasi-2D perovskite. (ii) The representation of quasi-2D perovskite crystal structure with various values of *n*. (iii) The strong and weak couplings in a layered structure depend on a different value of n where the weak coupling is observed for higher *n* values. (iv) The graph represents the exciton binding energies (Eb) and photoluminescence (PL) emission wavelength for various values of n in quasi-2D layered perovskite. *(i–iii) From L. Zhang, C. Sun, T. He, Y. Jiang, J. Wei, Y. Huang, M. Yuan, High-performance quasi-2D perovskite light-emitting diodes: from materials to devices. Light Sci. Appl. 10(1) (2021), https://doi.org/10.1038/s41377-021-00501-0, with permission and (iv) From L.N. Quan, F.P.G. de Arquer, R.P. Sabatini, E.H. Sargent, Perovskites for light emission, Adv. Mater. 30 (2018) 1801996, https://doi.org/10.1002/adma.201801996 with permission.*

structures are confined into quantum wells because of the potential barrier created by the organic spacer layer. As illustrated in Fig. 7(ii), the organic capping layer (spacer) serves as a "barrier," while the inorganic unit ($[BX_6]^{4-}$) acts as a "well." Therefore, the quasi-2D perovskite structure forms with an atomically fine and sharp interface between "wells" and "barriers." The dielectric constant between the organic barrier and inorganic wells strengthens the 2D confinement by image charge effect, also termed dielectric confinement, which makes the exciton binding energy (E_b) even larger (approx. a few hundred millivolts) [13]. Therefore, for layered structures, the selection of barrier (organic spacer) with different values of dielectric constant can significantly adjust the dielectric confinement effect and has a strong influence on controlling the value of E_b [69]. Due to both quantum confinement and dielectric confinement effect appearing from quantum well structure, the E_b of 2D and quasi-2D layered structure is much larger than their counterpart 3D structure. The large E_b makes it hard to produce free carriers via thermal activation. In these layered structures, the wave function is compressed in one direction because of quantum-well width limitation, which limits the charge carrier movement, and as a result, an increase in E_g and effective bandgap is observed. The various values of E_b and E_g against different values of n in a layered structure are shown in Fig. 7(iii). Thus, the carrier mobility in these structures is about two orders lower in magnitude as compared to 3D counterparts [68].

The degree of quantum confinement is also defined by the value of n, where the small value of n indicates the large bandgap (E_g) perovskites (Fig. 7(iii)). Thus, in 2D perovskites, the increasing value of n allows the E_g value to approach the value given by the 3D counterparts, which are suitable for photovoltaic applications whose value E_g (single junction) is ~1.1–1.4eV[37,70,71]. The quasi-2D single-crystal layered structure with $n=1, 2, 3$, and 4 has been successfully studied by Ishihara et al. [72]. Similarly, Stoumpos et al. synthesized and characterized the perovskite ($(CH_3(CH_2)_3NH_3)_2(CH_3NH_3)_4Pb_5I_{16}$) with $n=5$ [73]. The various values of n were systematically studied to profoundly understand the photophysical properties of a layered structure.

According to Table 1, the reported nanostructures of perovskite materials with the same chemical stoichiometry typically do not result in a significant change in the wavelength of emission peak. The significant changes in the E_g and therefore wavelength of the emission peak are observable for nanostructures with a size of less than 10nm due to quantum confinement effects. Yamauchi et al. [74] have utilized nanoporous silica with cylindrical pores (pore sizes of 3.3, 3.7, 4.2, 6.2, and 7.2nm (Fig. 7(iv)) as a scaffold for $CH_3NH_3Br_xI_{(3-x)}$ ($x=0, 1, 2$, and 3)). E_g values from 1.67 to 2.22eV, 1.86 to 2.28eV, 2.07 to 2.48eV, and 2.21 to 2.62eV were obtained by reducing pore size of the scaffold from 7.1 to 3.3nm for $x=0, 1, 2$, and 3, respectively, in their report. Applying a porous

TABLE 1 The optoelectronic properties of various common halide perovskites.

Formula	Angle length (°)	Angle bond (°)	Crystal system	Space group	Transition temperature (K)	Band gap (eV)	Lifetime (bulk, ns)	Trap density (cm^{-3})
$MAPbCl_3$	a = b = c = 5.666 a = 11.1763 b = 11.3409 c = 11.2804	α = β = γ = 90 α = β = γ =90	Cubic Orthorhombic	$Pm\bar{3}m$ *Pnma*	(Cubic, α-phase) 177.2 (Tetragonal, β-phase) 171.5 (Orthorhombic, γ-phase)	2.88	83±4 662±44	3.1×10^{10}
$MAPbBr_3$	a = b = c = 5.933	α = β = γ = 90	Cubic	$Pm\bar{3}m$	Cubic, α-phase) 236.3 (Tetragonal, β-phase) 155.1 (Tetragonal, γ-phase) 148.8 (Orthorhombic, δ-phase)	2.31 (Cubic) 2.22 (Tetragonal)	41±2 357±11	$(5.8 \pm 0.6) \times 10^9$
$MAPbI_3$	a = b = c = 6.391 a = 8.849 b = 8.849 c = 12.642	α = β = γ = 90 α = β = γ = 90	Cubic Tetragonal	$Pm\bar{3}m$ 14*cm*	(Cubic, α-phase) 330.4 (Tetragonal, β-phase) 161.4 (Orthorhombic, γ-phase)	1.54	22±6 1032±150	3.6×10^{10}
$CsPbCl_3$	a = b = c = 5.605	α = β = γ =90	Cubic	$Pm\bar{3}m$	(Cubic) 320 (Tetragonal) 315 (Orthorhombic)	2.85	–	–
$CsPbBr_3$	a = 8.2440 b = 11.7351 c = 8.1982	α = β = γ = 90	Orthorhombic	*Pnma*	(Cubic) 404 (Tetragonal) 325 (Orthorhombic)	2.26	4.4±0.1 30±3	3.6×10^{12}
$CsPbI_3$	a = 10.4342 b = 4.7905 c = 17.7610	α = β = γ = 90	Orthorhombic	*Pnma*	(Cubic) 634 (Orthorhombic, δ-phase)	1.76 (Cubic); 2.78 (Orthorhombic)	–	–

$FAPbBr_3$	a = b = c = 5.9944	α = β = γ = 90	Cubic	$Pm\bar{3}m$	–	2.27 (Cubic,*Pm3m*) 2.16 (Cubic, Im3) 2.10 (*Pm3m*)	687±23 2272±43	9.6×10^9
$FAPbI_3$	a = b = c = 6.3573 a = 8.6603 b = 8.6603 c = 7.9022 a = 17.7914 b = 17.7914 c = 10.9016 a = b = 8.98 c = 11.01	α = β = γ = 90 α = β = 90 γ = 120 α = β = 90 γ = 120 α = β = 90 γ = 120	Cubic Hexagonal Trigonal Trigonal	$Pm\bar{3}m$ $P6_3mc$ *P3* *P3m1*	(α-phase) 423 (δ-phase) (Tetragonal, β-phase) 140 (Trigonal, γ-phase)	1.51 (α-phase); 2.14 (δ-phase)	32 484	1.13×10^{10} 1.5×10^{11} (α-phase) 2.6×10^{12}

From K. Chen, S. Schünemann, S. Song, H. Tüysüz, Structural effects on optoelectronic properties of halide perovskites, Chem. Soc. Rev. 47(18) (2018) 7045–7077, https://doi.org/10.1039/c8cs00212f, with permission.

polydimethylsiloxane (PDMS) scaffold for perovskite [75] is also another approach. Utilizing scaffold for perovskite material can also protect it against environmentally destructive agents.

4. Thermal transport in halide perovskite

Thermal management plays a critical role in the design of solid-state materials for energy applications. Lead halide perovskites have emerged as promising candidates for thermoelectric applications, but their thermal properties are still poorly understood. Due to outstanding thermal and magnetic properties, recently perovskite has been extensively studied and successfully employed in various applications [75]. With the first report by Pisoni et al., on the thermal properties of $MAPbI_3$ [76], considerable efforts were directed toward the understanding of thermal properties, especially thermal transport and phonon behavior [77–79]. Interestingly, the different phonon- and electron-like dynamics in halide perovskites are classified as "phonon glass electron crystal" (PGEC) structures where charge transport is the same as found in an efficient semiconductor; however, similar to glass, a lower thermal conductivity is observed [80]. Similarly, Zhu et al. introduced a new name, "crystalline-liquid" to organic-inorganic halide perovskite material due to the coexistence of "crystalline" behavior in coherent transport of charge carriers also known as band-like transport and "liquid" like phonons activities have encouraged [81]. In PGECs, the transport of charge carriers and phonons is ballistic and diffusive, respectively, which ensures the highly efficient thermoelectric phenomenon of enhanced electric conductivity but lower thermal conductivity. The efficient electric conductivity is due to large electronic bandwidth; however, local vibrations interrupted the phonon bandwidth.

These days, the progress in the development of new thermal characterization tools is underway. Various kinds of custom-made setups for measuring thermal conductivities in perovskite are briefly summarized and shown in Fig. 8. The common thing among all techniques is the application of heat within the volume or at the surface of the material under investigation. However, it must be insured that the thermal measurement experiment must be carried out in dark to avoid the undesirable photo-induced effect. Also, it must be assured that the thermal conductivity measurement itself does not lead to the artifacts. As the thermal transport is greatly influenced by the presence of grain boundaries in polycrystalline materials, special care must be considered in this regard. Furthermore, in contact-type measuring technique, the glue applied to mount the thermocouples or heaters on the surface of sample causes degradation as well (Fig. 8(ii–iv)). Therefore, the noncontact-type optical flash measurement methods based on frequency-domain thermoreflectance (FDTR) were introduced (Fig. 8(v)). Interested readers may refer to these articles [82,83].

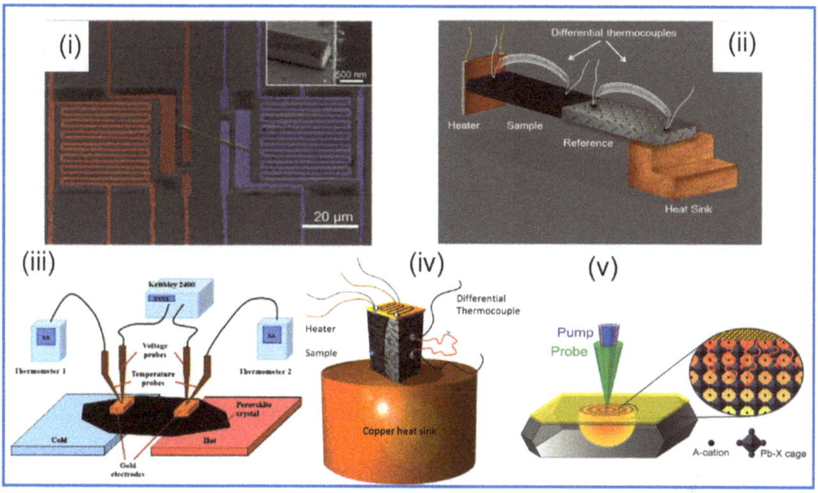

FIG. 8 The various techniques employed for the measurement of thermal conductivity of perovskite material. (i) Thermal conductivity of perovskite nanowires where the false color SEM image represents a microbridge device with a perovskite nanowire bridging between two suspended pads. A high-magnification SEM image of a typically measured nanowire can be seen in the inset, (ii) experimental setup for the measurement of thermal conductivity where the heater is connected at one side and *orange solid* represents the copper that acts as a heat sink. The *black* color denotes the samples, and the *gray* color represents the stainless steel that serves as a reference. The spiral springs are the thermocouples (type E) that are connected to the sample and reference surface through conducting epoxy, (iii) the schematics for custom-made setup for measuring the thermal conductivity of perovskite crystal. The DC power supply elements with electric resistance were used to produce a thermal gradient. Two thermometers are connected through copper wires to measure the temperature gradient, (iv) the home-made setup where the sample is placed on a copper heat sink and heated from the top through an electric resistance heater, and (v) the schematics of FDTR measurement on gold-coated perovskite single crystal. The crystal surface is heated by a laser pump *(blue)*, while the temperature-dependent measurement is taken through a probe laser *(green)* by reflectance from gold. *(i) From Y. Wang, R. Lin, P. Zhu, Q. Zheng, Q. Wang, D. Li, J. Zhu, Cation dynamics governed thermal properties of lead halide perovskite nanowires, Nano Lett. 18(5) (2018) 2772–2779, https://doi.org/10.1021/acs.nanolett.7b04437, with permission, (ii) From A. Pisoni, J. Jaćimović, O.S. Barišić, M. Spina, R. Gaál, L. Forró, E. Horváth, Ultra-low thermal conductivity in organic-inorganic hybrid perovskite* $CH_3NH_3PbI_3$*, J. Phys. Chem. Lett. 5(14) (2014) 2488–2492, https://doi.org/10.1021/jz5012109, with permission, (iii) From T. Ye, X. Wang, X. Li, A.Q. Yan, S. Ramakrishna, J. Xu, Ultra-high Seebeck coefficient and low thermal conductivity of a centimeter-sized perovskite single crystal acquired by a modified fast growth method, J. Mater. Chem. C, 5 (2017) 1255–1260, https://doi.org/10.1039/C6TC04594D, with permission, (iv) From A. Kovalsky, L. Wang, G.T. Marek, C. Burda, J.S. Dyck, Thermal conductivity of CH3NH3PbI3 and CsPbI3: measuring the effect of the methylammonium ion on phonon scattering, J. Phys. Chem. C. 121(6) (2017) 3228–3233, https://doi.org/10.1021/acs.jpcc.6b12231, with permission, and (v) From G.A. Elbaz, W.L. Ong, E.A. Doud, P. Kim, D.W. Paley, X. Roy, J.A. Malen, Phonon speed, not scattering, differentiates thermal transport in lead halide perovskites, Nano Lett. 17 (9) (2017) 5734–5739, https://doi.org/10.1021/acs.nanolett.7b02696, with permission.*

4.1 Compositional effect on thermal conductance

Temperature-dependent thermal conductivity of $MAPbI_3$ was reported by Xin Qian et al. using molecular dynamic simulations and density functional theory. A very low value of $0.59\,Wm^{-1}K^{-1}$ was observed for the tetragonal phase at temperature range of 160–300 K; however, the thermal conductivity increased up to $1.8\,Wm^{-1}K^{-1}$ at higher temperature (>330 K) where perovskite transforms to pseudocubic phase [84]. A strong anharmonicity and low velocity of acoustic phonons are attributed to lower thermal conductivity in the tetragonal phase. The first experimental report on thermal conductivity (k) of bulk single-crystal ($0.5\,Wm^{-1}K^{-1}$) and polycrystalline ($0.3\,Wm^{-1}K^{-1}$) $MAPbI_3$ was investigated by Pisoni et al., in which the author claimed that the phonon wave interaction and disorder of organic cation ($CH_3NH_3^+$) with its slow rotation are dominant mechanisms responsible for heat attenuation and for extremely low thermal conductivity at room temperature [76]. Later on, the report has attracted much research attention to the thermoelectric properties of perovskite materials. Compared to bulk single and polycrystalline growth, the solution-processed thin film contains different microstructures due to the presence of defects and grains, which leads to distinct thermal transport behavior. A similar value of thermal conductivity ($0.3–0.5\,Wm^{-1}K^{-1}$) of solution-processable $MAPbI_3$ thin film was calculated using the time-domain thermoreflectance (TDTR) pump-probe method [85]. Later, Chen et al. observed a higher value of thermal conductivity ($11.2 \pm 0.8\,Wm^{-1}K^{-1}$ at room temperature), in densely packed perovskite ($MAPbI_3$) film using the same TDTR technique [86]; however, this increased value has not been confirmed by anyone else. As compared to the previous reports, the higher value reported by Chen et al, was attributed to the loosely packed and higher roughness of the film. The thermal conductivity of three kinds of lead halide perovskite ($CH_3NH_3PbI_3$, $CH3NH_3PbBr_3$, and $CsPbBr_3$) single-crystal nanowires has been studied by Yuxi wang et al. The typical device design can be seen in Fig. 8(i) in which the perovskite nanowires have been bridged between two electrodes [87]. The color indication represents the heat flow from the heating membrane (red) to the sensor membrane (blue) through the bridged perovskite nanowire (green). The lower value of λ as compared to CsPbBr3 ($0.36\,Wm^{-1}\,K^{-1}$) is attributed to the dynamic disorder of organic cation, especially in lower temperature regime (<150 K); however, at higher temperatures, both lower group velocity and Umklapp scattering are possible reasons of lower thermal conductivity [81]. Table 2 represents the measurement technique used for measuring the thermal conductivity of perovskite materials and the value of thermal conductivity for various perovskite compositions.

In a working perovskite PV device, a large portion of sunlight is converted into heat in form of heat quanta "phonon" and subsequently activates phase transition in perovskite film in addition to the corresponding variation in their lattice parameters [91,92]. Such temperature or

TABLE 2 Values of thermal conductivities of the various compositions of perovskite along with their crystal structure and techniques used.

Sr. #	Materials	Crystal structure/ phase	Temperature range (K)	Thermal conductivity (λ) ($W m^{-1} K^{-1}$)	Measurement technique and instrument used	Reference
1	$CH_3NH_3PbI_3$	Single-crystal nanowires	35–320	0.22	Perovskite nanowires were bridged between two suspended membranes, both of which serve as a thermometer and heater (Fig. 8(i))	Wang et al. [87]
2	$CH_3NH_3PbBr_3$			0.32		
3	$CsPbBr_3$			0.36		
4						
5	$CH_3NH_3PbI_3$	Single crystal		0.34	Scanning near-field thermal microscopy (SThM) based on the 3ω-technique	Heiderhoff et al. [88]
6	$CH_3NH_3PbBr_3$	Single crystal		0.44		
7	$CH_3NH_3PbCl_3$	Single crystal		0.50		
8	$MAPbCl_3$ (cubic)	Single crystal	Room temperature	0.73 ± 0.18	Frequency-domain thermoreflectance (FDTR) measurements (Fig. 8(v))	Elbaz et al. [89]
9	$MAPbBr_3$ (cubic)			0.51 ± 0.12		
10	$MAPbI_3$ (tetragonal)			0.34 ± 0.08		
11	CsPbBr3 (orthorhombic)			0.46 ± 0.12		
12	$FAPbBr_3$ (cubic)			0.49 ± 0.12		
13	$CH_3NH_3PbI_3$	Single crystal	298–425	0.30–0.42	Custom-made setup; the sample was put on two substrates with a built-in resistive	Ye et al. [90]

Continued

TABLE 2 Values of thermal conductivities of the various compositions of perovskite along with their crystal structure and techniques used—cont'd

Sr. #	Materials	Crystal structure/ phase	Temperature range (K)	Thermal conductivity (λ) ($W m^{-1} K^{-1}$)	Measurement technique and instrument used	Reference
					heater, and the set temperature can be adjusted by the resistive heaters that were driven by a direct current power supply (Fig. 8(iii))	
14	$CH_3NH_3PbI_3$	Film	294	0.4	Noncontact time-domain thermoreflectance (TDTR) technique	Guo et al. [85]
15	$CH_3NH_3PbI_3$	Densely packed perovskite film	330	11.2±0.8	Noncontact time-domain thermoreflectance (TDTR) technique	Chen et al. [86]

temperature gradient initiates mechanical deformations in terms of residual stress and strain, which has a strong impact on optoelectronic properties of perovskite material, particularly in concentrator solar cells, LEDs, lasers, and photosensors; finally, structural and functional failure is observed over long device operating conditions [93]. Effective thermal management such as optimization of device geometry, doping in charge transport layers and connecting heat sink and spreaders and acquisition of material's thermal properties are imperative as both device performance and its lifetime are influenced by thermal gradients. Regarding perovskite PV devices, the lack of thermal evacuation causes mechanical stresses and decoupling of constituent layers of sandwiched structure, which results in a severe decline in device efficiency [76]. Similarly, the long-term operational stability of commercial PeLED is a daunting challenge where the device efficiency is notably reduced under the constant operating conditions at high current density, referred to as efficiency droop [94,95]. Under continuously operating conditions and high current density, the loss in device efficiency is attributed to luminescence quenching, severe auger recombination, as well as intensive joule heating [96,97].

Very recently, Xu et al. employed monocrystalline silicon as a substrate for PeLEDs and found enhanced device performance under continuous operation [98]. The Si has been proved as the best heat sink under continuous operational conditions, and the device configuration is shown in Fig. 9(i). The Si-based device temperature has merely increased up to 28–33°C at a high current density (300 mA cm^{-2}); however, the temperature of the device where ITO coated glass as the substrate has increased up to 85°C (Fig. 9(i)). Similarly, the time-dependent luminance (normalized) at various constant current densities of 100, 200, and 300 mA cm^{-2} for both Si-based and ITO-based devices was measured. The Si-based devices showed 113 times longer half-life (the time required to drop luminance to half of the original value) as compared to their counterpart (ITO-based devices) as shown in Fig. 9(iii and iv). Thus, due to less thermal effects and excellent heat dissipation, Si-based devices proved as efficient and stable PeLEDs for futuristic applications.

Lead halide perovskites decomposes even at moderate temperature (~60°C) therefore due to their low thermal conductivity the heat stress and the light illumination further accelerated the decomposition process. Therefore, high thermal conductivity is anticipated to dissipate effectively the heat and thus mitigate the thermal decomposition problem in perovskite-based optoelectronic devices.

5. Other outstanding properties

5.1 Ferroelectricity

Since first insight of halide perovskite in photovoltaics, one of the physical properties currently under strong investigation is a ferroelectric response in these materials. The role of ferroelectricity in approaching

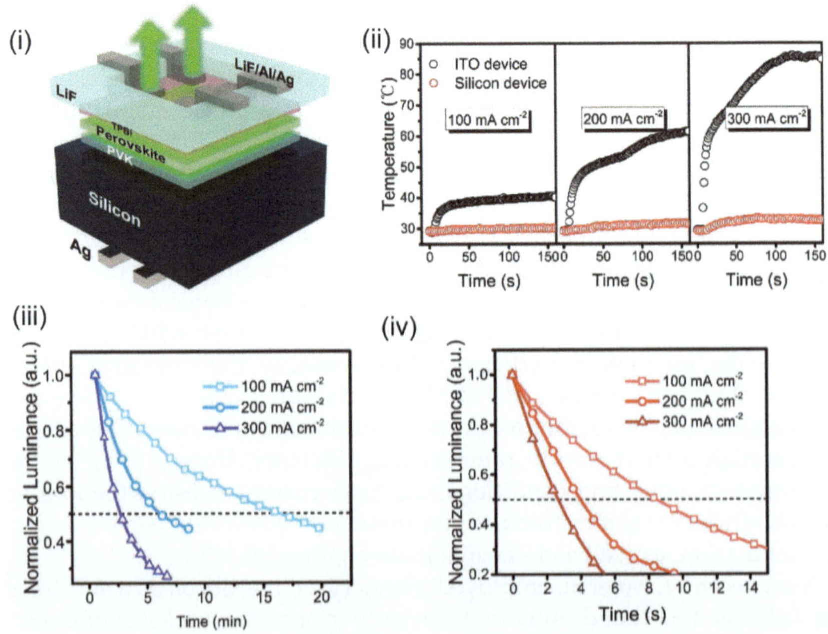

FIG. 9 The device architecture of PeLED on a silicon substrate with Ag/c-Si/perovskite/TBPI/LiF/Al/Ag configuration (i), variation in surface temperature of Si-based and ITO-based devices at various constant current densities ($100\,mA\,cm^{-2}$, $200\,mA\,cm^{-2}$, and $300\,mA\,cm^{-2}$, respectively) (ii), at different constant current densities, the luminance with respect to time for PeLED devices on ITO substrate (iii) and Si substrate (iv). *From H. Xu, X. Wang, Y. Li, L. Cai, Y. Tan, G. Zhang, Y. Wang, R. Li, D. Liang, T. Song, B. Sun, Prominent heat dissipation in perovskite light-emitting diodes with reduced efficiency droop for silicon-based display, J. Phys. Chem. Lett. 11(9) (2020) 3689–3698, https://doi.org/10.1021/acs.jpclett.0c00792, with permission.*

power conversion efficiency to its theoretical limit is still an enigma. Ferroelectricity defines by the spontaneous permanent polarization upon application of an external electric field and can also be defined as piezoelectric materials with reversible polarization [99]. Ferroelectric materials are a subclass of piezoelectric materials and are also known to be wide-bandgap polarizable materials. The ferroelectric nanodomains decrease the recombination of electron-hole pairs increasing carrier lifetime and hence intensifying the charge separation and its collection [100]. In organic-inorganic halide perovskites, while maintaining their semiconducting behavior, the organic part imparts ferroelectricity as well as chirality. In layered dimensional perovskite structure, the specific large "A" cation imparts an additional asymmetry to the structure, the alignment of which imparts ferroelectricity [101]. The weak interaction (hydrogen and halogen) between the A-site organic spacer and the octahedral layer (BX_6) imparts a certain level of dynamics in the A-site that results in A-site ordering of the dipole and resultant polarization [102]. At low temperature, the organic spacer in a layered structure aligns its dipole moment in a well-ordered configuration; however, at a higher temperature, the organic spacer becomes disordered, thus diminishing the polarization. Such phenomenon is similar to ferroelectric transition in many "ordered-disorder" types of oxides ($NaNO_2$) [103]. In 1D hybrid perovskite materials, the existing reports suggested that the ferroelectric polarization helps in band alignment favoring charge generation and charge transport affecting photovoltaic performances [100,104,105]. In a working PV device, the ferroelectricity in perovskite may also lead to the chemical segregation affecting the current-voltage curves of the solar cells [104,106].

Ferroelectricity in materials is characterized by switching of macroscopic polarization under an applied electrical field resulting in polarization-vs-electric (P-E) hysteresis loop. Although perovskite oxides show ferroelectric behavior, the presence of ferroelectricity in lead halide perovskite is still under debate. The ferroelectric behavior has been observed for the $MAPbI_3$ single crystals from dielectric and P-E loop measurements [107–109]. The measurement performed on $MAPbI_3$ device structure confirmed the absence of ferroelectricity in $MAPbI_3$ polycrystalline films [110]. Many other reports reported the presence of ferroelectric behavior in polycrystalline $MAPbI_3$ films using piezoresponse force microscopy (PFM) and quantify the ferroelectricity [106,111,112]. However, work performed by Liu et al. suggested that the ferroelastic nature or chemical segregation effect in halide perovskite results in the appearance of domain structure in the PFM measurements [113]. The use of conventional piezoresponse force microscopy technique for the determination of ferroelectricity in lead halide perovskite is limited by their high electronic and ionic conductivity. The conventional measurement, in materials which are not ferroelectric, could lead to a potential artifact such as ion migration, electrostatic forces, conducting states, and chemical compositions, as ferroelectric signals. Direct piezoelectric force microscopy (DPFM) has been employed to directly examine the ferroelectricity phenomenon in lead halide perovskite materials by Gomez et al. [99]. In DPFM, direct current signal is collected as a function of applied force and scan direction of the atomic force microscope tip in contact mode. A detailed experimental study has been performed by Gomez et al. on $MAPbI_3$ and triple cation perovskite films along with different ferroelectric materials varying the condition of the film. Their study confirms the origin of current profiles possibly due to ion migration or ferroelectricity. The ferroelectricity phenomenon has also been observed in many compositions tailored to hybrid halide perovskites due to the structural distortion and improved crystallinity [114,115].

In ferroelectric materials, the outstanding property of switchable polarization due to an applied electric field permits their employment in memory and logic device applications. Similarly, their extraordinary piezoelectric properties can be utilized in mechanical, optical, and energy transduction devices. Although these halide perovskite materials are extensively studied, the progress of halide perovskite ferroelectrics remains in its infancy stage as compared to their counterparts (inorganic perovskite). Various directions should be elucidated such as (i) to engineer the molecular structure, especially the organic cation parts, to increase the ferroelectric diploes and enhance the stability of a working device, (ii) insight into the mechanism that facilitates the design of material with wide bandgap, improved charged mobility, and large polarization, simultaneously, and (iii) in addition to PV devices, optical sensors, LEDs, and piezoelectric sensors, some sophisticated and high-quality devices like ferroelectric tunnel junctions and ferroelectric transistors should be developed urgently.

5.2 Ferroelasticity

Ferroelasticity is stress-induced crystallographic phase transformation resulting in spontaneous strain in the material by the displacement of domain walls separating ferroelastic domains. Ferroelasticity is the mechanical equivalent of ferroelectricity and ferromagnetism. It is defined by the elastic response of a material in the presence of applied external stress [116]. When stress is applied to the ferroelastic material, it results in the change in phase from one phase to another phase, which induces strain in the material.

A lot of studies have been made in recent years for the perovskite materials exhibiting ferroelasticity as a new variable for describing the high photovoltaic conversion efficiency. The coexistence of more than two phenomena such as piezoelectricity, ferroelasticity, and ferroelectricity in the perovskite materials makes it difficult to distinguish ferroelasticity from other phenomena. The study of ferroelasticity in perovskite materials provides an opportunity to control ferroelastic domains with applied stress exploited to tune the properties by strain engineering.

A study performed by Strelcov et al. confirms the existence of ferroelastic domains in both $CH_3NH_3PbI_3$ polycrystalline films and single crystals obtained using different fabrication methods in the pristine state and under applied stress by utilizing a combination of microscopic and nanoscale techniques [117]. Confirming the absence of ferroelectricity, ferroelastic domains in single crystals and polycrystalline films of $CH_3NH_3PbI_3$ can be controlled with applied stress. According to the report, $CH_3NH_3PbI_3$ exists in three crystallographic states, namely orthorhombic, tetragonal, and cubic phases. The lowering of symmetry from the cubic phase to the tetragonal phase can only be due to ferroelastic transition according to Aizu's classification [118–120].

Bari et al. found twin domains in an orthorhombic phase of $CH_3NH_3PbCl_3$ examined under uniaxial mechanical stress and electrical field proving ferroelasticity in the crystal [121]. The β and γ phases show changes in their domain structure on applying external uniaxial stress confirming the ferroelastic behavior of both phases. The absenteeism of ferroelectricity is confirmed by the absence of variation in domain wall positions on applying an external electric field of $15\,kV\,cm^{-1}$ for several minutes. Their study confirms that the $CH_3NH_3PbCl_3$ is not ferroelectric but ferroelastic in nature. The ab initio simulation was performed by Bristowe et al. in $CsPbI_3$ ferroelastic twins [122]. Their study provides a deep understanding of $CH_3NH_3PbI_3$ properties contributing to the result providing evidence on the ferroelectric nature of $CH_3NH_3PbI_3$.

The impact of the ferroelastic twin boundaries present in $CH_3NH_3PbI_3$ thin films and single crystals on the carrier transport and recombination in perovskites has been explored by the Huang group [123]. The electronically benign nature of the twin boundaries has been observed in their study and found to have an insignificant impact on the carrier transport along with acting as nonradiative recombination centers.

6. Conclusion

The chapter focused on the structural effect on halide perovskite properties. Various structural modifications such as ionic modifications, doping, and morphological impact on different optoelectronic properties in halide perovskite materials are discussed in detail. The quantum confinement and dielectric confinement in layered structures have a strong impact on controlling the values of exciton binding energy (E_b) and bandgap energy (E_g), thus bestowing astonishing optoelectronic properties. The effect of structural dimensionality on perovskite properties was thoroughly elaborated. Further, the thermal conductivity in halide perovskite materials along with their measurement techniques is explained. The ferroelectric and ferroelastic behavior in perovskite materials strongly enhances the efficiency of optoelectronic devices; however, the cause of such enhancement is still unclear and under investigation. Similarly, the last section indicates the presence of ferroelectric and ferroelastic effects in perovskite material. Thus, broad opportunities in this research direction can be explored by controlling the ferroelectric and ferroelastic crystal domains.

References

[1] H. Min, D.Y. Lee, J. Kim, G. Kim, K.S. Lee, J. Kim, M.J. Paik, Y.K. Kim, K.S. Kim, M.G. Kim, T.J. Shin, S. Il Seok, Perovskite solar cells with atomically coherent interlayers on SnO_2 electrodes, Nature 598 (7881) (2021) 444–450, https://doi.org/10.1038/s41586-021-03964-8.

[2] G. Pacchioni, Highly efficient perovskite LEDs, Nat. Rev. Mater. 6 (2) (2021) 108, https://doi.org/10.1038/s41578-021-00280-5.

[3] C. Li, Y. Ma, Y. Xiao, L. Shen, L. Ding, Advances in perovskite photodetectors, InfoMat 2 (6) (2020) 1247–1256, https://doi.org/10.1002/inf2.12141.

[4] H. Dong, C. Zhang, X. Liu, J. Yao, Y.S. Zhao, Materials chemistry and engineering in metal halide perovskite lasers, Chem. Soc. Rev. 49 (3) (2020) 951–982, https://doi.org/10.1039/c9cs00598f.

[5] Y. Zhou, J. Chen, O.M. Bakr, H.T. Sun, Metal-doped lead halide perovskites: synthesis, properties, and optoelectronic applications, Chem. Mater. 30 (19) (2018) 6589–6613, https://doi.org/10.1021/acs.chemmater.8b02989.

[6] Y. Fu, H. Zhu, J. Chen, M.P. Hautzinger, X.Y. Zhu, S. Jin, Metal halide perovskite nanostructures for optoelectronic applications and the study of physical properties, Nat. Rev. Mater. 4 (3) (2019) 169–188, https://doi.org/10.1038/s41578-019-0080-9.

[7] T.W. Lee, Emerging halide perovskite materials and devices for optoelectronics, Adv. Mater. 31 (47) (2019), https://doi.org/10.1002/adma.201905077.

[8] K. Wang, D. Yang, C. Wu, M. Sanghadasa, S. Priya, Recent progress in fundamental understanding of halide perovskite semiconductors, Prog. Mater. Sci. 106 (2019), 100580, https://doi.org/10.1016/j.pmatsci.2019.100580.

[9] X.K. Liu, F. Gao, Organic-inorganic hybrid Ruddlesden-popper perovskites: an emerging paradigm for high-performance light-emitting diodes, J. Phys. Chem. Lett. 9 (9) (2018) 2251–2258, https://doi.org/10.1021/acs.jpclett.8b00755.

[10] H. Tsai, W. Nie, J.-C. Blancon, C.C. Stoumpos, R. Asadpour, B. Harutyunyan, A.J. Neukirch, R. Verduzco, J.J. Crochet, S. Tretiak, L. Pedesseau, J. Even, M.A. Alam, G. Gupta, J. Lou, P.M. Ajayan, M.J. Bedzyk, M.G. Kanatzidis, A.D. Mohite, High-efficiency two-dimensional Ruddlesden–Popper perovskite solar cells, Nature 536 (7616) (2016) 312–316, https://doi.org/10.1038/nature18306.

[11] L. Mao, Y. Wu, C.C. Stoumpos, M.R. Wasielewski, M.G. Kanatzidis, White-light emission and structural distortion in new corrugated two-dimensional lead bromide perovskites, J. Am. Chem. Soc. 139 (14) (2017) 5210–5215, https://doi.org/10.1021/jacs.7b01312.

[12] M.E. Kamminga, H.H. Fang, M.R. Filip, F. Giustino, J. Baas, G.R. Blake, M.A. Loi, T.T.M. Palstra, Confinement effects in low-dimensional lead iodide perovskite hybrids, Chem. Mater. 28 (13) (2016) 4554–4562, https://doi.org/10.1021/acs.chemmater.6b00809.

[13] L. Pedesseau, D. Sapori, B. Traore, R. Robles, H.H. Fang, M.A. Loi, H. Tsai, W. Nie, J.C. Blancon, A. Neukirch, S. Tretiak, A.D. Mohite, C. Katan, J. Even, M. Kepenekian, Advances and promises of layered halide hybrid perovskite semiconductors, ACS Nano 10 (11) (2016) 9776–9786, https://doi.org/10.1021/acsnano.6b05944.

[14] M.I. Saidaminov, O.F. Mohammed, O.M. Bakr, Low-dimensional-networked metal halide perovskites: the next big thing, ACS Energy Lett. 2 (4) (2017) 889–896, https://doi.org/10.1021/acsenergylett.6b00705.

[15] J. Yin, P. Maity, M. De Bastiani, I. Dursun, O.M. Bakr, J.L. Brédas, O.F. Mohammed, Molecular behavior of zero-dimensional perovskites, Sci. Adv. 3 (12) (2017), https://doi.org/10.1126/sciadv.1701793.

[16] J. Chen, H. Xiang, J. Wang, R. Wang, Y. Li, Q. Shan, X. Xu, Y. Dong, C. Wei, H. Zeng, Perovskite white light emitting diodes: progress, challenges, and opportunities, ACS Nano 15 (11) (2021) 17150–17174, https://doi.org/10.1021/acsnano.1c06849.

[17] H. Wang, D.H. Kim, Perovskite-based photodetectors: materials and devices, Chem. Soc. Rev. 46 (17) (2017) 5204–5236, https://doi.org/10.1039/c6cs00896h.
[18] L. Lei, Q. Dong, K. Gundogdu, F. So, Metal halide perovskites for laser applications, Adv. Funct. Mater. 31 (16) (2021), https://doi.org/10.1002/adfm.202010144.
[19] K. Jesna George, V.V. Halali, C.G. Sanjayan, V. Suvina, M. Sakar, R.G. Balakrishna, Perovskite nanomaterials as optical and electrochemical sensors, Inorg. Chem. Front. 7 (14) (2020) 2702–2725, https://doi.org/10.1039/d0qi00306a.
[20] K. Chen, S. Schünemann, H. Tüysüz, Preparation of waterproof organometal halide perovskite photonic crystal beads, Angew. Chem. Int. Ed. 56 (23) (2017) 6548–6552, https://doi.org/10.1002/anie.201702556.
[21] Q. Dong, Y. Fang, Y. Shao, P. Mulligan, J. Qiu, L. Cao, J. Huang, Electron-hole diffusion lengths > 175 μm in solution-grown $CH_3NH_3PbI_3$ single crystals, Science 347 (6225) (2015) 967–970, https://doi.org/10.1126/science.aaa5760.
[22] J.H. Im, I.H. Jang, N. Pellet, M. Grätzel, N.G. Park, Growth of CH_3 NH_3 PbI_3 cuboids with controlled size for high-efficiency perovskite solar cells, Nat. Nanotechnol. 9 (11) (2014) 927–932, https://doi.org/10.1038/nnano.2014.181.
[23] H. Zhu, Y. Fu, F. Meng, X. Wu, Z. Gong, Q. Ding, M.V. Gustafsson, M.T. Trinh, S. Jin, X.-Y. Zhu, Lead halide perovskite nanowire lasers with low lasing thresholds and high quality factors, Nat. Mater. 14 (6) (2015) 636–642, https://doi.org/10.1038/nmat4271.
[24] V. Adinolfi, W. Peng, G. Walters, O.M. Bakr, E.H. Sargent, The electrical and optical properties of organometal halide perovskites relevant to optoelectronic performance, Adv. Mater. 30 (1) (2018), https://doi.org/10.1002/adma.201700764.
[25] G. Kieslich, S. Sun, A.K. Cheetham, An extended tolerance factor approach for organic-inorganic perovskites, Chem. Sci. 6 (6) (2015) 3430–3433, https://doi.org/10.1039/c5sc00961h.
[26] J.S. Manser, J.A. Christians, P.V. Kamat, Intriguing optoelectronic properties of metal halide perovskites, Chem. Rev. 116 (21) (2016) 12956–13008, https://doi.org/10.1021/acs.chemrev.6b00136.
[27] M.A. Green, A. Ho-Baillie, H.J. Snaith, The emergence of perovskite solar cells, Nat. Photonics 8 (7) (2014) 506–514, https://doi.org/10.1038/nphoton.2014.134.
[28] K. Korshunova, L. Winterfeld, W.J.D. Beenken, E. Runge, Thermodynamic stability of mixed Pb:Sn methyl-ammonium halide perovskites, Phys. Status Solidi B 253 (10) (2016) 1907–1915, https://doi.org/10.1002/pssb.201600136.
[29] V. D'Innocenzo, G. Grancini, M.J.P. Alcocer, A.R.S. Kandada, S.D. Stranks, M.M. Lee, G. Lanzani, H.J. Snaith, A. Petrozza, Excitons versus free charges in organo-lead tri-halide perovskites, Nat. Commun. 5 (2014), https://doi.org/10.1038/ncomms4586.
[30] K. Tanaka, T. Takahashi, T. Ban, T. Kondo, K. Uchida, N. Miura, Comparative study on the excitons in lead-halide-based perovskite-type crystals CH3NH3PbBr3 CH3NH 3PbI3, Solid State Commun. 127 (9–10) (2003) 619–623, https://doi.org/10.1016/S0038-1098(03)00566-0.
[31] F. Deschler, M. Price, S. Pathak, L.E. Klintberg, D.D. Jarausch, R. Higler, S. Hüttner, T. Leijtens, S.D. Stranks, H.J. Snaith, M. Atatüre, R.T. Phillips, R.H. Friend, High photoluminescence efficiency and optically pumped lasing in solution-processed mixed halide perovskite semiconductors, J. Phys. Chem. Lett. 5 (8) (2014) 1421–1426, https://doi.org/10.1021/jz5005285.
[32] S.D. Stranks, H.J. Snaith, Metal-halide perovskites for photovoltaic and light-emitting devices, Nat. Nanotechnol. 10 (5) (2015) 391–402, https://doi.org/10.1038/nnano.2015.90.
[33] Q. Lin, A. Armin, R.C.R. Nagiri, P.L. Burn, P. Meredith, Electro-optics of perovskite solar cells, Nat. Photonics 9 (2) (2015) 106–112, https://doi.org/10.1038/nphoton.2014.284.

[34] K. Gauthron, J.S. Lauret, L. Doyennette, G. Lanty, A.A. Choueiry, S.J. Zhang, A. Brehier, L. Largeau, O. Mauguin, J. Bloch, E. Deleporte, Optical spectroscopy of two-dimensional layered $(C_6H_5C_2H_4\text{-}NH_3)_2\text{-}PbI_4$ perovskite, Opt. Express 18 (6) (2010) 5912–5919, https://doi.org/10.1364/OE.18.005912.

[35] M.C. Gélvez-Rueda, M.B. Fridriksson, R.K. Dubey, W.F. Jager, W. van der Stam, F.C. Grozema, Overcoming the exciton binding energy in two-dimensional perovskite nanoplatelets by attachment of conjugated organic chromophores, Nat. Commun. 11 (1) (2020), https://doi.org/10.1038/s41467-020-15869-7.

[36] J. Tauc, A. Menth, States in the gap, J. Non Cryst. Solids 8 (1972) 569–585.

[37] W. Shockley, H.J. Queisser, Detailed balance limit of efficiency of p-n junction solar cells, J. Appl. Phys. 32 (3) (1961) 510–519, https://doi.org/10.1063/1.1736034.

[38] D. Kim, H.J. Jung, I.J. Park, B.W. Larson, S.P. Dunfield, C. Xiao, J. Kim, J. Tong, P. Boonmongkolras, S.G. Ji, F. Zhang, S.R. Pae, M. Kim, S.B. Kang, V. Dravid, J.J. Berry, J.Y. Kim, K. Zhu, D.H. Kim, B. Shin, Efficient, stable silicon tandem cells enabled by anion-engineered wide-bandgap perovskites, Science 368 (6487) (2020) 155–160, https://doi.org/10.1126/science.aba3433.

[39] K. Chen, S. Schünemann, S. Song, H. Tüysüz, Structural effects on optoelectronic properties of halide perovskites, Chem. Soc. Rev. 47 (18) (2018) 7045–7077, https://doi.org/10.1039/c8cs00212f.

[40] F.J. Iftikhar, Q. Wali, S. Yang, Y. Iqbal, R. Jose, S. Munir, I.A. Gondal, M.E. Khan, Structural and optoelectronic properties of hybrid halide perovskites for solar cells, Org. Electron. 91 (2021), https://doi.org/10.1016/j.orgel.2021.106077.

[41] S.D. Stranks, V.M. Burlakov, T. Leijtens, J.M. Ball, A. Goriely, H.J. Snaith, Recombination kinetics in organic-inorganic perovskites: excitons, free charge, and subgap states, Phys. Rev. Appl. 2 (3) (2014), https://doi.org/10.1103/PhysRevApplied.2.034007.

[42] Y. Dang, Y. Liu, Y. Sun, D. Yuan, X. Liu, W. Lu, G. Liu, H. Xia, X. Tao, Bulk crystal growth of hybrid perovskite material $CH_3NH_3PbI_3$, CrstEngComm 17 (3) (2015) 665–670, https://doi.org/10.1039/c4ce02106a.

[43] B. Suarez, V. Gonzalez-Pedro, T.S. Ripolles, R.S. Sanchez, L. Otero, I. Mora-Sero, Recombination study of combined halides (Cl, Br, I) perovskite solar cells, J. Phys. Chem. Lett. 5 (10) (2014) 1628–1635, https://doi.org/10.1021/jz5006797.

[44] J.H. Noh, S.H. Im, J.H. Heo, T.N. Mandal, S.I. Seok, Chemical management for colorful, efficient, and stable inorganic-organic hybrid nanostructured solar cells, Nano Lett. 13 (4) (2013) 1764–1769, https://doi.org/10.1021/nl400349b.

[45] G.E. Eperon, S.D. Stranks, C. Menelaou, M.B. Johnston, L.M. Herz, H.J. Snaith, Formamidinium lead trihalide: a broadly tunable perovskite for efficient planar heterojunction solar cells, Energ. Environ. Sci. 7 (3) (2014) 982–988, https://doi.org/10.1039/c3ee43822h.

[46] H. Zhu, A. Erbing, H. Wu, G.J. Man, S. Mukherjee, C. Kamal, M.B. Johansson, H. Rensmo, M. Odelius, E.M.J. Johansson, Tuning the bandgap in silver bismuth iodide materials by partly substituting bismuth with antimony for improved solar cell performance, ACS Appl. Energy Mater. 3 (8) (2020) 7372–7382, https://doi.org/10.1021/acsaem.0c00712.

[47] F. Hao, C.C. Stoumpos, D.H. Cao, R.P.H. Chang, M.G. Kanatzidis, Lead-free solid-state organic-inorganic halide perovskite solar cells, Nat. Photonics 8 (6) (2014) 489–494, https://doi.org/10.1038/nphoton.2014.82.

[48] A. Amat, E. Mosconi, E. Ronca, C. Quarti, P. Umari, M.K. Nazeeruddin, M. Grätzel, F. De Angelis, Cation-induced band-gap tuning in organohalide perovskites: interplay of spin-orbit coupling and octahedra tilting, Nano Lett. 14 (6) (2014) 3608–3616, https://doi.org/10.1021/nl5012992.

[49] M.R. Filip, G.E. Eperon, H.J. Snaith, F. Giustino, Steric engineering of metal-halide perovskites with tunable optical band gaps, Nat. Commun. 5 (2014), https://doi.org/10.1038/ncomms6757.
[50] S. Meloni, G. Palermo, N. Ashari-Astani, M. Grätzel, U. Rothlisberger, Valence and conduction band tuning in halide perovskites for solar cell applications, J. Mater. Chem. A 4 (41) (2016) 15997–16002, https://doi.org/10.1039/c6ta04949d.
[51] K.A. Bush, K. Frohna, R. Prasanna, R.E. Beal, T. Leijtens, S.A. Swifter, M.D. McGehee, Compositional engineering for efficient wide band gap perovskites with improved stability to photoinduced phase segregation, ACS Energy Lett. 3 (2) (2018) 428–435, https://doi.org/10.1021/acsenergylett.7b01255.
[52] Y. Zhou, Z.J. Yong, K.C. Zhang, B.M. Liu, Z.W. Wang, J.S. Hou, Y.Z. Fang, Y. Zhou, H.T. Sun, B. Song, Ultrabroad photoluminescence and electroluminescence at new wavelengths from doped organometal halide perovskites, J. Phys. Chem. Lett. 7 (14) (2016) 2735–2741, https://doi.org/10.1021/acs.jpclett.6b01147.
[53] Z.K. Tang, Z.F. Xu, D.Y. Zhang, S.X. Hu, W.M. Lau, L.M. Liu, Enhanced optical absorption via cation doping hybrid lead iodine perovskites, Sci. Rep. 7 (1) (2017), https://doi.org/10.1038/s41598-017-08215-3.
[54] A. De, N. Mondal, A. Samanta, Luminescence tuning and exciton dynamics of Mn-doped $CsPbCl_3$ nanocrystals, Nanoscale 9 (43) (2017) 16722–16727, https://doi.org/10.1039/c7nr06745c.
[55] E.R. Dohner, A. Jaffe, L.R. Bradshaw, H.I. Karunadasa, Intrinsic white-light emission from layered hybrid perovskites, J. Am. Chem. Soc. 136 (38) (2014) 13154–13157, https://doi.org/10.1021/ja507086b.
[56] D. Ma, Y. Fu, L. Dang, J. Zhai, I.A. Guzei, S. Jin, Single-crystal microplates of two-dimensional organic–inorganic lead halide layered perovskites for optoelectronics, Nano Res. 10 (6) (2017) 2117–2129, https://doi.org/10.1007/s12274-016-1401-6.
[57] I.C. Smith, E.T. Hoke, D. Solis-Ibarra, M.D. McGehee, H.I. Karunadasa, A layered hybrid perovskite solar-cell absorber with enhanced moisture stability, Angew. Chem. Int. Ed. 53 (42) (2014) 11232–11235, https://doi.org/10.1002/anie.201406466.
[58] Y.T. Li, L. Han, H. Liu, K. Sun, D. Luo, X.L. Guo, D.L. Yu, T.L. Ren, Review on organic-inorganic two-dimensional perovskite-based optoelectronic devices, ACS Appl. Electron. Mater. (2021), https://doi.org/10.1021/acsaelm.1c00781.
[59] H. Liang, F. Yuan, A. Johnston, C. Gao, H. Choubisa, Y. Gao, Y. Wang, L.K. Sagar, B. Sun, P. Li, G. Bappi, B. Chen, J. Li, Y. Wang, Y. Dong, D. Ma, Y. Gao, Y. Liu, M. Yuan, E.-H. Sargent, High color purity lead-free perovskite light-emitting diodes via Sn stabilization, Adv. Sci. 7 (8) (2020) 1903213, https://doi.org/10.1002/advs.201903213.
[60] J.A. Sichert, Y. Tong, N. Mutz, M. Vollmer, S. Fischer, K.Z. Milowska, R. García Cortadella, B. Nickel, C. Cardenas-Daw, J.K. Stolarczyk, A.S. Urban, J. Feldmann, Quantum size effect in organometal halide perovskite nanoplatelets, Nano Lett. 15 (10) (2015) 6521–6527, https://doi.org/10.1021/acs.nanolett.5b02985.
[61] L. Zhang, C. Sun, T. He, Y. Jiang, J. Wei, Y. Huang, M. Yuan, High-performance quasi-2D perovskite light-emitting diodes: from materials to devices, Light Sci. Appl. 10 (1) (2021), https://doi.org/10.1038/s41377-021-00501-0.
[62] N. Wang, L. Cheng, R. Ge, S. Zhang, Y. Miao, W. Zou, C. Yi, Y. Sun, Y. Cao, R. Yang, Y. Wei, Q. Guo, Y. Ke, M. Yu, Y. Jin, Y. Liu, Q. Ding, D. Di, L. Yang, W. Huang, Perovskite light-emitting diodes based on solution-processed self-organized multiple quantum wells, Nat. Photonics 10 (11) (2016) 699–704, https://doi.org/10.1038/nphoton.2016.185.
[63] M. Yuan, L.N. Quan, R. Comin, G. Walters, R. Sabatini, O. Voznyy, S. Hoogland, Y. Zhao, E.M. Beauregard, P. Kanjanaboos, Z. Lu, D.H. Kim, E.H. Sargent, Perovskite energy funnels for efficient light-emitting diodes, Nat. Nanotechnol. 11 (10) (2016) 872–877, https://doi.org/10.1038/nnano.2016.110.

[64] T. Qiu, Y. Hu, F. Xu, Z. Yan, F. Bai, G. Jia, S. Zhang, Recent advances in one-dimensional halide perovskites for optoelectronic applications, Nanoscale 10 (45) (2018) 20963–20989, https://doi.org/10.1039/c8nr05862h.

[65] L.C. Schmidt, A. Pertegás, S. González-Carrero, O. Malinkiewicz, S. Agouram, G. Mínguez Espallargas, H.J. Bolink, R.E. Galian, J. Pérez-Prieto, Nontemplate synthesis of $CH_3NH_3PbBr_3$ perovskite nanoparticles, J. Am. Chem. Soc. 136 (3) (2014) 850–853, https://doi.org/10.1021/ja4109209.

[66] A. Swarnkar, A.R. Marshall, E.M. Sanehira, B.D. Chernomordik, D.T. Moore, J.A. Christians, T. Chakrabarti, J.M. Luther, Quantum dot-induced phase stabilization of α-CsPbI3 perovskite for high-efficiency photovoltaics, Science 354 (6308) (2016) 92–95, https://doi.org/10.1126/science.aag2700.

[67] L. Polavarapu, B. Nickel, J. Feldmann, A.S. Urban, Advances in quantum-confined perovskite nanocrystals for optoelectronics, Adv. Energy Mater. 7 (16) (2017) 1700267, https://doi.org/10.1002/aenm.201700267.

[68] B. Cheng, T.Y. Li, P. Maity, P.C. Wei, D. Nordlund, K.T. Ho, D.H. Lien, C.H. Lin, R.Z. Liang, X. Miao, I.A. Ajia, J. Yin, D. Sokaras, A. Javey, I.S. Roqan, O.F. Mohammed, J.H. He, Extremely reduced dielectric confinement in two-dimensional hybrid perovskites with large polar organics, Commun. Phys. 1 (1) (2018), https://doi.org/10.1038/s42005-018-0082-8.

[69] A.O. El-Ballouli, O.M. Bakr, O.F. Mohammed, Structurally tunable two-dimensional layered perovskites: from confinement and enhanced charge transport to prolonged hot carrier cooling dynamics, J. Phys. Chem. Lett. 11 (14) (2020) 5705–5718, https://doi.org/10.1021/acs.jpclett.0c00359.

[70] D.H. Cao, C.C. Stoumpos, O.K. Farha, J.T. Hupp, M.G. Kanatzidis, 2D homologous perovskites as light-absorbing materials for solar cell applications, J. Am. Chem. Soc. 137 (24) (2015) 7843–7850, https://doi.org/10.1021/jacs.5b03796.

[71] A.H. Proppe, R. Quintero-Bermudez, H. Tan, O. Voznyy, S.O. Kelley, E.H. Sargent, Synthetic control over quantum well width distribution and carrier migration in low-dimensional perovskite photovoltaics, J. Am. Chem. Soc. 140 (8) (2018) 2890–2896, https://doi.org/10.1021/jacs.7b12551.

[72] T. Ishihara, J. Takahashi, T. Goto, Exciton state in two-dimensional perovskite semiconductor $(C_{10}H_{21}NH_3)_2PbI_4$, Solid State Commun. 69 (9) (1989) 933–936, https://doi.org/10.1016/0038-1098(89)90935-6.

[73] C.C. Stoumpos, C.M.M. Soe, H. Tsai, W. Nie, J.C. Blancon, D.H. Cao, F. Liu, B. Traoré, C. Katan, J. Even, A.D. Mohite, M.G. Kanatzidis, High members of the 2D Ruddlesden-Popper halide perovskites: synthesis, optical properties, and solar cells of (CH3(CH2)3-NH3)2(CH3NH3)4Pb5I16, Chem 2 (3) (2017) 427–440, https://doi.org/10.1016/j.chempr.2017.02.004.

[74] V. Malgras, S. Tominaka, J.W. Ryan, J. Henzie, T. Takei, K. Ohara, Y. Yamauchi, Observation of quantum confinement in monodisperse methylammonium lead halide perovskite nanocrystals embedded in mesoporous silica, J. Am. Chem. Soc. 138 (42) (2016) 13874–13881, https://doi.org/10.1021/jacs.6b05608.

[75] W. Cha, H.J. Kim, S. Lee, J. Kim, Size-controllable and stable organometallic halide perovskite quantum dots/polymer films, J. Mater. Chem. C 5 (27) (2017) 6667–6671, https://doi.org/10.1039/c7tc01562c.

[76] A. Pisoni, J. Jaćimović, O.S. Barišić, M. Spina, R. Gaál, L. Forró, E. Horváth, Ultra-low thermal conductivity in organic-inorganic hybrid perovskite $CH_3NH_3PbI_3$, J. Phys. Chem. Lett. 5 (14) (2014) 2488–2492, https://doi.org/10.1021/jz5012109.

[77] L. Dong, L. Ye, D. Zhao, L. Tong, F. Gao, W. Li, P. Lu, X.H. Bu, Thermal transport engineering in hybrid organic-inorganic perovskite phononic crystals, J. Phys. Chem. Lett. 11 (14) (2020) 5728–5733, https://doi.org/10.1021/acs.jpclett.0c01909.

[78] A.F. Akbulatov, V.M. Martynenko, L.A. Frolova, N.N. Dremova, I. Zhidkov, S.A. Tsarev, S.Y. Luchkin, E.Z. Kurmaev, S.M. Aldoshin, K.J. Stevenson, P.A. Troshin, Intrinsic thermal decomposition pathways of lead halide perovskites APbX3, Sol. Energy Mater. Sol. Cells 213 (2020), 110559, https://doi.org/10.1016/j.solmat.2020.110559.

[79] V.K. Sharma, V. Kanchana, M.K. Gupta, R. Mittal, Ultra-low thermal conductivity of orthorhombic $CH_3NH_3SnI_3$: a first principles investigation, J. Solid State Chem. 290 (2020), https://doi.org/10.1016/j.jssc.2020.121541.

[80] K. Miyata, T.L. Atallah, X.Y. Zhu, Lead halide perovskites: crystal-liquid duality, phonon glass electron crystals, and large polaron formation, Sci. Adv. 3 (10) (2017), https://doi.org/10.1126/sciadv.1701469.

[81] H. Zhu, K. Miyata, Y. Fu, J. Wang, P.P. Joshi, D. Niesner, K.W. Williams, S. Jin, X.Y. Zhu, Screening in crystalline liquids protects energetic carriers in hybrid perovskites, Science 353 (6306) (2016) 1409–1413, https://doi.org/10.1126/science.aaf9570.

[82] T. Haeger, R. Heiderhoff, T. Riedl, Thermal properties of metal-halide perovskites, J. Mater. Chem. C 8 (41) (2020) 14289–14311, https://doi.org/10.1039/D0TC03754K.

[83] M.A. Haque, S. Kee, D.R. Villalva, W. Ong, D. Baran, Halide perovskites: thermal transport and prospects for thermoelectricity, Adv. Sci. 7 (10) (2020) 1903389, https://doi.org/10.1002/advs.201903389.

[84] X. Qian, X. Gu, R. Yang, Lattice thermal conductivity of organic-inorganic hybrid perovskite $CH_3NH_3PbI_3$, Appl. Phys. Lett. 108 (6) (2016), https://doi.org/10.1063/1.4941921.

[85] Z. Guo, S.J. Yoon, J.S. Manser, P.V. Kamat, T. Luo, Structural phase- and degradation-dependent thermal conductivity of CH3NH3PbI3 perovskite thin films, J. Phys. Chem. C 120 (12) (2016) 6394–6401, https://doi.org/10.1021/acs.jpcc.6b00513.

[86] Q. Chen, C. Zhang, M. Zhu, S. Liu, M.E. Siemens, S. Gu, J. Zhu, J. Shen, X. Wu, C. Liao, J. Zhang, X. Wang, M. Xiao, Efficient thermal conductance in organometallic perovskite $CH_3NH_3PbI_3$ films, Appl. Phys. Lett. 108 (8) (2016), https://doi.org/10.1063/1.4942779.

[87] Y. Wang, R. Lin, P. Zhu, Q. Zheng, Q. Wang, D. Li, J. Zhu, Cation dynamics governed thermal properties of lead halide perovskite nanowires, Nano Lett. 18 (5) (2018) 2772–2779, https://doi.org/10.1021/acs.nanolett.7b04437.

[88] R. Heiderhoff, et al., Thermal conductivity of methylammonium lead halide perovskite single crystals and thin films: a comparative study, J. Phys. Chem. C 121 (51) (2017) 28306–28311. https://doi.org/10.1021/acs.jpcc.7b11495.

[89] G.A. Elbaz, W.L. Ong, E.A. Doud, P. Kim, D.W. Paley, X. Roy, J.A. Malen, Phonon speed, not scattering, differentiates thermal transport in lead halide perovskites, Nano Lett. 17 (9) (2017) 5734–5739. https://doi.org/10.1021/acs.nanolett.7b02696.

[90] T. Ye, X. Wang, X. Li, A.Q. Yan, S. Ramakrishna, J. Xu, Ultra-high Seebeck coefficient and low thermal conductivity of a centimeter-sized perovskite single crystal acquired by a modified fast growth method, J. Mater. Chem. C 5 (2017) 1255–1260. https://doi.org/10.1039/C6TC04594D.

[91] A. Mattoni, A. Filippetti, M.I. Saba, P. Delugas, Methylammonium rotational dynamics in lead halide perovskite by classical molecular dynamics: the role of temperature, J. Phys. Chem. C 119 (30) (2015) 17421–17428, https://doi.org/10.1021/acs.jpcc.5b04283.

[92] C.C. Stoumpos, C.D. Malliakas, J.A. Peters, Z. Liu, M. Sebastian, J. Im, T.C. Chasapis, A.-C. Wibowo, D.Y. Chung, A.J. Freeman, B.W. Wessels, M.G. Kanatzidis, Crystal growth of the perovskite semiconductor $CsPbBr_3$: a new material for high-energy radiation detection, Cryst. Growth Des. 13 (7) (2013) 2722–2727, https://doi.org/10.1021/cg400645t.

[93] C.C. Stoumpos, C.D. Malliakas, M.G. Kanatzidis, Semiconducting tin and lead iodide perovskites with organic cations: phase transitions, high mobilities, and near-infrared

photoluminescent properties, Inorg. Chem. 52 (15) (2013) 9019–9038, https://doi.org/10.1021/ic401215x.
[94] W. Zou, R. Li, S. Zhang, Y. Liu, N. Wang, Y. Cao, Y. Miao, M. Xu, Q. Guo, D. Di, L. Zhang, C. Yi, F. Gao, R.H. Friend, J. Wang, W. Huang, Minimising efficiency roll-off in high-brightness perovskite light-emitting diodes, Nat. Commun. 9 (1) (2018), https://doi.org/10.1038/s41467-018-03049-7.
[95] Y. Zou, Z. Yuan, S. Bai, F. Gao, B. Sun, Recent progress toward perovskite light-emitting diodes with enhanced spectral and operational stability, Mater. Today Nano 5 (2019), 100028, https://doi.org/10.1016/j.mtnano.2019.100028.
[96] J.C. Yu, D.W. Kim, D.B. Kim, E.D. Jung, J.H. Park, A.Y. Lee, B.R. Lee, D. Di Nuzzo, R.H. Friend, M.H. Song, Improving the stability and performance of perovskite light-emitting diodes by thermal annealing treatment, Adv. Mater. 28 (32) (2016) 6906–6913, https://doi.org/10.1002/adma.201601105.
[97] L. Zhao, K.M. Lee, K. Roh, S.U.Z. Khan, B.P. Rand, Improved outcoupling efficiency and stability of perovskite light-emitting diodes using thin emitting layers, Adv. Mater. 31 (2) (2019), https://doi.org/10.1002/adma.201805836.
[98] H. Xu, X. Wang, Y. Li, L. Cai, Y. Tan, G. Zhang, Y. Wang, R. Li, D. Liang, T. Song, B. Sun, Prominent heat dissipation in perovskite light-emitting diodes with reduced efficiency droop for silicon-based display, J. Phys. Chem. Lett. 11 (9) (2020) 3689–3698, https://doi.org/10.1021/acs.jpclett.0c00792.
[99] A. Gómez, Q. Wang, A.R. Goñi, M. Campoy-Quiles, A. Abate, Ferroelectricity-free lead halide perovskites, Energ. Environ. Sci. 12 (8) (2019) 2537–2547, https://doi.org/10.1039/c9ee00884e.
[100] V.G. Koukhar, N.A. Pertsev, R. Waser, Thermodynamic theory of epitaxial ferroelectric thin films with dense domain structures, Phys. Rev. B: Condens. Matter Mater. Phys. 64 (21) (2001) 2141031–21410315.
[101] M.E. Lines, A.M. Glass, Principles and Applications of Ferroelectrics and Related Materials, Oxford University Press, 2001.
[102] S. Tan, N. Zhou, Y. Chen, L. Li, G. Liu, P. Liu, C. Zhu, J. Lu, W. Sun, Q. Chen, H. Zhou, Effect of high dipole moment cation on layered 2D organic–inorganic halide perovskite solar cells, Adv. Energy Mater. 9 (5) (2019), https://doi.org/10.1002/aenm.201803024.
[103] Y. Ishibashi, H. Shiba, Successive phase transitions in ferroelectric $NaNO_2$ and SC$(NH_2)_2$, J. Physical Soc. Japan 45 (2) (1978) 409–413, https://doi.org/10.1143/JPSJ.45.409.
[104] J.M. Frost, K.T. Butler, F. Brivio, C.H. Hendon, M. Van Schilfgaarde, A. Walsh, Atomistic origins of high-performance in hybrid halide perovskite solar cells, Nano Lett. 14 (5) (2014) 2584–2590, https://doi.org/10.1021/nl500390f.
[105] J. Ma, L.W. Wang, Nanoscale charge localization induced by random orientations of organic molecules in hybrid perovskite CH3NH3PbI3, Nano Lett. 15 (1) (2015) 248–253, https://doi.org/10.1021/nl503494y.
[106] H. Röhm, T. Leonhard, M.J. Hoffmann, A. Colsmann, Ferroelectric domains in methylammonium lead iodide perovskite thin-films, Energ. Environ. Sci. 10 (4) (2017) 950–955, https://doi.org/10.1039/c7ee00420f.
[107] B. Chen, T. Li, Q. Dong, E. Mosconi, J. Song, Z. Chen, Y. Deng, Y. Liu, S. Ducharme, A. Gruverman, F.D. Angelis, J. Huang, Large electrostrictive response in lead halide perovskites, Nat. Mater. 17 (11) (2018) 1020–1026, https://doi.org/10.1038/s41563-018-0170-x.
[108] Q. Dong, J. Song, Y. Fang, Y. Shao, S. Ducharme, J. Huang, Lateral-structure single-crystal hybrid perovskite solar cells via piezoelectric poling, Adv. Mater. 28 (14) (2016) 2816–2821, https://doi.org/10.1002/adma.201505244.

[109] Y. Rakita, O. Bar-Elli, E. Meirzadeh, H. Kaslasi, Y. Peleg, G. Hodes, I. Lubomirsky, D. Oron, D. Ehre, D. Cahen, Tetragonal $CH_3NH_3PbI_3$ is ferroelectric, Proc. Natl. Acad. Sci. U. S. A. 114 (28) (2017) E5504–E5512, https://doi.org/10.1073/pnas.1702429114.

[110] M. Sajedi Alvar, M. Kumar, P.W.M. Blom, G.J.A.H. Wetzelaer, K. Asadi, Absence of ferroelectricity in methylammonium lead iodide perovskite, AIP Adv. 7 (9) (2017), https://doi.org/10.1063/1.4994957.

[111] M. Coll, A. Gomez, E. Mas-Marza, O. Almora, G. Garcia-Belmonte, M. Campoy-Quiles, J. Bisquert, Polarization switching and light-enhanced piezoelectricity in lead halide perovskites, J. Phys. Chem. Lett. 6 (8) (2015) 1408–1413, https://doi.org/10.1021/acs.jpclett.5b00502.

[112] Y. Kutes, L. Ye, Y. Zhou, S. Pang, B.D. Huey, N.P. Padture, Direct observation of ferroelectric domains in solution-processed $CH_3NH_3PbI_3$ perovskite thin films, J. Phys. Chem. Lett. 5 (19) (2014) 3335–3339, https://doi.org/10.1021/jz501697b.

[113] Y. Liu, L. Collins, R. Proksch, S. Kim, B.R. Watson, B. Doughty, T.R. Calhoun, M. Ahmadi, A.V. Ievlev, S. Jesse, S.T. Retterer, A. Belianinov, K. Xiao, J. Huang, B.G. Sumpter, S.V. Kalinin, B. Hu, O.S. Ovchinnikova, Chemical nature of ferroelastic twin domains in $CH_3NH_3PbI_3$ perovskite, Nat. Mater. 17 (11) (2018) 1013–1019, https://doi.org/10.1038/s41563-018-0152-z.

[114] B. Chen, J. Shi, X. Zheng, Y. Zhou, K. Zhu, S. Priya, Ferroelectric solar cells based on inorganic-organic hybrid perovskites, J. Mater. Chem. A 3 (15) (2015) 7699–7705, https://doi.org/10.1039/c5ta01325a.

[115] J. Xiao, J. Chang, B. Li, F.H. Isikgor, D. Wang, Z. Fan, Z. Lin, J. Ouyang, K. Zeng, J. Chen, Room temperature ferroelectricity of hybrid organic-inorganic perovskites with mixed iodine and bromine, J. Mater. Chem. A 6 (20) (2018) 9665–9676, https://doi.org/10.1039/c7ta09967c.

[116] V.K. Wadhawan, Ferroelasticity, Bull. Mater. Sci. 6 (4) (1984) 733–753, https://doi.org/10.1007/BF02744001.

[117] E. Strelcov, Q. Dong, T. Li, J. Chae, Y. Shao, Y. Deng, A. Gruverman, J. Huang, A. Centrone, CH3NH3PbI3 perovskites: ferroelasticity revealed, Sci. Adv. 3 (4) (2017), https://doi.org/10.1126/sciadv.1602165.

[118] K. Aizu, Possible species of "ferroelastic" crystals and of simultaneously ferroelectric and ferroelastic crystals, J. Physical Soc. Japan 27 (2) (1969) 387–396, https://doi.org/10.1143/JPSJ.27.387.

[119] K. Aizu, Possible species of ferromagnetic, ferroelectric, and ferroelastic crystals, Phys. Rev. B 2 (3) (1970) 754–772, https://doi.org/10.1103/PhysRevB.2.754.

[120] M.U. Rothmann, W. Li, Y. Zhu, U. Bach, L. Spiccia, J. Etheridge, Y.B. Cheng, Direct observation of intrinsic twin domains in tetragonal $CH_3NH_3PbI_3$, Nat. Commun. 8 (2017), https://doi.org/10.1038/ncomms14547.

[121] M. Bari, A.A. Bokov, Z.G. Ye, Ferroelasticity, domain structures and phase symmetries in organic-inorganic hybrid perovskite methylammonium lead chloride, J. Mater. Chem. C 8 (28) (2020) 9625–9631, https://doi.org/10.1039/d0tc02124e.

[122] A.R. Warwick, J. Íñiguez, P.D. Haynes, N.C. Bristowe, First-principles study of ferroelastic twins in halide perovskites, J. Phys. Chem. Lett. 10 (6) (2019) 1416–1421, https://doi.org/10.1021/acs.jpclett.9b00202.

[123] X. Xiao, W. Li, Y. Fang, Y. Liu, Y. Shao, S. Yang, J. Zhao, X. Dai, R. Zia, J. Huang, Benign ferroelastic twin boundaries in halide perovskites for charge carrier transport and recombination, Nat. Commun. 11 (1) (2020) 2215, https://doi.org/10.1038/s41467-020-16075-1.

CHAPTER

4

Synthesis techniques of metal halide perovskites

Ignacio Rosa-Pardo, Alejandro Cortés-Villena, Raquel E. Galian, and Julia Pérez-Prieto
Institute of Molecular Science, University of Valencia, Valencia, Spain

1. Introduction

Metal halide perovskites (MHPs) have emerged as one of the most promising materials for photovoltaic and light-emitting device (LED) technology due to their exceptional optoelectronic properties, specifically, their intrinsic capacity to transport charges and generate electricity and produce light because of their high defect tolerance and low nonradiative recombination. Their performance is superior when compared with related semiconductors in photovoltaic technology due to their high optical absorption coefficient, long carrier diffusion length, and high defect tolerance [1–3]. Pioneer examples were first focused on the use of MHPs as absorbent layers for photovoltaic applications and thin films prepared from the perovskite precursors. Since the first report by Kojima et al. in 2009 using a $CH_3NH_3PbI_3$ perovskite as a light harvester, thereby enabling a solar cell power conversion efficiency (PCE) of 3.8%, MHPs have become the most competitive and fast-growing material in photovoltaic history, not only because of their optical properties but also because of their low-cost solution processing [4]. Currently, solar cells with PCE values of up to 25.6% (certified 25.2%) based on cubic α-phase of formamidinium lead triiodide ($FAPbI_3$) perovskite films and long-term stability of ca. 450 h are achieved. Moreover, these authors demonstrated a promising application in LEDs with an external quantum efficiency of electroluminescence (EQE) higher than 10% [5].

In general, MHPs present the formula AMX_3, where A represents a monovalent cation such as methylammonium (MA), formamidinium (FA), or cesium (Cs^+); B represents a divalent cation such as lead (Pb^{2+}), tin (Sn^{2+}),

https://doi.org/10.1016/B978-0-323-88522-5.00001-6

or germanium (Ge^{2+}); and X is a halide anion (Cl^-, Br^-, and I^-) [6]. As most materials, LHPs can be classified according to their material dimension in bulk (3D), low-dimensional materials (2D, nanoplatelets, nanosheets; 1D nanowires), and nanoparticles (0D). Moreover, they can be classified in terms of their inorganic framework dimension as 3D (AMX_3), 2D (A_2MX_4), 1D (A_6MX_5), and 0D (AMX_6), as illustrated in Fig. 1A [7].

As mentioned above, a key advantage offered by halide perovskites, which makes them unique among other high-performance semiconductors, is their processability, evidenced by the easy and low-temperature

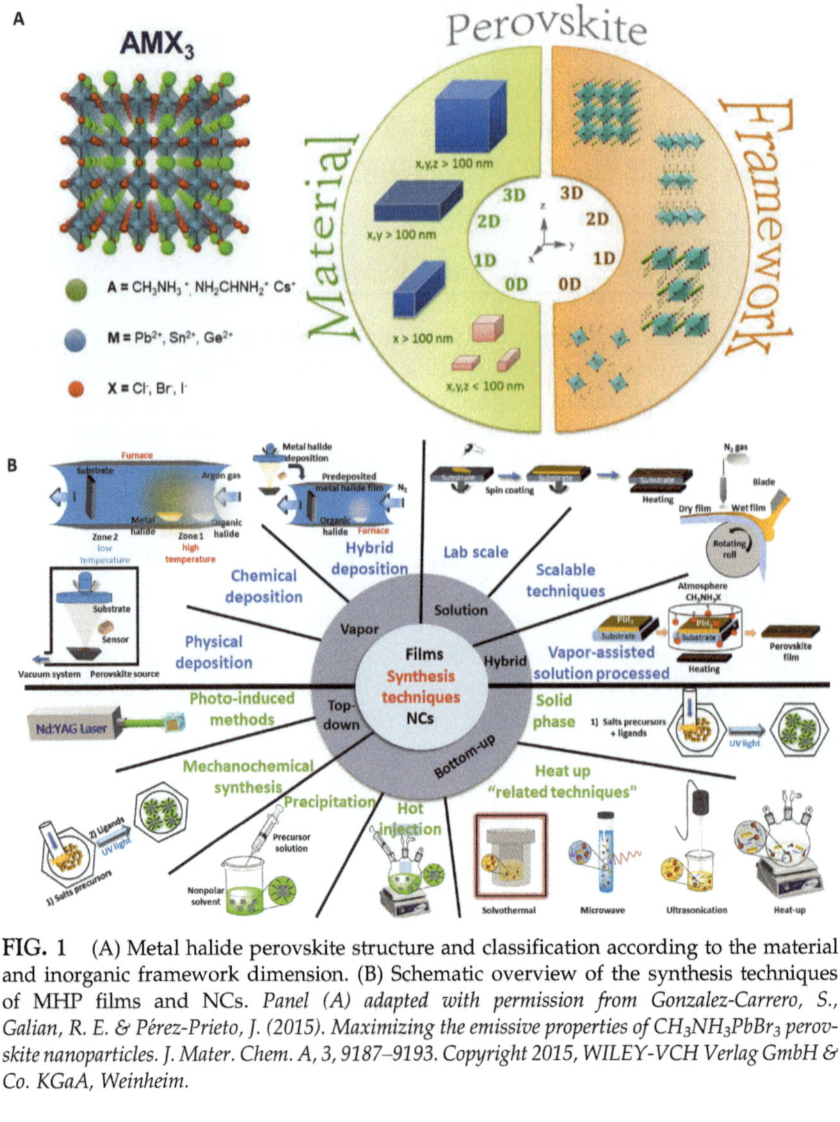

FIG. 1 (A) Metal halide perovskite structure and classification according to the material and inorganic framework dimension. (B) Schematic overview of the synthesis techniques of MHP films and NCs. *Panel (A) adapted with permission from Gonzalez-Carrero, S., Galian, R. E. & Pérez-Prieto, J. (2015). Maximizing the emissive properties of $CH_3NH_3PbBr_3$ perovskite nanoparticles. J. Mater. Chem. A, 3, 9187–9193. Copyright 2015, WILEY-VCH Verlag GmbH & Co. KGaA, Weinheim.*

routes used for their preparation. Synthesis approaches reported so far comprise perovskites with different dimensionality in terms of material and inorganic framework. Hence, it is highly desirable to make an illustrative classification, based on the fundamentals of the synthesis techniques, highlighting their advantages and drawbacks, with a view to perovskite-based solar cell and LED technology. The purpose of this chapter is to establish a comprehensive classification of different techniques used to prepare LHPs, focused on films and colloidal nanocrystals as shown in Fig. 1B. Thin-film technology is considered the backbone for advanced applications in several fields beyond optoelectronics, such as environmental applications and energy storage devices. The crucial issue for all applications of thin films depends strongly on their composition, morphology, and stability. A thin-film deposition involves preparing and depositing on a substrate a thin coating with thicknesses ranging from several nanometers to micrometers using different techniques and methodologies. The main advantage of using MHPs as thin-film absorbers, compared to classical organic compounds, is their high absorption coefficient and broad absorption band across the entire visible spectrum and near-infrared (NIR) window. Although LHP bulk/thin film presents an impressive response as absorbers in perovskite solar cells (PSCs), so far they have a limited application as emitting material for perovskite LEDs (PLEDs) due to their low photoluminescence efficiency. However, colloidal MHP nanocrystals are envisioned as promising materials in terms of photoluminescence and processability. The first report on the preparation of emissive $CH_3NH_3PbBr_3$ colloidal MHP nanocrystals was carried out by reverse microemulsion, using octylammonium bromide and oleic acid as organic ligands, octadecene as the solvent, and a mild temperature [8]. The resulting colloids presented a photoluminescence quantum yield of 20% and were successfully used to prepare thin films with good emission and an electroluminescent device with a better response than bulk LHPs. Organic ligands play a pivotal role in the preparation of colloidal LHP nanocrystals by controlling the growth size, determining their solubility, and providing colloidal stability, similar to chalcogenide semiconductor nanocrystals [9]. Since then, there have been many studies to gain deep insight into their inner structure to control the size, shape, and composition of the nanomaterial, which opened up doors for potential applications, such as photocatalysis [10] and optoelectronic applications behind light-emitting diodes [11] and photovoltaics [12] such as lasers [13], photodetectors [14], and amplified spontaneous emitters [15]. A comprehensive review on the state of the art of colloidal metal halide perovskite nanocrystals has recently been published to update the research in this area covering fundamental, shape-controlled synthesis and their evolution, surface chemistry, and their main applications [16].

This chapter is organized into two main sections according to the synthesis approaches reported for the preparation of LHP thin films (vacuum, chemical, hybrid deposition, etc.) and LHP colloidal nanocrystals (bottom-up and top-down methodologies). Moreover, we describe here pioneer examples of the use of these techniques, the key factors interfering in the preparation of good-quality thin films or colloidal nanocrystals with good optical properties and their impact on the power conversion efficiency of solar cell devices as well as the photoluminescence quantum yield (QY_{PL}) of the colloidal dispersion.

2. Techniques for the preparation of thin films

The techniques used to prepare LHP thin films are assorted according to the growth method used for the deposition of perovskites precursors: (i) vacuum-based methods, (ii) solution-processed methods, and (iii) hybrid methods, which combine vapor and solution approaches. It is worth mentioning that there is no ideal thin-film deposition technique for halide perovskite; therefore, the advantages and drawbacks of the different approaches will be discussed in this section (Table 1). Thus, the

TABLE 1 Comparison of thin-film preparation techniques and power conversion efficiency results.

Techniques		Advantages	Limitations	PCE (%) (up to)[a]
Vapor-based deposition	Physical deposition	Precise control. High reproducibility, uniform and coverage. Thinner films than solution-processed methods. High resistance. Not requirement of post-annealing treatments. Compatible with other deposition techniques. Avoid the use of solvents.	High cost. Need of extremely vacuum conditions. Complex stoichiometry control. Increased reaction times. Problems associated with interdiffusion aspects	20.8 [17]
	Chemical deposition	High performance. No need for high vacuum conditions. More cost-effective than physical methods. Conformal deposition. Atomic thickness and excellent control over stoichiometry for ALD.	Limitations in substrate size. Challenge for depositing dense and uniform films. The still limited number of reports for ALD.	18.2 [18]

TABLE 1 Comparison of thin-film preparation techniques and power conversion efficiency results—cont'd

Techniques		Advantages	Limitations	PCE (%) (up to)[a]
	Hybrid deposition	Easy fabrication. High quality. Better reproducibility and uniformity than one-step counterparts.	Takes longer than CVD counterparts. Interdiffusion problems.	14.7 [19]
Solution-processed deposition	Lab-scale deposition	Easy and low-temperature processing. Cost-effective technique. Good levels of reproducibility and morphology control. Two steps afford better results than one-step counterparts.	Different solubility of precursors. Need for annealing step. Incomplete coverage, pinholes, and cracks are frequently observed. High material waste. Use of hazardous solvents. Incompatible with large area manufacturing. Challenge of proper diffusion in two steps.	22.6 [20]
	Scalable deposition	Homogeneous, high coverage, large grains, and pinhole-free films. High material usage. Scalability and compatibility with R2R technology. Applicable to flexible substrates.	Multiple parameters need to be considered. Process engineering challenges.	20.3 [21]
Hybrid vapor-solution deposition	Vapor-assisted solution-processed	Improved morphology, better efficiencies compared to single-step spin-coating methods. Feasibility to deposit individually layers by whatever solution/vapor method.	Unreacted phases and residual solvent impede appropriate precursor diffusion.	19.1 [22]

[a] *The highest PCE value obtained by this technique is reported in this table.*

selected technique will depend on the performance and production requirements according to the final application of the LHP. The film formation technique, device architecture, and compositional engineering factors directly affect the efficiency of the solar cell. The fabrication of thin-film perovskites with large grain size, fewer grain boundaries, and high uniformity and surface coverage can reduce the recombination of the electrons and holes during their transport through the layer, so these parameters should determine the final performance of the thin film.

Most thin-film growth methods have been focused on solution-based processes, due to the attractive low cost and rapid deposition of these techniques compared to vacuum-based deposition. Moreover, they could be used for large-scale production and integration into complex devices. Focusing on perovskite tandem technologies for industry scalability and better efficiencies, the development of deposition methods for a full set of LHP compositions is becoming a high priority.

2.1 Vacuum-based methods

These are based on the deposition of the precursor materials in the vapor phase, often operated at pressures below the atmosphere, on a substrate. Compared to solution-processed approaches, these methods make perovskites with higher quality, crystallinity, and reproducibility. Moreover, they employ solid precursor materials, thereby avoiding the use of toxic solvents. The vacuum-based deposition can be roughly divided into (i) physical, (ii) chemical, and (iii) hybrid physical-chemical deposition methods.

2.1.1 Physical deposition methods

Underlying physical deposition methods involve physical processes that require first vaporizing a solid material, followed by condensation of the vaporized material on top of a substrate to form a thin film under vacuum or controlled atmospheric conditions. The atoms or molecules are guided in the form of vapor via a vacuum or low-pressure gaseous or plasma environment to the substrate surface for condensation to happen. These processes are used to deposit thin films with thicknesses in the range of a few nanometers to thousands of nanometers. Furthermore, films created by this method are highly durable and resistant to scratching and corrosion, although the main shortcoming is the high cost. Every physical deposition process can be attributed to three basic steps: (i) creation of the vapor phase by evaporation, sputtering or ion bombardment; (ii) transportation of the vapor phase from source to the substrate; and (iii) film growth on the substrate surface through nucleation and growth mechanisms. The most common physical processes are vacuum thermal evaporation, sputtering, and laser deposition.

Vacuum thermal evaporation

Vacuum thermal evaporation is an old and simple technique for thin-film and thick-film deposition, which is extensively applicable in the laboratory and industries. Basically, it refers to sublime materials into the gaseous phase by electrical heating under high vacuum conditions, followed by vapor condensation onto the cold substrate surface. Perovskites have been evaporated in four different ways: single-source vapor deposition, co-evaporation of different material components, sequential evaporation of the materials from single sources, and flash evaporation of a perovskite bulk material source.

Single-source vapor deposition. In single-source vapor deposition, a powder of the presynthesized perovskite or just a mixture of its precursors is added to a thermal crucible wrapped with a heating coil. The powder sublimates rapidly without a chemical reaction when a high electric current is applied to the coil (from 0 up to 100 A). Then, the evaporated material is transported and condensed atomistically onto the substrate surface to form the corresponding thin film. The only change in the whole process was the physical transformation of the solid powder to gaseous and back to solid again, being the material chemically unaltered. Owing to its simplicity and speed (only needs a few minutes), this technique is appealing to deposit films in a single step and does not even require post-annealing treatments, which makes it a low-temperature process that could be applied in the preparation of flexible plastic substrates. Nevertheless, due to a single-source process, it is considered poorly reproducible if compared to dual source and, since it is still a vacuum thermal process, that means it may present challenges associated with stoichiometry transfer and, no less important, the cost of the vacuum conditions.

Fan et al. [23] proposed this technique in 2016 for the deposition of organic-inorganic hybrid $CH_3NH_3PbI_3$ perovskite thin films at room temperature starting from small crystals or powders of high-quality, single-crystalline $CH_3NH_3PbI_3$ perovskite (Fig. 2A). This method was demonstrated to palliate the drawbacks of re-dissolution of the film, complete diffusion of the organic halide into the inner part of the metal halide layer, and the difficulty to balance the rate of both sources simultaneously obtained in the solution-processed methods, hybrid vapor-assisted solution-processed methods, and co-evaporation methods, respectively. A high level of tetragonal phase purity with no other phase peaks, full surface coverage of approximately 400 nm thickness, well-defined grain structure (extremely dense, compact, and uniform even on a large scale) and high crystallization was achieved with a PCE of 10.90% in a 9-mm^2 active area. Two years later, in order to overcome the intrinsic environmental instability of organic-inorganic hybrid perovskites due to the high volatility and moisture sensitivity of organic halide, Ajjouri et al. [24] demonstrated the deposition of fully inorganic cesium lead halide of pristine-mixed precursors by this technique. Whereas high-purity perovskite $CsPbX_3$ films were obtained for $X=Cl^-$ after a thermal annealing step, the $CsPb_2Br_5$ phase was produced to a large extent for $X=Br^-$. However, when they previously conducted dry

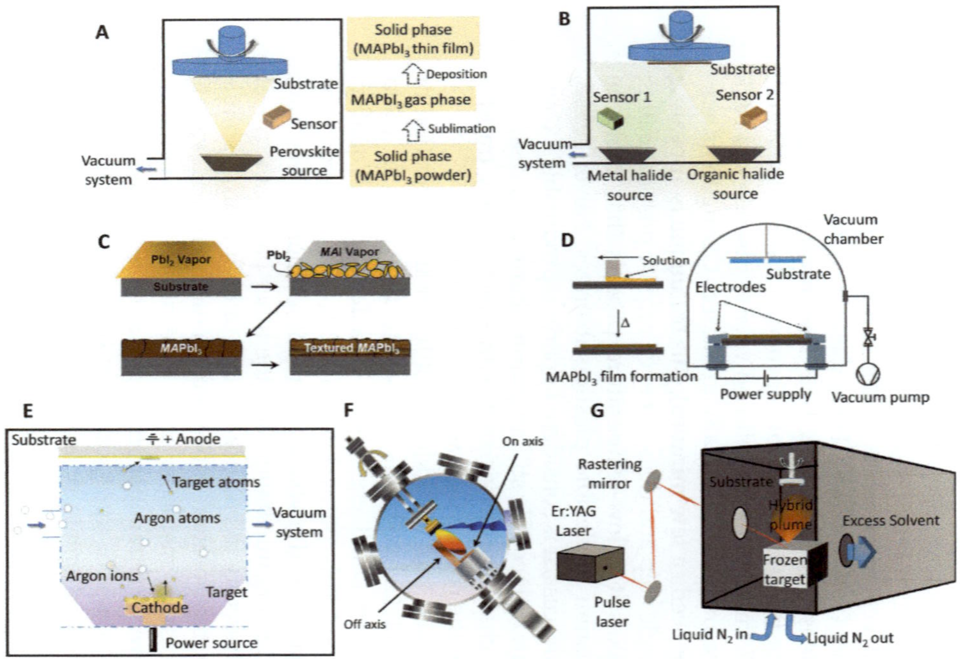

FIG. 2 Schematic illustrations of the (A) single-source vapor deposition. (B) Dual-source vapor deposition. (C) Sequential vapor deposition. (D) Flash evaporation. (E) Radiofrequency-magnetron sputtering. (F) Pulsed laser deposition. (G) Resonant-infrared matrix-assisted pulsed laser evaporation. *Panel (A) Adapted with permission under Creative Commons Attribution 4.0 International License from Fan, P., Gu, D., Liang, G.-X., Luo, J.-T., Chen, J.-L., Zheng, Z.-H., & Zhang, D.-P. (2016). High-performance perovskite CH3NH3PbI3 thin films for solar cells prepared by single-source physical vapour deposition. Sci. Rep., 6, 29910. Copyright 2016, The Authors, (B and C) adapted with permission from Hu, H., Wang, D., Zhou, Y., Zhang, J., Lv, S., Pang, S., Chen, X., Liu, Z., Padture, N. P., & Cui, G. (2014). Vapour-based processing of hole-conductor-free CH3NH3PbI3 perovskite/C60 fullerene planar solar cells. RSC Adv., 4, 28964–28967. Copyright 2014, Royal Society of Chemistry, (D) adapted with permission under Creative Commons Attribution 3.0 Unported License from Longo, G., Gil-Escrig, L., Degen, M. J., Sessolo, M., & Bolink, H. J. (2015). Perovskite solar cells prepared by flash evaporation. Chem. Commun., 51(34), 7376–7378. Copyright 2015, Royal Society of Chemistry, E and F adapted with permission from Bansode, U., Naphade, R., Game, O., Agarkar, S., & Ogale, S. (2015). Hybrid perovskite films by a new variant of pulsed excimer laser deposition: a room-temperature dry process. J. Phys. Chem. C, 119(17), 9177–9185. Copyright 2015, American Chemical Society, and (G) adapted from Barraza, E. T., Dunlap-Shohl, W. A., Mitzi, D. B., & Stiff-Roberts, A. D. (2018). Deposition of methylammonium lead triiodide by resonant infrared matrix-assisted pulsed laser evaporation. J. Electron. Mater., 47(2), 917–926.*

mechanochemical in a crucible of $CsPbX_3$ from precursors (CsX and PbX_2) by ball milling at room temperature, $CsPbX_3$ films of higher purity and morphology were formed (higher homogeneity with typical grain sizes of several hundreds of nanometers, with no observable pinholes).

Co-evaporation (also known as the dual-source vacuum deposition technique) occurs when solid precursors (usually a binary compound) are loaded into separate thermal crucibles and heated them electrically under vacuum. The same principle governs here as the previous strategy in which the corresponding precursors are sublimated, forming a film on the substrate surface facing the material sources as shown in Fig. 2B. Although the method exhibits high reproducibility due to the precise control of film thickness, uniform and smooth morphology, and compatibility with other vapor-deposited materials, it does not require an annealing step. The main challenge is the simultaneous control of evaporation rates of both material precursors to achieve the desired stoichiometry in the deposited films.

The first report using this technique by Era et al. [25] in 1997 demonstrated the success in a new self-organized growth method of layered $(RNH_3)_2PbI_4$ and also cubic $(CH_3NH_3)PbI_3$ perovskite thin films. Their results showed that a wide variety of organic ammonium halides can be used in the dual-source vacuum deposition technique to prepare a variety of perovskite films with tailored structures and properties. The first report on the successful vapor deposition of high-quality, pinhole-free methylammonium lead halide ($CH_3NH_3PbI_{3-x}Cl_x$) perovskite films was presented in mid-2013 by Liu et al. [26]. They created uniform flat films of the mixed-halide perovskite and demonstrated that dual-source evaporation can lead to efficient planar heterojunction architecture perovskite solar cells with a PCE of 15.4%, with an open-circuit voltage of 1.07 V over a small-area (0.076-cm^2) device. PCEs as high as 20.28% and 19% for 0.1- and 1-cm^2 PSCs, respectively, and the record value of 18.13% for a 21-cm^2 mini-module demonstrated their excellent scalability [27]. One of the main challenges for the successful perovskite film formation is the control over the organic halide evaporation rate because of its relatively high vapor pressure, so quartz crystal microbalance (QCM) was used to monitor the deposition rate of this precursor in a home-built instrument developed by Ono et al. [28] in 2014. CH_3NH_3I vapor partial pressure generated inside the vacuum chamber is precisely controlled and monitored by orientating the thickness monitor opposite to the evaporation direction of CH_3NH_3I to avoid the crossing from the metal halide source. By this method, perovskite films of ca. 50 and 135 nm thicknesses were prepared and showed PCEs of 6.3% and 9.9%, respectively.

Sequential evaporation process, also known as sequential vapor deposition, consists of the independent vapor deposition of multiple film layers on top of each other and subsequent conversion of the materials through diffusion/intercalation and recrystallization as shown in Fig. 2C. This method takes advantage of individual layers optimization, creates perovskite films with a much larger crystal domain size (micrometer grain size versus nanometers in the co-evaporation method), and prevents the contamination issues of CH_3NH_3I re-sublimation when higher sublimation temperature is employed. To palliate the interdiffusion and conversion inefficiencies, thin layers of the material are sequentially evaporated until the film has reached the desired thickness. By contrast, the CH_3NH_3I diffusion depth is limited, leaving the bottom metal halide layer unreacted.

The first sequential deposition was performed by Hu et al. [29] producing pinhole-free, uniform, and smooth films, and most of the textured $CH_3NH_3PbI_3$ grains within the films span the entire thickness of the film but with a significantly low PCE of 5.4%. Chen et al. [30] reported in the same year that this technique can also lead to high-performance perovskite solar cells of 15.4% PCE for small active areas. They used PEDOT: PSS [poly(3,4-ethylenedioxythiophene):poly(styrene sulfonate)] and C_{60}/Bphen (C_{60} fullerene/bathophenanthroline) as hole and electron-transporting layers, respectively, to form efficient and uniform planar-type perovskite solar cells.

Flash evaporation is also known as single-source thermal ablation, where a thermal sheet is initially coated with a perovskite layer with desirable stoichiometry or a thermal crucible is loaded with presynthesized perovskite powders, and then placed in a vacuum chamber and fed by passing a large current causing a fast sublimation of the entire compound without thermal decomposition of the organic constituents. Subsequently, the material is condensed onto the substrate surface, giving rise to the corresponding film (see Fig. 2D). The desired film thickness can be controlled by the volume of the precursor material and the substrate to heater distance. Flash evaporation offers the ability to produce high-quality films without the need to use harmful solvents or to balance deposition rates from multiple evaporation sources as occurs in co-evaporation. However, due to the high vapor pressure of CH_3NH_3I, high-quality films require not only high currents to make CH_3NH_3I and PbI_2 evaporation as simultaneous as possible, but also a CH_3NH_3I excess in the precursor (up to CH_3NH_3I to PbI_2 ratio of 2) in order to compensate the inevitable loss of CH_3NH_3I during the deposition.

Mitzi, et al. [31] were the first to report on this strategy for the fabrication of a range of layered 2D organic-inorganic hybrid lead and lead-free halide perovskite films from powders, resulting in optically active, small grain-sized polycrystalline thin films. They demonstrated that this technique can be used for both organic cations and inorganic frameworks. Further improvements in this technique were conducted by Longo et al. [32] in 2015 preparing a 3D tetragonal, crystalline organic-inorganic hybrid $CH_3NH_3PbI_3$ perovskite films from a preformed perovskite layer through a solution-processed method, which was used for building efficient, planar photovoltaic devices with a high PCE of 12.2%. Uniform and smooth polycrystalline perovskite thin films of the desired thickness were obtained by careful tuning of the deposition parameters (Fig. 2).

Sputtering

Sputtering is a nonthermal vaporization process whereby individual atoms escape from the target surface due to atomic collision cascades by high-energy ion bombardment. The glow discharge can be achieved by either using direct current (DC) or using radiofrequency (RF). There is a large variety of sputtering techniques, and their names are based on their source and the orientation of the process. Nevertheless, perovskite thin films have only been prepared by RF-magnetron sputtering techniques. Here, a magnetically confined plasma is created and the resulting positively charged ions collide with the target material producing ejected atoms from the surface. These atoms travel from the target to the substrate surface to create thin films (Fig. 2E). It constitutes a powerful technique for the deposition of both conductive (metals) and nonconductive (insulators) materials thin films, offering the possibility to grow consecutive layers of different materials with a nanometric control of the thickness and a reduced surface roughness. Other advantages are the homogeneity of the obtained films, the limited contamination from unwanted elements, and, conversely, the ease of doping. Despite its advantages, it has received very limited attention due to the slowness of the process and to the fact that the application of RF is not simple since it requires an expensive power supply and additional electronic accessories.

A successful route for the one-pot growth of $CH_3NH_3PbI_3$ films based on an RF-magnetron sputtering technique was first reported by Boroni et al. [33] in 2018. Highly reproducible single-phase hybrid perovskite films, full coverage of the substrate surface, and thickness ranging from less than 200 nm to more than 3 μm were obtained by starting from a single target made of a CH_3NH_3I and PbI_2 mixture with a 30% *w*/w excess of CH_3NH_3I. They presented a similar morphology to those reported from different vapor deposition methods, such as chemical vapor deposition (CVD) and thermal evaporation, but a more compact layer and better coverage of the substrate were obtained.

Laser deposition

This deposition method can further be classified as pulsed laser deposition and resonant-infrared matrix-assisted pulsed laser evaporation.

Pulsed laser deposition resembles flash evaporation in that the precursors are rapidly projected from a source onto a substrate in a vacuum chamber, but the energy required to vaporize them is provided by a UV laser striking the material rather than a more conventional heat source. Some advantages of this technique are related to conformal deposition and stoichiometric transfer; and the composition of the films grown by this technique is quite close to that of the target, even for a multicomponent target. Furthermore, it is a pollution-free growth (solvent-free) and accurately controls the deposition rate of the organic component. Nonetheless, the use of a UV laser introduces a complication, since the ablated particles are considerably more energetic than those that are thermally evaporated and can even sputter material (particularly volatile organic compounds such as CH_3NH_3I) from the substrate surface.

In 2015, Bansode et al. [34] proposed a new variant of the classical pulsed laser deposition process, which allowed the room-temperature growth and stoichiometry control of hybrid perovskite films through the use of nonstoichiometric single-target ablation and off-axis growth to compensate for the high volatility of CH_3NH_3I as shown in Fig. 2F. They achieved hybrid lead and lead-free mixed-halide perovskite film with a stoichiometry control that is not as easily accomplished by the other methods. Planar heterojunction perovskite solar cells achieved a maximum PCE of 7.66% for 0.09-cm^2 active area when the device was scanned from forwarding bias to short circuit, using ZnO nanoparticles as an electron-transporting layer without the requirement of a heating or sintering step, thus enabling its use for flexible substrates at room temperature.

Resonant-infrared matrix-assisted pulsed laser evaporation (RIR-MAPLE) is another laser-based method in which the target is composed of a solution of the precursors rather than a pellet. The laser energy is resonant with one of the bonds in the solvent (particularly –OH bonds), and hence, the ablation of the target solution should proceed in a gentler manner. The solvent, rather than the precursors, accepts most of the energy and vaporizes, carrying the precursors to the substrate on the solvent plume. This method offers a greater degree of control over the composition of the film compared to the pulsed laser counterpart since the stoichiometry of the precursors in the target solution should be transferred to the substrate. Consequently, stoichiometric films can be achieved without an excess of the organic halide component, which can be associated with other undesirable phases. Also, it overcomes the incompatible solubility and volatility of the different components of halide perovskites. On top of that, compared to vapor deposition techniques, RIR-MAPLE eliminates the risk of damaging sensitive organic materials through excessive heating (as in thermal evaporation) or by direct absorption of high laser energy (pulsed laser deposition). Despite that, this technique is still relatively unexplored for a variety of hybrid halide perovskite compositions. One complication is that while the target solvent should be pumped away by the dynamic vacuum, it can also co-deposit on the substrate along with the precursors. In such a case, a thermal annealing step is needed.

The first work using this technique was done by Barraza et al. [35] in 2018 based on the deposition of organic-inorganic hybrid perovskite films on multiple substrates, thus creating smooth, pinhole-free, compact grains, phase-pure films, by using stoichiometric targets (1:1 PbI_2: CH_3NH_3I molar ratio) in a double-solvent mixture, DMSO/MEG (monoethylene glycol). Post-annealing at gentle conditions (110°C for 10min) promoted solvent evaporation and crystal formation while minimizing thermal degradation of the film (Fig. 2G). These RIR-MAPLE-deposited $CH_3NH_3PbI_3$ films possessed comparable optical and crystallographic properties to those prepared by solution-prossessing methods. In the following year, Dunlap-Shohl et al. [36] investigated the optoelectronic properties of phase-pure, compact fine-grained microstructure RIR-MAPLE-deposited $CH_3NH_3PbI_3$ perovskite films and reported on the fabrication of perovskite solar cells, with the highest PCE of 12.2% so far for any perovskite solar cell in which the absorber layer was prepared by a laser ablation-based deposition process, highlighting the ability for being applied in devices with good-quality films.

2.1.2 Chemical deposition methods

Chemical deposition methods in vacuum processes involve chemical reactions between precursors, generally in the gas phase, to deposit the corresponding material onto a substrate surface. These coating methods do not involve purely physical processes, such as high-temperature vacuum evaporation with subsequent condensation or sputter bombardment that takes place in physical deposition methods, but involve a chemical reaction on the surface to be coated. These methods are also used to produce high-purity and high-performance films. The most known and widely used technique is chemical vapor deposition.

Chemical vapor deposition

Chemical vapor deposition methods are characterized by reactions between one or more volatile precursors in a vacuum chamber either in the gas phase or on the substrate surface, usually in the presence of some form of reaction initiator (either heat, high-energy radiation, or plasma), as well as the presence of specific chemicals that support the reaction. Often, these chemicals are directed to the substrate, and the volatile by-products are removed by an inert gas stream through the reaction chamber. The reaction process mainly includes three parts: (i) vaporization of the precursors and reactant transport process, (ii) chemical reaction and/or decomposition process between precursors, and (iii) elimination of reaction by-product. This perovskite deposition method is a promising alternative to solution-based methods of fabrication owing to high performance and resistance, relative ease of patterning, the ability to coat complex surfaces and using a range of pressure conditions even at ambient pressure (without the need for high or ultra-high vacuum conditions as in vacuum thermal evaporation), and therefore, it is more cost-effective than a physical deposition method. Furthermore, the ability to batch and scale up the process to enable mass production, together with the wide range of material compatibility, makes this technique a good candidate for uniform large-area deposition. However, chemical vapor deposition has some drawbacks, which may include limitations on size, due to the vacuum chamber size, as well as the high temperature required to drive the chemical reaction. It is worth mentioning that precursors for CVD processes must not only be volatile but also be stable enough to be able to be delivered to the reactor. In general, both lead halide and organic halide precursors are deposited onto a substrate in a single step. The advantage of using a single-step approach relapses on the uniformity across the batch.

In 2015, Tavakoli et al. [37] reported a facile one-step CVD method to fabricate efficient planar heterojunction perovskite solar cells with tuning compositions with a PCE of up to 11.1% and an open-circuit voltage of 0.97V. The perovskite layers were synthesized by co-sublimation of the two precursor materials placed in the high-temperature zone and at a given position of each of the sources in the furnace as shown in Fig. 3A. The sublimated vapors moved to the substrate that was positioned in a low-temperature zone where they reacted to produce the corresponding perovskite onto the substrate surface. By systematically optimizing the CVD parameters, such as temperature and growth time, high-quality films of $CH_3NH_3PbI_3$ and $CH_3NH_3PbI_{3-x}Cl_x$ were obtained.

Aerosol-assisted chemical vapor deposition (AACVD), shown in Fig. 3B, is one variation of the single-step CVD. Actually, this was the first report on a chemical vapor deposition method by Lewis et al. [38] in 2014. In this strategy, precursors are first dissolved in DMF in a stoichiometric ratio; then, they are nebulized in one step, followed by aerosol carrying into a hot tube furnace containing a glass substrate by an inert gas stream, such as argon, where thermal decomposition takes place to leave the $CH_3NH_3PbBr_3$ film. Compared with other deposition methods, it is a low-cost and scalable technique. The process occurs in onestep under ambient pressure and requires moderately volatile precursors, which should be a very attractive film fabrication technology and has a good potential for scale-up [39]. The films obtained are comparable in quality to those obtained by current methods used for the deposition of hybrid perovskites, such as spin coating, and do not require high vacuum conditions for their deposition, in contrast to physical vapor deposition methods. However, the ability to deposit dense and uniform perovskite films is still challenging.

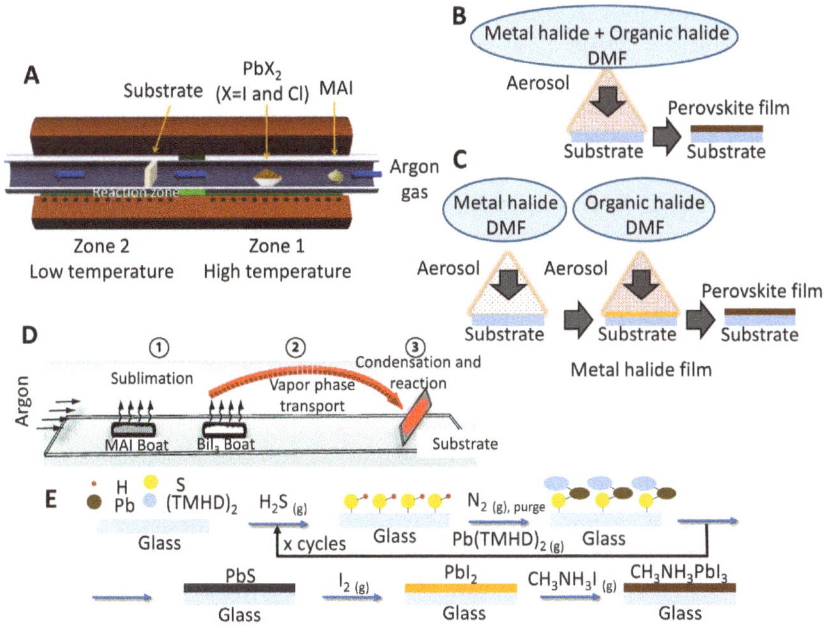

FIG. 3 Schematic illustration of the (A) one-step chemical vapor deposition; (B) one-step aerosol-assisted chemical vapor deposition; (C) two-step aerosol-assisted chemical vapor deposition; (D) ambient pressure chemical vapor deposition; (E) atomic layer deposition. *Panel (A) adapted with permission under Creative Commons Attribution 4.0 International License from Tavakoli, M. M., Gu, L., Gao, Y., Reckmeier, C., He, J., Rogach, A. L., Yao, Y., & Fan, Z. (2015). Fabrication of efficient planar perovskite solar cells using a one-step chemical vapor deposition method. Sci. Rep., 5, 14083. Copyright 2015, The Authors, (B–D) adapted with permission from Chen, X., Myung, Y., Thind, A., Gao, Z., Yin, B., Shen, M., Cho, S. B., Cheng, P., Sadtler, B., Mishra, R., & et al. (2017b). Atmospheric pressure chemical vapor deposition of methylammonium bismuth iodide thin films. J. Mater. Chem. A, 5, 24728–24739. Copyright 2017, Royal Society of Chemistry, and (E) adapted from Sutherland, B. R., Hoogland, S., Adachi, M. M., Kanjanaboos, P., Wong, C. T. O., McDowell, J. J., Xu, J., Voznyy, O., Ning, Z., Houtepen, A. J., & Sargent, E. H. (2015). Perovskite thin films via atomic layer deposition. Adv. Mater., 27(1), 53–58.*

Two-step aerosol-assisted chemical vapor deposition was developed as an improvement of one-step AACVD, which produces a less densely packed perovskite coating that might be originated from the relatively high deposition temperature. Hence, the two-steps AACVD allows for independent control of the deposition of each precursor layer, with better control over the perovskite film formation as illustrated in Fig. 3C.

One year after the development of the one-step AACVD, Chen et al. [39] prepared uniform, compact $CH_3NH_3PbI_3$ perovskite films on FTO (fluorine-doped tin oxide) glass substrates, which exhibited high uniformity, good phase, and micron-sized pinhole-free grains. Precursor solutions were prepared by dissolving lead halide and organic halides in

DMF and methanol, respectively. The mist was created using an ultrasonic humidifier, which was delivered to the reaction chamber through an inert nitrogen gas stream.

The atmospheric pressure chemical vapor deposition (APCVD) method was used by Chen et al. [40] to prepare lead-free $CH_3NH_3Bi_2I_9$ thin films at low temperatures (160°C) and atmospheric pressure (Fig. 3D). They co-sublimated CH_3NH_3I and BiI_3 (the precursors) in one step in a flow of argon stream, which condenses and then reacts thus obtaining films with better control over the morphology, purity, and coverage concerning than of solution-processed, undoped films and with the added advantage of rapid scale-up and integration with other solar manufacturing processes. The result is an *n*-type, dense, and polycrystalline $CH_3NH_3Bi_2I_9$ film with an optical bandgap of 1.80 eV, a room-temperature free-electron carrier density of $3.36 \cdot 10^{18}\,cm^{-3}$ and Hall mobility of $18\,cm^2\,V^{-1}\,s^{-1}$.

Atomic layer deposition

The atomic layer deposition (ALD) method is a variant of CVD that utilizes a series of successive, self-terminating, and irreversible gas-solid reactions of at least two compounds in low-vacuum and low-temperature conditions for the deposition of high-quality, thin, uniform, and conformal films over large areas with atomic thickness precision. Moreover, it does not produce pinholes and provides excellent stoichiometry control (Fig. 3E). This technique is also compatible with roll-to-roll manufacturing, enabling the deposition of a highly uniform coating. Specifically, ALD has been widely applied for the preparation of numerous layers of device fabrication prior to perovskite deposition, including electron transport layers (ETLs), hole transport layers (HTLs), charge carrier blocking, junction forming, or encapsulation layers. Although it was not used for the direct deposition of the perovskite absorber layer in PSCs, it worked successfully to grow a precursor seed layer that can then be converted to $CH_3NH_3PbI_3$ in a post-treatment step [41,42].

2.1.3 Hybrid physical-chemical vapor deposition

The hybrid physical-chemical vapor deposition (HPCVD) method consists of a metal halide deposition, usually, thermal evaporation, followed by heating an organic halide powder in the furnace in the presence of the lead halide film to yield the perovskite layer (Fig. 4). This method is used to achieve an easier fabrication, higher quality, and better reproducibility of perovskite films that can be used for scaling-up. Nonetheless, problems associated with proper diffusion are encountered in this technique.

Parallel to the first work in the single-step AACVD by Lewis et al. [38], Ha et al. [43] synthesized lead halide perovskite platelets with lateral dimensions in the 5-30 μm range and thicknesses from several atomic layers to several hundred nanometers; the longer diffusion lengths of ca. 200 nm were related to the high crystal quality of the platelet films.

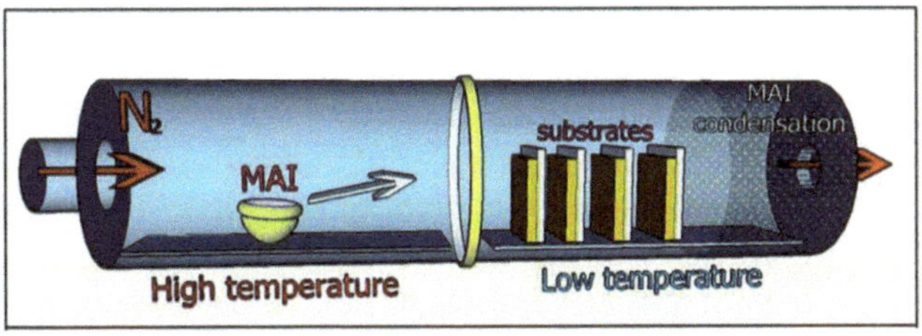

FIG. 4 Schematic illustration of the hybrid physical-chemical vapor deposition. *Adapted with permission under Creative Commons Attribution 3.0 Unported License from Leyden, M. R., Ono, L. K., Raga, S. R., Kato, Y., Wang, S., & Qi, Y. (2014). High performance perovskite solar cells by hybrid chemical vapor deposition. J. Mater. Chem. A, 2, 18742–18745. Copyright 2014, Royal Society of Chemistry.*

The method involved two steps: first, the growth of lead halide nanoplatelets on muscovite mica by van der Waals epitaxy. Then, the as-grown platelets placed downstream were converted to perovskites by a gas-solid hetero-phase reaction using methyl ammonium halide. Van der Waals epitaxy is a type of physical deposition method wherein a powder is evaporated and then deposited onto a substrate, whose surface is inert due to the absence of a high density of dangling bonds, in a quartz tube mounted on a single-zone furnace. This deposition method is often used for the growth of single-crystal 2D perovskite thin films. The first report demonstrating any PCE by chemical vapor deposition was carried out by Leyden et al. [44], months later from the first application to perovskites, through a low-pressure hybrid physical-chemical vapor deposition (LP-HPCVD) method. Lower pressures increase the rate of diffusion and the rate of sublimation of organic halides, thus permitting greater uniformity, faster deposition rates and the use of lower temperatures. First, $PbCl_2$ was deposited using thermal evaporation in a high vacuum, followed by sublimation of CH_3NH_3I through HPCVD in a furnace, thus resulting in efficiencies as high as 11.8% and stabilities of up to 1100h in a nitrogen environment. Shortening the deposition time by using formamidinium iodide (PCE of 14.2%) is of great interest to industrial processes.

2.2 Solution-processed methods

Solution-based deposition methods have been the most explored routes to deposit metal halide perovskites, especially in proof-of-concept studies in the preparation of optoelectronic devices, because they are easy and quick deposition approaches. The solution is used as a carrier of the perovskite components prepared from precursor salts dissolved in dimethylformamide (DMF), dimethylsulfoxide (DMSO), or a mixture of them, usually at concentrations in the 0.5–2M range. Solutions containing inorganic and organic salts have shown several bottlenecks during the crystallization of perovskite thin films, due to the different miscibility and solubility of the components, as well as the annealing step, thus leading to different crystallization rates and, as a consequence, lack of uniformity, and some pinholes and cracks formation. The solution process methods reported in this section are classified according to the size of the manufacturing area application, such as lab-scale and scalable techniques, and subclassified depending on the number of deposition steps. Several examples are presented, and the advantages and limitations of the different techniques are also discussed. It is noteworthy that some lab-scale techniques, such as dip coating, are compatible with other scalable techniques.

2.2.1 Lab scale

In general, single-step deposition methods have poor morphology, are less uniform, and present smaller grains compared to two-step deposition methods that exhibit better morphology with more uniform and larger grains.

One-step deposition

The one-step deposition methods are further classified into (i) spin coating, (ii) casting, (iii) dip coating, and (iv) capillarity thin-film growth as shown in Fig. 5.

Spin coating is considered the simplest and cost-effective method for preparing perovskite films. In the spin-coating strategy, a solution of perovskite precursors (metal and organic halides, generally stoichiometrically mixed) dissolved in a polar solvent, such as dimethylformamide (DMF), dimethylacetamide (DMA), dimethylsulfoxide (DMSO), gamma-butyrolactone (GBL), or their mixtures, is prepared. This solution is dropped onto a spinning substrate as shown in Fig. 5A. As the substrate spins (at hundreds to thousands of revolutions per minute), the liquid is removed from its surface, first by centrifugal force casting the excess solution off the substrate, and afterward by evaporation of the remaining solvent under a post-thermal treatment. As the solvent evaporates, the concentration of the perovskite precursors increases until it reaches the solubility limit (supersaturation), the point at which crystallization of the perovskite structure occurs. Generally, the as-formed film is transferred to a hot plate for relatively low-temperature annealing (typically 100–150°C), with the purposes to (i) remove any remaining solvent to cause supersaturation, (ii) assist in the nucleation and growth of the perovskite and (iii) setting the grains. The film performance is determined by several factors, such as precursor concentration, stoichiometry, solvent choice, as well as spin and environmental conditions, such as the presence of oxygen or moisture.

This technique was used by Kojima et al. [45] in 2009 to prepare the first successful perovskite solar cells of $CH_3NH_3PbBr_3$ and $CH_3NH_3PbI_3$, with a PCE of 3.13% and 3.81%, respectively. A stoichiometric amount of

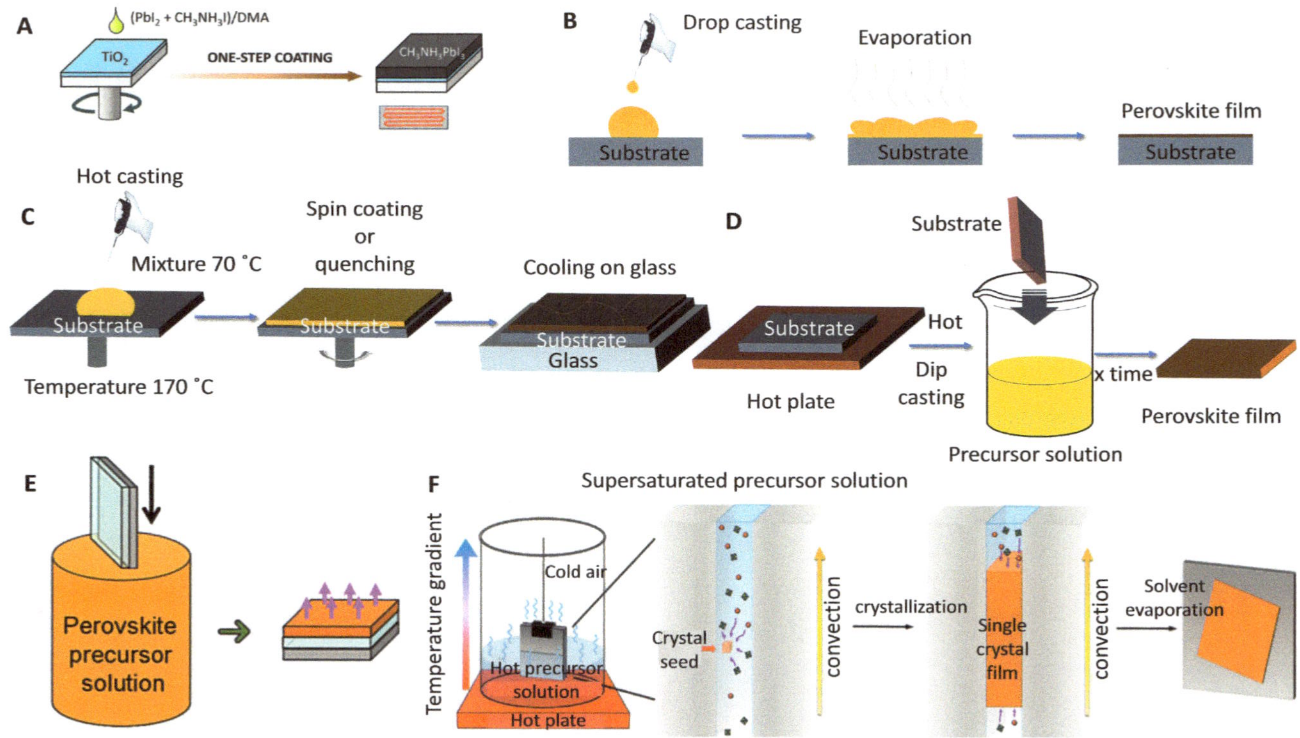

FIG. 5 Schematic illustration of the (A) one-step spin coating; (B) drop casting; (C) hot casting; (D) hot-dip casting; (E) one-step dip coating; (F) capillary thin-film growth. *Panel (A) Adapted with permission under Creative Commons Attribution (CC BY) license from Im, J. H., Kim, H. S., & Park, N. G. (2014). Morphology-photovoltaic property correlation in perovskite solar cells: one-step versus two-step deposition of CH3NH3PbI3. APL Mater., 2, 081510. Copyright 2014, Authors, (B–E) adapted with permission from Huang, L., Li, C., Sun, X., Xu, R., Du, Y., Ni, J., Cai, H., Li, J., Hu, Z., & Zhang, J. (2017). Efficient and hysteresis-less pseudo-planar heterojunction perovskite solar cells fabricated by a facile and solution-saving one-step dip-coating method. Org. Electron., 40, 13–23. Copyright 2016, Elsevier B.V. All rights reserved, and (F) adapted with permission from Chen, Y. X., Ge, Q. Q., Shi, Y., Liu, J., Xue, D. J., Ma, J. Y., Ding, J., Yan, H. J., Hu, J. S., & Wan, L. J. (2016b). General space-confined on-substrate fabrication of thickness-adjustable hybrid perovskite single-crystalline thin films. J. Am. Chem. Soc., 138(50), 16196–16199. Copyright 2016, American Chemical Society.*

CH_3NH_3X and PbX_2 was dissolved in DMF and GBL for $CH_3NH_3PbBr_3$ and $CH_3NH_3PbI_3$, respectively, and then spin-coated onto a TiO_2 film. Using a one-step preheating-assisted spin-coating method, a PCE of up to 18% was reported in 2015 [46]. This technique involved first casting a hot (70°C) precursor solution mixture of CH_3NH_3Cl and PbI_2 onto a substrate maintained at 180°C and, subsequently, spin-coating (15 s) to obtain a uniform perovskite film. The efficiency achieved using this technique can be high and with a good level of reproducibility and morphology control for small substrates. However, this approach tends to yield very rough films with poor substrate coverage, leading to incomplete pore filling when employing a mesoporous substrate. This fact, together with the generation of large material waste, limits this technique to lab scale. There are some different techniques from one-step spin coating, such as solvent engineering/antisolvent washing, solvent extraction, and gas quenching. It is noteworthy that, despite using variations in the spin-coating process, it is still a one-step process.

Solvent engineering or antisolvent washing consists of a spin-coating technique with the difference that while the substrate is spinning, a nonpolar antisolvent (e.g., toluene, chlorobenzene, or diethyl ether) is dripped onto the substrate that contained the precursor solution (Fig. 6A). The antisolvent effect promotes the supersaturation of the precursor solution by two mechanisms that lead to high nucleation density: (i) driving out the solvent much faster than it would evaporate or casting by the centrifugal force itself and (ii) reducing the precursor solubility in the remaining solution. This technique yields a flat and uniform perovskite thin film with a compact, pinhole-free microstructure, and it is adaptable to many different perovskite systems and many different device architectures. While solvent engineering is a reliable and portable process, the substrate washing step at the correct time should be carefully done.

Jeon et al. [47] were the first to report the solution-processed fabrication of efficient perovskite solar cells comprising a bilayer architecture and $CH_3NH_3Pb(I_{1-x}Br_x)_3$ perovskites formed by a solvent-engineering technique. When the perovskite was deposited using a mixture of GBL and DMSO, followed by a toluene drip while spinning, extremely uniform and dense layers were formed. However, precursor solutions that used DMF as the unique solvent produced films that often presented pinholes. This technique enabled a fully solution-processed perovskite solar cell with a certified PCE of 16.2%. The upcoming reports by the same group achieved certified efficiencies as high as 22.6% and 20.9% for small-area (ca. 0.094 cm^2) and medium-area (ca. 1 cm^2) cells, respectively [20].

Solvent extraction is a similar technique to solvent engineering but, in this case, the antisolvent applies through immersion in a bath (Fig. 6B). This technique offers some advantages over solvent engineering, specifically (i) it avoids the difficulty of precise timing and uniformly applying the antisolvent (better reproducibility), (ii) it reduces the risk of damage by

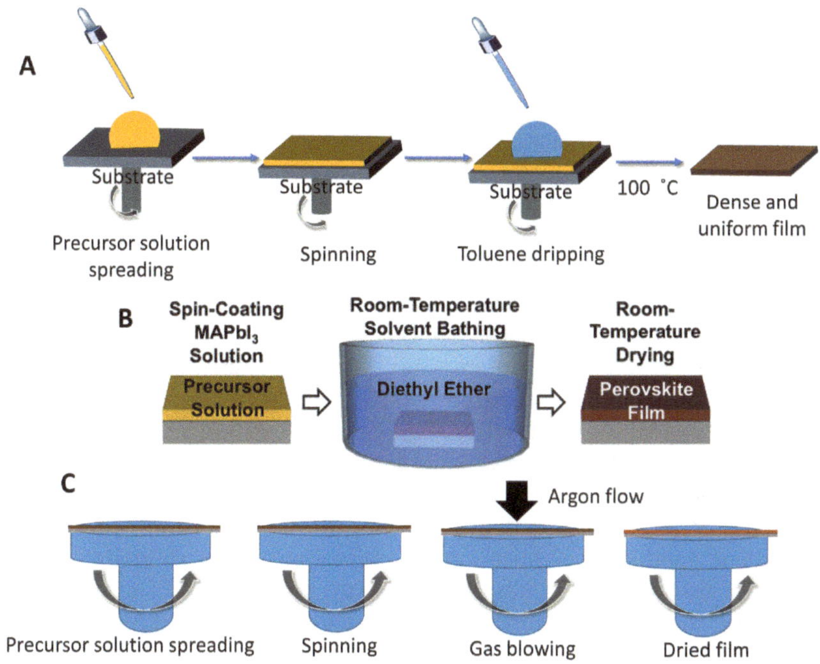

FIG. 6 Schematic illustration of the (A) solvent engineering/antisolvent washing; (B) solvent extraction; (C) gas quenching. *Panels (A and B) adapted with permission from Zhou, Y., Yang, M., Wu, W., Vasiliev, A. L., Zhu, K., & Padture, N. P. (2015). Room-temperature crystallization of hybrid-perovskite thin films via solvent-solvent extraction for high-performance solar cells. J. Mater. Chem. A, 3, 8178–8184. Copyright 2015, Royal Society of Chemistry and (C) adapted with permission from Huang, F., Dkhissi, Y., Huang, W., Xiao, M., Benesperi, I., Rubanov, S., Zhu, Y., Lin, X., Jiang, L., Zhou, Y., Gray-Weale, A., Etheridge, J., McNeill, C. R., Caruso, R. A., Bach, U., Spiccia, L., & Cheng, Y. B. (2014). Gas-assisted preparation of lead iodide perovskite films consisting of a monolayer of single crystalline grains for high efficiency planar solar cells. Nano Energy, 10, 10–18. Crown copyright 2014, Published by Elsevier Ltd. All rights reserved.*

the high-speed interaction of the solvent jet with the rapidly spinning film, (iii) it can be carried out at room temperature, and (iv) it can be extended to scalable deposition techniques and is compatible with roll-to-roll manufacturing processes. Moreover, the use of this technique contributes to the formation of smoother perovskite films. Special care must be taken when considering other perovskite compositions beyond $CH_3NH_3PbI_3$, as the transport properties of different ions can change the crystallization dynamics, thus leading to changes in morphology and creating pinholes in mixed-halide perovskite films.

Zhou et al. [48] demonstrated this technique enabled room-temperature solution processing of high-quality $CH_3NH_3PbI_3$ hybrid perovskite thin films, with excellent control over their thickness and smoothness, and consequently its application for fabricating perovskite solar cells. Planar perovskite solar cells using $CH_3NH_3PbI_3$ films displayed a PCE of up to 15.2%.

Gas quenching is a closely related technique to solvent engineering, in which the antisolvent is replaced with an inert gas stream such as argon or nitrogen during the one-step coating (Fig. 6C). The gas prompted rapid crystallization, leading to film morphologies like those produced by solvent engineering after annealing. The morphology and grain size of the perovskite film can be tuned by varying the gas pressure and quench timing. Gas quenching has been successfully applied to both hybrids (MA- and FA-based perovskite) systems. It is an attractive technique in terms of both material costs and safety concerns to remove the use of hazardous solvents (e.g., toluene, chlorobenzene) used in antisolvent washing. Furthermore, it is robust and readily scalable for large areas when combined with other deposition techniques such as blade coating and slot-die coating (see Section 2.2.2). After gas quenching, perovskite films are usually subjected to an annealing step to remove the remaining solvent.

Huang et al. [49] reported in 2014 the gas-quenching method to fabricate dense methylammonium lead iodide perovskite thin films of ca. 300 nm in thickness for high-efficiency planar solar cells. The films consisted of a monolayer of single-crystalline grains (300 and 500 nm) after an annealing step. Planar perovskite solar cells constructed from these films yielded a maximum efficiency of 17%.

Casting technique consists of a simple transfer of perovskite solution to a target substrate followed by natural evaporation of the solvent as shown in Fig. 5B. The advantages of this technique are that no special equipment is required and it is applied to obtain relatively thick and preferentially oriented large-grain perovskite films. It is possible to prepare large-area perovskite film, but the lack of control of film thickness, inhomogeneity, and accurate perovskite solution usage are the main drawbacks of this method. Casting includes drop casting and hot dip casting.

Drop casting consists of depositing the precursor solution onto a substrate and let the solvent evaporate under the influence of mild annealing as shown in Fig. 5C. This method has been successfully applied to mesoporous architectures in which the entire device consists of a mesoporous scaffold, where the crystallization of perovskite is fully constrained within the thick, multiple mesoporous layers, unlike the typical planar thin-film device structure. Although these cells possess the modest efficiencies, they enjoy high stability in ambient conditions, perhaps because the carbon electrode thickness (ca. 10 μm) offers a high degree of resistance to moisture. For instance, Mei et al. [50] reported a solar cell using this mesoporous architecture with a certified PCE of 12.8%, and it was stable for >1000 h in ambient air under full sunlight. The perovskite layer was infiltrated into the porous TiO_2/ZrO_2 scaffold by drop casting a solution of PbI_2, CH_3NH_3I, and 5-aminovaleric acid (5-AVA) in GBL, forming the new mixed-cation perovskite $(5\text{-AVA})_x(CH_3NH_3)_{1-x}PbI_3$ with lower defect concentration and better pore filling, as well as more complete contact with the TiO_2 scaffold. Moreover, a longer excitonic lifetime and a higher photoluminescence quantum yield as compared to $CH_3NH_3PbI_3$ were also achieved.

Hot dip casting is found to be one of the reliable approaches to fabricate halide perovskite photoactive layers for a large-throughput and ultrafast production with full surface coverage (Fig. 5D). Hot substrates facilitate film formation and crystallization, but they can also cause dewetting of the precursor solution on the substrate, thus altering the contact angle. Moreover, the films still require post-annealing steps. It can be done either by drop casting a precursor solution onto a preheated substrate or by dipping a hot substrate into a precursor solution, although the last one might be classified into a dip-coating technique as it will be discussed later.

This method, with some variations, was first reported by Nie et al. [46]. Although a high PCE of 18% was achieved, an unusual morphology (large, leaf-like structures of nearly a millimeter) was obtained. Later on, Liao et al. [51] prepared the film by dipping a hot substrate into a precursor solution and then letting it dry out. Perovskite crystallized on the substrate right after the substrate was withdrawn from the solution, and the PSC yielded a PCE of 12.4%. This is promising as it demonstrates the feasibility of the hot-casting process in large-throughput coating methodologies.

Dip coating is a simple and fast process that involves the deposition of a film by immersion of the substrate into a precursor solution (Fig. 5E). After the removal of the substrate from the solution and drying, a solid film is formed on the substrate surface. Dip coating is an ideal method to prepare thin films from chemical solutions since it is low cost and minimizes the waste of source materials, enabling scale-up and offering a good control on thickness. Although dip-coated films are free of pits and pinholes, they usually yield thicker films (from tens or hundreds of micrometers) with rougher surfaces than other coating techniques.

In 2017, Huang et al. [52] applied this strategy to sequentially prepare a dense sol-gel-derived titanium dioxide electron-transporting layer and a smooth perovskite film for hybrid perovskite solar cells. The precursor-covered substrate was blow-dried by a nitrogen stream and then annealed with a PCE of 11% by simply using spiro-MeOTAD as the hole transporter and Ag as the anode; its performance was comparable to that of a reference device (12%) fabricated by conventional spin coating. Both films have similar pinhole-free surface coverage, absorption properties, and crystalline structure.

Capillary thin-film growth is a technique in which two clean flat substrates are clipped together and then vertically dipped in a hot perovskite precursor solution to control the evaporation rate of the solvent as shown in Fig. 5F. The capillary pressure drove the solution onto the whole substrate. A temperature gradient was created, as the bottom solution maintained saturated conditions and the top solution film was supersaturated due to the cooler condition, thus inducing the perovskite crystallization in the clipped solution film. After a certain time, perovskite single-crystal thin films form in the confined space by the clipped substrates. The thickness was well defined by the distance between two substrates, which can be easily tuned from nano- to micrometers by clipping force. It is possible to synthesize bulk perovskite single crystals, which show significantly higher carrier mobility and longer diffusion length and carrier lifetime due to reduced trap-state densities (orders of magnitude compared with polycrystalline film caused by the considerable removal of grain boundaries). This technique affords the preparation of single-crystal perovskite thin films demanding high-quality and low-cost device applications. Though, the nature of the self-encapsulating process impedes the proper film characterization.

Chen et al. [53] reported this strategy in 2016 for the controllable fabrication of space-confined solution-processed growth on different substrates of various hybrid perovskite single-crystal submillimeter thin films with thicknesses from nano- to micrometers. These single-crystal thin films exhibited excellent air stability and comparable quality to bulk single crystals. This substrate-independent growth makes this technique attractive to the on-chip manufacture of perovskite devices for diverse applications.

Two-step deposition

The constrained solubility of lead halides can be an advantage because it stimulates the development of a wider range of two-step deposition approaches, without the fear that the second step will dissolve the lead halide layer. Two-step deposition processes rely on essentially the same steps: (i) a metal precursor is deposited on the substrate; (ii) the substrate is exposed to a vapor, liquid, or solid organic salt to drive the reaction (involving intercalated diffusion) that produces the perovskite; and (iii) a post-annealing step generally follows to drive off any excess of the organic salt and induces material crystallization. Lead halide perovskites are deposited from solvents such as DMF and/or DMSO, whereas the organic halides used alcohols, such as isopropyl alcohol (IPA) and ethanol (EtOH). This technique has been shown to be effective for $CH_3NH_3PbI_3$ films and multi-cation or multi-halide compositions, leading to good film morphology and high efficiencies. In comparison with direct coating of perovskite (e.g., $CH_3NH_3PbI_3$) precursors, it is generally easier to deposit uniform PbI_2 thin films over a large area, and the volume expansion from PbI_2 to perovskite can fill the pinholes to achieve compact perovskite films. Although it presents remarkably enhanced processing tolerance to humidity, better coverage, and morphology control than one-step deposition, its main drawback is the long and variable reaction time to form the perovskite.

Early works using two-step deposition usually soaked the lead halide films in organic halide solution for more than 10 minutes; mesoporous scaffolds were required to facilitate organic halide penetration. Two-step reactions on planar devices either result in an incomplete reaction or prolong the reaction time. The time consumption and indispensability of the mesoporous layer limited its scaling efforts.

The two-step deposition techniques are classified into (i) sequential spin coating, (ii) sequential dip coating, (iii) spin and dip coating, and (iv) electro(chemical bath) deposition and dip coating.

Sequential spin coating is based on spin coating of a metal halide solution in DMF or DMSO onto a substrate, followed by a second spin-coating process to deposit the organic halide solution in IPA or ethanol onto the metal halide layer as shown in Fig. 7A. A post-treatment step might be valid here to induce quick crystallization when blowing an inert gas or by dropping an antisolvent. The present two-step spin-coating process is expected to be more precise than the two-step dipping one because of its quantitatively managed procedure. Although this method results in a dense, uniform perovskite layer, the main challenge is the proper diffusion of the organic salt into a dense, compact PbI_2 film.

In 2014, Im et al. [54] reported the two-step spin-coating method to study the better morphology and higher photovoltaic performance of $CH_3NH_3PbI_3$ films in comparison with the one-step counterpart. The coverage, pore filling, and morphology of the deposited perovskite were found to be critical in the photovoltaic performance of the mesoporous TiO_2-based perovskite solar cells. Remarkably, the one-step coating showed a much better performance than the two-step coating with a PCE of 13.9% vs 7.5%. One year later, improvements in the two-step process were made by Yang et al. [55] through the introduction of an intramolecular exchange process, wherein a Lewis adduct $PbI_2{\cdot}DMSO$ solid film was first deposited instead of PbI_2, followed by spin coating of a solution of the organic ammonium halides CH_3NH_3X and thereafter annealing. CH_3NH_3X displaces the solvent molecules in the adduct, thereby forming the perovskite. The resulting films showed improved apparent grain size and exceptional performance, leading to the first certified PCE of a perovskite solar cell above 20%.

Sequential dip coating is based on the sequential dipping of a substrate into a lead and organic halide precursor solution in order to yield perovskite films from successive solid-state ion-exchange reactions (SSIER) as shown in Fig. 7B. It consists of a facile, cost-effective, and environmentally benign approach to preparing efficient perovskite films by simple dip-coating deposition, which may facilitate the production of large-area perovskite films using nonhalide lead precursors in an aqueous solution.

In 2018, Adnan et al. [56] reported an efficient methodology for preparing $MAPbI_3$ films using this technique, with layer adsorption of $Pb(NO_3)_2$

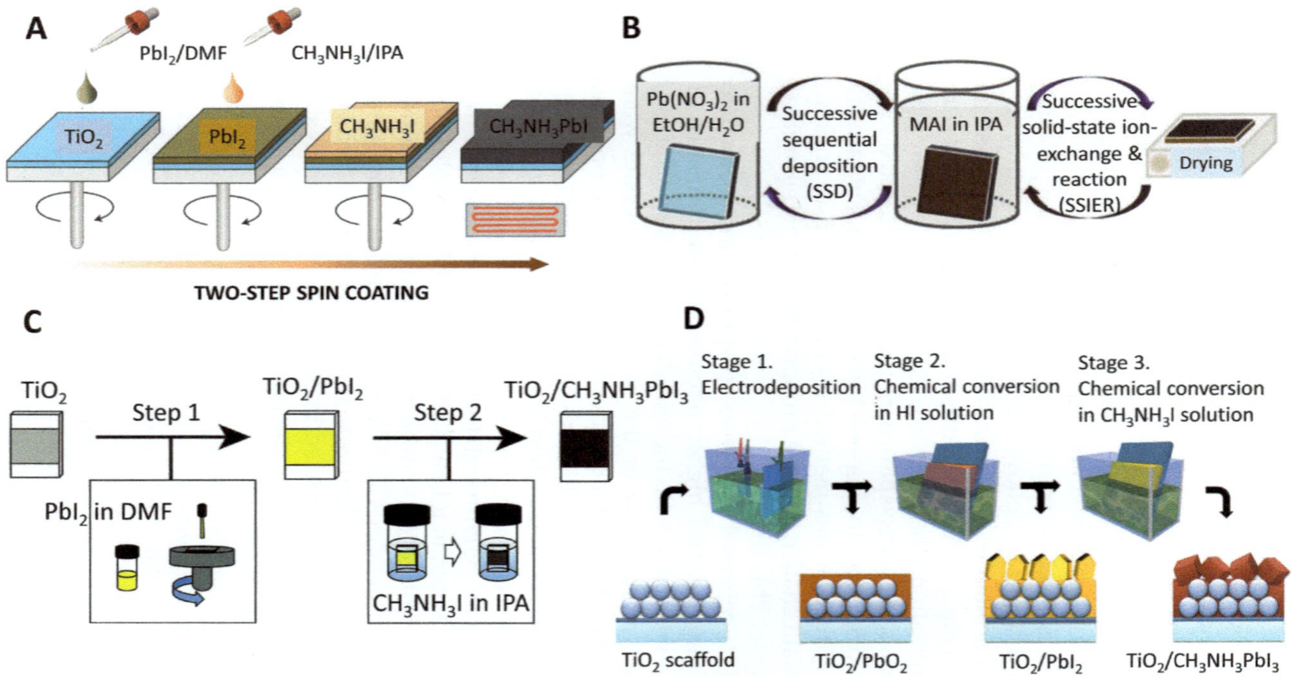

FIG. 7 Schematic illustration of the (A) two-step (sequential) spin coating; (B) two-step (sequential) dip coating; (C) sequential spin coating and dip coating; (D) sequential electrodeposition and dip coating. *Panel (A) Adapted with permission under Creative Commons Attribution (CC BY) license from Im, J. H., Kim, H. S., & Park, N. G. (2014). Morphology-photovoltaic property correlation in perovskite solar cells: one-step versus two-step deposition of CH3NH3PbI 3. APL Mater., 2, 081510. Copyright 2014, Authors, (B) adapted with permission under Creative Commons Attribution 4.0 International License from Adnan, M., & Lee, J. K. (2018). All sequential dip-coating processed perovskite layers from an aqueous lead precursor for high efficiency perovskite solar cells. Sci. Rep., 8(1), 2168. Copyright 2018, The Authors, (C) adapted from Harms, H. A., Tétreault, N., Pellet, N., Bensimon, M., & Grätzel, M. (2014). Mesoscopic photosystems for solar light harvesting and conversion: facile and reversible transformation of metal-halide perovskites. Faraday Discuss., 176, 251–269. Copyright 2014, Royal Society of Chemistry, and (D) adapted with permission from Chen, H., Wei, Z., Zheng, X., & Yang, S. (2015a). A scalable electrodeposition route to the low-cost, versatile and controllable fabrication of perovskite solar cells. Nano Energy, 15, 216–226. Copyright 2015, Elsevier Ltd. All rights reserved.*

in an aqueous solution followed by the reaction with MAI. Successive solid-state ion-exchange improved the morphology, crystallinity, surface coverage and stability of perovskite solar cells compared to materials produced by long-time dipping in CH_3NH_3I solution. The solar cells achieved a notable power conversion efficiency of 12.41%.

Spin coating and dip coating. In this technique, the metal halide layer is first introduced by spin coating a solution onto a substrate surface and subsequently transformed into the perovskite material by exposing it to a solution of the organic halide (Fig. 7C). This strategy enhanced the reproducibility and PCE (of ca. 15%) of the PSCs. The main drawback of this strategy is the incomplete conversion of PbI_2 to perovskite, due to the intercalation of organic halide precursors, which should penetrate completely from the surface to deeper layers of the film to guarantee full conversion.

In 1998, Liang et al. [57] were the first to report this technique in fabricating $CH_3NH_3PbI_3$ films, as well as a variety of 2D Pb- and Sn-based perovskites, for overcoming difficulties with morphology control and choice of solvent in one-step spin-coating techniques. Metal iodide films were first formed either by spin coating from a saturated methanol solution or by thermal vacuum evaporation and then dipping into an organic ammonium iodide solution, during which in situ formation of the perovskite occurred; while the 2D Pb-halide perovskites formed completely within minutes, a long time (up to hours) was required to complete the reaction of 3D systems. In addition to the kinetics of transformation, the main challenge associated with the conversion of planar PbI_2 films to perovskite is the transformation of a dense layered structure to a less dense 3D structure. Using a mesoporous rather than a planar architecture can accelerate the Pb halide-to-perovskite conversion significantly. This method was first implemented by Burschka et al. [58] in 2013 to form $CH_3NH_3PbI_3$ perovskite solar cells in which the perovskite completely infiltrates within the mesoporous TiO_2 scaffold, thus making the reaction between PbI_2 and CH_3NH_3I easier. The addition of additives, such as CH_3NH_3Cl to a CH_3NH_3I solution, has been reported to improve the photovoltaic performance of the device from 8.5% to ca. 15%.

Electrodeposition and dip coating. Electrodeposition, commonly known as electroplating, can be used to produce metals or semiconductors, such as metal oxides and metal chalcogenides. First, lead precursors (PbO_2, PbO or, PbI_2) are electrochemically deposited on a substrate. The lead precursor is then converted to perovskite by reacting with organic halides or by reacting with HI acid to form PbI_2, followed by perovskite formation as shown in Fig. 7D. This technique is versatile for producing simple, cost-effective surface coatings, and allows roll-to-roll production to large-area batch processing, good pore filling in mesoporous structures, and a dense and uniform perovskite film with good coverage. Some drawbacks of this method are that the solution is wasted after every deposition and post-thermal treatment is needed to form the final product.

Cui et al. [59] were the first in reporting this strategy for the fabrication of $CH_3NH_3PbI_3$ perovskites on a mesoscopic TiO_2 scaffold along with Chen et al. [60] in 2015. In the work reported by Cui et al, a PbO film was first electrochemically deposited on the TiO_2 film and then in situ transformed to PbI_2 by iodization and finally converted to $CH_3NH_3PbI_3$ by interdiffusion yielding a dense and flat upper layer on the surface with the perovskite well filled in the porous TiO_2 film to yield a high PCE of 12.5%. Chen et al. first fabricated PbO_2 films by electrodeposition, then converted to PbI_2 by immersion in a solution of HI in ethanol and finally to $CH_3NH_3PbI_3$ by immersion of the PbI_2 films into a CH_3NH_3I/isopropanol solution, thus producing a lower value of PCE of 10.31% but improving the cell area of electrodeposited perovskite solar cells up to 4 cm^2 with an excellent perovskite film uniformity. More recently, Wang et al. [61] have carried out the electrochemical deposition strategy on the direct fabrication of fully inorganic halide perovskites under air at room temperature, with highly controllable grain size, thickness, and fast fabrication within few minutes and an efficiency of up to 7.86%.

2.2.2 *Scalable techniques*

Even though all the above-mentioned techniques are fast, effective, and relatively reproducible methods in the laboratory, they are generally unsuitable for direct application in large-area manufacturing in terms of large substrate handling/throughput, film uniformity, coverage, and material usage. These challenges regarding large-area thin-film formation can be overcome by the introduction of scalable techniques. It is noteworthy that these techniques are compatible with either one- or two-step deposition approaches (Fig. 8).

Blade coating, also known as doctor blade or knife-over-edge coating, is a deposition method compatible with the roll-to-roll technique relying on a similar principle to spin coating. However, the excess solution is removed by sweeping a blade across the substrate surface, or vice versa in the case of roll-to-roll, coating at a precisely fixed height rather than by spinning rapidly as shown in Fig. 8A. The blade then spreads a predispensed ink and then the film dries, creating a solid thin film. This method is appealing to roll-to-roll manufacturing for optoelectronic device fabrication, as the solution can first be deposited onto a moving substrate and then swept under a stationary blade. Blade-coated perovskite films are typically polycrystalline, homogeneous, with higher coverage and larger crystalline grains and pinhole-free, showing long-lived photoluminescence. Compared to spin coating, there is almost no solution waste and it can operate under ambient conditions, showing resistance to humidity

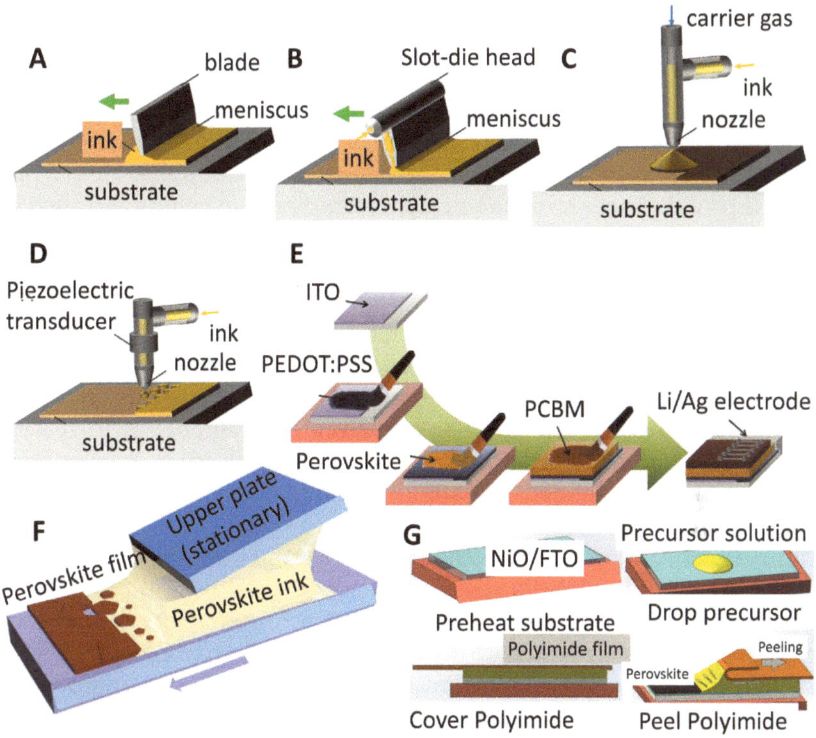

FIG. 8 Schematic illustration of the (A) blade coating; (B) slot-die coating; (C) spray coating; (D) inkjet printing; (E) brush painting; (F) meniscus-assisted solution printing; (G) soft-cover coating. *Panels (A, B, C, and D) are adapted with permission under Creative Commons (CC-BY-NC-ND) License from Howard, I. A., Abzieher, T., Hossain, I. M., Eggers, H., Schackmar, F., Ternes, S., Richards, B. S., Lemmer, U., & Paetzold, U. W. (2019). Coated and printed perovskites for photovoltaic applications. Adv. Mater., 31(26), 1806702. Copyright 2019, The Authors. Published by WILEY-VCH Verlag GmbH & Co. KGaA, Weinheim, (E) adapted with permission from ref Lee, J. W., Na, S. I., & Kim, S. S. (2017). Efficient spin-coating-free planar heterojunction perovskite solar cells fabricated with successive brush-painting. J. Power Sources, 339, 33–40. Copyright 2016, Elsevier B.V. All rights reserved, (F) adapted with permission under Creative Commons Attribution 4.0 International License from ref He, M., Li, B., Cui, X., Jiang, B., He, Y., Chen, Y., O'Neil, D., Szymanski, P., Ei-Sayed, M. A., Huang, J., & Lin, Z. (2017). Meniscus-assisted solution printing of large-grained perovskite films for high-efficiency solar cells. Nat. Commun., 8, 16045. Copyright 2017, The Authors, and (G) adapted with permission under Creative Commons Attribution 4.0 International (CC-BY-NC-ND) License from refYe, F., Chen, H., Xie, F., Tang, W., Yin, M., He, J., Bi, E., Wang, Y., Yang, X., & Han, L. (2016). Soft-cover deposition of scaling-up uniform perovskite thin films for high cost-performance solar cells. Energy Environ. Sci., 9, 2295–2301. Copyright 2016, Royal Society of Chemistry.*

and oxygen. In addition, it can be combined with another technique, such as solvent engineering, to further enhance efficiency.

In 2015, Kim et al. [62] reported the first preparation on $CH_3NH_3PbI_3$ perovskite films with a maximum PCE of 9.52% through the blade-coating technique. The formation of self-assembled large perovskite crystalline domains favored uniform film coverage and improved device ambient stability compared to that of the spin-coated film. This technique can also be applied to systems based on using the advanced solvent engineering technique reported elsewhere [38].

Slot-die coating is a process quite similar to blade coating but differs in respect that the precursor solution is dispensed onto the substrate at a precise rate by a die that serves as a print head rather than being controlled by the position of a blade held above the substrate (Fig. 8B). The evaporation rate of the solvent and the crystallization rate of the solutes are controlled by the solution flow rate as well as by substrate condition (if heat is applied) and the ambient environment (under vacuum). Although the process is attractive as there is no need to remove excess solution, it produces very rough and discontinuous films when coated on room-temperature substrates. It could be improved considerably by increasing the substrate temperature, inducing more rapid nucleation, or promoting the perovskite nucleation by applying antisolvent or gas-quenching techniques.

In 2015, Hwang et al. [63] demonstrated for the first time the two-step fabrication of fully slot-die-coated perovskite solar cells through a 3D printer in which all layers were fabricated through a scalable process. For the fully printed perovskite device, a hot slot-die coating of CH_3NH_3I was employed to promote the conversion reaction of PbI_2 to perovskite, yielding the best PCE of 11.96% under ambient conditions. They opened a new avenue for roll-to-roll perovskite manufacturing. Very recently, Fievez et al. [64] reported large-area, additive-free, low-temperature crystallization of such CH_3NH_3-free perovskite films via slot-die-coating technique with synergistic combination with the gas-quenching technique achieving $Cs_{0.16}FA_{0.84}Pb\ (I_{0.88}Br_{0.12})_3$ perovskite films over $10 \times 10\,cm^2$ substrates. They implemented this film into an (n-i-p) device configuration, reaching a maximum PCE of 18.03% over 0.09-cm^2 active area.

Spray coating has the potential for precise patterning of perovskite films. A perovskite ink or solution precursor is ejected through a nozzle, creating atomized droplets as small as possible to avoid film roughness that can then fall by gravity to a substrate surface or be directed by a carrier gas (Fig. 8C). The solvent evaporates, leading to the deposited perovskite precursors, which may then be annealed to form the final film. Atomization can be generated in a variety of ways, being ultrasonic stimulation the newest technology and showing more uniform droplet size, better airflow, and spray cone control. It is an attractive technique for high-volume manufacturing, as it allows large areas to be coated at high speed with only minimal loss of coating ink. Generally, one-step spray-coating methods are nonuniform, mostly dendritic or without full coverage films.

The first report on planar heterojunction $CH_3NH_3PbI_{3-x}Cl_x$ perovskite solar cells fabricated by a single-step ultrasonic spray-coating method from a DMF solution was made by Barrows et al. [65] in 2014. They found that while films coated on room-temperature substrates had poor surface coverage due to the formation of flower- or starburst-like structures; a better coverage was achieved by increasing the temperature, thus leading to PCE of up to 11%. Later, fully spray-coated triple-cation perovskite solar cells were prepared by Bishop et al. [66] in 2020 achieving a champion PCE value of 19.4% on small-area substrates (15.4 mm^2).

Inkjet printing is closely related to spray coating due to the similar mechanism and geometry of thin-film formation, and it is based on the ejection of fixed quantities of inks in the form of droplets from nozzles, but the resulting jet may be steered carefully in order to achieve precise patterns (Fig. 8D). The most common inkjet printing methods are continuous inkjet printing and drop-on-demand printing. The main processing advantages of this technique are the low-cost and mask-less on-demand patterning, which offers unmatched freedom of design, efficient use of materials, and scalability (from laboratory scale to larger-scale sheet-to-sheet or roll-to-roll manufacturing), without requiring contact to the substrate. It is difficult to get high-performance perovskite solar cells with uniform and compact crystallized perovskite films. Both the perovskite inks and printing parameters affect the ink-drop formation. Therefore, droplet wetting behavior on substrates and the subsequent film formation need to be optimized for the fabrication of defect-free uniform perovskite thin films. Moreover, this approach requires post-thermal treatment for 1–2 h.

The first report of inkjet printing by Li et al. [67] in early 2015 demonstrated that $CH_3NH_3PbI_3$ film was deposited onto a mesoscopic TiO_2 film in a single step using a 1:1 ink solution of $CH_3NH_3I:PbI_2$ in GBL. Appropriate optimization of CH_3NH_3Cl addition produced a flat and more uniform perovskite layer with a maximum PCE of 12.3% for a 0.04-cm^2 active area under AM 1.5 illumination conditions, and with high reproducibility. Further improvements were carried out by Liang et al. [68] using an innovative vacuum-assisted thermal annealing post-treatment for perovskite solar cells based on inkjet-printed $CH_3NH_3PbI_3$ films leading to a maximum power conversion efficiency of single-step deposition of ca. 17% on small-area (0.04 cm^2) and ca. 13% on large-area (4 cm^2) devices. In this case, Bag et al. [69] reported a two-step deposition method based on the use of *multi-channel* printing of organic cations onto prespin-coated PbI_2 films where the composition of the active layer can be easily and rapidly tuned by in situ mixing of cations from separate ink cartridges, for optimal performance and stability. They achieved a maximum PCE of 11.1%.

Brush painting is inspired by an old Chinese calligraphy tradition in which the "Maobi," also called "Chinese ink brush," is dipped into the ink, previously adsorbed and stored between hairs, which are pressed on the substrate surface to deliver the ink (Fig. 8E). Brush painting is an inexpensive method with advantages of high material utilization ratio, ease of patterning, high-speed fabrication, compatibility with roll-to-roll manufacturing, processability on various substrates (even on a nonflat surface), and suitability for large-area production. However, the quality and thickness of a perovskite film are difficult to control accurately, giving low reproducibility.

In 2017, Lee et al. [70] prepared all brush-painted heterojunction perovskite solar cells, except for an evaporated metal electrode, yielding a PCE of 9.08% on an ITO substrate and 7.75% on a flexible substrate.

Meniscus-assisted solution printing combines elements of both blade- and slot-die-coating techniques. Perovskite solution is loaded in a confined geometry composed of two nearly parallel plates (an upper stationary plate and a lower movable substrate separated by 300 μm). The lower substrate is mounted on a computer-programmed translational stage to expedite the evaporation of the solvent. Thus, the evaporative front progressively swept across the entire substrate to print a continuous micrometer-scale grain perovskite film by accelerating the solvent evaporation through the meniscus effect (Fig. 8F). Solutes gradually precipitated from the supersaturated phase and grow into large crystals as the outward flow continuously transports more solutes to the edge of the meniscus. In principle, large-grain-sized perovskite films can be achieved by this strategy since it presents advantages for flexible solar cells and other applications requiring a low-temperature continuous fabrication process. The low substrate speed reduces the possible implementation in a high-throughput production line; tuning the temperature and/or using a lower boiling-point solvent could produce more useful processing speeds.

He et al. [71] provided insights into the precise control over the crystal morphology and crystallinity of perovskite films. A high PCE approaching 20% was achieved in the planar $FA_{0.85}MA_{0.15}PbI_{2.55}Br_{0.45}$ perovskite solar cells, which can be ascribed to low trap-state density and long carrier lifetimes in large-grained perovskite films.

Soft-cover coating is a technique very similar to meniscus-assisted solution printing in which a film is deposited by first dropping a solution onto a heated substrate, and then, the solution is covered with a flexible sheet (such as polyimide or polytetrafluoroethylene) to spread the solution across the entire substrate. Later on, the top sheet is peeled back through a computer-controlled mechanical hand to expose the solution to ambient air, eventually leading to the perovskite crystallization by solvent evaporation (Fig. 8G). The thermally driven evaporation of the solvent leads to quick crystallization of the perovskite films under ambient conditions, thus resulting in larger apparent grain size and longer-lived photoluminescence than the spin-coating technique while maintaining a compact microstructure. The use of high-viscosity solvents smooths the rough borders in the perovskite film. This technique can be applied to the preparation of large-area solar cells. Though the efficiencies are not as high as meniscus-assisted solution printing, but the processing speed is faster.

In 2016, Ye et al. [72] reported this technique producing PCE up to 17.6% with a working area of 1 cm^2 with high reproducibility. Surface wettability, solution viscosity, and thermal crystallization are the key processing factors for the deposition of uniform perovskite films. As a result, a scaling-up, pinhole-free, large grain-sized, and rough-border-free perovskite film was presented with a large area of smooth surface of 51 cm^2 with a material utilization ratio of up to 80% higher than the widely used spin-coating method.

2.3 Hybrid vapor-solution methods

Hybrid methods consist of sequential deposition steps and rely on interdiffusion/intercalation processes. In all these strategies, the perovskite most likely nucleates on the surface of the solid precursor film and then propagates toward deeper parts of the film.

2.3.1 Vapor-assisted solution-processing methods

The deposition of the organic and inorganic precursors with a vapor-assisted solution process (VASP) is a promising method as metal halides can be individually deposited, either via a solution or via a vapor method, and then converted to a perovskite layer as shown in Fig. 9. Therefore, the reaction involved the transformation of the inorganic layer into a perovskite layer, which is a heterogeneous phase reaction between a solid layer of the inorganic part (PbX_2) and the organic part (CH_3NH_3X). The main advantage of this technique is the fast reaction between precursors, producing a well-defined grain structure (up to microscale), full surface coverage, and small surface roughness. Although high solar cell efficiency as compared with the single-step spin-coating method is obtained, the proper diffusion of the organic component is still challenging.

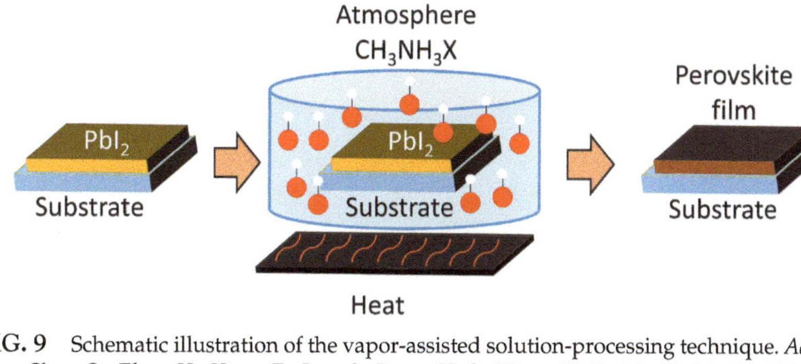

FIG. 9 Schematic illustration of the vapor-assisted solution-processing technique. *Adapted from Chen, Q., Zhou, H., Hong, Z., Luo, S., Duan, H. S., Wang, H. H., Liu, Y., Li, G., & Yang, Y. (2014b). Planar heterojunction perovskite solar cells via vapor-assisted solution process. J. Am. Chem. Soc., 136(2), 622–625.*

At the end of 2013, Chen et al. [73] used the VASP method to obtain polycrystalline perovskite thin films exhibiting compact grains spanning across the full-film thickness. A PbI_2 layer was first deposited on FTO glass coated with a compact layer of TiO_2 followed by exposing the PbI_2 film to CH_3NH_3I vapor at 150°C in N_2 atmosphere for 2h in a petri dish to form the perovskite film. The film consisted of fewer heterogeneous nucleation sites that eventually grew into large apparent perovskite grains but retained the original smooth PbI_2 film morphology. The corresponding film enabled a device power conversion efficiency of 12.2% on a planar architecture. Remarkably, Li et al. [22] employed the low-pressure vapor-assisted solution-processing method to fabricate a highly efficient 2D/3D $PEA_2MA_{n-1}Pb_nI_{3n+1}$ perovskite solar cells with a PCE of 19.10%, due to PEAI-modified perovskite grain size and surface potential. Another technique that involved a vapor-assisted solution-processing method is close space sublimation. This is a large-area and the low-cost technique that involves the sublimation of a powder material over a large area; positioning the substrate very close to the deposition source to facilitate a uniform deposition process onto a spin-coated PbX_2 film. Li et al. reported in 2015 the fabrication of $CH_3NH_3PbI_3$ perovskite film exhibiting high PCE of 16.2% and 13.8% in 0.01-cm^2 and 9-cm^2 active areas, respectively [74].

3. Techniques for the preparation of colloidal nanocrystals

Since the first report on the fabrication of colloidal MHP nanocrystals by Schmidt et al. in 2014, many strategies have been developed to obtain high-quality nanomaterials controlling the size, shape homogeneity, and optical properties. The different synthetic techniques for MHP NCs were coupled according to "top-down" and "bottom-up" approaches (Fig. 10). The top-down approach includes those strategies in which bulk material (3D material dimension) is fragmented mechanically, chemically, or using an irradiation source followed by a fast re-structuration process to produce nanomaterials (0D material dimension). Bottom-up methods consist

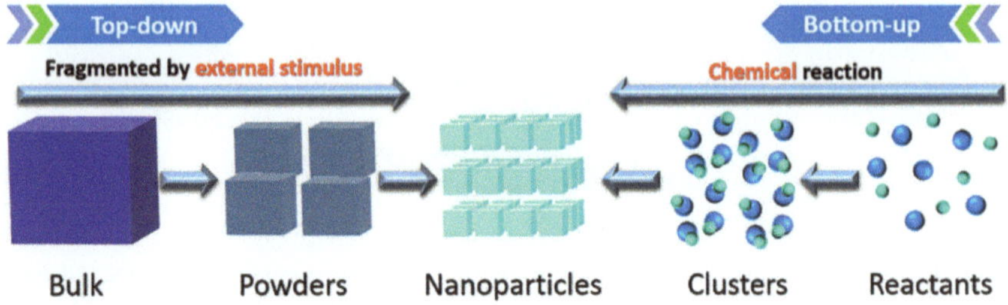

FIG. 10 Illustration of the top-down and bottom-up approaches for synthesizing nanocrystals (NCs).

of the reaction of perovskites precursors to form materials confined to the nanoscale, assisted by the presence of organic ligands. In this section, the colloidal MHP nanocrystals synthetic methods are presented and discussed in terms of size, morphology, and optical properties, pointing out their main advantages and drawbacks (Table 2).

TABLE 2 Summary of the MHP NCs preparation techniques

Techniques		Advantages	Limitations	QY_{PL} (%)
Bottom-up methods	Hot injection	Well-known nucleation growth process. Great control over the NCs size and shape. Suitable for tunable composition and doping.	Hard to scale-up. Synthesis performed in air-free conditions. Mainly studied the inorganic MHP NCs.	50%–90% [75]
	Precipitation methods	Simple strategy. Ambient conditions. Scalable. Suitable for the synthesis of all the $APbX_3$.	Requires polar solvents. Low control over the NC size and shape. Low reproducibility.	80%–92% [76]
	Heat-up-"related" techniques Solvothermal Microwave Ultrasonication Heat-up	Simpler than hot injection methods. Synthesis performed in air-free conditions.	Low control over the NC size and shape.	
		Rapid and homogeneous heating up rate. Good control over the NC size. Scalable.	Time-consuming process. More dangerous than alternatives.	up to 80% [77]
		Rapid and homogeneous heating up rate. Low energy consumption.	Scalable process limited by the capacity of microwave reactor. Requires absorbent molecules of microwave radiation.	up to 75% [78]

Continued

TABLE 2 Summary of the MHP NCs preparation techniques—cont'd

Techniques		Advantages	Limitations	QY_{PL} (%)
		Suitable for tunable composition. Scalable.	Low control over the NC size and shape.	75% [79]
		One-pot reaction. Good reproducibility. Scalable.	Lower NC morphology control and higher polydispersity than hot injection.	up to 80% [80]
	Solid phase	Simple strategy. Greener due to the lack of solvent and high product yield. Scalable.	Poor control over the NC size and shape. High polydispersity. Low QY_{PL}.	13%–39.8% [81]
Top-down methods	Grinding/ Milling Exfoliation	Suitable for tunable composition. Ambient conditions. Simple strategy. Good optical properties.	Poor control over the NC size and shape.	6%–91.7% [82]
	Photo-induced methods	Suitable for tunable composition ambient conditions.	Poor control over the NC size and shape. Low QY_{PL} for iodide and chloride MHP NCs. Few examples reported.	67% [83]

Some advantages of MHP nanocrystals are the easy tunability of the bandgap and higher radiative recombination rates than bulk perovskites, making them more suitable candidates for solar conversion systems like light harvesters and emitter materials in light-emitting diodes (LEDs).

3.1 Bottom-up methods

Although bottom-up methods include the templated and nontemplated approaches, we will consider in this section the nontemplated methodologies. However, a brief comment on the use of a template strategy to get the first MHP NCs will be mentioned below because those works inspired the preparation of novel methodologies or adapted from the knowledge of the synthesis approach of chalcogenide semiconductor nanocrystals.

Mesoporous materials were used to confine 3D perovskites to the nanometer scale, and as a result, improved uniform MHP films were achieved. In particular, the most used scaffolds were mesoporous metal oxides, such as alumina (Al_2O_3) and titanium dioxide (TiO_2) [58,84–86]; in the case of TiO_2, it also acts as electron transporter. Although the first use of perovskite as light-absorbing material in a solar cell device was reported in 2009 with a low PCE of 3.8%, it was not until 2012 when Grätzel and Snaith published the use of MHP NCs for solid-state solar cells with a good PCE of 9.7% and 10.9%, respectively [87,88].

It is well known that the presence of a high ratio of uncoordinated bonds on the nanocrystal semiconductor surface usually contributes to the formation of defect energy levels in the bandgap, thus producing exciton traps and lower radiative recombination rates. In particular, in MHP NCs, despite its high defect tolerance, the halide vacancies (Vx) located on the surface can have an important contribution in the nonradiative recombination. To avoid this drawback, nontemplate strategies were developed using different types of ligands to carry out complete surface passivation. Moreover, ligands play an important role in controlling crystal growth to regulate more accurately the size and shape, and for preventing NCs agglomeration, giving them dispersibility in different solvents and chemical stability. Briefly, ligands are classified into three groups: (i) X-type, involving ligands whose anchoring group has a single occupied orbital, such as ammoniums, carboxylates, phosphates, sulfonates, and thiols/thiolates; (ii) L-type, comprising ligands whose anchoring group presents a filled occupied orbital, such as phosphonates and amines; and (iii) Z-type, ligands whose anchoring group is empty [16]. Traditionally, the most used type ligands in MHP NCs are X- and L-type, especially primary aliphatic amines and carboxylic acid. In fact, surface chemistry studies have demonstrated that amines combined with carboxylic acids, commonly oleylamine (OLA) and oleic acid (OA), react via a simple acid-base reaction to form the ammonium (OLA^+) and carboxylate (OA^-) species in solution, which according to density functional theory (DFT) calculations is the most favorable surface ligand pair [89]. In this regard, even though phosphinic acids, secondary amines, zwitterionic acids have also been widely studied as ligand-anchoring groups for well-passivated perovskite NCs, it is necessary to gain more insight into its electrical conductivity to obtain NCs with efficient charge carrier transport for their applications in optoelectronic devices. In this section, we will focus on nontemplate strategies, taking into account that capping ligands provide more homogeneous and stable colloidal MHP NCs, with better optical properties and easier processability for optoelectronic devices applications.

3.1.1 *Liquid phase*

The vast majority of MHP NCs bottom-up synthesis methods are conducted in the liquid phase due to the achieved control in terms of shape, size, homogeneity, and composition. In these approaches, solvents usually play an important role in dissolving uniformly all the reactants, ions, and ligands to produce very homogeneous colloidal dispersions. The liquid-phase methods can be sorted as (i) hot injection methods, which often require injecting of one precursor at high temperatures and in an inert atmosphere; (ii) precipitation techniques, which require suitable and bad solvents in the same process and could be done at room temperature; and (iii) heat-up-related techniques, whose synthesis is carried out in one pot and needs some heat source.

Hot injection

Lamer and Dinegar demonstrated in 1950 that the production of NCs with a narrow size distribution requires fast nucleation followed by controlled growth of the formed nuclei [90]. In 1993, Murray et al. introduced a hot injection approach, which enabled an accurate control of these two phases for the synthesis of nearly monodisperse cadmium chalcogenide NCs [91]. Since then, the hot injection method is one of the mostly used strategies for synthesizing monodisperse semiconductor NCs. In particular, for colloidal AMX_3 NCs, this approach consists of the rapid injection of the cation precursor (A^+) into a hot high boiling solvent, which contains the ligands and the metal halide salt (MX_2), to favor the homogeneous nuclei formation (Fig. 11A). Once the fast nucleation process occurs, the growth stage can be easily controlled over time [92]. Generally, to obtain monodisperse MHP NCs, it is imperative to rapidly quench the reaction in an ice bath after 1–5 s of the injection. In fact, besides controlling the time, this technique allows a great number of modifications to fine-tune their size and morphology. The crucial parameters that must be accurately controlled are (a) injection temperature, (b) concentration and chain length of the organic ligands, (c) Cs/Pb ratio, and (d) concentration of precursors. By adjusting them properly, it is possible to obtain very homogeneous dispersions in different sizes of MHP NCs, such as nanocubes, nanoplatelets [93], and nanowires [94] with narrow PL FWHM and high photoluminescent quantum yields (QY_{PL}) (Fig. 11B). In addition, it is worth mentioning that this strategy enables a fine halide tunable composition engineering for managing the absorption and PL across the whole visible spectrum for both inorganic and hybrid LHP NCs. However, the main drawback is the scaling up for industrial applications, due to the difficulties of keeping the same injection conditions in a large-scale synthesis.

In 2015, Protesescu et al. demonstrated the potential of this approach for MHP NCs, preparing pure $CsPbX_3$ ($X{=}Cl^-$, Br^- and I^-) and mixed-halide perovskites combining PbX_2 salts [75]. Briefly, they injected the

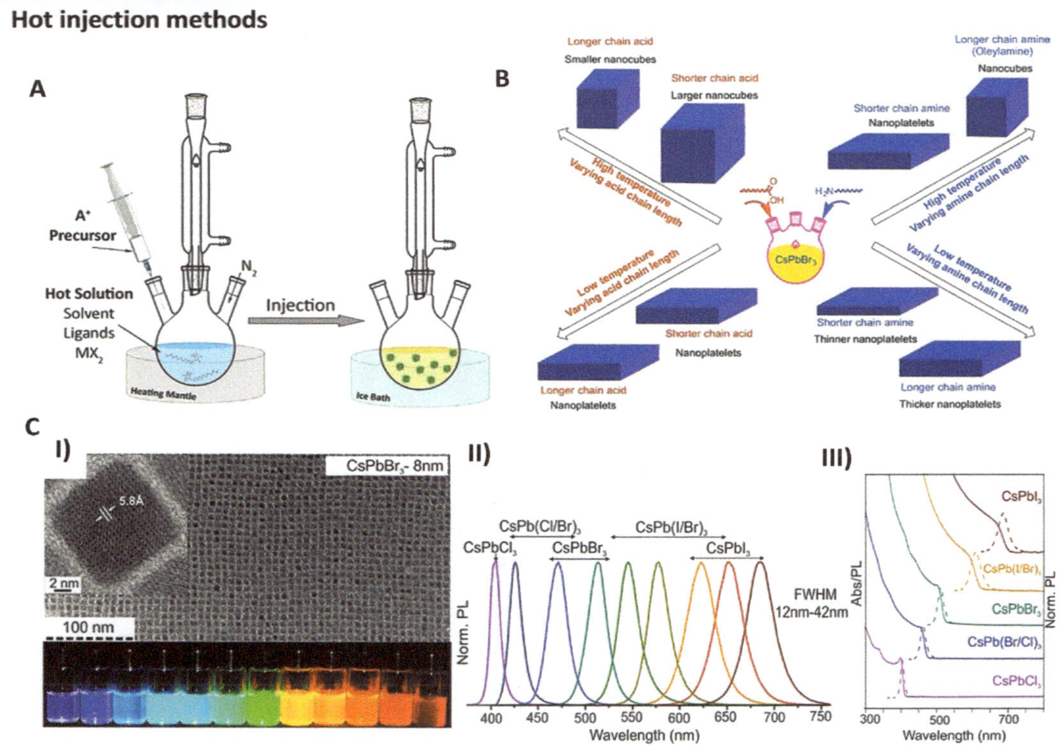

FIG. 11 (A) Scheme of the hot injection synthesis. (B) Representation of the dependence between the size-shape of the MHP NCs and the carboxylic acid-amine pair of ligands used. (C) (I) TEM of $CsPbBr_3$ NCs and colloidal dispersions of $CsPbX_3$ NCs under UV light ($\lambda = 365$ nm), (II) photoluminescence emission profiles, and (III) UV-vis absorption spectra of all inorganic $CsPbX_3$. *Panels (A) adapted with permission from Shamsi, J., Urban, A. S., Imran, M., De Trizio, L., & Manna, L. (2019). Metal halide perovskite nanocrystals: synthesis, post-synthesis modifications, and their optical properties. Chem. Rev., 119(5), 3296–3348. Copyright 2019, American Chemical Society, (B) adapted with permission from Pan, A., He, B., Fan, X., Liu, Z., Urban, J. J., Alivisatos, A. P., He, L., & Liu, Y. (2016). Insight into the ligand-mediated synthesis of colloidal CsPbBr3 perovskite nanocrystals: the role of organic acid, base, and cesium precursors. ACS Nano, 10(8), 7943–7954. Copyright 2016, American Chemical Society, and (C) adapted with permission from Protesescu, L., Yakunin, S., Bodnarchuk, M. I., Krieg, F., Caputo, R., Hendon, C. H., Yang, R. X., Walsh, A., & Kovalenko, M. V. (2015). Nanocrystals of cesium lead halide perovskites (CsPbX3, X = Cl, Br, and I): novel optoelectronic materials showing bright emission with wide color Gamut 15(6), 3692–3696. Copyright 2015, American Chemical Society.*

Cs-oleate precursor at temperatures between 140°C and 200°C in a mixture of ODE, PBX_2, OLA, and OA, and quenched the reaction after 5s to produce monodisperse NCs. Following this strategy, they obtained highly homogeneous NCs in the 4–15nm size range, depending on the injection temperature, and with optical properties that cover the visible spectra (400–700nm), high QY_{PL} (50% to 90%) and narrow FWHM (12 to 42nm) (Fig. 11C). One year later, the same group successfully extended this approach to hybrid NCs of $MAPbX_3$ ($X=Br^-$ and I^-) by replacing the Cs precursor with a methylamine solution but obtaining MHP NCs with a lower QY_{PL} (15% to 50%) than the inorganic ones [95].

Precipitation methods

Precipitation methods comprise those strategies where microcrystals are produced after a spontaneous crystallization and precipitation process by moving the initial precursor solutions into a supersaturated state, mainly reached by drastically solvent change (from a good solvent to dissolve the precursors to a bad one). One of the first reports using this strategy was reported in 1992 by Hitoshi et al. [96] about the preparation of organic chromophore microcrystals by a simple reprecipitation from an ethanol solutions of the precursors in water. Since then, a detailed research was done to elucidate the nanocrystallization process to have thorough control over the crystal size and shape by adjusting the concentration of the precursor solution, the temperature of the bad solvent, and especially the concentration and surfactant nature to confine the material stable to the nanoscale [97]. All the knowledge acquired has been used to easily prepare semiconductor NCs at room temperature [98]. In particular, during the last years, a large variety of precipitation methods have been developed for synthesizing MHP NCs, which can be classified according to the miscibility between the good polar and poor nonpolar solvent in (i) miscible solvents: ligand-assisted reprecipitation technique (LARP), and polar solvent-controlled ionization (PCI); and (ii) immiscible solvents or emulsion technique: emulsion LARP and reverse microemulsion.

Miscible solvents

Ligand-assisted reprecipitation technique (LARP) is a simple process occurring at room temperature, where all the precursor salts are dissolved in a good polar solvent, mainly dimethylformamide and dimethylsulfoxide, followed by its addition to a miscible and vigorously stirred poor solvent (usually toluene and hexane) in the presence of the ligands. In this miscible co-solvent, ion solubility is very low and, consequently, a spontaneous crystallization and precipitation process takes place (Fig. 12A). With this strategy, both hybrid and inorganic MHP NCs have been prepared with good control of the NC size, by adjusting the reaction temperature [99] and the morphology (quantum dots [8], nanocubes [100], nanorods nanowires, and nanoplatelets [101]) by changing the acid-amine ligand chain length.

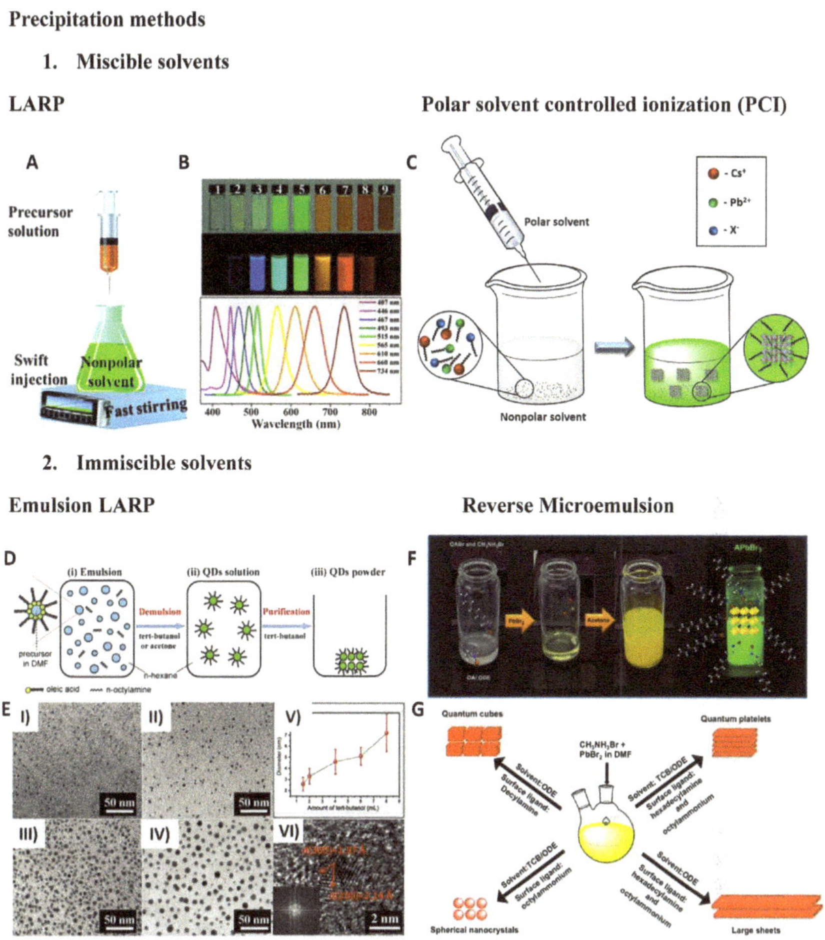

FIG. 12 (A–B) Scheme of the LARP technique, colloidal dispersion under UV light ($\lambda = 365$ nm), and PL emission profiles of all hybrid $MAPbX_3$. (C) Scheme of the PCI technique; (D–E) Scheme of the LARP emulsion and (I-IV) TEM image of $MAPbBr_3$ QDs with a diameter of 2.6 ± 0.6, 3.3 ± 0.7, 4.6 ± 1.1, and 7.2 ± 1.7 nm. (V) The plot of the average diameter of resultant QDs with the amount of *tert*-butanol. (VI) High-resolution TEM image of a single QD with lattice planes in two directions. Inset: FFT pattern transformed from the HRTEM image. (F–G) Schematic process of the reverse microemulsion technique and scheme of the solvent- and surface ligand dependence with the $MAPbBr_3$ NCs shape. *Panel (A) adapted with permission under a Creative Commons Attribution-NonCommercial 3.0 Unported Licence from Du, X., Wu, G., Cheng, J., Dang, H., Ma, K., Zhang, Y.-W., Tan, P.-F., & Chen, S. (2017). High-quality CsPbBr3 perovskite nanocrystals for quantum dot light-emitting diodes. RSC Adv., 7, 10391–10396. Copyright 2017, The Royal Society of Chemistry, (B) adapted with permission from Zhang, F., Zhong, H., Chen, C., Hu, X., Huang, H., Han, J., Zou, B., Dong, Y., (2015b). Brightly luminescent and color-tunable colloidal CH3NH3PbX3 (X = Br, I, Cl) quantum dots: potential alternatives for display technology. J. Mater. Chem. A, 9(4), 4533–4542. Copyright 2015, American Chemical Society, (C–E) adapted with permission from Huang, H., Susha, A. S., Kershaw, S. V., Hung, T. F., & Rogach, A. L. (2015a). Control of emission color of high quantum yield CH3NH3PbBr3 perovskite quantum dots by precipitation*

(Continued)

FIG. 12, CONT'D *temperature. Adv. Sci., 2(9), 1500194. Copyright 2015, American Chemical Society, (F) adapted with permission from Schmidt, L. C., Pertegás, A., González-Carrero, S., Malinkiewicz, O., Agouram, S., Mínguez Espallargas, G., Bolink, H. J., Galian, R. E., & Pérez-Prieto, J. (2014). Nontemplate synthesis of CH3NH3PbBr3 perovskite nanoparticles. J. Am. Chem. Soc., 136(3), 850–853. Copyright 2014, American Chemical Society, and (G) adapted with permission from Teunis, M. B., Johnson, M. A., Muhoberac, B. B., Seifert, S., & Sardar, R. (2017). Programmable colloidal approach to hierarchical structures of methylammonium lead bromide perovskite nanocrystals with bright photoluminescent properties. Chem. Mater., 29(8), 3526–3537. Copyright 2017, American Chemical Society.*

The simple laboratory material, the ambient conditions needed, and the great variety of precursors tested, make LARP technique one of the most popular and well-known procedures for preparing MHP NCs. In addition, it is easily scalable to obtain gram-scale synthesis [102]. However, one of the weak points of this technique is that MHP NCs are very sensitive to the polar solvents used to dissolve the precursors, thus reducing their stability during purification steps [103]. Moreover, as the nucleation and precipitation stages are fast processes and occur simultaneously, the process of the NC formation remains unclear and, in addition, it is not possible to obtain them as monodisperse MHP NCs, as in the case of hot injection approaches (Fig. 12).

The first precipitation method using miscible solvents was reported in 2012 by Papavassiliou et al. [104]. They described the preparation of $(CH_3NH_3)(CH_3C_6H_4CH_2NH_3)_2Pb_2Br_7$ or $(CH_3NH_3)(C_4H_9NH_3)_2Pb_2Br_7$ nano- and microcrystalline materials in solutions of toluene-PMMA with NC sizes between 30 and ca. 160 nm and narrow PL bands (20-35 nm) shifted to shorter wavelengths than that expected for bulk materials. However, it was not until 2015 that Zhang et al. [105] tuned the LARP technique by analyzing systematically all the steps involved in the production of $CH_3NH_3PbX_3$ ($X=Br^-, I^-, Cl^-$) NCs and studying the role of amines and carboxylic acids as ligands. With this strategy, they could obtain an array of QDs with narrow emission (20–50 nm), wide wavelength tunability of 400–750 nm, and absolute photoluminescence quantum yield (QY_{PL}) up to 70% (Fig. 12B).

Polar solvent-controlled ionization (PCI). This technique was developed in 2018 by Fang et al. [106] to overcome the drawbacks of using polar solvents. The precursors are not initially dissociated in polar solvents but metal oleates and alkylammonium halides are dispersed in nonpolar solvents. Then, a controlled volume of polar solvent is added until reaching a critical concentration of ions to induce the crystallization of the MHP NCs (Fig. 12C). This method enabled obtaining all inorganic $CsPbX_3$ NCs with good optical properties and controllable sizes. Thus, PCI can be considered as a greener approach than LARP, since it uses lower volumes of the usual polar solvents that usually have high boiling points and toxicity.

3.1.2 Immiscible solvents: Emulsion techniques

Precipitation approaches that use immiscible polar and nonpolar solvents are known as emulsion techniques [107]. In general, these approaches can be divided into two steps: emulsion formation and demulsion. In the first step, an emulsion is prepared after mixing a polar solvent that contains the precursors with an immiscible nonpolar solvent and ligands. In the second one, a demulsion process is carried out to precipitate the nanocrystals. We can classify these strategies into two groups: emulsion LARP when a demulsifier is employed, and reverse microemulsion when micelles are formed as nanoreactors.

Emulsion LARP is based on dissolving salt precursors in polar solvents and ligands, commonly carboxylic acids and amines, in nonpolar solvents. Then, both immiscible solutions are mixed to create an emulsion and, finally, a third solvent (known as a demulsifier), which is soluble in both, is added at room temperature, inducing the supersaturated state and the precipitation of the $APbX_3$ NCs.

The first emulsion LARP synthesis was reported by Huang et al. [76] in 2015, using *n*-hexane and DMF as immiscible solvents and *tert*-butanol as demulsifier (Fig. 12D). Monodisperse $MAPbBr_3$ NCs with a tunable size from 2 to 8 nm were achieved by varying the amount of demulsifier (Fig. 12E) and an absolute PL quantum yield of 80%–92% was obtained. This method provides better control of the crystallization and nucleation process because the emulsion generates droplets that act as microreactors for solvent mixing, thus improving the shape and size control of MHP NCs by varying the *tert*-butanol volume. The emulsion procedure is versatile and was extended by the same group for the preparation of $CsPbBr_3$ NCs.

Reverse microemulsion consists of mixing immiscible solvents by heating them up or by stirring vigorously the nonpolar solvent. After that, small, inverted micelles (with the hydrophilic head groups inward to stabilize the polar solvents which have dissolved the ions) are produced. These micelles act as nanoreactors where diffusion of the precursors between them can help to fine control the NC nucleation and growth. Eventually, once the MHP NCs are produced inside the micelles, a demulsifier is added to purify them.

The first synthesis of hybrid $MAPbBr_3$ NCs was carried out by reverse microemulsion approach (Fig. 12F) by Schmidt et al. in 2014, mixing the precursors in octadecene and using octylammonium bromide and OA as organic ligands and acetone as a polar solvent [8]. This strategy produced NCs with a typical emission centered at 525 nm and a QY_{PL} of 17% for the colloids and thin film. Soon after, the same group significantly enhanced the QY_{PL} to up to 83% by fine-tuning the molar ratio between the ammonium species (organic ligand and methylammonium) and $PbBr_2$ to 4:1. This enhancement in the PL was explained by the binding between the undercoordinated bromide ions (dangling bonds) and the ammonium cations [108]. Furthermore, in 2017 Teunis et al. demonstrated the suitability of this approach to control the shape of MHP NCs synthesizing nanoplatelets, nanosheets, nanocubes, and nanospheres by changing the nonpolar solvent, the temperature, and the ligand nature (Fig. 12G) [109].

Heat-up-"related" techniques

As we have shown in the hot injection section, this synthesis is arduous, demands controlled nitrogen or argon atmosphere, and requires the presynthesis of cesium precursors in rigorous conditions and at high temperatures for obtaining MHP NCs with high QY_{PL}. To overcome these limitations, some groups have tried to develop different strategies that could be simple, scalable, versatile, conducted in an air atmosphere, and especially in one pot. Therefore, heat-up-"related" techniques comprise those strategies in which mixing precursors, ligands, and solvents in a single one-pot step provides MHP NCs with at least the same quality as those obtained by hot injection synthesis.

3.1.3 *Solvothermal*

The solvothermal synthesis is a straightforward and scalable process that is based on mixing all the precursors with ligands in a high boiling solvent inside a sealed stainless vessel. Then, the reaction is conducted increasing the temperature above the boiling point of the solvent in a lab autoclave and kept in these conditions during the desired time. This strategy provides a homogeneous and fast phase of heating up, a fine controllable size, and a reproducible large-scale procedure. Nevertheless, the entire process consumes more time and is more hazardous than other alternatives (Fig. 13).

Although this method has been used for synthesizing other semiconductor nanomaterials, it was not until 2017 when Chen et al. [77] described for the first time a solvothermal strategy to synthesize high-quality MHP NCs controlling effectively the composition and morphology of $CsPbX_3$ with tunable and bright photoluminescence properties. To achieve these results, they placed cesium acetate (CsOAc), lead halides (PbX_2) as precursors, and the pair of oleic acid/oleylamine as ligands inside the reaction container. In addition, they added TOP (trioctylphosphine) and TOPO (trioctylphosphine oxide) to enhance the solubility of the salt precursor in the nonpolar solvents. Interestingly, they noticed that using a temperature of 160°C for 30 min and changing the halide composition without predissolving the precursors before the solvothermal process, monodisperse NCs can be obtained. These NCs presented absorption and emission spectra in the whole visible range, a QY_{PL} of up to 80%, and a narrow

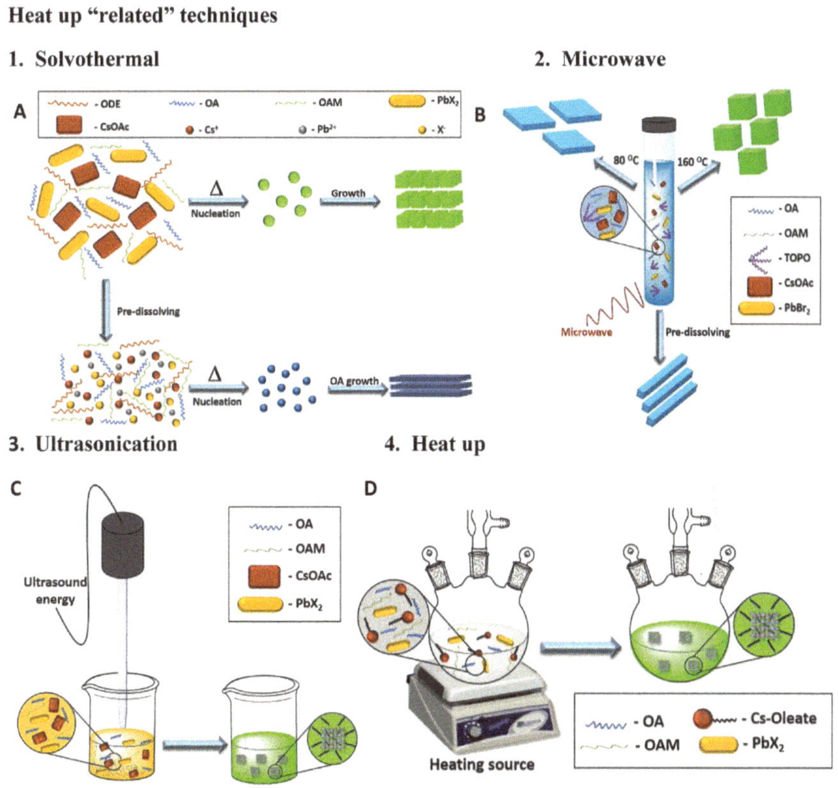

FIG. 13 (A) Schematic illustration of the solvothermal growth with and without predissolving the precursors. (B) Scheme of the microwave-assisted method under different temperatures. (C) Scheme of the synthesis of $CsPbX_3$ NCs by ultrasonication method. (D) Scheme of the single-step heat-up technique and emission of the MHP NCs obtained. *Panel (A) adapted from Chen, M., Zou, Y., Wu, L., Pan, Q., Yang, D., Hu, H., Tan, Y., Zhong, Q., Xu, Y., Liu, H., Sun, B., & Zhang, Q. (2017a). Solvothermal synthesis of high-quality all-inorganic cesium lead halide perovskite nanocrystals: from nanocube to ultrathin nanowire. Adv. Funct. Mater., 27(23), 1701121. Copyright 2017, WILEY-VCH Verlag GmbH & Co. KGaA, Weinheim, (B) adapted from Pan, Q., Hu, H., Zou, Y., Chen, M., Wu, L., Yang, D., Yuan, X., Fan, J., Sun, B., & Zhang, Q. (2017). Microwave-assisted synthesis of high-quality "all-inorganic" CsPbX3 (X = Cl, Br, I) perovskite nanocrystals and their application in light emitting diodes. J. Mater. Chem. C, 5, 10947–10954. Copyright 2017, The Royal Society of Chemistry, (C) adapted from Tong, Y., Bladt, E., Aygüler, M. F., Manzi, A., Milowska, K. Z., Hintermayr, V. A., Docampo, P., Bals, S., Urban, A. S., Polavarapu, L., & Feldmann, J. (2016). Highly luminescent cesium lead halide perovskite nanocrystals with tunable composition and thickness by ultrasonication. Angew. Chem. Int. Ed., 55(44), 13887–13892. Copyright 2016, Wiley-VCH Verlag GmbH & Co. KGaA, Weinheim, and (D) adapted from Chen, X., Peng, L., Huang, K., Shi, Z., Xie, R., & Yang, W. (2016a). Non-injection gram-scale synthesis of cesium lead halide perovskite quantum dots with controllable size and composition. Nano Res., 9, 1994–2006. Copyright 2016, Tsinghua University Press and Springer-Verlag Berlin Heidelberg.*

FWHM (12–36 nm). However, when the same conditions were maintained but the precursors were predissolved before treatment, ultrathin nanowires with a strong quantum confinement effect (Fig. 13A) and optical properties (absorption and emission) were shifted to shorter wavelengths in comparison with their bulk bandgap.

3.1.4 *Microwave*

The microwave-assisted methods require the use of a microwave as the heat source and polar solvents or organic ligands in the reaction medium able to absorb rapidly the energy of the microwave irradiation and reach the temperature required uniformly in a very short time. In this process, all the salt precursors and ligands are mixed simultaneously in an air microwave tube. Subsequently, the tube is placed in the microwave reactor to initiate the heating rate and carry out the reaction. This approach is also fast and simple and provides uniform MHP NCs from the size and shape point of view, and additionally, it allows a total and precise control of the heating rate with very low energy consumption. One of the main drawbacks of the procedure is that industrial applications are limited because of the volume capacity of the microwave reactor.

In 2017, Long et al. reported for the first time that high-quality $CsPbX_3$ NCs can be prepared through a practical, feasible, and rapid microwave-assisted method [78]. The size and morphology (nanocubes, nanorods, nanowires, hexagonal nanoplatelets, and nanocubes) can be easily tuned varying the reaction time, the molar ratio between the reactants, or the irradiation power. One month later, Pan et al. studied those reaction parameters (temperature, predissolution of precursors, and heating rates) that control the nucleation and growth of MHP NCs by using microwave irradiation, and obtained $CsPbX_3$ NCs with different shapes and morphologies, high QY_{PL} (75%) and narrow FWHM (from 9 to 34 nm) (Fig. 13B) [110]. The synthesis performed at 80°C yielded small and thin nanoplatelets (23.4 nm length and 3.3 nm width), while at 140°C monodisperse NCs (10.2 nm) were formed. However,not homogeneous MHP NCs were produced at temperatures above 180°C. Furthermore, keeping the same conditions, but predissolving the precursors and ligands and conducting the reaction at 160°C, thin nanorods (60 nm length and 2.7 nm width) were obtained. In addition, as the heating rate is one of the most important limitations of heat-up-related techniques because it is closely related to the nucleation event, they demonstrated that slow heating rates (8°C/min) gave inhomogeneous dispersions with broader distribution sizes in comparison with the very homogeneous MHP NCs dispersions obtained when heating rates were higher than 18°C/min.

3.1.5 Ultrasonication

The ultrasonication process is a very versatile method used to easily tune the MHP NCs composition by changing the halide and/or cation source without almost varying the reaction conditions. This strategy is carried out by adding all salt precursors and organic ligands in a lab container having the solvent (either nonpolar or polar solvents) and performing a high-density probe-type ultrasonication. The ultrasonic irradiation accelerates the dissolution of the precursors in the tested solvents, and the dissolution precursor rate determines the nucleation and growth rate of the nanocrystals. Despite that, the fundamentals of the growth process are not clear and required a detailed analysis of the steps to elucidate if MHP NCs are generated by a dissolution-recrystallization process and/or by a progressive fragmentation caused by the cavitation pathway. In general, it is a suitable candidate for MHP NCs scalable synthesis at room temperature, but the control over the size and morphology is reduced in comparison with other techniques.

Tong et al. in 2016 described the first approach for synthesizing MHP NCs by ultrasound treatment. In particular, they reported a simple, scalable, single-step, and polar-solvent-free method resulting in high-quality $CsPbX_3$ NCs and nanoplatelets (Fig. 13C) [79]. They show considerable air stability, high QY_{PL} (ca. 75%), a narrow emission across the visible spectrum, and a dependence between the Cs_2CO_3:PbX_2 ratio and the nanoplatelets thickness. Furthermore, the authors described for the first time the quantum size effect operating in the $CsPbI_3$ nanoplatelets. Soon after, Jang et al. extended satisfactorily this approach to hybrid MHP NCs, by using formamidinium and methylammonium as cations [111]. In addition, they prepared inorganic and hybrid MHP NCs with different halide compositions as $APbBr_{3-x}X_x$ ($A = Cs^+$, MA^+, and FA^+; $X = Cl^-, I^-$; and $x = 0$–1), thus tuning the bandgap over a wide range.

3.1.6 Heat-up

The heat-up methods, also known as convection or hot noninjection methods, are one of the simplest strategies to control the nucleation and growth of MHP NCs by controlled heating. In them, all the reagents (including salt precursors and ligands) are loaded at room temperature in a lab container with a high boiling nonpolar solvent, commonly octadecene. Subsequently, the mix is heated to the desired temperature in an air atmosphere to prepare the MHP NCs in a single-reaction pot. This is a synthetic procedure that can yield gram scales of high-quality NCs with only one batch reaction. However, despite good reproducibility and control over the size and composition, it tends to yield higher polydispersity and irregular shapes in comparison with other approaches mentioned in this section.

Chen et al. [80] in 2016 developed for the first time a low-cost heat-up approach (Fig. 13D) for the synthesis of $CsPbX_3$ (X=Cl^-, Br^-, and I^- or Cl^-/Br^- and I^-/Br^-) NCs with spherical morphology. They successfully tuned the particle size (1–10.4nm) and composition by varying only the reaction temperature (from 50°C to 170°C), enabling tunable PL across the entire visible region and QY_{PL} of >80% [80]. In addition, they succeeded in the preparation of large amounts of MHP NCs with high reproducibility and optical properties. Nevertheless, as the kinetics of this reaction are extremely fast and multiple variables are involved in the nucleation and growth process, an additional heat-up technique, called microfluidic system, was designed by Lignos et al. to gain insight into the MHP NCs kinetic produced by heat-up injection (hot injection approach) and noninjection techniques [112]. The system developed to carry out these precise studies contains T- and cross-junctions for mixing rapidly all the precursors and form droplets that can be quickly heated to the desired temperature. The synthesis can be monitored in real time by performing ultrafast kinetic measurements using efficient online PL and absorption measurements. With this setup, it has been able to study and optimize the synthesis parameters (such as molar ratios of Cs, Pb, and halide precursors, reaction temperatures, and reaction times), leading to the production of $CsPbX_3$ with superior PL performance and even, obtaining unique insights into the early stages of nucleation and growth process within the initial 0.1–5s. This approach was also used by the same group to study the formation process of pure and mixed $FAPbX_3$ (X=Br^- and I^-) aimed at tuning the bandgap between 700 and 800nm, which is not accessible with $CsPbBr_3$ NCs [113].

3.1.7 Solid phase

Even though NCs bottom-up methods are commonly conducted in liquid, some solvent-free studies have recently been published. The examples reported are mainly based on the use of a grinding approach, by ball milling or mortar-pestle, leading to the formation of NCs with a high product yield (more than 90%) [114]. Moreover, the drastic reduction in waste generation makes these methods "greener" and can also have more potential for being scalable in industrial applications [115]. In those examples, as the main aim is to prepare NCs for manufacturing solar cells and LEDs devices, they avoid the use of organic molecules as ligands to improve the efficiency in the transport charges. However, a few attempts have been described employing ligands to control the crystal growth and prevent NCs agglomeration for obtaining stable colloidal dispersions and for their use in applications that need processing solution methods.

In particular for perovskites, up to now, the unique strategy used to synthesize bottom-up stable MHP NCs with ligands in the solid phase and at room temperature was developed by Jana et al. [81]. In this study, a fixed amount of capping ligands (octylammonium bromide; OABr), $PbBr_2$, and ABr ($A=Cs^+/MA^+/FA^+$, where $MA^+=CH_3NH_3^+$ and $FA^+=CH(NH_2)_2{}^+$) were ground with a mortar and pestle, at room temperature for 10 min. Then, an orange solid was observed, and after a purification step by centrifugation, the precipitate was dried in a vacuum for further uses. With this strategy, they obtained $CsPbBr_3$, $MAPbBr_3$, and $FAPbBr_3$ NPs with square, rectangular, spherical, and parallelogram shapes, respectively, and with optical properties like those reported in the liquid phase but with the modest QY_{PL} and polydispersity. Finally, they proposed that the formation mechanism via bottom-up solid-state reaction of perovskite geometry ($APbBr_3$) works through solid-state diffusion of A-site cations into the channels and cracks created in crystal grains of $PbBr_2$ during the expansion of the crystal lattice. This proposal was confirmed by X-ray powder diffraction (PXRD) patterns collected at different time intervals where the faster disappearance of MABr peaks evidences the incorporation of the A-site cation into $PbBr_2$.

3.2 Top-down methods

The top-down methods to prepare MHP NCs consist of the fragmentation of bulk material by an external stimulus after exposure to a high-energy source. The main energy sources described in the synthesis of MHP NCs are mechanical (grinding or ultrasonication) and photophysical using laser irradiation exposure. In this section, synthetic strategies using ligands for preparing stable powder and/or colloidal NCs dispersions are discussed.

3.2.1 Mechanochemical synthesis

Mechanochemistry is the branch of chemistry that studies chemical transformations applying mechanical energy through forces of impacts, frictions, compressions, and shears. These operations are commonly implemented on materials by hand grinding, ball milling, and ultrasound energy. Although the mechanism of the mechanochemical synthesis is not thoroughly studied, it is accepted that the mechanical force generates either a homogenization or a size decreasing material process where the fragments are accessible to "fresh" reactive surfaces to support the formation of the desired products [116]. It is worth mentioning that these strategies are employed for reducing the size and for preparing homogeneous materials with mixed compositions.

Regarding the synthesis of smaller materials, the mechanochemical approach has been used for preparing oxide dispersions, alloying materials, metal oxides, and graphene layers [117]. However, its implementation as an alternative technique for preparing semiconductor NCs has been limited by the rigid crystal lattices of transition metal chalcogenides (such as CdSe or CdTe), which causes numerous structural defects that generate trap states within the semiconductor bandgap and, therefore, reduces the QY_{PL}. However, as MHP NCs present defect tolerance (structural defects are located outside the bandgap or close to the conduction and valence band), mechanochemical techniques are suitable candidates for preparing them [116]. In particular, these techniques present numerous advantages in comparison with other solution-processed syntheses because they provide finer control over the stoichiometry of the final MHP NCs, produce low amounts of side products are produced, they are simple to scale up and, in some approaches, avoid the use of organic solvents that influence the MHP NC stability.

Grinding: Milling

Grinding top-down techniques comprise those strategies in which the NCs are produced in two consecutive steps. In the first step, the bulk single perovskites crystals are formed and, in the second, ligands are added to confine the material to the nanoscale. This sequential approach, in light of the results reported, seems to be better for the preparation of high-quality MHP NCs because the FWHM is narrower than those obtained with the direct process, which leads to a larger polydispersion [117].

The grinding process can be classified as manual or electrical, if it is carried out by hand-grinding using mortar and pestle or automatically, using planetary ball mills. In general, electrically powered ball milling is predominantly employed because the mixture obtained by the manual approach is less uniform, producing some unreacted impurities in the sample. In the electrical process, the precursor salts and several beads (that exert on the sample the mechanical energy produced by the planetary rotation) are added in a sealed milling jar to form the metal halide bulk.

The ligands are placed inside the jar and the mixture is subjected to a second grinding round. In addition, the ball milling could be done in the absence (dry) or the presence (wet) of solvents, and in air conditions or in a controlled atmosphere, sealing it in a glovebox. With this approach, it has been possible to prepare MHP NCs with excellent phase purity, mixed compositions and crystalline structures which point out its high versatility and QY_{PL} of over 90%. All these values are very similar to the ones obtained with liquid-phase bottom-up techniques. However, as some important parameters still need to be optimized, such as time, the weight ratio between precursors beads and proper ligands, the MHP NCs obtained by grinding techniques are not as homogeneous in size and shape.

The first grinding top-down inorganic $CsPbX_3$ QDs ball-milling synthesis was reported by Zhu et al. in 2017. With this strategy and using oleylamine as a ligand, they obtained excellent optical properties with efficient QY_{PL} in the entire visible spectrum (Fig. 14A and B) [82]. In 2018, Protesescu et al. developed a wet ball milling for preparing easily $APbBr_3$

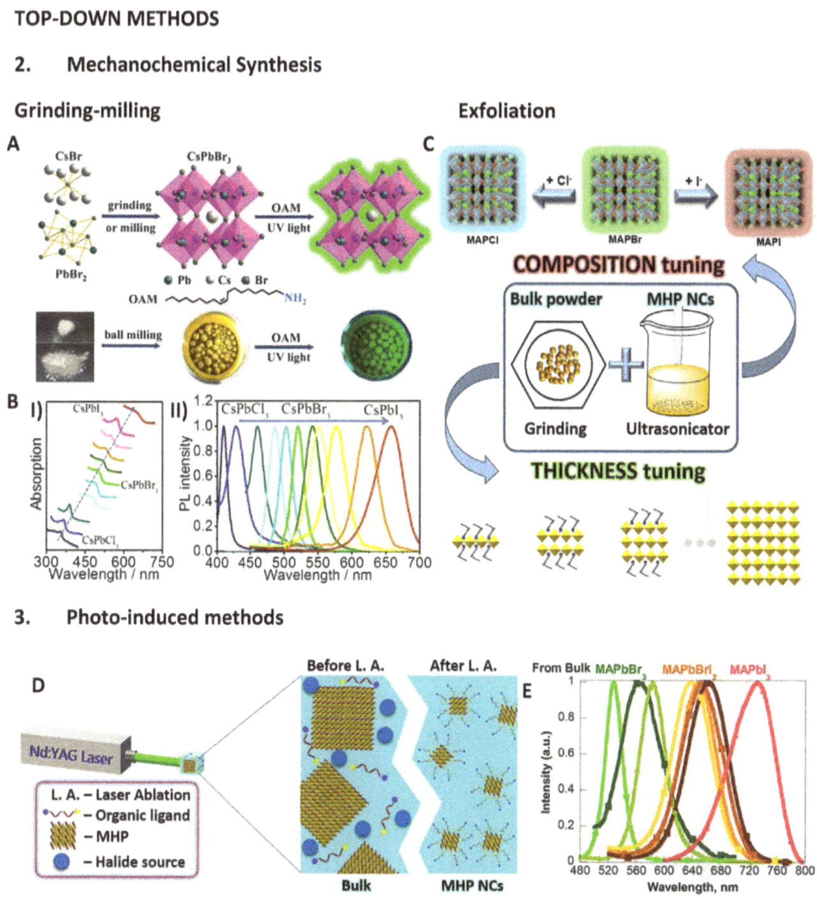

FIG. 14 (A–B) Schematic illustration of the two consecutive grinding steps and (I) UV-vis absorption, (II) emission. (C) Scheme of the ultrasonication synthesis and changes in the optical properties by varying the halide composition or separating the nanoplatelets according to their thickness. (D–E) Schematic representation of the synthesis by laser irradiation and fine-tuning of the optical properties of $MAPbX_3$ by varying the halide composition. *Panels (A and B) adapted with permission from Zhu, Z.-Y., Yang, Q.-Q., Gao, L.-F., Zhang, L., Shi, A.-Y., Sun, C.-L., Wang, Q., & Zhang, H.-L. (2017). Solvent-free mechanosynthesis of composition-tunable cesium lead halide perovskite quantum dots. J. Phys. Chem. Lett., 8(7), 1610–1614. Copyright 2017, American Chemical Society, (C) adapted from Hintermayr, V. A., Richter, A. F., Ehrat, F., Döblinger, M., Vanderlinden, W., Sichert, J. A., Tong, Y., Polavarapu, L., Feldmann, J., & Urban, A. S. (2016). Tuning the optical properties of perovskite nanoplatelets through composition and thickness by ligand-assisted exfoliation. Adv. Mater., 28(43), 9478–9485. Copyright 2016, WILEY-VCH Verlag GmbH & Co. KGaA, Weinheim, and (E) adapted with permission from Rosa-Pardo, I., Rando-Brotons, M., Pocoví-Martínez, S., Galian, R. E., & Prieto, J. (2019). Laser ablation of hybrid perovskite bulks into nanoparticles: adamantylammonium halides as ligands and halide sources. ChemNanoMat, 5(3), 328–333. Copyright 2019, Wiley-VCH Verlag GmbH & Co. KGaA, Weinheim.*

($A = Cs^+$ and FA^+) NCs by using mesitylene as solvent and oleylammonium bromide as capping ligand instead of the typical OA and OLA [117]. Furthermore, they described a decrease in the QY_{PL} and a blue-shift of $FAPbBr_3$ NCs with prolonged milling, probably due to the formation of trap states linked to crystal defects in wide bandgaps and the effect of quantum confinement, respectively. In 2019, to demonstrate the suitability of this approach for preparing an array of MHP NCs with pure and mixed composition, Chen et al. carried out a wet hand-grinding process to prepare satisfactorily the whole family of $APbX_3$ ($A = MA^+$, FA^+, and Cs; $X = Cl^-$, Br^-, and I^-) [118]. These NCs comprehended the whole visible range with good stability and halogen-rich surfaces with a QY_{PL} of up to 92%.

Exfoliation (Ultrasonication)

In this technique, the material is exfoliated by the frictions and shears generated by ultrasound energy. These strategies are very similar to the grinding/milling ones but place all the precursors in a container with a tip sonicator. Although a smaller number of publications have been reported using two-step ultrasonication techniques, it seems clear their suitability for preparing MHP NCs with different compositions and stoichiometries. Nevertheless, they enable very limited control over the size and shape, generating mainly nanoplatelets of different thicknesses.

In 2016, Hintermayr et al. synthesized 2D $MAPbX_3$ ($X = Cl^-$, Br^-, and I^-) nanoplatelets by a two-step approach in which the bulk was prepared by a dry step with a mortar and a pestle, and then, it was redispersed in toluene containing OLA for a sonicating step of 30 min. Afterward, a purification step was carried out, separating the nanoplatelets according to their thickness. With this procedure, they demonstrated the simple tuning of the absorption and emission spectra by varying the halide composition and separating nanoplatelets with a different quantum confinement (Fig. 14C) [119].

3.2.2 *Photo-induced methods*

Other less widespread alternatives for synthesizing MHP NCs by the top-down approach use the energy of photon irradiation to fragment bulk material in NCs. For that aim, it is necessary, as a light source, a pulse laser irradiation emitting in a wavelength where the bulk material can absorb the intense energy.

Although very few reports of photo-induced methods for MHP NCs have been published so far, because they require powerful pulsed lasers, this approach could allow the preparation of perovskite NCs with precise control of the composition thanks to their similarity with the nucleation and growth mechanism of the mechanochemical techniques. However, it seems much more difficult to regulate the specific conditions to modify uniformly the size and shape of the NCs.

In 2016, Amendola et al. published the first report of perovskite $MAPbBr_3$ colloids fragmenting perovskite bulk, prepared by a complex temperature crystallization process through laser pulse irradiation of the second harmonic of Nd:YAG laser at 532 nm [120]. Three years later, Rosa-Pardo et al. described the synthesis of stable $MAPbX_3$ ($X = Br^-$ and I^-) NCs based on the preparation of the bulk material by a simple ultrasound process followed by a controllable pulse laser irradiation step (Nd: YAG laser at 532 nm) (Fig. 14D). Besides, 2-adamantylammonium salts (ADAX) were used as ligand and as an iodine/bromine source to fine-tune the MHP NCs composition and,thus adjusting the optical properties from 582 to 700 nm (Fig. 14E) [83]. As the XRD spectra, $QY_{PL,}$ and FWHM showed, MHP NCs prepared by this technique have the same crystal structure and optical properties that the ones obtained by other common methods.

4. Challenges and perspectives

Challenges to be overcome for commercial applications are the lack of (i) large-area manufacturing processes; (ii) long-term chemical stability of the material; (iii) device packaging technology; and (iv) fabrication processes that allow a high level of control over film properties, such as crystal structure, composition, thickness, and conformality. Moreover, reduction of the toxicity of the perovskite film components is becoming a must.

Although high-quality thin films can be prepared by optimized solution-based processes, they can be incompatible with integration into complex devices (e.g., photonic circuits). There are a plethora of methods available to deposit halide perovskite thin films at the lab scale, mostly founded on solution-based methods, but several challenges remain for the controlled formation of halide perovskite-based thin films to fully investigate their exceptional optoelectronic properties. Nevertheless, this decade is witnessing an abundance of studies addressing these issues.

Key challenges of halide perovskite thin-film preparation and potential strategies for succeeding in this task are (i) control of grain size formation to tackle phase segregation in multi-cation-multi-halide perovskites. Studies in this field aim at resulting in an in-depth understanding of growth mechanisms and formation kinetics andat defining the role of grain boundaries in phase segregation, and its final influence on conversion efficiency; (ii) computational studies of physical and chemical processes occurring at all stages of thin-film fabrication by in situ monitoring to guide the synthesis of crystalline halide perovskite thin films; (iii) integration of halide perovskite top cells in tandem solar cells by using solvent-free thin-film synthesis approaches; (iv) conformal deposition to facilitate integration on the bottom cells with high roughness.

Further applications of perovskites rely on succeeding in epitaxial growth as a strategy to manufacture monocrystalline, grain boundary-free films, which will enable fundamental understanding of material properties, and new physical phenomena that increase their perspective.

5. Conclusion

Perovskite solar cells are the most promising alternative to silicon-based solar cells, but they are still facing some limitations, in particular reaching greater power conversion efficiency and stability. The interface of a perovskite polycrystalline film is a defect-rich area, this being the main factor that reduce its efficiency. Engineering interfaces have created expectations for achieving efficient PSCs with long-term chemical stability; in addition, passivation at the interface can provide a smooth and dense perovskite layer. Interface contact reduces the formation of interfacial defects and recombination of carriers and eases the transport of interfacial carriers. Analysis of the interface passivation materials will allow the identification of suitable materials and therefore the acquisition of fundamental knowledge for designing PSCs with improved efficiency and stability.

Acknowledgments

This work was supported by the Spanish MINECO (PID2020-115710GB, I.R.P. FPU grant FPU17/05564), partially co-financed with FEDER funds, and the "Maria de Maeztu" Programme for Units of Excellence in R&D (CEX2019-000919-M; A.C.V. PRE2018-084294). The authors also thank the financial support by Generalitat Valenciana (PROMETEO/2019/080 and IDIFEDER/2018/064, co-financed by FEDER).

References

[1] A. Kojima, M. Ikegami, K. Teshima, T. Miyasaka, Highly luminescent lead bromide perovskite nanoparticles synthesized with porous alumina media, Chem. Lett. 41 (4) (2012) 397–399.

[2] D.B. Mitzi, C.A. Feild, Z. Schlesinger, R.B. Laibowitz, Transport, optical, and magnetic properties of the conducting halide perovskite ch3nh3sni3, J. Solid State Chem. 114 (1) (1995) 159–163.

[3] H.J. Snaith, Perovskites: the emergence of a new era for low-cost, high-efficiency solar cells, J. Phys. Chem. Lett. 4 (21) (2013) 3623–3630.

[4] V. Blay, R.E. Galian, L.M. Muresan, D. Pankratov, P. Pinyou, G. Zampardi, Research frontiers in energy-related materials and applications for 2020–2030, Adv. Sustain. Syst. 4 (2) (2020) 1900145.

[5] J. Jeong, M. Kim, J. Seo, H. Lu, P. Ahlawat, A. Mishra, Y. Yang, M.A. Hope, F.T. Eickemeyer, M. Kim, et al., Pseudo-halide anion engineering for α-FAPbI3 perovskite solar cells, Nature 59 (2021) 381–385.

[6] D.B. Mitzi, Progress in Inorganic Chemistry, vol. 48, 1999.

[7] S. González-Carrero, R.E. Galian, J. Pérez-Prieto, Organometal halide perovskites: bulk low-dimension materials and nanoparticles, Part. Part. Syst. Charact. 32 (7) (2015) 709–720.

[8] L.C. Schmidt, A. Pertegás, S. González-Carrero, O. Malinkiewicz, S. Agouram, G. Mínguez Espallargas, H.J. Bolink, R.E. Galian, J. Pérez-Prieto, Nontemplate synthesis of CH3NH3PbBr3 perovskite nanoparticles, J. Am. Chem. Soc. 136 (3) (2014) 850–853.

[9] R.E. Galian, J. Pérez-Prieto, Synergism at the nanoscale: photoactive semiconductor nanoparticles and their organic ligands, in: Research Perspectives on Functional Micro- and Nanoscale Coatings, vol. 3, 2016.

[10] A. Dandia, P. Saini, R. Sharma, V. Parewa, Visible light driven perovskite-based photocatalysts: a new candidate for green organic synthesis by photochemical protocol, Curr. Res. Green Sustain. Chem. 3 (2020), 100031.

[11] Y.-H. Kim, C. Wolf, Y.-T. Kim, H. Cho, W. Kwon, S. Do, A. Sadhanala, C.G. Park, S.-W. Rhee, S.H. Im, et al., Highly efficient light-emitting diodes of colloidal metal-halide perovskite nanocrystals beyond quantum size, ACS Nano 11 (7) (2017) 6586–6593.

[12] C. McDonald, C. Ni, P. Maguire, P. Connor, J.T.S. Irvine, D. Mariotti, V. Svrcek, Nanostructured perovskite solar cells, Nanomaterials 9 (10) (2019) 1481.

[13] Q. Wei, X. Li, C. Liang, Z. Zhang, J. Guo, G. Hong, G. Xing, W. Huang, Recent progress in metal halide perovskite micro- and nanolasers, Adv. Opt. Mater. 17 (2019) 1900080.

[14] D. Liu, Y. Guo, M. Que, X. Yin, J. Liu, H. Xie, C. Zhang, W. Que, Metal halide perovskite nanocrystals: application in high-performance photodetectors, Mater. Adv. 2 (2021) 856–879.

[15] M.L. De Giorgi, F. Krieg, M.V. Kovalenko, M. Anni, Amplified spontaneous emission threshold reduction and operational stability improvement in CsPbBr3 nanocrystals films by hydrophobic functionalization of the substrate, Sci. Rep. 9 (2019) 17964.

[16] A. Dey, J. Ye, A. De, E. Debroye, S.K. Ha, E. Bladt, A.S. Kshirsagar, Z. Wang, J. Yin, Y. Wang, L.N. Quan, F. Yan, M. Gao, X. Li, J. Shamsi, T. Debnath, M. Cao, M.A. Scheel, S. Kumar, J.A. Steele, M. Gerhard, L. Chouhan, K. Xu, X.-g. Wu, Y. Li, Y. Zhang, A. Dutta, C. Han, I. Vincon, A.L. Rogach, A. Nag, A. Samanta, B.A. Korgel, C.-J. Shih, D.R. Gamelin, D.H. Son, H. Zeng, H. Zhong, H. Sun, H.V. Demir, I.G. Scheblykin, I. Mora-Seró, J.K. Stolarczyk, J.Z. Zhang, J. Feldmann, J. Hofkens, J.M. Luther, J. Pérez-Prieto, L. Li, L. Manna, M.I. Bodnarchuk, M.V. Kovalenko, M.B.J. Roeffaers, N. Pradhan, O.F. Mohammed, O.M. Bakr, P. Yang, P. Müller-Buschbaum, P.V. Kamat, Q. Bao, Q. Zhang, R. Krahne, R.E. Galian, S.D. Stranks, S. Bals, V. Biju, W.A. Tisdale, Y. Yan, R.L.Z. Hoye, L. Polavarapu, State of the art and prospects for halide perovskite nanocrystals, ACS Nano 15 (7) (2021) 10775–10981.

[17] Y.-H. Chiang, M. Anaya, S.D. Stranks, Multisource vacuum deposition of methylammonium-free perovskite solar cells, ACS Energy Lett. 5 (8) (2020) 2498–2504.

[18] G. Tong, L. Huan, G. Li, T. Zhang, C. Li, L. Yu, J. Xu, Y. Jiang, K. Chen, Mixed cation perovskite solar cells by stack-sequence chemical vapor deposition with self-passivation and gradient absorption layer, Nano Energy 48 (2018) 536–542.

[19] Y. Peng, G. Jing, T. Cui, A hybrid physical–chemical deposition process at ultra-low temperatures for high-performance perovskite solar cells, J. Mater. Chem. A 3 (2015) 12436–12442.

[20] N.J. Jeon, H. Na, E.H. Jung, T.-Y. Yang, Y.G. Lee, G. Kim, H.-W. Shin, I., Seok, S., Lee, J., & Seo, J., A fluorene-terminated hole-transporting material for highly efficient and stable perovskite solar cells, Nat. Energy 3 (2018) 682–689.

[21] Y. Deng, X. Zheng, Y. Bai, Q. Wang, J. Zhao, J. Huang, Surfactant-controlled ink drying enables high-speed deposition of perovskite films for efficient photovoltaic modules, Nat. Energy 3 (2018) 560–566.

[22] M.H. Li, H.H. Yeh, Y.H. Chiang, U.S. Jeng, C.J. Su, H.W. Shiu, Y.J. Hsu, N. Kosugi, T. Ohigashi, Y.A. Chen, P.S. Shen, P. Chen, T.F. Guo, Highly efficient 2D/3D hybrid

perovskite solar cells via low-pressure vapor-assisted solution process, Adv. Mater. 30 (30) (2018) 1801401.
[23] P. Fan, D. Gu, G.-X. Liang, J.-T. Luo, J.-L. Chen, Z.-H. Zheng, D.-P. Zhang, High-performance perovskite CH3NH3PbI3 thin films for solar cells prepared by single-source physical vapour deposition, Sci. Rep. 6 (2016) 29910.
[24] Y. El Ajjouri, F. Palazon, M. Sessolo, H.J. Bolink, Single-source vacuum deposition of mechanosynthesized inorganic halide perovskites, Chem. Mater. 30 (21) (2018) 7423–7427.
[25] M. Era, T. Hattori, T. Taira, T. Tsutsui, Self-organized growth of PbI-based layered perovskite quantum well by dual-source vapor deposition, Chem. Mater. 9 (1) (1997) 8–10.
[26] M. Liu, M.B. Johnston, H.J. Snaith, Efficient planar heterojunction perovskite solar cells by vapour deposition, Nature 501 (7467) (2013) 395–398.
[27] J. Li, H. Wang, X.Y. Chin, H.A. Dewi, K. Vergeer, T.W. Goh, J.W.M. Lim, J.H. Lew, K.P. Loh, C. Soci, et al., Highly efficient thermally co-evaporated perovskite solar cells and mini-modules, Joule 5 (2020) 1035–1053.
[28] L.K. Ono, S. Wang, Y. Kato, S.R. Raga, Y. Qi, Fabrication of semi-transparent perovskite films with centimeter-scale superior uniformity by the hybrid deposition method, Energy Environ. Sci. 7 (12) (2014) 3989–3993.
[29] H. Hu, D. Wang, Y. Zhou, J. Zhang, S. Lv, S. Pang, X. Chen, Z. Liu, N.P. Padture, G. Cui, Vapour-based processing of hole-conductor-free CH3NH3PbI3 perovskite/C60 fullerene planar solar cells, RSC Adv. 4 (2014) 28964–28967.
[30] C.-W. Chen, H.-W. Kang, S.-Y. Hsiao, P.-F. Yang, K.-M. Chiang, H.-W. Lin, Efficient and uniform planar-type perovskite solar cells by simple sequential vacuum deposition, Adv. Mater. 26 (38) (2014) 6647–6652.
[31] D.B. Mitzi, M.T. Prikas, K. Chondroudis, Thin film deposition of organic-inorganic hybrid materials using a single source thermal ablation technique, Chem. Mater. 11 (3) (1999) 542–544.
[32] G. Longo, L. Gil-Escrig, M.J. Degen, M. Sessolo, H.J. Bolink, Perovskite solar cells prepared by flash evaporation, Chem. Commun. 51 (34) (2015) 7376–7378.
[33] S. Bonomi, D. Marongiu, N. Sestu, M. Saba, M. Patrini, G. Bongiovanni, L. Malavasi, Novel physical vapor deposition approach to hybrid perovskites: growth of MAPbI3 thin films by RF-Magnetron sputtering, Sci. Rep. 8 (1) (2018) 15388.
[34] U. Bansode, R. Naphade, O. Game, S. Agarkar, S. Ogale, Hybrid perovskite films by a new variant of pulsed excimer laser deposition: a room-temperature dry process, J. Phys. Chem. C 119 (17) (2015) 9177–9185.
[35] E.T. Barraza, W.A. Dunlap-Shohl, D.B. Mitzi, A.D. Stiff-Roberts, Deposition of methylammonium lead triiodide by resonant infrared matrix-assisted pulsed laser evaporation, J. Electron. Mater. 47 (2) (2018) 917–926.
[36] W.A. Dunlap-Shohl, E.T. Barraza, A. Barrette, K. Gundogdu, A.D. Stiff-Roberts, D.B. Mitzi, MAPbI3 solar cells with absorber deposited by resonant infrared matrix-assisted pulsed laser evaporation, ACS Energy Lett. 3 (2) (2018) 270–275.
[37] M.M. Tavakoli, L. Gu, Y. Gao, C. Reckmeier, J. He, A.L. Rogach, Y. Yao, Z. Fan, Fabrication of efficient planar perovskite solar cells using a one-step chemical vapor deposition method, Sci. Rep. 5 (2015) 14083.
[38] D.J. Lewis, P. O'Brien, Ambient pressure aerosol-assisted chemical vapour deposition of (CH3NH3)PbBr3, an inorganic-organic perovskite important in photovoltaics, Chem. Commun. 50 (48) (2014) 6319–6321.
[39] S. Chen, J. Briscoe, Y. Shi, K. Chen, R.M. Wilson, S. Dunn, R. Binions, A simple, low-cost CVD route to high-quality CH3NH3PbI3 perovskite thin films, CrystEngComm 17 (2015) 7486–7489.
[40] X. Chen, Y. Myung, A. Thind, Z. Gao, B. Yin, M. Shen, S.B. Cho, P. Cheng, B. Sadtler, R. Mishra, et al., Atmospheric pressure chemical vapor deposition of methylammonium bismuth iodide thin films, J. Mater. Chem. A 5 (2017) 24728–24739.

[41] G. Popov, M. Mattinen, T. Hatanpää, M. Vehkamäki, M. Kemell, K. Mizohata, J. Räisänen, M. Ritala, M. Leskelä, Atomic layer deposition of PbI2 thin films, Chem. Mater. 31 (3) (2019) 1101–1109.
[42] B.R. Sutherland, S. Hoogland, M.M. Adachi, P. Kanjanaboos, C.T.O. Wong, J.J. McDowell, J. Xu, O. Voznyy, Z. Ning, A.J. Houtepen, E.H. Sargent, Perovskite thin films via atomic layer deposition, Adv. Mater. 27 (1) (2015) 53–58.
[43] S.T. Ha, X. Liu, Q. Zhang, D. Giovanni, T.C. Sum, Q. Xiong, Synthesis of organic-inorganic lead halide perovskite nanoplatelets: towards high-performance perovskite solar cells and optoelectronic devices, Adv. Opt. Mater. 2 (9) (2014) 838–844.
[44] M.R. Leyden, L.K. Ono, S.R. Raga, Y. Kato, S. Wang, Y. Qi, High performance perovskite solar cells by hybrid chemical vapor deposition, J. Mater. Chem. A 2 (2014) 18742–18745.
[45] A. Kojima, K. Teshima, Y. Shirai, T. Miyasaka, Organometal halide perovskites as visible-light sensitizers for photovoltaic cells, J. Am. Chem. Soc. 131 (17) (2009) 6050–6051.
[46] W. Nie, H. Tsai, R. Asadpour, J.-C. Blancon, A.J. Neukirch, G. Gupta, J.J. Crochet, M. Chhowalla, S. Tretiak, M.A. Alam, et al., High-efficiency solution-processed perovskite solar cells with millimeter-scale grains, Science 347 (6221) (2015) 522–525.
[47] N.J. Jeon, J.H. Noh, Y.C. Kim, W.S. Yang, S. Ryu, S.I. Seok, Solvent engineering for high-performance inorganic-organic hybrid perovskite solar cells, Nat. Mater. 13 (9) (2014) 897–903.
[48] Y. Zhou, M. Yang, W. Wu, A.L. Vasiliev, K. Zhu, N.P. Padture, Room-temperature crystallization of hybrid-perovskite thin films via solvent-solvent extraction for high-performance solar cells, J. Mater. Chem. A 3 (2015) 8178–8184.
[49] F. Huang, Y. Dkhissi, W. Huang, M. Xiao, I. Benesperi, S. Rubanov, Y. Zhu, X. Lin, L. Jiang, Y. Zhou, A. Gray-Weale, J. Etheridge, C.R. McNeill, R.A. Caruso, U. Bach, L. Spiccia, Y.B. Cheng, Gas-assisted preparation of lead iodide perovskite films consisting of a monolayer of single crystalline grains for high efficiency planar solar cells, Nano Energy 10 (2014) 10–18.
[50] A. Mei, X. Li, L. Liu, Z. Ku, T. Liu, Y. Rong, M. Xu, M. Hu, J. Chen, Y. Yang, M. Grätzel, H. Han, A hole-conductor-free, fully printable mesoscopic perovskite solar cell with high stability, Science 345 (6194) (2014) 295–298.
[51] H.C. Liao, P. Guo, C.P. Hsu, M. Lin, B. Wang, L. Zeng, W. Huang, C.M.M. Soe, W.-F. Su, M.J. Bedzyk, M.R. Wasielewski, A. Facchetti, R.P.H. Chang, M.G. Kanatzidis, T.J. Marks, Enhanced efficiency of hot-cast large-area planar perovskite solar cells/modules having controlled chloride incorporation, Adv. Energy Mater. 7 (8) (2017) 1601660.
[52] L. Huang, C. Li, X. Sun, R. Xu, Y. Du, J. Ni, H. Cai, J. Li, Z. Hu, J. Zhang, Efficient and hysteresis-less pseudo-planar heterojunction perovskite solar cells fabricated by a facile and solution-saving one-step dip-coating method, Org. Electron. 40 (2017) 13–23.
[53] Y.X. Chen, Q.Q. Ge, Y. Shi, J. Liu, D.J. Xue, J.Y. Ma, J. Ding, H.J. Yan, J.S. Hu, L.J. Wan, General space-confined on-substrate fabrication of thickness-adjustable hybrid perovskite single-crystalline thin films, J. Am. Chem. Soc. 138 (50) (2016) 16196–16199.
[54] J.H. Im, H.S. Kim, N.G. Park, Morphology-photovoltaic property correlation in perovskite solar cells: one-step versus two-step deposition of CH3NH3PbI 3, APL Mater. 2 (2014), 081510.
[55] W.S. Yang, J.H. Noh, N.J. Jeon, Y.C. Kim, S. Ryu, J. Seo, S.I. Seok, High-performance photovoltaic perovskite layers fabricated through intramolecular exchange, Science 348 (6240) (2015) 1234–1237.
[56] M. Adnan, J.K. Lee, All sequential dip-coating processed perovskite layers from an aqueous lead precursor for high efficiency perovskite solar cells, Sci. Rep. 8 (1) (2018) 2168.

[57] K. Liang, D.B. Mitzi, M.T. Prikas, Synthesis and characterization of organic-inorganic perovskite thin films prepared using a versatile two-step dipping technique, Chem. Mater. 10 (1) (1998) 403–411.
[58] J. Burschka, N. Pellet, S.-J. Moon, R. Humphry-Baker, P. Gao, M.K. Nazeeruddin, M. Grätzel, Sequential deposition as a route to high-performance perovskite-sensitized solar cells, Nature 499 (2013) 316–319.
[59] X.P. Cui, K.J. Jiang, J.H. Huang, X.Q. Zhou, M.J. Su, S.G. Li, Q.Q. Zhang, L.M. Yang, Y.L. Song, Electrodeposition of PbO and its in situ conversion to CH3NH3PbI3 for mesoscopic perovskite solar cells, Chem. Commun. 51 (8) (2015) 1457–1460.
[60] H. Chen, Z. Wei, X. Zheng, S. Yang, A scalable electrodeposition route to the low-cost, versatile and controllable fabrication of perovskite solar cells, Nano Energy 15 (2015) 216–226.
[61] X. Wang, S. Abbasi, D. Zhang, J. Wang, Y. Wang, Z. Cheng, H. Liu, W. Shen, Electrochemical deposition of CsPbBr3 perovskite for photovoltaic devices with robust ambient stability, ACS Appl. Mater. Interfaces 12 (45) (2020) 50455–50463.
[62] J.H. Kim, S.T. Williams, N. Cho, C.C. Chueh, A.K.Y. Jen, Enhanced environmental stability of planar heterojunction perovskite solar cells based on blade-coating, Adv. Energy Mater. 5 (4) (2015) 1401229.
[63] K. Hwang, Y.S. Jung, Y.J. Heo, F.H. Scholes, S.E. Watkins, J. Subbiah, D.J. Jones, D.Y. Kim, D. Vak, Toward large scale roll-to-roll production of fully printed perovskite solar cells, Adv. Mater. 27 (7) (2015) 1241–1247.
[64] M. Fievez, P.J. Singh Rana, T.M. Koh, M. Manceau, J.H. Lew, N.F. Jamaludin, B. Ghosh, A. Bruno, S. Cros, S. Berson, S.G. Mhaisalkar, W.L. Leong, Slot-die coated methylammonium-free perovskite solar cells with 18% efficiency, Sol. Energy Mater. Sol. Cells 230 (2021), 111189.
[65] A.T. Barrows, A.J. Pearson, C.K. Kwak, A.D.F. Dunbar, A.R. Buckley, D.G. Lidzey, Efficient planar heterojunction mixed-halide perovskite solar cells deposited via spray-deposition, Energy Environ. Sci. 7 (9) (2014) 2944–2950.
[66] J.E. Bishop, C.D. Read, J.A. Smith, T.J. Routledge, D.G. Lidzey, Fully spray-coated triple-cation perovskite solar cells, Sci. Rep. 10 (2020) 6610.
[67] S.-G. Li, K.-J. Jiang, M.-J. Su, X.-P. Cui, J.-H. Huang, Q.-Q. Zhang, X.-Q. Zhou, L.-M. Yang, Y.-L. Song, Inkjet printing of CH3NH3PbI3 on a mesoscopic TiO2 film for highly efficient perovskite solar cells, J. Mater. Chem. A 3 (2015) 9092–9097.
[68] C. Liang, P. Li, H. Gu, Y. Zhang, F. Li, Y. Song, G. Shao, N. Mathews, G. Xing, One-step inkjet printed perovskite in air for efficient light harvesting, Solar RRL 2 (2) (2018) 1700217.
[69] M. Bag, Z. Jiang, L.A. Renna, S.P. Jeong, V.M. Rotello, D. Venkataraman, Rapid combinatorial screening of inkjet-printed alkyl-ammonium cations in perovskite solar cells, Mater. Lett. 164 (2016) 472–475.
[70] J.W. Lee, S.I. Na, S.S. Kim, Efficient spin-coating-free planar heterojunction perovskite solar cells fabricated with successive brush-painting, J. Power Sources 339 (2017) 33–40.
[71] M. He, B. Li, X. Cui, B. Jiang, Y. He, Y. Chen, D. O'Neil, P. Szymanski, M.A. Ei-Sayed, J. Huang, Z. Lin, Meniscus-assisted solution printing of large-grained perovskite films for high-efficiency solar cells, Nat. Commun. 8 (2017) 16045.
[72] F. Ye, H. Chen, F. Xie, W. Tang, M. Yin, J. He, E. Bi, Y. Wang, X. Yang, L. Han, Soft-cover deposition of scaling-up uniform perovskite thin films for high cost-performance solar cells, Energy Environ. Sci. 9 (2016) 2295–2301.
[73] Q. Chen, H. Zhou, Z. Hong, S. Luo, H.S. Duan, H.H. Wang, Y. Liu, G. Li, Y. Yang, Planar heterojunction perovskite solar cells via vapor-assisted solution process, J. Am. Chem. Soc. 136 (2) (2014) 622–625.
[74] G. Li, J.Y.L. Ho, M. Wong, H.-S. Kwok, Low cost, high throughput and centimeter-scale fabrication of efficient hybrid perovskite solar cells by closed space vapor transport, Phys. Status Solidi (RRL) 10 (2) (2016) 153–157.

[75] L. Protesescu, S. Yakunin, M.I. Bodnarchuk, F. Krieg, R. Caputo, C.H. Hendon, R.X. Yang, A. Walsh, M.V. Kovalenko, Nanocrystals of cesium lead halide perovskites (CsPbX3, X = Cl, Br, and I): novel optoelectronic materials showing bright emission with wide color, Gamut 15 (6) (2015) 3692–3696.

[76] H. Huang, F. Zhao, L. Liu, F. Zhang, X.G. Wu, L. Shi, B. Zou, Q. Pei, H. Zhong, Emulsion synthesis of size-tunable CH3NH3PbBr3 quantum dots: an alternative route toward efficient light-emitting diodes, ACS Appl. Mater. Interfaces 7 (51) (2015) 28128–28133.

[77] M. Chen, Y. Zou, L. Wu, Q. Pan, D. Yang, H. Hu, Y. Tan, Q. Zhong, Y. Xu, H. Liu, B. Sun, Q. Zhang, Solvothermal synthesis of high-quality all-inorganic cesium lead halide perovskite nanocrystals: from nanocube to ultrathin nanowire, Adv. Funct. Mater. 27 (23) (2017) 1701121.

[78] Z. Long, H. Ren, J. Sun, J. Ouyang, N. Na, High-throughput and tunable synthesis of colloidal CsPbX3 perovskite nanocrystals in a heterogeneous system by microwave irradiation, Chem. Commun. 53 (71) (2017) 9914–9917.

[79] Y. Tong, E. Bladt, M.F. Aygüler, A. Manzi, K.Z. Milowska, V.A. Hintermayr, P. Docampo, S. Bals, A.S. Urban, L. Polavarapu, J. Feldmann, Highly luminescent cesium lead halide perovskite nanocrystals with tunable composition and thickness by ultrasonication, Angew. Chem. Int. Ed. 55 (44) (2016) 13887–13892.

[80] X. Chen, L. Peng, K. Huang, Z. Shi, R. Xie, W. Yang, Non-injection gram-scale synthesis of cesium lead halide perovskite quantum dots with controllable size and composition, Nano Res. 9 (2016) 1994–2006.

[81] A. Jana, M. Mittal, A. Singla, S. Sapra, Solvent-free, mechanochemical syntheses of bulk trihalide perovskites and their nanoparticles, Chem. Commun. 53 (21) (2017) 3046–3049.

[82] Z.-Y. Zhu, Q.-Q. Yang, L.-F. Gao, L. Zhang, A.-Y. Shi, C.-L. Sun, Q. Wang, H.-L. Zhang, Solvent-free mechanosynthesis of composition-tunable cesium lead halide perovskite quantum dots, J. Phys. Chem. Lett. 8 (7) (2017) 1610–1614.

[83] I. Rosa-Pardo, M. Rando-Brotons, S. Pocoví-Martínez, R.E. Galian, J. Prieto, Laser ablation of hybrid perovskite bulks into nanoparticles: adamantylammonium halides as ligands and halide sources, ChemNanoMat 5 (3) (2019) 328–333.

[84] B. Cai, Y. Xing, Z. Yang, W.H. Zhang, J. Qiu, High performance hybrid solar cells sensitized by organolead halide perovskites, Energy Environ. Sci. 6 (6) (2013) 1480–1485.

[85] H.-S. Kim, C.-R. Lee, J.-H. Im, K.-B. Lee, T. Moehl, A. Marchioro, S.-J. Moon, R. Humphry-Baker, J.-H. Yum, J.E. Moser, et al., Lead iodide perovskite sensitized all-solid-state submicron thin film mesoscopic solar cell with efficiency exceeding 9%, Sci. Rep. 2 (591) (2012).

[86] J.H. Noh, S.H. Im, J.H. Heo, T.N. Mandal, S.I. Seok, Chemical management for colorful, efficient, and stable inorganic-organic hybrid nanostructured solar cells, Nano Lett. 13 (4) (2013) 1764–1769.

[87] L. Etgar, P. Gao, Z. Xue, Q. Peng, A.K. Chandiran, B. Liu, M.K. Nazeeruddin, M. Grätzel, Mesoscopic CH3NH3PbI3/TiO2 heterojunction solar cells, J. Am. Chem. Soc. 134 (42) (2012) 17396–17399.

[88] M.M. Lee, J. Teuscher, T. Miyasaka, T.N. Murakami, H.J. Snaith, Efficient hybrid solar cells based on meso-superstructured organometal halide perovskites, Science 338 (6107) (2012) 643–647.

[89] C.K. Ng, C. Wang, J.J. Jasieniak, Synthetic evolution of colloidal metal halide perovskite nanocrystals, Langmuir 35 (36) (2019) 11609–11628.

[90] V.K. LaMer, R.H. Dinegar, Theory., Production and mechanism of formation of monodispersed hydrosols, J. Am. Chem. Soc. 72 (11) (1950) 4847–4854.

[91] C.B. Murray, D.J. Norris, M.G. Bawendi, Synthesis and characterization of nearly monodisperse CdE (E = sulfur, selenium, tellurium) semiconductor nanocrystallites, J. Am. Chem. Soc. 115 (19) (1993) 8706–8715.

[92] J. Shamsi, A.S. Urban, M. Imran, L. De Trizio, L. Manna, Metal halide perovskite nanocrystals: synthesis, post-synthesis modifications, and their optical properties, Chem. Rev. 119 (5) (2019) 3296–3348.
[93] A. Pan, B. He, X. Fan, Z. Liu, J.J. Urban, A.P. Alivisatos, L. He, Y. Liu, Insight into the ligand-mediated synthesis of colloidal CsPbBr3 perovskite nanocrystals: the role of organic acid, base, and cesium precursors, ACS Nano 10 (8) (2016) 7943–7954.
[94] D. Zhang, S.W. Eaton, Y. Yu, L. Dou, P. Yang, Solution-phase synthesis of cesium lead halide perovskite nanowires, J. Am. Chem. Soc. 137 (29) (2015) 9230–9233.
[95] O. Vybornyi, S. Yakunin, M.V. Kovalenko, Polar-solvent-free colloidal synthesis of highly luminescent alkylammonium lead halide perovskite nanocrystals, Nanoscale 8 (12) (2016) 6278–6283.
[96] H. Kasai, H.S. Nalwa, H. Oikawa, S. Okada, H. Matsuda, N. Minami, A. Kakuta, K. Ono, A. Mukoh, H. Nakanishi, A novel preparation method of organic microcrystals, Jpn. J. Appl. Phys. 31 (8) (1992) L1132–L1134.
[97] H. Nakanishi, H. Oikawa, Single Organic Nanoparticles, 2003.
[98] B.S. Santos, P.M.A. Farias, A. Fontes, Handbook of Self Assembled Semiconductor Nanostructures for Novel Devices in Photonics and Electronics, vol. 26, 2008.
[99] H. Huang, A.S. Susha, S.V. Kershaw, T.F. Hung, A.L. Rogach, Control of emission color of high quantum yield CH3NH3PbBr3 perovskite quantum dots by precipitation temperature, Adv. Sci. 2 (9) (2015) 1500194.
[100] X. Du, G. Wu, J. Cheng, H. Dang, K. Ma, Y.-W. Zhang, P.-F. Tan, S. Chen, High-quality CsPbBr3 perovskite nanocrystals for quantum dot light-emitting diodes, RSC Adv. 7 (2017) 10391–10396.
[101] S. Seth, A. Samanta, A facile methodology for engineering the morphology of CsPbX3 perovskite nanocrystals under ambient condition, Sci. Rep. 6 (2016) 37693.
[102] S. Wei, Y. Yang, X. Kang, L. Wang, L. Huang, D. Pan, Room-temperature and gram-scale synthesis of CsPbX3 (X = Cl, Br, I) perovskite nanocrystals with 50-85% photoluminescence quantum yields, Chem. Commun. 52 (45) (2016) 7265–7268.
[103] X. Li, D. Yu, F. Cao, Y. Gu, Y. Wei, Y. Wu, J. Song, H. Zeng, Healing all-inorganic perovskite films via recyclable dissolution-recyrstallization for compact and smooth carrier channels of optoelectronic devices with high stability, Adv. Funct. Mater. 26 (32) (2016) 5903–5912.
[104] G.C. Papavassiliou, G. Pagona, N. Karousis, G.A. Mousdis, I. Koutselas, A. Vassilakopoulou, Nanocrystalline/microcrystalline materials based on lead-halide units, J. Mater. Chem. 22 (2012) 8271–8280.
[105] F. Zhang, H. Zhong, C. Chen, X. Hu, H. Huang, J. Han, B. Zou, Y. Dong, Brightly luminescent and color-tunable colloidal CH3NH3PbX3 (X = Br, I, Cl) quantum dots: potential alternatives for display technology, J. Mater. Chem. A 9 (4) (2015) 4533–4542.
[106] F. Fang, W. Chen, Y. Li, H. Liu, M. Mei, R. Zhang, J. Hao, M. Mikita, W. Cao, R. Pan, K. Wang, X.W. Sun, Employing polar solvent controlled ionization in precursors for synthesis of high-quality inorganic perovskite nanocrystals at room temperature, Adv. Funct. Mater. 28 (10) (2018) 1706000.
[107] M.A. Malik, M.Y. Wani, M.A. Hashim, Microemulsion method: a novel route to synthesize organic and inorganic nanomaterials: 1st nano update, Arab. J. Chem. 5 (4) (2012) 397–417.
[108] S. Gonzalez-Carrero, R.E. Galian, J. Pérez-Prieto, Maximizing the emissive properties of CH3NH3PbBr3 perovskite nanoparticles, J. Mater. Chem. A 3 (2015) 9187–9193.
[109] M.B. Teunis, M.A. Johnson, B.B. Muhoberac, S. Seifert, R. Sardar, Programmable colloidal approach to hierarchical structures of methylammonium lead bromide perovskite nanocrystals with bright photoluminescent properties, Chem. Mater. 29 (8) (2017) 3526–3537.

[110] Q. Pan, H. Hu, Y. Zou, M. Chen, L. Wu, D. Yang, X. Yuan, J. Fan, B. Sun, Q. Zhang, Microwave-assisted synthesis of high-quality "all-inorganic" CsPbX3 (X = Cl, Br, I) perovskite nanocrystals and their application in light emitting diodes, J. Mater. Chem. C 5 (2017) 10947–10954.

[111] D.M. Jang, D.H. Kim, K. Park, J. Park, J.W. Lee, J.K. Song, Ultrasound synthesis of lead halide perovskite nanocrystals, J. Mater. Chem. C 4 (2016) 10625–10629.

[112] I. Lignos, S. Stavrakis, G. Nedelcu, L. Protesescu, A.J. Demello, M.V. Kovalenko, Synthesis of cesium lead halide perovskite nanocrystals in a droplet-based microfluidic platform: fast parametric space mapping, Nano Lett. 16 (3) (2016) 1869–1877.

[113] R.M. MacEiczyk, K. Dümbgen, I. Lignos, L. Protesescu, M.V. Kovalenko, A.J. Demello, Microfluidic reactors provide preparative and mechanistic insights into the synthesis of formamidinium lead halide perovskite nanocrystals, Chem. Mater. 29 (19) (2017) 8433–8439.

[114] P. Sadhukhan, S. Kundu, A. Roy, A. Ray, P. Maji, H. Dutta, S.K. Pradhan, S. Das, Solvent-free solid-state synthesis of high yield mixed halide perovskites for easily tunable composition and band gap, Cryst. Growth Des. 18 (6) (2018) 3428–3432.

[115] B.A. Rosales, L. Wei, J. Vela, Synthesis and mixing of complex halide perovskites by solvent-free solid-state methods, J. Solid State Chem. 271 (2019) 206–215.

[116] F. Palazon, Y. Ajjouri, H.J. Bolink, Making by grinding: mechanochemistry boosts the development of halide perovskites and other multinary metal halides, Adv. Energy Mater. 13 (2020) 1902499.

[117] L. Protesescu, S. Yakunin, O. Nazarenko, D.N. Dirin, M.V. Kovalenko, Low-cost synthesis of highly luminescent colloidal lead halide perovskite nanocrystals by wet ball milling, ACS Appl. Nano Mater. 1 (3) (2018) 1300–1308.

[118] D. Chen, J. Li, X. Chen, J. Chen, J. Zhong, Grinding synthesis of APbX3 (A = MA, FA, Cs; X = Cl, Br, I) perovskite nanocrystals, ACS Appl. Mater. Interfaces 11 (10) (2019) 10059–10067.

[119] V.A. Hintermayr, A.F. Richter, F. Ehrat, M. Döblinger, W. Vanderlinden, J.A. Sichert, Y. Tong, L. Polavarapu, J. Feldmann, A.S. Urban, Tuning the optical properties of perovskite nanoplatelets through composition and thickness by ligand-assisted exfoliation, Adv. Mater. 28 (43) (2016) 9478–9485.

[120] V. Amendola, I. Fortunati, C. Marega, A.L. Abdelhady, M.I. Saidaminov, O.M. Bakr, High-purity hybrid organolead halide perovskite nanoparticles obtained by pulsed-laser irradiation in liquid, ChemPhysChem 18 (9) (2017) 1047–1054.

CHAPTER

5

Ab initio studies on perovskites

Tudor Luca Mitran[a], Rachel Elizabeth Brophy[b], Marina Cuzminschi[a,c], Nicolae Filipoiu[a,c], Movaffaq Kateb[d], Ioana Pintilie[e], Andrei Manolescu[d], and George Alexandru Nemnes[a,c]

[a]Horia Hulubei National Institute for Physics and Nuclear Engineering, Magurele, Ilfov, Romania, [b]Reykjavik University, Icelandic School of Energy, Reykjavik, Iceland, [c]Faculty of Physics, MDEO Research Center, Research Institute of the University of Bucharest (ICUB), University of Bucharest, Magurele, Ilfov, Romania, [d]Reykjavik University, Department of Engineering, Reykjavik, Iceland, [e]National Institute of Materials Physics, Magurele, Ilfov, Romania

1. Band gap and other electronic properties

Halide perovskites, with formula ABX_3, where A and B are organic or inorganic metallic cations and X is a halide anion, have recently proven to be of high technological interest, mostly for solar-cell applications. Their large chemical diversity and peculiar electronic, optoelectronic, structural, and dynamical characteristics require sound theoretical understanding and modeling procedures in order to efficiently guide experimental investigations throughout this vast family of materials. Several recent efforts to structure the theoretical results in this field have been put forth [1–6]. One of the most pressing issues in simulating perovskite materials, especially those that incorporate heavy elements, is finding a suitable theoretical framework that is able to closely approximate the experimental results while also being computationally efficient. Furthermore, there are several issues that are still under debate, as will be indicated [3].

Usual DFT approaches using the exchange-correlation (XC) functional based on local-density approximations (LDA) or generalized gradient approximations (GGA) result in the underestimation (LDA) or overestimation

Low-Dimensional Halide Perovskites
https://doi.org/10.1016/B978-0-323-88522-5.00012-0

(GGA) of lattice parameters in static calculations, mostly because they do not include van der Waals (vdW) and hydrogen bonding [5,7]. In order to improve the binding estimate in perovskite compounds, a solution is to mix LDA and GGA (in order to compensate the under/overestimates) [5] or to include dispersive interactions (DFT-D2) [7]. In a comparative study of different approximation levels in DFT, Beck et al. [7] have also indicated that hybrid functionals, like Heyd-Scuseria-Ernzerhof (HSE), show good structural estimates, but at the cost of being more computationally expensive. This level of accuracy is considered well suited for lead-based compounds, and even relativistic effects, such as the spin-orbit coupling (SOC), can be safely ignored when only estimating their structural properties [4]. On the other hand, the accurate treatment of defects and their different equilibrium configurations require the use of hybrid functionals with SOC and self-interaction correction [8].

The correct treatment of the band gap, conduction, and valence band edges also needs to be treated with care, especially for lead-based perovskites. For example, semilocal functionals (such as Perdew-Burke-Ernzerhof—PBE) can appear to accurately estimate the band gap when compared to experimental values. This is just a fortuitous canceling out the GGA underestimate with the overestimate caused by the lack of SOC [1,5,6]. The problem of correcting the XC term was treated in a recent benchmark study by Hernández-Haro et al. by comparing 24 different local, gradient-corrected, metagradient-corrected, and nonlocal XC functionals, with SOC and dispersion corrections [9]. They suggest using RPBE (gradient-corrected XC) as a good compromise between accuracy and computational performance.

Tao et al. [10] and Valadares et al. [11] have also suggested that an approximate quasiparticle method with corrected self-interaction error (DFT-1/2) can reach the accuracy of the much more demanding Green's function method with screened Coulomb interaction (GW) for several ABX_3 Pb- and Sn-based perovskites. Tao et al. also reported a significant improvement of the lattice parameters when including vdW corrections and indicated some general trends of the band gap values: a positive correlation with the electronegativity of the B ion and a negative correlation with the symmetry of the crystal. There are also reports of a near-GW accuracy in simulating different lead halide compounds with a modified Becke-Johnson GGA (mBJ-GGA) XC functional [12,13] which is known to be a good choice for band gap determination in crystalline solids [14].

Leppert et al. [15] have also found that even some popular GW implementations, namely one-shot $G_0 W_0$ and eigenvalue self-consistent $G W_0$, inherit the nonsystematic errors of their semilocal DFT starting points (GGA) and are not suitable for halide perovskites. They conclude that band gap values computed with $G W_0$ based on a hybrid functional starting point are in good agreement with experimental values and that such an approach is suitable for materials with large SOC and dispersive bands. Filip et al. [2] described a correction scheme for band gap and effective mass estimations in the GW approximation.

Because of their compositional flexibility, halide perovskites also lend themselves to band gap tuning, which is especially important in optoelectronic applications. Berger [1] indicated that substitutions on the B and X sites have the most significant impact on the band gap, but unit cell reduction and quantum confinement can also be used toward the same effect. It was suggested [16] that the band gap value can be adjusted by applying pressure on the order of tens of GPa, although this would be challenging from an engineering standpoint and also because of the well-known phase transition that is characteristic of halide perovskites [16–18]. A study of layered hybrid organic-inorganic cells [19] indicates that the band gap can, alternatively, be adjusted by the choice of the organic spacer.

One of the main efforts of recent years has been the transition to environmentally friendly lead-free perovskites [20,21] while still preserving their record efficiency. Lead-free halide perovskites also require a high level of simulation accuracy since correlated d-valence electrons are ineffectively represented by standard LDA or GGA. In a study of Cs2NaMCl6 (M=Ti, V), Mir et al. [22] pointed out the need to include SOC and also mBJ and Hubbard potentials in order to have a good enough estimate of the exchange-correlation energy in these compounds. Zhao et al. [23] used a hybrid functional (HSE+SOC) approach in the study of quaternary halides based on cation transmutation (replacing two divalent Pb^{2+} ions with one monovalent and one trivalent ion). The proposed quaternary halides show good phase stability and a wide range of tunable band gaps. A similar method (HSE0+SOC) is used by Liu et al. [24,25] for the study of Ge-based perovskites, indicating $CsGe_{2/3}Si_{1/3}I_3$, with a band gap of 1.34eV, as a potential lead-free candidate.

The discovery of new perovskite materials has also been of high interest recently. DFT simulations can be used for screening out perovskites by their stability and band gap values [26], but these methods are still very computationally expensive. More efficient high-throughput screening can be performed by combining DFT simulations with machine learning (ML). Yılmaz and Yıldırım [27] have compiled a comprehensive list of recent studies that use ML algorithms and DFT-generated data. Jain et al. [28] used DFT results as training data for a support vector regression algorithm, which, in turn, is used to predict the formability of 454 inorganic halides. The screened candidates are compared to results from known databases and subjected to further DFT investigations in order to select two promising compounds for photovoltaic applications: $RbSnCl_3$ and $RbSnBr_3$.

In a recent study, Li et al. [24] used five different ML algorithms (Bayesian bridge regression, linear regression, elastic nets, support vector regression, and gradient boosting regression) to predict the transmission function for $MaPbI_3$ and 48 different metal electrodes. The ML algorithms used 13 easily available descriptors and their predictions were compared to values obtained by a DFT nonequilibrium Green's function method. They show that gradient boosted regression has the best performance ($R^2 = 0.98$) and can be used in fast high-throughput screening even for complex transport phenomena. In another interesting ML study, Park et al. [29] used DFT-generated data and a random forest regression (RFR) algorithm to predict the octahedral distortion parameters and band gap of ABC_3 chalcogenide-halide and hybrid perovskites. From the RFR algorithm, they are also able to extract the importance of each feature used in the prediction and show their correlation with each predicted parameter as described in Fig. 1.

Schmidt et al. [30] performed a large benchmark study on several machine learning methods (ridge regression, random forests, extremely randomized trees, and neural networks) trained with up to 119 features on 20.000 systems and tested on 250.000 cubic perovskite systems. Extremely randomized trees give the best performance when predicting the distance to the convex hull of stability and could be used to restrict the search space and speed-up high-throughput DFT simulations. In order to show the flexibility of ML methods, we also mentioned the fact that they were used to guide or accelerate several experimental studies focused on

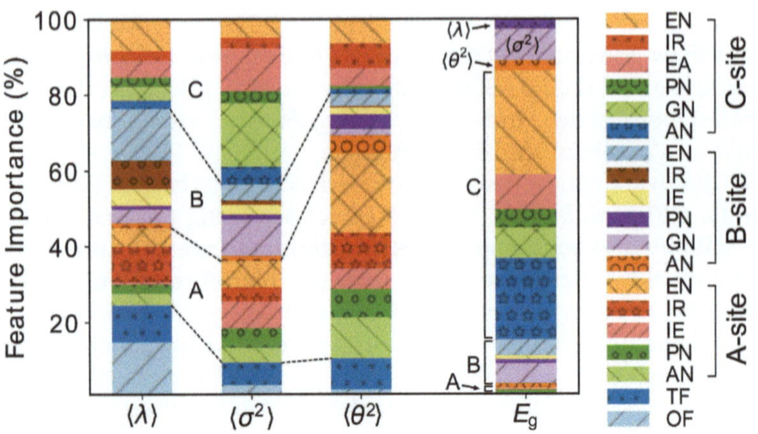

FIG. 1 Feature importance and correlation with predicted parameter. *Reproduced from Park, H., Mall, R., Alharbi, F. H., Sanvito, S., Tabet, N., Bensmail, H., & El-Mellouhi, F. (2019). Exploring new approaches towards the formability of mixed-ion perovskites by DFT and machine learning. Physical Chemistry Chemical Physics, 21(3), 1078–1088. https://doi.org/10.1039/c8cp06528d by permission of the PCCP Owner Societies*

finding capping layers that suppress perovskite degradation [31], guiding robotic systems that synthesize perovskite single crystals [32], and optimizing the composition of mixed cation perovskite thin films [33].

2. Ferroelectricity of hybrid perovskites

Using ferroelectric materials in solar cell applications is a rather old and appealing idea [34]. The built-in electric field naturally occurring inside the ferroelectric material is supposed to provide the conditions for separating electrons and holes. However, the practical realization of a ferroelectric-based solar cell remained elusive due to the complications related to the presence of contact electrodes as well as finding a good ferroelectric absorber.

The appearance of the $CH_3NH_3PbI_3$ (MAPI) perovskite as a photoactive material in the context of unusually high power conversion efficiencies (PCEs) refueled these ideas and it was regarded as a potential ferroelectric absorber. The very high mobility of the free carriers with mean free paths larger than 1 μm was, at the time, regarded as potential evidence for the ferroelectric character. The electric dipole of methylammonium (MA) molecules has been considered a potential source of ferroelectricity in hybrid perovskites and experiments were designed to measure the local polarization. Using piezoforce microscopy (PFM), ferroelectric domains of ~100 nm were identified, which are comparable to the grain size in the MAPI layer [35].

The ferroelectric contributions leading to anomalous hysteretic effects were investigated by Frost et al. using a combined DFT and MD approach [36]. The model implies that electron-hole recombination is influenced by the local ferroelectric domains, which impacts the dynamic J-V characteristics. Using DFT and symmetry mode analysis, Stroppa et al. investigated the ferroelectric polarization of MAPI and determined a small ferroelectric polarization in the tetragonal phase of $\sim 4.42\,\mu C/cm^2$ [37]. The authors also proposed routes to enhance the polarization. Using first-principles calculations, Fan et al. also found a rather small ferroelectric polarization of $\sim 8\,\mu C/cm^2$ in the MAPI perovskite [38].

However, a detailed study concerning a possible correlated motion of electric dipoles in MAPI performed by Goehry et al. showed that the typical correlation time is of the order of tens of picoseconds, which excludes the hypothesis of macroscopic ferroelectricity [39]. In fact, the rotation of the organic cations is correlated with the deformation of the Pb-I cages and this may lead to microscopic ferroelectric domains, as it is suggested in Fig. 2. These, however, may have a significant impact on the carrier recombination: even if the ferroelectric domains are restricted to very narrow time and space domains, they may induce an essential separation of the electron and hole pathways, reducing the charge recombination.

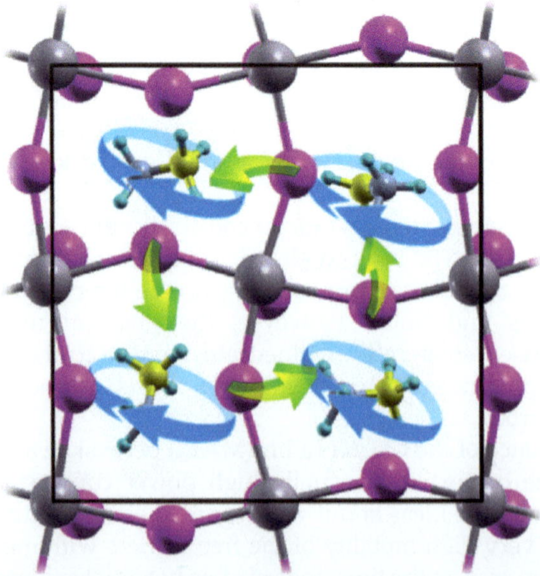

FIG. 2 The motions of MA cations are correlated with the tilting of the Pb-I octahedra, leading to the appearance of microscopic ferroelectric domains. *Reprinted with permission from Goehry, C., Nemnes, G. A., & Manolescu, A. (2015). Collective Behavior of Molecular Dipoles in CH3NH3PbI3. J Phys Chem C, 119(34), 19674–19680. https://doi.org/10.1021/acs.jpcc.5b05823. Copyright (2015) American Chemical Society*

The microscopic ferroelectric polarization may explain the high performance of the PSCs [40], as suggested by Frost et al. Thus, a microstructure with randomly oriented ferroelectric domains can render channels for efficient charge transport, which decrease the electron-hole recombination, as depicted in Fig. 3. This picture was also revealed by drift-diffusion equations coupled to a mesoscale polarization model [41], which may explain the relatively large fill factors in the J-V characteristics.

Furthermore, other first principles studies revealed ferroelectric effects, e.g., in lead-free $CsSnI_3$ films as a function of epitaxial strain [42]. Here it was established that both compressive and tensile strain induces the polar phases *P4bm* and *Pmc21-(I)*. Also, a direct-to-indirect band gap transition was observed in the case of *P4bm*, while the *Pmc21-(I)* has a direct band gap, making it suitable as light-harvesting material.

The importance of ferroelectric ordered domains on $MAPbX_3$ solar cell performance was demonstrated based on simulations of drift-diffusion equations coupled with piezoresponse force micrographs [43]. More recently, ferroelectricity was identified in perovskite-related structures like $[TMAEA]Pb_2Cl_6$ [44].

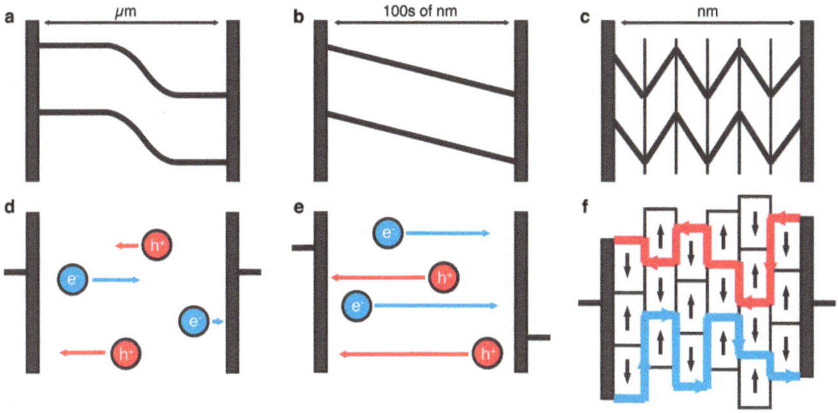

FIG. 3 Suggested pathways for electrons and holes in the presence of microscopic ferroelectric domains. *Reprinted with permission from Frost, J. M., Butler, K. T., Brivio, F., Hendon, C. H., Van Schilfgaarde, M., & Walsh, A. (2014). Atomistic origins of high-performance in hybrid halide perovskite solar cells. Nano Lett, 14(5), 2584–2590. https://doi.org/10.1021/nl500390f. Copyright (2014) American Chemical Society.*

3. Point defects

A better theoretical understanding of native defects in perovskite materials is still an open issue because of the challenges posed by their complex structure. There are several recent reviews that cover the theoretical and computational state of the art of the field [8,25,45]. Although a large number of native point defects and defect complexes have been studied with DFT methods, many show conflicting results and different estimates for the defect levels [46]. It was suggested that, when estimating defect formation energies with DFT methods, it is critical to take into account self-interaction error and use more precise hybrid functionals (like HSE06 or PBE0) with SOC in order to get an accurate picture [45,47]. The actual intrinsic soft nature of metal halide perovskite gives rise to complex dynamic effects, such as ion diffusion and a dynamic density of trap states, that are difficult to simulate efficiently. Although displaying intrinsic structural disorder and defects, which are usually detrimental to electronic and optoelectronic materials, lead halide perovskites are highly defect-tolerant and show surprisingly long lifetimes of the photogenerated carriers and high conversion efficiencies for solar cells and light-emitting devices [8].

Defects also strongly influence ion diffusion, which leads to problems particular to perovskite materials, like the current-voltage hysteresis, moisture [48], and light exposure-related degradation [6,49,50]. Ion migration, more precisely methylammonium and Pb vacancies, is also indicated as the cause for slow electrical material response under light irradiation [51]. Supported by experimental results that indicate exceptionally long charge carrier lifetimes, theoretical investigations also point to the fact that defects in general [25], and grain boundaries in particular [5], could be classified as electrically benign for perovskites. Most stable types of defects usually appear outside the band gap or very close to the band edges [1] and only give rise to short trapping times at room temperature [47].

More detailed investigations on $MAPbI_3$ have shown that grain boundaries are important since they enhance electron-phonon coupling and reduce the band gap, but can be mitigated by chlorine doping [52]. Surfaces can also lead to a number of point defects such as vacancies, interstitials, and antisites which can lead to carrier trapping, in which case Pb-rich conditions are preferable to I-rich [53]. The investigation of trap states and nonradiative charge recombination processes can be efficiently performed with nonadiabatic MD in the framework of time-domain density functional theory [25] which simulates the dynamics of charge carriers coupled to lattice vibrations and effectively mimics time-resolved experiments. In a study by English [54], hybrid DFT and Born-Oppenheimer MD were used to compute the vibrational density of states and electronic structure of $CH_3NH_3PbX_3$ (where X is Cl, Br, or I) and TiO_2 interfaces. They find that acoustic phonon modes lead to coupling between Pb and halide atoms and between Ti and O atoms, which indicates possible modulation of the dynamical properties of the perovskite by host-substrate vibrations.

Employing first principle calculations performed using the VASP code in combination with the standard frozen-core projector augmented wave (PAW) method, Yin et al. [55] conducted electronic structure calculations on tetragonal MAPI. They studied also the point defects properties, using the formation energies of twelve neutral defects under three different conditions: (Pb-poor and I-rich), (moderate), and (Pb-rich and I-poor), as described in Fig. 4. The following point defects were investigated: vacancies (V_{MA}, V_{Pb}, V_I), cation substitutions (MA_{Pb}, Pb_{MA}), interstitial (MA_i, Pb_i, I_i), and antisite substitutions (MA_I, Pb_I, I_{MA}, I_{Pb}). It was revealed that the dominant contributions to the band structure come from the Pb vacancies (V_{Pb}) and MA interstitials (MA_i), showing the changing character of the electrical conductivity. By representing the formation energy as a function of Fermi level position, it was concluded that halide perovskites like MAPI exhibits unusual defect physics, being quite different from common p-type thin-film solar cell absorbers. Depending on the growth mechanism (Pb-rich or I-rich), it can offer flexible conductivity, changing from good p-type to good n-type, which may be the key to realize thin-film solar cells with high efficiency.

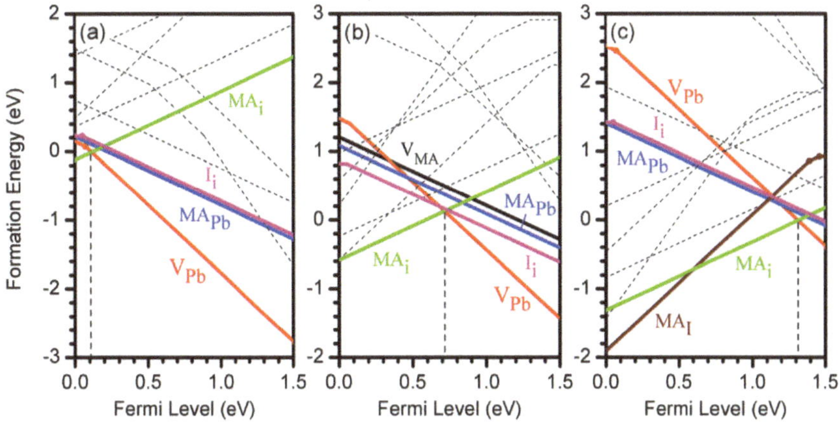

FIG. 4 The defect formation energies as a function of Fermi level position for different chemical potentials: (A) Pb-poor and I-rich, (B) moderate, and (C) Pb-rich and I-poor. Displayed with dashed lines are the defects with much higher formation energies. *Reproduced From Yin, W. J., Shi, T., & Yan, Y. (2014). Unusual defect physics in CH3NH3PbI3 perovskite solar cell absorber. Appl. Phys. Lett. 104(6). https://doi.org/10.1063/1.4864778, with the permission of AIP publishing.*

4. 2D-3D perovskites

More recently, many computational works are based on layered 2D-3D organohalide perovskites, since this family of perovskite materials offers improved features when compared with 3D perovskites. One main advantage is the possibility for tuning the optical [56] and electrical [57,58] properties, by replacing the organic or inorganic components. Also, when reducing the dimensionality, from 3D to 2D, the layered halide perovskites are expected to have a more stable [59] structure and an increased exciton binding energy. Moreover, these layered systems are regarded as natural multiple quantum wells where the insulating organic layers act as potential "barriers" and the semiconducting inorganic layers act as potential "wells" [60].

The most commonly used 2D halide perovskites that have emerged by cutting layers along the crystallographic plane (100) [61] are classified into two main categories: Ruddlesden-Popper (RP) [56,58] and Dion-Jacobson (DJ) [62] perovskites. Having pairs of interdigitated interlayer spacers between the two offset layers, Ruddlesden-Popper structures are dominating the halide perovskite applications.

In Stoumpos et al. [63], Ruddlesden-Popper halide perovskites with variable numbers of layers are investigated, having the chemical formula $(BA)_2(MA)_{n-1}Pb_nI_{3n+1}$ where n takes the values 1, 3, and 4 and BA stands for $CH_3(CH_2)_3NH_3$ (butylammonium). The DFT simulations were performed by Vienna Ab initio Simulation Package using the projector augmented wave (PAW) and potentials within the PBEsol exchange-correlation functional. Although underestimated by DFT at the PBEsol level, the band gap simulations prove one of the most important aspects of 2D halide perovskites, namely the band gap tunability when reducing the dimensionality of the layered perovskites, as one can see from Fig. 5.

Band gap simulations confirm that all structures are semiconductor types, having a clear direct band gap value ranging from 1.99 eV for $n=1$ to 0.96 eV for $n=4$. The large bandwidths and small effective masses for both electrons and holes show that they can easily play an important role in achieving tunable electronic properties. As reported by Mao et al. [64], the change in the band gap could be a result of the indirect effect of the organic cation on the inorganic framework. Therefore, the intercalation of different cations between perovskite layers can improve the material stability [65], but also can be a powerful tool for band gap engineering [66].

In a recent study, Mao et al. [62] used DFT implemented by SIESTA package with the Perdew-Burke-Ernzerhof (PBE) functional and generalized gradient approximation (GGA) to calculate the electronic properties of a 7-layered DJ phase perovskite in comparison with an RP phase perovskite with the same number of layers [67]. The DJ phase was formed from the $(4AMP)MA_6Pb_7I_{22}$, where 4AMP is 4-aminomethylpyridinium and

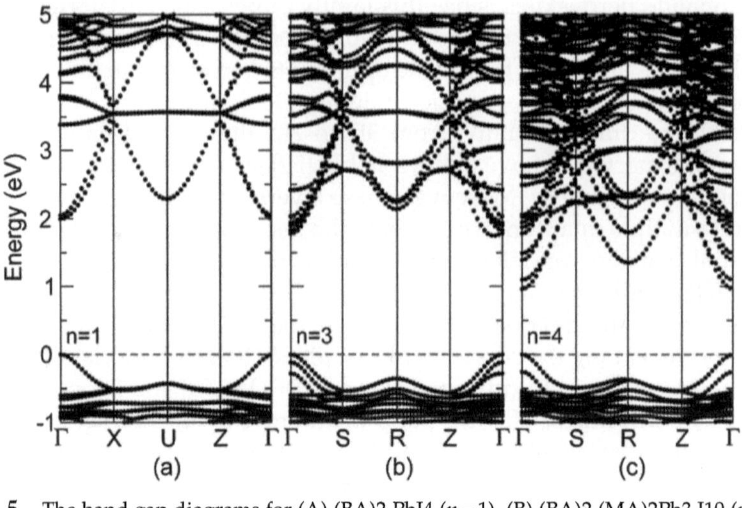

FIG. 5 The band gap diagrams for (A) (BA)2 PbI4 ($n=1$), (B) (BA)2 (MA)2Pb3 I10 ($n=3$), and (C) (BA)2 (MA)3Pb4I13 ($n=4$). Fermi level is set to zero (*red* line). *Reprinted with permission from Stoumpos, C. C., Cao, D. H., Clark, D. J., Young, J., Rondinelli, J. M., Jang, J. I., Hupp, J. T., & Kanatzidis, M. G. (2016). Ruddlesden-Popper Hybrid Lead Iodide Perovskite 2D Homologous Semiconductors. Chemistry of Materials, 28(8), 2852–2867. https://doi.org/10.1021/acs.chemmater.6b00847. Copyright (2016) American Chemical Society.*

the RP phase, formed from $(BA)_2MA_6Pb_7I_{22}$ (BA = butylammonium). They concluded that in the DJ phase, separate channels are formed in the structure, which corresponds to different pathways for electrons and holes.

5. Perovskite/ETL and perovskite/HTL interfaces

In solar cell applications, the perovskite material is usually sandwiched between two buffer layers, known as the electron transporter layer (ETL) and hole transporter layer (HTL). In standard (inverted) solar cell configuration, the ETL (HTL) layers should be transparent so that the sunlight reaches the perovskite layer, where the generation of free electrons and holes predominantly occurs. Subsequently, aided by a proper sequence of band offsets, the electrons and holes are separated at the interfaces between perovskite and ETL and HTL. The requirements for a material to become a good ETL of HTL include high carrier mobility, low production costs, absence of toxicity or encapsulation, chemical stability, and lack of trap states [68]. Another necessary condition is the alignment of the energy levels.

Both the ETL and HTL operate as filters for photogenerated charge carriers. For instance, the ETL prevents the flow of photogenerated holes by an energy barrier at the interface and, at the same time, it favors the electron flow. In order for this to happen, the ETL conduction band minimum and the valence band maximum energies, respectively, have to be equal or less than the ones of the perovskite. This type of alignment is considered to be Type II [68,69]. The energy barrier at the HTL/perovskite interface is set to block the photoexcited electron flow while favoring the hole flow, while the opposite is found for holes at the ETL/perovskite interface. A proper selection of semiconductors as ETLs and HTLs can further increase the power conversion efficiency and increase the stability of the organohalide perovskites. TiO_2 and ZnO can be used as ETL, while Cu_2O, CuSCN, NiO, CuI, spiro-OMeTAD [70], and spiro-MeOTAD [71] as HTL. However, even though using spiro-MeOTAD higher efficiency devices were fabricated, it also promoted the degradation of the device [72].

5.1 Titanium dioxide

The crystal phase of TiO_2 [73] has a significant impact on the electron collection. On the other hand, the chlorine substitution of iodine may enhance the performance of the PSC [74]. A reduced chlorine inclusion in the $CH_3NH_3PbI_{3-x}Cl_3$ was shown by theoretical and experimental investigations, which lends the idea that the chlorine species are located mostly in the vicinity of the TiO_2-perovskite interface. In Nemnes et al. [75], the band structure and electron collection at the rutile-TiO_2/halide perovskite $CH_3NH_3PbI_{3-x}Cl_3$ interface were investigated (Fig. 6).

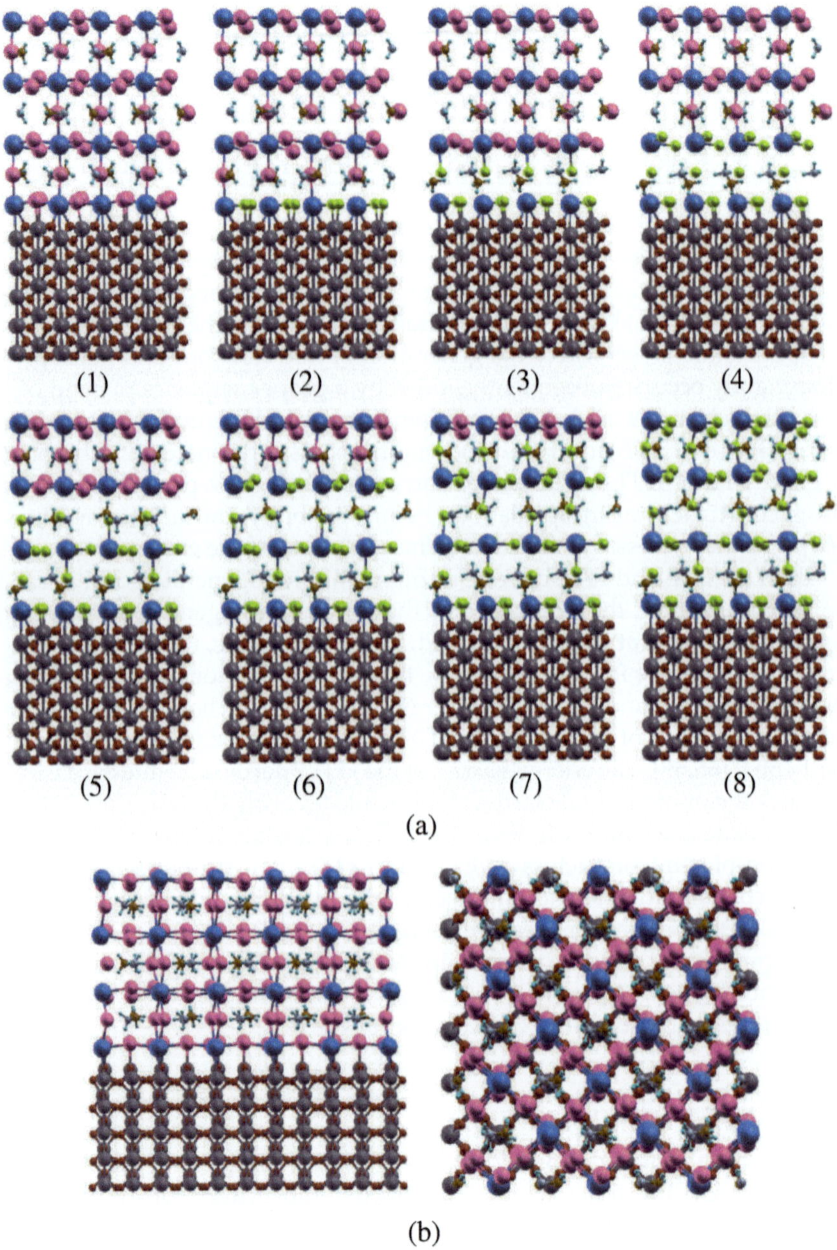

FIG. 6 (A) TiO2 (*bottom layer*) and MAPbI3 (*top layer*) heterostructures with increasing concentration of Cl doping. (B) The 45o rotation and top view of the interface. *Reproduced from Nemnes, G. A., Goehry, C., Mitran, T. L., Nicolaev, A., Ion, L., Antohe, S., Plugaru, N., & Manolescu, A. (2015). Band alignment and charge transfer in rutile-TiO2/CH3NH3PbI3-xClx interfaces. Physical Chemistry Chemical Physics, 17(45), 30417–30423. https://doi.org/10.1039/c5cp05466d by permission of the PCCP Owner Societies.*

The presence of chlorine in the vicinity of the junction highly affects the conduction and valence band offsets and also increases the perovskite band gap [75]. The increased band gap introduces a larger conduction band offset and enhances the electron transfer in the ETL layer. However, with the increase of the chlorine concentration in the compound, the valence band offset reduces, leading to lowered capacity of the rutile-TiO_2 layer as the hole blocking material. This allows one to conclude that there exists an optimal chlorine concentration for the mixed halogen perovskite.

The charge transfer analysis via the displacement charge technique demonstrated that the chlorine systems exhibit more electron charge in the ETL near the interface. Moreover, it was determined that the charge transfer saturated with the increase of the chlorine concentration. The system consisting of the first layer with complete Cl substitution and second layer 50% substitutions of the chlorine on iodine was found to exhibit the optimal performance.

Anatase phase of TiO_2 is often found as ETL. In Long and Prezhdo [76], the TiO_2(001)/perovskite(100) interface was investigated using ab initio nonadiabatic (NA) MD for both pristine and doped material. Iodine was replaced with chlorine or bromine and lead with tin. $CH_3NH_3PbI_3$ ($MAPbI_3$) is an efficient light-absorbing material and is also very cost-effective. Its band gap favors light absorption from visible to near-infrared light spectrum. Its charge diffusion lengths are $\sim$100 nm, larger than in other materials [77]. Moreover, a way to enhance the electron and hole diffusion length is achieved by doping. Diffusion lengths higher than 1 μm can be reached by chlorine doping of the perovskite. Moreover, Cl and Br doping can increase the photo conversion efficiency [76,78]. In contrast, tin-doped $MAPbI_3$ has a lower conversion efficiency; however, it increases the range of the absorption solar spectrum [79].

Long and Prezhdo [76] have reached the conclusion that doping with chlorine or bromide significantly reduces the recombination rate while doping with tin does the contrary. The findings are surprising because all of the dopants are considerably lighter than the substituted atoms. The lighter atoms were anticipated to accelerate the relaxation by rendering higher vibration frequencies. Meanwhile, it should be noted that tin dopants increase the bonding between the perovskite and the anatase TiO_2 interface, while chlorine and bromide reduce the bonding. Tin atoms strengthen the donor-acceptor coupling, while chlorine and bromide weaken it. Furthermore, tin dopants do not influence the quantum decoherence times significantly, while chlorine and bromide dopants considerably reduce the quantum coherence times.

Also, the behavior of the doped systems with the undoped material at room temperature conditions was compared. Additional distortion is present in the doped systems in comparison with the pristine material. When the perovskite is doped with chlorine or bromide, the number of the I-Ti bonds reduces. Therefore, it decreases the interaction between donor and acceptor species. On the other hand, tin doping not only keeps the I-Ti bonds but also promotes new bonding, for example, TiO_2-$CH_3NH_3^+$, which, as a consequence, leads to increased donor-acceptor coupling.

5.2 Zinc oxide

Another candidate for the ETL is zinc oxide (ZnO). It has a similar band gap to TiO_2, lower processing temperature and higher electron mobility. In addition, the open circuit voltage and power conversion efficiency can be further improved by interfacial modification with AZO (Al-doped ZnO) [80]. Improved carrier transport and diminished charge recombination at the interface can also be achieved by the Al doping of ZnO [81].

Interfaces made of $MAPbI_3$ and nanocolumnar ZnO and AZO were studied by ab initio density functional theory (DFT) calculations in Nicolaev et al. [82]. Spikes that block the electron transfer into the oxide layer were observed in the conduction and valence band edges using the partial density of states (PDOS) analysis at the ZnO/perovskite interface (Fig. 7). In order to eliminate the spikes, it was suggested that the Al doping of the ZnO at the vicinity of the interface is beneficial. In this way, the band gap of the AZO layer is being shifted to the lower energies relative to the $MAPbI_3$ band gap. Moreover, the required band offset can be obtained by different Al doping concentrations.

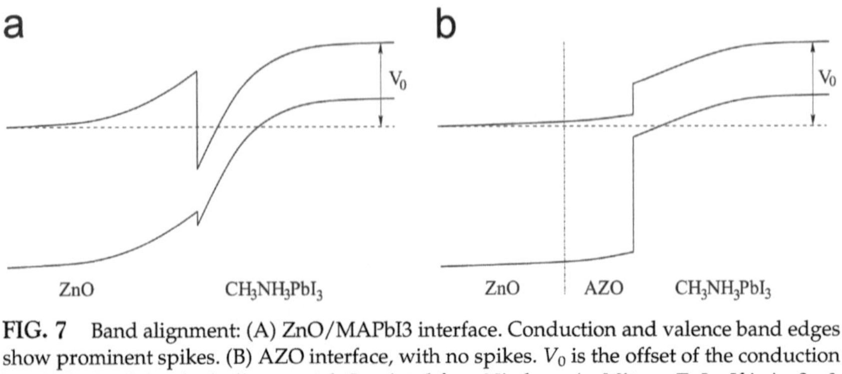

FIG. 7 Band alignment: (A) ZnO/MAPbI3 interface. Conduction and valence band edges show prominent spikes. (B) AZO interface, with no spikes. V_0 is the offset of the conduction band obtained for the bulk material. *Reprinted from Nicolaev, A., Mitran, T. L., Iftimie, S., & Nemnes, G. A. (2016). Optimization of halide perovskite solar cells based on nanocolumnar ZnO. Sol. Energy Mater. Sol. Cells, 158, 202–208. https://doi.org/10.1016/j.solmat.2015.10.023 with permission from Elsevier.*

5.3 Copper oxides

Experimental results show that CuO, Cu_2O, or CuO_x materials can be used as HTLs [83–88], the maximum efficiency obtained reaching 17%. The analysis of $Cu_2O/MAPbI_3$ interfaces parallel to the (001) plane was presented in Castellanos-Águila et al. [68]. The authors have investigated four atomic-scaled models of the $Cu_2O/MaPbI_3$ interface: O/PbI, O/MAI, Cu/PbI and Cu/MAI (Fig. 8). Band alignment, local density of states, projected density of states, energy of adhesion, interface states, and charge transfer have been here analyzed.

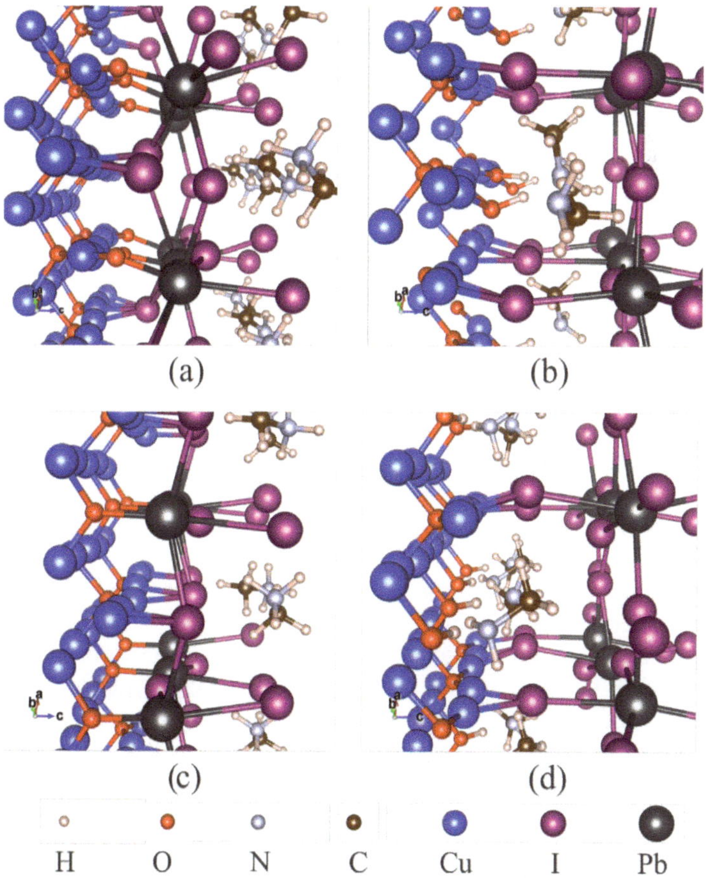

FIG. 8 The optimized interfaces obtained after the relaxation procedure: (A) O/PbI, (B) O/MAI, (C) CuPbI, and (D) Cu/MAI. *Reprinted (adapted) with permission From Castellanos-Águila, J. E., Lodeiro, L., Menéndez-Proupin, E., Montero-Alejo, A. L., Palacios, P., Conesa, J. C., & Wahnón, P. (2020). Atomic-Scale Model and Electronic Structure of Cu2O/CH3NH3PbI3Interfaces in Perovskite Solar Cells. ACS Applied Materials and Interfaces, 12(40), 44648–44657. https://doi.org/10.1021/acsami.0c11187. Copyright 2020 American Chemical Society.*

The band alignment obtained during the density functional theory calculations suggests that Cu_2O is advantageous for photovoltaic applications as HTL. Simulations showed that O/PbI and Cu/PbI type interfaces present the best choice for solar cells. Nevertheless, for both terminations interface states are present, with energies similar to the conduction band minimum, which have the potential of becoming recombination centers. The authors concluded that an alteration of the interface composition or the addition of a buffer layer might be necessary to boost the efficiency of the solar cells.

5.4 Nickel oxides

NiO_x is also considered a suitable HTL material due to its energy level matching with $MAPbI_3$ and stability [89–92]. The best power conversion efficiency with the $NiO_x/MAPbI_3$ heterojunction can reach up to 20% [93]. However, nickel oxide lacks the necessary conductivity. To solve this problem, the surface modification has been proposed, as well as doping with other elements and pretreatment of NiO_x. Doping with Cu [94,95], Mg, Li [96], and Cs [97] were implemented.

NiO_x doped with silver was investigated as a promising HTL in the study [98] both theoretically by the means of DFT and experimentally. The samples were prepared by the one-step spin-coating method. The calculations showed that Ag is a p-type dopant in NiO_x and usually occupied the substitutional Ni site. Moreover, doping with silver increased electrical conductivity, hole mobility, and optical transmittance. From the experimental perspective, the crystallinity improved and the surface coverage was revealed. Also, reduced hysteresis and increased efficiency were observed.

Another way to enhance the hole-transport, the stability of the device and the reduction of the number of surface defects is the modification of the surface of NiO_x. Appropriate surface modifiers for nickel oxide were found to be alkali metal salts, such as KCl [99], and small organic molecules, such as pyridine-terminated conjugated small molecules [100]. In Xie et al. [101], the performance of the adenine-coated NiO_x was investigated as the HTL for a perovskite interface. Vienna Ab initio Simulation Package (VASP) was used for these calculations. The NiO (111) surface was obtained by relaxing bulk NiO, while $MAPbI_3$ was used in the cubic configuration. Adenine coating was taken into account by the insertion of only one layer of $C_5N_5H_3$ between the nickel oxide and the perovskite. The calculations showed that adenine aids in holding the perovskite in its crystalline structure. Also, improved charge transport was obtained. A drop of 0.1 eV of the highest occupied molecular orbital was observed using the adenine coating.

5.5 Graphene

Moreover, graphene-based materials have promising applications in perovskite-based solar cells as both ETL and HTL [102] because of their extraordinary electronic and optical properties [103–106]. Graphene nanoflakes have been used together with TiO_2 as ETL and, due to the addition of graphene, the efficiency of the solar cell was increased from 10% to 15.6% [107]. Moreover, reduced graphene oxide (rGO) can substitute spiro-OMeTAD as an HTL, at the same time inhibiting the recombination of photogenerated charges and enhancing the efficiency [108].

A systematic computational analysis concerning the interface between the oxidized graphene sheet and $MAPbI_3$ with different oxygen concentrations (Fig. 9) and its influence on the electronic properties of the heterojunction was performed by Zibouche et al. [102]. Energy band alignment suggests that the rGO is more suitable for charge carrier extraction in comparison with the highly or fully oxidized graphene sheets. Smaller valence band offsets in comparison with the conduction ones are formed for heterostructures with oxygen concentrations up to 33% in the rGO. In this way, rGO can be used to extract both holes and electrons. However, in this instance, a blocking layer is needed to select the requested charge carriers. The concentration of electrons becomes predominant for concentrations

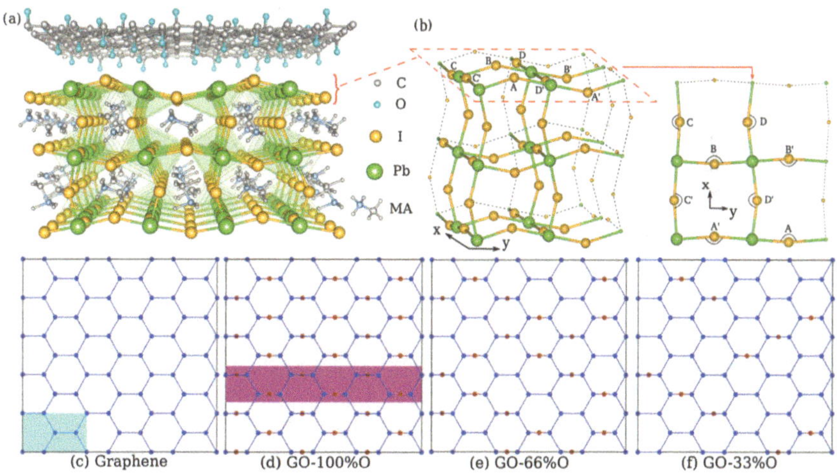

FIG. 9 (A) MAPbI3/graphene oxide heterojunction. (B) Perovskite unit cell excluding the methylammonium organic molecule. Along x-direction are taken angles A, A′, B, and B′ and along y-direction angles C, C′, D, and D′. (C) Graphene supercell, the cyan rectangle represents the unit cell of undoped graphene. (D) Fully oxidized graphene. (E) Graphene oxide with 66% oxygen concentration. (F) Graphene oxide with 33% oxygen concentration. *Reprinted (adapted) with permission from Zibouche, N., Volonakis, G., & Giustino, F. (2018). Graphene Oxide/Perovskite Interfaces for Photovoltaics. J. Phys. Chem. C, 122(29), 16715–16726. https://doi.org/10.1021/acs.jpcc.8b03230. Copyright 2018 American Chemical Society.*

between 33% and 66%, making these devices more suitable as ETL. The concentrations above 66% are not appropriate for charge carrier extraction.

Moreover, the work function and the band offsets were investigated for different oxygen rearrangements in rGO. Even though in the case of rGO the rearrangements matter, it does not really influence the heterojunction performance. The applications of graphene oxide with different oxidation levels are promising on their own as ETL or HTL or as a complement for TiO_2 as ETL.

6. MD studies in hybrid perovskites

Understanding molecular motion, like the rotation of methylammonium or ionic transport inside the perovskite material which leads to material degradation, is of great importance. Shortly after the initial success, it became clear that perovskite materials have stability-related problems. Several degradation factors have been identified, such as humidity, or moisture, temperature variations, exposure to oxygen, and the UV component of light [109]. There were attempts, however with little progress, to enhance its stability by doping [25,110] or other approaches, like replacing the MA molecule with other inorganic cations like Cs and other small molecules [111], or halogen substitutions [112]. MD was used to estimate the implication of these degradation factors on the stability of the perovskite and of the entire solar cell device as well.

The computational methodology can be classified as ab initio methods, when the electronic distribution of the valence states is taken into account with a DFT approach, i.e., quantum mechanically, or as force field methods when effective interaction forces between pairs of atoms are initially calculated and used to integrate in time the classical Newton's second law. The major difference between these methods is the number of atoms that can be realistically treated, much smaller in the first case, but potentially with better accuracy. Ab initio methods can only treat short timescale phenomena, whereas force fields allow the study of much longer timescales but neglect the fine details of the dynamics, which sometimes have a critical role. Depending on the application or effect of interest, one or both of these types of methods have been used.

6.1 Ab initio MD

Ab initio DFT methods can be used to simulate short-scale MD and investigate a large range of processes that take place in halide perovskites. We already mentioned a couple of examples, like the correlation of the methylammonium dipoles [35], or vibrational density of states [54]. Another example is the simulation of fast oxygen diffusion in iodine vacancy, forming superoxide species [113].

Using real-time time-dependent DFT and nonadiabatic MD, Li et al. [114] found that compression increases solar cell efficiency by decreasing the band gap and increasing the carrier lifetime in $MAPbI_3$. Zhou et al. [115] studied the origin of the density of states broadening and found that this effect is reduced in the lower temperature phases (tetragonal, orthorhombic) when compared to the cubic phase. They also traced back the broadening to anharmonic vibrations of the iodine atoms and the vibrational modes of the MA cations. Investigating the effect of anharmonic nuclear motion in $CsPbBr_3$, $CsPbI_3$, and $MAPbI_3$, MD simulations indicate a dynamic shortening of the disorder potential for electrons and holes, narrow band gap distributions, and sharp optical absorption edges [116,117].

McKechnie et al. [118] used ab initio MD, DFT, and GW methods to investigate the disorder-induced spin-splitting and band gap widening that appears in both organic and inorganic halides. Long carrier lifetimes have been attributed to low nonadiabatic coupling and short charge recombination, characteristics which are positively correlated with the excess of PbI2 or MAI-poor perovskites [119]. In a complementary study, Ambrosio et al. [120] attribute long charge carrier lifetimes to the spatial separation of electrons and holes into polaronic states. This is caused by thermal vibrations in the inorganic sublattice and it is only moderately influenced by the orientation disorder of the organic cations. High precision DFT methods can even be used in conjunction with classical MD (CMD) by developing fully transferable force fields. Following such an approach, Balestra et al. [121] used a genetic algorithm to fit force field parameters to DFT energies of a large set of configurations. Then, using CMD simulations, MD simulations of heat transport and ion migration were accessible at larger timescales (ns, as opposed to ps in the case of DFT-MD) and larger system sizes ($6 \times 6 \times 6$).

6.2 Force field methods

Force field MD simulations have been widely used to study the thermodynamic and mechanical properties of solids [122]. In particular, force field MD simulations are of interest when the system size becomes important and an ab initio simulation becomes computationally prohibitive. However, it is known that the precision of the results of an MD simulation depends on the accuracy of the force field. Motivated by its hybrid nature, a combination of different force fields can be utilized to model MAPI [123]. Apparently, Mattoni et al. [124] were the first who proposed a hybrid model for MAPI, which is widely known as MYP. This model separates the interactions taking place in hybrid perovskites between the constituting components as organic-organic (describes by generalized AMBER force fields—GAFF), inorganic-inorganic (modeled by a Buckingham-Coulomb potential), and hybrid inorganic-organic (given as a sum of Buckingham-Coulomb and Lennard-Jones terms). More precisely, the cross interaction between MA and Pb/I can be described by Lennard-Jones and Buckingham force fields along with the Coulomb term. In the original MYP model, Mattoni et al. [124] used Buckingham-Coulomb for Pb/I and C/N interactions, as it gives more numerical stability, and Lennard-Jones was only used for Pb/I and H interactions. However, there exists an MYP model that defines MA and Pb/I interactions only by Lennard-Jones force field [125] which gives less numerical stability for free surfaces and defected lattices.

Using MYP, Mattoni et al. studied the rotational behavior of the MA cation in MAPI. They have been able to explain the experimental observation of the freezing cation in orthorhombic phase, as well as a quasidisordered cation (C-N), angles with respect to the inorganic lattice. They could also demonstrate the temperature dependence of the cation relaxation time that was a source of debate in ab initio and experimental studies.

Since a high concentration of defects (vacancies, interstitials, and grain boundaries) is obtained in solution processing of MAPI, force field MD can be used to study related phenomena at a timescale beyond ab initio MD. As discussed later in the subsection on hysteresis, these defects play an important role in the hysteric J-V behavior of MAPI. Delugas et al. [126] proposed a random walk defect dynamics and utilized the mean-squared displacement (MSD), which is linearly proportional to the diffusion coefficient, and they predicted the diffusion of iodine interstitial. It has been shown that the diffusion of the iodine vacancy is dominating over iodine interstitial in the temperature range of 300 to 450°C. However, the diffusion coefficient of iodine interstitial showed a larger variation with the temperature that might surpass the one corresponding to the iodine vacancy at elevated temperatures.

7. More applications of MD methods

7.1 Effects of water molecules

When water is added to the system, it acts as a catalyst causing the breakdown of the chemical order. By incorporating in situ measurements of the varying relative humidity (RH) at 98% into the system, the absorbance layer and band edge are reduced. This reduction causes the absorption factor to drop by one half [109].

DFT methods can model the perovskite degradation due to humidity exposure [52], or adsorption of water on surfaces, and indicate routes to increase stability. A detailed MD study [127] simulates the actual percolation of water molecules in the $MAPbI_3$ slab, where it forms a hydrated bulk phase by interacting with the Pb sites and releasing I ions and desorbing MA molecules. The same study also indicates almost negligible changes to the band gap, even under 4:1 $MAPbI_3$:water stoichiometry, and similarly negligible volume expansion, which would indicate as to why the presence of water is hard to detect for usual characterization techniques.

NPT and NVT statistical ensembles were orchestrated with periodic boundary conditions to reproduce the density of the water solvent. As well as the Ewald method to account for the long-range electrostatic forces. The simulations show that the water solvent is fully dispersed within the system due to the polarization of the water molecules. The MSD extracted from the simulations showed that the lead atoms diffuse slower than the rest. This was attributed to the solvation of the lead ions by the water molecules, the conclusion being that the degradation is likely from a chemical reaction and not due to the polarity [125]. IR spectroscopy showed that the water molecules interfere with the bonding of MA and I-molecules and join to form NH_3 and OH. This reaction causes the NH molecules to decrease in stretching vibration [128]. Overall, the presence of water and humidity within the system causes unfavorable bond reactions and results in the degradation of the cell.

7.2 Temperature effects

The maximum efficiency of perovskite solar cells occurs at room temperature and temperatures over 423 K lead to enhanced degradation. The structural changes of the perovskite material with varying temperatures were observed while incorporating in situ X-ray diffraction (XRD) [129]. The XRD measurement showed peak shifts when exposed to temperatures above 373 K. This can be due to the modifications of lattice spacing, meaning the structure of the perovskite material changed from tetragonal to cubic structure, believed to happen at 328 K.

Actually, one of the most interesting facts about the perovskite material is that it is able to change its overall structure when exposed to different temperatures. At the ground state, the MAPI structure is found to be in the orthorhombic phase. When in this phase, the material is set to stay stable with confirmation from DFT calculations [130]. This structure is in supercell expansion of the cubic perovskite lattice.

The effects of temperature on the perovskite system depend deeply on the overall trajectories of the molecules when in a relaxed state. In order to understand this reaction, first principle calculations were used to construct the MYP force field. MYP was used to model the angular trajectories as well as the time-correlation functions for each molecule within the system. Through the use of MYP with set parameters based on Lennard-Jones and Buckingham terms, it is possible to conceptually understand the trajectories of the molecules as they undergo phase transitions. It is also important to note that DFT calculations were also used in this simulation to ensure an agreement between the two calculations. A zero temperature system was used as a baseline for the effects of temperature. The stability of the system was achieved between 100 K and 150 K where the orthorhombic structure was obtained [124]. There is a phase change from the orthorhombic structure to the tetragonal structure when 150 K is passed. This phase transition is due to the vibration and reorientation of the molecules. At temperatures above 180 K, the degradation and disorganization of the material increase. This study concludes that the molecules below 180 K do not disrupt the lattice.

7.3 Ionic transport

The perovskite material has low activation energies for the migration of halide-ion vacancies, and consequently a high ionic conductivity. The migration and conductivity are the main cause of the degradation. Many of the studies regarding this ionic mobility relate it to the hysteresis of the J-V curves. To understand the defect migration, first principle techniques with kinetic experiments were used to monitor the relaxation of the system.

In order to set realistic parameters for the perovskite system, classical MD along with DFT was utilized to simulate the ionic transport and effects on the lattice. The force field to be used in perovskite simulations was defined through a combination of Lennard-Jones, Buckingham, and Coulombic forces [121]. This includes the use of short- and long-range forces that affect the overall migration of the atoms within the system. The change in geometry due to the migration was simulated using force field adjustment algorithm (FLAMA), and the energy of each geometry was calculated. An NPTPR ensemble was incorporated to find the equilibrium density and temperature effects on the geometry, which was outlined earlier. Once all parameters were set, the MSD was exported to locate the position of the atoms in real time.

Eames et al. [131] considered three mechanisms to describe the ionic mobility in perovskite solar cells: iodide migration on the octahedron edge, lead migration on a diagonal, and CH_3NH_3 migration into a vacancy in the A-site cage. In the MD approach, depending on the size of the material, vacancies can be created either by displacing or by deleting a single atom in the system. It is important to ensure that net charge neutrality stays constant in order to reduce extrinsic effects. The mentioned study concluded that the iodide migrates with the lowest activation energy of

0.58 eV and the methylammonium with activation energy of 0.85 eV. Whereas the lead atoms seem to remain immobile, having high levels of activation energy, suggesting high stability, the iodide atoms thus have the highest mobility rates and represent the ionic carriers. These ions take the shortest linear path to fill a vacancy, with a slight curve to surpass other iodine atoms. The migration rate depends on environmental factors such as temperature or electric field. For example, at higher temperatures, the migration is much faster than at room temperature [131].

7.4 Hysteresis

One of the most significant phenomena related to the time response of the solar cell, and implicitly to the degradation, is attributed to the hysteresis of the current-voltage measurements that is present in perovskite solar cells when exposed to light [132]. In perovskite solar cells, the current density measured during the forward scan, i.e., by increasing the voltage in time (from short circuit to open circuit conditions), is different than during the backward scan, i.e., by decreasing the voltage. The magnitude of the hysteresis effect depends on how fast the scan is performed. The reason is that the voltage is changing faster than the time needed by the space charge to relax.

In general, the charge collection in the forward and backward scan is determined by a measurement protocol [133]. In addition, depending on the structure of the solar cells and initial conditions, different forms of hysteresis can be obtained. Several factors are involved in this phenomenon: ionic migration, charge traps, ferroelectric polarization, or charge accumulation at the interfaces between the components of the solar cell device [109].

First principle molecular dynamic simulations were constructed at varying temperatures to identify the specific processes causing hysteresis. Simulations were done in order to understand the ferroelectric effects, by focusing on the rearrangement of the atoms as they went through phase changes from orthorhombic to tetragonal [134]. The results of this MD simulation were consistent with experimental results, concerning the tilting angle reduction. In a 30-ps simulation, the effects of ferroelectric hysteresis were contested. However, in conjunction with previous classical MD simulations, the presence of external bias and electric field causes the MA to produce a counterfield, which results in a larger time constant for reorganization. However, this reaction occurs within the picosecond scale and does not account for the hysteresis that takes place on the millisecond scale.

Further simulations were done to test the ionic migration effects of hysteresis. These first-principles MD simulations were initiated by placing a single vacancy within the structure. Initial testing showed that no ionic migration occurs within the first 10 ps, making it plausible that the migration is the main cause of hysteresis. To confirm this hypothesis, a third set of simulations were done using string simulations, which allowed for longer computation time. The string simulation focused on the vacancy-driven ionic migration and the energy barrier. The results confirmed that halides have the lowest activation energy and are more likely to migrate in the presence of vacancies, which is congruent with other experimental data. The migration ultimately affects the polarization of the system and it is a factor in the hysteresis [134].

8. Concluding remarks

We have discussed some of the currently active issues behind perovskite solar cell simulations and development. The intrinsic chemical flexibility and complex physical effects that are behind halide perovskite materials are responsible not only for their surprising performance but also for the difficulty of accurately modeling and optimizing their properties. We have indicated some of the caveats and current best practices for DFT simulations of their electronic properties and the influence of defects. A computationally efficient molecular dynamics model tailored for reproducing the main dynamical properties of MAPI was described. Defect migration, temperature, and moisture degradation were also presented in relation to the performance and durability of solar cell structures. A better understanding of the structural properties, such as the dimensionality effects of 2D/3D materials and thermal-induced phase transitions or the complex interface properties, charge dynamics, and hysteresis effects, must be further investigated in order to obtain more realistic models for perovskites and perovskite-based devices.

Acknowledgments

This work was supported by a grant of the Romanian Ministry of Education and Research, CCCDI-UEFISCDI, project number PN-III-P2-2.1-PED-2019-1567, within PNCDI III, and by the EEA Grants 2014–2021, under Project Contract No. 36/2021. REB was also funded by the Icelandic Student Innovation Fund and by the National Power Company of Iceland (Landsvirkjun) via the Sustainability Institute and Forum of Reykjavík University.

References

[1] R.F. Berger, Design principles for the atomic and electronic structure of halide perovskite photovoltaic materials: insights from computation, Chem. Eur. J. 24 (35) (2018) 8708–8716, https://doi.org/10.1002/chem.201706126.

[2] M.R. Filip, G. Volonakis, F. Giustino, Hybrid Halide Perovskites: Fundamental Theory and Materials Design, arXiv, 2018. https://arxiv.org.

[3] J. Shi, S. Yun, First-principles DFT calculations for perovskite solar cells, in: S. Yun, A. Hagfeldt (Eds.), Counter Electrodes for Dye-Sensitized and Perovskite Solar Cells, John Wiley & Sons, Ltd, 2018, pp. 487–509, https://doi.org/10.1002/9783527813636.ch19.

[4] L.D. Whalley, J.M. Frost, Y.K. Jung, A. Walsh, Perspective: theory and simulation of hybrid halide perovskites, J. Chem. Phys. 146 (22) (2017), https://doi.org/10.1063/1.4984964.

[5] W.J. Yin, J.H. Yang, J. Kang, Y. Yan, S.H. Wei, Halide perovskite materials for solar cells: a theoretical review, J. Mater. Chem. A 3 (17) (2015) 8926–8942, https://doi.org/10.1039/c4ta05033a.

[6] C.-J. Yu, Advances in modelling and simulation of halide perovskites for solar cell applications, J. Phys.: Energy 1 (2) (2019), 022001, https://doi.org/10.1088/2515-7655/aaf143.

[7] H. Beck, C. Gehrmann, D.A. Egger, Structure and binding in halide perovskites: analysis of static and dynamic effects from dispersion-corrected density functional theory, APL Mater. 7 (2) (2019), https://doi.org/10.1063/1.5086541.

[8] D. Meggiolaro, F. De Angelis, First-principles modeling of defects in lead halide perovskites: best practices and open issues, ACS Energy Lett. 3 (9) (2018) 2206–2222, https://doi.org/10.1021/acsenergylett.8b01212.

[9] N. Hernández-Haro, J. Ortega-Castro, Y.B. Martynov, R.G. Nazmitdinov, A. Frontera, DFT prediction of band gap in organic-inorganic metal halide perovskites: an exchange-correlation functional benchmark study, Chem. Phys. 516 (2019) 225–231, https://doi.org/10.1016/j.chemphys.2018.09.023.

[10] S.X. Tao, X. Cao, P.A. Bobbert, Accurate and efficient band gap predictions of metal halide perovskites using the DFT-1/2 method: GW accuracy with DFT expense, Sci. Rep. 7 (1) (2017), https://doi.org/10.1038/s41598-017-14435-4.

[11] F. Valadares, I. Guilhon, L.K. Teles, M. Marques, Electronic structure panorama of halide perovskites: approximated DFT-1/2 quasiparticle and relativistic corrections, J. Phys. Chem. C 124 (34) (2020) 18390–18400, https://doi.org/10.1021/acs.jpcc.0c03672.

[12] H.M. Ghaithan, Z.A. Alahmed, A. Lyras, S.M.H. Qaid, S.A. Ald-Wayyan, Computational investigation of the folded and unfolded band structure and structural and optical properties of cspb(i1-xbrx)3 pernovskites, Crystals 10 (2020).

[13] R.A. Jishi, Modified Becke-Johnson exchange potential: improved modeling of lead halides for solar cell applications, AIMS Mater. Sci. 3 (1) (2016) 149–159, https://doi.org/10.3934/matersci.2016.1.149.

[14] T. Rauch, M.A.L. Marques, S. Botti, Local modified Becke-Johnson exchange-correlation potential for interfaces, surfaces, and two-dimensional materials, J. Chem. Theory Comput. 16 (2020).

[15] L. Leppert, T. Rangel, J.B. Neaton, Towards predictive band gaps for halide perovskites: lessons from one-shot and eigenvalue self-consistent GW, Phys. Rev. Mater. 3 (10) (2019), https://doi.org/10.1103/PhysRevMaterials.3.103803.

[16] R. Rajeswarapalanichamy, A. Amudhavalli, R. Padmavathy, K. Iyakutti, Band gap engineering in halide cubic perovskites cspbbr3-yiy (y = 0), Mater. Sci. Eng. B 258 (2020), 114560, https://doi.org/10.1016/j.mseb.2020.114560.

[17] M. Faghihnasiri, J. Beheshtian, F. Shayeganfar, R. Shahsavari, Phase transition and mechanical properties of cesium bismuth silver halide double perovskites (Cs2AgBiX6, X = Cl, Br, I): A DFT approach, Phys. Chem. Chem. Phys. 22 (10) (2020) 5959–5968, https://doi.org/10.1039/c9cp05342e.

[18] J.H. Lee, A. Jaffe, Y. Lin, H.I. Karunadasa, J.B. Neaton, Origins of the pressure-induced phase transition and metallization in the halide perovskite (CH3NH3)PbI3, ACS Energy Lett. 5 (7) (2020) 2174–2181, https://doi.org/10.1021/acsenergylett.0c00772.

[19] Y. Cho, T.C. Berkelbach, Optical properties of layered hybrid organic-inorganic halide perovskites: a tight-binding GW-BSE study, J. Phys. Chem. Lett. 10 (20) (2019) 6189–6196, https://doi.org/10.1021/acs.jpclett.9b02491.

[20] M. Roknuzzaman, K.K. Ostrikov, H. Wang, A. Du, T. Tesfamichael, Towards lead-free perovskite photovoltaics and optoelectronics by ab-initio simulations, Sci. Rep. 7 (2017) 14025.

[21] M. Wang, W. Wang, B. Ma, W. Shen, L. Liu, K. Cao, S. Chen, W. Huang, Lead-free perovskite materials for solar cells, Nano-Micro Lett. 13 (2021) 62.

[22] S.A. Mir, D.C. Gupta, Analysis of cage structured halide double perovskites Cs_2NaMCl_6 (M = Ti, V) by spin polarized calculations, J. Alloys Compd. 854 (2021) 156000.

[23] X.G. Zhao, J.H. Yang, Y. Fu, D. Yang, Q. Xu, L. Yu, S.-H. Wei, L. Zhang, Design of lead-free inorganic halide perovskites for solar cells via cation-transmutation, J. Am. Chem. Soc. 139 (2017) 2630–2638.

[24] L. Li, Y. You, S. Hu, Y. Shi, G. Zhao, C. Chen, Y. Wang, A. Stroppa, W. Ren, Electronic transport of organic-inorganic hybrid perovskites from first-principles and machine learning, Appl. Phys. Lett. 114 (8) (2019), https://doi.org/10.1063/1.5045512.

[25] W. Li, R. Long, J. Tang, O.V. Prezhdo, Influence of defects on excited-state dynamics in lead halide perovskites: time-domain ab initio studies, J. Phys. Chem. Lett. 10 (13) (2019) 3788–3804, https://doi.org/10.1021/acs.jpclett.9b00641.

[26] M. Kar, T. Körzdörfer, Computational high throughput screening of inorganic cation based halide perovskites for perovskite only tandem solar cells, Mater. Res. Exp. 7 (5) (2020), https://doi.org/10.1088/2053-1591/ab8c0d.

[27] B. Yılmaz, R. Yıldırım, Critical review of machine learning applications in perovskite solar research, Nano Energy 80 (2021), 105546, https://doi.org/10.1016/j.nanoen.2020.105546.

[28] D. Jain, S. Chaube, P. Khullar, S. Goverapet Srinivasan, B. Rai, Bulk and surface DFT investigations of inorganic halide perovskites screened using machine learning and materials property databases, Phys. Chem. Chem. Phys. 21 (35) (2019) 19423–19436, https://doi.org/10.1039/c9cp03240a.

[29] H. Park, R. Mall, F.H. Alharbi, S. Sanvito, N. Tabet, H. Bensmail, F. El-Mellouhi, Exploring new approaches towards the formability of mixed-ion perovskites by DFT and machine learning, Phys. Chem. Chem. Phys. 21 (3) (2019) 1078–1088, https://doi.org/10.1039/c8cp06528d.

[30] J. Schmidt, J. Shi, P. Borlido, L. Chen, S. Botti, M.A.L. Marques, Predicting the thermodynamic stability of solids combining density functional theory and machine learning, Chem. Mater. 29 (12) (2017) 5090–5103, https://doi.org/10.1021/acs.chemmater.7b00156.

[31] N.T.P. Hartono, J. Thapa, A. Tiihonen, F. Oviedo, C. Batali, J.J. Yoo, Z. Liu, R. Li, D.F. Marrón, M.G. Bawendi, T. Buonassisi, S. Sun, How machine learning can help select capping layers to suppress perovskite degradation, Nat. Commun. 11 (1) (2020), https://doi.org/10.1038/s41467-020-17945-4.

[32] J. Kirman, A. Johnston, D.A. Kuntz, M. Askerka, Y. Gao, P. Todorović, D. Ma, G.G. Privé, E.H. Sargent, Machine-learning-accelerated perovskite crystallization, Matter 2 (4) (2020) 938–947, https://doi.org/10.1016/j.matt.2020.02.012.

[33] A. Ali, H. Park, R. Mall, B. Aïssa, S. Sanvito, H. Bensmail, A. Belaidi, F. El-Mellouhi, Machine learning accelerated recovery of the cubic structure in mixed-cation perovskite thin films, Chem. Mater. 32 (7) (2020) 2998–3006, https://doi.org/10.1021/acs.chemmater.9b05342.

[34] Y. Yuan, Z. Xiao, B. Yang, J. Huang, Arising applications of ferroelectric materials in photovoltaic devices, J. Mater. Chem. A 2 (17) (2014) 6027–6041, https://doi.org/10.1039/c3ta14188h.

[35] Y. Kutes, L. Ye, Y. Zhou, S. Pang, B.D. Huey, N.P. Padture, Direct observation of ferroelectric domains in solution-processed CH3NH3PbI3 perovskite thin films, J. Phys. Chem. Lett. 5 (19) (2014) 3335–3339, https://doi.org/10.1021/jz501697b.
[36] J.M. Frost, K.T. Butler, A. Walsh, Molecular ferroelectric contributions to anomalous hysteresis in hybrid perovskite solar cells, APL Mater. 2 (8) (2014), https://doi.org/10.1063/1.4890246.
[37] A. Stroppa, C. Quarti, F. De Angelis, S. Picozzi, Ferroelectric polarization of CH3NH3PbI3: a detailed study based on density functional theory and symmetry mode analysis, J. Phys. Chem. Lett. 6 (12) (2015) 2223–2231, https://doi.org/10.1021/acs.jpclett.5b00542.
[38] Z. Fan, J. Xiao, K. Sun, L. Chen, Y. Hu, J. Ouyang, K.P. Ong, K. Zeng, J. Wang, Ferroelectricity of CH3NH3PbI3 perovskite, J. Phys. Chem. Lett. 6 (7) (2015) 1155–1161, https://doi.org/10.1021/acs.jpclett.5b00389.
[39] C. Goehry, G.A. Nemnes, A. Manolescu, Collective behavior of molecular dipoles in CH3NH3PbI3, J. Phys. Chem. C 119 (34) (2015) 19674–19680, https://doi.org/10.1021/acs.jpcc.5b05823.
[40] J.M. Frost, K.T. Butler, F. Brivio, C.H. Hendon, M. Van Schilfgaarde, A. Walsh, Atomistic origins of high-performance in hybrid halide perovskite solar cells, Nano Lett. 14 (5) (2014) 2584–2590, https://doi.org/10.1021/nl500390f.
[41] T.S. Sherkar, L. Jan Anton Koster, Can ferroelectric polarization explain the high performance of hybrid halide perovskite solar cells? Phys. Chem. Chem. Phys. 18 (1) (2016) 331–338, https://doi.org/10.1039/c5cp07117h.
[42] G. Song, B. Gao, G. Li, J. Zhang, First-principles study on the electric structure and ferroelectricity in epitaxial CsSnI3 films, RSC Adv. 7 (65) (2017) 41077–41083, https://doi.org/10.1039/c7ra07735a.
[43] D. Rossi, A. Pecchia, M.A.D. Maur, T. Leonhard, H. Röhm, M.J. Hoffmann, A. Colsmann, A.D. Carlo, On the importance of ferroelectric domains for the performance of perovskite solar cells, Nano Energy 48 (2018) 20–26, https://doi.org/10.1016/j.nanoen.2018.02.049.
[44] H.Y. Zhang, X.J. Song, H. Cheng, Y.L. Zeng, Y. Zhang, P.F. Li, W.Q. Liao, R.G. Xiong, A three-dimensional lead halide perovskite-related ferroelectric, J. Am. Chem. Soc. 142 (10) (2020) 4604–4608, https://doi.org/10.1021/jacs.0c00375.
[45] S.G. Motti, D. Meggiolaro, S. Martani, R. Sorrentino, A.J. Barker, F. De Angelis, A. Petrozza, Defect activity in lead halide perovskites, Adv. Mater. 31 (47) (2019), https://doi.org/10.1002/adma.201901183.
[46] M.H. Du, Density functional calculations of native defects in CH3NH3PbI3: effects of spin-orbit coupling and self-interaction error, J. Phys. Chem. Lett. 6 (8) (2015) 1461–1466, https://doi.org/10.1021/acs.jpclett.5b00199.
[47] H. Jin, E. Debroye, M. Keshavarz, I.G. Scheblykin, M.B.J. Roeffaers, J. Hofkens, J.A. Steele, It's a trap! on the nature of localised states and charge trapping in lead halide perovskites, Mater. Horiz. 7 (2) (2020) 397–410, https://doi.org/10.1039/c9mh00500e.
[48] Q. Li, Z. Chen, I. Tranca, S. Gaastra-Nedea, D. Smeulders, S. Tao, Compositional effect on water adsorption on metal halide perovskites, Appl. Surf. Sci. 538 (2021), 148058, https://doi.org/10.1016/j.apsusc.2020.148058.
[49] J.M. Ball, A. Petrozza, Defects in perovskite-halides and their effects in solar cells, Nat. Energy 1 (11) (2016), https://doi.org/10.1038/nenergy.2016.149.
[50] A. Walsh, A. Zunger, Instilling defect tolerance in new compounds, Nat. Mater. 16 (10) (2017) 964–967, https://doi.org/10.1038/nmat4973.
[51] J.M. Azpiroz, E. Mosconi, J. Bisquert, F. De Angelis, Defect migration in methylammonium lead iodide and its role in perovskite solar cell operation, Energy Environ. Sci. 8 (7) (2015) 2118–2127, https://doi.org/10.1039/c5ee01265a.

[52] R. Long, W. Fang, O.V. Prezhdo, Moderate humidity delays electron-hole recombination in hybrid organic-inorganic perovskites: time-domain ab initio simulations rationalize experiments, J. Phys. Chem. Lett. 7 (16) (2016) 3215–3222, https://doi.org/10.1021/acs.jpclett.6b01412.
[53] H. Uratani, K. Yamashita, Charge carrier trapping at surface defects of perovskite solar cell absorbers: a first-principles study, J. Phys. Chem. Lett. 8 (4) (2017) 742–746, https://doi.org/10.1021/acs.jpclett.7b00055.
[54] N.J. English, Dynamical properties of organo lead-halide perovskites and their interfaces to titania: insights from Density Functional Theory, Heliyon 6 (3) (2020), e03427, https://doi.org/10.1016/j.heliyon.2020.e03427.
[55] W.J. Yin, T. Shi, Y. Yan, Unusual defect physics in CH3NH3PbI3 perovskite solar cell absorber, Appl. Phys. Lett. 104 (6) (2014), https://doi.org/10.1063/1.4864778.
[56] Z. Zhang, W.H. Fang, R. Long, O.V. Prezhdo, Exciton dissociation and suppressed charge recombination at 2D perovskite edges: key roles of unsaturated halide bonds and thermal disorder, J. Am. Chem. Soc. 141 (39) (2019) 15557–15566, https://doi.org/10.1021/jacs.9b06046.
[57] A. Fraccarollo, V. Cantatore, G. Boschetto, L. Marchese, M. Cossi, Ab initio modeling of 2D layered organohalide lead perovskites, J. Chem. Phys. 144 (16) (2016), https://doi.org/10.1063/1.4947305.
[58] Z. Zhang, W.H. Fang, M.V. Tokina, R. Long, O.V. Prezhdo, Rapid decoherence suppresses charge recombination in multi-layer 2D halide perovskites: time-domain ab initio analysis, Nano Lett. 18 (4) (2018) 2459–2466, https://doi.org/10.1021/acs.nanolett.8b00035.
[59] B.E. Cohen, M. Wierzbowska, L. Etgar, High efficiency and high open circuit voltage in Quasi 2D perovskite based solar cells, Adv. Funct. Mater. 27 (5) (2017), https://doi.org/10.1002/adfm.201604733.
[60] E.A. Muljarov, S.G. Tikhodeev, N.A. Gippius, T. Ishihara, Excitons in self-organized semiconductor/insulator superlattices: PbI-based perovskite compounds, Phys. Rev. B 51 (20) (1995) 14370–14378, https://doi.org/10.1103/PhysRevB.51.14370.
[61] C. Ortiz-Cervantes, P. Carmona-Monroy, D. Solis-Ibarra, Two-dimensional halide perovskites in solar cells: 2D or not 2D? ChemSusChem 12 (8) (2019) 1560–1575, https://doi.org/10.1002/cssc.201802992.
[62] L. Mao, R.M. Kennard, B. Traore, W. Ke, C. Katan, J. Even, M.L. Chabinyc, C.C. Stoumpos, M.G. Kanatzidis, Seven-layered 2D hybrid lead iodide perovskites, Chemistry 5 (10) (2019) 2593–2604, https://doi.org/10.1016/j.chempr.2019.07.024.
[63] C.C. Stoumpos, D.H. Cao, D.J. Clark, J. Young, J.M. Rondinelli, J.I. Jang, J.T. Hupp, M.G. Kanatzidis, Ruddlesden-popper hybrid lead iodide perovskite 2D homologous semiconductors, Chem. Mater. 28 (8) (2016) 2852–2867, https://doi.org/10.1021/acs.chemmater.6b00847.
[64] L. Mao, W. Ke, L. Pedesseau, Y. Wu, C. Katan, J. Even, M.R. Wasielewski, C.C. Stoumpos, M.G. Kanatzidis, Hybrid Dion-Jacobson 2D lead iodide perovskites, J. Am. Chem. Soc. 140 (10) (2018) 3775–3783, https://doi.org/10.1021/jacs.8b00542.
[65] L.N. Quan, M. Yuan, R. Comin, O. Voznyy, E.M. Beauregard, S. Hoogland, A. Buin, A.-R. Kirmani, K. Zhao, A. Amassian, D.H. Kim, E.H. Sargent, Ligand-stabilized reduced-dimensionality perovskites, J. Am. Chem. Soc. 138 (8) (2016) 2649–2655, https://doi.org/10.1021/jacs.5b11740.
[66] A. Fraccarollo, L. Marchese, M. Cossi, Ab initio design of low band gap 2D tin organohalide perovskites, J. Phys. Chem. C 122 (7) (2018) 3677–3689, https://doi.org/10.1021/acs.jpcc.7b08928.
[67] C.M.M. Soe, G. Nagabhushana, R. Shivaramaiah, H. Tsai, W. Nie, J.-C. Blancon, F. Melkonyan, D.H. Cao, B. Traoré, L. Pedesseau, M. Kepenekian, C. Katan, J. Even, T.J. Marks, A. Navrotsky, A.D. Mohite, Structural and thermodynamic limits of layer thickness in 2D halide perovskites, Proc. Natl. Acad. Sci. U. S. A. 116 (2019) 58.

[68] J.E. Castellanos-Águila, L. Lodeiro, E. Menéndez-Proupin, A.L. Montero-Alejo, P. Palacios, J.C. Conesa, P. Wahnón, Atomic-scale model and electronic structure of Cu2O/CH3NH3PbI3 interfaces in perovskite solar cells, ACS Appl. Mater. Interfaces 12 (40) (2020) 44648–44657, https://doi.org/10.1021/acsami.0c11187.
[69] M. Cardona, Y.Y. Peter, Fundamentals of semiconductors, vol. 38, 2005.
[70] G.A. Casas, M.A. Cappelletti, A.P. Cédola, B.M. Soucase, Peltzer y Blancá, E. L., Analysis of the power conversion efficiency of perovskite solar cells with different materials as Hole-Transport Layer by numerical simulations, Superlattice. Microst. 107 (2017) 136–143, https://doi.org/10.1016/j.spmi.2017.04.007.
[71] E. Mosconi, T. Etienne, F. De Angelis, First-principles modeling of organohalide thin films and interfaces, in: Organic-Inorganic Halide Perovskite Photovoltaics: From Fundamentals to Device Architectures, Springer International Publishing, 2016.
[72] H. Wang, A. Guerrero, A. Bou, A.M. Al-Mayouf, J. Bisquert, Kinetic and material properties of interfaces governing slow response and long timescale phenomena in perovskite solar cells, Energy Environ. Sci. 12 (7) (2019) 2054–2079, https://doi.org/10.1039/c9ee00802k.
[73] J.W. Lee, T.Y. Lee, P.J. Yoo, M. Grätzel, S. Mhaisalkar, N.G. Park, Rutile TiO2-based perovskite solar cells, J. Mater. Chem. A 2 (24) (2014) 9251–9259, https://doi.org/10.1039/c4ta01786b.
[74] M. Liu, M.B. Johnston, H.J. Snaith, Efficient planar heterojunction perovskite solar cells by vapour deposition, Nature 501 (7467) (2013) 395–398, https://doi.org/10.1038/nature12509.
[75] G.A. Nemnes, C. Goehry, T.L. Mitran, A. Nicolaev, L. Ion, S. Antohe, N. Plugaru, A. Manolescu, Band alignment and charge transfer in rutile-TiO2/CH3NH3PbI3-xClx interfaces, Phys. Chem. Chem. Phys. 17 (45) (2015) 30417–30423, https://doi.org/10.1039/c5cp05466d.
[76] R. Long, O.V. Prezhdo, Dopants control electron-hole recombination at perovskite-TiO2 interfaces: ab initio time-domain study, ACS Nano 9 (11) (2015) 11143–11155, https://doi.org/10.1021/acsnano.5b05843.
[77] G. Xing, N. Mathews, S. Sun, S.S. Lim, Y.M. Lam, M. Gratzel, S. Mhaisalkar, T.C. Sum, Long-range balanced electron- and hole-transport lengths in organic-inorganic CH3NH3PbI3, Science 342 (6156) (2013) 344–347, https://doi.org/10.1126/science.1243167.
[78] J.H. Noh, S.H. Im, J.H. Heo, T.N. Mandal, S.I. Seok, Chemical management for colorful, efficient, and stable inorganic-organic hybrid nanostructured solar cells, Nano Lett. 13 (4) (2013) 1764–1769, https://doi.org/10.1021/nl400349b.
[79] Y. Ogomi, A. Morita, S. Tsukamoto, T. Saitho, N. Fujikawa, Q. Shen, T. Toyoda, K. Yoshino, S.S. Pandey, T. Ma, S. Hayase, CH3NH3SnxPb(1-x)I3 perovskite solar cells covering up to 1060 nm, J. Phys. Chem. Lett. 5 (6) (2014) 1004–1011, https://doi.org/10.1021/jz5002117.
[80] J. Dong, Y. Zhao, J. Shi, H. Wei, J. Xiao, X. Xu, J. Luo, J. Xu, D. Li, Y. Luo, Q. Meng, Impressive enhancement in the cell performance of ZnO nanorod-based perovskite solar cells with Al-doped ZnO interfacial modification, Chem. Commun. 50 (87) (2014) 13381–13384, https://doi.org/10.1039/C4CC04908J.
[81] K. Mahmood, B.S. Swain, H.S. Jung, Controlling the surface nanostructure of ZnO and Al-doped ZnO thin films using electrostatic spraying for their application in 12% efficient perovskite solar cells, Nanoscale 6 (15) (2014) 9127–9138, https://doi.org/10.1039/c4nr02065k.
[82] A. Nicolaev, T.L. Mitran, S. Iftimie, G.A. Nemnes, Optimization of halide perovskite solar cells based on nanocolumnar ZnO, Sol. Energy Mater. Sol. Cells 158 (2016) 202–208, https://doi.org/10.1016/j.solmat.2015.10.023.
[83] S. Chatterjee, A.J. Pal, Introducing Cu2O thin films as a hole-transport layer in efficient planar perovskite solar cell structures, J. Phys. Chem. C 120 (3) (2016) 1428–1437, https://doi.org/10.1021/acs.jpcc.5b11540.

[84] A.M. Elseman, M.S. Selim, L. Luo, C.Y. Xu, G. Wang, Y. Jiang, D.B. Liu, L.P. Liao, Z. Hao, Q.L. Song, Efficient and stable planar n-i-p perovskite solar cells with negligible hysteresis through solution-processed Cu2O nanocubes as a low-cost hole-transport material, ChemSusChem 12 (16) (2019) 3808–3816, https://doi.org/10.1002/cssc.201901430.

[85] X. Miao, S. Wang, W. Sun, Y. Zhu, C. Du, R. Ma, C. Wang, Room-temperature electrochemical deposition of ultrathin CuOx film as hole transport layer for perovskite solar cells, Scr. Mater. 165 (2019) 134–139, https://doi.org/10.1016/j.scriptamat.2019.02.032.

[86] W. Sun, Y. Li, S. Ye, H. Rao, W. Yan, H. Peng, Y. Li, Z. Liu, S. Wang, Z. Chen, L. Xiao, Z. Bian, C. Huang, High-performance inverted planar heterojunction perovskite solar cells based on a solution-processed CuOx hole transport layer, Nanoscale 8 (20) (2016) 10806–10813, https://doi.org/10.1039/c6nr01927g.

[87] W. Yu, F. Li, H. Wang, E. Alarousu, Y. Chen, B. Lin, L. Wang, M.N. Hedhili, Y. Li, K. Wu, X. Wang, O.F. Mohammed, T. Wu, Ultrathin Cu2O as an efficient inorganic hole transporting material for perovskite solar cells, Nanoscale 8 (11) (2016) 6173–6179, https://doi.org/10.1039/c5nr07758c.

[88] C. Zuo, L. Ding, Solution-processed Cu2O and CuO as hole transport materials for efficient perovskite solar cells, Small 11 (41) (2015) 5528–5532, https://doi.org/10.1002/smll.201501330.

[89] J.H. Park, J. Seo, S. Park, S.S. Shin, Y.C. Kim, N.J. Jeon, H.W. Shin, T.K. Ahn, J.H. Noh, S.-C. Yoon, C.S. Hwang, S.I. Seok, Efficient CH3NH3PbI3 perovskite solar cells employing nanostructured p-type NiO electrode formed by a pulsed laser deposition, Adv. Mater. 27 (27) (2015) 4013–4019, https://doi.org/10.1002/adma.201500523.

[90] X. Yin, P. Chen, M. Que, Y. Xing, W. Que, C. Niu, J. Shao, Highly efficient flexible perovskite solar cells using solution-derived NiOx hole contacts, ACS Nano 10 (3) (2016) 3630–3636, https://doi.org/10.1021/acsnano.5b08135.

[91] X. Yin, J. Liu, J. Ma, C. Zhang, P. Chen, M. Que, Y. Yang, W. Que, C. Niu, J. Shao, Solvothermal derived crystalline NiOx nanoparticles for high performance perovskite solar cells, J. Power Sources 329 (2016) 398–405, https://doi.org/10.1016/j.jpowsour.2016.08.102.

[92] H. Zhang, J. Cheng, F. Lin, H. He, J. Mao, K.S. Wong, A.K.Y. Jen, W.C.H. Choy, Pinhole-free and surface-nanostructured niox film by room-Temperature solution process for high-performance flexible perovskite solar cells with good stability and reproducibility, ACS Nano 10 (1) (2016) 1503–1511, https://doi.org/10.1021/acsnano.5b07043.

[93] L. Xu, X. Chen, J. Jin, W. Liu, B. Dong, X. Bai, H. Song, P. Reiss, Inverted perovskite solar cells employing doped NiO hole transport layers: a review, Nano Energy 63 (2019), https://doi.org/10.1016/j.nanoen.2019.103860.

[94] J.W. Jung, C.C. Chueh, A.K.Y. Jen, A low-temperature, solution-processable, Cu-doped nickel oxide hole-transporting layer via the combustion method for high-performance thin-film perovskite solar cells, Adv. Mater. 27 (47) (2015) 7874–7880, https://doi.org/10.1002/adma.201503298.

[95] J.H. Kim, P.W. Liang, S.T. Williams, N. Cho, C.C. Chueh, M.S. Glaz, D.S. Ginger, A.K.Y. Jen, High-performance and environmentally stable planar heterojunction perovskite solar cells based on a solution-processed copper-doped nickel oxide hole-transporting layer, Adv. Mater. 27 (4) (2015) 695–701, https://doi.org/10.1002/adma.201404189.

[96] W. Chen, Y. Wu, Y. Yue, J. Liu, W. Zhang, X. Yang, H. Chen, E. Bi, I. Ashraful, M. Gratzel, L. Han, Efficient and stable large-area perovskite solar cells with inorganic charge extraction layers, Science 350 (6263) (2015) 944–948, https://doi.org/10.1126/science.aad1015.

[97] W. Chen, F.Z. Liu, X.Y. Feng, A.B. Djurišić, W.K. Chan, Z.B. He, Cesium doped NiOx as an efficient hole extraction layer for inverted planar perovskite solar cells, Adv. Energy Mater. 7 (19) (2017), https://doi.org/10.1002/aenm.201700722.

[98] Y. Wei, K. Yao, X. Wang, Y. Jiang, X. Liu, N. Zhou, F. Li, Improving the efficiency and environmental stability of inverted planar perovskite solar cells via silver-doped nickel oxide hole-transporting layer, Appl. Surf. Sci. 427 (2018) 782–790, https://doi.org/10.1016/j.apsusc.2017.08.184.

[99] W. Chen, Y. Zhou, G. Chen, Y. Wu, B. Tu, F.Z. Liu, L. Huang, A.M.C. Ng, A.B. Djurišić, Z. He, Alkali chlorides for the suppression of the interfacial recombination in inverted planar perovskite solar cells, Adv. Energy Mater. 9 (19) (2019), https://doi.org/10.1002/aenm.201803872.

[100] H. Cheng, Y. Li, G. Zhao, K. Zhao, Z.S. Wang, Pyridine-terminated conjugated organic molecules as an interfacial hole transfer bridge for NiOx-based perovskite solar cells, ACS Appl. Mater. Interfaces 11 (32) (2019) 28960–28967, https://doi.org/10.1021/acsami.9b09530.

[101] L. Xie, Z. Cao, J. Wang, A. Wang, S. Wang, Y. Cui, Y. Xiang, X. Niu, F. Hao, L. Ding, Improving energy level alignment by adenine for efficient and stable perovskite solar cells, Nano Energy 74 (2020), https://doi.org/10.1016/j.nanoen.2020.104846.

[102] N. Zibouche, G. Volonakis, F. Giustino, Graphene oxide/perovskite interfaces for photovoltaics, J. Phys. Chem. C 122 (29) (2018) 16715–16726, https://doi.org/10.1021/acs.jpcc.8b03230.

[103] K.P. Loh, S.W. Tong, J. Wu, Graphene and graphene-like molecules: prospects in solar cells, J. Am. Chem. Soc. 138 (4) (2016) 1095–1102, https://doi.org/10.1021/jacs.5b10917.

[104] J.C. Yu, J.I. Jang, B.R. Lee, G.W. Lee, J.T. Han, M.H. Song, Highly efficient polymer-based optoelectronic devices using PEDOT:PSS and a go composite layer as a hole transport layer, ACS Appl. Mater. Interfaces 6 (3) (2014) 2067–2073, https://doi.org/10.1021/am4051487.

[105] A. Zubarev, D. Dragoman, Ballistic charge carrier transmission through graphene multi-barrier structures in uniform magnetic field, J. Phys. D. Appl. Phys. 47 (42) (2014), 425302, https://doi.org/10.1088/0022-3727/47/42/425302.

[106] A. Zubarev, D. Dragoman, Effect of Schrödinger-Dirac transformations on electron transport in graphene-based structures surrounded by conducting regions, J. Phys. D. Appl. Phys. 46 (2013).

[107] J.T.W. Wang, J.M. Ball, E.M. Barea, A. Abate, J.A. Alexander-Webber, J. Huang, M. Saliba, I. Mora-Sero, J. Bisquert, H.J. Snaith, R.J. Nicholas, Low-temperature processed electron collection layers of graphene/TiO2 nanocomposites in thin film perovskite solar cells, Nano Lett. 14 (2) (2014) 724–730, https://doi.org/10.1021/nl403997a.

[108] A.L. Palma, L. Cinà, S. Pescetelli, A. Agresti, M. Raggio, R. Paolesse, F. Bonaccorso, A. Di Carlo, Reduced graphene oxide as efficient and stable hole transporting material in mesoscopic perovskite solar cells, Nano Energy 22 (2016) 349–360, https://doi.org/10.1016/j.nanoen.2016.02.027.

[109] D. Wang, M. Wright, N.K. Elumalai, A. Uddin, Stability of perovskite solar cells, Sol. Energy Mater. Sol. Cells 147 (2016) 255–275, https://doi.org/10.1016/j.solmat.2015.12.025.

[110] J. Liu, O.V. Prezhdo, Chlorine doping reduces electron-hole recombination in lead iodide perovskites: time-domain ab initio analysis, J. Phys. Chem. Lett. 6 (22) (2015) 4463–4469, https://doi.org/10.1021/acs.jpclett.5b02355.

[111] T. Zhou, M. Wang, Z. Zang, L. Fang, Stable dynamics performance and high efficiency of ABX3-type super-alkali perovskites first obtained by introducing H5O2 cation, Adv. Energy Mater. 9 (29) (2019), https://doi.org/10.1002/aenm.201900664.

[112] A. Aziz, N. Aristidou, X. Bu, R.J.E. Westbrook, S.A. Haque, M.S. Islam, Understanding the enhanced stability of bromide substitution in lead iodide perovskites, Chem. Mater. 32 (1) (2020) 400–409, https://doi.org/10.1021/acs.chemmater.9b04000.

[113] N. Aristidou, C. Eames, I. Sanchez-Molina, X. Bu, J. Kosco, M. Saiful Islam, S.A. Haque, Fast oxygen diffusion and iodide defects mediate oxygen-induced degradation of perovskite solar cells, Nat. Commun. 8 (2017), https://doi.org/10.1038/ncomms15218.
[114] W. Li, Z. Chen, J. Tang, O.V. Prezhdo, Anti-correlation between band gap and carrier lifetime in lead halide perovskites under compression rationalized by ab initio quantum dynamics, Chem. Mater. 32 (11) (2020) 4707–4715, https://doi.org/10.1021/acs.chemmater.0c01287.
[115] L. Zhou, A.J. Neukirch, D.J. Vogel, D.S. Kilin, L. Pedesseau, M.A. Carignano, A.D. Mohite, J. Even, C. Katan, S. Tretiak, Density of states broadening in CH3NH3PbI3 hybrid perovskites understood from ab initio molecular dynamics simulations, ACS Energy Lett. 3 (4) (2018) 787–793, https://doi.org/10.1021/acsenergylett.8b00166.
[116] C. Gehrmann, D.A. Egger, Dynamic shortening of disorder potentials in anharmonic halide perovskites, Nat. Commun. 10 (1) (2019), https://doi.org/10.1038/s41467-019-11087-y.
[117] M.Z. Mayers, L.Z. Tan, D.A. Egger, A.M. Rappe, D.R. Reichman, How lattice and charge fluctuations control carrier dynamics in halide perovskites, Nano Lett. 18 (12) (2018) 8041–8046, https://doi.org/10.1021/acs.nanolett.8b04276.
[118] S. McKechnie, J.M. Frost, D. Pashov, P. Azarhosh, A. Walsh, M. van Schilfgaarde, Dynamic symmetry breaking and spin splitting in metal halide perovskites, Phys. Rev. B 98 (2018) 085108.
[119] C.-J. Tong, L. Li, L.-M. Miu, O.V. Prezhdo, Long carrier lifetimes in PbI_2-rich perovskites rationalized by *ab initio* nonadiabatic molecular dynamics, ACS Energy Lett. 3 (2018) 1868–1874.
[120] F. Ambrosio, J. Wiktor, F. De Angelis, A. Pasquarello, Origin of low electron–hole recombination rate in metal halide perovskites, Energy Environ. Sci. 11 (2018) 101–105.
[121] S.R.G. Balestra, J.M. Vicent-Luna, S. Calero, S. Tao, J.A. Anta, Efficient modelling of ion structure and dynamics in inorganic metal halide perovskites, J. Mater. Chem. A 8 (23) (2020) 11824–11836, https://doi.org/10.1039/D0TA03200J.
[122] M.P. Allen, D.J. Tildesley, Computer Simulation of Liquids, Oxford University Press, 1989.
[123] A. Mattoni, A. Filippetti, C. Caddeo, Modeling hybrid perovskites by molecular dynamics, J. Phys. Condens. Matter 043001 (2017), https://doi.org/10.1088/1361-648x/29/4/043001.
[124] A. Mattoni, A. Filippetti, M.I. Saba, P. Delugas, Methylammonium rotational dynamics in lead halide perovskite by classical molecular dynamics: the role of temperature, J. Phys. Chem. C 119 (30) (2015) 17421–17428, https://doi.org/10.1021/acs.jpcc.5b04283.
[125] J.J. Gutierrez-Sevillano, S. Ahmad, S. Calero, J.A. Anta, Molecular dynamics simulations of organohalide perovskite precursors: solvent effects in the formation of perovskite solar cells, Phys. Chem. Chem. Phys. 17 (35) (2015) 22770–22777, https://doi.org/10.1039/c5cp03220b.
[126] P. Delugas, C. Caddeo, A. Filippetti, A. Mattoni, Thermally activated point defect diffusion in methylammonium lead trihalide: anisotropic and ultrahigh mobility of iodine, J. Phys. Chem. Lett. 7 (13) (2016) 2356–2361, https://doi.org/10.1021/acs.jpclett.6b00963.
[127] E. Mosconi, J.M. Azpiroz, F. De Angelis, Ab initio molecular dynamics simulations of methylammonium lead iodide perovskite degradation by water, Chem. Mater. 27 (13) (2015) 4885–4892, https://doi.org/10.1021/acs.chemmater.5b01991.
[128] C. Müller, T. Glaser, M. Plogmeyer, M. Sendner, S. Döring, A.A. Bakulin, C. Brzuska, R. Scheer, M.S. Pshenichnikov, W. Kowalsky, A. Pucci, R. Lovrinčić, Water infiltration in methylammonium lead iodide perovskite: fast and inconspicuous, Chem. Mater. 27 (22) (2015) 7835–7841, https://doi.org/10.1021/acs.chemmater.5b03883.

[129] L. Cojocaru, S. Uchida, Y. Sanehira, V. Gonzalez-Pedro, J. Bisquert, J. Nakazaki, T. Kubo, H. Segawa, Temperature effects on the photovoltaic performance of planar structure perovskite solar cells, Chem. Lett. 44 (11) (2015) 1557–1559, https://doi.org/10.1246/cl.150781.

[130] J.M. Frost, A. Walsh, Molecular motion and dynamic crystal structures of hybrid halide perovskites, in: Organic-Inorganic Halide Perovskite Photovoltaics: From Fundamentals to Device Architectures, Springer International Publishing, 2016, pp. 1–17, https://doi.org/10.1007/978-3-319-35114-8_1.

[131] C. Eames, J.M. Frost, P.R.F. Barnes, B.C. O'Regan, A. Walsh, M.S. Islam, Ionic transport in hybrid lead iodide perovskite solar cells, Nat. Commun. 6 (2015), https://doi.org/10.1038/ncomms8497.

[132] A.K. Jena, T. Miyasaka, Hysteresis characteristics and device stability, in: Organic-Inorganic Halide Perovskite Photovoltaics: From Fundamentals to Device Architectures, Springer International Publishing, 2016, pp. 255–284, https://doi.org/10.1007/978-3-319-35114-8_10.

[133] G.A. Nemnes, C. Besleaga, A.G. Tomulescu, A. Palici, L. Pintilie, A. Manolescu, I. Pintilie, How measurement protocols influence the dynamic J-V characteristics of perovskite solar cells: theory and experiment, Sol. Energy 173 (2018) 976–983, https://doi.org/10.1016/j.solener.2018.08.033.

[134] S. Meloni, T. Moehl, W. Tress, M. Franckeviius, M. Saliba, Y.H. Lee, P. Gao, M.K. Nazeeruddin, S.M. Zakeeruddin, U. Rothlisberger, M. Graetzel, Ionic polarization-induced current-voltage hysteresis in CH3NH3PbX3 perovskite solar cells, Nat. Commun. 7 (2016), https://doi.org/10.1038/ncomms10334.

CHAPTER

6

Lead-free halide perovskites

Mohammad Hatamvand[a], *Somayeh Gholipour*[b], *Mozhgan Yavari*[c], *Mahboubeh Hadadian*[d], *Mohammad Sajedi Alvar*[e], *Bart Roose*[f], *Yaser Abdi*[g], *Yiqiang Zhan*[h], *Yonghua Chen*[a], *and Wei Huang*[a]

[a]Key Laboratory of Flexible Electronics (KLoFE) & Institute of Advanced Materials (IAM), Nanjing Tech University (Nanjing Tech), Nanjing, Jiangsu, China, [b]METU GUNAM Center, Middle East Technical University, Ankara, Turkey, [c]Department of Electrical and Electronic Engineering, Advanced Technology Institute, University of Surrey, Guildford, Surrey, United Kingdom, [d]Department of Mechanical and Materials Engineering, Faculty of Technology, University of Turku, Turku, Finland, [e]Max Planck Institute for Polymer Research, Mainz, Germany, [f]Department of Physics, Cavendish Laboratory, University of Cambridge, Cambridge, United Kingdom, [g]Department of Physics, Nanophysics Research Laboratory, University of Tehran, Tehran, Iran, [h]State Key Laboratory of ASIC and System, Centre of Micro-Nano System, School of Information Science and Technology, Fudan University, Shanghai, China

1. Introduction

Metal halide perovskite materials have attracted enormous interest in the photovoltaic (PV) field in the last decade. The popularity of metal halide perovskites is largely due to their unique PV properties; long carrier lifetimes and diffusion lengths, suitable (tunable) bandgap, and high absorption coefficient [1,2]. Adding to this is the possibility to process these materials using cost-effective, large-scale solution-processing methods, like spray-coating and ink-based printing techniques [3]. Halide perovskite materials have an ABX_3 structure. A is typically a large cation such as methylammonium ($MA = CH_3NH_3$), formamidinium ($FA = CH$

Low-Dimensional Halide Perovskites
https://doi.org/10.1016/B978-0-323-88522-5.00009-0

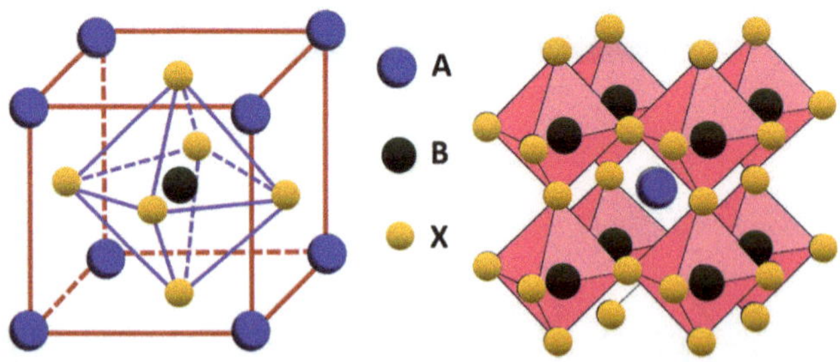

FIG. 1 Schematic of the ABX_3 perovskite structure. A cations sit in the middle of a cube formed by B cations, which in turn are surrounded by corner-sharing octahedra of X anions. *Reproduced with permission from reference Z. Yi, N. H. Ladi, X. Shai, H. Li, Y. Shen, M. Wang, Will organic-inorganic hybrid halide lead perovskites be eliminated from optoelectronic applications?, Nanoscale Adv. 1 (4) (2019) 1276–1289. Copyright 2021, Royal Society of Chemistry.*

$(NH_2)_2)$, or Cs, B is a divalent metal such as Pb^{+2}, and X is a halide anion (Fig. 1). However, the toxicity of lead (Pb) is a significant challenge toward the commercialization of perovskite solar cells (PSCs). Pb-free perovskite compositions are now gaining ground to address this environmental issue [4]. In addition, Pb-free perovskites have demonstrated lower bandgaps (~1.2eV) compared to Pb-based perovskites (>1.45eV), making them promising materials for tandem solar cells [5]. Besides solar cells, Pb-free perovskites also find use in light-emitting devices [6], sensors [7], energy storage applications [8], and piezoelectric materials [9].

Pb-free perovskites should similarly have excellent optoelectronic properties and, in addition, low toxicity if they are to fully replace Pb-based perovskites [10]. Fig. 2A shows elements that can potentially replace Pb, based on their charge and ionic radius [10]. The bandgap of various Pb-free perovskites is shown in comparison to $MAPbI_3$ and $FAPbI_3$ in Fig. 2B [10].

A whole host of divalent [11–15], or alloys of monovalent and trivalent metal ions, has been suggested as alternatives to Pb. However, many of these compounds suffer from suboptimal bandgap and optoelectronic properties. The most encouraging alternative so far is tin (Sn), which is abundant and nontoxic to humans. Sn-based perovskites can achieve lower bandgaps than Pb-based perovskites and have comparable carrier lifetimes and diffusion lengths [3,16]. Therefore, Sn-based perovskites have attracted much attention [17,18]. However, Sn-based perovskites come with their own set of challenges. Sn oxidizes readily from Sn^{+2} to Sn^{+4} [19], which results in increased carrier recombination. The crystallization of Sn-based perovskites is fast and therefore harder to control, leading to crystals with smaller grain size and lower quality than Pb-based perovskites [20,21]. Therefore, Sn-based PSCs generally have lower performance in comparison with Pb-based PSCs. The highest power

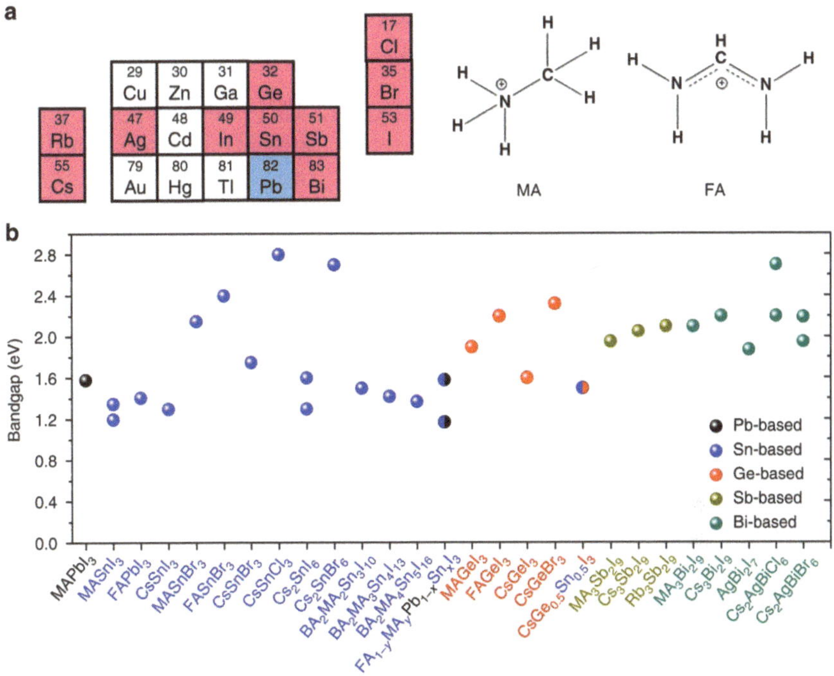

FIG. 2 (A) Potential replacements for Pb (marked in *pink*). (B) Bandgap for different halide perovskite materials. *From Prospects for low-toxicity lead-free perovskite solar cells (n.d.). Nature Communications Volume. https://doi.org/10.1038/s41467-019-08918-3. Reproduced with permission from reference W. Ke, M.G. Kanatzidis, Prospects for low-toxicity Pb-free perovskite solar cells, Nat. Commun. 10 (2019) 1–4. Copyright 2021, Nature.*

conversion efficiency (PCE) for Pb-free, Sn-based PSCs is 13.24% [22], while Pb-based PSCs have shown much higher (>25%) efficiencies [23]. Thus, stability and efficiency are the two main properties that need to be addressed for Sn to replace Pb in PSCs.

This chapter will investigate the potential of Pb-free halide perovskite materials, highlighting recent advances: providing a new perspective on the limitations and challenges, providing fresh horizons, and inspiring novel research toward improved Pb-free halide perovskites. Section 2 will give an overview of all Pb-free perovskite compositions reported to date. In Section 3, the optical properties of Pb-free perovskites are discussed. Section 4 investigates the electronic properties of Pb-free perovskites. Progress on the implementation of Pb-free perovskites in optoelectronic devices is described in Section 5. Section 6 discusses the stability of Pb-free perovskites. Finally, Section 7 describes the future and outlook of the Pb-free perovskite materials.

2. Different Pb-free perovskite composition and crystalline structures

Organic-inorganic Pb-based halide perovskite ($APbX_3$) has a cubic, tetragonal, or orthorhombic crystal structure (depending on temperature). The A-site cation is at the center of a cube formed by corner-sharing BX_6 octahedra (Fig. 3). The Goldschmidt tolerance factor is used to predict if an ABX_3 compound will form a stable perovskite structure [24]. The B site is not necessarily a divalent metal; an alloy of monovalent and trivalent metals can be used to yield the so-called double perovskites. The structural dimensionality of perovskites can be reduced from 3D to two-dimensional (2D), one-dimensional (1D), or even zero-dimensional (0D) [25], by increasing the size of the A-site cations. This section will give an overview of different Pb-free compositions and crystalline structures.

2.1 Pb-free perovskites

There are five types of (quasi-)3D perovskites, $AB^{2+}X_3$, $A_2B^+B^{3+}X_6$, $A_2B^{4+}X_6$, $AB^{4+}Ch_3$, and $A_4B^{3+}B^{5+}X_{12}$, as shown in Fig. 3. These materials generally exhibit better charge transport than their lower-dimensional counterparts. A logical effort while replacing Pb is thus to preserve the £D structure.

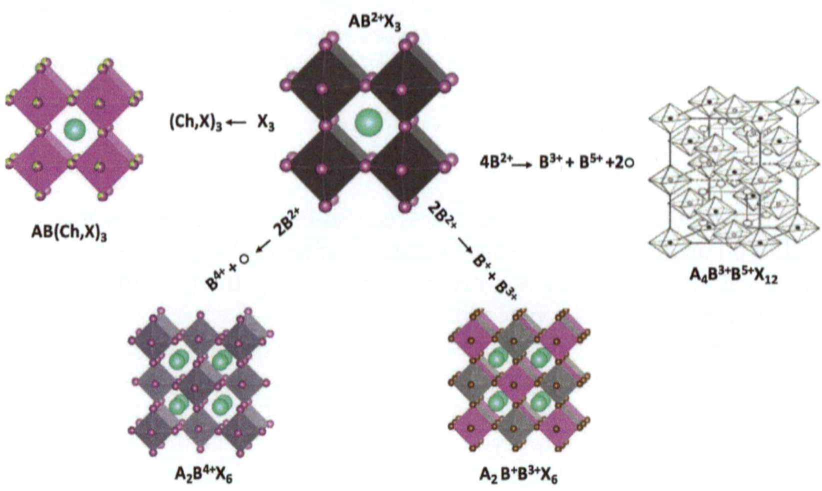

FIG. 3 Crystal structures of 3D lead-free perovskites and their derivative diagrams between different types of 3D (quasi-3D) perovskite structures. *From Lead Halide Perovskites to Lead-Free Metal Halide Perovskites and Perovskite Derivatives. (n.d.). Advanced Materials. https://doi.org/10.1002/adma.201803792. Redesigned with permission from references Z. Xiao, Z. Song, Y. Yan, From lead halide perovskites to Pb-free metal halide perovskites and perovskite derivatives, Adv. Mater. 31 (47) (2019) 1–22 and C.R. Hubbard, R.A. Jacobson, Proceedings of the Iowa Academy of Science the crystal structure of Rb4SbIIISbvBr12, 75 (1968) 85–96. Copyright 2021, Wiley-VCH.*

2.1.1 $AB^{2+}X_3$ type perovskites

Pb can be readily replaced by other divalent metals of suitable size, such as group 14 elements (Sn and germanium (Ge)), alkaline earth metals (magnesium (Mg), calcium (Ca), strontium (Sr), and barium (Ba)), or other metals with stable divalent states (copper (Cu), iron (Fe), palladium (Pd), and europium (Eu)). Most studies on Pb substitution for photovoltaic applications are Sn- and Ge-based. These metals have an ionic radius and electronic configuration comparable to Pb. However, Sn^{2+} and Ge^{2+} are easily oxidized into Sn^{4+} and Ge^{4+} in the presence of oxygen or moisture, which has a detrimental effect on their optoelectronic properties [26]. Mg, Ca, Sr, and Ba yield moisture-sensitive perovskites with a large bandgap. This makes them unattractive candidates for solar cells but may be of use in other optoelectronic devices.

2.1.2 $A_2B^+B^{3+}X_6$ type double perovskites

In $A_2B^+B^{3+}X_6$ double perovskites, also named elpasolite, the divalent (B^{2+}) metal is replaced with both a monovalent (B^+; Cu, silver (Ag), gold (Au), sodium (Na), potassium (K), thallium (Tl), and indium (In)) and a trivalent (B^{3+}; antimony (Sb), bismuth (Bi), In, terbium (Tb), and Eu) metallic cation [27]. As various elements can be used as B^+ and B^{3+}, the halide double perovskites exhibit rich combinatorial chemistry, with a wide variety of optoelectronic properties [27–33].

Double perovskites based on nontoxic Bi and Ag are generally more stable against heat, light, and moisture than $MAPbI_3$ [34]. Cs_2AgBiX_6 (X: Br, Cl) belongs to the cubic space group *Fm-3m* and is more stable than Pb-based perovskite. $Cs_2AgBiBr_6$ has an indirect bandgap of 1.95 eV and could be suitable for use in tandem solar cells [27]. But for single-junction solar cells it needs to be reduced. Mixing Bi with Sb ($Cs_2AgBi_{1-x}Sb_xBr_6$) can bring the bandgap down to 1.6 eV, which is comparable to that of the most efficient Pb-based perovskites [35]. The synthesis of pure Sb-based double perovskites is challenging because nonperovskite side products are easily formed. Due to the oxidation of In^I to In^{III}, $Cs_2InBiCl_6$ and $Cs_2InSbCl_6$ double perovskites were found to be unsuitable for optoelectronic devices [36]. But $Cs_2AgInCl_6$, where In is already in the 3+ state, is stable and has a direct bandgap [34]. The Ga-based compounds $Cs_2LiGaBr_6$ and $Cs_2NaGaBr_6$ are similar to the In-based double perovskites, with direct bandgaps [37]. According to theoretical calculations, using Na(I) or Cu(I) as the B^+-cation leads to Cu/Na(I)-based double perovskites with optoelectronic properties suitable for solar cells. However, $[Cu(I)X_6]$ octahedra are thermodynamically unstable and decompose to nonperovskite-containing $[Cu(I)X_4]$ tetrahedra [38]. Lanthanide-based double perovskites with the crystal structures of $Cs_2NaTbCl_6$ and $Cs_2NaEuCl_6$ have been studied as early as 1970 [39,40] but have not found an optoelectronic application as of yet. Mixed-valence double perovskites have the same element as the monovalent and trivalent cations. $Cs_2Au^+Au^{3+}I_6$ has a direct bandgap of 1.31 eV [41], although it is stoichiometrically a single perovskite.

2.1.3 $A_2B^{4+}X_6$ *type vacancy-ordered double perovskites*

$A_2B^{4+}X_6$ type vacancy-ordered double perovskites have one tetravalent (B^{4+}; titanium (Ti), palladium (Pd), platinum (Pt), niobium (Nb)) cation and a vacancy (e.g., $A_2B^{4+}\circ X_6$, where ○ is a vacancy) instead of two B^{2+}. The crystal structure of these perovskite derivatives is composed of a face-centered lattice of isolated BX_6 units with A-site cations occupying the cuboctahedral voids. The distorted quasi-3D structure is sometimes considered to be a 0D structure. A_2SnI_6 (A = Cs, MA), Cs_2PdBr_6, and Cs_2TiBr_6 [42–44] have been investigated for photovoltaic applications. For Cs_2SnI_6, the conventional double perovskite ($A_2B^+B^{3+}X_6$), cubic *Fm-3m* space group has been reported [45]. Depending on doping, Cs_2SnI_6 can be either an n- or a p-type semiconductor with excellent carrier mobility and a bandgap of ≈1.3 eV [45,46]. Due to the high metallic conductivity, efficient devices based on tin perovskite have not been reported yet. Therefore, to decrease the conductivity, Sn has been replaced by other metals, such as Pd^{4+}. The resulting compound is solution-processable, shows long photoluminescence (PL) lifetimes, a bandgap of 1.6 eV, and is resistant to moisture and light [47]. The nontoxic Ti-based Cs_2TiX_6 [48] has a suitable bandgap (1.38–1.78 eV) and band energies, as well as excellent intrinsic and environmental stability. Semimetal ion Te^{4+}-based vacancy-ordered double perovskites ($A_2Te^{4+}X_6$) have a tunable bandgap (1.42–2.02 eV), high mobility, and strong optical absorption [49]. The confinement effect, due to the isolated BX_6 octahedra, is favorable for the PL of vacancy-ordered double perovskites. PL can be further enhanced by impurity doping (e.g., manganese (Mn^{2+}), Bi^{3+}, etc.), which is a promising way to further enhance the optoelectronic properties of these materials [50].

2.1.4 $A_4B^{3+}B^{5+}X_{12}$ *type vacancy-ordered perovskites*

$A_4B^{3+}B^{5+}X_{12}$ type vacancy-ordered perovskites can be obtained by replacing four B^{2+} with B^{3+}, B^{5+}, and two vacancies (○). In 1963, a pure antimony compound ($Cs_4Sb^{3+}Sb^{5+}Cl_{12}$) with this structure was first reported [51]. The mixed-valence (Sb^{3+} and Sb^{5+}) compound $Rb_4Sb^{3+}Sb^{5+}X_{12}$ crystallizes in the tetragonal space group 141/amd (D_{4h}^{19}) [52]. Compounds using In and Bi have since been reported [53]. Tin substitution can affect the charge transfer band and electrical conductivity [54]. As there are few reports on this type of perovskite, their optoelectronic properties are still largely unexplored.

2.1.5 Chalcogenide-substitution: ($AB^{IV}Ch_3$, $AB^{III}(X, Ch)_3$) type perovskites

Halides can be substituted with chalcogenides (Ch; oxygen (O), sulfur (S), selenium (Se), and tellurium (Te)). These materials exhibit good stability as the transition metals and the chalcogenides form a strong covalent bond. The direct bandgap of zirconium (Zr)-based chalcogenide perovskites ($AZrCh_3$) can be changed by using different inorganic cations or chalcogenides. This results in bandgaps ranging from 1.73eV to 2.87eV for $CaTiS_3$, $SrZrS_3$, $SrZrS_3$, and $BaZr(O_xS_{1-x})_3$ [55]. The only Ch-based perovskite photovoltaic device reported to date with direct bandgap is $BiFeO_3$ [56]. Partial substitution of halides with chalcogenides seems to yield the best materials in terms of stability and bandgap. Density functional theory (DFT) calculations show that the direct bandgaps of partially substituted Bi perovskite compounds ($MABiChX_2$) vary between 1.24 and 2.00eV [57]. However, $MASbSI_2$ is thermodynamically unstable and decomposes into a nonperovskite phase. Although a 3% efficient solar cell was made, it is unclear whether the light-absorbing layer contains $MASbSI_2$ or a mixture of nonperovskite phases such as Sb_2S_3 and $MA_3Sb_2I_9$ [58,59].

2.2 Low-dimensional Pb-free perovskites

Although 3D perovskites have exhibited the best performance in solar cells, lower-dimensional perovskites also hold much potential due to their high stability. By using a larger A-site

cation, the BX_6 can be confined to lower dimensionality. Mixed dimensional perovskite structure can be assembled by using A-site cations of varying lengths. These compounds are usually referred to using a $n=$"number of corner-sharing 2D layers" notation, where $n=1$ is the true 2D perovskite and $n=\infty$ is the 3D perovskite. Like in 3D perovskites, double perovskites can be made by using metals with other valences, yielding the following low-dimensional perovskite types (from 2D to 0D): 2D $A_2B^{2+}X_4$, 2D $A_4B^{+}B^{3+}X_8$, 2D $A_4B^{2+}B_2^{3+}X_{12}$, 2D $A_3B_2^{3+}X_9$, 1D $AB^{2+}X_3$, 1D $A_2B^{3+}X_5$, and 0D $A_3B_2^{3+}X_9$ perovskites.

2.2.1 2D $A_2B^{2+}X_4$ type perovskites

In this structure, 2D perovskite sheets are separated by space or large organic monovalent cations. This is the 2D equivalent of the 3D $AB^{2+}X_3$ perovskite, with a common formula of $A_2B^{2+}X_4$. Butylammonium (BA: $C_4H_9NH_3^+$) is commonly used as an A-site cation with Sn or Ge as B-site metals. Alternatively, a transition metal with a small ionic radius can be used [60,61]. This type of perovskites is more stable in ambient air than their 3D structure because the organic BA layers form a penetration barrier for moisture and oxygen [62,63]. Different transition metal cation-based perovskites have been reported; $MA_2CuCl_x Br_{4-x}$, MA_2FeCl_4, and MA_2PdCl_4. However, only the Cu(II)-based perovskites have been used for solar cells [26]. For Ge, $(PEA)_2GeI_4$ (PEA = phenyl ethyl ammonium) was reported with a direct bandgap of 2.12 eV [64]. By partially substituting Ge with Sn, the bandgap can be narrowed to 1.95 eV [65]. All Sn- and Ge-based 2D perovskites exhibit better stability than their 3D counterparts [64]. By mixing in a smaller organic cation, such as MA, a mixed 2D/3D dimensionality can be achieved, which shows improved optoelectronic properties. $(BA)_2(MA)_{n-1}Sn_nI_{3n+1}$ has a bandgap of 1.42–1.5 eV for $n = 3$ or 4. The efficiency of $(BA)_2(MA)_3Sn_4I_{13}$ ($n = 4$) devices reached up to 2.5% and showed better stability toward moisture for unencapsulated devices. 90% of initial PCE was retained after one month of storage and 50% PCE after four months [66,67]. To date, the best photovoltaic efficiency of 12.4% was achieved for $PEA_{0.15}FA_{0.75}SnI_3$. Devices retained 90% of initial efficiency after storage in an inert atmosphere for 3800 h [68].

2.2.2 2D $A_4B^+B^{3+}X_8$ type double perovskites

The 3D double Pb-free perovskites $A_2B^+B^{3+}X_6$ can be converted to a 2D structure with a common formula of $A_4B^+B^{3+}X_8$, by employing a large organic A-site cation. Most of the 2D double perovskite structures are based on Ag^+, Bi^{3+}, Br^-, and the large organic cation BA. Karunadasa et al. reported $(BA)_4AgBiBr_8$ with a direct bandgap of 2.77 eV [69]. By incorporating Cs, the bandgap decreases to 2.4 eV for the mixed 2D/3D $(BA)_2CsAgBiBr_7$ [69]. The hexylammonium compounds $(C_6H_{16}N_2)_2CuBiI_8{\cdot}0.5H_2O$ and $(C_6H_{16}N_2)_2AgBiI_8{\cdot}H_2O$ were reported to have bandgaps of 1.68 and 1.93 eV [70]. Using alkyldiammonium cations in mixed valence double perovskite leads to narrow bandgap materials $[NH_3(CH_2)_8NH_3]_2Au^IAu^{III}I_6(I_2)_3$ and $[NH_3(CH_2)_7NH_3]_2Au^IAu^{III}I_6(I_2)_3$ (containing I_3^-) with a bandgap of 1.14 eV and 0.95 eV, respectively [71]. I_3 and the AuI_6 octahedra have a strong electronic interaction, leading to narrower bandgaps compared to the 3D mixed valence perovskite [72,73].

2.2.3 2D $A_3B_2^{3+}X_9$ type perovskites

By substitution of B^{2+} in $AB^{2+}X_3$ with trivalent B^{3+}(Bi, Sb), the lower energy structure of perovskite with the formula of $A_3B^{3+}{}_2X_9$ can be obtained. This composition can yield both 0-D dimers (with space group $P6_3/mmc$, which will be considered later) and the 2-D layered form ($P3$-$m1$) with corner-sharing $[BX_6]^{3-}$ octahedra (Fig. 4). For example, $Cs_3Sb_2I_9$ has a 2-D layered form and a 0-D dimer. The structure and dimensions of Sb/Bi-based perovskites can be affected by the choice of cations and anions. Small cations (potassium, rubidium, or ammonium) lead to the 2D layered perovskite structure. Whereas for larger cations (e.g., Cs, MA), the 0D dimers are produced. 2D $Cs_3Sb_2I_9$ perovskite exhibits a

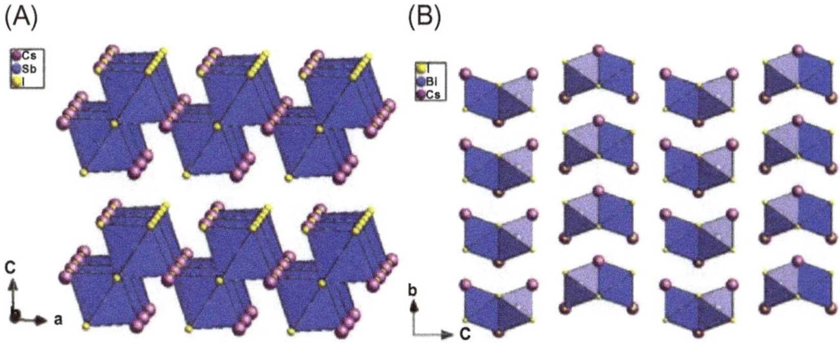

FIG. 4 Crystal structures of low-dimensional lead-free perovskites. (A) 2D layered form of Cs3Sb23+I9. (B) 0D dimer form of Cs3Bi23+I9. *From Structural and Functional Diversity in Lead-Free Halide Perovskite Materials. (n.d.). Advanced Materials. https://doi.org/10.1002/adma.201900326. Reproduced with permission from references W. Ning, F. Gao, Structural and functional diversity in Pb-free halide perovskite, Materials 1900326 (2019) 1–21. Copyright 2021, Wiley-VCH.*

bandgap of 2.05eV and increased stability in ambient air compared to $MAPbI_3$ [74]. A-site substitution of $Cs_3Sb_2I_9$ with a small cation (Rb^+, NH_4^+, and K^+) leads to 2D derivatives such as $Rb_3Sb_2I_9$ or $(NH_4)_3Sb_2I_9$, stabilizing the layered structure and improving physical properties [75,76]. An experimental and theoretical study demonstrated that an increasing amount of Cl transforms the 0D phase to the 2D layered phase for $MA_3Sb_2Cl_x I_{9-x}$ [77].

2.2.4 2D $A_4B^{2+}B^{3+}{}_2X_{12}$ type perovskites

Upon introduction of B^{2+} into 2D double-layered $A_3B^{3+}{}_2X_9$ type perovskites, the lattice changes to 2D triple-layered $A_4B^{2+}B^{3+}{}_2X_{12}$ type perovskites. Solis-Ibarra et al. incorporated Cu^{2+} into the 2D large bandgap perovskite $Cs_3Sb_2Cl_9$ and reported the first 2D $A_4B^{2+}B^{3+}{}_2X_{12}$ type perovskite ($Cs_4CuSb_2Cl_{12}$) with a direct bandgap of 1.0eV, higher conductivity than $MAPbI_3$ and negligible heat, light, and moisture-induced degradation. However, poor solubility in polar solvents makes thin-film fabrication challenging [78]. This can be overcome by using a carbon matrix [79]. When Cu is substituted with germanium, the resulting $FA_4GeSb_2Cl_{12}$ is highly soluble in polar solvents, enabling thin-film fabrication. Exhibiting excellent conductivity and stability and a direct bandgap of ~1.3eV make this an interesting solar cell material [80].

2.2.5 1D $AB^{2+}X_3$ type perovskites

1D $AB^{2+}X_3$ type perovskites can be obtained by introducing potential order-to-disorder (OTD) A-site cations into $AB^{2+}X_3$ type perovskites. Common examples of OTD cations are flexible alkyl amines, saturated heterocycle-based amines, saturated naphthalene-based amines, and their derivatives. The dielectric, ferroelectric, and piezoelectric properties are affected by the A-site cation. (Pyrrolidinium)$MnCl_3$ is a 1D $AB^{2+}X_3$ type perovskite exhibiting ferroelectric properties [81]. When using 3-pyrrolinium instead of pyrrolidinium, a new isostructural 1D $AB^{2+}X_3$ type perovskite was developed with high Curie temperature and high fatigue resistance [82]. The ferroelectric and photoluminescent properties of TMCM-$MnCl_3$ (TMCM = trimethylchloromethyl ammonium) [83] and TMCM-$CdBr_3$ have also been reported [84].

2.2.6 1D $A_2B^{3+}X_5$ type perovskites

By using large A-site cations and trivalent B^{3+}, 1D $A_2B^{3+}X_5$type perovskites can be achieved. Luo et al. reported $(C_6H_{13}N)_2BiI_5$ perovskite with narrow direct bandgap and high moisture-stability compared to Pb compounds (Fig. 5) [85].

2.2.7 0D $A_3B_2{}^{3+}X_9$ type perovskites (dimer phases)

By using B^{3+} with a lone pair, 0D $A_3B_2{}^{3+}X_9$ type perovskites can be obtained. This is a hexagonal phase consisting of 0D face-sharing bioctahedra $[B_2I_9]^{3-}$ (Fig. 4B). $Cs_3Bi_2I_9$ and $MA_3Bi_2I_9$ both exhibit a 0D structure with face-sharing bismuth iodide octahedra. A wide range of A-site cations (MA, Cs, Rb, K, guanidinium, cyclohexylammonium, or imidazolium) can be used to form a 0D perovskite structure [86]. The single-crystal structure of 0D bismuth $MA_3Bi_2I_9$ was investigated by Kaskel et al. In this

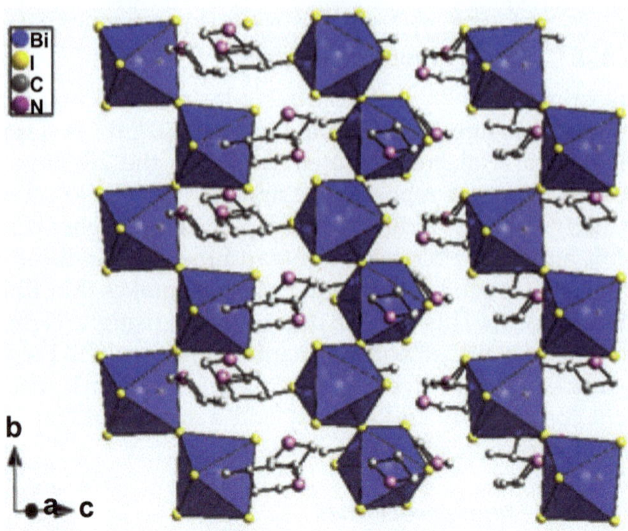

FIG. 5 Crystal structures of 1D lead-free perovskites of (C6H13N)2Bi23+I5. *From Structural and Functional Diversity in Lead-Free Halide Perovskite Materials. (n.d.). Advanced Materials. https://doi.org/10.1002/adma.201900326. Reproduced with permission from references W. Ning, F. Gao, Structural and functional diversity in Pb-free halide perovskite, Materials 1900326 (2019) 1–21. Copyright 2021, Wiley-VCH.*

direct bandgap material, the $[BiI_6]^{3-}$ octahedra are surrounded by randomly oriented MA^+ cations [87]. Another 0D perovskite is $Cs_3Bi_2I_9$. This is an indirect bandgap semiconductor with a lower bandgap compared to $MA_3Bi_2I_9$ [88]. The atomic radius of trivalent Sb^{3+} (76pm) is lower than that of Pb^{2+} (119pm). Due to the high oxidation state and small ionic radius, the $A_3Sb_2I_9$ compounds possess a 0-D dimer crystal structure [89,90]. $Cs_3Sb_2I_9$ is a highly stable material with an indirect bandgap [86].

As this section illustrates, there is an enormous parameter space to explore for Pb-free perovskite materials, of which current work has barely scratched the surface. More research is needed in order to replace their Pb-based counterparts and to uncover the wide range of physical and optoelectronic properties exhibited by these materials.

3. Optical properties

3.1 Absorption and photoluminescence

Both absorption and PL are highly important optical properties of semiconductors in optoelectronic devices. In a semiconductor, absorption will occur when an electron is excited from the valence band (VB) into the conduction band (CB) when the energy of an incident photon is greater than the bandgap of the semiconductor. When the excited charge carriers are transferred to the electrodes, a photocurrent is generated. PL occurs when the excited electron relaxes back to the VB before it is extracted, thereby emitting a photon with an energy equal to the bandgap [91]. Investigating the absorption and PL spectra thus reveals essential optoelectronic properties of semiconductors. Understanding these properties will allow further improvements for Pb-free perovskite materials [92–98].

In this section, we highlight Sn-based perovskites and Ag/Bi double perovskites. These are two of the most promising Pb-free halide perovskites. This is exemplary of what can be learned about a material by studying the optoelectronic properties.

3.2 Absorption and photoluminescence of Sn-based Pb-free perovskites

The bandgap of Pb-free $CsSnX_3$ perovskite materials can be tuned from 2.8 to 1.3eV by varying the halide ratio (X=Cl, Br, I) [99,100]. As Fig. 6A, C shows, increasing the average atomic radius of the halide shifts the PL peak to lower energies. The reason for this shift is the decreased electronegativity with increased radius, which weakens ionic interactions. Fig. 6B depicts slow and fast PL decay of Pb-free perovskite $CsSnBr_3$, which are 8.8 and 0.7ns, respectively. Slow and fast PL decay for $CsSnBr_{0.5}I_{0.5}$ are 3.3 and

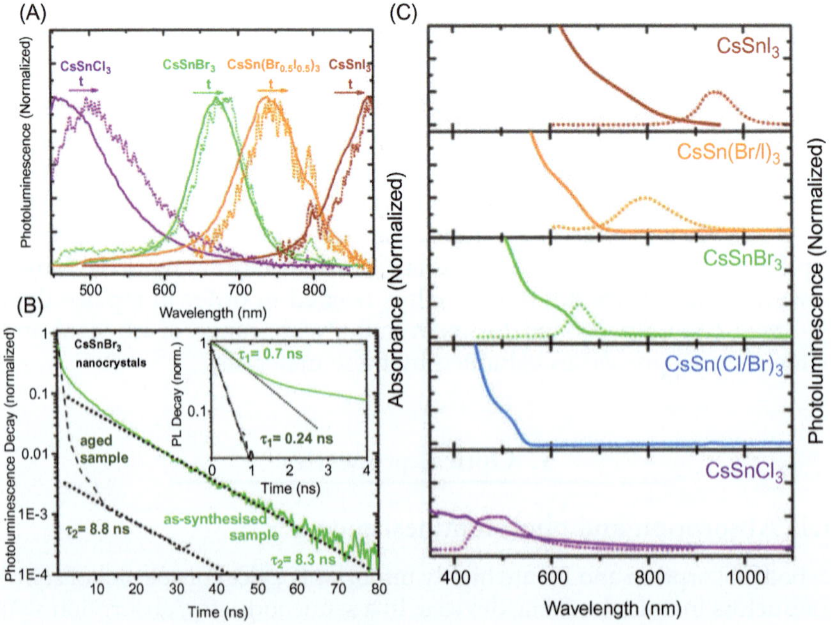

FIG. 6 (A) PL diagrams of lead-free $CsSnX_3$ (X=Cl, Br, Br0.5I0.5, I) with different compositions of halides. (B) Slow and fast (inset) PL decay kinetics of as-synthesized and aged CsSnBr3 excited at 490nm, and detected at 590–900nm. (C) Absorbance and steady-state PL of $CsSnX_3$ containing pure and mixed halides. *From Synthesis and Optical Properties of Lead-Free Cesium Tin Halide Perovskite Nanocrystals. (n.d.). J. Am. Chem. Soc. Reproduced with permission from reference T.C. Jellicoe, et al., Synthesis and optical properties of Pb-free cesium tin halide perovskite nanocrystals, J. Am. Chem. Soc. 138 (9) (2016) 2941–2944. Copyright 2021, American Chemical Society.*

0.3ns; meanwhile slow and fast PL decay for $CsSnI_3$ are 2.6 and 0.3ns, respectively. Indicating that the size of the halide has a profound effect on defect chemistry and as a result on PL decay lifetime.

3.3 Absorption and photoluminescence of Bi-based Pb-free perovskites

The bandgaps for the halide double perovskite Cs_2AgBiX_6(X=Cl, Br, I) were calculated by DFT and spin-orbit coupling. The calculated band gaps for $Cs_2AgBiCl_6$, $Cs_2AgBiBr_6$, and Cs_2AgBiI_6 were 1.91, 1.42, and 0.89eV, respectively, which is in agreement with experimental results [101]. Changing the halide from Cl to I leads to lower bandgaps and sharper absorption peaks. The bandgap of Pb-free perovskite structures can be adjusted most readily using different halides, as summarized in Table 1. This can be used to dial in the desired bandgap for a specific optoelectronic device application.

Doping of double perovskite $Cs_2NaBiCl_6$ lattice has been shown to be a second route to adjust optical properties. Doping of $Cs_2NaBiCl_6$ with Ag^+ shifts the excitonic absorption energy from 3.82 to 3.48 eV and improves PL quantum yield (PLQY) from 1.6% to 20% (Fig. 7A, B). Mn^{2+}-doped $Cs_2NaBiCl_6$ exhibits a broad emission peak at 585 nm (2.12 eV) and a PLQY of 3% (Fig. 7C, D). Eu^{3+}-doped $Cs_2NaBiCl_6$ also shows a strong emission at 591 and 615 nm (2.10 and 2.02 eV) (Fig. 7E, F) [108].

TABLE 1 Optical properties of different lead-free perovskite structures.

Chemical structure	Absorption peak (nm)	Emission peak (nm)	PLQY (%)	Ref.
Cs_2AgBiI_6	500	575	0.1	[102]
$Cs_2AgBiBr_6$	440	465	0.7	[102]
$Cs_2AgBiCl_6$	360	395	6.7	[102]
$Cs_3Bi_2Br_9$	439	468	0.2	[103]
$Cs_3Bi_2Br_9$	439	460	4.5	[103]
$Cs_3Bi_2Br_9$	396	410	19.4	[104]
$Cs_3Bi_2Cl_9$	290	393	26.4	[104]
$Cs_3Bi_2I_9$	395	545	0.018	[104]
$MA_3Bi_2Cl_9$	360	360	15	[105]
$MA_3Bi_2Br_9$	376	423	12	[105]
$MA_3Bi_2I_9$	400	540	0.03	[105]
$Cs_3Sb_2Br_9$	375	410	46	[106]
$Cs_3Sb_2Cl_9$	360	370	11	[106]
$Cs_3Sb_2I_9$	520	560	23	[106]
$CsSnCl_3$	420	490	$\leq$0.14	[99]
$CsSnBr_3$	610	660	$\leq$0.14	[99]
$CsSnl_3$	750	945	$\leq$0.14	[99]
Cs_2SnI_6	–	620	$\leq$0.48	[107]

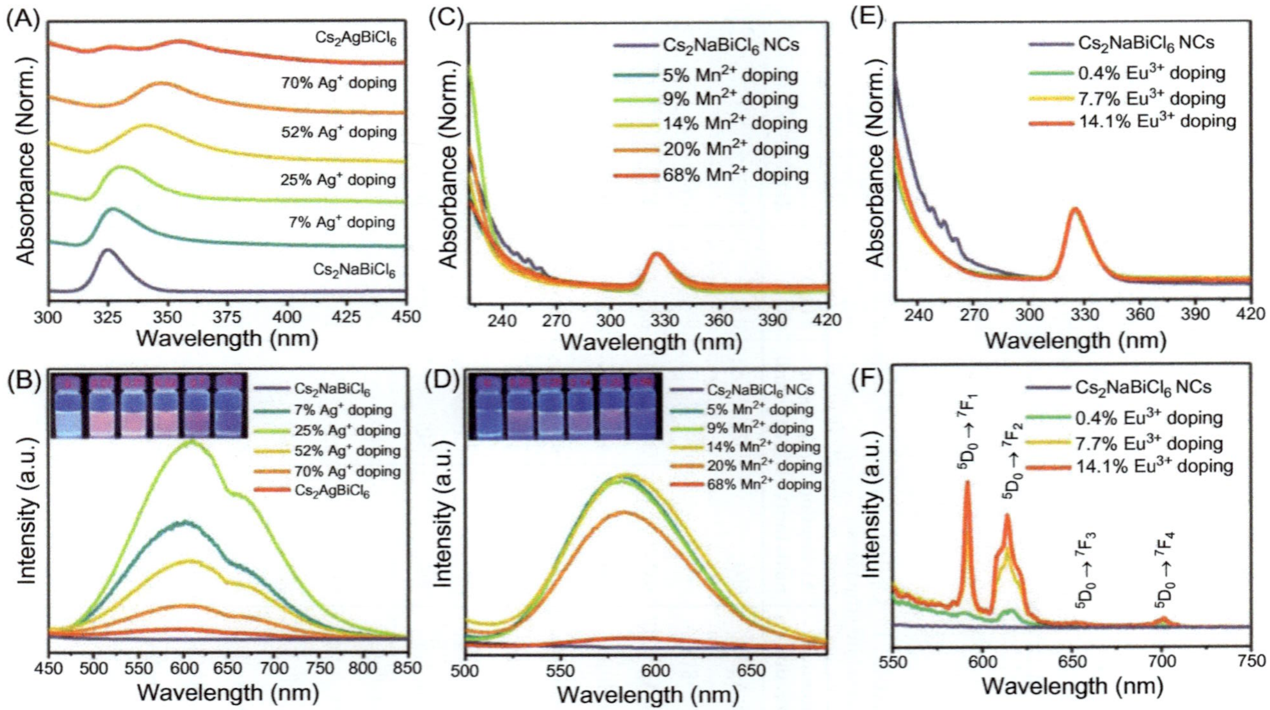

FIG. 7 (A, B) UV–vis absorption and PL spectra (excitation at 325 nm, inset) for Ag+-doped Cs2NaBiCl6. (C, D) UV–vis absorption and PL spectra (excitation at 354 nm, inset) for Mn^{2+}-doped Cs2NaBiCl6. (E, F) UV–vis absorption and PL spectra (excitation at 300 nm) for Eu^{3+}-doped Cs2NaBiCl6. *From Improving Lead-Free Double Perovskite Cs2NaBiCl6 Nanocrystal Optical Properties via Ion Doping. (n.d.). Adv. Opt. Mater. https://doi.org/10.1002/adom.201901919. Reproduced with permission from reference M.M. Yao, et al., Improving Pb-free double perovskite Cs2NaBiCl6 nanocrystal optical properties via ion doping, Adv. Opt. Mater. 8 (8) (2020) 2–9. Copyright 2021, Wiley-VCH.*

4. Electronic properties

4.1 Transport of charge carriers in Pb-free perovskites

4.1.1 Ionic conduction in organic-inorganic perovskites

The transport of ions in halide perovskites has been widely investigated [109,110]. The motion of ions in halide perovskites was initially reported three decades ago by Mizusaki et al. [109]. It has since been found to play a profound role in device operation. Unger et al. [111] and Snaith et al. [112] introduced ion migration as one of the major culprits for the hysteresis in the *J-V* characteristics of PSCs. The nature and the mechanism of ion conduction in halide perovskites, known as solid ionic-electronic conductors, have been intensively investigated [112–135]. Electric fields are the main driver behind the migration of ions, irrespective of the perovskite crystal structure. Numerous direct and indirect indications of ion migration and their implications on the device functionality of PSCs have been reported. This includes *J-V* hysteresis, phase segregation, aging, and long-term instability of PSCs [118,128,136–139].

The migration of mobile ionic species in solids is a defect-mediated phenomenon [110]. The migration speed of mobile ions depends on a number of variables. Available interstitial space, vacancies, hopping distance, and the dimension and charge of the ions all play a role. Ion migration is thermally activated; ion motion in perovskite films thus depends on the activation energy of ions in the film. This implies that ions with different properties migrate at different rates. For instance, in $MAPbI_3$ perovskites, Pb^{2+}, MA^+, and I^- (or their vacancies) have different values of migration activation energies [115,124,134]. Among them, migration of the heavy Pb^{2+} is negligible due to its large activation energy [115]. However, migration of MA^+ and I^- ions (or their vacancies) has been commonly observed and reported [134,140]. Due to the lower activation energy of the I^-, it is the most easily migrating ionic species in $MAPbI_3$ [115,134,141,142]. Slow migration of MA^+, with higher activation energy than that of I^- ion, has also been confirmed [115,134,141,142].

It has to be noted that from theoretical and experimental studies on I^- (or V_I^+) migration in $MAPbI_3$, a broad range of values have been reported for activation energy (~0.1 to 0.6 eV) [113,115,116,119,122,123,126,129,130,143–145] and diffusion coefficient (10^{-16} to $10^{-7}\,m^2\,s^{-1}$) [115,119,120,122,131,134,141,142,146,147]. Similarly, for MA^+, a range of activation energies (~0.4 to 1.1 eV) and diffusion coefficients (10^{-15} to $10^{-20}\,m^2\,s^{-1}$) have been reported [113,115,116,120,126,130,134,141,142,147].

Moreover, the reported values for ion concentrations in $MAPbI_3$ from simulations and theoretical models are rather high in the range of 10^{24} to $10^{25}\,m^{-3}$ [115,121,128,147,148], whereas the experimentally measured values are often a few order of magnitude lower (in the range of 10^{21} to $10^{23}\,m^{-3}$) [123,134,137,146,149–151].

The presence of mobile ionic charges in halide perovskite devices, and their accumulation at the interfaces of perovskite-transport layers, has a big effect on the electric-field distribution within the perovskite layer [115,117,121,137,145,147,149,152]. The distribution of the electric field strongly affects the recombination, transport, and extraction of charge carriers in PSCs [117,137]. To what degree mobile ionic charges influence the stability and device functionality of PSCs depends on the concentration and diffusivity of the ions.

The mechanism of ion diffusion in Pb-free halide double perovskites is much less understood [153–155], although it is suspected that in-migration plays an important role here as well. Xu et al. suggested that the fast ion migration of the dominant defects in Cs_2AgInX_6 (X=Cl, Br) may play a crucial role in the stability of double perovskites [155]. They concluded that the most dominant acceptors are V_{Ag}^-, while the most dominant donors are Ag_i^+ and V_X^+. This is true in both $Cs_2AgInBr_6$ and $Cs_2AgInCl_6$ under Ag-rich conditions, with concentrations of $>10^{18}\,cm^{-3} > 10^{16}\,cm^{-3}$, respectively [155].

The calculated ion diffusion barriers and coefficients for different ionic species in Cs_2AgInX_6 (X=Br, Cl) are summarized in Table 2 [155]. In addition to vacancy-mediated mechanism, the dominant mechanism of ion diffusion in $MAPbI_3$ [115], ion diffusion through the interstitial mechanism is also included. It was found that Ag diffusion through the second mechanism occurs at a much faster rate. Due to the higher freedom for structural relaxation in $Cs_2AgInBr_6$, the diffusion of vacancies in $Cs_2AgInBr_6$ is faster than that in $Cs_2AgInCl_6$ [155].

In Cs_2AgInX_6 (X=Cl, Br) halide double perovskites, the defects with lower formation energies all have low activation energy. This causes fast diffusion and leads to poor device stability [155].

TABLE 2 Theoretically calculated ion diffusion barriers (ΔEa) and the ion diffusion coeffcients (*D*) of defect migrations in Cs_2AgInX_6 (X=Br, Cl).

			D **($cm^2\,s^{-1}$)**	
Defect	**X=Br**	**X=Cl**	**X=Br**	**X=Cl**
Ag_i^+	0.15	0.17	1.70×10^{-5}	7.87×10^{-6}
V_{Ag}^-	0.31	0.55	7.11×10^{-8}	4.71×10^{-12}
V_X^+	0.24	0.41	2.88×10^{-7}	2.30×10^{-10}
V_{Cs}^-	0.99	1.12	9.43×10^{-20}	6.38×10^{-22}
V_{In}^0	1.68	1.88	1.15×10^{-30}	4.3×10^{-34}

For V_{Cs}^- and V_{In}^0 intrinsic defects, which have higher formation energies, the diffusion coeffcients are below $<10^{-20}$ $cm^2 s^{-1}$. This suggests that these defects are almost immobile and unlikely to cause hysteresis in the current-voltage curve or play a role in the instability [155]. This is similar to $MAPbI_3$, where the Pb^{2+} ions are approximately immobile, and the degradation is caused by the migration of MA^+ and I^- ions [115,116,156].

Using first-principle calculations, Lan et al. studied halide diffusion in Pb-free double perovskites $Cs_2Ag(Bi, Sb, In)X_6$ (X=Cl, Br) [153]. Their calculations predicted that formation energies of halide vacancies are related to the electronic configurations of B-site cations. It was found that $Cs_2AgInCl_6$ exhibited the lowest Cl vacancy formation energy. The formation energy of Br vacancies is lower than that of Cl vacancies, indicating that the concentration of halide vacancies is higher in $Cs_2AgBiBr_6$. Similarly, the migration barrier of Br in $Cs_2AgBiBr_6$ was predicted to be lower than that of Cl in $Cs_2AgBiCl_6$ [116,153]. The lower predicted migration barrier can explain the origin of the significant hysteresis observed in $Cs_2AgBiBr_6$ solar cells [153]. To improve optoelectronic properties of Pb-free double perovskites, the suppression of halide ion diffusion is critical [153,157].

Theoretical defect calculations on $MASnI_3$ perovskite indicated that the Sn-I bond is easily breakable. As a consequence, an excessively high defect density ($>10^{17}$ cm^{-3}) is found [158]. Similarly, theoretical investigations reveal that the dominant defects in Ge-based perovskite compounds are Ge vacancies. A relatively low formation enthalpy leads to a high density of holes [159]. Sun et al. reported that $MAGeI_3$ is more stable than $MASnI_3$ based on calculations of the defect formation energy [160].

4.2 Electronic charge transport properties

The transportation of electrons and holes in a semiconductor material is directly dependent on the electronic band structure of the semiconductor that describes the available and unavailable range of energy states [161,162]. DFT is a commonly used method for predicting the electronic band structure of a solid. The effective mass of electronic charge carriers is defined as the apparent mass of charge carriers in the crystal and can be determined from the curvature of the CB and VB in the electronic band structure [161–163]. A physical quantity that determines the quality of charge carrier transport within a semiconductor is the charge carrier mobility μ, which is inversely proportional to the effective mass and the scattering rate of the carriers. Theoretically, for the holes and electrons in cubic $MAPbI_3$, effective masses of $m_h^* \sim 0.1\, m_0$ and $m_e^* \sim 0.15\, m_0$ have been calculated, comparable to those of the classical inorganic semiconductors [161,164,165]. For orthorhombic and the tetragonal phases of $MAPbI_3$, the effective masses obtained by first-principles calculations [161,165,166] are in good agreement with the reduced mass measured by magneto-absorption in thin films ($\sim 0.1m_0$) [167]. These relatively low effective masses suggest that high charge carrier mobilities in the range of several hundreds of $cm^2\,V^{-1}\,S^{-1}$ can be obtained for electrons and holes in $MAPbI_3$. However, the experimentally obtained mobilities and diffusion coefficients are not fully consistent with these theoretical predictions and experimental measurements [164,168,169]. Despite the low effective masses of the electronic charge carriers in hybrid perovskite compounds that were either theoretically predicted or experimentally measured, the values for mobilities (or diffusion coefficients) are lower than those of their classical inorganic counterparts. Over the past decade, a broad range of mobility values from 10^{-7} to $10^3\,cm^2\,V^{-1}\,S^{-1}$ have been reported for single crystals and polycrystalline thin films of hybrid perovskite compounds [164,168,169]. The significant discrepancy between the values of charge carrier mobility measured by different techniques on various samples may be the result of a difference in perovskite composition and fabrication conditions, resulting in different material properties. In addition, different experimental techniques to measure mobility can give rise to different values [168].

Considering such modest values of charge carrier mobility, long carrier [170–172] lifetimes of $\sim$0.1- to >1 μs seem to be the origin of the efficient carrier collection in PSCs [164,170]. Superposition of the two effects, namely, the modest charge carrier mobility and the long lifetime of the charge carriers, results in long diffusion lengths of $\sim$0.1- to >1 μm in thin films [167,171–173] and $\sim$2–15 μm in perovskite single crystals [170,174,175]. The efficient charge carrier collection in PSCs is a direct consequence of long diffusion lengths that are several times longer than the absorption depth [164,170,174,175].

From relativistic GW calculations, Umari et al. showed that $MASnI_3$ exhibits excellent optoelectronic properties similar to $MAPbI_3$. Electron transport is more efficient than in $MAPbI_3$ ($\mu = 10^2$–$10^3\,cm^2\,V^{-1}\,s^{-1}$ vs $\mu = 10$–$10^2\,cm^2\,V^{-1}\,s^{-1}$) [165]. From temperature-dependent Hall mobility measurements on $MASnI_3$, Takahashi et al. measured μ_p (250 K) = $200\,cm^2\,V^{-1}\,s^{-1}$, which is comparable to silicon ($500\,cm^2\,V^{-1}\,s^{-1}$) and GaAs ($320\,cm^2\,V^{-1}\,s^{-1}$) [176].

Ran et al. observed short carrier diffusion length, high trap density, and space charge limited current (SCLC) mobility of $3.47 \times 10^{-7}\,cm^2\,V^{-1}\,s^{-1}$ for $MA_3Bi_2I_9$ thin films [177]. Using time-resolved THz photoconductivity analysis, Milot et al. showed that $FASnI_3$ has charge-carrier mobility of $22\,cm^2\,V^{-1}\,s^{-1}$. Moreover, $FASnI_3$ exhibits strong radiative bimolecular recombination and low Auger recombination rates similar to GaAs, which is promising for light-emitting devices [178]. Hall effect measurements of the black orthorhombic perovskite phase of $CsSnI_3$ indicated a p-type direct bandgap semiconductor with a hole mobility of $\sim 585\,cm^2\,V^{-1}\,s^{-1}$ [179].

These observations indicate that similar to Pb-based perovskites, the reported values of charge carrier mobilities in Pb-free perovskites are also covering a broad range of values. Additionally, one can recognize that the transport properties of Pb-free perovskites are comparable to their Pb-based counterparts, which makes them promising candidates for replacing the Pb-based perovskites in the future.

5. Pb-free perovskite for optoelectronic devices

5.1 Solar cells

As described in Section 2, host cations can potentially be employed to replace Pb. The most promising candidates are Sn^{2+}, Ge^{2+}, Ni^{2+}, Mn^{2+}, Mg^{2+}, and Co^{2+} [180]. Sn-based perovskites were the first to be explored as Pb-free perovskites compounds because Sn^{2+} has an ionic radius (1.35 Å) similar to Pb^{2+} (1.49 Å). Their lower bandgap and higher charge carrier mobility (10^2–10^3 $cm^2\ V^{-1}\ s^{-1}$) make them potentially even better candidates for photovoltaics (PVs) [181]. A large amount of work has been done on Sn-based organic-inorganic hybrid PSCs. Table 3 summarizes the architecture used and performance parameters for Sn-based PSCs to date.

Despite the promising optoelectronic properties of Sn-based perovskites, employing them in PSCs has been challenging due to the ease of oxidation of Sn^{2+} to Sn^{4+} in ambient conditions. Sn^{4+} is a p-type dopant, causing self-doping of perovskite [176] and limiting PCE. The other factor that has impeded progress in Sn-based perovskites is their poor film morphology. Rapid crystallization results in films with many pinholes. However, compositional engineering and novel deposition techniques have greatly improved the film quality of Sn-based perovskites.

Several methods have been implemented to stop the process of oxidation, but as of now, no strategy could fully prevent the oxidation of Sn^{2+} to Sn^{4+}. Using an excess of Sn^{2+} precursors (SnI_2 [210], SnF_2 [185], and $SnCl_2$ [201]) in the precursor solution has been used to prevent the formation of Sn^{2+} vacancies. Of these, SnF_2 is the most effective and most widely used. It has been shown to prevent oxidation and improve film morphology. This leads to reduced background carrier density and intrinsic defects [18,179,193]. Reducing agents, such as hydrazine (N_2H_4) and hypophosphorous acid (H_3PO_2), have been used in combination with SnF_2, resulting in improved stability [191]. $SnCl_2$ has been found to concentrate on the surface of $CsSnI_3$, helping to prevent the penetration of water and oxygen. When exposed to an ambient atmosphere, $SnCl_2$ gets converted to SnO_2, which acts as an electron transport layer [201,211]. Although progress is made on Sn-based PSCs, their efficiency remains much lower than

TABLE 3 Various organic/inorganic cation-based Sn-perovskite and their corresponding device architecture and parameters.

Sn-perovskites	Device-structure	J_{SC} (mA cm^{-2})	V_{OC} (V)	FF	PCE (%)	Ref.
$MASnI_3$	Inverse	17.8	0.60	30	3.2	[182]
$MASnI_3$	Regular	16.8	0.88	42	6.4	[183]
$MASnI_3$ (hydrazine)	Regular	22.9	0.49	64	7.1	[158]
$MASnI_3$ (SnF_2)	Inverse	20.7	0.57	66	7.8	[184]
$MASnI_3$ ($SnF_2+N_2H_4$)	Regular	19.9	0.38	52	3.9	[185]
$MASnI_{2.6}Br_{0.4}$	Inverse	1.0	0.15	51	0.1	[186]
$MASnI_{1.5}Br_{1.5}$	Regular	5.0	0.45	48	1.1	[187]
$MASnI_2Br$	Regular	11.7	0.82	57	5.7	[17]
$MASnBr_3$	Regular	7.9	0.88	59	4.3	[188]
$FASnI_3$	Inverse	11.7	0.04	23	0.1	[189]
$FASnI_3$ (SnF_2)	Inverse	22.1	0.47	61	6.2	[190]
$FASnI_3$ (SnF_2+pyrazine)	Regular	23.7	0.32	63	4.8	[191]
TG-$FASnI_3$	Inverse	22.4	0.73	72	11.8	[192]
$FASnI_3$ (SnF_2)	Inverse	24.5	0.24	36	2.1	[193]
$FASnI_3$ (SnF_2+EDAI_2)	Inverse	21.3	0.58	72	8.9	[194]
$FASnI_3$ (5-AVAI)	Inverse	18.9	0.59	62	7.0	[195]
$FASnI_3$ (SnF_2+TMA)	Inverse	22.45	0.47	68	7.1	[196]
$FASnI_3$ (CDTA)	Inverse	21.83	0.64	74	10.3	[197]
$FASnI_3$ (THDH)	Inverse	22.1	0.54	71	8.48	[198]
$FASnI_2Br$	Inverse	6.8	0.47	54	1.7	[199]
$CsSnI_3$	Inverse	9.5	0.52	61	3	[200]
$CsSnI_3$ ($SnCl_2$)	Inverse	9.9	0.50	68	3.6	[201]
$CsSnI_3$ (SnI_2)	Inverse	8.5	0.47	54	2.13	[202]
$CsSnBr_3$ (SnF_2)	Regular	9.00	0.41	58	2.1	[203]
$FA_{0.75}MA_{0.25}SnI_3$ (SnF_2)	Inverse	21.2	0.61	63	8.12	[204]
$FA_{0.75}MA_{0.25}SnI_{2.75}Br_{0.25}$ (SnF_2+MACl)	Inverse	22.3	0.52	0.7	8.07	[205]
$FA_{0.8}Cs_{0.2}SnI_3$	Inverse	16.1	0.24	36	1.4	[189]

TABLE 3 Various organic/inorganic cation-based Sn-perovskite and their corresponding device architecture and parameters—cont'd

Sn-perovskites	Device-structure	J_{SC} (mA cm^{-2})	V_{OC} (V)	FF	PCE (%)	Ref.
$CsPb_{0.9}Sn_{0.1}IBr_2$	Regular	14.3	1.26	63	11.33	[206]
$CsSn_{0.3}Pb_{0.7}I_3$	Inverse	20.96	0.64	70	9.4	[207]
$PEA_{0.15}FA_{0.85}SnI_3$	Inverse	17.4	0.94	75	12.4	[68]
$(BEA)FA_2Sn_3I_{10}$	Inverse	18.9	0.62	55	6.43	[208]
$(FA_{0.9}EA_{0.1})_{0.98}EDA_{0.01}SnI_3$	Inverse	20.32	0.84	78	13.24	[22]
$CsSn_{0.5}Ge_{0.5}I_3$	Inverse	18.61	0.63	60	7.11	[209]

Regular and inverse device structures represent n-i-p (regular) and p-i-n (inverse) architectures. *5-AVAI*, 5-ammonium valeric acid iodide; *CDTA*, 2-cyano-3-[5-[4-(diphenylamino)phenyl]-2-thienyl]-propenoic acid; *EDAI$_2$*, ethylenediammonium diiodide; *TG*, templated growth; *THDH*, trihydrazine dihydriodide; *TMA*, trimethylamine.

Pb-based perovskites. One critical issue with Sn-PSCs is the large open-circuit voltage (V_{OC}) loss resulting from a high trap density [10]. Reducing additives [191,201,212] and manipulation of perovskite dimensionality [213–215] have improved the V_{OC} of the Sn-based devices from ~0.3 V to ~0.6 V [216]. However, the V_{OC} loss for Sn-based PSCs is still >0.7 V while that of Pb-based perovskite is only ~0.4 V. Hence, one obvious way for raising the PCE of Sn-based perovskites would be to minimize the voltage loss. Recently, the use of indene-C60 bisadduct (ICBA) as ETL in $PEA_xFA_{1-x}SnI_3$ PSCs has demonstrated a V_{OC} of over 0.9 V [68]. Better energy level alignment of ICBA (as compared to C60) improves attainable voltage and reduces interfacial recombination. Addition of NH_4SCN and using PEDOT:PSS as hole transporting layer further reduce the trap density, resulting in a high V_{OC} of 0.94 V with a PCE of 12.4%. Moreover, the device shows excellent shelf life, losing only 10% of the initial efficiency after 5 months of storage (Fig. 8). In mixed Pb/Sn perovskites, adding Sn powder to the precursor solution has been shown to reduce the amount of Sn^{4+} in the solution through comproportionation [217]. Further improvements were made by adding a zwitterionic molecule, which prevents the oxidation of Sn^{2+} even in the ambient atmosphere [218]. These strategies can potentially be used in pure Sn-based perovskites as well to further boost device efficiency.

Sb and Bi, both close to Pb in the periodic table, have ionic radii and electronic configurations similar to Pb. This allows them to be effectively incorporated into the perovskite lattice. Sb and Bi form lower dimensional $A_3M_2X_9$ perovskites and have emerged as promising Pb-free candidates as they are nontoxic and stable under ambient conditions. $MA_3Bi_2I_9$ shows

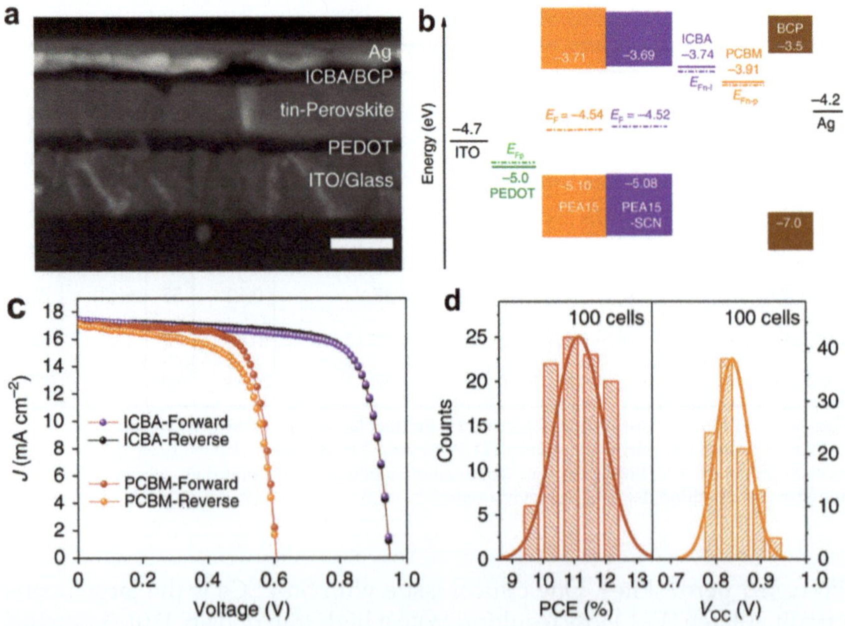

FIG. 8 (A) Cross-section SEM image (scale bar = 200 nm) of $PEA_{0.15}FA_{0.85}SnI_3$-based PSC (PEA15-SCN) with ICBA as ETL and PEDOT as HTL device (NH_4SCN was used as an additive in the perovskite). (B) Schematic illustration of energy levels of materials in the device. Dashed lines represent the quasi-Fermi level of ICBA (E_{Fn-I}), PCBM (E_{Fn-P}), and PEDOT (E_{Fp}). (C) *J–V* curves of the PEA15-SCN device with ICBA and best device of PEA15-SCN film with PCBM as ETL. (D) Histograms for PCE and V_{OC} of PEA15-SCN device. *Reproduced with permission from reference X. Jiang, et al., Ultra-high open-circuit voltage of tin perovskite solar cells via an electron transporting layer design, Nat. Commun. 11 (1) (2020). Copyright 2021, Nature.*

an indirect bandgap of 2.04 eV and better long-term stability in an environment with a relative humidity of 61% in comparison to $MAPbI_3$. $MA_3Bi_2I_9$, $Cs_3Bi_2I_9$, and $MA_3Bi_2I_{9-x}Cl_x$, with optical bandgaps of 2.1, 2.2, and 2.4 eV, respectively, have been found to yield 1%, 0.1%, and 0.003% PCE [219]. The relatively low PCE is in part caused by the rapid crystallization from solution, resulting in low-quality films. A 2-step approach to obtain $MA_3Bi_2I_9$ films from evaporated BiI_3 film under high vacuum results in compact, pinhole-free layers with large grains. Still, the PCE of just ~1.6% is still far behind that of $MAPbI_3$ [220]. As the wide bandgap of $MA_3Bi_2I_9$ (2.1 eV) is nonideal, attempts have been made to lower the bandgap by doping. For instance, sulfur doping of $MA_3Bi_2I_9$ perovskites at a relatively low temperature (120°C) lowers the bandgap to 1.45 eV, which is even lower than $MAPbI_3$ [221].

$A_3Sb_2X_9$-based perovskites form lower dimensional structures [222]. 2D $Cs_3Sb_2I_9$ has a high absorption coefficient that is comparable to Pb-based perovskites. However, the less preferable dimer structure is formed when deposited via solution processes [13]. This can be prevented by replacing Cs with Rb; the resulting $Rb_3Sb_2I_9$ forms a stable 2D layered material yielding a PCE of 0.66% [14]. The morphology of double cation $A_3Sb_2I_9$ (A=Cs, MA) can be improved using additives, resulting in a PCE of 2.04% [223]. For $MA_3Sb_2I_9$, dimer formation can be prevented by incorporating Cl^-, yielding a PCE of 2% [224]. Efforts employing Bi and Sb perovskites are summarized in Table 4.

Bi and Sb form double perovskites ($A_2B^+B^{3+}X_6$) in combination with monovalent metals (Ag, Au, Cu, K) [27]. $Cs_2BiAgBr_6$ PSCs have shown PCEs in the range of 1%–3% [231–233]. Theoretical investigations suggest that Bi/In and Bi/Tl double perovskite materials have direct bandgaps of around ~2eV [234]. However, as Tl is even more toxic than Pb, its use in PSCs is not attractive [235]. The performance of Ag/Bi perovskites is largely limited by their wide and indirect bandgap. The development of a narrower direct bandgap double perovskite material (through doping) is thus an important goal to further improve the performance of these materials. Nevertheless, with a V_{OC} of ~1V achieved in $Cs_2AgBiBr_6$

TABLE 4 Optical band gap and photovoltaic parameters of various perovskite, perovskite-like, and nonperovskite materials based on Bi, Sb, Ag/Bi-halide, and Cu.

Active layer	E_g (eV)	J_{SC} (mA cm^{-2})	V_{OC} (V)	FF	PCE (%)	Ref.
$MA_3Bi_2I_9$	2.1	1.00	0.67	62.5	0.42	[225]
$MA_3Bi_2I_9$	2.22	1.39	0.83	34	0.39	[177]
$MA_3Bi_2I_9$	2.1	1.10	0.65	0.50	0.36	[226]
$MA_3Bi_2I_9$	–	0.94	0.51	0.61	0.31	[227]
$MA_3Bi_2I_9$	–	0.83	0.56	49	0.26	[228]
$C_5H_6NBiI_4$	1.98	2.71	0.62	0.54	0.9	[229]
$(H_3NC_6H_{12}NH_3)BiI_5$	2.1	0.12	0.40	43	0.03	[230]
$Cs_3Bi_2I_9$	2.03	2.15	0.85	60	1.09	[219]
$CsBi_3I_{10}$	1.77	3.4	0.31	38	0.4	[230]
$MA_3Sb_2I_9$	2.14	1.0	0.9	55	0.49	[12]
$MA_3Sb_2I_9$	1.95	5.41	0.62	60.	2.04	[223]
$Rb_3Sb_2I_9$	2.1	2.11	0.55	57	0.66	[14]
$Cs_3Sb_2I_9$	2.05	<0.1	0.31	–	<0.1	[13]
$Cs_3Sb_2I_9$	2.0	2.91	0.6	48.1	0.84	[223]

double-perovskite and $MA_3Bi_2I_9$-based devices [231], these materials may be suitable for tandem solar cells.

The overall combination of A, B, and X leads to distinct electro-optical properties with different levels of structural stability. There are many promising Pb-free perovskites for solar cell applications. However, Sn, Bi, and Sb-based PSCs have not yet shown the same performances as Pb-based or double metal Pb/Sn devices. It is thus evident that, despite the massive potential of Pb-free perovskites, much work remains to be done.

5.2 LED

Lead halide perovskites are also promising materials for light-emitting diodes (LEDs) [236]. Perovskite LEDs can yield high luminance at a low driving voltage, a feat that is challenging for existing LEDs [237,238]. Mo et al. made centimeter-sized Cu(I)-based 1D $CsCu_2I_3$ single crystals (SCs) crystals (Fig. 9A), where Cu(I) fully substitutes Pb. Their high quality was confirmed by high-resolution transient electron microscope (HRTEM) (Fig. 9B). The thermal analysis demonstrates high thermal stability as high as 480°C. $CsCu_2I_3$ SCs also exhibit high ambient air PL stability (>2000h). A record PLQY of 50.4% was obtained by using oleic acid as an additive during growth (see Fig. 9C); this is comparable to the best yellow-emitting Pb halide perovskites. PLQY values for Sn and Cu(I)-based perovskites are summarized in Table 5.

White LEDs using $CsCu_2I_3$ SCs as yellow phosphors (Fig. 9D) have a half-life of ~150h in ambient air at a continuous driving current of 50mA (Fig. 9E). These promising results show that Pb-free perovskites can be applied in solid-state lighting applications.

5.3 Detectors

The main metric for assessing the performance of photodetectors is the signal to noise (S/N) ratio, which in effect is the ratio between photocurrent and dark current. A high photocurrent is thus desirable in order to increase the S/N ratio. High responsivity photo-detecting devices have been made using $MAPbI_3$ [247–249]. However, Pb-free photodetectors are promising to be a good alternative due to their wider range of bandgaps, spanning from the near-infrared to the UV. Color image sensing in digital cameras is one area where the tunability and low-temperature processing of perovskites may provide an advantage over existing technology [250].

Device performances of some UV photodetectors are presented in Table 6. The response rate and detection limit so far have not been able

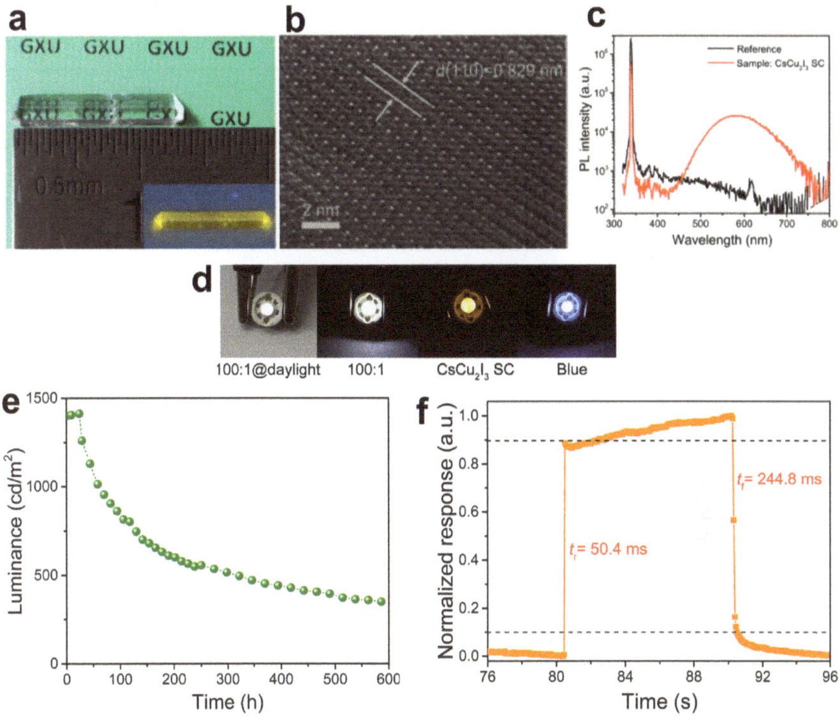

FIG. 9 (A) Photographs of typical 1D CsCu2I3 SC under ambient light and 365 nm UV light (inset). (B) HRTEM of the 1D CsCu2I3 SC. (C) Excitation and emission spectra of the Reference and 1D CsCu2I3 SC collected by Edinburgh Instruments FLS980 through a 120 mm integrating sphere. The excitation wavelength is set to be 340 nm. Absolute PLQY was determined by the ratio photons emitted and absorbed by the 1D CsCu2I3 SC. (D) Photographs of the LEDs with lighting up at driving current of 50 mA, showing bright blue-, yellow-, and white-light emissions. For the WLEDs, the weight ratio of the yellow to blue phosphors is used as 100:1. (E) Lifetime measurement of the WLED at a continuous driving current of 50 mA in air atmosphere. (F) Rising and falling edges of the PD to estimate the rise time (tr) and fall time (tf). *Reproduced with permission from reference X. Mo, et al., Highly-efficient all-inorganic Pb-free 1D CsCu2I3 single crystal for white-light emitting diodes and UV photodetection, Nano Energy 81 (2021). Copyright 2021, Elsevier.*

to match those of their Pb-based perovskite counterparts [256–258]. However, the response rate is strongly influenced by the device structure and big improvements can be made by moving away from the planar device configuration and large channel width that has been used so far. The film quality can also be further improved to prevent the formation of pinholes and grain boundaries. CsCu2I3 UV photodetectors exhibit a relatively fast response time as low as 50.4 ms (at 340 nm, see Fig. 9F). Using innovative deposition methods and vertical device architecture, device performance can be greatly enhanced, as has been shown in $Cs_2AgBiBr_6$ visible photodetectors [251].

TABLE 5 Summarization of PLQYs for all-inorganic light-emitting lead-free halide compounds.

Materials	Morphology	Dimension	Emission peak (nm)	PLQY (%)	Ref.
PEA_2SnI_4	Thin film	2D	633	1.35	[239]
$(OAm)_2SnBr_4$	Thin film	2D	621	68	[240]
$CsSnBr_3$	Thin film	3D	672	–	[241]
$CsCu_2I_3$	Nanorod	1D	561	11	[242]
$CsCu_2I_3$	Thin film	1D	548	20.6	[243]
$CsCu_2I_3$	Nanocrystal	1D	553	5.0	[244]
$CsCu_2I_3$	Powder	1D	576	3.23	[245]
$CsCu_2I_3$	SC	1D	580	50.4	[246]

SC, single crystal.

TABLE 6 Summary of the device parameters of lead-free perovskite UV photodetectors.

Photodetectors	Detectivity (Jones)	Responsivity	Response time	Detection limit	Ref.
$ITO/Cs_3Cu_2I_5/ITO$	1.10×10^9	$0.11\,mA\,W^{-1}$	–	$111\,\mu W\,cm^{-2}$	[251]
$Au/Cs_3Cu_2I_5/Au$	1.12×10^{12}	$17.8\,A\,W^{-1}$	465/897 μs	$0.07\,mW\,cm^{-2}$	[252]
$Au/Cs_3Cu_2I_5/GaN/In$	1.4×10^{12}	$0.28\,A\,W^{-1}$	95/130 ms	$1\,\mu W\,cm^{-2}$	[253]
$Cs_3Cu_2I_5$/Si-core/shell nanowires	3.10×10^{12}	$130\,mA\,W^{-1}$	92.5/189.2 μs	$12\,\mu W\,cm^{-2}$	[254]
$Cs_2AgInCl_6$ SCs	10^{12}	$0.013\,A\,W^{-1}$	1 ms	$0.1\,\mu W\,cm^{-2}$	[255]

SC, single crystal.
Reproduced with permission from reference F. Cao, X. Xu, D. Yu, H. Zeng, Pb-free halide perovskite photodetectors spanning from near-infrared to X-ray range: a review, Nanophotonics (2021). Copyright 2021, De Gruyter.

Besides light detection, Pb-free perovskites can also be used to detect X-rays. Bi-based perovskites exhibit strong X-ray absorption, large bulk resistance, suppressed ion migration, and efficient charge collection. This makes them promising candidates for X-ray detectors with high sensitivities and low detection limits.

5.4 Memristor

Cheng et al. used $Cs_2AgBiBr_6$ to fabricate an environmentally robust memristor (Fig. 10A, B) with a sandwich-like structure of $ITO/Cs_2AgBiBr_6/Au$. The device exhibited good memory performance, including 1000 "writing" and "erasing" cycles [259]. The device can operate under

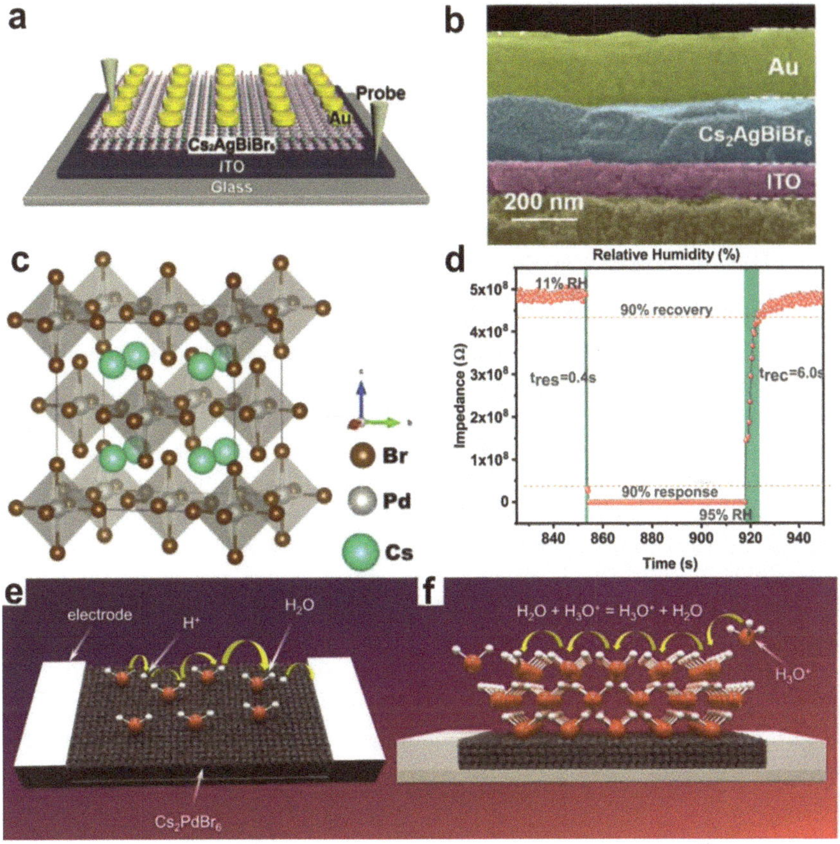

FIG. 10 (A) Typical sandwiched structure. (B) Scanning electron microscopic cross-section view of ITO/Cs2AgBiBr6/Au memory device. (C) 3D crystal structure of Cs2PdBr6 showing the unit cell. (D) Response and recovery time. (E, F) Schematic diagram of humidity sensing mechanism for Cs2PdBr6 humidity sensor. *(B) Reproduced with permission from reference X.F. Cheng, et al., Environmentally robust memristor enabled by Pb-free double perovskite for high-performance information storage, Small 15 (49) (2019). Copyright 2021, Wiley-VCH. (E, F) Reproduced with permission from reference W. Ye, Q. Cao, X.F. Cheng, C. Yu, J.H. He, J.M. Lu, A Pb-free Cs2PdBr6 perovskite-based humidity sensor for artificial fruit waxing detection, J. Mater. Chem. A 8 (34) (2020). Copyright 2021, Royal Society of Chemistry.*

harsh conditions, such as high humidity of 80% (3 months, 25°C), temperatures up to 180°C, and γ-ray irradiation for a dose of 5×10^5 rad. The excellent stability was attributed to the strong Ag-Br bond and good crystallinity of $Cs_2AgBiBr_6$.

5.5 Humidity sensor

Pb-Free Cs_2PdBr_6 (see Fig. 10C) can be used as humidity, which was investigated by Ye et al. As shown in Fig. 10D, the sensor exhibited a response time of 0.7 s and recovery time of 1.7 s with an impedance variation range of 105 Ω under relative humidity ranging from 11% to 95%. At low humidity conditions, only a small amount of water molecules are adsorbed. The water molecules dissociate to form free protons, which hop to adjacent water molecules to form H_3O^+. When an external electric field is applied, the protons migrate toward one side of the device (Fig. 10E). For high humidity conditions, a continuous water film is formed. In this case, the migration of H_3O^+ is the dominant process (Fig. 10F). One possible application is in detecting waxing on apples and oranges [260].

To conclude, Sn-based perovskites are promising materials for solar cells, but poor stability and film quality are two main issues that need to be addressed in order to achieve high-efficiency PSCs. Ternary perovskites based on Bi and Sb have shown promising environmental stability. Efforts should be focused here on reducing the bandgap and improving film morphology. Double perovskites based on the Ag/Bi combination are promising materials; understanding the mechanism of defects passivation of these compounds can help to improve the device performance.

Due to the distinct optoelectronic properties of Pb-free perovskites, they can be applied beyond solar cells, including LEDs, photodetectors, X-ray detectors, memristors, and humidity sensors. For example, LEDs based on $CsCu_2I_3$ SC showed superior stability under thermal stress with a record PLQY of 50.4% and photodetectors fabricated by the 1D $CsCu_2I_3$ SC show distinct UV response with a fast response time as low as 50.4 ms. Moreover, the strong Ag-Br bond and good crystallinity of $Cs_2AgBiBr_6$, result in a memristor with excellent memory performance including 1000 writing and erasing cycles.

6. Stability of Pb-free perovskites

Pb-free perovskites have found their functions in different applications, including photodetectors, lasers, and photovoltaic applications. Nevertheless, due to their poor stability against heat, light, oxygen, and humidity, these materials are still far from commercial implementation. In the past few years, many efforts have been made to overcome this problem and improve the long-term stability of Pb-free perovskites. It is worth noting that most approaches for stabilizing Pb-free perovskites are still inspired by photovoltaics [261]. Various stabilities have been reported for different types of Pb-free hybrid perovskites. The following discussion has a focus on previous efforts in stabilizing the Pb-free perovskites.

6.1 Sn-based halide perovskites

Sn-based halide perovskites own promising optoelectronic properties such as higher charge-carrier mobility, tunable bandgap, and long carrier diffusion length [262]. But their stability under ambient atmosphere is poor. This originates from the facile oxidation of Sn^{2+} to Sn^{4+} even in the presence of trace amounts of oxygen and water [185].

Many strategies have been employed to improve the stability of Sn-based perovskite and inhibit the oxidation of Sn^{2+} [263]. One proven approach to suppress the instability of Sn^{2+} is incorporating various kinds of additives, such as tin halide, hypophosphorous acid (HPA), ethylenediammonium diiodide, and Lewis bases in the precursor solution [18,190,264]. One of the most effective additives that have been used for stabilizing the Sn-based perovskite is ethylenediammonium [262]. Incorporation of this cation enhanced the air stability of the perovskite absorber and improved its photoelectric properties. Encapsulated PSCs retained 96% of the initial efficiency after aging over 1000 h. Jokar et al. used guanidinium (GA^+), a nonpolar organic cation, in combination with ethylenediammonium diiodide as an additive. The initial PCE of 8.5% increased to achieve a record PCE of 9.6% after storage in a glovebox environment for 2000 h. Moreover, the unencapsulated devices were stable under continuous illumination for 1 h [263].

As discussed in Section 5, the incorporation of tin halides (i.e., SnF_2, $SnCl_2$, SnI_2, $SnBr_2$) in the system is a general approach for enhancing the stability of Sn-based perovskites [261,264]. In 2014, Kumar et al. first reported that adding SnF_2 into the precursor solution notably improved the performance and stability of the perovskite. SnF_2 is stable in air and moisture, and the fluoride ion in SnF_2 acts as a stabilizer against oxidation [18]. Liao et al. used SnF_2 along with antisolvent (diethyl ether) dripping to prepare high-quality, uniform, and pinhole-free $FASnI_3$ thin films. The encapsulated devices indicated high long-term stability under illumination, retaining over 94% of the initial efficiency and current density after 2700 s. Moreover, the encapsulated devices were stable for 30 days in the glove box, retaining 85% of the initial efficiency [190].

Although the incorporation of tin halides such as SnF_2 in perovskite is a commonly applied method for preventing Sn^{2+} oxidation and stabilizing the Sn-based perovskites [191,261], the excess tin halide could potentially cause severe phase separation in the film leading to poor film morphologies [185,191]. Many strategies have been applied to address this issue. For instance, pyrazine as a mediator has a good binding affinity to SnF_2 which could hinder the phase separation caused by excess SnF_2. A homogeneous dispersion of SnF_2 was made via the formation of the SnF_2-pyrazine complex in $FASnI_3$ perovskite [191]. The PSCs maintained 98% of the initial efficiency over 100 days. The $SnCl_2$-additive complex formation for eliminating phase separation was also achieved using hydroxybenzene sulfonic acid or its salt in $FASnI_3$ [212]. Fig. 11A shows the schematic illustration of the fabrication of $FASnI_3$ film in the presence of excess $SnCl_2$ and the potassium salt of hydroquinone sulfonic acid. Fig. 11B, C show the aggregates on the perovskite film caused by $SnCl_2$. Fewer aggregates form in the presence of the additive (Fig. 11C) in

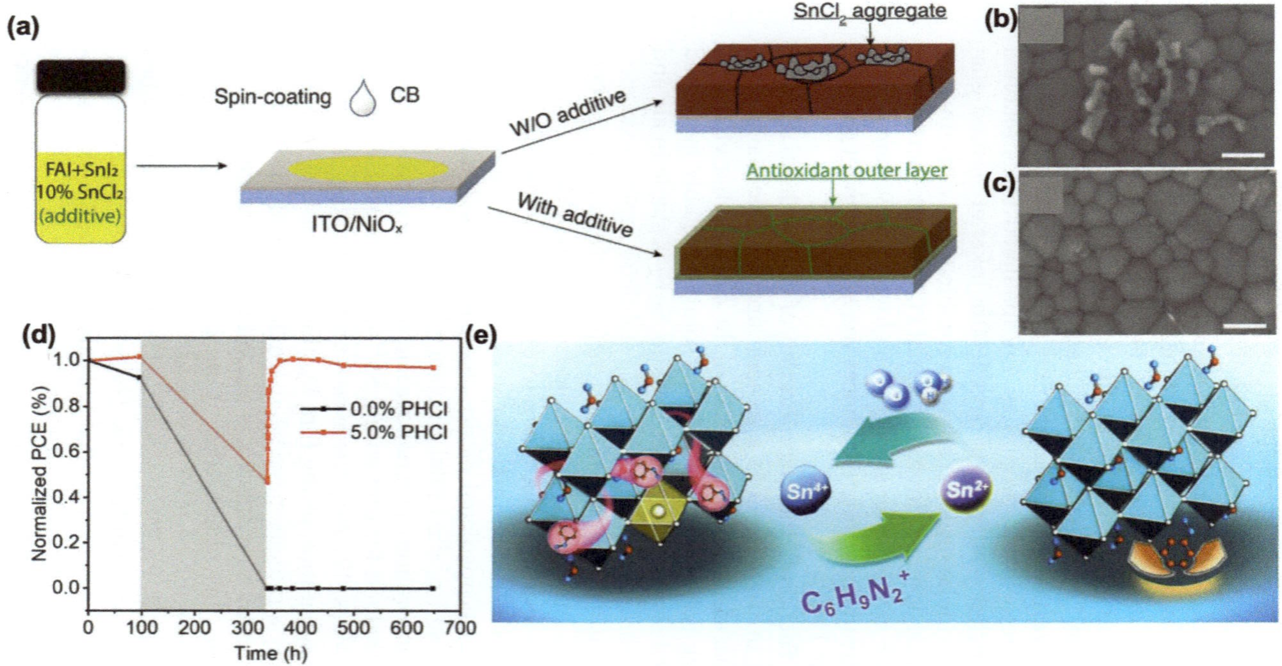

FIG. 11 (A) The scheme of the method for perovskite film preparation and the morphologies of the perovskite films in the presence of excess $SnCl_2$ with and without additive. SEM images of the (B) pristine and (C) additive modified FASnI3 films (the scale bar is 500nm). (D) The stability of the devices with PHCl for the self-repairing process. (E) Schematic illustration of possible SRTSP mechanism for PHCl. *(B, C) Reproduced with permission from reference Q. Tai, et al., Antioxidant grain passivation for air-stable tin-based perovskite solar cells, Angew. Chem. Int. Ed. 58 (3) (2019) 806–810. Copyright 2021, Wiley-VCH. (E) Reproduced with permission from reference C. Wang, et al., Self-repairing tin-based perovskite solar cells with a breakthrough efficiency over 11%, Adv. Mater. 32 (31) (2020) 1907623. Copyright 2021, Wiley-VCH.*

comparison to the perovskite film without additive in the precursor solution (Fig. 11B). Furthermore, the reducing hydroxybenzene group of the additive molecule served as an O_2 scavenger or antioxidant to further protect the perovskite. The resulting nonencapsulated $FASnI_3$ solar cells maintained 80% of the efficiency for over 500h in air.

Considering that the inferior instability of Sn-based perovskites is mainly originated from the oxidation of Sn^{2+}, reducing the Sn^{4+} to Sn^{2+} provides a practical approach for enhancing stability. Various reducing agents such as hypophosphorous acid (HPA) [265], metallic tin powder [266], hydrazinium chloride [267], hydrazine vapor [185], and phenylhydrazine-containing compound [268] have been used for this purpose. Hydrazine is widely known as a strong reducing agent and a strong base, potentially preventing or suppressing Sn^{4+} formation. Song et al. [185] introduced a reducing hydrazine vapor atmosphere approach during the preparation of tin-based halide perovskites. The reducing vapor atmosphere process resulted in a more than 20% reduction of Sn^{4+}/Sn^{2+} ratios.

Phenylhydrazine hydrochloride (PHCl) is an additive with a reductive hydrazino group and a hydrophobic phenyl group. It was incorporated into $FASnI_3$ perovskite film to reduce the existing Sn^{4+} and prevent further degradation of perovskite. The PSCs showed superior stability in the glovebox for 110days [268]. After an efficiency drop by exposing the device to air, a remarkable recovery in efficiency was observed when the device was stored back in the glove box (Fig. 11D). A mechanism for self-repairing capability was proposed using PHCL as a self-repairing trap state passivator (SRTSP). PHCL showed both the passivation and self-repairing of trap states. The reductive hydrazino group of PHCl induced reduction of Sn^{4+} and passivated Sn-vacancy, while the large hydrophobic phenyl group of PHCl blocks water and oxygen (Fig. 11E).

Recently, ligand modification strategies with various large organic molecules have been developed for improving the stability of Sn-based perovskites. In this strategy, the perovskite is protected against moisture and oxygen by functionalizing the grain boundaries and film surface using the hydrophobic organic ligands attached to the surface. Li et al. [269] developed a film-formation strategy for the simultaneous restriction of Sn^{2+} oxidation and regulation of crystallization using fluoro-aniline isomers. Due to their hydrophobic nature, fluorinated ligands enhanced the structure stability, by passivating defect-related trap states and protecting the perovskite from moisture. The efficiency of the device with 2-F-PEAI additives exhibited only 15% PCE loss after 1600h stored in a nitrogen-filled glove box under continuous illumination. Enhancing the stability of Sn-based perovskite has also been achieved by applying the π-conjugated Lewis base 2-cyano-3-[5-[4-(diphenylamino)phenyl]-2-thienyl]-propenoic acid (CDTA) [212]. In addition to the formation of uniform CDTA-treated $FASnI_3$ films, the introduction of the π-conjugated systems also enhanced the stability of the perovskite layer by retarding moisture penetration.

An in-situ encapsulation of Sn-based perovskite was reported by Chen et al. [270] using the bulky organic cation 4-(aminomethyl)piperidinium (4AMP), a bulky cation also used to obtain low-dimensional Dion-Jacobson perovskites. Fig. 12A shows the schematic illustration of the 4AMP functionalization of both grain surfaces and grain boundaries in $FASnI_3$ perovskite polycrystalline thin film. In addition to enhancing device performance, 4AMP was efficient in enhancing the stability of PSCs. The devices maintained 77% of the initial efficiency for 500 h under illumination. The solid ionic bonding caused by 4AMP provided a functionalization of perovskite surface and grain boundaries, hindering oxygen and water ingress and suppressing the generation of Sn-defects.

An essential factor in the stability of Pb-free perovskites, including Sn-based perovskite, is the appropriate choice of A-site cation to ensure adequate protection against water penetration [190]. It has been reported that $FASnI_3$ shows a lower rate of oxidation of Sn^{2+} and better stability than $MASnI_3$. Replacing the organic cations (e.g., MA^+ and FA^+) with Cs^+ to form all-inorganic Sn-based halide perovskites has further shown great potential for achieving superior thermal stability [265].

Inspired by low-dimensional Pb-based perovskites, quasi-2D or hybrid 2D/3D Pb-free perovskites were synthesized by replacing the small hydrophilic cations with bulkier hydrophobic ammonium cations such as butylammonium (BA^+) or phenylethylammonium (PEA^+). Low-dimensional Sn-based perovskites have shown higher resilience against oxidation, humidity, and light than their 3D counterparts [208]. 1,4-Butanediamine (BEA) was incorporated into $FASnI_3$ to prepare low-dimensional Dion-Jacobson-phase $(BEA)FA_{n-1}Sn_nI_{3n+1}$ perovskite. The introduction of BEA stabilized the perovskite structure and inhibited the oxidation of Sn^{2+}. The obtained unencapsulated solar cells retained 90% of the initial efficiency over 200 h under ambient environment, while the $FASnI_3$ devices fully degraded within 72 h [208].

For addressing the stability issues of Sn-based perovskites ($MASnI_3$ nanowires) in photodetectors, a strategy based on the application of porous alumina templates was presented by Waleed et al. [271]. It was shown that by embedding nanowires in mechanically and chemically robust templates, the device lifetime was increased by up to three orders of magnitude in comparison with a planar thin film. The superior stability is a result of the effective protection from water and oxygen provided by the templates.

Perovskite morphology is also an important parameter in stability. Wang et al. [272] showed that hollow $CsSnBr_3$ perovskite cubic nanocages are more stable than solid cubes. The change in the absorbance of the perovskite films over time is shown in Fig. 12B, C. Moreover, using perfluorooctanoic acid (PFOA) as a strong electron-withdrawing capping agent further stabilizes the Sn^{2+} on the surface of hollow nanocubes (Fig. 12D).

6.2 Bi and Sb-based halide perovskites

Bi and Sb-based perovskites are promising candidates for nontoxic Pb-free perovskites due to their stability. Due to their trivalent oxidation state, they form $A_3B_2X_9$ perovskites [273,274].

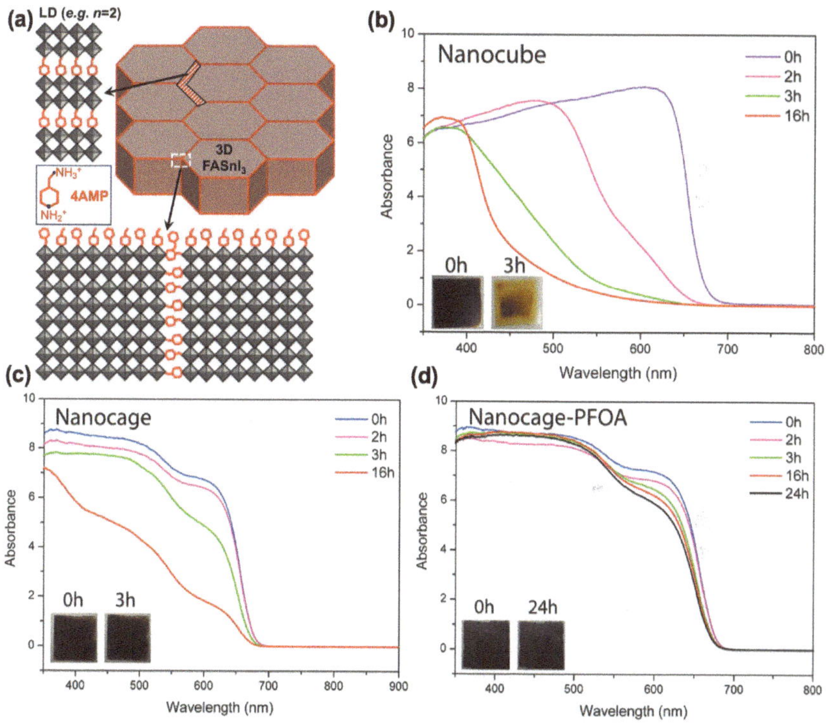

FIG. 12 (A) Schematic illustration of 4AMP functionalization of grain surfaces and grain boundaries in FASnI3 perovskite polycrystalline thin film, together with occasional low-dimensional nanoplatelet. (B) Absorbance spectra of the CsSnBr3 nanocubes at different times. (C) Absorbance spectra of the CsSnBr3 cubic nanocages. (D) Absorbance spectra of CsSnBr3 nanocages after treatment with PFOA. Insets in B, C, and D are the as-prepared films on a glass slide and after air exposure for 3h. *(A) Reproduced with permission from reference M. Chen, et al., High-performance Pb-free solar cells based on tin-halide perovskite thin films functionalized by a divalent organic cation, ACS Energy Lett. 5, (7) (2020) 2223–2230. Copyright 2021, American Chemical Society. (D) Reproduced with permission from reference A. Wang, Y. Guo, F. Muhammad, Z. Deng, Controlled synthesis of Pb-free cesium tin halide perovskite cubic nanocages with high stability, Chem. Mater. 29 (15) (2017) 6493–6501. Copyright 2021, American Chemical Society.*

Typically, Bi-based perovskites such as $MA_3Bi_2I_9$ and $Cs_3Bi_2I_9$ exhibit superior intrinsic stability to $MAPbI_3$ [219,275]. Park et al. [219] showed that Bi-based halide perovskite with Cs^+ and MA^+ organic cations was stable for one month in the dark and dry air. Kulkarni et al. [227] fine-tuned the morphology of the $MA_3Bi_2I_9$ perovskite by using *N*-methyl-2-pyrrolidone (NMP) as a morphology controller. The prepared perovskite showed excellent stability over 30 days under ambient conditions (without encapsulation). Lyu et al. [273] showed that $(MA)_3Bi_2I_9$ perovskite is stable in ambient air (~50% RH) for more than 40 days. Excellent long-term stability was presented by Jain et al. [276] using a vapor-assisted solution process for the controlled formation of uniform $MA_3Bi_2I_9$ perovskite films. The solar cells exhibited a negligible drop in efficiency under continuous illumination. The high stability of this material is attributed to its layered nature.

Similar to the Sn-based perovskites, the choice of A-site cation of perovskite affects the stability in Bi-based perovskites. Superior tolerance to humid conditions and heat can be achieved through the use of bulkier organic cations. A dense and pinhole-free $(C_6H_5NH_3)BiI_4$ perovskite was prepared using a gas pump treatment technique. The resulting film was stable under constant moisture exposure for one year [275].

Inorganic cations also enhance the stability of the Bi-based perovskites. All inorganic $Cs_3Bi_2X_9$ nanocrystals are promising materials due to their high stability. $Cs_3Bi_2Br_9$ quantum dots showed high thermal stability with only a slight reduction in PL intensity after being heated to 180 °C for 1 h. The enhanced thermal stability was attributed to a combination of the solid framework and chemical bonds in all-inorganic perovskite, and the strong ligand binding [277].

Analogous to Bi, Sb is also a popular trivalent substitution for Pb in perovskite that exhibits high stability [278]. $Cs_3Sb_2Br_9$ quantum dots were synthesized via a ligand-assisted reprecipitation technique [106]. $Cs_3Sb_2Br_9$ quantum dots maintained 70% of the initial PLQY when stored in air for 35 days without encapsulation (Fig. 13A). Furthermore, $Cs_3Sb_2Br_9$ quantum dots showed high photostability retaining 50% of their initial relative intensity under illumination for 108 h (Fig. 13B).

6.3 Germanium perovskites

Several theoretical studies have indicated that germanium halide perovskites are promising for optoelectronic applications such as photovoltaics [160]. Nevertheless, these materials have been rarely investigated experimentally because of the spontaneous oxidation of Ge^{2+} to Ge^{4+} [279,280]. Modifying the chemical composition can lead to enhanced

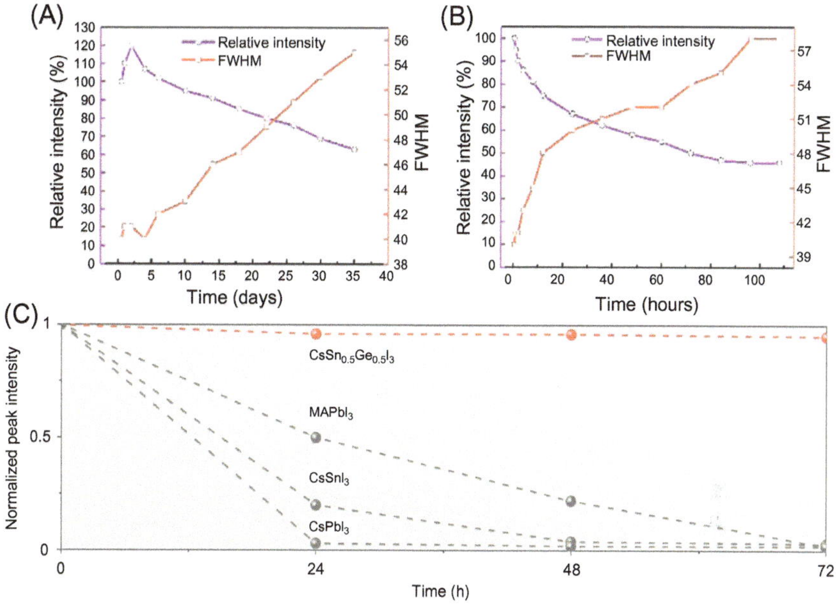

FIG. 13 (A) Air stability test. (B) Photostability test of Cs3Sb2Br9 quantum dots. (C) Plots of relative XRD peak intensities as a function of time. *(B) Reproduced with permission reference J. Zhang, et al., High quantum yield blue emission from Pb-free inorganic antimony halide perovskite colloidal quantum dots, ACS Nano 11 (9) (2017) 9294–9302. Copyright 2021, American Chemical Society. (C) Reproduced with permission from reference M. Chen, et al., Highly stable and efficient all-inorganic Pb-free perovskite solar cells with native-oxide passivation, Nat. Commun. 10 (1) (2019) 16. Copyright 2021, Nature.*

stability of the Ge-based perovskite. For example, partly substituting the X-site anion from iodide to bromide of $MAGeI_3$ perovskite resulted in improved ambient stability [280].

6.4 Alloyed perovskites

The rapid oxidation of Ge is a major concern in the Ge-based perovskites. Alternatively, Ge has been applied for alloying tin-based perovskites; $ASn_{1-x}Ge_xX_3$. Alloying Ge in Sn-based perovskite results in better stability under ambient conditions. The superior stability is due to the formation of a stable oxide layer on the surface of the perovskite because of the extremely high oxidation activity of Ge(II) [209].

Chen et al. [209] achieved high stability in $CsSn_{0.5}Ge_{0.5}I_3$ perovskites compared with Sn-based and even $MAPbI_3$ perovskite. The relative intensities of the main peaks in XRD patterns of $CsSn_{0.5}Ge_{0.5}I_3$, $CsSnI_3$, $CsPbI_3$, and $MAPbI_3$ as a function of time are shown in

Fig. 13C. $CsSn_{0.5}Ge_{0.5}I_3$ perovskite showed the highest stability because of the ultrathin uniform passivating layer on the perovskite surface.

In general, the use of B-site alloyed perovskites is an effective strategy for substituting toxic Pb atoms in perovskite materials [281]. Other elements have also been utilized in preparing stable double perovskite. For example, Zhang et al. [282] prepared a novel Pb-free double perovskite material with Na^+ and Bi^{3+} cations, Cs_2NaBiI_6, which exhibited good tolerance against moisture and oxygen in ambient air.

Obtaining perovskite materials with high stability and resistance to harsh working conditions is vital for practical applications. Pb-free perovskites have great potential for providing stable perovskite materials. However, work is still needed to develop this type of perovskites with long-term durability suitable for commercialization.

7. Future outlooks and remarks

The significant progress of halide perovskite structures in different fields including solar cells, LEDs, and sensors shows their huge potential for replacing existing technologies. Most halide perovskite materials contain Pb elements due to the superior material properties this yields. However, the main issue for Pb-based halide perovskites is the toxicity that causes environmental and health consequences. Therefore, replacing Pb with other elements is essential to tackle toxicity issues. As solar cells will make large-area solar modules, producing Pb-free devices is much more important than other optical devices with a small area such as LEDs. Therefore, much research has been carried out on Pb-free perovskites for solar cells. Despite these efforts, the PCE of Pb-free Sn-based PSCs is still only ~13% [283], as compared to ~25% for Pb-based PSCs. In addition, the poor stability (the oxidation of Sn^{2+} to Sn^{4+}) and film quality are two main challenges that need to be addressed to develop competitive Pb-free PSCs. Thus, incorporation of tin halides such as SnF_2 in perovskite is a commonly applied method for preventing Sn^{2+} oxidation and stabilizing the Sn-based perovskites, where the fluorinated ligands enhanced the structure stability, by passivating defect-related trap states and protecting the perovskite from moisture. Double perovskites, composed of an alloy of monovalent and trivalent metals, and low dimensional perovskites have shown good environmental stability. But the diffusion of halide ions needs to be further suppressed to increase the performance of these materials. This can be achieved by modifying the morphology and increasing the structural stability.

Pb-free perovskites have a wide bandgap range from the near infra-red to the UV. This opens up novel applications for Pb-free halide perovskites, such as LEDs and photodetectors. $CsCu_2I_3$ is a promising material for

LEDs with high PL stability and a record PLQY of 50.4%. $FASnI_3$-based LEDs exhibit strong radiative bimolecular recombination and low Auger recombination rates similar to GaAs. A $Cs_2AgBiBr_6$ perovskite memristor can operate under harsh conditions with excellent stability due to the strong Ag-Br bond and good crystallinity of $Cs_2AgBiBr_6$.

From these many potential applications, it is evident that there is a bright future for Pb-free perovskites. Continuing research efforts may well lead to materials that are capable of surpassing and replacing Pb-based devices, as well as finding new uses for this extraordinary class of materials.

References

[1] N. Arora, et al., Intrinsic and extrinsic stability of formamidinium lead bromide perovskite solar cells yielding high photovoltage, Nano Lett. 16 (11) (2016) 7155–7162.

[2] B. Wang, T. Chen, Exceptionally stable CH3 NH3 Pbi3 films in moderate humid environmental condition, Adv. Sci. 3 (2) (2015) 1–5.

[3] F. Giustino, H.J. Snaith, Toward Pb-free perovskite solar cells, ACS Energy Lett. 1 (6) (2016) 1233–1240.

[4] W. Ke, C.C. Stoumpos, M.G. Kanatzidis, Unleaded' perovskites: status quo and future prospects of tin-based perovskite solar cells, Adv. Mater. 31 (47) (2019).

[5] C. Wang, Z. Song, C. Li, D. Zhao, Y. Yan, Low-bandgap mixed tin-lead perovskites and their applications in all-perovskite tandem solar cells, Adv. Funct. Mater. 29 (47) (2019) 1–30.

[6] J. Sun, J. Yang, J.I. Lee, J.H. Cho, M.S. Kang, Pb-free perovskite nanocrystals for light-emitting devices, J. Phys. Chem. Lett. 9 (7) (2018) 1573–1583.

[7] Z. Weng, et al., Pb-free Cs2BiAgBr6 double perovskite-based humidity sensor with superfast recovery time, Adv. Funct. Mater. 29 (24) (2019).

[8] L. Yang, et al., Perovskite Pb-free dielectrics for energy storage applications, Prog. Mater. Sci. 102 (December 2018) (2019) 72–108.

[9] T. Zheng, J. Wu, D. Xiao, J. Zhu, Recent development in Pb-free perovskite piezoelectric bulk materials, Prog. Mater. Sci. 98 (January) (2018) 552–624.

[10] W. Ke, M.G. Kanatzidis, Prospects for low-toxicity lead-free perovskite solar cells, Nat. Commun. 10 (2019) 1–4.

[11] D. Cortecchia, et al., Pb-free MA2CuClxBr4-x hybrid perovskites, Inorg. Chem. 55 (3) (2016) 1044–1052.

[12] J.C. Hebig, I. Kühn, J. Flohre, T. Kirchartz, Optoelectronic properties of (CH3NH3) 3Sb2I9 thin films for photovoltaic applications, ACS Energy Lett. 1 (1) (2016) 309–314.

[13] B. Saparov, et al., Thin-film preparation and characterization of Cs3Sb2I9: a Pb-free layered perovskite semiconductor, Chem. Mater. 27 (16) (2015) 5622–5632.

[14] P.C. Harikesh, et al., Rb as an alternative cation for templating inorganic Pb-free perovskites for solution processed photovoltaics, Chem. Mater. 28 (2016) 7496–7504.

[15] M.N. Islam, M.A. Hadi, J. Podder, Influence of Ni doping in a lead-halide and a Pb-free halide perovskites for optoelectronic applications, AIP Adv. 9 (12) (2019) 1–9.

[16] A. Toshniwal, V. Kheraj, Development of organic-inorganic tin halide perovskites: a review, Sol. Energy 149 (2017) 54–59.

[17] F. Hao, C.C. Stoumpos, D.H. Cao, R.P.H. Chang, M.G. Kanatzidis, Pb-free solid-state organic-inorganic halide perovskite solar cells, Nat. Photonics 8 (6) (2014) 489–494.

[18] M.H. Kumar, et al., Pb-free halide perovskite solar cells with high photocurrents realized through vacancy modulation, Adv. Mater. 26 (41) (2014) 7122–7127.

[19] T. Leijtens, R. Prasanna, A. Gold-Parker, M.F. Toney, M.D. McGehee, Mechanism of tin oxidation and stabilization by lead substitution in tin halide perovskites, ACS Energy Lett. 2 (9) (2017) 2159–2165.
[20] F. Hao, et al., Solvent-mediated crystallization of CH3NH3SnI3 films for heterojunction depleted perovskite solar cells, J. Am. Chem. Soc. 137 (35) (2015) 11445–11452.
[21] T. Yokoyama, et al., Overcoming short-circuit in Pb-free CH3NH3SnI3 perovskite solar cells via kinetically controlled gas-solid reaction film fabrication process, J. Phys. Chem. Lett. 7 (5) (2016) 776–782.
[22] K. Nishimura, et al., Pb-free tin-halide perovskite solar cells with 13% efficiency, Nano Energy 74 (January) (2020), 104858.
[23] National Renewable Energy Laboratory, NREL_efficiencies_chart.pdf, 2020.
[24] D. Gesetze, et al., Die Gesetze der Krystallochemie, Naturwissenschaften, 1926.
[25] R. Wang, J. Wang, S. Tan, Y. Duan, Z. Wang, Y. Yang, Opportunities and challenges of Pb-free perovskite optoelectronic devices, Trends Chem. 1 (4) (2019) 368–379.
[26] Z. Xiao, Z. Song, Y. Yan, From lead halide perovskites to Pb-free metal halide perovskites and perovskite derivatives, Adv. Mater. 31 (47) (2019) 1–22.
[27] A.H. Slavney, T. Hu, A.M. Lindenberg, H.I. Karunadasa, A bismuth-halide double perovskite with long carrier recombination lifetime for photovoltaic applications, J. Am. Chem. Soc. 138 (7) (2016).
[28] E.T. McClure, M.R. Ball, W. Windl, P.M. Woodward, Cs2AgBiX6 (X = Br, Cl): new visible light absorbing, Pb-free halide perovskite semiconductors, Chem. Mater. 28 (5) (2016) 1348–1354.
[29] W. Deng, et al., Synthesis of Cs2AgSbCl6 and improved optoelectronic properties of Cs2AgSbCl6/TiO2 heterostructure driven by the interface effect for Pb-free double perovskites solar cells, Appl. Phys. Lett. 111 (15) (2017) 2–7.
[30] F. Wei, et al., Synthesis and properties of a Pb-free hybrid double perovskite: (CH3NH3) 2AgBiBr6, Chem. Mater. 29 (3) (2017) 1089–1094.
[31] F. Wei, et al., The synthesis, structure and electronic properties of a Pb-free hybrid inorganic-organic double perovskite (MA)2KBiCl6 (MA = methylammonium), Mater. Horiz. 3 (4) (2016) 328–332.
[32] Z. Deng, F. Wei, S. Sun, G. Kieslich, A.K. Cheetham, P.D. Bristowe, Exploring the properties of Pb-free hybrid double perovskites using a combined computational-experimental approach, J. Mater. Chem. A 4 (31) (2016) 12025–12029.
[33] X.G. Zhao, et al., Design of Pb-free Inorganic Halide Perovskites for Solar Cells via Cation-transmutation, arXiv (2017).
[34] G. Volonakis, et al., Cs2InAgCl6: a new Pb-free halide double perovskite with direct band gap, J. Phys. Chem. Lett. 8 (4) (2017) 772–778.
[35] E.M. Hutter, C. Ge, D. Bartesaghi, F.C. Grozema, T.J. Savenije, Band-like charge transport in Cs 2 AgBiBr 6 and mixed antimony—bismuth Cs 2 AgBi 1—x Sb x Br 6 halide double perovskites, CS Omega 3 (2018) 11655–11662.
[36] Z. Xiao, K.Z. Du, W. Meng, J. Wang, D.B. Mitzi, Y. Yan, Intrinsic instability of Cs2In(I)M (III)X6 (M = Bi, Sb; X = Halogen) double perovskites: a combined density functional theory and experimental study, J. Am. Chem. Soc. 139 (17) (2017) 6054–6057.
[37] Y. Saeed, B. Amin, H. Khalil, F. Rehman, H. Ali, Correction: Cs2NaGaBr6: a new lead-free and direct band gap halide double perovskite, RSC Adv. 10 (2020) 17444–17451.
[38] Z. Xiao, K.-Z. Du, W. Meng, D.B. Mitzi, Y. Yan, Chemical origin of the stability difference between copper(I)- and silver(I)-based halide double perovskites, Angew. Chem. 129 (40) (2017) 12275–12279.
[39] V. Poblete, G. Navarro, V. Martin, M. Alvarez, Synthesis and structural characterization of luminescent materials. The elpasolite Cs 2 NaTbCl 6, Powder Diffract. 17 (1) (2002) 10–12.

[40] L.R. Morss, M. Siegal, L. Stenger, N. Edelstein, Preparation of cubic chloro complex compounds of trivalent metals: Cs2NaM Cl61a, Inorg. Chem. 9 (7) (1970) 1771–1775.
[41] L. Debbichi, et al., Mixed valence perovskite Cs2Au2I6: a potential material for thin-film Pb-free photovoltaic cells with ultrahigh efficiency, Adv. Mater. 30 (12) (2018) 1–6.
[42] M.M. Elsenety, A. Kaltzoglou, M. Antoniadou, I. Koutselas, A.G. Kontos, P. Falaras, Synthesis, characterization and use of highly stable trimethyl sulfonium tin(IV) halide defect perovskites in dye sensitized solar cells, Polyhedron 150 (Iv) (2018) 83–91.
[43] M.G. Brik, I.V. Kityk, Modeling of lattice constant and their relations with ionic radii and electronegativity of constituting ions of A2XY6 cubic crystals (A=K, Cs, Rb, Tl; X=tetravalent cation, Y=F, Cl, Br, I), J. Phys. Chem. Solid 72 (11) (2011) 1256–1260.
[44] Y. Cai, et al., Computational study of halide perovskite-derived A2BX6 inorganic compounds: chemical trends in electronic structure and structural stability, Chem. Mater. 29 (18) (2017) 7740–7749.
[45] B. Lee, et al., Air-stable molecular semiconducting iodosalts for solar cell applications: Cs2SnI6 as a hole conductor, J. Am. Chem. Soc. 136 (43) (2014) 15379–15385.
[46] B. Lee, A. Krenselewski, S. Il Baik, D.N. Seidman, R.P.H. Chang, Solution processing of air-stable molecular semiconducting iodosalts, Cs2SnI6-: XBrx, for potential solar cell applications, Sustain. Energy Fuels 1 (4) (2017) 710–724.
[47] N. Sakai, et al., Solution-processed cesium hexabromopalladate(IV), Cs2PdBr6, for optoelectronic applications, J. Am. Chem. Soc. 139 (17) (2017) 6030–6033.
[48] M. Chen, et al., Cesium titanium(IV) bromide thin films based stable Pb-free perovskite solar cells, Joule 2 (3) (2018) 558–570.
[49] D. Ju, et al., Tellurium-based double perovskites A2TeX6 with tunable band gap and long carrier diffusion length for optoelectronic applications, ACS Energy Lett. 4 (1) (2019) 228–234.
[50] Z. Tan, et al., Highly efficient blue-emitting Bi-doped Cs2SnCl6 perovskite variant: photoluminescence induced by impurity doping, Adv. Funct. Mater. 28 (29) (2018) 1–10.
[51] W. Ning, F. Gao, Structural and functional diversity in Pb-free halide perovskite materials, Adv. Mater. 31 (22) (2019) 1–21.
[52] C.R. Hubbard, R.A. Jacobson, The crystal structure of Rb4SbIIISbvBr12, Proc. Iowa Acad. Sci. 75 (1968) 85–96.
[53] K. Prassides, P. Day, A.K. Cheetham, Crystal structures of mixed-valency and mixed-metal salts A2MIII0.5SbV0.5X6 (A = Rb, Cs; M = Sb, Bi, In, Tl, Fe, Rh; X = Cl, Br). A powder neutron diffraction study, Inorg. Chem. 24 (1985) 545–552, https://doi.org/10.1021/ic00198a023.
[54] L. Atkinson, P. Day, Charge transfer in mixed-valence solids. Part V. Semiconductivity of hexachloroantimonates(III,V), J. Chem. Soc. A. 2432–2436 (1969), https://doi.org/10.1039/J19690002432.
[55] S. Perera, et al., Chalcogenide perovskites—an emerging class of ionic semiconductors, Nano Energy 22 (2016) 129–135.
[56] D. Tiwari, D.J. Fermin, T.K. Chaudhuri, A. Ray, Solution processed bismuth ferrite thin films for all-oxide solar photovoltaics, J. Phys. Chem. C 119 (11) (2015) 5872–5877.
[57] Y.Y. Sun, et al., Discovering Pb-free perovskite solar materials with a split-anion approach, Nanoscale 8 (12) (2016) 6284–6289.
[58] F. Hong, B. Saparov, W. Meng, Z. Xiao, D.B. Mitzi, Y. Yan, Viability of Pb-free perovskites with mixed chalcogen and halogen anions for photovoltaic applications, J. Phys. Chem. C 120 (12) (2016) 6435–6441.
[59] R. Nie, A. Mehta, B.W. Park, H.W. Kwon, J. Im, S. Il Seok, Mixed sulfur and iodide-based Pb-free perovskite solar cells, J. Am. Chem. Soc. 140 (3) (2018) 872–875.
[60] S.H. Park, I.H. Oh, S. Park, Y. Park, J.H. Kim, Y.D. Huh, Canted antiferromagnetism and spin reorientation transition in layered inorganic-organic perovskite (C 6H 5CH 2CH 2NH 3) 2MnCl 4, Dalton Trans. 41 (4) (2012) 1237–1242.

[61] P. Zhou, J.E. Drumheller, B. Patyal, R.D. Willett, Magnetic properties and critical behavior of quasi-two-dimensional systems [C6H5(CH2)nNH3]2CuBr4 with n=1, 2, and 3, Phys. Rev. B 45 (21) (1992) 12365–12376.

[62] H. Tsai, et al., High-efficiency two-dimensional Ruddlesden–Popper perovskite solar cells, Nature 536 (7616) (2016) 312–316.

[63] D.H. Cao, C.C. Stoumpos, O.K. Farha, J.T. Hupp, M.G. Kanatzidis, 2D homologous perovskites as light-absorbing materials for solar cell applications, J. Am. Chem. Soc. 137 (24) (2015) 7843–7850.

[64] P. Cheng, T. Wu, J. Zhang, Y. Li, J. Liu, L. Jiang, X. Mao, R.-F. Lu, W.-Q. Deng, K. Han, (C6H5C2H4NH3)2GeI4: A Layered Two-Dimensional Perovskite with Potential for Photovoltaic Applications, J. Phys. Chem. Lett. 8 (2017) 4402–4406, https://doi.org/10.1021/acs.jpclett.7b01985.

[65] P. Cheng, T. Wu, J. Liu, W.Q. Deng, K. Han, Pb-free, two-dimensional mixed germanium and tin perovskites, J. Phys. Chem. Lett. 9 (10) (2018) 2518–2522.

[66] L. Ma, M.G. Ju, J. Dai, X.C. Zeng, Tin and germanium based two-dimensional Ruddlesden-Popper hybrid perovskites for potential Pb-free photovoltaic and photoelectronic applications, Nanoscale 10 (24) (2018) 11314–11319.

[67] D.H. Cao, et al., Thin films and solar cells based on semiconducting two-dimensional Ruddlesden-Popper (CH3(CH2)3NH3)2(CH3NH3)n-1SnnI3n+1 perovskites, ACS Energy Lett. 2 (5) (2017) 982–990.

[68] X. Jiang, et al., Ultra-high open-circuit voltage of tin perovskite solar cells via an electron transporting layer design, Nat. Commun. 11 (1) (2020).

[69] B.A. Connor, L. Leppert, M.D. Smith, J.B. Neaton, H.I. Karunadasa, Layered halide double perovskites: dimensional reduction of Cs2AgBiBr6, J. Am. Chem. Soc. 140 (15) (2018) 5235–5240.

[70] L.Y. Bi, et al., Two-dimensional Pb-free iodide-based hybrid double perovskites: crystal growth, thin-film preparation and photocurrent responses, J. Mater. Chem. A 7 (34) (2019) 19662–19667.

[71] L.M. Castro-Castro, A.M. Guloy, Organic-based layered perovskites of mixed-valent gold(I)/gold(III) iodides, Angew. Chem. Int. Ed. 42 (2003) 2771–2774, https://doi.org/10.1002/anie.200350929.

[72] V. Pecunia, et al., Perovskite-inspired Pb-free Ag2BiI5 for self-powered NIR-blind visible light photodetection, Nano-Micro Lett. 12 (1) (2020) 1–12.

[73] L. Castro-Castro, A. Guloy, Organic-based layered perovskites of mixed-valent gold(I)/gold(III) iodides, Angew. Chem. 115 (24) (2003) 2877–2880.

[74] H. Zhang, L. Fang, R.-Z. Yuan, Triammonium nonaiododiantimonate(III), (NH 4) 3 [Sb 2 I 9], Acta Crystallogr., Sect. E: Struct. Rep. Online 61 (5) (2005) i70–i72.

[75] C. Zuo, L. Ding, Pb-free perovskite materials (NH 4) 3 Sb 2 I x Br 9 − x, Angew. Chem. 129 (23) (2017) 6628–6632.

[76] L. Absorbers, A. Based, T. Structural, A review on Pb-free hybrid halide perovskites as, Molecules 25 (5039) (2020) 1–40.

[77] F. Jiang, D. Yang, Y. Jiang, T. Liu, X. Zhao, Y. Ming, Chlorine-incorporation-induced formation of the layered phase for antimony-based Pb-free perovskite solar cells, J. Am. Chem. Soc. 140 (2017) 1019–1027.

[78] B. Vargas, E. Ramos, E. Pérez-Gutiérrez, J.C. Alonso, D. Solis-Ibarra, A direct bandgap copper-antimony halide perovskite, J. Am. Chem. Soc. 139 (27) (2017) 9116–9119.

[79] P.M. Jayasankar, A.K. Pathak, S.P. Madhusudanan, S. Murali, S.K. Batabyal, Double perovskite Cs4CuSb2Cl12 microcrystalline device for cost effective photodetector applications, Mater. Lett. (2019) 127200.

[80] W.B. Dai, S. Xu, J. Zhou, J. Hu, K. Huang, M. Xu, Solar energy materials and solar cells Pb-free, stable, and e ff ective double FA 4 Ge II Sb III Cl 12 perovskite for photovoltaic applications, 192 (December 2018) (2019) 140–146.

[81] Y. Zhang, W.Q. Liao, D.W. Fu, H.Y. Ye, Z.N. Chen, R.G. Xiong, Highly efficient red-light emission in an organic-inorganic hybrid ferroelectric: (Pyrrolidinium)MnCl3, J. Am. Chem. Soc. 137 (15) (2015) 4928–4931.
[82] H.Y. Ye, et al., High-temperature ferroelectricity and photoluminescence in a hybrid organic-inorganic compound: (3-pyrrolinium)MnCl3, J. Am. Chem. Soc. 137 (40) (2015) 13148–13154.
[83] Y.M. You, et al., An organic-inorganic perovskite ferroelectric with large piezoelectric response, Science 357 (6348) (2017) 306–309.
[84] W.Q. Liao, Y.Y. Tang, P.F. Li, Y.M. You, R.G. Xiong, Competitive halogen bond in the molecular ferroelectric with large piezoelectric response, J. Am. Chem. Soc. 140 (11) (2018) 3975–3980.
[85] W. Zhang, et al., (C6H13N)2BiI5: a one-dimensional Pb-free perovskite-derivative photoconductive light absorber, Inorg. Chem. 57 (8) (2018) 4239–4243.
[86] R. Kour, et al., Potential substitutes for replacement of lead in perovskite solar cells: a review, Global Chall. 3 (11) (2019) 1900050.
[87] K. Eckhardt, V. Bon, J. Getzschmann, J. Grothe, F.M. Wisser, S. Kaskel, Crystallographic insights into (CH3NH3)3(Bi2I9): a new Pb-free hybrid organic-inorganic material as a potential absorber for photovoltaics, Chem. Commun. 52 (14) (2016) 3058–3060.
[88] A.J. Lehner, et al., Crystal and electronic structures of complex bismuth iodides A3Bi2I9 (A = K, Rb, Cs) related to perovskite: aiding the rational design of photovoltaics, Chem. Mater. 27 (20) (2015) 7137–7148.
[89] P. Karuppuswamy, K. Moorthy, A. Mohapatra, Nano Energy Role of a hydrophobic sca ff old in controlling the crystallization of methylammonium antimony iodide for e ffi cient Pb-free perovskite solar cells, Nano Energy 45 (October 2017) (2018) 330–336.
[90] S. Bhaumik, S. Ray, S.K. Batabyal, Recent advances of Pb-free metal halide perovskite single crystals and nanocrystals: synthesis, crystal structure, optical properties, and their diverse applications, Mater. Today Chem. 18 (2020), 100363.
[91] F. Wang, X.K. Liu, F. Gao, Fundamentals of Solar Cells and Light-Emitting Diodes, Elsevier Inc., 2019.
[92] J. Zhou, Z. Xia, M.S. Molokeev, X. Zhang, D. Peng, Q. Liu, Composition design, optical gap and stability investigations of Pb-free halide double perovskite Cs2AgInCl6, J. Mater. Chem. A 5 (29) (2017) 15031–15037.
[93] A. Natik, Y. Abid, R. Moubah, M. Abid, H. Lassri, Ab-initio investigation of the structural, electronic and optical properties of Pb-free halide Cs2TiI6 double perovskites, Solid State Commun. 319 (May) (2020), 114006.
[94] M. Roknuzzaman, et al., Electronic and optical properties of Pb-free hybrid double perovskites for photovoltaic and optoelectronic applications, Sci. Rep. 9 (1) (2019) 1–7.
[95] S. Zhao, K. Yamamoto, S. Iikubo, S. Hayase, T. Ma, First-principles study of electronic and optical properties of Pb-free double perovskites Cs2NaBX6 (B = Sb, Bi; X = Cl, Br, I), J. Phys. Chem. Solid 117 (February) (2018) 117–121.
[96] M. Palummo, E. Berrios, D. Varsano, G. Giorgi, Optical properties of Pb-free double perovskites by Ab initio excited-state methods, ACS Energy Lett. (2020) 457–463.
[97] M.A. Ghebouli, T. Chihi, B. Ghebouli, M. Fatmi, Study of the structural, elastic, electronic and optical properties of lead free halide double perovskites Cs2AgBiX6(X = Br, Cl), Chin. J. Phys. 56 (1) (2018) 323–330.
[98] A. Yan, K. Li, Y. Zhou, Y. Ye, X. Zhao, C. Liu, Tuning the optical properties of Cs2SnCl6: Bi and Cs2SnCl6:Sb Pb-free perovskites via post-annealing for white LEDs, J. Alloys Compd. 822 (2020), 153528.
[99] T.C. Jellicoe, et al., Synthesis and optical properties of Pb-free cesium tin halide perovskite nanocrystals, J. Am. Chem. Soc. 138 (9) (2016) 2941–2944.

[100] L. Chen, C. Lee, Y. Chuang, Z. Wu, C. Chen, Synthesis and optical properties of Pb-free cesium tin halide perovskite quantum rods with high-performance solar cell application, J. Phys. Chem. Lett. 7 (2016) 5028–5035.

[101] N. Rajeev Kumar, R. Radhakrishnan, Electronic, optical and mechanical properties of Pb-free halide double perovskites using first-principles density functional theory, Mater. Lett. 227 (2018) 289–291.

[102] B. Yang, et al., Pb-free silver-bismuth halide double perovskite nanocrystals, Angew. Chem. Int. Ed. 57 (19) (2018) 5359–5363.

[103] B. Yang, et al., Lead-Free, air-stable all-inorganic cesium bismuth halide perovskite nanocrystals, Angew. Chem. 129 (41) (2017) 12645–12649.

[104] M. Leng, et al., All-inorganic bismuth-based perovskite quantum dots with bright blue photoluminescence and excellent stability, Adv. Funct. Mater. 28 (1) (2018) 1–11.

[105] M. Leng, et al., Pb-free, blue emitting bismuth halide perovskite quantum dots, Angew. Chem. 128 (48) (2016) 15236–15240.

[106] J. Zhang, et al., High quantum yield blue emission from Pb-free inorganic antimony halide perovskite colloidal quantum dots, ACS Nano 11 (9) (2017) 9294–9302.

[107] A. Wang, et al., Controlled synthesis of Pb-free and stable perovskite derivative Cs2SnI6 nanocrystals via a facile hot-injection process, Chem. Mater. 28 (22) (2016) 8132–8140.

[108] M.M. Yao, et al., Improving Pb-free double perovskite Cs2NaBiCl6 nanocrystal optical properties via ion doping, Adv. Opt. Mater. 8 (8) (2020) 2–9.

[109] J. Mizusaki, K. Arai, K. Fueki, Ionic conduction of the perovskite-type halides, Solid State Ion. 11 (3) (1983).

[110] N.-G. Park, M. Grätzel, T. Miyasaka (Eds.), Organic-Inorganic Halide Perovskite Photovoltaics, Original Text: From Fundamentals to Device Architectures, Springer International Publishing, 2016. https://www.springer.com/de/book/9783319351124. (Accessed 20 May 2019).

[111] E.L. Unger, et al., Hysteresis and transient behavior in current-voltage measurements of hybrid-perovskite absorber solar cells, Energ. Environ. Sci. 7 (11) (2014).

[112] H.J. Snaith, et al., Anomalous hysteresis in perovskite solar cells, J. Phys. Chem. Lett. 5 (9) (2014).

[113] J.M. Azpiroz, E. Mosconi, J. Bisquert, F. De Angelis, Defect migration in methylammonium lead iodide and its role in perovskite solar cell operation, Energ. Environ. Sci. 8 (7) (2015).

[114] M. Bag, et al., Kinetics of ion transport in perovskite active layers and its implications for active layer stability, J. Am. Chem. Soc. 137 (40) (2015).

[115] C. Eames, J.M. Frost, P.R.F. Barnes, B.C. O'Regan, A. Walsh, M.S. Islam, Ionic transport in hybrid lead iodide perovskite solar cells, Nat. Commun. 6 (2015).

[116] J. Haruyama, K. Sodeyama, L. Han, Y. Tateyama, First-principles study of ion diffusion in perovskite solar cell sensitizers, J. Am. Chem. Soc. 137 (32) (2015).

[117] W. Tress, N. Marinova, T. Moehl, S.M. Zakeeruddin, M.K. Nazeeruddin, M. Grätzel, Understanding the rate-dependent J-V hysteresis, slow time component, and aging in CH3NH3PbI3 perovskite solar cells: the role of a compensated electric field, Energ. Environ. Sci. 8 (3) (2015).

[118] S. Van Reenen, M. Kemerink, H.J. Snaith, Modeling anomalous hysteresis in perovskite solar cells, J. Phys. Chem. Lett. 6 (19) (2015).

[119] T.Y. Yang, G. Gregori, N. Pellet, M. Grätzel, J. Maier, The significance of ion conduction in a hybrid organic-inorganic lead-iodide-based perovskite photosensitizer, Angew. Chem. Int. Ed. 54 (27) (2015).

[120] Y. Yuan, et al., Photovoltaic switching mechanism in lateral structure hybrid perovskite solar cells, Adv. Energy Mater. 5 (15) (2015).

[121] P. Calado, et al., Evidence for ion migration in hybrid perovskite solar cells with minimal hysteresis, Nat. Commun. 7 (2016).

[122] P. Delugas, C. Caddeo, A. Filippetti, A. Mattoni, Thermally activated point defect diffusion in methylammonium lead trihalide: anisotropic and ultrahigh mobility of iodine, J. Phys. Chem. Lett. 7 (13) (2016).
[123] D.W. DeQuilettes, et al., Photo-induced halide redistribution in organic-inorganic perovskite films, Nat. Commun. 7 (2016).
[124] J.M. Frost, A. Walsh, What is moving in hybrid halide perovskite solar cells, Acc. Chem. Res. 49 (3) (2016).
[125] C. Li, et al., Iodine migration and its effect on hysteresis in perovskite solar cells, Adv. Mater. 28 (12) (2016).
[126] S. Meloni, et al., Ionic polarization-induced current-voltage hysteresis in CH3NH3PbX3 perovskite solar cells, Nat. Commun. 7 (2016).
[127] E. Mosconi, D. Meggiolaro, H.J. Snaith, S.D. Stranks, F. De Angelis, Light-induced annihilation of Frenkel defects in organo-lead halide perovskites, Energ. Environ. Sci. 9 (10) (2016).
[128] G. Richardson, et al., Can slow-moving ions explain hysteresis in the current-voltage curves of perovskite solar cells? Energ. Environ. Sci. 9 (4) (2016).
[129] J. Xing, Q. Wang, Q. Dong, Y. Yuan, Y. Fang, J. Huang, Ultrafast ion migration in hybrid perovskite polycrystalline thin films under light and suppression in single crystals, Phys. Chem. Chem. Phys. 18 (44) (2016).
[130] O.S. Game, G.J. Buchsbaum, Y. Zhou, N.P. Padture, A.I. Kingon, Ions matter: description of the anomalous electronic behavior in methylammonium lead halide perovskite devices, Adv. Funct. Mater. 27 (16) (2017).
[131] C. Li, A. Guerrero, S. Huettner, J. Bisquert, Unravelling the role of vacancies in lead halide perovskite through electrical switching of photoluminescence, Nat. Commun. 9 (1) (2018).
[132] D. Meggiolaro, et al., Iodine chemistry determines the defect tolerance of lead-halide perovskites, Energ. Environ. Sci. 11 (3) (2018).
[133] F. Ebadi, N. Taghavinia, R. Mohammadpour, A. Hagfeldt, W. Tress, Origin of apparent light-enhanced and negative capacitance in perovskite solar cells, Nat. Commun. 10 (2019) 1574.
[134] M.H. Futscher, et al., Quantification of ion migration in CH3NH3PbI3 perovskite solar cells by transient capacitance measurements, Mater. Horizons 6 (7) (2019).
[135] D. Moia, et al., Ionic-to-electronic current amplification in hybrid perovskite solar cells: ionically gated transistor-interface circuit model explains hysteresis and impedance of mixed conducting devices, Energ. Environ. Sci. 12 (4) (2019).
[136] A. Riquelme, et al., Identification of recombination losses and charge collection efficiency in a perovskite solar cell by comparing impedance response to a drift-diffusion model, Nanoscale 12 (33) (2020).
[137] S.A.L. Weber, et al., How the formation of interfacial charge causes hysteresis in perovskite solar cells, Energ. Environ. Sci. 11 (9) (2018).
[138] B. Chen, M. Yang, S. Priya, K. Zhu, Origin of J-V hysteresis in perovskite solar cells, J. Phys. Chem. Lett. 7 (5) (2016).
[139] B. Roose, Ion migration drives self-passivation in perovskite solar cells and is enhanced by light soaking, RSC Adv. 11 (20) (2021).
[140] A. Dualeh, et al., Impedance spectroscopic analysis of lead iodide perovskite-sensitized solid-state solar cells, ACS Nano 8 (1) (2014) 362–373.
[141] A. Senocrate, et al., The nature of ion conduction in methylammonium lead iodide: a multimethod approach, Angew. Chem. Int. Ed. 56 (27) (2017).
[142] A. Senocrate, et al., Slow CH3NH3+ diffusion in CH3NH3PbI3 under light measured by solid-state NMR and tracer diffusion, J. Phys. Chem. C 122 (38) (2018).
[143] H. Yu, H. Lu, F. Xie, S. Zhou, N. Zhao, Native defect-induced hysteresis behavior in organolead iodide perovskite solar cells, Adv. Funct. Mater. 26 (9) (2016).

[144] A. Pockett, G.E. Eperon, N. Sakai, H.J. Snaith, L.M. Peter, P.J. Cameron, Microseconds, milliseconds and seconds: deconvoluting the dynamic behaviour of planar perovskite solar cells, Phys. Chem. Chem. Phys. 19 (8) (2017).
[145] M. Sajedi Alvar, P.W.M. Blom, G.J.A.H. Wetzelaer, Space-charge-limited electron and hole currents in hybrid organic-inorganic perovskites, Nat. Commun. 11 (1) (2020).
[146] L. Bertoluzzi, R.A. Belisle, K.A. Bush, R. Cheacharoen, M.D. McGehee, B.C. O'Regan, In situ measurement of electric-field screening in hysteresis-free PTAA/FA0.83Cs0.17Pb (I0.83Br0.17)3/C60 perovskite solar cells gives an ion mobility of ~3 × 10–7 cm2/(V s), 2 orders of magnitude faster than reported for metal-oxide-contacted perovskite cells with hysteresis, J. Am. Chem. Soc. 140 (40) (2018).
[147] M. Sajedi Alvar, P.W.M. Blom, G.J.A.H. Wetzelaer, Device model for methylammonium lead iodide perovskite with experimentally validated ion dynamics, Adv. Electron. Mater. 6 (6) (2020).
[148] N.E. Courtier, G. Richardson, J.M. Foster, A fast and robust numerical scheme for solving models of charge carrier transport and ion vacancy motion in perovskite solar cells, App. Math. Model. 63 (2018).
[149] R.A. Belisle, et al., Interpretation of inverted photocurrent transients in organic lead halide perovskite solar cells: proof of the field screening by mobile ions and determination of the space charge layer widths, Energ. Environ. Sci. 10 (1) (2017).
[150] O. Almora, I. Zarazua, E. Mas-Marza, I. Mora-Sero, J. Bisquert, G. Garcia-Belmonte, Capacitive dark currents, hysteresis, and electrode polarization in lead halide perovskite solar cells, J. Phys. Chem. Lett 6 (9) (2015).
[151] S.T. Birkhold, J.T. Precht, R. Giridharagopal, G.E. Eperon, L. Schmidt-Mende, D.S. Ginger, Direct observation and quantitative analysis of mobile frenkel defects in metal halide perovskites using scanning kelvin probe microscopy, J. Phys. Chem. C 122 (24) (2018).
[152] M. Sajedi Alvar, M. Kumar, P.W.M. Blom, G.J.A.H. Wetzelaer, K. Asadi, Absence of ferroelectricity in methylammonium lead iodide perovskite, AIP Adv. 7 (9) (2017).
[153] C. Lan, S. Zhao, J. Luo, P. Fan, First-principles study of anion diffusion in Pb-free halide double perovskites, Phys. Chem. Chem. Phys. 20 (37) (2018).
[154] M. Wang, et al., Lead—Free Perovskite Materials for Solar Cells, no. 0123456789, Springer Singapore, 2021.
[155] J. Xu, J.B. Liu, B.X. Liu, B. Huang, Unique ion diffusion properties in Pb-free halide double perovskites: a first-principles study, J. Power Sources 412 (2019).
[156] J.H. Yang, W.J. Yin, J.S. Park, S.H. Wei, Fast self-diffusion of ions in CH3NH3PbI3: the interstiticaly mechanism: versus vacancy-assisted mechanism, J. Mater. Chem. A 4 (34) (2016).
[157] M. Pantaler, et al., Hysteresis-free Pb-free double-perovskite solar cells by interface engineering, ACS Energy Lett. 3 (8) (2018).
[158] F. Li, et al., A cation-exchange approach for the fabrication of efficient methylammonium tin iodide perovskite solar cells, Angew. Chem. Int. Ed. 58 (20) (2019).
[159] W. Ming, H. Shi, M.H. Du, Large dielectric constant, high acceptor density, and deep electron traps in perovskite solar cell material CsGeI3, J. Mater. Chem. A 4 (36) (2016).
[160] P.-P. Sun, Q.-S. Li, L.-N. Yang, Z.-S. Li, Theoretical insights into a potential Pb-free hybrid perovskite: substituting Pb2+ with Ge2+, Nanoscale 8 (3) (2016) 1503–1512.
[161] F. Brivio, K.T. Butler, A. Walsh, M. Van Schilfgaarde, Relativistic quasiparticle self-consistent electronic structure of hybrid halide perovskite photovoltaic absorbers, Phys. Rev. B: Condens. Matter Mater. Phys. 89 (15) (2014).
[162] S.M. Sze, K.K. Ng, Photodetectors and solar cells, in: Physics of Semiconductor Devices, John Wiley & Sons, Ltd, 2006, pp. 663–742. https://doi.org/10.1002/9780470068328.ch13.

[163] T. Zhao, W. Shi, J. Xi, D. Wang, Z. Shuai, Intrinsic and extrinsic charge transport in CH3 NH3 PbI3 perovskites predicted from first-principles, Sci. Rep. 7 (2016).

[164] T.M. Brenner, D.A. Egger, L. Kronik, G. Hodes, D. Cahen, Hybrid organic—inorganic perovskites: low-cost semiconductors with intriguing charge-transport properties, Nat. Rev. Mater. 1 (1) (2016).

[165] P. Umari, E. Mosconi, F. De Angelis, Relativistic GW calculations on CH3 NH3 PbI 3 and CH3 NH3 SnI3 perovskites for solar cell applications, Sci. Rep. 4 (2014).

[166] E. Menéndez-Proupin, P. Palacios, P. Wahnón, J.C. Conesa, Self-consistent relativistic band structure of the CH3 NH3 PbI3 perovskite, Phys. Rev. B: Condens. Matter Mater. Phys. 90 (4) (2014).

[167] A. Miyata, et al., Direct measurement of the exciton binding energy and effective masses for charge carriers in organic-inorganic tri-halide perovskites, Nat. Phys. 11 (7) (2015).

[168] J. Peng, Y. Chen, K. Zheng, T. Pullerits, Z. Liang, Insights into charge carrier dynamics in organo-metal halide perovskites: from neat films to solar cells, Chem. Soc. Rev. 46 (19) (2017).

[169] L.M. Herz, Charge-carrier mobilities in metal halide perovskites: fundamental mechanisms and limits, ACS Energy Lett. 2 (7) (2017).

[170] D. Shi, et al., Low trap-state density and long carrier diffusion in organolead trihalide perovskite single crystals, Science 347 (6221) (2015).

[171] S.D. Stranks, et al., Electron-hole diffusion lengths exceeding 1 micrometer in an organometal trihalide perovskite absorber, Science 342 (6156) (2013).

[172] G. Xing, et al., Long-range balanced electron-and hole-transport lengths in organic-inorganic CH3NH3PbI3, Science 342 (6156) (2013).

[173] C. Wehrenfennig, G.E. Eperon, M.B. Johnston, H.J. Snaith, L.M. Herz, High charge carrier mobilities and lifetimes in organolead trihalide perovskites, Adv. Mater. 26 (10) (2014).

[174] Q. Dong, et al., Electron-hole diffusion lengths $>$ 175 μm in solution-grown CH3NH3PbI3 single crystals, Science 347 (6225) (2015) 967–970.

[175] M.I. Saidaminov, et al., High-quality bulk hybrid perovskite single crystals within minutes by inverse temperature crystallization, Nat. Commun. 6 (2015).

[176] Y. Takahashi, H. Hasegawa, Y. Takahashi, T. Inabe, Hall mobility in tin iodide perovskite CH3NH3SnI 3: evidence for a doped semiconductor, J. Solid State Chem. 205 (2013).

[177] C. Ran, et al., Construction of compact methylammonium bismuth iodide film promoting Pb-free inverted planar heterojunction organohalide solar cells with open-circuit voltage over 0.8 V, J. Phys. Chem. Lett. 8 (2) (2017).

[178] R.L. Milot, G.E. Eperon, T. Green, H.J. Snaith, M.B. Johnston, L.M. Herz, Radiative monomolecular recombination boosts amplified spontaneous emission in HC(NH2) 2SnI3 perovskite films, J. Phys. Chem. Lett. 7 (20) (2016).

[179] I. Chung, et al., CsSnI 3: Semiconductor or metal? High electrical conductivity and strong near-infrared photoluminescence from a single material. High hole mobility and phase-transitions, J. Am. Chem. Soc. 134 (20) (2012).

[180] S.F. Hoefler, G. Trimmel, T. Rath, Progress on Pb-free metal halide perovskites for photovoltaic applications: a review, Monatsh. Chem. 148 (5) (2017) 795–826.

[181] C.C. Stoumpos, C.D. Malliakas, M.G. Kanatzidis, Semiconducting tin and lead iodide perovskites with organic cations: phase transitions, high mobilities, and near-infrared photoluminescent properties, Inorg. Chem. 52 (15) (2013).

[182] B. Zhao, et al., High open-circuit voltages in tin-rich low-bandgap perovskite-based planar heterojunction photovoltaics, Adv. Mater. 29 (2) (2017).

[183] N.K. Noel, et al., Pb-free organic-inorganic tin halide perovskites for photovoltaic applications, Energ. Environ. Sci. 7 (9) (2014).

[184] P. Wang, et al., Ion exchange/insertion reactions for fabrication of efficient methylammonium tin iodide perovskite solar cells, Adv. Sci. 7 (9) (2020).
[185] T.-B. Song, et al., Importance of reducing vapor atmosphere in the fabrication of tin-based perovskite solar cells, J. Am. Chem. Soc. 139 (2) (2017) 836–842.
[186] Z. Yang, et al., Stabilized wide bandgap perovskite solar cells by tin substitution, Nano Lett. 16 (12) (2016).
[187] T. Yokoyama, T. Bin Song, D.H. Cao, C.C. Stoumpos, S. Aramaki, M.G. Kanatzidis, The origin of lower hole carrier concentration in methylammonium tin halide films grown by a vapor-assisted solution process, ACS Energy Lett. 2 (1) (2017).
[188] F. Hao, C.C. Stoumpos, R.P.H. Chang, M.G. Kanatzidis, Anomalous band gap behavior in mixed Sn and Pb perovskites enables broadening of absorption spectrum in solar cells, J. Am. Chem. Soc. 136 (22) (2014).
[189] X. Liu, et al., Improved efficiency and stability of Pb-Sn binary perovskite solar cells by Cs substitution, J. Mater. Chem. A 4 (46) (2016).
[190] W. Liao, et al., Pb-free Inverted planar formamidinium tin triiodide perovskite solar cells achieving power conversion efficiencies up to 6.22%, Adv. Mater. 28 (42) (2016) 9333–9340.
[191] S.J. Lee, et al., Fabrication of efficient formamidinium tin iodide perovskite solar cells through SnF2-pyrazine complex, J. Am. Chem. Soc. 138 (12) (2016) 3974–3977.
[192] X. Liu, et al., Templated growth of FASnI3crystals for efficient tin perovskite solar cells, Energ. Environ. Sci. 13 (9) (2020).
[193] T.M. Koh, et al., Formamidinium tin-based perovskite with low Eg for photovoltaic applications, J. Mater. Chem. A 3 (29) (2015) 14996–15000.
[194] E. Jokar, C.H. Chien, A. Fathi, M. Rameez, Y.H. Chang, E.W.G. Diau, Slow surface passivation and crystal relaxation with additives to improve device performance and durability for tin-based perovskite solar cells, Energ. Environ. Sci. 11 (9) (2018).
[195] M.E. Kayesh, et al., Coadditive engineering with 5-ammonium valeric acid iodide for efficient and stable Sn perovskite solar cells, ACS Energy Lett. 4 (1) (2019).
[196] Z. Zhu, C.C. Chueh, N. Li, C. Mao, A.K.Y. Jen, Realizing efficient Pb-free formamidinium tin triiodide perovskite solar cells via a sequential deposition route, Adv. Mater. 30 (6) (2018).
[197] T. Wu, et al., Efficient and stable tin-based perovskite solar cells by introducing π-conjugated Lewis base, Sci. China Chem. 63 (1) (2020) 107–115.
[198] F. Li, et al., Trihydrazine dihydriodide-assisted fabrication of efficient formamidinium tin iodide perovskite solar cells, Sol. RRL 3 (9) (2019).
[199] M. Zhang, et al., Low-temperature processed solar cells with formamidinium tin halide perovskite/fullerene heterojunctions, Nano Res. 9 (6) (2016).
[200] T.K. Todorov, K.B. Reuter, D.B. Mitzi, High-efficiency solar cell with earth-abundant liquid-processed absorber, Adv. Mater. 22 (20) (2010).
[201] K.P. Marshall, M. Walker, R.I. Walton, R.A. Hatton, Enhanced stability and efficiency in hole-transport-layer-free CsSnI 3 perovskite photovoltaics, Nat. Energy 1 (12) (2016).
[202] K.P. Marshall, R.I. Walton, R.A. Hatton, Tin perovskite/fullerene planar layer photovoltaics: improving the efficiency and stability of Pb-free devices, J. Mater. Chem. A 3 (21) (2015).
[203] S. Gupta, T. Bendikov, G. Hodes, D. Cahen, CsSnBr3, A Pb-free halide perovskite for long-term solar cell application: insights on SnF2 addition, ACS Energy Lett. 1 (5) (2016) 1028–1033.
[204] Z. Zhao, et al., Mixed-organic-cation tin iodide for Pb-free perovskite solar cells with an efficiency of 8.12%, Adv. Sci. 4 (11) (2017).
[205] Z. Jin, et al., Enhanced efficiency and stability in Sn-based perovskite solar cells with secondary crystallization growth, J. Energy Chem. 54 (2021).

[206] J. Liang, et al., CsPb0.9Sn0.1IBr2 based all-inorganic perovskite solar cells with exceptional efficiency and stability, J. Am. Chem. Soc. 139 (40) (2017).

[207] Z. Yang, X. Zhang, W. Yang, G.E. Eperon, D.S. Ginger, Tin–lead alloying for efficient and stable all-inorganic perovskite solar cells, Chem. Mater. 32 (7) (2020).

[208] P. Li, et al., Low-dimensional dion–jacobson-phase lead-free perovskites for high-performance photovoltaics with improved stability, Angew. Chem. Int. Ed. 59 (17) (2020) 6909–6914.

[209] M. Chen, et al., Highly stable and efficient all-inorganic Pb-free perovskite solar cells with native-oxide passivation, Nat. Commun. 10 (1) (2019) 16.

[210] T. Bin Song, T. Yokoyama, S. Aramaki, M.G. Kanatzidis, Performance enhancement of Pb-free tin- based perovskite solar cells with reducing atmosphere-assisted dispersible additive, ACS Energy Lett. 2 (4) (2017).

[211] P.R. Moses, H.O. Finklea, J.R. Lenhard, R.W. Murray, L.M. Wier, J.C. Lennox, X-ray photoelectron spectroscopy of alkylamine-silanes bound to metal oxide electrodes, Anal. Chem. 50 (4) (1978).

[212] Q. Tai, et al., Antioxidant grain passivation for air-stable tin-based perovskite solar cells, Angew. Chem. Int. Ed. 58 (3) (2019) 806–810.

[213] J. Qiu, Y. Xia, Y. Chen, W. Huang, Management of crystallization kinetics for efficient and stable low-dimensional Ruddlesden–Popper (LDRP) Pb-free perovskite solar cells, Adv. Sci. 6 (1) (2019).

[214] S. Shao, et al., Enhancing the crystallinity and perfecting the orientation of formamidinium tin iodide for highly efficient Sn-based perovskite solar cells, Nano Energy 60 (2019).

[215] F. Wang, et al., 2D-quasi-2D-3D hierarchy structure for tin perovskite solar cells with enhanced efficiency and stability, Joule 2 (12) (2018).

[216] Y. Liao, et al., Highly oriented low-dimensional tin halide perovskites with enhanced stability and photovoltaic performance, J. Am. Chem. Soc. 139 (19) (2017).

[217] R. Lin, et al., Monolithic all-perovskite tandem solar cells with 24.8% efficiency exploiting comproportionation to suppress Sn(ii) oxidation in precursor ink, Nat. Energy 4 (10) (2019).

[218] K. Xiao, et al., All-perovskite tandem solar cells with 24.2% certified efficiency and area over 1 cm2 using surface-anchoring zwitterionic antioxidant, Nat. Energy 5 (11) (2020).

[219] B.-W. Park, B. Philippe, X. Zhang, H. Rensmo, G. Boschloo, E.M.J. Johansson, Bismuth based hybrid perovskites A3Bi2I9 (A: Methylammonium or Cesium) for solar cell application, Adv. Mater. 27 (43) (2015) 6806–6813.

[220] Z. Zhang, et al., High-quality (CH3NH3)3Bi2I9 film-based solar cells: pushing efficiency up to 1.64%, J. Phys. Chem. Lett. 8 (17) (2017).

[221] M. Vigneshwaran, et al., Facile synthesis and characterization of sulfur doped low bandgap bismuth based perovskites by soluble precursor route, Chem. Mater. 28 (18) (2016).

[222] H. Hu, B. Dong, W. Zhang, Low-toxic metal halide perovskites: opportunities and future challenges, J. Mater. Chem. A 5 (23) (2017).

[223] K.M. Boopathi, et al., Solution-processable antimony-based light-absorbing materials beyond lead halide perovskites, J. Mater. Chem. A 5 (39) (2017) 20843–20850.

[224] F. Jiang, et al., Chlorine-incorporation-induced formation of the layered phase for antimony-based Pb-free perovskite solar cells, J. Am. Chem. Soc. 140 (3) (2018).

[225] X. Zhang, et al., Active-layer evolution and efficiency improvement of (CH3NH33Bi2I9)-based solar cell on TiO2-deposited ITO substrate, Nano Res. 9 (10) (2016).

[226] S.S. Mali, H. Kim, D.H. Kim, C. Kook Hong, Anti-solvent assisted crystallization processed methylammonium bismuth iodide cuboids towards highly stable Pb-free perovskite solar cells, ChemistrySelect 2 (4) (2017).

[227] A. Kulkarni, T. Singh, M. Ikegami, T. Miyasaka, Photovoltaic enhancement of bismuth halide hybrid perovskite by N-methyl pyrrolidone-assisted morphology conversion, RSC Adv. 7 (16) (2017) 9456–9460.
[228] T. Singh, A. Kulkarni, M. Ikegami, T. Miyasaka, et al., ACS Appl. Mater. Interfaces 8 (23) (2016).
[229] T. Li, Y. Hu, C.A. Morrison, W. Wu, H. Han, N. Robertson, Pb-free pseudo-three-dimensional organic-inorganic iodobismuthates for photovoltaic applications, Sustain. Energy Fuels 1 (2) (2017).
[230] M.B. Johansson, H. Zhu, E.M.J. Johansson, Extended photo-conversion spectrum in low-toxic bismuth halide perovskite solar cells, J. Phys. Chem. Lett. 7 (17) (2016).
[231] E. Greul, M.L. Petrus, A. Binek, P. Docampo, T. Bein, Highly stable, phase pure Cs2AgBiBr6 double perovskite thin films for optoelectronic applications, J. Mater. Chem. A 5 (37) (2017).
[232] C. Wu, et al., The dawn of Pb-free perovskite solar cell: highly stable double perovskite Cs2AgBiBr6 Film, Adv. Sci. 5 (3) (2018).
[233] W. Ning, et al., Long electron–hole diffusion length in high-quality Pb-free double perovskite films, Adv. Mater. 30 (20) (2018).
[234] C.N. Savory, A. Walsh, D.O. Scanlon, Can Pb-free halide double perovskites support high-efficiency solar cells, ACS Energy Lett. 1 (5) (2016).
[235] C. Zhou, H. Lin, S. Lee, M. Chaaban, B. Ma, Organic-inorganic metal halide hybrids beyond perovskites, Mater. Res. Lett. 6 (10) (2018).
[236] Z.K. Tan, et al., Bright light-emitting diodes based on organometal halide perovskite, Nat. Nanotechnol. 9 (9) (2014).
[237] X. Yang, et al., Efficient green light-emitting diodes based on quasi-two-dimensional composition and phase engineered perovskite with surface passivation, Nat. Commun. 9 (1) (2018).
[238] J. Wang, et al., Interfacial control toward efficient and low-voltage perovskite light-emitting diodes, Adv. Mater. 27 (14) (2015).
[239] H. Liang, et al., High color purity Pb-free perovskite light-emitting diodes via Sn stabilization, Adv. Sci. 7 (8) (2020).
[240] X. Zhang, et al., Bright orange electroluminescence from Pb-free two-dimensional perovskites, ACS Energy Lett. 4 (1) (2019).
[241] F. Yuan, et al., All-inorganic hetero-structured cesium tin halide perovskite light-emitting diodes with current density over 900 A cm^{-2} and its amplified spontaneous emission behaviors, Phys. Status Solidi Rapid Res. Lett. 12 (5) (2018).
[242] P. Vashishtha, et al., Cesium copper iodide tailored nanoplates and nanorods for blue, yellow, and white emission, Chem. Mater. 31 (2019) 9003–9011.
[243] Z. Ma, et al., Stable yellow light-emitting devices based on ternary copper halides with broadband emissive self-trapped excitons, ACS Nano 14 (4) (2020).
[244] P. Cheng, et al., Colloidal synthesis and optical properties of all-inorganic low-dimensional cesium copper halide nanocrystals, Angew. Chemie 131 (45) (2019).
[245] R. Roccanova, et al., Bright luminescence from nontoxic $CsCu_2X_3$ (X = Cl, Br, I), ACS Mater. Lett. 1 (4) (2019).
[246] X. Mo, et al., Highly-efficient all-inorganic Pb-free 1D CsCu2I3 single crystal for white-light emitting diodes and UV photodetection, Nano Energy 81 (2021).
[247] R. Dong, et al., High-gain and low-driving-voltage photodetectors based on organolead triiodide perovskites, Adv. Mater. 27 (11) (2015).
[248] T. Moehl, et al., Strong photocurrent amplification in perovskite solar cells with a porous TiO2 blocking layer under reverse bias, J. Phys. Chem. Lett. 5 (21) (2014).
[249] D. Li, G. Dong, W. Li, L. Wang, High performance organic-inorganic perovskite-optocoupler based on low-voltage and fast response perovskite compound photodetector, Sci. Rep. 5 (2015).

[250] S. Yakunin, Y. Shynkarenko, D.N. Dirin, I. Cherniukh, M.V. Kovalenko, Non-dissipative internal optical filtering with solution-grown perovskite single crystals for full-colour imaging, NPG Asia Mater. 9 (9) (2017).
[251] Z.X. Zhang, et al., Sensitive deep ultraviolet photodetector and image sensor composed of inorganic Pb-free Cs3Cu2I5 perovskite with wide bandgap, J. Phys. Chem. Lett. 10 (18) (2019).
[252] W.Q. Liang, et al., A solution-processed ternary copper halide thin films for air-stable and deep-ultraviolet-sensitive photodetector, Nanoscale 12 (33) (2020).
[253] Y. Li, et al., Highly stable and spectrum-selective ultraviolet photodetectors based on Pb-free copper-based perovskites, Mater. Horizons 7 (2) (2020).
[254] W. Liang, et al., Strategy of all-inorganic Cs3Cu2I5/Si-core/shell nanowire heterojunction for stable and ultraviolet-enhanced broadband photodetectors with imaging capability, ACS Appl. Mater. Interfaces 12 (33) (2020).
[255] J. Luo, et al., Cs2AgInCl6 double perovskite single crystals: parity forbidden transitions and their application for sensitive and fast UV photodetectors, ACS Photonics 5 (2) (2018).
[256] L. Dou, et al., Solution-processed hybrid perovskite photodetectors with high detectivity, Nat. Commun. 5 (2014).
[257] C. Bao, et al., Low-noise and large-linear-dynamic-range photodetectors based on hybrid-perovskite thin-single-crystals, Adv. Mater. 29 (39) (2017).
[258] C. Bao, et al., High performance and stable all-inorganic metal halide perovskite-based photodetectors for optical communication applications, Adv. Mater. 30 (38) (2018).
[259] X.F. Cheng, et al., Environmentally robust memristor enabled by Pb-free double perovskite for high-performance information storage, Small 15 (49) (2019).
[260] W. Ye, Q. Cao, X.F. Cheng, C. Yu, J.H. He, J.M. Lu, A Pb-free Cs2PdBr6perovskite-based humidity sensor for artificial fruit waxing detection, J. Mater. Chem. A 8 (34) (2020).
[261] J. Li, J. Duan, X. Yang, Y. Duan, P. Yang, Q. Tang, Review on recent progress of Pb-free halide perovskites in optoelectronic applications, Nano Energy 80 (2021), 105526.
[262] W. Ke, et al., Enhanced photovoltaic performance and stability with a new type of hollow 3D perovskite {en}FASnI3, Sci. Adv. 3 (8) (2017), e1701293.
[263] E. Jokar, C.-H. Chien, C.-M. Tsai, A. Fathi, E.W.-G. Diau, Robust tin-based perovskite solar cells with hybrid organic cations to attain efficiency approaching 10%, Adv. Mater. 31 (2) (2019) 1804835.
[264] K.P. Marshall, M. Walker, R.I. Walton, R.A. Hatton, Enhanced stability and efficiency in hole-transport-layer-free CsSnI3 perovskite photovoltaics, Nat. Energy 1 (12) (2016) 16178.
[265] W. Li, J. Li, J. Li, J. Fan, Y. Mai, L. Wang, Addictive-assisted construction of all-inorganic CsSnIBr2 mesoscopic perovskite solar cells with superior thermal stability up to 473 K, J. Mater. Chem. A 4 (43) (2016) 17104–17110.
[266] F. Gu, et al., Improving performance of Pb-free formamidinium tin triiodide perovskite solar cells by tin source purification, Sol. RRL 2 (10) (2018) 1800136.
[267] M.E. Kayesh, et al., Enhanced photovoltaic performance of FASnI3-based perovskite solar cells with hydrazinium chloride coadditive, ACS Energy Lett. 3 (7) (2018) 1584–1589.
[268] C. Wang, et al., Self-repairing tin-based perovskite solar cells with a breakthrough efficiency over 11%, Adv. Mater. 32 (31) (2020) 1907623.
[269] P. Li, et al., Ligand orientation-induced lattice robustness for highly efficient and stable tin-based perovskite solar cells, ACS Energy Lett. 5 (7) (2020) 2327–2334.
[270] M. Chen, et al., High-performance Pb-free solar cells based on tin-halide perovskite thin films functionalized by a divalent organic cation, ACS Energy Lett. 5 (7) (2020) 2223–2230.
[271] A. Waleed, et al., Pb-free perovskite nanowire array photodetectors with drastically improved stability in nanoengineering templates, Nano Lett. 17 (1) (2017) 523–530.

[272] A. Wang, Y. Guo, F. Muhammad, Z. Deng, Controlled Synthesis of Pb-free cesium tin halide perovskite cubic nanocages with high stability, Chem. Mater. 29 (15) (2017) 6493–6501.
[273] M. Lyu, et al., Organic–inorganic bismuth (III)-based material: a Pb-free, air-stable and solution-processable light-absorber beyond organolead perovskites, Nano Res. 9 (3) (2016) 692–702.
[274] S. Öz, et al., Zero-dimensional (CH3NH3)3Bi2I9 perovskite for optoelectronic applications, Sol. Energy Mater. Sol. Cells 158 (2016) 195–201.
[275] X.-L. Li, L.-L. Gao, B. Ding, Q.-Q. Chu, Z. Li, G.-J. Yang, (C6H5NH3)BiI4: a Pb-free perovskite with >330 days humidity stability for optoelectronic applications, J. Mater. Chem. A 7 (26) (2019) 15722–15730.
[276] S.M. Jain, et al., An effective approach of vapour assisted morphological tailoring for reducing metal defect sites in Pb-free, (CH3NH3)3Bi2I9 bismuth-based perovskite solar cells for improved performance and long-term stability, Nano Energy 49 (2018) 614–624.
[277] Y. Lou, M. Fang, J. Chen, Y. Zhao, Formation of highly luminescent cesium bismuth halide perovskite quantum dots tuned by anion exchange, Chem. Commun. 54 (30) (2018) 3779–3782.
[278] S. Ghosh, B. Pradhan, Pb-free metal halide perovskite nanocrystals: challenges, applications, and future aspects, ChemNanoMat 5 (3) (2019) 300–312.
[279] C.C. Stoumpos, et al., Hybrid germanium iodide perovskite semiconductors: active lone pairs, structural distortions, direct and indirect energy gaps, and strong nonlinear optical properties, J. Am. Chem. Soc. 137 (21) (2015) 6804–6819.
[280] I. Kopacic, et al., Enhanced performance of germanium halide perovskite solar cells through compositional engineering, ACS Appl. Energy Mater. 1 (2) (2018) 343–347.
[281] L. Chu, et al., Pb-free halide double perovskite materials: a new superstar toward green and stable optoelectronic applications, Nano-Micro Lett. 11 (1) (2019) 16.
[282] C. Zhang, et al., Design of a novel and highly stable Pb-free Cs2NaBiI6 double perovskite for photovoltaic application, Sustain. Energy Fuels 2 (11) (2018) 2419–2428.
[283] V. Pecunia, L.G. Occhipinti, A. Chakraborty, Y. Pan, Y. Peng, et al., APL Mater. 8 (10) (2020).

Further reading

[284] Z. Yi, N.H. Ladi, X. Shai, H. Li, Y. Shen, M. Wang, Will organic-inorganic hybrid halide lead perovskites be eliminated from optoelectronic applications? Nanoscale Adv. 1 (4) (2019) 1276–1289.
[285] W. Ning, F. Gao, Structural and functional diversity in Pb-free halide perovskite, Materials 1900326 (2019) 1–21.
[286] Y. Kim, et al., Pure cubic-phase hybrid iodobismuthates AgBi 2 I 7 for thin-film photovoltaics, Angew. Chemie 128 (33) (2016).
[287] H. Zhu, M. Pan, M.B. Johansson, E.M.J. Johansson, High photon-to-current conversion in solar cells based on light-absorbing silver bismuth iodide, ChemSusChem 10 (12) (2017).
[288] K.W. Jung, et al., Silver bismuth iodides in various compositions as potential Pb-free light absorbers for hybrid solar cells, Sustain. Energy Fuels 2 (1) (2018).
[289] I. Turkevych, et al., Photovoltaic rudorffites: Pb-free silver bismuth halides alternative to hybrid lead halide perovskites, ChemSusChem 10 (19) (2017).
[290] X.P. Cui, et al., Cupric bromide hybrid perovskite heterojunction solar cells, Synth. Met. 209 (2015).
[291] X. Li, et al., Organic-inorganic copper(II)-based material: a low-toxic, highly stable light absorber for photovoltaic application, J. Phys. Chem. Lett. 8 (8) (2017).

[292] R. Nie, et al., Efficient solar cells based on light-harvesting antimony sulfoiodide, Adv. Energy Mater. 8 (7) (2018).
[293] A. Kulkarni, T. Singh, A.K. Jena, P. Pinpithak, M. Ikegami, T. Miyasaka, Vapor annealing controlled crystal growth and photovoltaic performance of bismuth triiodide embedded in mesostructured configurations, ACS Appl. Mater. Interfaces 10 (11) (2018).
[294] U.H. Hamdeh, R.D. Nelson, B.J. Ryan, U. Bhattacharjee, J.W. Petrich, M.G. Panthani, Solution-processed BiI3 thin films for photovoltaic applications: improved carrier collection via solvent annealing, Chem. Mater. 28 (18) (2016).
[295] A. Kulkarni, A.K. Jena, M. Ikegami, T. Miyasaka, Performance enhancement of AgBi2I7 solar cells by modulating a solvent-mediated adduct and tuning remnant BiI3 in one-step crystallization, Chem. Commun. 55 (28) (2019).
[296] J. Jeong, et al., Ternary halide perovskites for highly efficient solution-processed hybrid solar cells, ACS Energy Lett. 1 (4) (2016).
[297] P.W. Liang, et al., High-performance planar-heterojunction solar cells based on ternary halide large-band-gap perovskites, Adv. Energy Mater. 5 (1) (2015).
[298] F. Cao, X. Xu, D. Yu, H. Zeng, Pb-free halide perovskite photodetectors spanning from near-infrared to X-ray range: a review, Nanophotonics. 10 (2021) 2221–2247.

CHAPTER

7

Low-dimensional halide perovskite for solar cell applications

M. Atikur Rahman[a] *and Faisal Islam Chowdhury*[b]

[a]Department of Electrical & Electronic Engineering, University of Chittagong, Chittagong, Bangladesh, [b]Department of Chemistry, University of Chittagong, Chittagong, Bangladesh

1. Introduction

Because of having distinctive features with reduced dimensions, low-dimensional materials are crucial in optoelectronic, electrical, and modern photonics fields. Materials with a minimal number of dimensions in the nanometer range (1–100nm) are termed low-dimensional materials. Generally, the halide perovskite material that adopts 0-2D dimension (below 3D) is referred to as low-dimensional halide perovskite [1]. However, 3D perovskite can also act as low-dimensional perovskite by satisfying related requirements of dimensionality. In 0D, 1D, or 2D halide perovskite, massive organic-inorganic cations divide metal halide anions, and at the cellular level, they are bulk arrays of molecular entities or clusters, perovskite quantum-wells, and quantum-wires. As a result of quantum size effects, low-dimensional halide perovskites have dramatically variable optical characteristics. Low-dimensional nanostructure halide perovskites have distinct physical features, including strong luminescence and a larger optical bandgap due to the quantum confinement effect, which have not been found in their conventional materials [2]. Varied low-dimensional halide perovskites have been described with various synthesis techniques, including 0D structures like quantum dots and nanocrystals, 1D nanostructures like nanowires and nanorods, 2D nanostructures like nanoplatelets and layered structures [1]. In recent years, organic-inorganic hybrid low-dimensional halide perovskite materials have received much attention

Low-Dimensional Halide Perovskites
https://doi.org/10.1016/B978-0-323-88522-5.00006-5

for photovoltaic (PV) applications, particularly in solar cells. According to data from Clarivate Analytics in 2018, more than 1000 institutes across the globe are actively working on photovoltaics and optoelectronics using halide perovskite, resulting in more than 8000 scholarly articles on the subject [3]. In the past, oxide perovskites were widely used in the PV field. Metal oxide perovskites have poor semiconducting characteristics [4], making them unsuitable for photovoltaic applications.

In addition, a family of oxide perovskites separate from halide perovskites has oxide anions rather than halide anions. These halide perovskites demonstrate the semiconductive characteristics that are needed for PV applications. Since 2009 when the initial fabrication of dye-sensitized solar cells, including perovskites with halide ions, were investigated, halide perovskites have been studied extensively as light-collecting material to ease the looming energy shortfall. A solar cell formed of halide perovskites demonstrated a maximum power conversion efficiency (PCE) of 22.7%, higher than previously reported for dye-sensitized solar cells, organic photovoltaics, and traditional thin-film solar cells, demonstrating significant advancements and breakthroughs in halide perovskite solar cell (PSC) research [5]. Because of extensive research, performance and efficiencies have increased. Fig. 1 shows the increasing trend of Halide PSCs efficiencies since 2008 to 2021. Efficiencies of PSCs are affected by energy loss in the core and junction of perovskite layers, as well as charge extraction and transit in devices, among other factors. In the last 2 years, various methods have been used to increase the working efficiency of PSCs, such as absorber layer modification, interface engineering, charge transport layer design, etc. Acquired PCEs have been listed and demonstrated in Table 1.

In 2021, Seo et al. [7] published a solar cell based on perovskite with a validated PCE of 25.2%. Outstanding charge transport features and optical properties, including higher external quantum efficiency and better light absorptivity, result in their remarkable PCE, which has not been observed in other materials. In 2020, another work by Kim et al. [13] with $FAPbI_3$ layered solar cell involved substituting methylenediammonium (MDA), and cesium (Cs) cations were found in tiny amounts (0.03 M fraction) in the solution. For the FA cations, the α-phase $FAPbI_3$ was stabilized without modifying its intrinsic bandgap. However, the UV-vis absorption of the perovskite layer was improved, resulting in better efficiency of close to 24.4%. In 2021, For 1 cm^2 PSCs, Peng et al. [14] developed a confined nanoscale interaction at the polymer-passivated interface of perovskite/TiO_2 in order to get a higher fill factor. They also achieved a validated PCE of 21.6% for the cell. Another study in 2020 based on hole transport material (HTM) design was carried out by Jeong et al. [15] To make use of fluorination's advantages in terms of noncovalent interactions, hydrophobicity, and energy levels, two fluorinated analogs of Spiro-OMeTAD were created by Jeong et al. [15], who employed HTMs over PSC-based solar cells that result in PCE of 24.8% along with better stability.

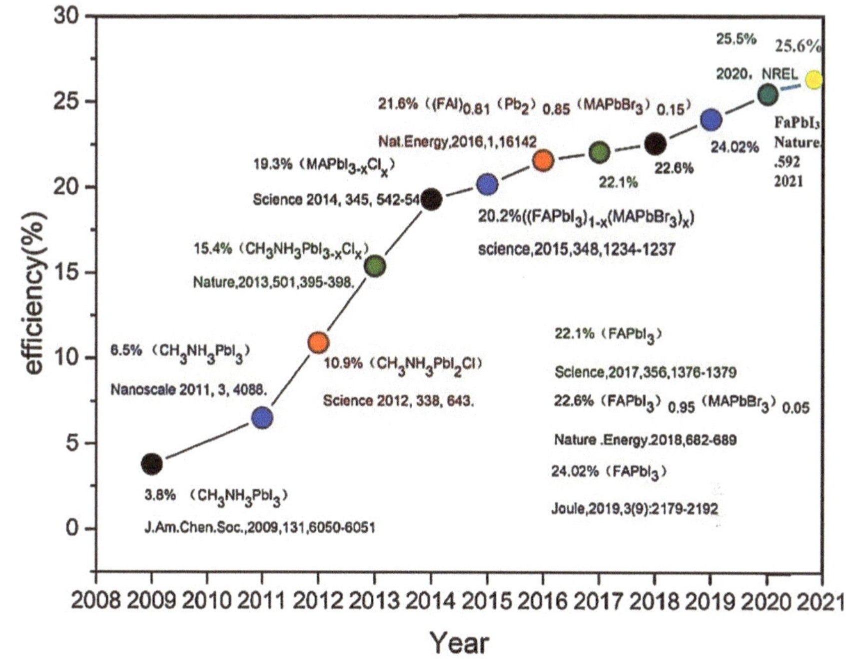

FIG. 1 Over the period 2008–21, increment and progress of the efficiency of perovskite solar cells (PSCs). *Reproduced from T. Dai, Q. Cao, L. Yang, M. Aldamasy, M. Li, Q. Liang, H. Lu, Y. Dong, Y. Yang, Strategies for high-performance large-area perovskite solar cells toward commercialization, Crystals 11 (2021). Reproduced from J. Jeong, M. Kim, J. Seo, H. Lu, P. Ahlawat, A. Mishra, Y. Yang, M. Hope, F. Eickemeyer, M. Kim, Y. Yoon, I. Choi, B. Darwich, S. Choi, Y. Jo, J. Lee, B. Walker, S. Zakeeruddin, L. Emsley, U. Rothlisberger, A. Hagfeldt, D. Kim, M. Grätzel, J. Kim, Pseudo-halide anion engineering for α-FAPbI$_3$ perovskite solar cells, Nature 592 (2021).*

TABLE 1 Perovskite solar cell efficiency based on the literature of 2020–21.

Device	Year	Power conversion efficiency (PCE) (%)	Reference
α-Phase $FAPbI_3$-based PSC	2020	24.4	[13]
α-Phase $FAPbI_3$-based PSC	2021	25.6	[10]
$MAPbI_3$-based PSC	2020	21.36	[11]
Mixed halide CsFAMA-based PSC	2020	22.3	[12]
Polymer passivated halide PSC	2021	21.6	[14]
Spiro-OMeTAD HTM fabricated FAPbI3 PSC	2020	24.8	[15]
SnO2 layered PSC	2021	25.2%	[7]

Unlike typical semiconductor nanomaterials, low-dimensional halide perovskites can be made via low-cost techniques using solution-processable processes. Furthermore, as thin films of halide perovskite could be made using various solution-based methods like facile spin coating, these thin films of halide perovskite with various compositions may be easily fabricated using limited cost, large-scale, and easy approaches. Low-dimensional perovskites have been developed in a variety of morphologies, including quantum dots (zero-dimensional), perovskite nanorods, or nanowires (one-dimensional) perovskite nanoplatelets and nanosheets, and perovskite nanoplatelets and nanosheets (two-dimensional). However, unfortunately, some limitations are included with those solution-processed thin films of low-dimensional halide perovskites, which impede the industrial implementation of optoelectronic devices such as layered solar cells [8]. In low-dimensional perovskite, the insulated organic cations obstruct carrier movement, lowering the PCE of PSCs. Lack of homogenous stability, long-term performance reduction, and current-voltage hysteresis may be considered hindrances to the commercialization of halide perovskites [9]. It has been discovered that by decreasing the three-dimensionality of bulk perovskites, it is possible to generate another class of perovskite materials, termed layered perovskites. A layered perovskite structure comprises an endless number of two-dimensional slabs with the structure ABX_3 that are divided by an element, which can be any size. Here, $A_{n\text{-}1}B_nX_{3n+1}$ is the general formula for perovskites with a layered structural arrangement. For layered perovskites, the number n specifies the size of the two-dimensional slabs in the structural formula (Fig. 4).

When compared to three-dimensional perovskites, the layered perovskites exhibit a variety of physical and chemical characteristics that vary depending on the number of layers present. Quan et al. claimed the best performance ever achieved by a 2D halide perovskite-based solar cell in their research. A study was conducted in this paper to determine the influence of layer number on the solar cell performance of halide perovskite $n=1$ (considered to be pure 2D) to $n=40$–60 (considered to be quasi-2D) was used to adjust the number of layers in the model [6]. They had found the best performance when the value of n was higher. There have been many studies recently on low-dimensional halide perovskites and single-crystalline bulk perovskites. For example, the fabrication of halide perovskites with good crystallinity as well as the analysis of their inherent features and applications in order to overcome the limitation included with the implementation of halide PSCs.

In addition, low-dimensional halide perovskites solar cells may be fine-tuned and adjusted to match the required qualities for stable solar cell application because of their chemical flexibility and electrical and optical properties. They have shown exceptional electrical, dielectric, optical, and optoelectronic capabilities, and they are a popular study issue in novel semiconducting and optoelectronic devices. PSCs of inorganic-organic lead halide perovskite with low dimensions have made significant advancements in photovoltaic devices. The enormous potential of low-dimensional lead halide PSCs as an upcoming PV technology is attested to not only by their outstanding performance but also by the low-cost, solution-based procedures employed for device manufacturing. This chapter provides a brief overview of the various halide perovskite materials for solar cell applications, their limitations, and how to overcome these with the aid of low-dimensional halide perovskite.

2. Perovskite solar cells of low-dimensional halide perovskite

2.1 Perovskite solar cells with 0D halide perovskite

The interesting features of 0D perovskites from their unusual structure with solitary metal halide octahedra have recently attracted much attention in the PSC field. Metal-halide octahedron anions or clusters can form 0D perovskite and are isolated by other organic or inorganic cations. Sometimes, metal halide wires and individual octahedra are cut out of 2D perovskites to form 0D or 1D perovskite. Low-dimensional (0D and 1D) perovskites convey excellent stability and hydrophobicity by which the 3D active layer is well protected under adverse environmental conditions, resulting in higher mixed PSC performance. 0D metal halide perovskites are very stable in the environment. According to the information provided in Table 2, 0D perovskite shows high stability compared to other perovskites. This is because the organic shells that maintain metal halide octahedron units/clusters act as an ideal host-guest system. The maximum degree of stability is very advantageous for industrial applications. However, a significant problem is that with a 0D structure, we do not get good power conversion efficiency. A planar heterojunction structure with 0D

TABLE 2 Stability comparison among perovskite of various dimensions.

Perovskite type	Stability	References
0D	High	[16]
1D	Medium	[16]
Quasi-2D	Medium	[16]
Corrugated 2D	Medium	[16]
3D	Low	[16]

perovskite (ITO/PEDOT/$(CH_3NH_3)_3Bi_2I_9$/PCBM/Ca/Al) was fabricated to form a solar cell by Öz et al. [17] In this cell, they found poor efficiency of 0.1%, which may be reasonable for energy level mismatch of the transport layer, bandgap, and higher exciton binding energy [17]. Charge trapping at grain boundaries and interfaces and ion transport is the primary cause of reported hysteresis in 3D perovskites, impacting the devices' performance and the long-term stability of the PSCs. This is why fabricating a solar cell by forming the heterojunction of the different dimensional perovskite elements with other elements has received attention. Quantum dots, nanosheets, and 3D-2D-0D dimensionally profiled interface structures that contain $CsPbBrI_2$ material fabricated by Zhang et al. [18] can be shown as an example of heterojunction. Both $CsPbBrI_2$ and the HEL layers are tuned for energy alignment between the conduction and valence band. A constant upshifting of energy levels is caused by the profiled structure, leading to an optimum value of J_{SC} of up to 12.93 mA cm^{-2} and an 80.5% fill factor value, and consequently a higher power conversion efficiency (PCE) of 12.39% [18]. Again adjusting the stoichiometric ratio of CsI and PbI precursors led to the formation of a 0D-3D heterostructure with 0D Cs_4PbI_6 and—3D $CsPbI_3$ designated by Bai et al. [19]. As a result of this heterostructure, it is possible to realize a stable all-inorganic PSC with a high PCE of 16.39%, which is extremely efficient. Because isolated octahedrons' metal halides are in 0D perovskites rather than in 3D perovskites, ion movement between octahedrons is inhibited, resulting in reduced hysteresis. An initial planar heterojunction solar cell with ITO/PEDOT/$(CH_3NH_3)_3Bi_2I_9$/PCBM/Ca/Al structure, 0D perovskite $(CH_3NH_3)_3Bi_2I_9$ as absorber layer was developed by Öz et al. [17] and Chen et al. [20] presented 0D/3D mixed perovskite. However, due to the low PCE of 0D fabricated solar cells, they recommended more research on increasing the performance of 0D PSCs by decreasing bandgaps and exciton binding energy.

2.2 PSC with 1D halide perovskite

Large exciton binding energies in 1D halide perovskites are typically considered undesirable for solar systems at the molecular level. 1D perovskites have not been studied as active material as much as 3D and 2D perovskites for solar cells. That is why poor literature exists on this topic. Theoretically, 1D perovskites can be more stable than 2D and 3D equivalents. Ma et al. [21] reported a high-performance solar cell based on 1D halide perovskite. They reported solar cells formed of 1D $BDAPbI_4$ (1,4-benzene diammonium lead iodide) along with $MAPbI_3$. A thin film of composite perovskite was made for solar cells following the formula $BDAPbI_4 \cdot n$-1$MAPbI_3$ with $n=2$–4 (when $n=2,3,4$, it is referred to as N2,

N3, N4 samples). PSCs with absorber layers of N2, N3, and N4 samples were built and tested under normal AM1.5G light intensity. The testing solar cell device was created using an ITO/PEDOT: PSS/perovskite/C60 (20 nm)/BCP (8 nm)/Ag(70 nm) configuration. The prepared cells were analyzed using a reference solar cell of silicon due to calibration and examined under a 100 mW cm^{-2} light source and an air mass of 1.5 irradiance. To produce maximum performance, the thicknesses of the N2, N3, and N4 samples were kept to 350, 350, and 400 nm, respectively. The sample produced PCE of up to 14% and showed large carrier stability, higher mobility, and lower quantum confinement. This hybrid model PSC with low-dimensional showed better stability. The PCE for the N4 specimen was 14.1%, and the PCEs for the N3 and N2 specimens were 9.0% and 3.8%, respectively, with PSCs exhibiting negligible J-V hysteresis as shown in Fig. 2. In addition, embedded N4 PSC stability was investigated along with 3D $MAPbI_3$ to compare the results. Fig. 3D shows the comparison of normalized PCE of 3D and N4 samples that have been presented as a function of storage duration. After 1000 h, the power conversion efficiency of the 3D PSC reduced to 70% from the original value, whereas PSC with N4 remained at 95% after 1000 h from the tested performance value.

The impact of 1D halide perovskite and their understanding of solar cell applications will fuel ongoing research to widen the use of low-dimensional 1D halide perovskite materials. In terms of enhancing the stability of PSCs, a hybrid dimension structure composed of one-dimensional and three-dimensional perovskites (1D-3D) is an effective method to employ. Organic-inorganic hybrid perovskites with 1D and 2D organic-inorganic perovskites have been discovered to alter the 3D perovskite absorbent layer to increase the long-term ambient stability of PSCs by creating mixed heterojunctions and crystal boundaries passivation. For example, adding a 1D $TAPbI_3$ perovskite layer on top of a 3D $FAMAPbI_3$ layer resulted in a maximum PCE of 18.97%, as reported by Gao et al. [22] Furthermore, after 60 days of storage in ambient air, the constructed device retained 92% of its original efficiency. To increase efficiency, Zardari et al. [23] reported a new strategy. According to this study, PSCs can be made cheaply and more reliably using a self-assembly approach that grows the 1D-perovskite nanowires network. Following their report, a small amount of Urotropin (UTP) was added to the precursor of the perovskite solution used during the two-step sample preparation technique. The 1D perovskite nanowires exhibited passivated grain boundaries, fewer defects, and better film crystallinity.

The optimum device's PCE rose from 11.47% to 19.15% with minimal hysteresis, and improved stability was noticed in this device, losing just 5.08% of its PCE during a 42-day period. Similarly, to create the 1D/3D perovskite heterostructure, Yang et al. [24] cross-link the 3D perovskite

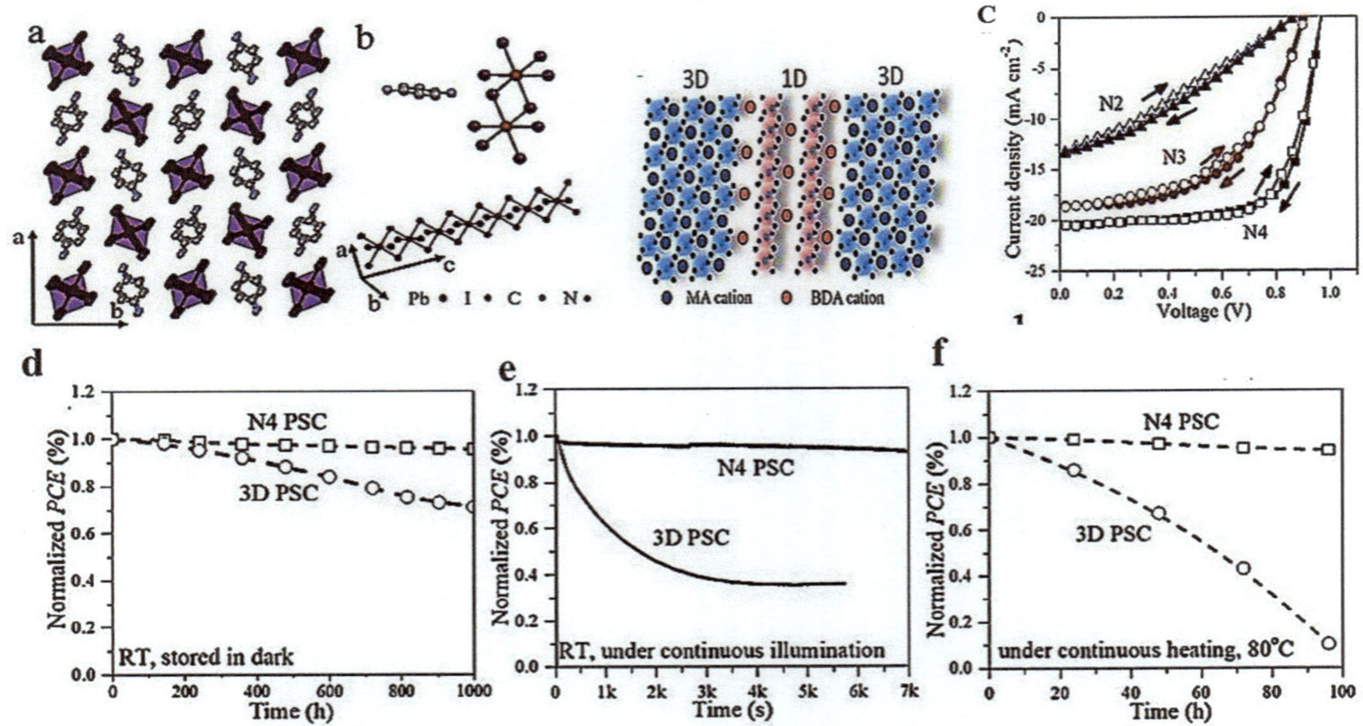

FIG. 2 (A) $BDAPbI_4$ (1D perovskite) structure. (B) Unit and notation of different atoms of $BDAPbI_4$. (C) Graph of different PSC with scan sample J-V curves with their scanning directions. (D–F) Operational and storage stability of N4 sample with PSC formed by 3D perovskite under different conditions shown in the figure. *Reproduced from C. Ma, D. Shen, B. Huang, X. Li, W.C. Chen, M.F. Lo, P. Wang, M. Hon-Wah Lam, Y. Lu, B. Ma, C.S. Lee, High performance low-dimensional perovskite solar cells based on a one dimensional lead iodide perovskite, J. Mater. Chem. A 7 (2019) 8811–8817.*

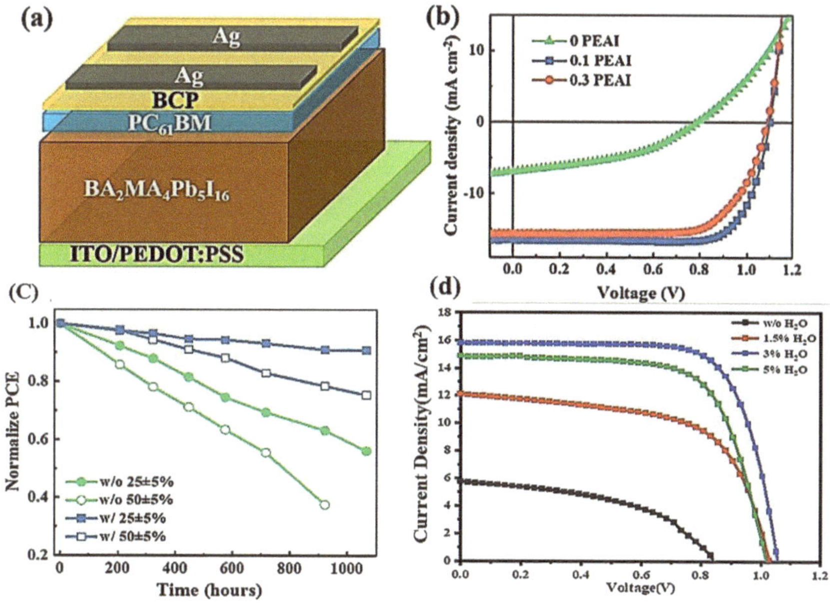

FIG. 3 (A) The PSC planar structure formed of 2D perovskite. (B) A series of J-V curves for PSCs made from $BA_2MA_4Pb_5I_{16}$ perovskite films and varied PEAI concentrations are shown. (C) Stability of open systems both with and without 0.1 PEAI in various humidities. (D) $BA_2MA_3Pb_4I_{13}$-based perovskite solar cell film's J-V curve, along with varied H_2O additive, demonstrates improved results. *Reproduced from X. Lian, J. Chen, M. Qin, Y. Zhang, S. Tian, X. Lu, G. Wu, H. Chen, The second spacer cation assisted growth of a 2D perovskite film with oriented large grain for highly efficient and stable solar cells, Angew. Chem. 131 (2019). Reproduced from Z. Liu, H. Zheng, D. Liu, Z. Liang, W. Yang, H. Chen, L. Ji, S. Yuan, Y. Gu, S. Li, Controllable two-dimensional perovskite crystallization via water additive for high-performance solar cells, Nanoscale Res. Lett. 15 (2020).*

film using polymerizable propargylammonium (PA^+) in situ at surfaces and grain boundaries. That increased the interfacial carrier transport and relieved residual tensile strain in prepared films. During 3055 h of constant illumination, the comparable devices attained a maximum PCE of 21.19% and were able to keep 93% efficiency [24].

2.2.1 *Perovskite solar cells with 2D halide perovskite*

By regulating the light absorption qualities and demonstrating a quick exciton production process, 2D perovskites with their direct bandgaps demonstrate exceptional power conversion efficiencies (PCEs) [25]. When it comes to solar cell applications, 2D perovskite structures show very high stability, but they have poor charge transport capability and a wide bandgap. Many efforts have been made to achieve high-performance 2D PSCs and improved stability, including precursor solvent engineering, additive engineering, hot-coasting, composition engineering, and interface engineering [26]. To understand the improved stability of 2D perovskite, a group fabricated a 2D perovskite cell and found that after 46 days of exposure to up to 52% relative humidity, 2D$(PEA)_2(MA)_2Pb_3I_{10}$ perovskite thin films did not disintegrate. A comparison revealed that the reference device with 3D $MAPbI_3$ held under the same circumstances degraded in just 4 days, demonstrating that $MAPbI_3$ is completely transformed into PbI_2 after 40 days [27]. It proves the outstanding stability of 2D halide perovskite. To design 2D halide perovskite materials, large, organic cations, like PEA (phenyl-ethyl ammonium) and $C_4H_9NH_3$ (BA) usually need to be put together. These organic cations are added to the octahedron Pb-I repeating units to make 2D halide perovskites [28]. Smith et al. [29] reported the first PSCs produced by 2D halide perovskites. The team explored the roots of instability in typical 3D $CH_3NH_3PbI_3$ and presented a novel idea of 2D layer halide perovskites, which was created utilizing phenyl ethylammonium iodide (PEAI) molecules instead of methylammonium iodide (MAI), as illustrated in Fig. 4A. High-performance PSCs may be made using perovskite films with optimal grain size and film orientation. The power conversion efficiency of solar cells based on Ruddlesden-Popper (RP) 2D perovskite relies heavily on manufacturing high-grade films with big grains aligned toward the thickness of films [30]. Due to the obvious features of high formation energy and enhanced hydrophobicity, 2D RP perovskite with chemical formula $A_2B_{n-1}M_nX_{3n+1}$ exhibits superior moisture resistance to its 3D equivalent. Lian et al. [30] reported a new 2D $BA_2MA_{n-1}Pb_nI_{3n+1}$ ($n=5$) perovskite film having a grain size of 1 μm where, by injecting a second spacer cation (SSC^+) into the solution, preferred orientation growth can be achieved. For the optimized fine-grained PSCs mentioned above, the highest PCE of 14.09% was achieved using phenylethylammonium as SSC^+. This may be observed from Fig. 3A–C. Furthermore, after 1000 h of air exposure, the uncapped device maintains roughly 90% of its original efficiency, demonstrating excellent moisture stability. Their study paves the way for the commercialization of PSCs that are both highly efficient and extremely stable. Due to the effect of dielectric confinement when the organic layer and its inorganic framework are mixed, the exciton binding energy of the layered 2D perovskite is better because it makes the organic layer more stable. However, to prevent charges from moving between adjacent inorganic plates, thick coatings of massive organic ligands will act as insulators. Hence, the PCE of 2D layered perovskite is lower than that of its 3D counterpart. Therefore, various methods and techniques are currently being adopted in order to increase the cell efficiency of designed 2D perovskite-based solar cells.

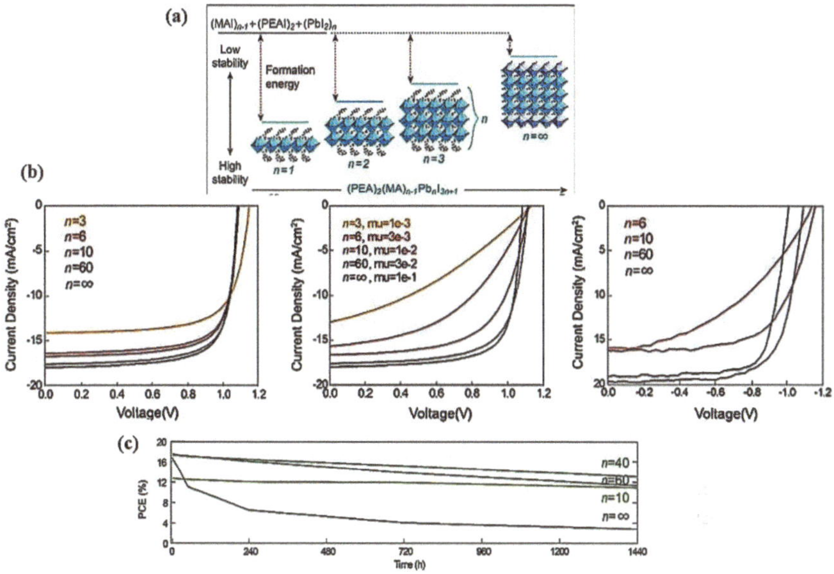

FIG. 4 (A) Stability and formation energetics of perovskite with unit cell layered perovskite structure of $(C_8H_9NH_3)_2(CH_3NH_3)_{n\text{-}1}Pb_nI_{3n+1}$. (B) J-V hysteresis curve of different n values. (C) The development of device performance as a function of time for various *n* values. *Reproduced from L.N. Quan, M. Yuan, R. Comin, O. Voznyy, E.M. Beauregard, S. Hoogland, A. Buin, A.R. Kirmani, K. Zhao, A. Amassian, Ligand-stabilized reduced-dimensionality perovskites, J. Am. Chem. Soc. 138 (2016) 2649–2655.*

Improving PCE of 2D PSCs also requires controlling the crystallization of 2D perovskite films. Liu et al. [26] introduced water additive technique for crystallization of the film in the precursor solution in order to improve the efficiency of 2D PSC. Improved crystallinity and orientation order were seen in the perovskite film they made ($BA_2MA_3Pb_4I_{13}, n=4$) after water additive treatment. As a result, PSCs with this high-quality 2D perovskite layer had reduced trap-state density and thus higher photovoltaic efficiency. They reported an additive of 3% H_2O with a fabricated device to improve the PCE from 2.29% to 12.15% compared to perovskite film without additive, following Fig. 3D. On the other hand, Fu et al. [31] used NH_4SCN and NH_4Cl additives and achieved an ideal PCE of 14.1% for 2D PSCs. Zhao and Zhu [32] used a thermal decomposition technique to create $MAPbI_2Br$ nanosheets having a bandgap value of 1.8 eV from a precursor comprising PbI_2, MABr, and MACl. Plane solar cells constructed on an interlayer of $MAPbI_2Br$ nanosheets demonstrated a PCE of 10% [32]. Quasi-2D perovskites with halide ions follow the chemical formula $R(NH_3)_2A_{n\text{-}1}\ B_nX_{3n+1}$. They have the same layered structure as 3D halide perovskite containing long aliphatic chain cations, which ultimately prevent moisture from adsorbing and increase the stability of halide perovskite [33]. Representatives of the quasi-2D family integrate the enhanced stability of 2D halide perovskites with the superior optoelectronic properties of 3D perovskites, including wide-range photocarrier diffusion.

Many researchers have lately concentrated on creating alternate and suitable spacer cations for regulating PCEs and maintaining PSC stability, inspired by Smith's work. When compared to the 3D structure of $CH_3NH_3PbI_3$, the moisture stability of 2D layered perovskites with a structure of $(PEA)_2(MA)_2Pb_3I_{10}$ ($n=3$) was considerably improved. Resulting in small absorption coefficient and weak carrier transport of $(PEA)_2(MA)_2Pb_3I_{10}$ ($n=3$), PCEs were lower than 5% from PSCs with a device structure of FTO/compact TiO_2/$(PEA)_2(MA)_2Pb_3I_{10}$ ($n=3$)/spiro-OMeTAD/Au, as shown in Fig. 4B and C.

Cohen et al. [34] discovered that the 2D layered perovskite with $(PEA)_2(MA)_2Pb_nI_{3n+1}$ ($n=60$) had improved stability and that PSCs made from this unique 2D perovskite had good PCEs. Solar cell performance from PSCs by 3D $CH_3NH_3PbI_3$was decreased from 16.47% to 0.72% after 2 weeks in humid air, whereas PCEs from PSCs by PEA-based 2D layered were reduced from 17.21% to 12.80% after 2 weeks where relative humidity was 55%. Furthermore, because the insulator OS cations hampered charge mobility, these devices performed poorly.

Quan and colleagues claimed the best performance and efficiency of a solar cell implemented with 2D halide perovskite [34,35]. The influence of layer number on the efficiency of solar cells made with halide perovskite was explored in this study. The layer number was varied with range $n=1$ (denoting pure 2D) to $n=40$–60 (denoting quasi-2D) and $n=40$–60 (3D), as shown in Fig. 4. Compared to pure 2D halide perovskites, the halide perovskite with layer number $n=40$–60 (quasi-2D) demonstrated better exciton diffusion length, carrier mobility, and absorption [34]. In addition, its stability was better compared to 3D high-dimensional halide perovskites. The van der Walls couplings between the capped organic molecules were believed to be responsible for the enhanced stability of solar cells made of quasi-2D and low-dimensional 2D halide perovskites, which was also validated by DFT simulations [35]. Due to the toxicity issue of Pb, research has been triggered on lead-free PSCs. Later, nontoxic tin-based perovskite appeared as a new generation of PSCs [36]. However, due to the poor performance of PSCs built from these Pb-less materials, attempts have been undertaken to reduce the content to decrease the toxicity of lead rather than eliminate it. This is why the partial replacement of Pb by Sn in Pb-based halide perovskite materials, resulting in Pb-Sn hybrid perovskites, has received much attention [37].

3. Current challenges regarding commercialization of halide perovskites solar cells and prospects

While most major scientists are already achieving PCEs of more than 20% in a testing device, stability and Pb toxicity remain two significant barriers to implementing PSCs [38]. According to a recent study, the interfacing state of PSCs is the most important aspect determining the devices' stability and performance. For practical uses and industrial implementation within 25 years, PSCs must provide steady power output under operational circumstances of actual solar radiation, increased temperature owing to the heat, and ambient oxygen and moisture. With perovskite, stability/instability for a long time involves both extrinsic and intrinsic stability concerns. Both internal and structural stability, as well as stability during various external environmental factors, including heat, oxygen, light, and humidity, are vital [39].

Under the continuous operation mode of the cells, performance degradation or material degradation concerns require immediate attention and should be resolved for the next generation of halide PSCs. Previous studies have shown that the distance ratio determines the formability of halide perovskite material among akylammonia (CH_3NH_3) cation-halide anions and metal cation-halide anions (A:X and M:X). This ratio is referred to as the tolerance factor [40], and the formula is as follows:

$$TF = \frac{R_A + R_B}{\sqrt{2(R_A + R_B)}}$$

The Goldschmidt tolerance factor (τ), an experimental metric frequently calculated to forecast the development of distinct crystal structures of ABX_3, may be used to assess the structural stability of perovskite compounds [41]. A stable perovskite will be formed if it is near the intermediate range between 1 and 0.8, distant from both zones of non-perovskite (Fig. 5). Perovskite structures are less likely to occur with values of 0.8 and greater than 1 of the Goldschmidt tolerance factor (τ). With a value of 0.9, 2D halide $MAPbI_3$ produces a dark photoactive perovskite phase because of stands in the middle of the perovskite zone. For the perovskite structure to be more stable, cations such as K, Na, Rb, and Cs with significant differences in ionic sizes between them and the matrix system (FA/MA mixture) are the best until no nonperovskite secondary structure used. In other words, phase purity is important for the intrinsic long-term stability of mixed perovskites.

After improving the internal stability of halide perovskites, cell stability to environmental factors such as moisture, oxygen (air), heat, light, external bias, and so on will be significant. Perovskite's moisture sensitivity has been a source of concern for its long-term stability [42]. Water molecules

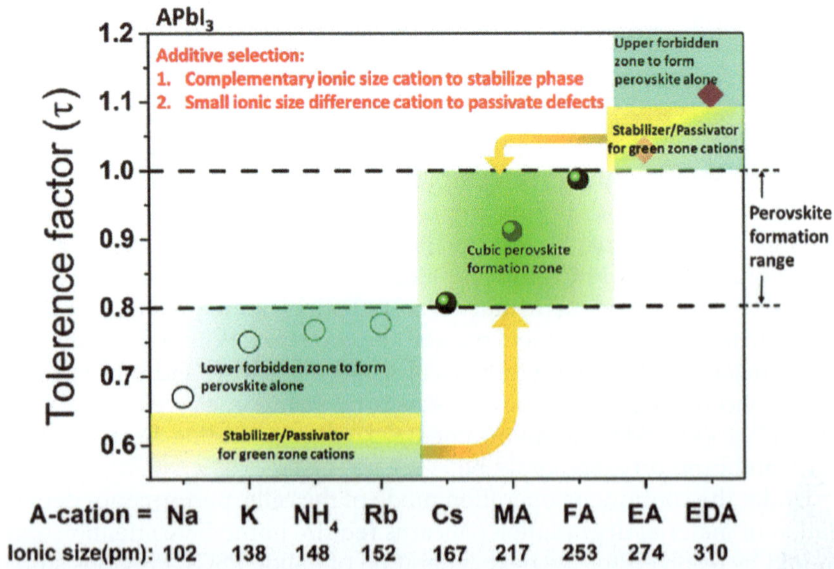

FIG. 5 Tolerance factors (τ) in the $APbI_3$ perovskite system for various cations (A). Cations such as Cs, MA, and FA provide a value in the 0.8–1.0 range, suggesting the creation of the cubic perovskite phase structure. *Reproduced from T. Dai, Q. Cao, L. Yang, M. Aldamasy, M. Li, Q. Liang, H. Lu, Y. Dong, Y. Yang, Strategies for high-performance large-area perovskite solar cells toward commercialization, Crystals 11 (2021).*

are known to form the hydrogen bond to perovskite easily in order to produce hydrated compounds that change the features of perovskite locally [43]. As a result, appropriate insulation of PSCs becomes a requirement for preventing moisture-induced deterioration. Moisture-resistant halide perovskite PSCs can be created by technically changing the perovskite layer with small molecules or integrating 2D perovskites with hydrophobic organic groups into 3D perovskites. Chaudhary et al. demonstrated that when $MAPbI_3$ is immersed in water for up to 100 s, a polyvinylpyridine (PVP) ultra-thin layer placed on the surface of $MAPbI_3$ 2D halide perovskite and also in combination with spiroOMeTAD enhances the resistivity of water in the solar cell (Fig. 6). In addition, the PVP-modified device has a higher V_{oc} and efficiency [44].

3.1 Perovskite solar cells with toxic lead-free low-dimensional halide perovskite

3.1.1 Tin-based low-dimensional halide perovskite for solar cells

Tin-based PSC received attention as the most acceptable substitute for Pb replacement is likely to be Sn [45]. Because of their comparable ionic

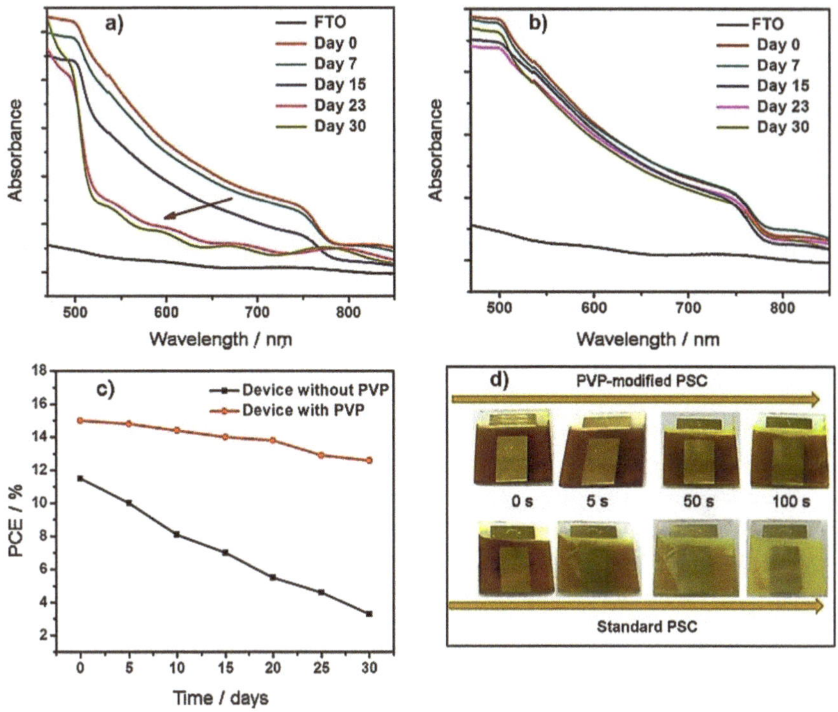

FIG. 6 (A) Absorbance of light in device without PVP. (B) Absorbance in with PVP modified device. (C) Power conversion efficiency for systems with and without PVP variation during days held at room temperature in 50% relative humidity. (D) Images of conventional PVP-modified PSCs and without PVP after they have been submerged in water show that they have changed (discoloration). *Reproduced from B. Chaudhary, A. Kulkarni, A.K. Jena, M. Ikegami, Y. Udagawa, H. Kunugita, K. Ema, T. Miyasaka, Poly (4-vinylpyridine)-based interfacial passivation to enhance voltage and moisture stability of lead halide perovskite solar cells, ChemSusChem 10 (2017) 2473–2479.*

structure, they are among the most promising contenders to replace Pb. Several lead-free low-dimensional perovskites have been proposed [46]. However, compared to others, tin-based PSC is one of the lead-free perovskites that have been reported so far, with a power conversion efficiency of 9.6% [47]. Chen et al. [48] showed that their Sn-based fabricated solar cell system was adjustable in terms of photoluminescence by using hybrid materials $CsSnX_3$(X=Cl, Br, and I) with various halides. Under AM 1.5G, a PCE of 12.96% was earned by substituting the halide of the $CsSnX_3$ quantum rods [48]. But challenges are affiliated with Sn-based PSCs because of different oxidation states such as Sn^{2+} to Sn^{4+} and stability issues [49]. $SnCl_2$, SnF_2, and H_3PO_2 have made Sn perovskites more stable. Their stability, however, lags far behind that of perovskite made of Pb.

Such additives balance the lacking Sn^{2+} in the film by supplying extra Sn^{2+}. Further knowledge of the effect of such additions would undoubtedly aid in discovering more appropriate additives to stabilize Sn^{2+} and, as a result, improve performance and stability. Snaith et al. developed a solar cell with Pb-free perovskite based on mesoporous TiO_2 substrates employing $MASnI_3$, an Sn-based perovskite as the PSC with PCE of more than 6% [50]. However, stability under ambient conditions appeared to be a major issue. Therefore, low-dimensional 2D halide perovskite appeared to solve the stability issue. Marshall et al. created Sn-based PSCs following the structure ITO/CuI/$CsSnI_3$/C_{60}/BCP/Al, as shown in Fig. 7. The findings show that additional added parts of the SnI_2 precursor aid in increasing the performance of photovoltaic applications and solar cell stability [51]. Individually, research on Sn-based PSCs with a reversed planar structure and the findings was presented by two research teams. As illustrated in Fig. 7, Marshall et al. [51] developed a model with an ITO/CuI/$CsSnI_3$/C_{60}/BCP/Al structure. Their findings showed that having too much SnI_2 precursor might improve PV performance and system stability. In other work, as illustrated in Fig. 7B, an antisolvent technique was devised by Zhang et al. [52] to produce $FASnI_2Br$ films that were incredibly consistent and compact systems, having a reversed ITO/PEDOT:PSS/$FASnI_2Br$/C_{60}/Ca/Al planar structure. With reduced temperature annealing at 75°C, the $FASnI_2Br$ perovskite with a bandgap value of 1.68 eV demonstrated excellent PV performance. After 30 h of storage in a nitrogen atmosphere without wrapping, the gadget prepared with this material retained more than 60% of its efficiency.

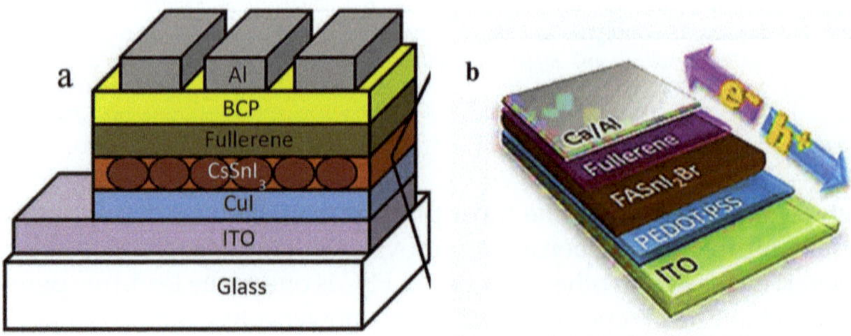

FIG. 7 PSCs based on Sn that have an reversed planar structure in device configurations: (A) ITO/CuI/$CsSnI_3$/C_{60}/BCP/Al and (B) ITO/PEDOT:PSS/$FASnI_2Br$/C_{60}/Ca/Al. *Reproduced from K.P. Marshall, R.I. Walton, R.A. Hatton, Tin perovskite/fullerene planar layer photovoltaics: improving the efficiency and stability of lead-free devices, J. Mater. Chem. A 3 (2015) 11631–11640 and Reproduced from M. Zhang, M. Lyu, J.-H. Yun, M. Noori, X. Zhou, N.A. Cooling, Q. Wang, H. Yu, P.-C. Dastoor, L. Wang, Low-temperature processed solar cells with formamidinium tin halide perovskite/fullerene heterojunctions, Nano Res. 9 (2016) 1570–1577.*

3.1.2 Copper-based perovskites

Transitional metals like Cu have a relatively stable divalent condition compared to Sn and Ge in the air. As a result, a few scientific studies have looked into copper-based perovskites to replace Pb-based 2D halide perovskite. Halide perovskites based on Cu typically have a 2D structure instead of the 3D structure of traditional perovskites based on Pb. The general chemical formula followed for Cu-based PSCs is $(RNH_3)_2CuX_4$, with RNH^{3+} denoting various aromatic or aliphatic ammonium cations and the sign X denoting halogens compound [43]. As previously stated, Cu-based perovskites are substantially more stable than Sn- and Ge-based perovskites and can be synthesized utilizing a simple solution procedure in ambient conditions. Cu-based PSCs with power conversion efficiencies of 0.51% and 0.63%, respectively, utilizing $(F\text{-}C_8H_8NH_3)_2CuBr_4$ and $(CH_3CH_2CH_2CH_2NH_3)_2CuBr_4$ were manufactured by Cui et al. [53] under ambient conditions. $MA_2CuCl_xBr_{4x}$ was investigated as PSC by Cortecchia et al. [54], as shown in Fig. 8. The Cu-based perovskite light absorber's absorption range may be modified by adjusting the Br/Cl ratio. Unfortunately, their $MA_2CuCl_xBr_{4x\text{-}1}$ based device PSCs did not perform as well as intended, with a power conversion efficiency of just 0.017%.

3.1.3 Germanium-based perovskite

Because of their extraordinary optical characteristics for solar applications, germanium halide perovskites are an appealing alternative to lead halide perovskites. On the other hand, the PCE of solar cells based on germanium perovskites was consistently below 0.2% up to this point, and device stability has also been a concern. However, altering the chemical composition of the germanium perovskite, for example, by mixing bromide ions into the methylammonium germanium iodide perovskite, results in a significant improvement in solar cell performance, as well as a modest improvement in the stability of the low-dimensional germanium perovskite. The first report of germanium halide PSCs was published in 2015 by Krishnamoorthy et al. [55], as shown in Fig. 8D. Some theoretical studies calculated following the density functional theory have been conducted on Ge-based perovskites, for example, $AGeI_3$ (A=Cs, FA, MA) and $MAGeX_3$ (X=Br, I Cl), and other related compounds [40,41].

Following the theoretical calculations and materials analysis, perovskites based on Ge have somewhat wider bandgaps than the lead versions, which makes them more appropriate for solar applications, according to researchers. The Ge^{2+}, on the other hand, is even more unstable than the Sn^{2+}. A low V_{oc} was seen in the device made of Ge-based perovskites, and the maximum PCE measured was just 0.2% (compared to the other devices).

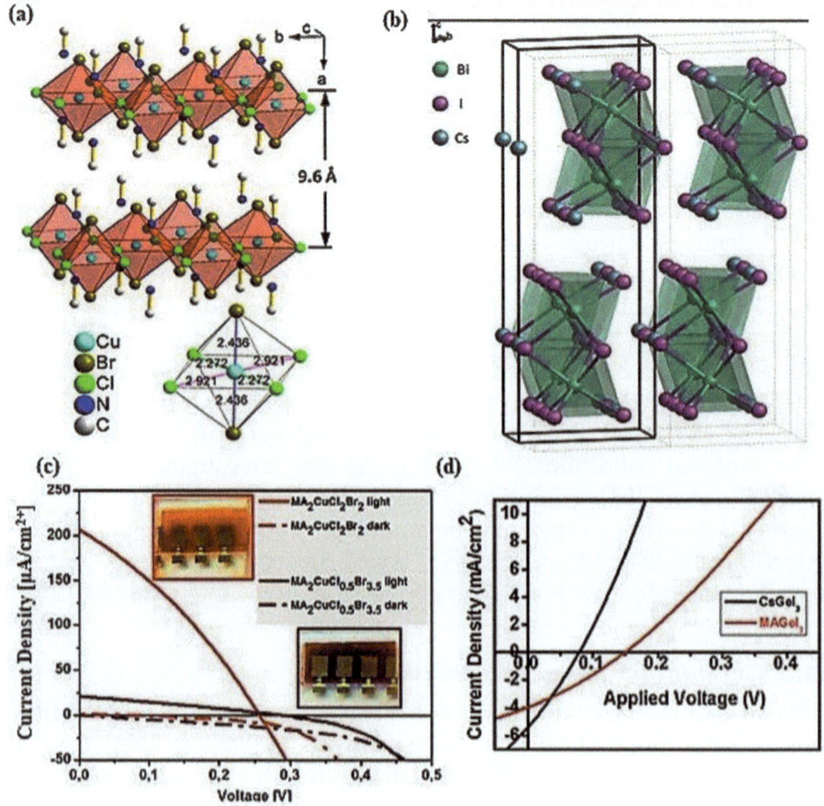

FIG. 8 (A) Cu based $MA_2CuCl_2Br_2$ structure. (B) Lead-free bismuth-based $Cs_3Bi_2I_9$ structure. (C) Under one sunlight irradiation, the J-V curves of solar cells sensitized using $MA_2CuCl_2Br_2$ *(red)*, $MA_2CuCl_{0.5}Br_{3.5}$ *(brown)* were measured. The dark current is represented by the *dashed red and brown lines.* (D) A comparison of the J-V graph of PV devices made using various Ge-based perovskites. *(A and C) Reproduced from D. Cortecchia, H.A. Dewi, J. Yin, A. Bruno, S. Chen, T. Baikie, P.P. Boix, M. Grätzel, S. Mhaisalkar, C. Soci, Lead-free $MA_2CuCl_xBr_{4-x}$ hybrid perovskites, Inorg. Chem. 55 (2016) 1044–1052. (B) Reproduced from B. Park, B. Philippe, X. Zhang, H. Rensmo, G. Boschloo, E.M.J. Johansson, Bismuth based hybrid perovskites $A_3Bi_2I_9$ (A: methylammonium or cesium) for solar cell application, Adv. Mater. 27 (2015) 6806–6813. (D) Reproduced from T. Krishnamoorthy, H. Ding, C. Yan, W.L. Leong, T. Baikie, Z. Zhang, M. Sherburne, S. Li, M. Asta, N. Mathews, Lead-free germanium iodide perovskite materials for photovoltaic applications, J. Mater. Chem. A 3 (2015) 23829–23832.*

3.1.4 *Bithmus-based perovskites*

Lower ($A_3M_2I_9$; $M = Bi^{3+}_{,}$ Sb^{3+}) dimensional perovskite materials based on bismuth and antimony have remarkable stability under environmental factors like heat, humidity, and constant light radiation. $A_3Bi_2X_9$, a promising Bi-based halide perovskite, has recently gained a lot of interest.

Bi-based halide perovskites are made up of bi-octahedral $(Bi_2I_9)_3$ clusters with cations like Cs and MA that create a low-dimensional 0D perovskite structure. Perovskites based on Bi not only have lower toxicity but also have far higher stability than Pb-based perovskites, which is a significant advantage. $MA_3Bi_2I_9$ has exhibited outstanding stability for more than 40 days when exposed to immanent air and humidity at room temperature for extended periods. Consequently, when maintained in an ambient atmosphere with a humidity of 55%, the performance of photovoltaic cells using $MA_3Bi_2I_9$ is able to remain steady for a minimum of 3 weeks. Moreover, because of an indirect and broad bandgap (2 eV) in most studies, their efficiency has been confined to 2%. Unfortunately, the efficiency of $MA_3Bi_2I_9$-based devices is insufficient, with just 0.19% on the normal structure [46] and 0.1% on the inverted structure [49]. In contrast, a device based on $Cs_3Bi_2I_9$ has a substantially greater proficiency of more than 1%. Although the efficiency of PSCs based on Bi is still inferior to that of Pb-based PSCs, given that the devices follow the ideal structure like Pb-based perovskite, future progress is envisaged for Bi-based PSCs.

3.2 Metal halide perovskite for solar cell applications

3.2.1 *Organic metal halide perovskite solar cells*

In recent years, organometal halide perovskites have emerged as a very promising family of functional materials with potential applications in a wide range of fields. The optical absorption coefficients of organic metal halide PSCs are extremely high, and their charge carrier conveyance is excellent. Additionally, there are low-cost, high-efficiency fabrication methods available. Because of their extraordinary structural tunability, these materials can have three-dimensional (3D), two-dimensional (2D), one-dimensional (1D), and zero-dimensional (0D) structures [56]. The characteristics of 1D and 0D perovskites are surprising and valuable, and they have significantly better characteristics than 3D and 2D perovskites, for which low-dimensional perovskite is considered a promising candidate for solar cell application nowadays. Following an investigation by Im et al. [57], low-dimensional nanowires have recently been shown to have improved transport capacities comparable to three-dimensional nanocuboid polycrystals by constructing solar cells based on $CH_3NH_3PbI_3$ nanowires having a solar cell performance of 14.71%, utilizing an invariant two-step spin coating process as shown in Fig. 9. The grain boundaries in nanocrystalline materials have been found to obstruct carrier transportation in thin films of 3D perovskite, [58] which may be avoided by using low-dimensional nanocrystals in the thin film. According to A. Jen and his colleagues, the influence of organic substituents on cations such as phenethylammonium (PEA), 4-methoxyphenethylammonium (MeO-PEA), and 4-fluorophenethylammonium (F-PEA) in quasi-2D perovskites has been studied recently [59]. PSCs crafted with F-PEA-based quasi-2D perovskite ($n = 5$) presented a high PCE of 14.5% alongside exciting stability by maintaining a 90% PCE reduction from its initial PCE after 40 days in an air medium, which could be due to the smaller bandgap and higher charge transport properties of the perovskite cell. In addition, exciton binding energy (E_b) is an essential factor in PSCs for generating photovoltaic activities [60]. Wu et al. [61] demonstrated that producing significant interfacial dipoles among both the working organic perovskite layer and the interfacial layer impacted the exciton binding energy in organic solar cells. Due to their extraordinary inherent properties like greater dielectric constants, low exciton binding energies, inherent ferroelectric polarization, and a wide range of ambipolar transport properties, solar cells based on organo-metal halide perovskite have demonstrated significant advances in solar cell performance in solar applications in the last 5 years [62]. Organo-metal halide perovskites have therefore emerged as one of the most promising photovoltaic materials for developing next-generation solar cells. Their combined iodide-chloride counterpart ($CH_3NH_3PbI_{3-x}Cl_x$) is believed to be one of the most promising options for developing high-efficiency solar cells that enable the display of enhanced charge transport capabilities while also exhibiting ambipolar behavior [63]. Furthermore, recent research has demonstrated that charge transport characteristics may be further enhanced using advanced chloride doping, electrode-interface design, single crystal growth techniques, and structural development, among other techniques [62].

3.2.2 *Inorganic and hybrid metal halide perovskite*

Organic hybrid perovskites such as $MAPbI_3$, $FAPbI_3$, and mixture perovskites (FA/MA, MA/Cs, FA/Cs, FA/MA/Cs, etc.) shows better efficiency. However, the environmental and poor thermal stability of these materials is thought to be caused by the organic cation component (MA or FA) [64]. For example, $MAPbI_3$ degrades readily at 120°C and even at 80°C if heated over an extended time [64]. Consequently, it is believed that substituting an organic cation with an inorganic cation is an effective technique for improving stability. Power conversion efficiency (PCE) has increased dramatically in recent years, from 3.8% to 25.5%, thanks to the discovery of organic-inorganic metal halide (OMH) PSCs. Due to their rapid gains in PCE, the hybrid, all-inorganic lead halide perovskites have received a lot of interest in the solar world. Yang et al. [65] synthesized an inorganic halide 2D perovskite of $Cs_2PbI_2Cl_2$ nanosheets (NSs), including a Ruddlesden-Popper (RP) framework. They used an NSs/$CsPbI_2Br$/NSs layered heterostructure to improve stability as well as the PCE of $CsPbI_2Br$ solar cells [65].

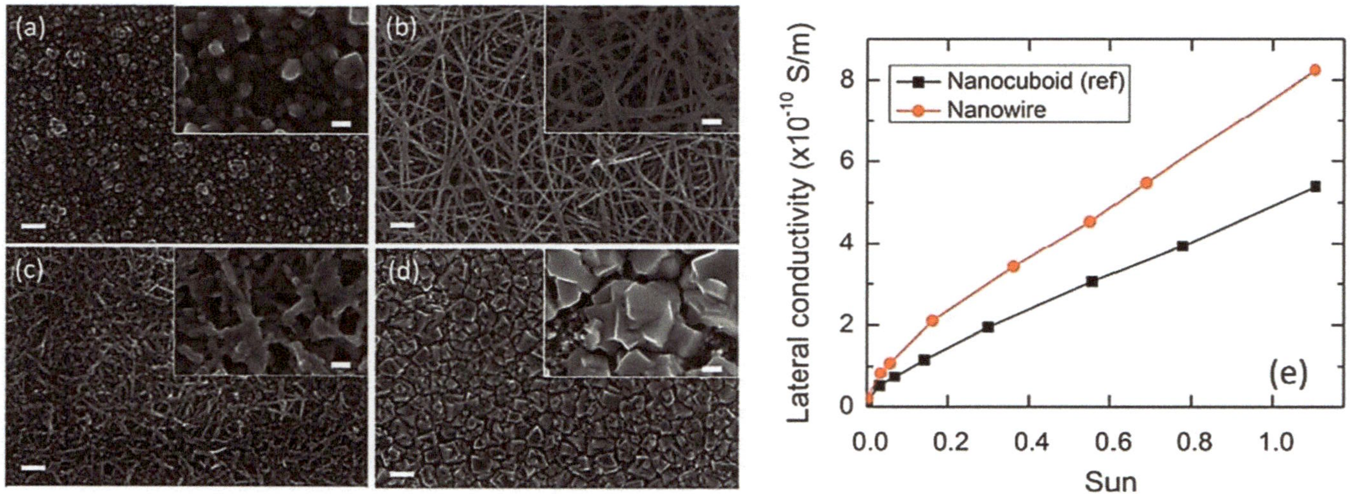

FIG. 9 Photographs of $CH_3NH_3PbI_3$ were obtained with an electron microscope after a two-step spin coating method utilizing a CH_3NH_3I solution in IPA: (A) for 10 μL of dimethylformamide; (B) for 50 μL of dimethylformamide; (C) for 100 μL of dimethylformamide; and (D) without dimethylformamide. (200 nm in insets). (E) The in-plane perovskite's DC conductivity was measured by exhaling two gold contacts of 0.14 m separated onto a layer. *Reproduced from J.-H. Im, I.-H. Jang, N. Pellet, M. Grätzel, Growth of* $CH_3NH_3PbI_3$ *cuboids with controlled size for high-efficiency perovskite solar cells, Nat. Nanotechnol. 9 (2014) 927–932.*

$Cs_2PbI_2Cl_2$ is a new inorganic 2D perovskite with a standard RP phase, extreme high carrier mobilities with a value of $1.39 \times 10^4 cm^2 V^{-1} s^{-1}$ for holes and $9.39 \times 10^3 cm^2 V^{-1} s^{-1}$ for electrons, lower effective masses, outstanding ambient/thermal stability, a reduced exciton binding energy, and a direct bandgap, indicating its possible applications as a solar cell. The efficiency of $CsPbI_2Br$ PSCs has increased from 15.02% to 16.65% of the nanostructured NSs/$CsPbI_2Br$/NSs inorganic perovskite film, which has been developed. Furthermore, despite being exposed to ambient conditions for an extended time, the encapsulate-less $CsPbI_2Br$ devices with the NSs/$CsPbI_2Br$/NSs nanostructures retain more than 90% of their initial efficiencies. Inorganic halide perovskite nanocrystals (NCs) with plentiful elements, variable energy levels, outstanding crystallinity, and high stability are particularly appealing for chemical doping and interface design in inorganic $CsPbX_3$-based PSCs. The resultant $CsSnX_3$ quantum rods (X=Cl, Br, and I) exhibit composition-tunable photoluminescence (PL), which was explored by Chen et al. [48] Under AM 1.5G, a power conversion efficiency of 12.96% was attained by substituting the halide of the $CsSnX_3$ quantum rods. In its perovskite phase, these lead-free quantum rods have proven to be an excellent way to create absorber layers that boost light-harvesting and charge collecting for solar applications. Gupta et al. [66] investigated the possibility of using cesium tin bromide ($CsSnBr_3$), an all-inorganic lead-free perovskite material, for solar cell applications. Under one sun ($100 mW cm^{-2}$) photo luminance, $CsSnBr_3$-based solar cells demonstrated PCEs of 2.1%, a short circuit current (J_{SC}) of $9 mA cm^{-2}$, an open circuit potential (V_{OC}) of 0.41 V, and a fill factor (FF) of 58%, which, while modest when compared to Pb-based cells, represented one of the optimal outcomes reported to date, as shown in Fig. 10 [66]. Gupta et al. discovered that the addition of tin fluoride (SnF_2) was useful in attaining high device performance; this was considered to be due either to the lowering of ambient carrier density by neutralizing trapping or to the replacement of cation vacancies in the semiconductor.

4. Conclusions

Because of excellent performance in solar cell applications, halide perovskites of low dimension have been regarded as innovative, effective materials in recent years. They emerged in relate to solving some issues with high-performance PSCs. The optical characteristics of perovskite materials have recently been proved by a large number of organizations to be highly dependent on their dimensions; hence, the capacity to modify their dimensions completely will be critical for understanding the relationships between property, device behavior, and structure. Although low-dimensional halide perovskite nanoparticles have been widely studied in recent years, numerous issues must be answered before they can be utilized in practical applications. PSC technology has come a long way in

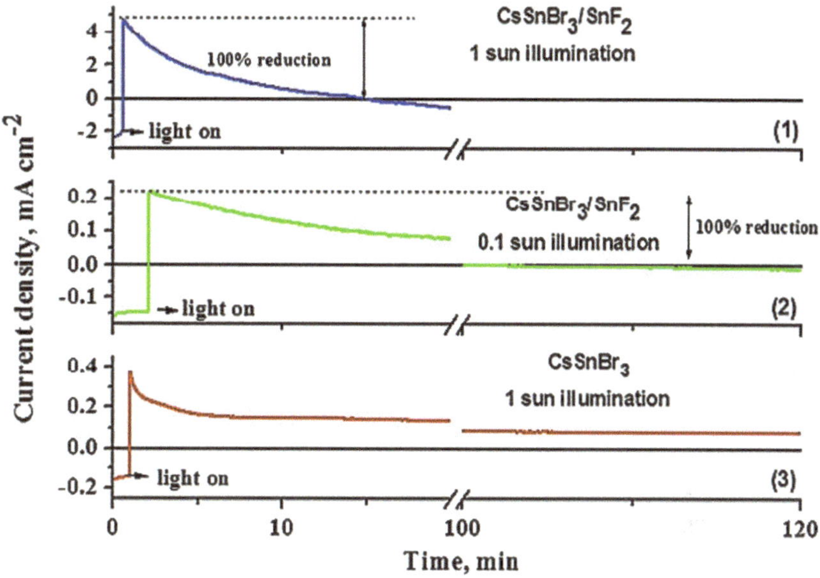

FIG. 10 With continuous illumination, current density as a function of time was calculated when the greatest Power Point was in the ambient environment: (1) for 0.2 V applied bias; (2) at 0.08 V applied bias; and (3) at 0.08 V applied bias when only $CsSnBr_3$ with ~0.1 sun intensity. *Reproduced from S. Gupta, T. Bendikov, G. Hodes, D. Cahen, $CsSnBr_3$, a lead-free halide perovskite for long-term solar cell application: insights on SnF_2 addition, ACS Energy Lett. 1 (2016).*

the last few years, but the biggest problem is that the material is very unstable when it comes into contact with various environmental factors such as heat, vapor, and oxygen. As a result, device performance has become less effective over time. Different strategies are being utilized to eliminate these issues. Lead toxicity is another issue for commercialization. This chapter includes the literature on PSC research up to 2021. Insights and advancements in halide PSCs were discussed along with commercialization issues. Although considerable resources have been invested in exploring low-dimensional perovskites, many obstacles remain to be addressed. One of the most critical factors in commercializing halide perovskite-based devices is stability. This chapter provided an overview of the many low-dimensional halide perovskite materials of various sizes and forms, as well as their controlled production techniques for utilizing in PSCs.

References

[1] K. Hong, Q. Van Le, S.Y. Kim, H.W. Jang, Low dimentional halide perovskite: review and issues, J. Mater. Chem. C 6 (2018) 2189–2209.

[2] J. Zhang, X. Yang, H. Deng, K. Qiao, U. Farooq, M. Ishaq, F. Yi, H. Liu, J. Tang, H. Song, Low-dimensional halide perovskites and their advanced optoelectronic applications, Nanomicro Lett. 9 (2017) 1–26.

[3] V.N. Veettil, V. Chitra, Lantibiotics of milk isolates: a short review on characterization and potential applications, J. Microbiol. Biotechnol. Food Sci. (2021), e3702.

[4] W. Ouyang, F. Teng, J. He, X. Fang, Enhancing the photoelectric performance of photodetectors based on metal oxide semiconductors by charge-carrier engineering, Adv. Funct. Mater. 29 (2019), 1807672.

[5] J.H. Rhee, C.-C. Chung, E.W.-G. Diau, A perspective of mesoscopic solar cells based on metal chalcogenide quantum dots and organometal-halide perovskites, NPG Asia Mater. 5 (2013) e68.

[6] L. Na Quan, D. Ma, Y. Zhao, O. Voznyy, H. Yuan, E. Bladt, J. Pan, F. García de Arquer, R. Sabatini, Z. Piontkowski, A. Emwas, P. Todorović, R. Quintero-Bermudez, G. Walters, J. Fan, M. Liu, H. Tan, M. Saidaminov, L. Gao, Y. Li, D. Anjum, N. Wei, J. Tang, D. McCamant, M. Roeffaers, S. Bals, Sargent, Edge stabilization in reduced-dimensional perovskites, Nat. Commun. 11 (2020).

[7] J.J. Yoo, G. Seo, M.R. Chua, T.G. Park, Y. Lu, F. Rotermund, Y.-K. Kim, C.S. Moon, N.J. Jeon, J.-P. Correa-Baena, V. Bulović, S. Shin, M. Bawendi, J. Seo, Efficient perovskite solar cells via improved carrier management, Nature 590 (2021) 587–593.

[8] A.R.B.M. Yusoff, M.K. Nazeeruddin, Low-dimensional perovskites: from synthesis to stability in perovskite solar cells, Adv. Energy Mater. 8 (2018), 1702073.

[9] A.H. Slavney, R.W. Smaha, I.C. Smith, A. Jaffe, D. Umeyama, H.I. Karunadasa, Chemical approaches to addressing the instability and toxicity of lead–halide perovskite absorbers, Inorg. Chem. 56 (2017) 46–55.

[10] J. Jeong, M. Kim, J. Seo, H. Lu, P. Ahlawat, A. Mishra, Y. Yang, M. Hope, F. Eickemeyer, M. Kim, Y. Yoon, I. Choi, B. Darwich, S. Choi, Y. Jo, J. Lee, B. Walker, S. Zakeeruddin, L. Emsley, U. Rothlisberger, A. Hagfeldt, D. Kim, M. Grätzel, J. Kim, Pseudo-halide anion engineering for α-$FAPbI_3$ perovskite solar cells, Nature 592 (2021).

[11] H. Fan, F. Li, P. Wang, Z. Gu, J. Huang, K. Jiang, B. Guan, L. Yang, X. Zhou, Y. Song, Methylamine-assisted growth of uniaxial-oriented perovskite thin films with millimeter-sized grains, Nat. Commun. 11 (2020).

[12] X. Zheng, Y. Hou, C. Bao, J. Yin, F. Yuan, Z. Huang, K. Song, J. Liu, J. Troughton, N. Gasparini, C. Zhou, Y. Lin, D. Xue, B. Chen, A. Johnston, N. Wei, M. Hedhili, M. Wei, A. Alsalloum, P. Maity, B. Turedi, C. Yang, D. Baran, T. Anthopoulos, Y. Han, Z. Lu, O. Mohammed, F. Gao, E. Sargent, O. Bakr, Managing grains and interfaces via ligand anchoring enables 22.3%-efficiency inverted perovskite solar cells, Nat. Energy 5 (2020).

[13] G. Kim, H. Min, K. Lee, D. Lee, S. Yoon, S. Seok, Impact of strain relaxation on performance of α-formamidinium lead iodide perovskite solar cells, Science 370 (2020).

[14] J. Peng, D. Walter, Y. Ren, M. Tebyetekerwa, Y. Wu, T. Duong, Q. Lin, J. Li, T. Lu, M. Mahmud, O. Lem, S. Zhao, W. Liu, Y. Liu, H. Shen, L. Li, F. Kremer, H. Nguyen, D. Choi, K. Weber, K. Catchpole, Nanoscale localized contacts for high fill factors in polymer-passivated perovskite solar cells, T. White, Science 371 (2021).

[15] M. Jeong, I. Choi, E. Go, Y. Cho, M. Kim, B. Lee, S. Jeong, Y. Jo, H. Choi, J. Lee, J. Bae, S. Kwak, D. Kim, C. Yang, Stable perovskite solar cells with efficiency exceeding 24.8% and 0.3-V voltage loss, Science 369 (2020).

[16] S. Sun, M. Lu, X. Gao, Z. Shi, X. Bai, W. Yu, Y. Zhang, 0D perovskites: unique properties, synthesis, and their applications, Adv. Sci. 8 (2021).

[17] S. Öz, J. Hebig, E. Jung, T. Singh, A. Lepcha, S. Olthof, F. Jan, Y. Gao, R. German, P. van Loosdrecht, K. Meerholz, T. Kirchartz, S. Mathur, Zero-dimensional $(CH_3NH_3)\ 3Bi_2I_9$ perovskite for optoelectronic applications, Sol. Energy Mater. Sol. Cells 158 (2016).

[18] J. Zhang, D. Bai, Z. Jin, H. Bian, K. Wang, J. Sun, Q. Wang, S. Liu, 3D–2D–0D interface profiling for record efficiency all-inorganic $CsPbBrI_2$ perovskite solar cells with superior stability, Adv. Energy Mater. 8 (2018).

[19] F. Bai, J. Zhang, Y. Yuan, H. Liu, X. Li, C. Chueh, H. Yan, Z. Zhu, A. Jen, A 0D/3D Heterostructured all-inorganic halide perovskite solar cell with high performance and enhanced phase stability, Adv. Mater. 31 (2019).

[20] X. Chen, F. Zhang, Y. Ge, L. Shi, S. Huang, J. Tang, Z. Lv, L. Zhang, B. Zou, H. Zhong, Centimeter-sized Cs_4PbBr_6 crystals with embedded $CsPbBr_3$ nanocrystals showing superior photoluminescence: nonstoichiometry induced transformation and light-emitting applications, Adv. Funct. Mater. 28 (2018).

[21] C. Ma, D. Shen, B. Huang, X. Li, W.C. Chen, M.F. Lo, P. Wang, M. Hon-Wah Lam, Y. Lu, B. Ma, C.S. Lee, High performance low-dimensional perovskite solar cells based on a one dimensional lead iodide perovskite, J. Mater. Chem. A 7 (2019) 8811–8817.

[22] L.L. Gao, I. Spanopoulos, W.J. Ke, S. Huang, I. Hadar, L. Chen, X.L. Li, G.J. Yang, M.G. Kanatzidis, An in situ cross-linked 1D/3D perovskite heterostructure improves the stability of hybrid perovskite solar cells for over 3000 h operation, ACS Energy Lett. 4 (2019).

[23] P. Zardari, A. Rostami, Construction of 1D perovskite nanowires by urotropin passivation towards efficient and stable perovskite solar cell, Sol. Energy Mater. Sol. Cells 227 (2021).

[24] N. Yang, C. Zhu, Y. Chen, H. Zai, C. Wang, X. Wang, H. Wang, S. Ma, Z. Gao, X. Wang, J. Hong, Y. Bai, H. Zhou, B. Cui, Q. Chen, An in situ cross-linked 1D/3D perovskite heterostructure improves the stability of hybrid perovskite solar cells for over 3000 h operation, Energy Environ. Sci. 13 (2020).

[25] M. Pandey, K.W. Jacobsen, K.S. Thygesen, Band gap tuning and defect tolerance of atomically thin two-dimensional organic–inorganic halide perovskites, J. Phys. Chem. Lett. 7 (2016) 4346–4352.

[26] Z. Liu, H. Zheng, D. Liu, Z. Liang, W. Yang, H. Chen, L. Ji, S. Yuan, Y. Gu, S. Li, Controllable two-dimensional perovskite crystallization via water additive for high-performance solar cells, Nanoscale Res. Lett. 15 (2020).

[27] G. Xing, N. Mathews, S. Sun, S.S. Lim, Y.M. Lam, M. Grätzel, S. Mhaisalkar, T.C. Sum, Long-range balanced electron- and hole-transport lengths in organic-inorganic $CH_3NH_3PbI_3$, Science 342 (2013) 344–347.

[28] L. Dou, Emerging two-dimensional halide perovskite nanomaterials, J. Mater. Chem. C 5 (2017) 11165–11173.

[29] I.C. Smith, E.T. Hoke, D. Solis-Ibarra, M.D. McGehee, H.I. Karunadasa, A layered hybrid perovskite solar-cell absorber with enhanced moisture stability, Angew. Chem. Int. Ed. 53 (2014) 11232–11235.

[30] X. Lian, J. Chen, M. Qin, Y. Zhang, S. Tian, X. Lu, G. Wu, H. Chen, The second spacer cation assisted growth of a 2D perovskite film with oriented large grain for highly efficient and stable solar cells, Angew. Chem. 131 (2019).

[31] W. Fu, J. Wang, L. Zuo, K. Gao, F. Liu, D. Ginger, A. Jen, Two-dimensional perovskite solar cells with 14.1% power conversion efficiency and 0.68% external radiative efficiency, ACS Energy Lett. 3 (2018).

[32] Y. Zhao, K. Zhu, Efficient planar perovskite solar cells based on 1.8 eV band gap $CH_3NH_3PbI_2Br$ nanosheets via thermal decomposition, J. Am. Chem. Soc. 136 (2014) 12241–12244.

[33] O. Weber, D. Ghosh, S. Gaines, P. Henry, A. Walker, M. Islam, M. Weller, Phase behavior and polymorphism of formamidinium lead iodide, Chem. Mater. 30 (2018).

[34] B. Cohen, M. Wierzbowska, L. Etgar, High efficiency and high open circuit voltage in quasi 2D perovskite based solar cells, Adv. Funct. Mater. 27 (2017), 1604733.

[35] L.N. Quan, M. Yuan, R. Comin, O. Voznyy, E.M. Beauregard, S. Hoogland, A. Buin, A.R. Kirmani, K. Zhao, A. Amassian, Ligand-stabilized reduced-dimensionality perovskites, J. Am. Chem. Soc. 138 (2016) 2649–2655.

[36] U. Mandadapu, S. Vedanayakam, K. Thyagarajan, B. Babu, Optimisation of high efficiency tin halide perovskite solar cells using SCAPS-1D, Int. J. Simul. Process. Model. 13 (2018).

[37] N. Kumari, S.R. Patel, J.V. Gohel, Superior efficiency achievement for $FAPbI_3$-perovskite thin film solar cell by optimization with response surface methodology technique and partial replacement of Pb by Sn, Optik (Stuttg) 176 (2019) 262–277.

[38] Z. Shi, A.H. Jayatissa, Perovskites-based solar cells: a review of recent progress, materials and processing methods, Materials 11 (2018) 729.

[39] A.K. Jena, A. Kulkarni, T. Miyasaka, Halide perovskite photovoltaics: background, status, and future prospects, Chem. Rev. 119 (2019) 3036–3103.

[40] M. Abd Mutalib, N.A. Ludin, N.A.A.N. Ruzalman, V. Barrioz, S. Sepeai, M.A.M. Teridi, M.S. Su'ait, M.A. Ibrahim, K. Sopian, Progress towards highly stable and lead-free perovskite solar cells, Mater. Renew. Sustain. Energy 7 (2018) 1–13.

[41] C.J. Bartel, C. Sutton, B.R. Goldsmith, R. Ouyang, C.B. Musgrave, L.M. Ghiringhelli, M. Scheffler, New tolerance factor to predict the stability of perovskite oxides and halides, Sci. Adv. 5 (2019) 0693.

[42] S.G. Hashmi, A. Tiihonen, D. Martineau, M. Ozkan, P. Vivo, K. Kaunisto, V. Ulla, S.M. Zakeeruddin, M. Grätzel, Long term stability of air processed inkjet infiltrated carbon-based printed perovskite solar cells under intense ultra-violet light soaking, J. Mater. Chem. A 5 (2017) 4797–4802.

[43] E. Mosconi, J.M. Azpiroz, F. De Angelis, *Ab Initio* molecular dynamics simulations of methylammonium lead iodide perovskite degradation by water, Chem. Mater. 27 (2015) 4885–4892.

[44] B. Chaudhary, A. Kulkarni, A.K. Jena, M. Ikegami, Y. Udagawa, H. Kunugita, K. Ema, T. Miyasaka, Poly (4-vinylpyridine)-based interfacial passivation to enhance voltage and moisture stability of lead halide perovskite solar cells, ChemSusChem 10 (2017) 2473–2479.

[45] M. Awais, R.L. Kirsch, V. Yeddu, M.I. Saidaminov, Tin halide perovskites going forward: frost diagrams offer hints, ACS Mater. Lett. 3 (2021) 299–307.

[46] A. Abate, Perovskite solar cells go lead free, Joule 1 (2017) 659–664.

[47] E. Jokar, C. Chien, C. Tsai, A. Fathi, E.W. Diau, Robust tin-based perovskite solar cells with hybrid organic cations to attain efficiency approaching 10%, Adv. Mater. 31 (2019), 1804835.

[48] L. Chen, C. Lee, Y. Chuang, Z. Wu, C. Chen, Synthesis and optical properties of lead-free cesium tin halide perovskite quantum rods with high-performance solar cell application, J. Phys. Chem. Lett. 7 (2016).

[49] T. Mahmoudi, W.-Y. Rho, M. Kohan, Y.H. Im, S. Mathur, Y.-B. Hahn, Suppression of Sn^{2+}/Sn^{4+} oxidation in tin-based perovskite solar cells with graphene-tin quantum dots composites in active layer, Nano Energy 90 (2021), 106495.

[50] H.J. Snaith, A. Abate, J.M. Ball, G.E. Eperon, T. Leijtens, N.K. Noel, S.D. Stranks, J.T.-W. Wang, K. Wojciechowski, W. Zhang, Anomalous hysteresis in perovskite solar cells, J. Phys. Chem. Lett. 5 (2014) 1511–1515.

[51] K.P. Marshall, R.I. Walton, R.A. Hatton, Tin perovskite/fullerene planar layer photovoltaics: improving the efficiency and stability of lead-free devices, J. Mater. Chem. A 3 (2015) 11631–11640.

[52] M. Zhang, M. Lyu, J.-H. Yun, M. Noori, X. Zhou, N.A. Cooling, Q. Wang, H. Yu, P.C. Dastoor, L. Wang, Low-temperature processed solar cells with formamidinium tin halide perovskite/fullerene heterojunctions, Nano Res. 9 (2016) 1570–1577.

[53] X. Cui, K. Jiang, J. Huang, Q. Zhang, M. Su, L. Yang, Y. Song, X. Zhou, Cupric bromide hybrid perovskite heterojunction solar cells, Synth. Met. 209 (2015).

[54] D. Cortecchia, H.A. Dewi, J. Yin, A. Bruno, S. Chen, T. Baikie, P.P. Boix, M. Grätzel, S. Mhaisalkar, C. Soci, Lead-free $MA_2CuCl_xBr_{4-x}$ hybrid perovskites, Inorg. Chem. 55 (2016) 1044–1052.
[55] T. Krishnamoorthy, H. Ding, C. Yan, W.L. Leong, T. Baikie, Z. Zhang, M. Sherburne, S. Li, M. Asta, N. Mathews, Lead-free germanium iodide perovskite materials for photovoltaic applications, J. Mater. Chem. A 3 (2015) 23829–23832.
[56] H. Lin, C. Zhou, Y. Tian, T. Siegrist, B. Ma, Low-dimensional organometal halide perovskites, ACS Energy Lett. 3 (2017).
[57] J.-H. Im, I.-H. Jang, N. Pellet, M. Grätzel, Growth of $CH_3NH_3PbI_3$ cuboids with controlled size for high-efficiency perovskite solar cells, Nat. Nanotechnol. 9 (2014) 927–932.
[58] J. Yun, A. Ho-Baillie, S. Huang, S. Woo, Y. Heo, J. Seidel, F. Huang, Y. Cheng, M. Green, Benefit of grain boundaries in organic–inorganic halide planar perovskite solar cells, J. Phys. Chem. Lett. 6 (2015).
[59] W. Fu, H. Liu, X. Shi, L. Zuo, X. Li, A.K. Jen, Tailoring the functionality of organic spacer cations for efficient and stable quasi-2D perovskite solar cells, Adv. Funct. Mater. 29 (2019), 1900221.
[60] J. Wu, H. Cha, T. Du, Y. Dong, W. Xu, C. Lin, J.R. Durrant, A comparison of charge carrier dynamics in organic and perovskite solar cells, Adv. Mater. (2021), 2101833.
[61] T. Wu, Y.-C. Hsiao, M. Li, N.-G. Kang, J.W. Mays, B. Hu, Dynamic coupling between electrode interface and donor/acceptor interface via charge dissociation in organic solar cells at device-operating condition, J. Phys. Chem. C 119 (2015) 2727–2732.
[62] Y.-C. Hsiao, T. Wu, M. Li, Q. Liu, W. Qin, B. Hu, Fundamental physics behind high-efficiency organo-metal halide perovskite solar cells, J. Mater. Chem. A 3 (2015) 15372–15385.
[63] Z. Arain, C. Liu, Y. Yang, M. Mateen, Y. Ren, Y. Ding, X. Liu, Z. Ali, M. Kumar, S. Dai, Elucidating the dynamics of solvent engineering for perovskite solar cells, Sci. China Mater. 62 (2019) 161–172.
[64] T. Miyasaka, A. Kulkarni, G.M. Kim, S. Öz, A.K. Jena, Perovskite solar cells: can we go organic-free, lead-free, and dopant-free, Adv. Energy Mater. 10 (2020), 1902500.
[65] S. Yang, W. Liu, Y. Han, Z. Liu, W. Zhao, C. Duan, Y. Che, H. Gu, Y. Li, S. Liu, 2D $Cs_2PbI_2Cl_2$ nanosheets for holistic passivation of inorganic $CsPbI_2Br$ perovskite solar cells for improved efficiency and stability, Adv. Energy Mater. 10 (2020) 2002882.
[66] S. Gupta, T. Bendikov, G. Hodes, D. Cahen, $CsSnBr_3$, a lead-free halide perovskite for long-term solar cell application: insights on SnF_2 addition, ACS Energy Lett. 1 (2016).

CHAPTER

8

Halide perovskite for light-emitting diodes

Rajan Kumar Singh[a], *Chung-Hsin Lu*[a,b,c], *Radha Tamrakar*[d], *Neha Jain*[e], *Anupriya Singh*[f], *Mohan Lal Meena*[b], *and Sudipta Som*[a]

[a]Department of Chemical Engineering, National Taiwan University (NTU), Taipei, Taiwan, [b]Department of Chemical Engineering, National Taiwan University Science and Technology (NTUST), Taipei, Taiwan, [c]Advanced Research Centre of Green Materials Science & Technology, Taipei, Taiwan, [d]Department of Physics, Govt. K. N. M. M., Damoh, Madhya Pradesh, India, [e]Department of Physics, Dr. H. S. Gour Central University, Sagar, Madhya Pradesh, India, [f]Research Center for Applied Science, Academia Sinica, Taipei, Taiwan

1. Introduction

Inorganic organic halide (OIHP) perovskites are the good light harvesters due to their outstanding optical and electrical properties, for example, high photoluminescence quantum yield (PLQY), adjustable optical energy gap, high color purity, and captivating electrons which aroused great interest of scientists in perovskite materials nowadays [1,2]. Besides, their low cost and easy to synthesize technology are also favorable for device formation. As high-efficiency light-gathering materials are used in photovoltaic (PV) devices, LEDs, lasers, bio-imaging, and other photoelectric detecting devices, they are on the rise [3]. The history of these perovskite materials is very interesting. The German mineralogist Gustav Rose (Gustav Rose) discovered the firstly perovskite-type oxide of perovskite ($CaTiO_3$) in 1839 while Perovski conducted extensive research on its structure. Perovskite-based oxide materials have thermoelectric applications and attract ferroelectric, magnetic, and superconducting properties.

Low-Dimensional Halide Perovskites
https://doi.org/10.1016/B978-0-323-88522-5.00002-8

In 1958, Moller's research team first reported the structure of halide-based perovskite, 119 years after the discovery of oxide-based perovskite materials [3–5]. The photoconductivity of these materials can be attuned by exchanging the composition of the halides. Furthermore, Weber and Naturforsch first reported the first organic cationic methylammonium (MA) perovskite material in 1978 [6]. On the other hand, Mitzi et al. conveyed bright and intense photoluminescence (PL) in a visible spectrum in the 1990s [7].

Low-dimensional-based perovskite materials have a larger surface-to-volume ratio, quantum confinement effect, and anisotropic geometry, so their optical properties are better than polycrystalline or bulk thin-film perovskites. At the same time, low-dimensional materials also have wide and adjustable band gaps, high luminescence performance and composition flexibility, and excellent charge-carrying capabilities, which lead to the behavior of photo-generated vectors and contribute to excellent performance and reliability optoelectronic devices. Therefore, many exertions have recently been made to develop low-dimensional perovskite materials ranging from 0-D to 3-D. 0D (zero-dimensional) perovskite structures are quantum dots, 1D (one-dimensional) calcium nanotubes and nanowires, and 2D (two-dimensional) hybrid perovskite structures, such as nanosheets and nanoplates as shown in Fig. 1A [8].

The change in the size of the perovskite molecules can be achieved by changing the fabrication approaches and some other ligands. Because of the size and chemical composition, band gap of perovskites also is tuned as shown in Fig. 1B. The applications of these perovskite materials vary according to their size. The 3D perovskite application is for solar cell equipment. Perovskite 2D and 0D materials are best suited for LED and photovoltaic display devices (see Fig. 1C and D). It is a one-dimensional halide perovskite light collector for photoelectric sensing, and the applications of these perovskite materials vary according to their size. [8].

In our daily life, industry, science, and technology, light plays a significant part in modern civilization. The first batch of light-emitting diode was produced in the 1960s, and since then, they have been supplying oxygen to the human body like oxygen in everyday life [11]. LED replaces traditional lighting and display equipment and consumes less energy and money. Record-breaking improvements are observed every day with the introduction of new materials and technologies for LEDs. In general, choosing the right light collection material is one of the significant features of LED-based devices. Silicon and gallium arsenide semiconductor LEDs are the most commercially available LEDs, and these LEDs are expensive and have poor performance. On the other hand, compared to traditional LEDs, semiconductor-based solution-processing LEDs are rapidly

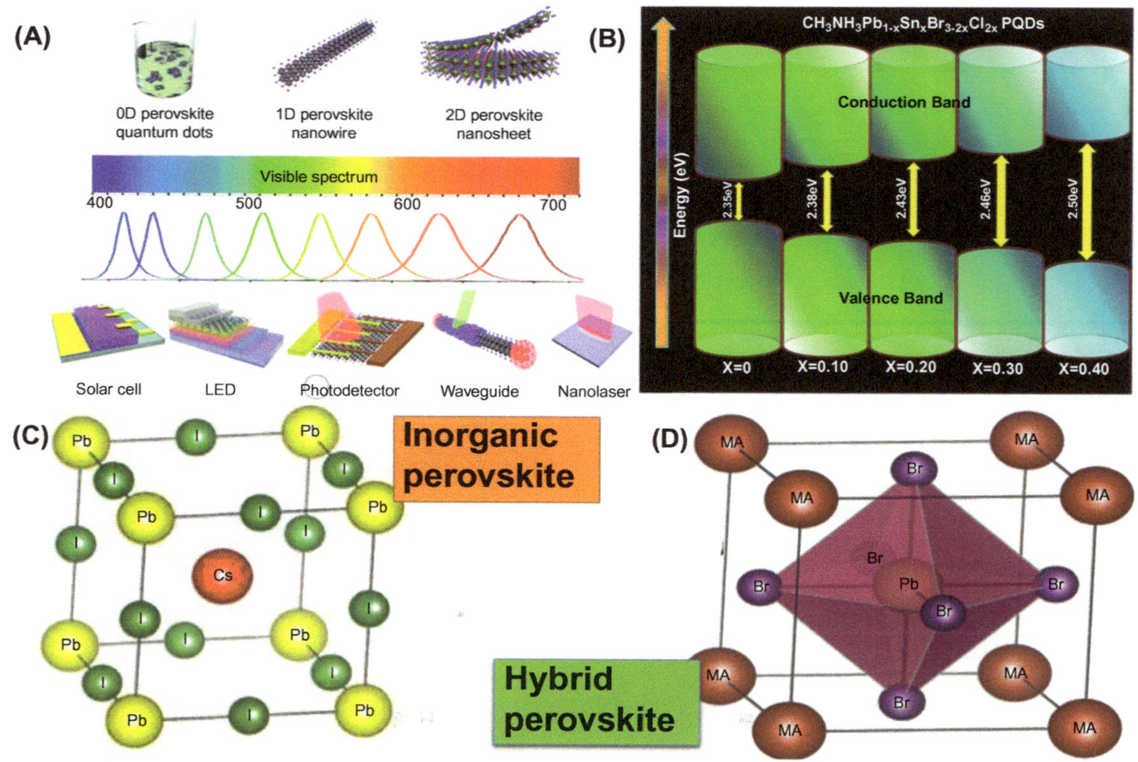

FIG. 1 (A) Various dimensions of the perovskite materials and their applications [8], (B) band-gap tuning with the incorporation of $SnCl_2$ into $MAPbBr_3$ perovskite [9], and (C, D) cubic perovskite structure of $CsPbI_3$ inorganic perovskite and $MAPbBr_3$ hybrid perovskite, respectively [10].

improved due to their high performance and low cost traditional organic light-emitting diodes (OLED). Perovskite-based LEDs are better than traditional LEDs and OLEDs due to their negligible stroke changes and higher PLQE. The color-adaptable and narrow bandwidth characteristics of halide perovskite LEDs can achieve high color purity to match ultrahigh luminescence displays (such as QLED TV). In addition, the color gamut of the perovskite LED can reach 140% of National Television System Committee (NTSC) standard, which is much higher than OLED and traditional QDLED. Depending on its structure and application, perovskite-type LED can be divided into two parts, namely, electroluminescent LED (active LED) and photoluminescent LED (passive backlight LED), and can achieve high color purity that matches ultrahigh luminescent displays (such as QLED TV). Depending on its structure and application, perovskite-type LED can be divided into two parts, namely, electroluminescent LED (active LED) and photoluminescent LED (passive backlight LED), and can achieve high color purity that matches ultrahigh luminescence displays (such as QLED TV). In addition, the color gamut of the perovskite LED can reach 140% of the NTSC standard. This value is much higher than OLED and traditional QDLED. Depending on its structure and application, perovskite-type LED can be divided into two parts, namely, electroluminescent LED (active LED) and photoluminescent LED (passive backlight LED).

1.1 Why perovskite LEDs are good?

Perovskites have good light absorption and bandgap tunability, long diffusion length, high vector mobility, bipolar charge transport, and high defect tolerance. Halide perovskite has shown outstanding enactment in PV, PQDs materials, and perovskite sensing devices, so it has been a hot spot for scientists and industry since the last few years. Perovskite halide materials have many applications in the field of display and backlight devices, but they are becoming promising materials, especially for LEDs as shown in Fig. 1B. With the incorporation of $SnCl_2$ into $MAPbBr_3$ perovskite, the band-gap gradually increases with an increase in the concentration of dopant [9]. The emitted color can be controlled by the composition and size of the material, which makes them more feasible for LEDs. The PL and PLQY emission of perovskite materials can be improved by exchanging 3D perovskite nanoparticles or quantum dots into 0D [12]. Recently, the PCE of halide-based solar cells has increased to 24%, and the external quantum efficiency of perovskite LEDs has reached 22%. Therefore, in recent years, the perovskite material received researchers' interests due to their high EQE, PLQY, narrow emission spectrum (<15 nm), high color purity, low cost of production, synthetic route of solution treatment, pure and adjustable luminescence. Most forms of perovskite (3-D) and nanoparticles/quantum dots (0-D) are used in LED applications [12].

2. The application of LED in halide perovskite

Current LEDs' display technology is mostly established on two main stratagem assemblies. This can be alienated into photo-luminescence (PL), i.e., passive LEDs type and electro-luminescence (EL), i.e., active LEDs type according to their excitation approach [13].

2.1 Electroluminescence LEDs (Active LEDs)

EL LEDs are a device in which electrical energy (EE) is converted into optical energy (light energy, LE). The perovskite LEDs are formed by forming a P-N junction [14]. The carriers, *viz,* holes and electrons from p-type and n-type perovskite semiconductors pass through the barrier. The graphic diagram of LEDs' operation is illustrated in Fig. 2A and B, respectively. The energy level diagrams later than contact among p-type and n-type semiconductors are demonstrated in Fig. 2(B1). The barrier height is reduced when a forward bias is applied. The electrons and holes are recombined and release photons. It is the working principle of homo-junction LE which is formed by two semiconductors having the same bandgap [16]. Here, photons reabsorption takes place due to the same bandgap of semiconductors. To resolve this problem, another type of junction that used LEDs is heterojunction, in which two semiconductors having different band gaps has been used. Fig. 2(C1) displays the band energy diagram of heterojunction semiconductor which shows the migration of hole and electron. Fig. 2(C2) represents that in forward bias, electron-hole recombination gives light energy emission [15]. From the research observation, it was observed that a heterojunction semiconductor has advanced carrier insertion EQE than the homo-junction.

The ratio of a hole electrical current density to an electron electrical density is called the injection ratio. The injection ratio is expressed by Eq. (1) [17]

$$\frac{J_\mathrm{p}}{J_\mathrm{n}} = \frac{D_{\mathrm{p}2}p_{10}L_{\mathrm{n}1}}{D_{\mathrm{n}1}n_{20}L_{\mathrm{p}2}} \exp\left(\frac{\Delta E}{k_0 T}\right) \tag{1}$$

where n_{10} is the n-type semiconductor electron concentration, p_{20} is the p-type semiconductor hole concentration, D_{n1} stands for the diffusion coefficients of the n-type, D_{p2} stands for the diffusion coefficients of the p-type semiconductors, L_{n1} is the electron diffusion length, L_{p2} is the hole

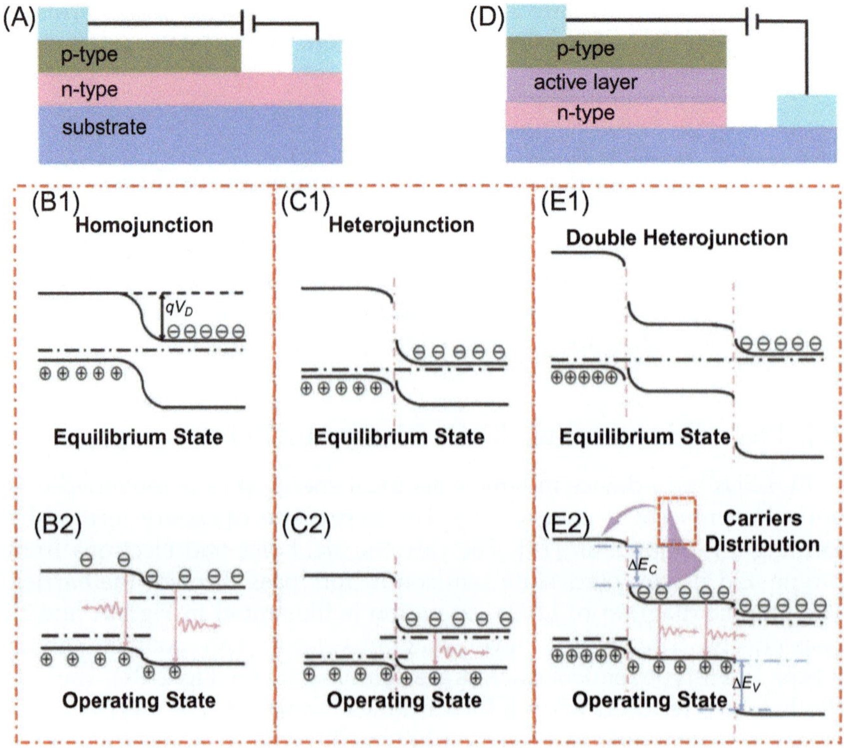

FIG. 2 The basic operation of light-emitting diodes: (A) Device construction of p–n junction-based LEDs, with homojunction and heterojunction LEDs. (B1, B2) Energy level representation of homo-junction LEDs at stable and operating states, respectively. (C1, C2) Energy level representation of hetero-junction LEDs at stable and operating states, respectively. (D) Device structure of double heterojunction LED. (E1, E2) An energy level diagram of double heterojunction LEDs [15].

diffusion length, and ΔE stands for bandgap difference between n and p-type semiconductors. When the bandgap is zero, the semiconductor is homojunction, $\Delta E = 0$, whereas, for the heterojunction, $\Delta E \gg k_{0T}$. The hole transport layer (HTL) and an electron transport layer (ETL) are correspondingly created by using p- and n-type semiconductors. To create double heterojunction, a semiconductor has a narrow band placed between HTL and ETL (Fig. 2(E1)) [17]. Fig. 2(E2) represents injected electron-hole confined in the active layer where ΔE_C stands for the gap in the conduction band between the hole transport layer and the active layer, and ΔE_V stands for the gap in the valence band (VB) between the electron transport layer (ETL) and the active layer. A value of ΔE_C and ΔE_v should be high so that electron-hole carrier loss would be minimum [18].

2.1.1 Charge carrier recombination

The LED works on the principle in which the electron-hole recombination process generates photons. In forward bias, many carriers (electrons and holes) present in an active layer are migrated and the thermal-equilibrium of hetero-junction has been destroyed. Again, electrons and holes are recombining so that a new equilibrium reaches. In this phenomenon, the radiative and nonradiative process has been taking place. The photons are emitted due to the radiative process and thermal loss occurs due to the nonradiative process. There are three nonradiative emission processes, which are monomolecular trap-assisted recombination, bimolecular, and the trimolecular Auger recombination. The 2D perovskites film has stronger anisotropic absorption, emission, and improved chemical stabilities than the corresponding 3D perovskite crystal. The carrier recombination rate has been improved after modification in the dimension [19].

2.1.2 Various schemes toward improving the performance of PeLEDs

The solar cells are generally fabricated through perovskite materials because they have small exciton binding energy (EBE) and a long lifetime of charge carriers [20]. However, these properties of perovskite limit applications of perovskite in LEDs. Thereby, a range of approaches is applied to get better performance of perovskite LEDs which are summarized in this section:

(a) Perovskite crystal defect management

In perovskite, crystal defects have been created in the crystal by modifying grain boundaries and creating quenching centers [21]. The roll of defects is to improve luminescence performance or sometimes it increases nonradiative recombination losses. There are several ways to create defects in crystal by creating vacancies of methylammonium (MA), lead (Pb) or iodine (I), and interstitial addition of MA, Pb, or I. Perovskite materials have high defect tolerance so that large radiative recombination has been taking place [22]. Due to this property of perovskite solar cells, small open voltage has been observed in perovskite solar cells. The defects created by ions Br^-, K^+, CH_3COO^-, and NH_4^+ are shown in Fig. 3 [15]. In addition, ethylenediamine (EDA), poly(ethylene oxide), 4,4′-diaminodiphenyl sulfone, LiBr, and KBr are used to move Pb or halide ion. Similarly, ammonium-based cation such as methylammonium and fluorophenylmethylammonium aroused passive Pb-based halides. Ammonium cations are also introduced by Lewis bases, together with halide anions ($X^- = F^-$, Cl^-, Br^-, I^-), trifluoroacetate, carboxyl derivatives, etc. Some organic ligands, for example, oleic acid and oleylamine reduce surface traps, nanocrystal aggregation, and poor

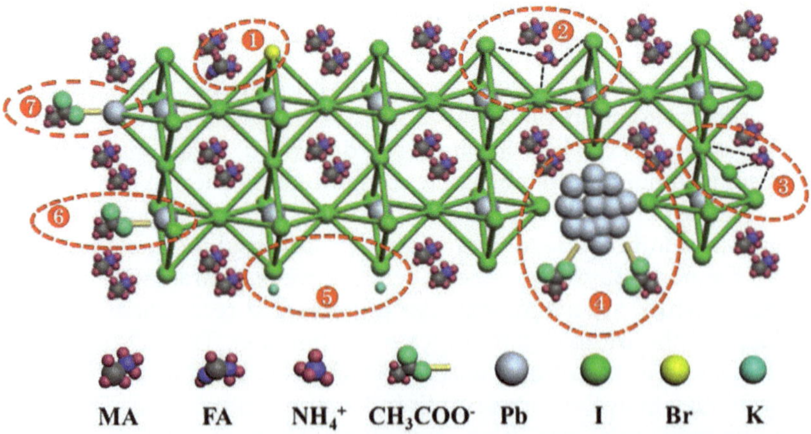

FIG. 3 General vacancies of perovskite and passivation functional molecules [4].

photoluminescence emission. In addition, colloidal synthesis of perovskites indicates that ligands are present on the surface and thin film crystallization.

(b) Perovskite crystallization

The crystallization and nucleation of perovskites are very important to prepare efficient LEDs. The perovskite films used as active layers should have a high crystalline size. The crystallinity can be improved by introducing some conjugates so that compact crystals have been formed and the grain size of the perovskite has been modified. The additive impurities enhance radiative recombination rate and it does not affect the conductivity of carriers, and it has negligible confinement of perovskites. The grain boundaries of perovskites affect the performance of LEDs. The grain size was modified by using 1,3,5-Tris(1-phenyl-1H-benzimidazole-2-yl) benzene (TPBi) and it enhanced the coverage of the perovskite film [23]. The residual amount of dimethyl sulfoxide (DMSO) is removed by using chloroform antisolvent, and after this step, perovskite crystal growth has been improved [23].

(c) Balancing charge carrier rate

The charge carriers' electrons and holes play a vital role in LEDs so balancing of charge carrier is an operational way to improve its performance. If the charge carriers are not balanced, then certain nonreductive processes like trimolecular and Auger recombination have been taken place. To balance charge carriers between HTL and ETL, optimization of charge carriers between the carriers' transport layer and perovskite layers is important. Fig. 4 shows a list of materials that were reported to improve the performance of ETL and HTL [15].

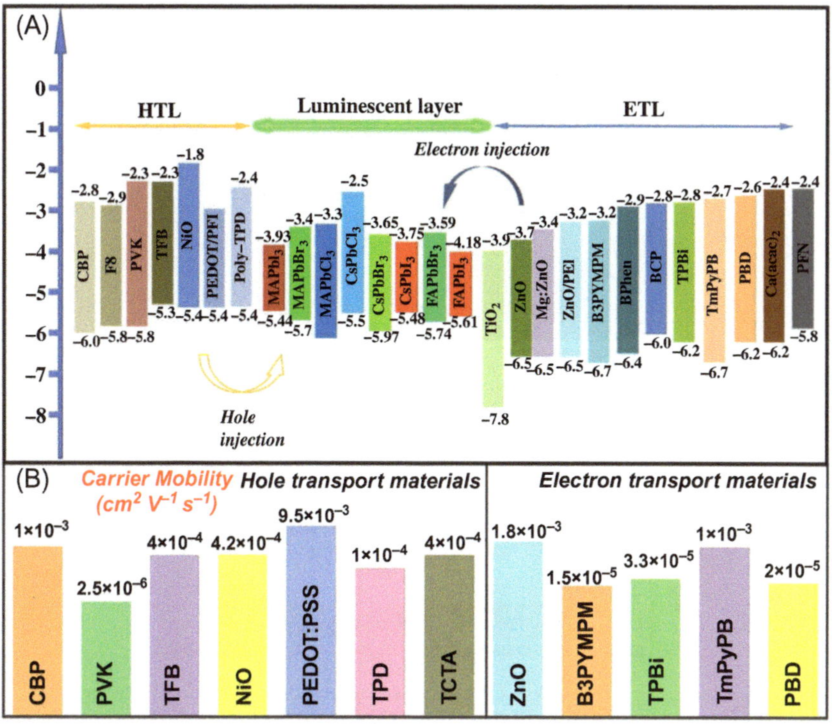

FIG. 4 Fundamental characteristics of the carrier injection materials. (A) Energy level, (B) carrier mobility of the many materials [15].

The carrier mobility of holes is slower than the electrons carrier mobility. Thereby, to balance carrier transformation, injection of the holes should be higher than the electrons so that migration of both carriers can be balanced. It can be done by increasing the bandgap of hole transport material and perovskite to increase the hole injection amount. Similarly, electron injection can be decreased by increasing the bandgap of electron transport material.

2.2 Photoluminescent LED (passive LED backlight)

For backlight applications, nano/quantum dot materials are the best choice. The principle of luminescence quantum dots is like nanoscale semiconductor crystals. Since electrons are limited by the quantum scale, quantum dots differ from traditional semiconductor materials. When the (electron-hole) e-h pair is cramped in the QDs, the size of the quantum dot is close to the wavelength of the material wave of the quantum dot material, the energy level of the electron in the quantum dot is quantified, and

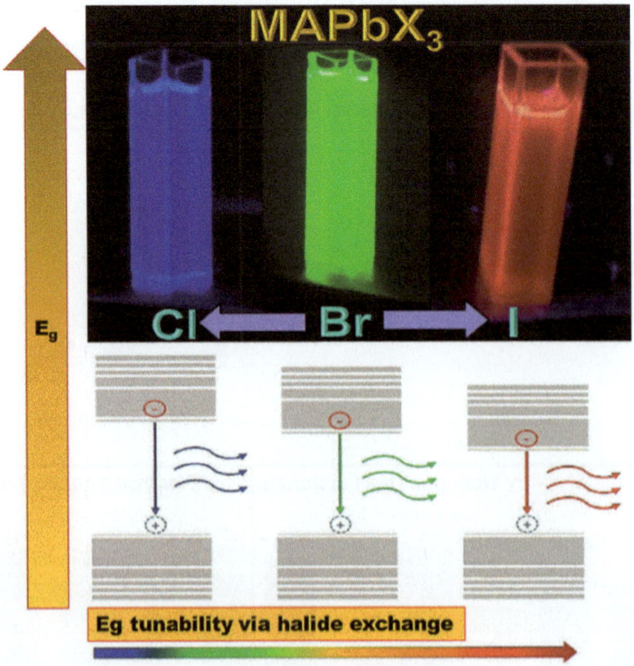

FIG. 5 Color tuning of perovskite QDs with exchanging the halides and their corresponding energy bandgap.

the bandgap can be changed by the magnitude of the QDs. By changing the energy bandgap of the semiconductors, the color of the light emission through the QDs can be adjusted. This phenomenon is called the quantum confinement effect. Additionally, the bandgap of the perovskite kind of quantum dots can be tuned by changing their materials composition as shown in Fig. 5. By changing the halide part from Cl to I, the bandgap of QDs gradually decreases. The smaller the quantum dot, the wider the energy gap and the shorter the wavelength of light. In addition, by combining these two points, it is possible to accurately and systematically apply quantum dot materials with different chromaticities to full-color displays [24].

Common quantum dot materials are colloidal suspensions (colloidal suspension), which can be synthesized by wet chemical methods to precisely control the size, shape, and purity of quantum dots and obtain high photoluminescence quantum yield (PLQY) [25]. In addition, because of the characteristics of colloidal suspension, it is inexpensive to use in large scale, large-scale solution processing, easy to process, and easy to manufacture. QDs are also applicable to building highly efficient and stable LEDs. LEDs are generally based on inorganic, organic-inorganic hydride perovskite materials, and have QDs instead of luminous layers, nonluminous layers, or both.

The unique characteristics of quantum dots can benefit two different display technologies, i.e., LCD (liquid crystal displays) and (OLED) organic material-based light-emitting displays [26]. These two technologies are based on different operating mechanisms. The LCD screen has a backlight, usually, a white LED. The polarizer and backlight filter can control the amount of light passing over the stack. Presently, several display industrialists are using QD to attain better backlighting than the initial w-LEDs. OLED screens do not use for a backlight. Each R (red), G (green), and B (blue) pixel emits light, resulting in an image with higher contrast as well as extra vivid colors.

Although OLED has become an established display technology alongside LCD technologies, the short-term strategies of display industrialists and scientists suggest the use of OLED in conjunction with QD. This can solve several problems faced by LCD and OLED technologies and combine the best of both worlds. As the picture is shown in Fig. 6, according to the optical shape of QDs, QLCD devices have two unique packaging options. The first route is a QD film with green QDs and red QD emission films (QDEF) embedded in the polymer. This QD polymer composite is sandwiched between two protective insulating films [27]. At the same

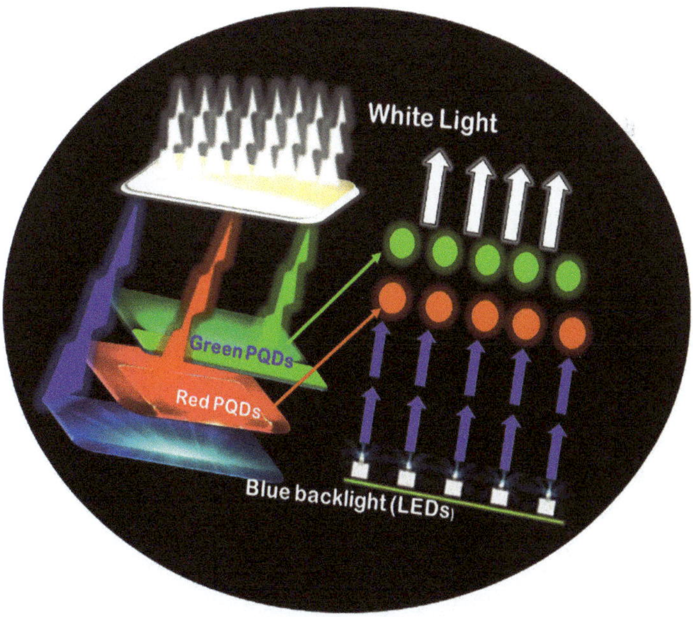

FIG. 6 Two backlighting approaches: Perovskite QDs enhanced films with different layering and white light emission and mixed perovskite QDs for white-LEDs.

time, the essential brightness enhancement film (BEF) is also used here to increase the brightness of an LCD panel and regulate the viewing angle by concentrating light. Among the products released by Samsung, QD Vision, and Nanosys, this QD movie option has been successfully commercialized in QLCD applications [28]. A 46-inch QLCD TV panel based on the QD movie option was recently reported by Samsung. However, complex manufacturing and high costs limit their further applications. The second option is to use QD directly on the LED chip, as shown in Fig. 6. Compared with QD film, this method has greater stability, lesser cost, better adaptability for large-area processing, higher absorption and emission adjustability, as well as greater ease of modification, so it has many advantages Like high cost-effectiveness and received wide attention [29].

The white LED prototype was fabricated using $CH_3NH_3Pb_{1-x}Mn_xI_3$ PQD and green QD on the blue LED chip by Singh et al. [30]. The perovskite light body was integrated as a color conversion layer united with the (B) blue LED, and the different color parameters of an obtained W-LED were shown in Fig. 7. According to the diagram of CIE (Fig. 7A), it was clear that the color point of the blue LED and the connecting line of the green and red PQD will always emit white light so that the combination can consistently produce white light. The CCT and CRI standards of the analogous white emission were evaluated to draw the CCT-CRI patterning diagrams, Fig. 7B.

The imitation outcomes showed that using the above amalgamation can easily achieve warm W-light with CCT < 5000 K and CRI > 80. A sequence of the W-LEDs was fabricated by packaging blue InGaN LED chips through altered numbers of green PQDs and red perovskite QDs to verify the mock-up results. The electroluminescence (EL) spectrum of the manufactured WLED is shown in Fig. 7C, demonstrating the high brightness of WLED. Fig. 7D shows the different color rendering parameters of the manufactured WLED. The CCT value can be adjusted by changing the ratio of R to G from cold light (11,000 K) to warm light (3800 K) as shown in the picture. Fig. 7E shows the combination of blue LED chip and green and red PQD effectively produces warm white light (4000 K) and high CRI (91), improving stability because only under faster aging conditions (up to 85°C and relative humidity up to 85%), 10% of the early strength had disappeared after 200 h as shown in Fig. 7F. These results proved the successful production of highly efficient passive white LEDs by using PQD emitters as the color conversion layer. Later, in Chen et al., $CsPbX_3$ @ glass @PDMS yellow film was used as an LCD backlight to improve the color gamut and reproducibility [31]. Due to the flexibility of the PDMS matrix, $CsPbX_3$ @ glass @ PDMS film has the potential for flexible display. To prove its successful practical application, the traditional backlit LCD combined the traditional blue-chip and $CsPbX_3$ @ glass

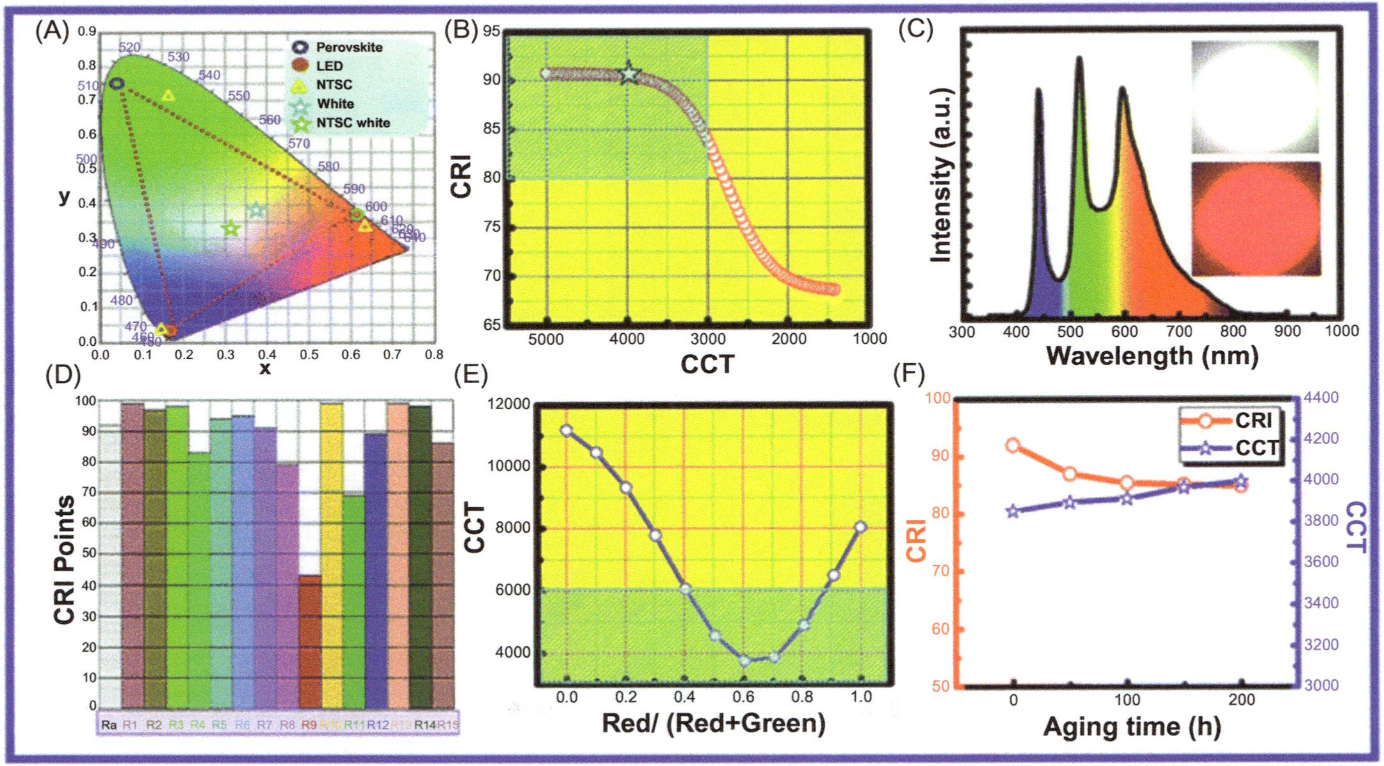

FIG. 7 (A) The color harmonizing of blue LED and green PQDs and red perovskite QD in the CIE diagram (XYZ color space). (B) CCT-CRI values. (C) EL. (D) Different CRI values of warm W-LEDs. (E, F) Amend the CCT and CRI values of warm white LEDs prepared of blue LED chips and green and red PQDs with aging time (h) [30].

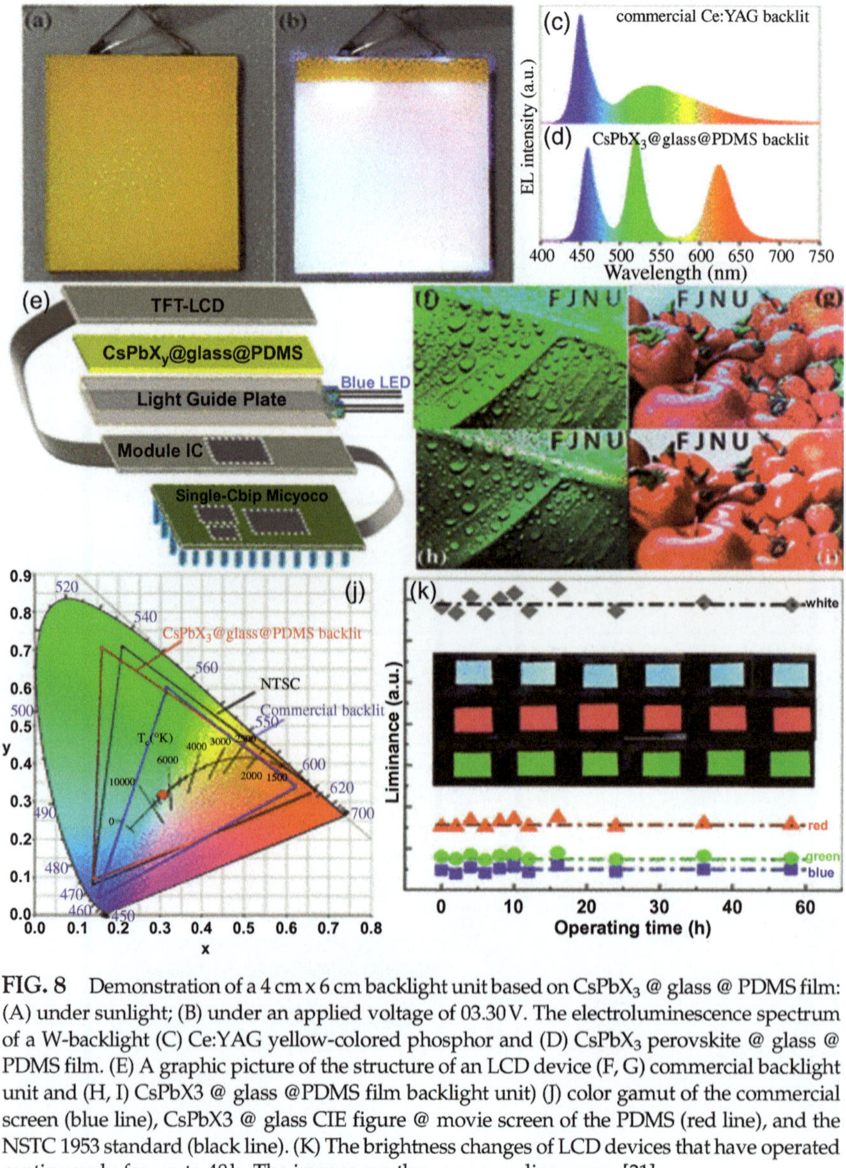

FIG. 8 Demonstration of a 4 cm x 6 cm backlight unit based on $CsPbX_3$ @ glass @ PDMS film: (A) under sunlight; (B) under an applied voltage of 03.30 V. The electroluminescence spectrum of a W-backlight (C) Ce:YAG yellow-colored phosphor and (D) $CsPbX_3$ perovskite @ glass @ PDMS film. (E) A graphic picture of the structure of an LCD device (F, G) commercial backlight unit and (H, I) CsPbX3 @ glass @PDMS film backlight unit) (J) color gamut of the commercial screen (blue line), CsPbX3 @ glass CIE figure @ movie screen of the PDMS (red line), and the NSTC 1953 standard (black line). (K) The brightness changes of LCD devices that have operated continuously for up to 48 h. The images are the corresponding green [31].

@PDMS yellow film instead of using yellow Ce:YAG phosphor. A functioning bias voltage of 3.30 V was applied to a blue LED, which additional excites the $CsPbX_3$ @ glass @PDMS yellow film, resulting in a bright white LED, as shown in Fig. 8A and B. The corresponding EL spectrum showed 450 nm (blue-LED chip), 518 nm (green perovskite

$CsPbBr_3$@glass), and 630 nm (red perovskite $CsPbBr_{1.5}I_{1.5}$ @ glass) tricolor narrowband emission (Fig. 8D). Traditional (commercial) backlight with blue-LED chips and yellow-colored Ce:YAG phosphors displays broadband emission (Fig. 8C). Therefore, the proposed $CsPbX_3$ @ glass @PDMS offers a better alternative to LCD backlight applications.

Fig. 8 also showed a TFT-LCD commercial panel, which was illuminated through the current yellow film of the $CsPbX_3$@glass@PDMS and combined with the blue-LED chip on a light guided plate (Fig. 8E). Compared to commercial LCDs, LCD displays with CsPbX3 @ glass @ PDMS component backlight display more object color details, with pure green, more pronounced color rendering, and pure red with high saturation (Fig. 8F, I). The color gamut of an assembled backlit display was found to be 103% of NTSC 1953, and for the commercial backlight display was 152% (Fig. 8J), respectively. The stability of the working $CsPbX_3$ @ glass @ PDMS backlight was measured by recording the EL spectrum, and after the backlight was lit for 48 h, almost 100% EL intensity can be maintained. Therefore, after 48 h of continuous use (the LCD insertion position), the brightness of the LCD device (green, red, and white) remains unchanged (Fig. 8K). Finally, the flexibility of the $CsPbX_3$ glass PDMS film was tested. Even after severe deformation and stretching, it can still maintain its bright luminescence (Fig. 9A and B). The tensile strength of $CsPbBr_3$

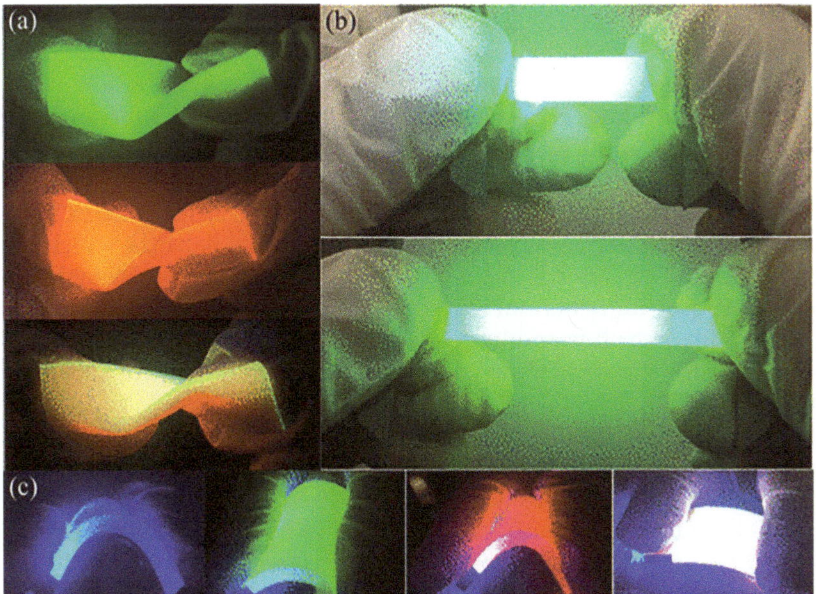

FIG. 9 (A) Distorted green (G), red (R), and yellow (Y) luminescent images of $CsPbX_3$@ glass @ PDMS film. (B) Stretched luminous image of green perovskite $CsPbBr_3$@ glass @ PDMS film. (C) Flexible RGB and white backlight set when 3.30V is applied [31].

@ glass @PDMS at break can increase ~180% of the original material and has high mechanical durability due to the good elasticity of the PDMS matrix as shown in Fig. 9C. Using $CsPbBr_3$ @ glass @ PDMS, $CsPbBr_{1.5}I_{1.5}$ @ glass @PDMS, and mixed @ glass @PDMS, the curved backlight unit produces green, red, and white light, respectively, were fabricated. Thus it can be concluded that PQDs @glass can be a potential candidate for flexible displays as well as for additional innovative optoelectronic devices.

QLED devices can be used to obtain the widest color gamut from QD-based EL [32]. The industrialization of QLED is a future goal because its adolescent technology and expensive cost limit its practical application. Comparing the performance and maturity of these 4 display technologies, it can be seen that QLCD (i.e., QD-based backlighting) occupies a dominant position in the current and near future.

3. Lead-free perovskite LED

Perovskite materials have attracted a lot of attention due to their development in displays in recent years. In addition to displays, perovskite materials are also advancing cutting-edge applications such as electroluminescent modules (EL), PV, image sensors, and IR light detectors. Currently, commercial quantum dot materials mainly based on Pb QD films (QDEF, 3M) used in LCD screens perform well, but cadmium is a toxic substance whether synthetic or applied. In terms of disposal, there are potential dangers and high costs, so scientists are looking for alternative methods. Perovskite is a newly developed semiconductor material. Regardless of its application in photoluminescence (PL), it has the features of easy synthesis, high efficiency, and uniform size distribution. Perovskite is used in solar cells with bright eyes, but the material contains heavy metals, and there are also toxicity and environmental hazards, which cannot replace the meaning of cadmium. In addition, the surface instability of perovskite quantum dots makes it compatible with resins solvents and the compatibility, which is another challenge.

Many studies are underway to moderate the harmfulness of perovskite ingredients [33]. Scientists hope to obtain Pb-free perovskite light harvesters with great potential in practical applications. In this section, a brief overview of the lack of leadership in the field of optoelectronics and the use of lead-free perovskites will be a valuable allusion to the development of LED-based nontoxic perovskite light collectors. Many lead-free perovskite LEDs have been developed so far at the research level. Fig. 10 provides an overview of possible replacements for Pb^{2+} in perovskite and perovskite counterparts. Since the strict meaning of perovskite is limited to 3D-connected BX_6 octahedral, certain lead-free perovskites should be called perovskite communication objects [34–36].

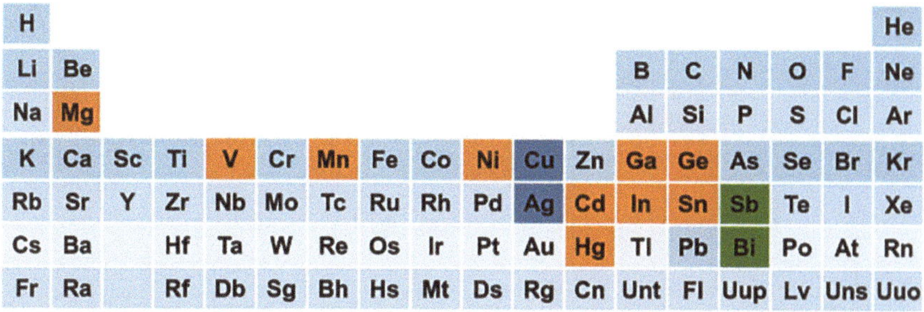

FIG. 10 Candidates for lead substitution in halide perovskites in the periodic table. *Reprinted with permission M. Wang, W. Wang, B. Ma, W. Shen, L. Liu, K. Cao, S. Chen, W. Huang, Lead-free perovskite materials for solar cells, Nano-Micro Lett. 13 (2021) 62. https://doi.org/10.1007/s40820-020-00578-z.*

Chaoqun first conveyed lead-free perovskite LED in 2016, which was based on solution-processed $CsSnI_3$ inorganic perovskite [37]. The LED was used for infrared lighting, the PL emission wavelength was 950 nm, and the maximum EQE was as high as 3.8%. Unfortunately, Sn^{2+} has oxidation problems under environmental conditions and degrades quickly.

Instead, Tan's et al. also transferred a series of Sn-based EL LEDs (structure shown in Fig. 11A). It is based on ITO/PEDOT device structure: PSS/ $MASn(Br_{1-x}I_x)_3$/F8/Ca/Ag and $MASn(Br_{1-x}I_x)_3$ ($x=0.5$–1) thin film. When the Br/I halide composition was different, the EL peak spectrum also changes between 6667 and 945 nm. When the emission material is $CH_3NH_3SnI_3$ film, the maximum EQE was reached up to 0.72% as shown in Fig. 11B. [39] The quantum well structure has shown worthy stability, high PLQY, and outstanding enactment in perovskite LEDs. The unfamiliar PLQY can be attributed to energy transfer and exciton restriction [44]. 2D perovskites based on tin elements were first used to fabricate EL LEDs and showed valuable results. By adjusting different precursor quotients, the solution-treated $(PEAI)_x(CsI)_y(SnI_2)_z$ stable and uniform film shows a range of EL spectra. For such a structure (Fig. 11C and D), $(PEAI)_{3.5}(CsI)_5(SnI_2)_{4.5}$ achieved the highest EQE (up to 3%) with $40\,W\,sr^{-1}m^{-2}$ radiations at 920 nm. In addition, Zhang's team also reported on IR-based EL LEDs, which are constructed on 2D Sn-based perovskites [45]. A visible EL red LED is reported with the structure, ITO/PEDOT: PSS/ $(PEA)_2SnI_4$/F8/LiF/Al, where PEDOT:PSS and F8 are hole and electron inoculation layers, respectively, as shown in Fig. 11E. This visible EL LED exhibited an EL peak at 618 nm, a current efficiency (CE) of $0.003\,cd\,A^{-1}$, and a maximum brightness of $0.15\,cd\,m^{-2}$ [41].

Chu et al. first reports Sb-based perovskite film with ITO/PEDOT: PSS/$Cs_3Sb_2I_9$ film/TPBi/LiF/Al device structure. It has been proven that at a voltage of 2 V, the current density increases rapidly and the radiation at 6 V is $0.012\,W\,sr^{-1}m^{-2}$ [156]. In addition, many researchers have also

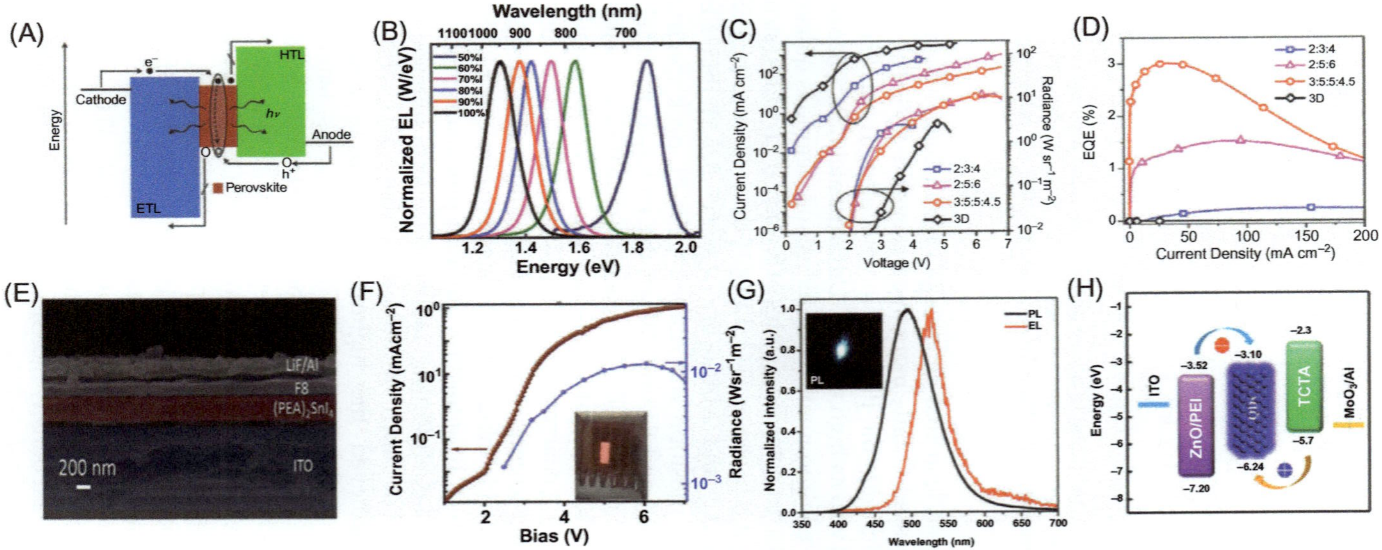

FIG. 11 (A) Representation diagram of the construction of a typical electroluminescent device [38]. (B) ITO/PEDOT: normalized EL spectrum of PSS/CH_3NH_3Sn $(Br_{1-x}I_x)_3$/F8/Ca/Ag LED [39]. (C) $(PEAI)_x(CsI)_y(SnI_2)_z$ current density and emissivity in relation to the control voltage. (D) The relationship between the EQE value of $(PEAI)_x(CsI)_y(SnI_2)_z$ and the current density. For $(PEAI)_{3.5}(CsI)_5(SnI_2)_{4.5}$ MQW LED, the maximum EQE reaches 3.0% [40]. (E) ITO/PEDOT: PSS/$(PEA)_2SnI_4$/F8/LiF/Al device cross section [41]. (F) Glass/ITO/PEDOT: current density vs. voltage (JV) and emissivity vs. tension characteristics of PSS/$Cs_3Sb_2I_9$ film/TPBi/LiF/Al [42]. (G) PL spectrum of $CsCuBr_2$ microcross on Si substrate and EL spectrum of LED at 2.8 V. Image: PL photo of $CsCuBr_2$ microcross. Reprinted with permission [43]. (H) Simplified energy levels of each layer of the heterostructure in the LED based on $Cs_3Sb_2Br_9$ nanocrystals. *Reprinted with permission Z. Ma, Z. Shi, D. Yang, F. Zhang, S. Li, L. Wang, D. Wu, Y. Zhang, G. Na, L. Zhang, X. Li, Y. Zhang, C. Shan, Electrically-driven violet light-emitting devices based on highly stable lead-free perovskite $Cs_3Sb_2Br_9$ quantum dots, ACS Energy Lett. 5 (2020) 385–394. https://doi.org/10.1021/acsenergylett.9b02096.*

reported red, green, and blue EL lead-free perovskite LEDs in recent years. By integrating $CsCuBr_2$ micro-cross (MC) into Ag/$CsCuBr_2$/pSi/In-Ga device structure, a 527 nm green EL LED is displayed. Shi et al. produced LEDs based on EL lead-free perovskite nanomaterial. The EL LED using $Cs_3Sb_2Br_9$ quantum dots emits purple light at 408 nm and the EQE is about 0.206% [46].

Wang et al. reported a highly efficient, bright lead-free warm white LED based on fully inorganic perovskite and cesium-cesium halide. Fig. 12A represents the device structure of the lead-free warm white LED display, namely, ITO/PEDOT: PSS/metal halide/TmPyPB/LiF/Al. The LED device showed a wide-ranging EL spectrum of 565 nm peak and an FWHM of 121 nm (see Fig. 12B). The CIE color coordinates are (0.44, 0.53). Fig. 12C predicted the dependence of current density and brightness on driving voltage. Before the device was turned on, it showed a low current density, indicating that the leakage current from the organic thin layer between the crystals was suppressed. After that the brightness increased rapidly and reached a brightness of 1570 cd m^{-2} at a low voltage of 5.4 V. Fig. 12D shows the relationship between EQE and current density. The LED achieves a peak EQE of 3.1%. Fig. 12E displays the histogram of peak EQE. Statistics from 91 devices show that the mean peak EQE is 1.8% and the relative standard deviation was 23.5% [47]. Hey and group reported the room temperature fabrication of $CsPb_{1-x}Sn_xBr_3$ PQDs by solution-processing by a modified LARP method. The structure of the EL LED based on the lead-deficient perovskite QD was shown in Fig. 13A, and the energy level of each layer in the LED was given in Fig. 13A and B. The brightness of the best material composition was reported at 1600 cd m^{-2} at 3.6 V. The highest CE, EQE, and energy efficiency (PE) were recorded at 4.89 cd A^{-1}, respectively, 1.8% and 6.41 lmw^{-1}, as shown in Fig. 13C [48].

Given the noxiousness of Pb-based PQDs, lead-free $Cs_3Bi_2Br_9$ PQD was renowned as an ideal attractant [49]. However, the lower PLQY hindered its practical application. Pan and his team reported that Sm^{3+} ions were incapacitated with $Cs_3Bi_2Br_9$ PQD with effective white-light emission by an improved LART method. More specifically, 10.9%–20.8% increment was recorded in Sm^{3+}-doped PQDs rather than undoped ones. Additionally, stability with water resistance was also expressively enhanced [49,50]. As keeping in the knowledge of the best performance of Sm doped PQDs, W-LEDs were fabricated with Sm^{3+}:$Cs_3Bi_2Br_9$ lead-free perovskite with 12.6lm/W luminous efficiency. Thus, this farmable performance of W-LEDs showed that lead-free PQDs have potential perspectives in W-LEDs-based applications.

Singh and his team worked hard to announce instantaneous cation and anion exchange with Co^{2+} and Cl^- ions in the host $MAPb_{1-x}Co_xBr_{3-2x}Cl_{2x}$ PQD, and then use polymethyl methacrylate (PMMA) for matrix

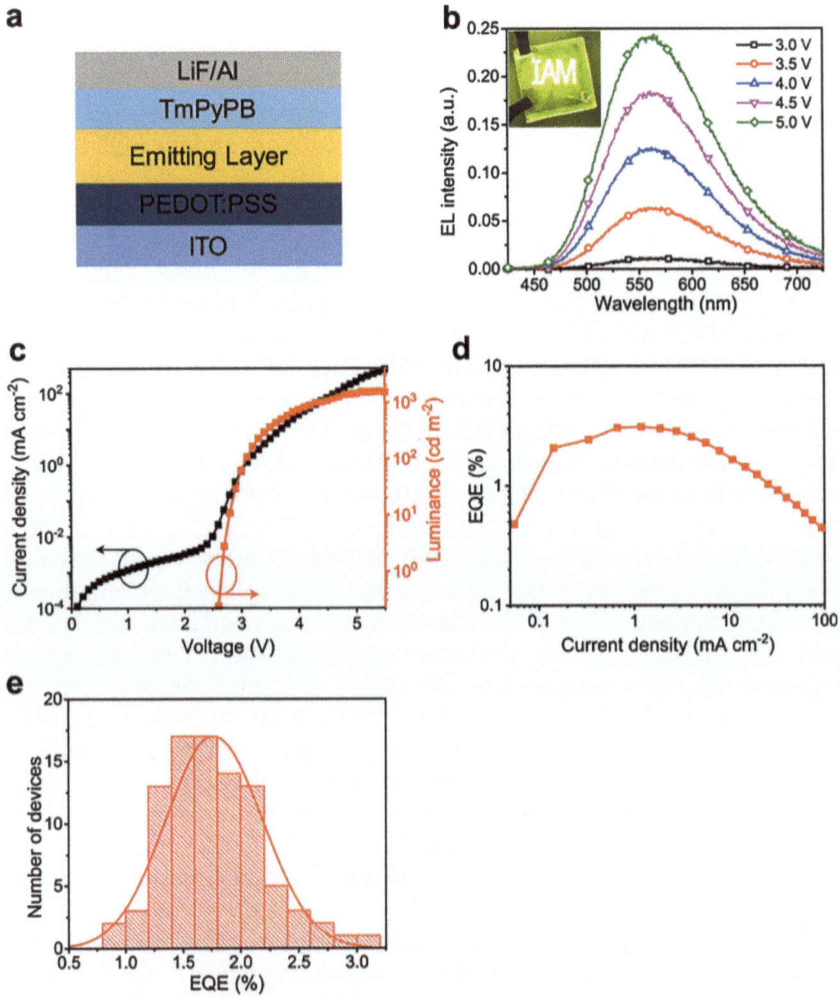

FIG. 12 (A) Schematic diagram of the device structure of Cs-Cu-I LED. (B) EL spectra of the device under various voltages. Picture, LED picture with IAM logo. (C) Relationship between current density and brightness and control voltage. (D) Relationship between EQE and current density. (E) Histogram of peak EQE [47].

encapsulation, which was a stable PQD with a favorable method of improved PLQY [51,52]. Even when Co^{2+} and Cl^{-} ions were incorporated into the perovskite QD, 95% high PLQY was observed. They showed a prototype of WLEDs with PQDs/PMMA composite materials on blue LED chips. The backlight used PQDs/PMMA composite materials, which exhibit excellent environmental stability and stable atmosphere, and provide NTSC standard 124% wide color gamut. This research paved the way

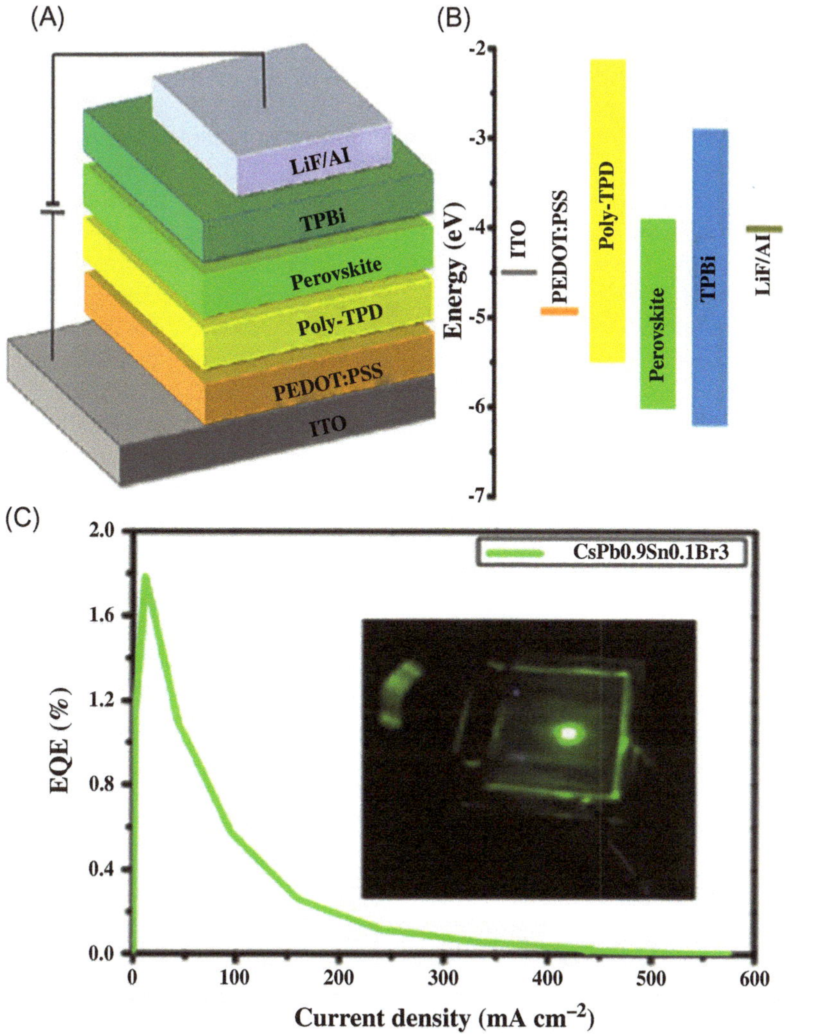

FIG. 13 (A) Lead-deficient perovskite QDs EL LED device structure. (B) Schematic diagram of each layer of perovskite QDs EL LED. (C) EQE. The insert part shows an image of a control device with a transmission area of 2mm × 2mm. Production of $CsPb_{0.9}Sn_{0.1}Br_3$ perovskite QDs EL LED devices.

for the research community to create efficient optoelectronic devices. In another work, Singh et al. reported a simple method to generate nontoxic $CH_3NH_3PbBr_3$ perovskite by replacing lead (Pb^{2+}) and chloride ions (Cl^-) with tin (Sn2 +) ions instead of bromide ions (Br) quantum dots (PQDs) ions [9].

$MAPb_{1-x}Sn_xBr_{3-2x}Cl_{2x}$ PQDs showed strong band-gap tunability from a dark green color to the cyan, and the change of composition (x) was marked as a black circle in the CIE diagram, as given in Fig. 14A. A change in color emission predicted excellent hue and saturation, making the color visible to the naked eye, as shown in Fig. 14B. The figure presented an image of $MAPb_{1-x}Sn_xBr_{3-2x}Cl_{2x}$ PQDs from ultraviolet radiation as a function of composition (x). Because of these inherently excellent optical characteristics, the resulting PQD can be a good choice for the optoelectronic display industry. Therefore, in their research, the synthesized $MAPb_{1-x}Sn_xBr_{3-2x}Cl_{2x}$ PQDs were used as the conversion layer to fabricate a prototype white light-emitting diode (WLED) to demonstrate its usefulness in display technology with a wide range of color gamut applications as shown in Fig. 14C. The resulting $MAPb_{1-x}Sn_xBr_{3-2x}Cl_{2x}$ PQD exhibits a moderately pure color gamut and the preferred triangles for red, green, and blue luminescent materials, as shown in the CIE color coordinate map (see 14c (A species)). The change of composition (x) in

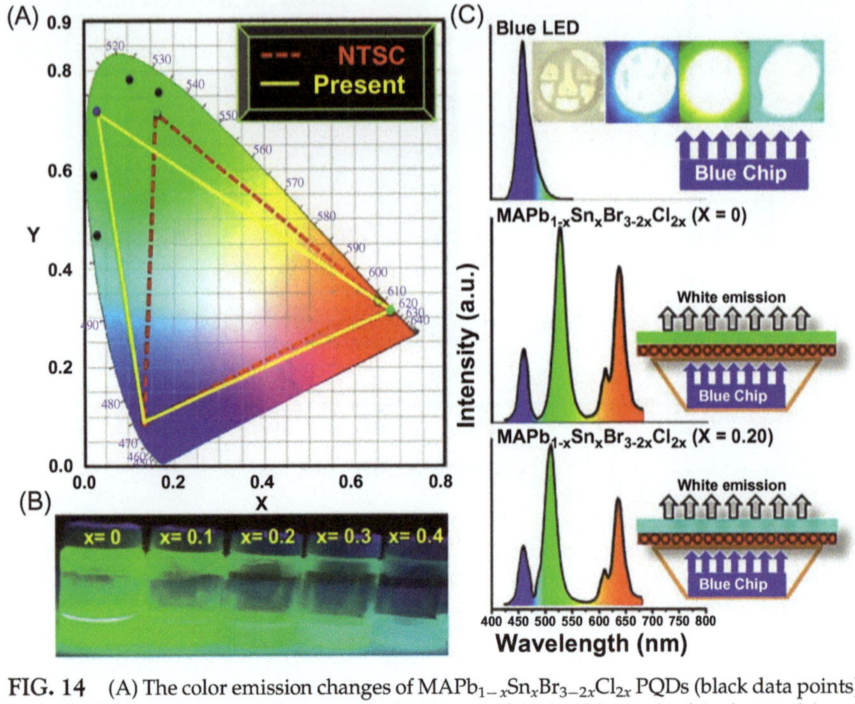

FIG. 14 (A) The color emission changes of $MAPb_{1-x}Sn_xBr_{3-2x}Cl_{2x}$ PQDs (black data points) obtained in the diagram of CIE and the color gamut of the NTSC standard (red triangle) and current situation (yellow triangle). (B) Photos of PQDs solutions under UV light. (C) For blue and white light emission, EL spectrum and graphic scheme include the number prefabricated prototype LEDs of $MAPb_{1-x}Sn_xBr_{3-2x}Cl_{2x}$ PQD. The image shows a graphic image of the blue LED chip and the accumulated W-LED [9].

$MAPbBr_3$:Sn PQDs expanded the color gamut in the green area. The color gamut enclosed up to 117% and 123% of the NTSC TV color standards. The enlarged color gamut, high PLQY, narrow FWHM, and the effective synthesis of Sn-doped PQD offered great prospects for exploring its application in display applications.

4. Perovskite LED stability

For the successful commercialization of perovskites light sources, their long-term stability is of prodigious importance [52]. The electrically pumped perovskite light sources, particularly LEDs, suffer from self-degradation and hydrolysis in the presence of atmospheric water [10]. It has been reported that the normalized EQE of the hybrid perovskite ($MAPbI_3$) film-based LEDs declined to half its original value within around 9 min [53]. Replacing the organic A site, i.e., MA cation by Cs, i.e., inorganic cations, is postulated to be the most effective way to solve the problem of instability of perovskite devices. Unfortunately, the emission intensity without an encapsulated all-inorganic $CsPbBr_3$ perovskite LED drops to half its maximum intensity in just 4 min. This fast degradation in inorganic Cs-based perovskite was attributed to the high defect and trap density [54]. These short-lived but highly efficient devices advocated the requirement of developing new strategies to extend the life of perovskite-based LEDs. The latest advances and efforts made for the stabilization of perovskite-based LED stabilization have been outlined in this section.

The stability of a perovskite-based LED depends primarily on the selection of stable perovskite material. Using all-inorganic perovskites (such as $CsPbX_3$) was considered as compared to organic-inorganic hybrid perovskites (such as $MAPbX_3$ and $FAPbX_3$) to solve the low thermal and light stability of perovskite films. Protesescu et al. synthesized nanocrystals (NC) of perovskite with a mixture of Cs and FA cation at A site ($FA_{0.1}Cs_{0.9}PbI_3$) to obtain a stable structure with effective emissions [55]. One of the developing and effective approaches to improve the stability of perovskite films is the use of 2D-RP phase perovskites. These perovskite films exhibited better stability because the reaction of the perovskite film with outside moisture is inhibited by the occurrence of long-chain hydrophobic organic ammonium cations in its structure [56]. In addition to the prolonged stability, 2-dimensional perovskites possess higher exciton binding energy (>300 meV) that resulted in improvement of the electroluminescent efficiency of perovskite-based LEDs [57]. Cai et al. synthesized a 2D perovskite light harvester $(BA)_2(MA)_{n-1}Pb_nI_{3n+1}$ and suggested that the stability of the LED was closely associated with the phase purity of 2D perovskite [58]. The phase pure 2D perovskite LEDs exhibited a high degree of stability of more than 14h. This outstanding stability as compared to 2D/3D hybrid of 3D perovskite (degradable within minutes) paved new pathways for stable perovskite-based LEDs. The formation energy of perovskite is very low which facilitates its easy production but leads to its poor stability too. To increase the formation energy and thus the stability of the perovskite, an effective and widely accepted strategy is to dope the perovskite film with a smaller divalent cation. Zou et al. used Mn^{2+} as the doping ion to form $CsPbX_3$:Mn NCs with higher formation energy and showed better stability for these doped NCs [59].

Doping of the smaller ionic radii Mn^{2+} ion trace (~0.97 Å) than Pb^{2+} (~1.33 Å) (38) increases the formation energy of $CsPbX_3$ QD perovskite formation and thus increase the thermal stability of the perovskite film [60]. They demonstrated the effect of exposing the films of $CsPbX_3$ QD doped with different Mn^{2+} to ambient atmosphere. Films doped with Mn^{2+} exhibited increased stability in the air as shown in Fig. 15A. Subsequently, three types of perovskite LEDs were fabricated via pure $CsPbBr_3$ perovskite and $CsPbBr_3$:Mn (2.6 mol%) and $CsPbBr_3$:Mn (3.8 mol%)

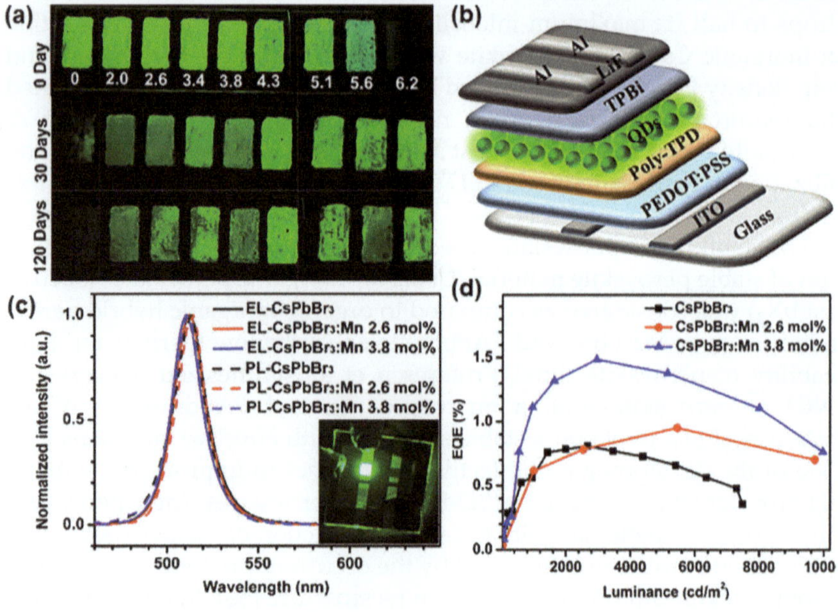

FIG. 15 (A) Photo of PL $CsPbBr_3$: Mn QD emission on a glass coating under 365 nm UV light. (B) (A) via undoped $CsPbBr_3$ perovskite and $CsPbBr_3$:Mn (2.6 mol%) and $CsPbBr_3$: Mn (3.8 mol%) perovskite QD was used as a green light emitter. (C) Normalized EL spectra of $CsPbBr_3$ PQDs and $CsPbBr_3$: Mn (2.6 mol%) and $CsPbBr_3$:Mn (3.8 mol%) PQDs under functional voltage of 6.0 V and their analogous comparisons of photoluminescence emission spectra. The inset shows an EL photograph of a representative PLED device. (D) The relationship between EQE and the brightness of these devices [58,59].

perovskite material as the active green emitter. Device architecture of all PLED devices was ITO/PEDOT:PSS (as HTL)/poly-TPD/$CsPbBr_3$:Mn or undoped $CsPbBr_3$ layers QD/TPBi/LiF/Al (Fig. 15B). Fig. 15C and D shows that although doping Mn^{2+} had no substantial impact on the current density and voltage characteristics (Fig. 15C) but the PLED-$Mn_{2.60}$ and PLED-$Mn_{3.80}$ devices were brighter than pure PLED devices (~1.30 times higher).

Bi et al. recommended the use of copper (Cu) metal as an appropriate replacement element for Pb to synthesize $(PEA)_4Cu_4I_4$ material [60]. The high-quality single crystal of $(PEA)_4Cu_4I_4$ demonstrated an excited state band in the orange-red emission and yielded a high PLQY of up to 68%. Additionally, it exhibited exceptional stability which is due to the hydrophobic nature of PEA and the stable Cu–N bond. An analogous strategy of doping bivalent cations was presented by Chen et al., NCs of $CsPbI_3$ doped with Ni^{2+} ions were synthesized [61]. Doped NCs of Ni^{2+}(3.3mol%): $CsPbI_3$ perovskite showed 95.0%–100% PLQY due to reduction in nonradiative decay rate and increase in radiative transition decay rate in the presence of Ni^{2+}. Additionally, these Ni^{2+}-doped films demonstrated outstanding stability of 7months in ambient air as compared to only 15 days for pristine films. These doped NCs were used as high-efficiency red light emitters in high-performance perovskite LEDs with EQE of ~7%. The stability of undoped and Ni^{2+}-doped $CsPbI_3$ films was measured by directly exposing them to the atmosphere. The undoped film undergoes a phase change which was evident from the color change of the film from red to yellow in three days, while the doped film maintains its red color (Fig. 16A). They concluded that the undoped cubic $CsPbI_3$ briskly transformed to the nonperovskite orthorhombic phase in three continuing days. Nevertheless, Ni^{2+}-doped $CsPbI_3$ perovskite retained its cubic phase for up to seven months. Further, Ni^{2+}:CsPbI3 perovskite retained its 100% PLQY even after storing for 100 days, while undoped $CsPbI_3$ completely lost its PL emission within only 10 days (Fig. 16B). Briefly, the addition of a suitable amount of Ni^{2+} significantly improved the stability of PeNCs.

Xiang et al. reported $CsPbBrI_2$ NC@glass doped with Gd and demonstrated that the bandgap of $CsPbBrI_2$@glass QDs can be attuned via varying the doping amount of Gd and treating temp [62]. As-prepared Gd-doped $CsPbBrI_2$@glass PQDs showed vivid optical as well as hydrothermal properties and also demonstrated excellent light stability. The outstanding luminescent performance and stability of the prepared $CsPbBrI_2$ NCs@glass were efficient enough to replace the commercially available red phosphors. They further demonstrated W-LED device using Gd:$CsPbBrI_2$ QDs@glass and found that the CRI of these W-LED devices was greater than that of commercially available red powder (Fig. 16C–E). The color coordinates of the best WLEDs were found (0.3340; 0.3386),

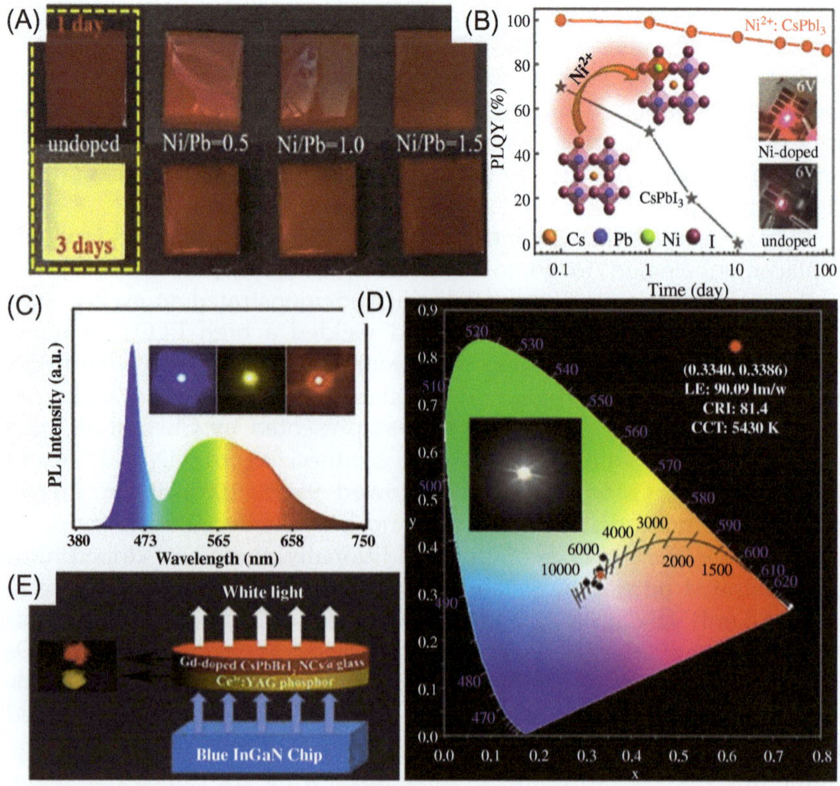

FIG. 16 (A) Compares the relationship between pure $CsPbI_3$ film and Ni-incorporated $CsPbI_3$ perovskite film exposed to the atmosphere for 3 days. (B) The relationship between PLQY of undoped and Ni-incorporated $CsPbI_3$ perovskite and storage time [61]. (C) PL spectrum of blue-chip InGaN, Ce^{3+}:YAG phosphorus, and Gd doped with $CsPbBrI_2$ QDs@glass and corresponding photographs. (D) Photographs of modified coordinates of CIE and white-LED devices. (E) Graphic representation of a structure of W-LED and Ce^{3+}:YAG samples of phosphorus and Gd induced with $CsPbBrI_2$ QDs underneath UV irradiation [62].

which is approximately the same as the color coordinates of the prototype white light emission (0.33; 0.33), LE was 90.09 lm/W, CCT was 5430 K, and CRI was 81.40. Gd:$CsPbBrI_2$ PQDs@glass was also found to be stable even at great currents. These results recommended good scenarios for Gd: $CsPbBrI_2$ PQDs@glass in commercialized white-LED devices and can be used as an auxiliary for plant lighting.

For a highly efficient LED, nonradiative recombination should be less. Nonradiative recombination usually takes place due to charge trapping in defects states of perovskite leading to poor performance of PeLED [63].

Further, these defect states are active states and more defects lead to the lower stability of perovskite. Additionally, the interface between the different layers of LEDs also plays an important role in the efficient charge transfer. So, the development of stable and efficient LEDs requires efficient surface and interface engineering to reduce perovskite film defects and optimize the charge transfer performance of the device.

In this regard, the Kido group synthesized QD by exchange of anion from primeval $CsPbBr_3$ perovskite using halide anion covering alkyl ammonium (OAM-I) and aryl ammonium salts (An-HI) [63]. Anion swapped PQD exhibited a strong red-shift from green emission to red emission at 649.0nm due to the replacement of bromine anion to iodine-based ammonium salts and also exhibited higher PLQY. An anion swapped PQD was used to manufacture LEDs using the structure with the band alignment as given in Fig. 17A. The normalized electroluminescence spectra of perovskite QD LEDs based on crude $CsPbBr_3$ QD, OAM-I QD, and An-HI QD were shown in the figure. Similar to the PL spectrum, the LED EL spectrum that was based on the anion-exchanged red PQD showed a strong red shift from wavelength 511nm (crude) to 653nm (OAM-I) and 645nm (An-HI). The current density and EQE of the anion anion-exchanged red perovskite QD LED were found higher than that of the original $CsPbBr_3$ PQD-LED as shown in Fig. 17C and D. OAM-I, the anionic red perovskite QD LED exhibited a lower switching voltage, while An-HI exhibits a lower switching voltage of 2.7V as shown in Fig. 17E and F and the original $CsPbBr_3$ QD LED has a greater switching voltage of 5.60V. This can be attributed to lowering the barrier to inject holes between poly-TPD and exchange anion-based PQD. In contrast, the high VB (up to 6.10 eV) of the original $CsPbBr_3$ perovskite leads to a higher orifice injection barrier. Due to the anionic defect in PQD, the EQE of the original $CsPbBr_3$ PQD-LED was only 0.17% and for anion-exchanged counterparts, OAM-I- and An-HI-based PQDs-LEDs were 21.3% and 14.1%, respectively. Further, functioning luminance stability and voltage shift of the OAM-I- as well as An-HI-based PQD-LEDs were measured at a current density of 1.25 $mA\,cm^{-2}$ with the preliminary luminance of ~100 $cd.m^{-2}$. An OAM-I-based PQD-LEDs demonstrated a short functioning lifetime (LT_{50}) of 5min and an operative voltage shift (ΔV) of 0.70V. An-HI-based PQD-LEDs possessed a longer operating time of 180min for the LT50, which was 36 times longer than the OAM-I-based PQD-LEDs. Also, the operating voltage drift was significantly reduced for An-HI-based PQD-LEDs. Therefore, for the highly efficient and longer operational stability of perovskite LEDs, composition and surface ligands of the anion exchange PQD are necessary to control.

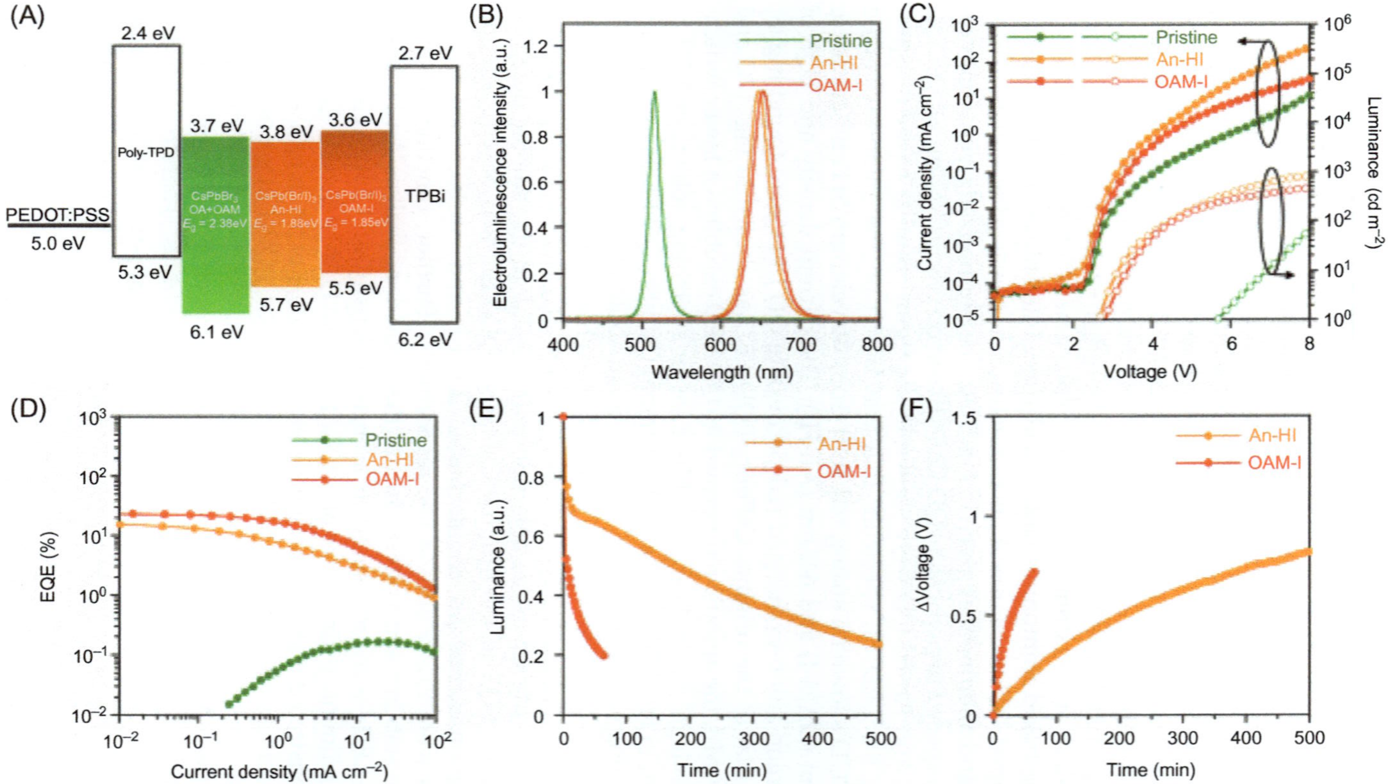

FIG. 17 (A) Energy illustration of PQD LED. (B) EL spectrum of perovskite QD LED at 25.0 mA cm^{-2}. (C) The C-V-V-brightness curve of perovskite QD-LED. (D) EQE curve of PQD LED. (E, F) Lifetime and voltage drift of PQD-LEDs at a constant current density of 1.25 mA cm^{-2} [63].

5. Current challenge and perspective of perovskite LEDs

In conclusion, in this chapter, perovskite-based LEDs with different compassions and their properties are described. Perovskite materials have been invested in large amounts of research in academic research units around the world. The most promising applications are currently in luminescent materials and components, such as QD-LEDs and solar cells. In recent years, more funds have been invested in perovskite-based materials for optoelectronic applications because of their excellent and unique properties. Day-by-day scientific and technological achievements rely on the support of the state and industry to implement but still there are some challenges to industrializing the perovskite LEDs.

(1) EL-based LEDs are useful for self-light generation optoelectronics but still their efficiency is only up to 21%. So there is a need to improve the efficiency of the EL-PLEDs. In the case of white and backlight LEDs, perovskite QDs are very effective and used for flexible optoelectronic devices. But highly intense blue light emission and the blinking problem of such kinds of LEDs are still not solved.

(2) Perovskite LEDs have challenges because of the toxicity and instability of the perovskite materials. In the past, perovskite/traditional quantum dot materials were suspected of toxic risks due to Pb/Cd presence. Besides these environmental problems result in sales and acceptance of products in the market. Today's largest market is estimated to be used in the LCD industry and can be widely used in biomedical, sensing, lighting, and display technologies. The research and development of Pb/Cd-free quantum dot materials has been going on. Thus, lead-deficient/-free stable perovskite-based LEDs can be industrialized in the coming future.

(3) Stability issues with perovskite LEDs are another concern about the commercializing the perovskite LEDs. Perovskite LEDs are very sensitive to oxygen, humidity, heat, UV-light, etc. Thus because of their factors, the optical, electrical, structural stability, and device performance are still poor which is also restricting its application in LEDs.

(4) Up to now, most of the perovskite materials and LEDs are still fabricated on the laboratory scale. The synthesis of perovskite materials is sophisticated and needs more care. Thus, large-scale production methods of perovskite-based LEDs are essential to commercialize the perovskite LEDs.

References

[1] J. Shamsi, A.S. Urban, M. Imran, L. DeTrizio, L. Manna, Metal halide perovskite nanocrystals: synthesis, post-synthesis modifications, and their optical properties, Chem. Rev. 119 (2019) 3296–3348, https://doi.org/10.1021/acs.chemrev.8b00644.

[2] R.K. Singh, P. Sharma, C.-H. Lu, R. Kumar, N. Jain, J. Singh, Structural, morphological and thermodynamic parameters investigation of tunable $MAPb_{1-x}Cd_xBr_{3-2x}I_{2x}$ hybrid perovskite, J. Alloys Compd. 866 (2021), 158936, https://doi.org/10.1016/j.jallcom.2021.158936.

[3] A. Singh, P.-T. Lai, A. Mohapatra, C.-Y. Chen, H.-W. Lin, Y.-J. Lu, C.W. Chu, Panchromatic heterojunction solar cells for Pb-free all-inorganic antimony based perovskite, Chem. Eng. J. 419 (2021), 129424, https://doi.org/10.1016/j.cej.2021.129424.

[4] R.K. Singh, N. Jain, P. Singh, R. Kumar, J. Singh, S. Dutta, T.-M. Chen, S. Som, H.C. Swart, 2-Development in the innovation of lead halide-based perovskite quantum dots from rare earth-doped garnet-based phosphors for light-emitting diodes, in: S.J. Dhoble, V.B. Pawade, H.C. Swart, V.B.T.-S. of L.D.O.M. Chopra (Eds.), Woodhead Publishing Series in Electronic and Optical Materials, Woodhead Publishing, 2020: pp. 21–56. doi:https://doi.org/10.1016/B978-0-08-102935-0.00002-2.

[5] D. Weber, CH3NH3PbX3, ein Pb(II)-System mit kubischer Perowskitstruktur/CH3NH3PbX3, a Pb(II)-System with Cubic Perovskite Structure, Zeitsch. Naturforsch. B 33 (1978) 1443–1445, https://doi.org/10.1515/znb-1978-1214.

[6] R.K. Singh, S.R. Dash, R. Kumar, N. Jain, J. Singh, Role of organic and inorganic cations on thermal behavior of lead iodide perovskites, in: AIP Conf. Proc, American Institute of Physics Inc., 2018, https://doi.org/10.1063/1.5031736.

[7] M.A. Peña, J.L.G. Fierro, Chemical structures and performance of perovskite oxides, Chem. Rev. 101 (2001) 1981–2017, https://doi.org/10.1021/cr980129f.

[8] M.M. Lee, J. Teuscher, T. Miyasaka, T.N. Murakami, H.J. Snaith, Efficient hybrid solar cells based on meso-superstructured organometal halide perovskites, Science (80-) 338 (2012) 643–647, https://doi.org/10.1126/science.1228604.

[9] R.K. Singh, S. Som, S. Dutta, N. Jain, J. Singh, R. Kumar, C.-H. Lu, Rapid and room temperature synthesis of $MAPb_{1-x}Sn_xBr_{3-2x}Cl_{2x}$ perovskite quantum dots with enhanced lifetime in warm WLEDs: A step towards environmental friendly perovskite light harvester, Chem. Eng. J. 391 (2020), 123629, https://doi.org/10.1016/j.cej.2019.123629.

[10] R.K. Singh, R. Kumar, N. Jain, S.R. Dash, J. Singh, A. Srivastava, Investigation of optical and dielectric properties of CsPbI3 inorganic lead iodide perovskite thin film, J. Taiwan Inst. Chem. Eng. 96 (2019) 538–542, https://doi.org/10.1016/j.jtice.2018.11.001.

[11] N. Zheludev, The life and times of the LED—a 100-year history, Nat. Photon. 1 (2007) 189–192, https://doi.org/10.1038/nphoton.2007.34.

[12] P. Zhu, J. Zhu, Low-dimensional metal halide perovskites and related optoelectronic applications, InfoMat. 2 (2020) 341–378, https://doi.org/10.1002/inf2.12086.

[13] F. Liu, C. Ding, Y. Zhang, T.S. Ripolles, T. Kamisaka, T. Toyoda, S. Hayase, T. Minemoto, K. Yoshino, S. Dai, M. Yanagida, H. Noguchi, Q. Shen, Colloidal synthesis of air-stable alloyed $CsSn_{1-x}Pb_xI_3$ perovskite nanocrystals for use in solar cells, J. Am. Chem. Soc. 139 (2017) 16708–16719, https://doi.org/10.1021/jacs.7b08628.

[14] L. Yan, Z. Li, T. Niu, X. Xu, S. Xie, G. Dong, Q. Xue, H.-L. Yip, Effects of ZnI2 doping on the performance of methylammonium-free perovskite solar cells, J. Appl. Phys. 128 (2020) 43102, https://doi.org/10.1063/5.0012370.

[15] A. Liu, C. Bi, R. Guo, M. Zhang, X. Qu, J. Tian, Electroluminescence principle and performance improvement of metal halide perovskite light-emitting diodes, Adv. Opt. Mater. n/a (2021) 2002167. doi:https://doi.org/10.1002/adom.202002167.

[16] M. Saba, F. Quochi, A. Mura, G. Bongiovanni, Excited state properties of hybrid perovskites, Acc. Chem. Res. 49 (2016) 166–173, https://doi.org/10.1021/acs.accounts.5b00445.

[17] H. Kroemer, Theory of a wide-gap emitter for transistors, Proc. IRE 45 (1957) 1535–1537, https://doi.org/10.1109/JRPROC.1957.278348.

[18] T. Kim, J. Lim, S. Song, Recent progress and challenges of electron transport layers in organic–inorganic perovskite solar cells, Energies 13 (2020), https://doi.org/10.3390/en13215572.

[19] Y.-H. Kim, C. Wolf, H. Kim, T.-W. Lee, Charge carrier recombination and ion migration in metal-halide perovskite nanoparticle films for efficient light-emitting diodes, Nano Energy 52 (2018) 329–335, https://doi.org/10.1016/j.nanoen.2018.07.030.
[20] M. Yuan, L.N. Quan, R. Comin, G. Walters, R. Sabatini, O. Voznyy, S. Hoogland, Y. Zhao, E.M. Beauregard, P. Kanjanaboos, Z. Lu, D.H. Kim, E.H. Sargent, Perovskite energy funnels for efficient light-emitting diodes, Nat. Nanotechnol. 11 (2016) 872–877, https://doi.org/10.1038/nnano.2016.110.
[21] J. Liang, X. Han, J.-H. Yang, B. Zhang, Q. Fang, J. Zhang, Q. Ai, M.M. Ogle, T. Terlier, A.-A. Martí, J. Lou, Defect-engineering-enabled high-efficiency all-inorganic perovskite solar cells, Adv. Mater. 31 (2019) 1903448, https://doi.org/10.1002/adma.201903448.
[22] D. Luo, R. Su, W. Zhang, Q. Gong, R. Zhu, Minimizing non-radiative recombination losses in perovskite solar cells, Nat. Rev. Mater. 5 (2020) 44–60, https://doi.org/10.1038/s41578-019-0151-y.
[23] C. Wang, F. Wang, Z. Liu, Y. Zhao, Y. Liu, Q. Yue, H. Zhu, Y. Deng, Y. Wu, D. Zhao, N-doped carbon hollow microspheres for metal-free quasi-solid-state full sodium-ion capacitors, Nano Energy 41 (2017) 674–680, https://doi.org/10.1016/j.nanoen.2017.10.025.
[24] S. Shi, L. Cao, H. Gao, Z. Tian, W. Bi, C. Geng, S. Xu, Solvent- and initiator-free fabrication of efficient and stable perovskite-polystyrene surface-patterned thin films for LED backlights, Nanoscale (2021), https://doi.org/10.1039/D0NR08759A.
[25] M. Shi, R. Li, C. Li, Halide perovskites for light emission and artificial photosynthesis: opportunities, challenges, and perspectives, Eco Mat. 3 (2021), e12074, https://doi.org/10.1002/eom2.12074.
[26] G.J. Lee, J.-G. Lee, Y. Kim, T. Park, Y.W. Ko, J.-H. Ko, The effect of the reflective property of a reflection film on the performance of backlight units with quantum-dot films for LCD applications, J. Inf. Disp. 22 (2021) 55–61, https://doi.org/10.1080/15980316.2020.1813822.
[27] Z. Yang, M. Gao, W. Wu, X. Yang, X.W. Sun, J. Zhang, H.-C. Wang, R.-S. Liu, C.-Y. Han, H. Yang, W. Li, Recent advances in quantum dot-based light-emitting devices: challenges and possible solutions, Mater. Today 24 (2019) 69–93, https://doi.org/10.1016/j.mattod.2018.09.002.
[28] E. Jang, S. Jun, H. Jang, J. Lim, B. Kim, Y. Kim, White-light-emitting diodes with quantum dot color converters for display backlights, Adv. Mater. 22 (2010) 3076–3080, https://doi.org/10.1002/adma.201000525.
[29] J.S. Steckel, J. Ho, C. Hamilton, J. Xi, C. Breen, W. Liu, P. Allen, S. Coe-Sullivan, Quantum dots: the ultimate down-conversion material for LCD displays, J. Soc. Inf. Disp. 23 (2015) 294–305, https://doi.org/10.1002/jsid.313.
[30] R.K. Singh, S. Som, S. Dutta, N. Jain, M.-T. Kuo, J. Singh, R. Kumar, T.-M. Chen, Rapid synthesis of hybrid methylammonium lead iodide perovskite quantum dots and rich MnI 2 substitution favouring Pb-free warm white LED applications, Nanoscale Adv. 1 (2019) 2999–3008, https://doi.org/10.1039/c9na00330d.
[31] J. Lin, Y. Lu, X. Li, F. Huang, C. Yang, M. Liu, N. Jiang, D. Chen, Perovskite quantum dots glasses based backlit displays, ACS Energy Lett. 6 (2021) 519–528, https://doi.org/10.1021/acsenergylett.0c02561.
[32] X. Dai, Z. Zhang, Y. Jin, Y. Niu, H. Cao, X. Liang, L. Chen, J. Wang, X. Peng, Solution-processed, high-performance light-emitting diodes based on quantum dots, Nature 515 (2014) 96, https://doi.org/10.1038/nature13829.
[33] Y. Peng, T.N. Huq, J. Mei, L. Portilla, R.A. Jagt, L.G. Occhipinti, J.L. Mac Manus-Driscoll, R.L.Z. Hoye, V. Pecunia, Lead-free perovskite-inspired absorbers for indoor photovoltaics, Adv. Energy Mater. 11 (2021) 2002761, https://doi.org/10.1002/aenm.202002761.
[34] X. Li, X. Gao, X. Zhang, X. Shen, M. Lu, J. Wu, Z. Shi, V.L. Colvin, J. Hu, X. Bai, W.W. Yu, Y. Zhang, Lead-free halide perovskites for light emission: recent advances and perspectives, Adv. Sci. 8 (2021) 2003334, https://doi.org/10.1002/advs.202003334.

[35] M. Wang, W. Wang, B. Ma, W. Shen, L. Liu, K. Cao, S. Chen, W. Huang, Lead-free perovskite materials for solar cells, Nano-Micro Lett. 13 (2021) 62, https://doi.org/10.1007/s40820-020-00578-z.

[36] J. Li, J. Duan, X. Yang, Y. Duan, P. Yang, Q. Tang, Review on recent progress of lead-free halide perovskites in optoelectronic applications, Nano Energy 80 (2021), 105526, https://doi.org/10.1016/j.nanoen.2020.105526.

[37] W.-L. Hong, Y.-C. Huang, C.-Y. Chang, Z.-C. Zhang, H.-R. Tsai, N.-Y. Chang, Y.-C. Chao, Efficient low-temperature solution-processed lead-free perovskite infrared light-emitting diodes, Adv. Mater. 28 (2016) 8029–8036, https://doi.org/10.1002/adma.201601024.

[38] B.R. Sutherland, E.H. Sargent, Perovskite photonic sources, Nat. Photon. 10 (2016) 295–302, https://doi.org/10.1038/nphoton.2016.62.

[39] M.L. Lai, T.Y.S. Tay, A. Sadhanala, S.E. Dutton, G. Li, R.H. Friend, Z.-K. Tan, Tunable near-infrared luminescence in tin halide perovskite devices, J. Phys. Chem. Lett. 7 (2016) 2653–2658, https://doi.org/10.1021/acs.jpclett.6b01047.

[40] Y. Wang, R. Zou, J. Chang, Z. Fu, Y. Cao, L. Zhang, Y. Wei, D. Kong, W. Zou, K. Wen, N. Fan, N. Wang, W. Huang, J. Wang, Tin-based multiple quantum well perovskites for light-emitting diodes with improved stability, J. Phys. Chem. Lett. 10 (2019) 453–459, https://doi.org/10.1021/acs.jpclett.8b03700.

[41] L. Lanzetta, J.M. Marin-Beloqui, I. Sanchez-Molina, D. Ding, S.A. Haque, Two-dimensional organic tin halide perovskites with tunable visible emission and their use in light-emitting devices, ACS Energy Lett. 2 (2017) 1662–1668, https://doi.org/10.1021/acsenergylett.7b00414.

[42] A. Singh, N.-C. Chiu, K.M. Boopathi, Y.-J. Lu, A. Mohapatra, G. Li, Y.-F. Chen, T.-F. Guo, C.-W. Chu, Lead-free antimony-based light-emitting diodes through the vapor–anion-exchange method, ACS Appl. Mater. Interfaces 11 (2019) 35088–35094, https://doi.org/10.1021/acsami.9b10602.

[43] T. Li, X. Mo, C. Peng, Q. Lu, C. Qi, X. Tao, Y. Ouyang, Y. Zhou, Distinct green electroluminescence from lead-free CsCuBr2 halide micro-crosses, Chem. Commun. 55 (2019) 4554–4557, https://doi.org/10.1039/C8CC09265F.

[44] N. Wang, L. Cheng, R. Ge, S. Zhang, Y. Miao, W. Zou, C. Yi, Y. Sun, Y. Cao, R. Yang, Y. Wei, Q. Guo, Y. Ke, M. Yu, Y. Jin, Y. Liu, Q. Ding, D. Di, L. Yang, G. Xing, H. Tian, C. Jin, F. Gao, R.H. Friend, J. Wang, W. Huang, Perovskite light-emitting diodes based on solution-processed self-organized multiple quantum wells, Nat. Photon. 10 (2016) 699–704, https://doi.org/10.1038/nphoton.2016.185.

[45] X. Zhang, C. Wang, Y. Zhang, X. Zhang, S. Wang, M. Lu, H. Cui, S.V. Kershaw, W.W. Yu, A.L. Rogach, Bright orange electroluminescence from lead-free two-dimensional perovskites, ACS Energy Lett. 4 (2019) 242–248, https://doi.org/10.1021/acsenergylett.8b02239.

[46] A. Wang, X. Yan, M. Zhang, S. Sun, M. Yang, W. Shen, X. Pan, P. Wang, Z. Deng, Controlled synthesis of lead-free and stable perovskite derivative Cs_2SnI_6 nanocrystals via a facile hot-injection process, Chem. Mater. 28 (2016) 8132–8140, https://doi.org/10.1021/acs.chemmater.6b01329.

[47] H. Chen, L. Zhu, C. Xue, P. Liu, X. Du, K. Wen, H. Zhang, L. Xu, C. Xiang, C. Lin, M. Qin, J. Zhang, T. Jiang, C. Yi, L. Cheng, C. Zhang, P. Yang, M. Niu, W. Xu, J. Lai, Y. Cao, J. Chang, H. Tian, Y. Jin, X. Lu, L. Jiang, N. Wang, W. Huang, J. Wang, Efficient and bright warm-white electroluminescence from lead-free metal halides, Nat. Commun. 12 (2021) 1421, https://doi.org/10.1038/s41467-021-21638-x.

[48] J. Deng, H. Wang, J. Xun, J. Wang, X. Yang, W. Shen, M. Li, R. He, Room-temperature synthesis of excellent-performance $CsPb_{1-x}Sn_xBr_3$ perovskite quantum dots and application in light emitting diodes, Mater. Des. 185 (2020), 108246, https://doi.org/10.1016/j.matdes.2019.108246.

[49] Y. Zhu, J. Zhu, H. Song, J. Huang, Z. Lu, G. Pan, Samarium doping improves luminescence efficiency of $Cs_3Bi_2Br_9$ perovskite quantum dots enabling efficient white light-emitting diodes, J. Rare Earths. 39 (2021) 374–379, https://doi.org/10.1016/j.jre.2020.06.007.

[50] Q. Xing, X. Zhang, D. Wu, Y. Han, M. Nirmali Wickramaratne, H. Dai, X. Wang, Ultrasound-assisted synthesis and characterization of heparin-coated Eu3+ doped hydroxyapatite luminescent nanoparticles, Colloid Interface Sci. Commun. 29 (2019) 17–25, https://doi.org/10.1016/j.colcom.2019.01.001.

[51] R.K. Singh, P. Sharma, R. Kumar, S. Som, S. Dutta, N. Jain, R. Chaurasiya, M.L. Meena, J.-S. Ho, S.-W. Dai, J. Singh, C.-H. Lu, H.-W. Lin, $CH_3NH_3Pb_{1-x}Co_xBr_{3-2x}Cl_{2x}$ perovskite quantum dots for wide-color backlighting, ACS Appl. Nano Mater. (2021), https://doi.org/10.1021/acsanm.0c03019.

[52] G. Li, K. Chen, Y. Cui, Y. Zhang, Y. Tian, B. Tian, Y. Hao, Y. Wu, H. Zhang, Stability of perovskite light sources: status and challenges, Adv. Opt. Mater. 8 (2020) 1902012, https://doi.org/10.1002/adom.201902012.

[53] Z. Xiao, R.A. Kerner, L. Zhao, N.L. Tran, K.M. Lee, T.-W. Koh, G.D. Scholes, B.P. Rand, Efficient perovskite light-emitting diodes featuring nanometre-sized crystallites, Nat. Photon. 11 (2017) 108, https://doi.org/10.1038/nphoton.2016.269.

[54] Z. Wang, Z. Luo, C. Zhao, Q. Guo, Y. Wang, F. Wang, X. Bian, A. Alsaedi, T. Hayat, Z. Tan, Efficient and stable pure green all-inorganic perovskite $CsPbBr_3$ light-emitting diodes with a solution-processed NiO_x interlayer, J. Phys. Chem. C 121 (2017) 28132–28138, https://doi.org/10.1021/acs.jpcc.7b11518.

[55] L. Protesescu, S. Yakunin, S. Kumar, J. Bär, F. Bertolotti, N. Masciocchi, A. Guagliardi, M. Grotevent, I. Shorubalko, M.I. Bodnarchuk, C.-J. Shih, M.V. Kovalenko, Dismantling the "Red Wall" of colloidal perovskites: highly luminescent formamidinium and formamidinium–cesium lead iodide nanocrystals, ACS Nano 11 (2017) 3119–3134, https://doi.org/10.1021/acsnano.7b00116.

[56] T. Niu, J. Lu, M.-C. Tang, D. Barrit, D.-M. Smilgies, Z. Yang, J. Li, Y. Fan, T. Luo, I. McCulloch, A. Amassian, S.(.F.). Liu, K. Zhao, High performance ambient-air-stable $FAPbI_3$ perovskite solar cells with molecule-passivated Ruddlesden–Popper/3D heterostructured film, Energ. Environ. Sci. 11 (2018) 3358–3366, https://doi.org/10.1039/C8EE02542H.

[57] B.-E. Cohen, M. Wierzbowska, L. Etgar, High efficiency and high open circuit voltage in quasi 2D perovskite based solar cells, Adv. Funct. Mater. 27 (2017) 1604733, https://doi.org/10.1002/adfm.201604733.

[58] H. Tsai, W. Nie, J.-C. Blancon, C.C. Stoumpos, C.M.M. Soe, J. Yoo, J. Crochet, S. Tretiak, J. Even, A. Sadhanala, G. Azzellino, R. Brenes, P.M. Ajayan, V. Bulović, S.D. Stranks, R.H. Friend, M.G. Kanatzidis, A.D. Mohite, Stable light-emitting diodes using phase-pure Ruddlesden–Popper layered perovskites, Adv. Mater. 30 (2018) 1704217, https://doi.org/10.1002/adma.201704217.

[59] S. Zou, Y. Liu, J. Li, C. Liu, R. Feng, F. Jiang, Y. Li, J. Song, H. Zeng, M. Hong, X. Chen, Stabilizing cesium lead halide perovskite lattice through Mn(II) substitution for air-stable light-emitting diodes, J. Am. Chem. Soc. 139 (2017) 11443–11450, https://doi.org/10.1021/jacs.7b04000.

[60] L. Wang, H. Sun, C. Sun, D. Xu, J. Tao, T. Wei, Z. Zhang, Y. Zhang, Z. Wang, W. Bi, Lead-free, stable orange-red-emitting hybrid copper based organic–inorganic compounds, Dalton Trans. 50 (2021) 2766–2773, https://doi.org/10.1039/D0DT04413J.

[61] M. Liu, N. Jiang, H. Huang, J. Lin, F. Huang, Y. Zheng, D. Chen, Ni^{2+}-doped $CsPbI_3$ perovskite nanocrystals with near-unity photoluminescence quantum yield and superior structure stability for red light-emitting devices, Chem. Eng. J. 413 (2021), 127547, https://doi.org/10.1016/j.cej.2020.127547.

[62] Q. He, Y. Zhang, Y. Yu, Y. Chen, M. Jin, E. Mei, X. Liang, L. Zhai, W. Xiang, Ultrastable Gd^{3+} doped $CsPbBrI_2$ nanocrystals red glass for high efficiency WLEDs, Chem. Eng. J. 411 (2021), 128530, https://doi.org/10.1016/j.cej.2021.128530.
[63] T. Chiba, Y. Hayashi, H. Ebe, K. Hoshi, J. Sato, S. Sato, Y.-J. Pu, S. Ohisa, J. Kido, Anion-exchange red perovskite quantum dots with ammonium iodine salts for highly efficient light-emitting devices, Nat. Photon. 12 (2018) 681–687, https://doi.org/10.1038/s41566-018-0260-y.

CHAPTER

9

Other applications of halide perovskites

Shivam Porwal[a], Dinesh Kumar[a], Subrata Ghosh[a], Sakshi Kansal[a], Surbhi Priya[a], Amreesh Chandra[b], and Trilok Singh[a]

[a]Functional Materials and Device Laboratory, School of Energy Science and Engineering, Indian Institute of Technology Kharagpur, Kharagpur, 721302, India, [b]Department of Physics, Indian Institute of Technology Kharagpur, Kharagpur, 721302, India

1. Introduction

Halide perovskite materials have attracted tremendous attention since 2009, when some of the path-breaking studies reported their high carrier mobility, excellent optoelectronic properties, low recombination rates, tunable bandgaps, and relatively long diffusion lengths [1–3]. All these unique characteristics opened up new vistas for halide perovskite materials in the areas of energy conversion, storage, piezoelectric, optoelectronic devices, etc. These are individually described below.

2. Halide perovskite for batteries and supercapacitors

With the growing worldwide efforts toward climate change mitigation, the use of renewable energy sources has seen a phenomenal rise over the last few decades. Most renewable energy generation sources suffer from the intrinsic limitation of "intermittency." Therefore, the renewable energy-based future landscape will have energy storage devices as an integral component. In addition, energy storage devices are already indispensable in mobile, electric vehicles, and off-grid distribution networks.

Low-Dimensional Halide Perovskites
https://doi.org/10.1016/B978-0-323-88522-5.00007-7

As one can see, the range of applications for storage devices is vast with varying requirements in the performance parameters. No single energy storage device has been able to cater the needs of all the applications. But a combination of two devices, viz., Li-batteries and supercapacitors, has superseded the other solutions. The ion-based batteries are considered suitable for high energy density applications while the supercapacitors cater for the needs of high-power requirements. Lithium-ion batteries offer an energy density in the range of 100–250 $Wh\,kg^{-1}$, with low self-discharge, high operating voltage, and appreciable cycle life in comparison to conventional alkaline or dry cell batteries. Fig. 1 depicts the working mechanism of a typical Li-ion battery. More recently, metal-air batteries (where Faradic reactions prevail between the pure metal anode and air cathode) have become competitive on account of their high theoretical energy density, ranging from 1090 $Wh\,kg^{-1}$ for Zn-air battery to 5210 $Wh\,kg^{-1}$ for lithium-air batteries. In comparison, supercapacitors or ultracapacitors [4] exhibit higher specific capacity than the conventional parallel-plate capacitance ($<0.1\,Wh\,kg^{-1}$). They store charge either electrostatically or electrochemically, have fast charge-discharge rates, long cycle life, and high-power densities. Initially, the use of halide perovskite materials was mainly focused in the area of photovoltaics on account of their outstanding power conversion efficiencies, which is as high as 25.5% [5].

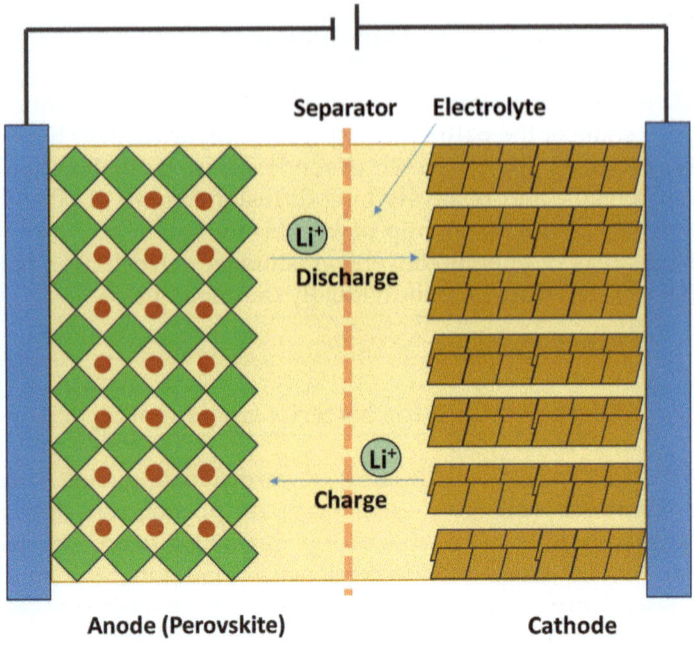

FIG. 1 Working mechanism of Li-ion battery.

Recently, it has been realized that the halide perovskite materials possess an explicit potential of ion diffusion within the perovskite lattice, which is considered to be one of the reasons for the major disadvantage of power conversion efficiency in solar cells. However, this characteristic of halide perovskites makes them promising candidates for energy storage applications. Halide perovskites display the ionic motion of A^+ (MA^+, Cs^+, K^+, FA^+), B^{+2} (Pb^{+2}, Sn^{+2}), and X^- (Br^-, I^-, Cl^-) ions, vacancies, or interstitials owing to their low activation energy values required for migration [6,7]. The ion migration is not only limited to the smaller size of inorganic cations or anions. Even the organic ions, with much larger sizes, can become mobile within the halide perovskite lattice [8,9].

Additionally, halide perovskite nanostructures exhibit an appreciable catalytic ability for oxygen reduction, tunable functionality, facile carrier generation, and relatively higher specific capacity than their counterparts. These properties make them useful for metal-air batteries, supercapacitors, and Li-ion batteries. Peng et al. [10] reported the use of halide perovskite materials for Li-ion batteries in 2015, where hydrothermally synthesized $CH_3NH_3PbBr_3$ microcrystals displayed a charge-discharge capacity of nearly $330\,mAh\,g^{-1}$, with a retained capacity of $200\,mAh\,g^{-1}$ after 200th cycles. These values were comparable to those of the commercial graphite-based batteries ($372\,mAh\,g^{-1}$). Moreover, the group further explored the compositional effect of halogens on battery performance by performing a compendious comparison between the 3-D $CH_3NH_3PbBr_3$ and 3-D $CH_3NH_3PbI_3$ as anode materials for the Li-ion battery. The first discharge capacity for $CH_3NH_3PbBr_3$ ($331.8\,mAh\,g^{-1}$) was found to be much larger than that for $CH_3NH_3PbI_3$ ($43.6\,mAh\,g^{-1}$). However, the capacities of both the halide perovskites drastically fell after 30 cycles owing to the structural decomposition with successive charge-discharge cycles. A comprehensive study of the mechanism of the interaction between Li-ions and the halide perovskites was successfully carried out by Dawson et al. [11], where the group studied the $CH_3NH_3PbBr_3$ material and suggested the occurrence of both intercalation and conversion mechanisms. However, the high first discharge capacity of halide perovskite was found to be approximately six times more than the maximum theoretical capacity of $CH_3NH_3PbBr_3$ ($55.96\,mAh\,g^{-1}$) when one Li-ion per formula unit was intercalated. These results indicated the dominance of the conversion mechanism over the intercalation; otherwise, six intercalating Li-ions per formula unit would have been required. Further, they illustrated that the charge transfer during the first discharge cycle was equal to 2.4 Li^+ ions per mole of $CH_3NH_3PbBr_3$, which exceeded the limit of the vacant site per mole available in primary $CH_3NH_3PbBr_3$ necessary for an intercalation mechanism. They proposed two plausible mechanisms for conversion reaction (Eq. 1 and 2) [11]:

$$Li + C\,H_3NH_3PbX_3 \rightarrow CH_3NH_3X + LiPbX_2 \quad (1)$$

$$2Li + CH_3NH_3PbX_3 \rightarrow CH_3NH_3X + 2LiX + Pb \quad (2)$$

The first reaction here involves the phase separation of perovskite with Li efficiently being intercalated in the PbX_2 ($X=Br^-$, I^-, Cl^-) structure. However, the second reaction is similar to the Faradic reactions observed in conventional Li-ion battery materials, where the metal cation of the corresponding oxide or fluoride on the cathode is reduced by Li metal to form the metal and the lithium oxide or fluoride (for instance, CoO+2Li Co +Li_2O). Experimental data suggested that the second one is the thermodynamically more favorable reaction owing to the formation of stable lithium halides. This result was further supported by PXRD patterns, which indicated the appearance of new Pb peaks, and appreciable suppression in the intensities associated with halide perovskite peaks. This clearly indicated a significant change in the crystal lattice, which directly modifies the response characteristics. Another approach toward understanding the mechanism of Li-ion intake in $CH_3NH_3PbBr_3$ has been proposed by Vicente et al. [12] by analyzing the **operando-XRD** data. The group deduced that, initially, when Li molar content is less than 1, the pure and lithiated phases coexist without structural rearrangement. However, when the Li molar content exceeds the value of 1 but remains less than 2, conversion reaction dominates and that leads to the formation of solid Pb. Beyond the value of 2, an alloying occurs, leading to the formation of a lithium-lead alloy.

Tathavadekar et al. recognized that lowering the dimension eventually improves the electrochemical performances of the halide perovskite-based Li-ion batteries, as the effective surface area for Li-ion accommodation increases with simultaneous increment in the accessibility of Li-ions for the Pb atoms [13]. The group demonstrated the electrochemical response of 1-D benzidine lead iodide, 2-D butyl ammonium lead iodide, and 3-D methylammonium lead iodide with respective first discharge capacities of $1605\,mAh\,g^{-1}$, $1580\,mAh\,g^{-1}$, and $476\,mAh\,g^{-1}$ for the 1-D, 2-D, and 3-D perovskites, respectively, at a current density of $100\,mA\,g^{-1}$.

To address the environmental concerns associated with the use of lead, Li-ion batteries utilizing Pb-free perovskites are also being proposed. Pandey et al. [14] reported that the 2-D $(CH_3NH_3)_2CuBr_4$ and 3-D Cs_2CuBr_4 possess a reversible capacity of $\sim 480\,mAh\,g^{-1}$ and $420\,mAh\,g^{-1}$, respectively. Interestingly, it was observed that the capacity of the 2-D material increases slowly with subsequent charge-discharge cycles, displaying an increased capacity of $630\,mAh\,g^{-1}$ after the 140th cycle. This observation is attributed to the slow percolation of electrolyte into the electrode and also the appearance of Li-ion intercalation sites as a function of cycling. Similarly, the use of $CH_3NH_3NiCl_3$ perovskite has been reported by Lopez et al. This Pb-free perovskite is able to deliver a high first-discharge capacity of $650\,mAh\,g^{-1}$, with a capacity retention of more than 50% after the 19th cycle at the current density of $32\,mA\,g^{-1}$ [15]. Additionally, Kostopoulou et al. [16] have illustrated the electrochemical performance of all inorganic Cs_4PbBr_6 perovskite for Li-ion batteries. The material can manage a discharge capacity of $377\,mAh\,g^{-1}$ with a 75% capacity retention after 100 cycles. Moreover, it has been found that the crystal structure remains intact till ~40 cycles, which indicates a reversible intercalation mechanism. On further cycling, an irreversible conversion reaction dominates, which leads to the formation of CsBr and $PbBr_2$. Furthermore, as lithium-sulfur batteries exhibit higher specific energy density ($500\,Wh\,kg^{-1}$) than Li-ion batteries ($100–250\,Wh\,kg^{-1}$), various research on probing halide perovskite for Li-sulfur batteries are coming to the forefront. It has been indicated that there are strong interactions present between the halide perovskite and lithium polysulfides such as S_8, Li_2S_6, and Li_2S_2, which may be able to immobilize the sulfur species and can enhance the overall electrochemical performance of the batteries [17,18].

The usage of halide perovskites is not limited to batteries. They are also finding applications in supercapacitors. It has been observed that the electrode and electrolyte materials derived from halide perovskites have a high surface area and appreciable ionic conductivity, which can be beneficially harvested in supercapacitors. The ionic conduction of halide perovskite material was firstly investigated by Zhou et al. by integrating a perovskite thin film in a symmetrical electrochemical capacitor (EC). The electrochemical studies showed that the perovskite layer serves both as a polarized electrode and solid electrolyte, where iodine-ion migration takes place. They further reported that the EC exhibits cyclic stability even beyond 10,000 cycles. Furthermore, the lower density for mobile carriers in the perovskite films than the perovskite electronic devices indicates that the ion formation and transport mechanisms are mostly activated by the movement of free charge carriers and may also be interface-dependent [19]. Snaith et al. constructed a 350 nm thick $MAPbI_3$ film in a parallel-plate capacitor that achieved a capacitance of $\sim 1.8\,mF\,cm^{-2}$ with an energy density of $\sim 2 \times 10^{-6}\,Wh\,kg^{-1}$ in the operating voltage window of 1 V [20]. However, the detailed mechanism for halide-perovskite-based supercapacitors remains to be extensively explored. Recently, Popoola et al. assembled a supercapacitor based on the composite electrodes of metal halide hybrid perovskites ($MAPbI_3$ and $MABi_2I_9$)-nickel oxide (NiO) nanoparticles via wet chemical and spin coating technique [21]. High specific capacitance values of $407\,F\,g^{-1}$ and $368\,F\,g^{-1}$ with energy densities of $56.5\,Wh\,kg^{-1}$ and $51.1\,Wh\,kg^{-1}$ were observed for NiO@ $MAPbI_3$- and NiO@$MABi_2I_9$-based supercapacitors, respectively. Herein, they reported that both the diffusion-controlled intercalation of electrolyte ions and

Faradic reactions influence the charge-storage process. It is well known that the halide perovskite materials undergo phase transition with varying temperatures from orthorhombic at low temperatures to tetragonal at room temperature to finally attain a cubic phase at elevated temperatures. It was experimentally observed by Slonopas et al. [22] that the temperature-dependent phase transition in halide perovskites can impact the charging time and also the overall electrochemical performance of the supercapacitors. The group revealed the specific capacitance value of ~523 mF cm^{-2}, 432 mF cm^{-2}, and 384 mF cm^{-2} at 350 K, 300 K, and 100 K, respectively. The increase in capacitance values with temperature could be attributed to the phase transition of perovskite from orthogonal to tetragonal to cubic geometry. Recently, a new application of halide perovskites materials has been introduced in photobatteries and photorechargeable supercapacitors, where energy generation, viz., solar energy and storage, will be potentially obtained in a single perovskite material. In such systems, firstly, under irradiation, the photoactive material produces electron-hole pairs. The electrons move toward the energy storage electrode and helps in storing the charges electrostatically. Volder et al. [23] developed a single-layer multifunctional 2D halide perovskite $(C_6H_9C_2H_4NH_3)PbI_4$ integrated into a photobattery, which simultaneously accomplished photo-charging and Li-ion storage. The overall capacity achieved by the photobattery was found to be 100 mAh g^{-1}. Another notable work by Wang et al. suggested incorporation of the $MAPbI_3$ material with a conjugated polymer P3HT in a photorechargeable battery [24]. The photobattery exhibits a high power conversion efficiency of nearly 9.94%, with specific capacities varying from 46.1 to 77.5 mAh g^{-1}. Further, in the photo-supercapacitor, a halide perovskite-based solar cell is connected to the energy storage electrode, viz., sharing a common electrode [18]. Huang et al. [25] connected an inorganic $CsPbBr_{2.9}I_{0.1}$-based perovskite solar cell with a supercapacitor to assemble a photosupercapacitor that delivers capacitance of 30 mF cm^{-2}. Another photocapacitor has been designed by Liu and the group, where they incorporated the polyaniline polymer and halide perovskite. This device shows a high specific areal capacitance of 422 mF cm^{-2} with a coulombic efficiency of 96% [26]. Liang et al. fabricated a novel "solar capacitor" comprising $CsPbBr_3$-based inorganic perovskite solar cell (PSC) and an all-inorganic silica-gel-electrolyte-based supercapacitor [27]. It has been observed that as-synthesized solar capacitor works in a wide potential window of 1.2 V and possesses energy and power density of 6.8 μWh cm^{-2} and 450 μW cm^{-2}, respectively. Zhou et al. [28] vertically combined a perovskite solar cell (PSC), MoO_3/Au/MoO_3 transparent electrode, and a WO_3-based electrochromic electrode to develop the color-tunable photovoltachromic supercapacitors. Although halide perovskite materials face toxicity and stability issues, they may still serve as a remarkable material for energy storage devices and integrated energy generation and storage systems.

3. Gas sensing application of organic-inorganic hybrid halide perovskite material

In the recent past, various sensing techniques, along with different mechanisms, have been used in gas sensors for the identification of hazardous pollutants. But, low selectivity and high operating temperatures limit their large-scale utilization. The organic-inorganic halide perovskites (OIHPs) have shown promising applications in gas sensing with improved selectivity [29]. With crystal structure ABX_3, they belong to a family of crystalline materials that have a hybrid structure. In terms of structural and optical properties, these materials are extremely adaptable [30]. The crystal structure of OIHPs can be conveniently modified by altering the structure of A, B, and X, and therefore, the properties of these materials can be tunned accordingly. These greatly tunable properties of inorganic-organic hybrid perovskites lead to high sensitivity, selectivity, and room temperature operation, along with high response and recovery time. Perovskite semiconductors have recently shown their ability to detect a variety of chemicals, temperature changes, metal ions, and explosives. This section will provide an analysis of the most significant and current studies on inorganic-organic metal halide perovskite as a gas sensor.

3.1 Gas sensing mechanism

The basic theory of hybrid metal halide perovskite gas sensors is based on a change in resistance caused by the chemically active substance interacting with the analyte gas. When the surface of hybrid metal halide perovskite comes in contact with the atmosphere, oxygen molecules from the vicinity get adsorbed over the surface, and the adsorbed molecules then capture the conducting electron present in the active material, inevitably forming a depletion layer, which alters the electrical resistance of the sensor. Adsorption of oxygen species over the active material can be described as [31]:

$$O_2(gas) \rightarrow O_2(ads) \tag{3}$$

$$O_2(ads) + e^- \rightarrow O_2^-(ads) \tag{4}$$

$$O_2^-(ads) + e^- \rightarrow 2O^- \tag{5}$$

$$O^-(ads) + e^- \rightarrow O^{2-}(ads) \tag{6}$$

These adsorbed species act as an active place for the attachment of target gas molecules. As the target gas is made incident over the active material, depending upon whether the gas is oxidizing or reducing, the electrical resistance increases or decreases. If the analyte gas is reducing in nature, it will contribute to the numbers of majority charge carriers

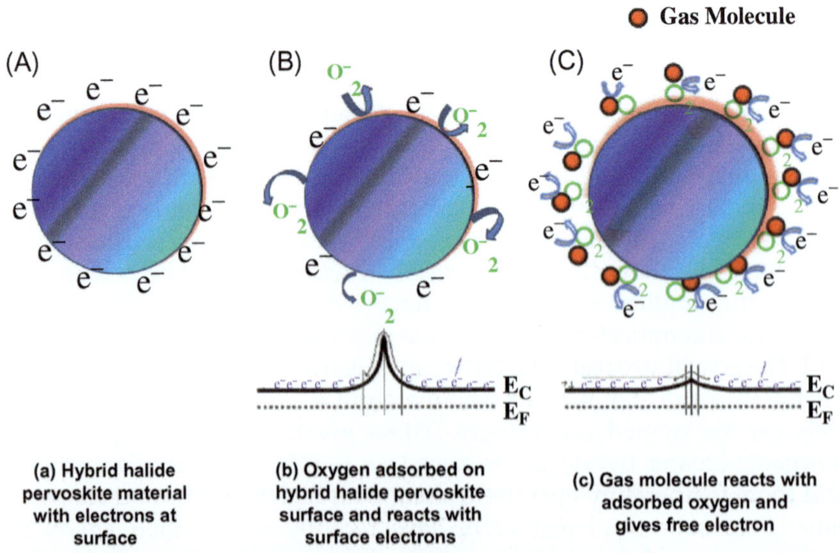

FIG. 2 Schematic representation of gas sensing mechanism.

for an *n*-type sensor leading to an increase in conductivity or decrease in resistance, and if the hybrid metal halide perovskite sensor material is of *p*-type then, on exposure to a reducing gas, the conductivity will decrease and the electrical resistance will increase for the entire system. Just opposite results are recorded in the case of oxidizing analyzed gas. Fig. 2 explains the gas sensing mechanism of hybrid metal halide perovskite.

3.2 Gas sensor based on inorganic and hybrid metal halide perovskite

Relatively high gas sensing properties are exhibited by organic and inorganic metal halide perovskites. The factors that influence the ability of perovskite-based sensor devices to detect specific analyte gas are (a) chemical composition of analyte gas and (b) hybrid perovskite's composition. A solid-state gas sensor based on paper electronics that detects NH_3 gas selectively with a detection capability better than 1 ppm was reported by Maity et al. [32] The active sensor content of $CH_3NH_3PbI_3$ was deposited on paper and operated at room temperature. When exposed to just 10 ppm NH_3 steam, the current through the paper sensor rises by one order of magnitude. For 1 ppm of NH_3 gas in the air, the measured sensitivity is 55%. This research sets up perovskite halide as a modern solid-state gas sensing material with sub-ppm sensitivity using basic paper electronics [32]. Jiao et al. synthesized $MAPbBr_{3-x}I_x$ using a one-step solution process for sensing acetone, ethanol, and ammonia. It has been observed that the active perovskite materials showed favorable response and selectivity for ammonia, whereas the faster response and recovery time for ammonium sensing are attributed to cation exchange, adsorption, and reversible weak bonding between MA^+ and NH^{4+}. Moreover, long-term exposure to the high concentration of ammonia gas has accelerated the irreversible decomposition of $MAPbBr_{3-x}I_x$ perovskite, resulting in a decrease in the response value. In the case of ethanol and acetone, slightly short response and recovery time along with good stability were observed as a result of the reversible physical transition from gas molecules to intrinsic bromide vacancies, and uncoordinated Pb^{2+} cations in $MAPbBr_{3-x}I_x$ perovskite [33]. For the identification of aliphatic amines such as diethylamine, monomethyl amine, and trimethylamine along with volatile organic compounds, Kim et al. fabricated a fluorescence chemosensor using $CH_3NH_3PbBr_3$ perovskite nanoparticles. The sensor exhibited excellent selectability of ~89% for amine vapor along with a very fast reversible/irreversible fluorescence quenching rate (<1 s). Quantum chemical calculations and X-ray diffraction (XRD) measurements were used to confirm the detection mechanism that took place due to fluorescence quenching. This phenomenon of fluorescence quenching is originated by the change of the basic structure of perovskite into its component by permanent or reversible interactions of hydrogen bonds between aliphatic amines and perovskite [34].

Zero-dimensional hybrid organic metal halides have emerged as an attractive class in the field of the latest light-emitting substance. Li et al., in their research, reported a novel lead-free compound $(C_9NH_{20})_2MnBr_4$ and the fabricated sensor devices were highly selective and exhibited fast response/recovery time [35]. Zhang et al. used the solution method to synthesize organotin halide perovskite $HC(NH_2)_2SnI_3(FASnI_3)$ and later Pt nanoparticles were decorated over the perovskite layer to develop a sensor element. The effect of morphological and structural properties of Pt coated $HC(NH_2)_2SnI_3(FASnI_3)$ electrode was studied for the formaldehyde gas sensor. The sensor exhibited high sensing performance, such as the limit of detection for 10 ppm formaldehyde at 80°C was calculated to be 47.5 ppb, and also the response and recovery time was recorded to be 40 and 37 s. This superior result was attributed to the formation of *n-n* heterojunction due to the interaction of $HC(NH_2)_2SnI_3$ with SnO_2. The combination of lead-free perovskite and photocatalytic Pt nanoparticles improved the operating temperature and gas sensor response of the sensor [36]. An extremely responsive hydrogen sulfide gas sensor made of methylammonium lead perovskite was fabricated by Lee et al. In the active layer of the perovskite-based sensor, pseudohalide molecular anion (SCN) was incorporated as a dopant. An excellent limit of detection in the range of 200 ppb of hydrogen sulfide gas was recorded for both $MAPbI_{3-x}Br_x$-based and $MAPbI_{3-x}(SCN)_x$-based gas sensors. Among both the sensors, the $MAPbI_{3-x}(SCN)_x$-based gas sensor exhibited improved sensing performance with better stability [37]. A layered $(C_4H_9NH_3)_2PbI_4$ perovskite material has been used as an active gas sensor device whereas the device showed a quick response (11.8 s at 140°C)/recovery time, high selectivity, and repeatability due to the physical adsorption [38]. Chen et al. used $Au/MASnI_3/SnO_2$ material to detect NO_2 gas under UV illumination conditions at room temperature. The sensor performed best at room temperature, which included excellent response, i.e., 240 for 5 ppm NO_2, along with quick recovery (~12 s), better selectivity than Au/SnO_2- or SnO_2-based sensors under UV illumination at room temperature. Further limit of detection was calculated to be 55 ppb at an operating voltage of 1.1 V. The improved sensing performance of the device was attributed to the light absorption ability of $MASnI_3$, efficient generation, and conduction of photo-induced electrons along with catalytic effect [39]. $CH_3NH_3PbI_{3-x}Cl_x$ perovskite sensor was operated for the various concentrations of ozone gas; due to the passivation of surface electrons, a change in resistance was observed [40]. Further, Hu et al. studied the effect of humidity of $CH_3NH_3PbI_{3-x}Cl_x$ perovskite films for possible application in humidity sensing. They observed a difference in electrical resistance of exposed (humidity) films due to shrinkage of the depletion region on the surface of films [41]. Furthermore, $CH_3NH_3PbBr_3$ humidity sensor was developed and subjected to different concentrations of moisture (from 7% to 98%), sensor exhibited high sensitivity even ~0.04% humidity was detected along with a fast response and recovery time of 250 s and 30–70 s, respectively. The enhancement in sensing properties was attributed to the formation of hydrogen bonds between iodine and water molecules. Instability in the sensor performance was observed after the sensor was exposed to more than 50% of humidity concentration for 24 h. The recent developments in the field of gas sensing using perovskite are summarized in Table 1 [42].

3.3 Future prospects

Gas sensors have a wide range of applications, from controlling what living beings breathe to detecting toxic gases to determining clinical gases. The entire sensor depends upon the performance of the sensing element, and the unique physical properties of metal halide perovskite such as electrical conductivity, wavelength, and photoluminescence emission intensity, along with an exceptional ability to reverse the changes upon removal of foreign stimuli, make it a perfect and attractive choice. Unfortunately, according to the present scenario, only a few inorganic and

Analyte gas	Active material	Type of sensor	Result	Year of publication	Reference
NH_3	$CH_3NH_3PbI_3$	Paper sensor	Sensitivity <55%	2019	[32]
NH_3	$MAPbBr_{3-x}I_x$	Chemosensor	Response time: <20 s	2020	[33]
VOC, aliphatic amines	$CH_3NH_3PbBr_3$ perovskite nanoparticles	Fluorescence chemosensor	Response time: <1 s Selectivity: 89%	2017	[34]
Acetone	$(C_9NH_{20})_2$ MnBr4	Fluorescence sensor	Photoluminescence quantum efficiency (PLQE) of 81.08%	2019	[35]
Formaldehyde	$HC(NH_2)_2SnI_3$ ($FASnI_3$)	Chemosensor	Response time:40 s Sensitivity: 47.5	2019	[36]
Hydrogen sulfide	$MAPbI_{2-x}$ $(SCN)_{x-}$	Chemosensor	Response time: 30 s LOD: 200 ppb	2020	[37]
p-Xylene gas	Layered $(C_4H_9NH_3)$ $_2PbI_4$	Chemiresistive Sensor	Response time:4.6 s Recovery time: 5.9 s	2018	[38]
NO_2	$Au/MASnI_3/SnO_2$	Resistive Sensor	Recovery time: 12 s LOD: 55 ppb	2019	[39]
Ozone	$CH_3NH_3PbI_{3-x}Cl_x$	Chemiresistive Sensor	Response time: 13 s	2018	[40]
Humidity	$CH_3NH_3PbI_{3-x}Cl_x$	Chemiresistive Sensor	Response time: 74 s Resistance decrease: from 98% to 25%	2015	[41]
Humidity	$CH_3NH_3PbB_3$	Optical sensor	Response time:250 s Sensitivity: 0.04%	2016	[42]

hybrid metal halide perovskites have been studied for sensing applications. With ongoing attempts to develop and prepare stable perovskites with new compositions and better structural characteristics, it is expected that the potential of metal halide perovskites to translate any input into an optical or electric signal will be addressed in the upcoming time to enable sturdy and strong platforms for sensing.

4. Resistive switching random access memory (*ReRAM*) devices using metal halide perovskite

Metal halide perovskites have recently found their application in nanoscale electrical memory storage devices where the perovskite as an active material can be switched between a low-resistance state (ON state) and a high-resistance state (OFF state) under the current-voltage (*I-V*) scan. Various structures of perovskite such as 3-D, 2-D, and 0-D quantum dot-based [43] have been explored for resistive random access memory (*ReRAM*) storage devices [44–46].

The electromagnetic interaction between the A-site cation and $[BX_6]^{-4}$ octahedra site is mainly responsible for the bandgap formation in these materials. The rotational motion of A-site cation provides microscopic electrical properties [47]. Solution-processed perovskites have free carrier mobility typically in the range of 5–10 $cm^2\ V^{-1}$ s and diffusion length in the order of a few micrometers [48,49]. Perovskite semiconducting materials are ionic crystals with electrical transport from free charge carriers as well as from the ionic motion of mobile ions. Perovskite semiconducting materials also offer tunable optical, chemical, and electrical properties that make these materials highly suitable for resistive switching devices described later in the chapter [50]. Further low cost, low temperature, and solution processability also make these materials a great candidate for next-generation memory storage devices.

The resistive switching phenomenon was first discovered by Hickmott in the year 1964 [51], with an aluminum oxide thin film sandwiched between two different metal electrodes. Resistive switching is defined as the change in the resistance value from high to low and low to high value under applied periodic input. The resistive switching phenomenon is extensively studied in various oxide-based insulators and wide bandgap semiconductors such as TiO_2, SiO_x, Al_2O_3, $SrTiO_3$, and NiO_X thin films [52]. Typically, *ReRAM* devices have a device structure of metal-insulator-metal (*MIM*) or metal-semiconductor-metal (*MSM*) type (Fig. 3A) configuration. Semiconductors with low-cost fabrication, solution processability, and tunable electrical parameters were always desired for the fabrication of *ReRAM* devices. Recently, with the advent of perovskite semiconducting materials alongside their exceptional properties

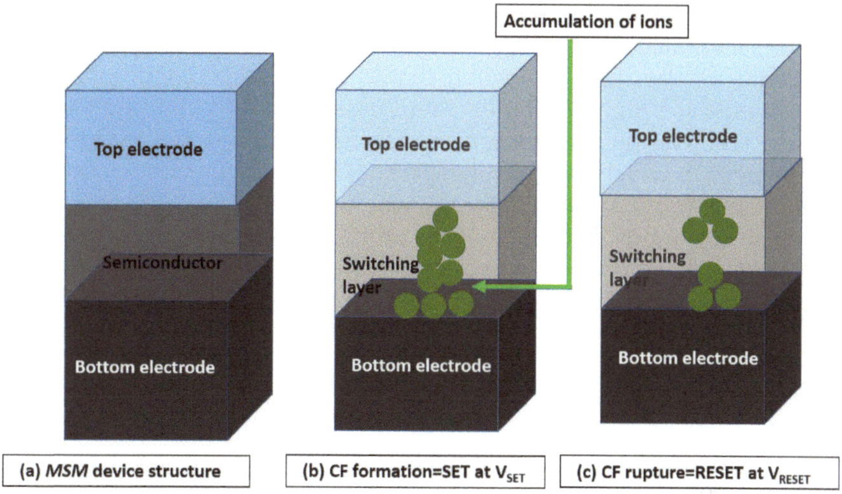

FIG. 3 Schematic of *ReRAM* (A) MSM (metal-semiconductor-metal) device structure. (B and C) Conduction filament formation.

mentioned above, these materials have become a powerful choice for *ReRAM* applications. Besides this, perovskite-based *ReRAMs* have also shown ultrahigh switching speed [53], large storage density, low power consumption [54], also the possibility of integration into flexible memory devices [45]. In the next few sections, we will briefly describe device fundamentals, working principles, recent progress, and associated quality factors of *ReRAM* devices.

The concept of a resistive switching device originated from the "Memristors." [55] Memristors are considered to be a fourth passive circuit element after capacitor (dv/dq), resistance (dv/di), inductor ($d\Phi/di$), which links charge with the magnetic flux ($d\Phi/dq$) and is also known as memristance. The concept was first coined by Leon Chua by preserving symmetry between current, charge, flux, voltage, and completed a theoretical quartet. In an ideal memristor, memristance should only be a function of charge $q(t)$ and flux $\Phi(t)$. Practical *ReRAM* devices have more dynamic behavior than the ideal memristor. Therefore, we need to understand various other models to explain resistive switching behavior in the *ReRAM* devices [56]. Strukov et al. [57] confirmed that all practical memristors could be regarded as *ReRAM* devices.

Nonvolatile *ReRAM* is defined as the device's resistance retention capability even after removing the applied input. *ReRAM* devices are capable of resistance switching; this resistance switching is established by external perturbations such as voltage, current, magnetic field, and temperature. Like all other memory storage devices, *ReRAM* devices follow the same writing, read and erase sequence for data storage, data interpretation, and data deletion. The writing process is done through the binary operation of using binary bits "1" or high and "0" or low. In *ReRAM* devices, a set stands for low resistance value, that is, a set (HIGH) condition, and a reset stands for high resistance value, a reset (LOW) condition of the device. The reading process is the interpretation of stored information on the device, which is again done through the application of external voltage or current sweep. Typically, the magnitude of voltage needed for reading is much smaller than the writing voltage, and it is up to the order of 1/10th of the writing voltage.

4.1 Conduction filament formation

Conduction filament (CF) formation (Fig. 3B and C) is regarded as a primary working principle for *ReRAM* devices [58]. *CFs* are the high conduction (metallic conductivity) pathways formed during applied periodic input on the device, and this is mediated by the migration of cations or anions or defects through the active semiconducting layer. The switching behavior in the *ReRAM* device is mainly divided into three different categories, namely, electrochemical switching memory (ECM), valence change memory (VCM), and thermochemical change (TCM) memory. All three different switching behaviors can also simultaneously occur in the same device. CF formation due to migration of cations (A site perovskite cations or metal electrode cations) through the active semiconducting layer known as ECM-type resistive switching and migration of anions (halides, oxides, vacancies) is known as VCM-type resistive switching. Sometimes ECM and VCM effects are simultaneously observed in the different *ReRAM* devices. These switching processes are mutually competitive with the dominance of one process over other types of processes inside the device is widely observed in different cases [55]. Sun et al. compared mutual competition of cation-mediated (ECM) and vacancy-mediated (VCM) conduction filament formation [59] in *ReRAM* devices. Authors have discovered that metallic CF dominates whenever the active perovskite semiconductor layer thickness is small (<100 nm), and vacancy CF dominates whenever the active layer thickness is large (>300 nm). The dominance of CF of one type can also be established through the temperature dependence of the overall device resistance value. Whenever a metallic CF form (ECM type), device resistance shows a strong dependence on temperature, a similar behavior observed in different metallic materials. In contrast, whenever anionic CF (VCM type) forms weak device resistance, temperature dependence is observed. This is very identical to the temperature dependence of the resistance value as observed in various semiconductor materials.

The other important parameter for realizing the resistive switching process is the electroforming (poling) voltage, wherein a high electric field is applied to the active material, joule heating can produce sufficient energy for vacancies to overcome their thermal energy barrier, which leads to the formation of CF. The magnitude of poling voltage is much larger than the writing voltage/current, and this process is also known as a soft dielectric breakdown of the semiconducting layer. This provides essential conductivity to the sandwiched semiconductor layer so that CF formation can take place. To avoid hard breakdown while performing the electroforming process, a compliance current in reverse is applied to protect the overall device. The other advanced model to understand resistive switching behavior is space charge limiting current (SCLC) [60], Poole-Frenkel [61] method, trap assisted tunneling [62] (*TAT*) model, and statistical approach for CF interconnections [56].

4.2 Importance of *I-V* hysteresis

Metal halide semiconducting perovskite materials fall in the category of ionic conductors. Besides the motion of free charge carriers (i.e., electrons and holes), halide perovskites are also prone to the large numbers (10^{14}–$10^{17}\,cm^{-3}$) [48] of intrinsic point defects. These materials suffer from large numbers of point defects such as atomic vacancies, interstitial atoms, antisite occupations, Frenkel and Schottky defects. Point defects in semiconducting perovskite mainly contribute to shallow energy levels by having low defect formation energy [63]. These materials also suffer from 1-D defects (such as screw axis, edge, and partial dislocations) and 2-D defects (such as surface states and grain boundaries) [64]. Transport in metal halide perovskite is defined by hoping, multiple trapping, and de-trapping of charge carriers due to these defects. These ionic defects possess heavy mass, and hence, a slower response (smaller mobility) compared to the smaller effective mass of free charge carrier electrons and holes [65]. Because of this slow response and localized band transport of the ionic defects, there exist a hysteresis in the forward scan and reverse scan of *I-V* characteristics of perovskite-based devices [52,55].

ReRAM devices have pinched hysteresis *I-V* loop as their characteristic signature (Fig. 4), and these hysteresis loops are typically found to have a shape of "8" in the *I-V* response. The hysteresis feature appears due to nanoscale movement of the defects and ionic species. If the periodic input frequency is comparable with nanoscale process frequencies, effects on the hysteresis loop are much more pronounced. At the higher frequencies compared to characteristics, nanoscale process frequency hysteresis effects start to diminish. Though hysteresis is lethal in perovskite-based solar cells [65], it is the decisive mechanism in perovskite-based *ReRAM* devices.

4.3 Device working principle

Resistive switching in *ReRAM* devices can be explored under various operating conditions (modes), such as periodic voltage sweeps, pulse sequential, single voltage sweep, and *I-V* sweep mode. *I-V* sweep mode is

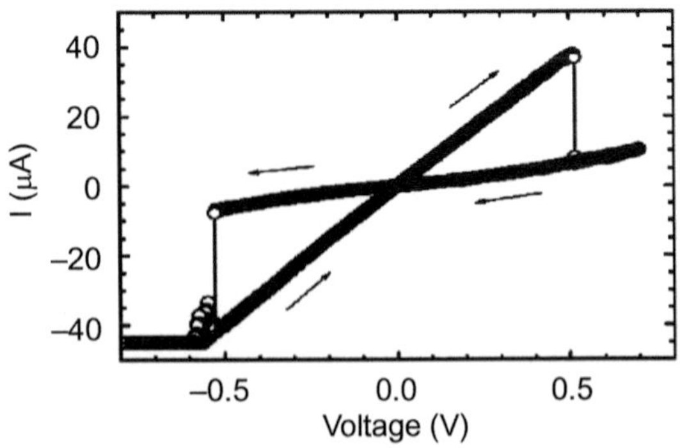

FIG. 4 Linear scale I-V response of *ReRAM* device showing large hysteresis [66]. *From A. Beck, J.G. Bednorz, C. Gerber, C. Rossel, D. Widmer, Reproducible switching effect in thin oxide films for memory applications. Appl. Phys. Lett. 77 (139) (2000) 4, https://doi.org/10.1063/1.126902. © 2000 American Institute of Physics.*

highly informative, and we can easily determine V_{set}, V_{reset}, low-resistance, and high-resistance values from the hysteresis response itself (Fig. 4). The VCM type of switching is mainly observed in bipolar devices (when resistive switching from high to low value is done through switching of the voltage polarity from a positive to a negative value). The majority of the perovskite-based *ReRAM* devices have shown bipolar switching behavior.

I-V response of the *ReRAM* devices can be measured in bipolar as well as unipolar configuration, whereas based on the applied input, *ReRAM* devices can be further classified into three categories, namely, unipolar, bipolar, and threshold switching. During unipolar switching, the set and reset of resistance states happen at the same voltage polarity. But in the bipolar configuration, set and reset happen at different voltage polarities and both set and reset states are the metastable states. In threshold voltage switching, only one metastable state is present. As the name suggests, bipolar switching is polarity-dependent switching; different polarities are required for different types of charge carriers, such that one type of polarity of the voltage is required for the anion migration and the other for the cation migration. Through bipolar switching, one can establish the presence of ambipolar transport present in the metal halide perovskite semiconductors.

4.4 Role of metal electrodes

Under a bipolar operation, both of the electrodes take part in resistive switching while only one is during the unipolar switching. The switching type ECM device is realized through oxidation of the active metal electrodes (Ag, Cu) rather than the inert metal electrodes (Pt). The oxidized cation then migrates toward the other electrode through the active semiconducting perovskite layer and gets reduced at the other electrode. The collection of reduced cations at the opposite polarity electrodes forms a conduction filament as shown above in Fig. 3B). This provides high conduction pathways for charge carriers through the active semiconductor layer. The formation of CF leads to the set condition, and when the polarity is reversed, rupturing of CF leads to the reset condition of the device. Under the VCM type of switching, both of the electrodes can be active or inert type. Various *ReRAM* devices have shown resistance switching phenomena only while using active metal electrodes, but some researchers also showed resistance switching for inert metal electrodes as well. Wu et al. [67] studied the impact of top metal electrodes on the performance of *ReRAM* device by varying the top electrode material types, such as InGa, Ag, Al. The resistive switching mechanism [67] was induced by the substitution of top electrode metal cation with PB-I octahedra cage in perovskite. Ag-based metal electrodes have mainly shown an interface (surface) type of switching, while other metal electrodes have also shown bulk CF formation switching as well. A detailed study of the influence of metal electrodes on device behavior and unipolar and bipolar switching can be found somewhere else [68,69].

4.5 Quality factors and their limitations

Memory storage devices have several quality factors which need to be taken into consideration while designing these devices. The resistance ratio of set and reset state indicates nondestructive reading from the device. The device having a low set/reset resistance ratio value does not allow a change in written information by the application of a reading voltage which is much smaller in magnitude than the writing voltage. This is also established through the current ratio of low-resistance state (LRS) to the high-resistance state (HRS) of the device. Devices having a high current ratio (smaller current in the HRS state of the device) are more efficient for nondestructive reading. The value of set and reset voltage and write current is related to the power consumption of the device. In order to employ *ReRAM* devices in complementary metal-oxide-semiconductor (CMOS) and other integrated circuits, the value of power consumption by the device should be minimal. Endurance of the *ReRAM* device indicates recyclability of the device, while retention time is concerned with the durability of stored data on the device even after the removal of external input. Retention time is indicated by the capability of the device to remain in set or reset conditions. After removal of the input signal, ionic species start to lose their mobility and devices try to retain their memorized state. Further, a smaller read voltage also ensures a long retention time because larger reading voltages can force unwanted resistive switching in the devices. *ReRAM* devices should also have high-temperature tolerance because the temperature can also cause resistive switching due to the nanoscale movement of defects inside the device. The writing speed of the device is measured through the pulse width of the input signal. Lower write speed indicates poor switching capability of the device from HRS to LRS and vice versa. Perovskite-based *ReRAM* devices are observed to have very high writing (smaller input pulse width) speed because of the nanoscale movement of the ionic species.

4.6 Progress in perovskite-based ReRAM devices

The most commonly used device structure for *ReRAM* fabrication is glass substrate usually coated with transparent conducting oxide, perovskite active layer, and active metal electrodes. 2-D layered perovskite structure [70] and 0-D quantum dot-based devices are also studied by various researchers [71]. The majority of the reported devices are found to be bipolar in nature [56]. Li et al. used a device stack of FTO/TiO_2/Al_2O_3/$MAPbI_3$/Au and studied the impact of the hysteresis curve using planar device configuration [72]. Charge trapping and de-trapping due to the localized states were observed to be the responsible mechanism behind the resistive switching. The retention time of 3.5 s and $V_{set,\ reset}$ at (−4, 4 V) with a bipolar resistance switching behavior was observed. *MSM* device structure Ag/$MAPbI_3$/Pt [47] reported by Choi et al. showed low set voltage (0.15 V) with a set/reset current ratio greater than 10^6 and retention time over 1000 s. Further, Choi et al. studied the impact of defect densities on the set voltage and hysteresis behavior of the device and observed that defect density mainly impacts the operating voltage of the device [73]. Heo et al. [74] studied the effect of crystal grain size from 60 to 600 nm of $MAPbI_3$-based thin film and obtained a storage capacity >0.1 TB in^{-2}, cycles endurance >600, data retention time > 104 s, $V_{set} \sim 0.7$ V, and $V_{reset} \sim 0.61$ V bias voltage. Park et al. investigated temperature dependence [75] of the various parameters of the *ReRAM* device and developed a thermal-assisted hopping model, and made an accurate determination of temperature dependence of the *I-V* response of the *ReRAM* device.

Xiaoning et al. [76] proposed a photo-assisted electroforming process in the device stack FTO/$MAPbI_3$/Al(Au) and obtained an ultralow-power consumption of the device having a magnitude of 0.12 mW. Light irradiation lowered the ion migration energy, and this leads to a significant reduction in the required poling power. Photoconductivity of the active semiconducting material also helped in overshooting of the current by CF formation, which facilitates the efficient transport of charge carriers. Guan et al. [77] proposed a novel artificial iconic memory having stacked perovskite multilayers in the configuration ITO/perovskite/Au/perovskite/Ag. This device was capable of sensing photo memory information and simultaneous detection of incident irradiation. The single device was capable of acting as a photodetector as well as a *ReRAM* device. $MAPbI_3$-based devices are fragile to small variations in external parameters of the device, so in the next section, we have discussed the issue of stability of *ReRAM* devices in little more detail.

4.7 Stability and low-dimensional perovskite-based devices

Metal halide perovskites are prone to degrade under exposure to irradiation, humidity, and temperature stress, and hence, the device's performance deteriorates with time under harsh external conditions. Various techniques have been adopted to make perovskite-based *ReRAM* devices stable, such as device encapsulation [78], surface passivation [79], bandgap engineering (halide mixing), and the use of all inorganic perovskite materials. Further stability of 3-D perovskite can also be enhanced by using the lower-dimensional structures such as mixed dimensional (3D/2D), 2-D, and 0-D perovskites. Hwang et al. [80] studied the impact of halide mixing on the switching behavior of the active layer of mixed halide perovskite ($MAPbI_{3-x}Br_x$). Authors have observed a reduction in V_{set} due to a decrease in the energy barrier to the motion of anion species with the increase in Br content of the active layer. Br-anion-based perovskites are observed to have a lower activation energy barrier for ionic migration than I-anion-based perovskites. Hence, by increasing the Br content in the active layer of perovskite, a smaller writing voltage and faster switching speed can be achieved. A mixed 2D/3D (*n*-butylammonium iodide on top of the 3-D $MAPbI_{3-x}Cl_x$) perovskite device deposited by Xia et al. [81] showed better stability than its 3-D counterpart. The device was also observed to have a smaller overall V_{set} (than its 3-D counterparts) and a significantly higher set/reset current ratio ($\sim 10^3$) was obtained. The improved performance in this configuration was attributed to the surface passivation effect of 2-D perovskite on top of the 3-D layer. A 2-D perovskite-based device fabricated by Seo et al. [70] with the active layer $BA_2MA_{N-1}PB_NI_{3N+1}$ (BA = butylammonium, MA = methylammonium) showed resistance switching at an electric field of magnitude $0.12 \times 10^6\,V\,m^{-1}$. Upon changing the dimensionality from 3-D to 2-D, the set/reset current ratio increased by a factor of 10^5, which also showed the high retention capability of 2-D perovskite-based *ReRAM* devices. The apparent increase in the current ratio was observed due to the enhancement in Schottky barrier height and reduction in the current flows during the high-resistance state of the device. Unipolar resistive switching with large grain decomposition into smaller during the CF formation was observed in the device stack of ITO/PEDOT: PSS/quasi-2D perovskite/Bphen (20 nm)/Ag proposed by

TABLE 2 Summary of best ReRAM device structures and their performance parameters.

Device structure	V_{set}/V_{reset} (V)	Current ratio	Retention (s)	Reference
Au/$MAPbI_{3-x}Cl_x$/FTO	1.47/−1.41	10^4	4×10^4	[84]
Ag/$CsPbI_3$/ITO glass	−0.1/0.8	10^2	10^3	[85]
ITO/PEDOT:PSS/$MAPbBr_3$/Al	−0.2/3	3.6×10^6	10^4	[86]
ITO/$MA_3Sb_2Br_9$/PMMA/Ag	−0.18/0.37	10^2	10^4	[87]
ITO/$Cs_2AgBiBr_6$/Au	0.65/−3.1	10^3	10^5	[88]
Ag/$CsPbBr_3$/Pt/PET	−0.98/0.70	10^3	10^5	[89]
ITO/PEDOT:PSS/$(PEA)_2(MA)_{n-1}PbnI_{3n+1}$/PCBM/Ag for $n=5$ (Ruddlesden-Popper)	0.2/−1	10^4	10^5	[90]

Yu et al. [82] Quantum dot-based *ReRAM* devices are also widely explored due to their unique properties. [44,71,83] Chen et al. [44] studied $CsPbBr_3$ based on perovskite quantum dots (0-D perovskite) and obtained a set/reset current ratio of 10^7 in a bipolar resistive switching device with a reading voltage −0.3 V, operation voltage as −2.4 V/1.55 V, and data retention of over 1000 s. The uses of low-dimensional perovskite structures not only enhance the material library for device fabrication but have also shown great promising device parameters. Table 2 summarizes the parameters of some of the best-performing perovskite-based *ReRAM* devices. Further efforts are still needed to overcome the stability and scalability challenges of perovskite-based *ReRAM* devices.

5. Piezoelectric energy harvesting using halide perovskites

The piezoelectric effect is primarily divided into two basic mechanisms; the direct piezoelectric effect and the converse piezoelectric effect. In direct piezoelectric, the materials produce an electric field under applied stress, and in the converse piezoelectric, the piezoelectric materials are deformed under an electric field. The electrical behavior of piezoelectric materials can be described by using Hooke's law as [91]

$$D = \varepsilon E \tag{7}$$

For defining a system,

$$S = s\sigma \tag{8}$$

where D is displacement, ε.

The piezoelectric effects are governed by the following equations [92]:

$$\begin{bmatrix} S \\ D \end{bmatrix} = \begin{bmatrix} s^E & d^t \\ d & \varepsilon^\sigma \end{bmatrix} \begin{bmatrix} \sigma \\ E \end{bmatrix} \tag{9}$$

where S, σ.

The piezoelectric coefficients are defined as e_{ij}

$$d_{ij} = \left(\frac{\partial D_i}{\partial \sigma_j}\right)^E$$

$$e_{ij} = \left(\frac{\partial D_i}{\partial S_j}\right)^E$$

$$g_{ij} = -\left(\frac{\partial E_i}{\partial \sigma_j}\right)^D$$

$$h_{ij} = -\left(\frac{\partial E_i}{\partial S_j}\right)^D$$

The first two equations are related to the direct piezoelectric effect and the last two equations are related to the indirect piezoelectric effect.

6. Importance of halide perovskite for piezoelectricity

Recently, perovskite materials have attracted attention toward piezoelectric energy harvesting applications because of their good piezoelectric, ferroelectric, and dielectric properties. The other important characteristics of halide perovskite are structural, electrical, and optical properties that can be tuned by changing the composition and synthesis roots for desired applications. When perovskite devices are subjected to mechanical stress, the charges within the active layer are induced and create a potential difference between the top and bottom electrodes, allowing electrons to flow. This phenomenon is achieved in halide perovskites due to ferroelectric polarization. Although ceramic perovskites are more promising candidates for piezoelectricity with high piezoelectric coefficients, however, due to high processing temperature and rigid nature, these materials cannot be used for wider applications. So, researchers have focused on developing good-performing and flexible energy harvesters through halide perovskites with device structure as shown in Fig. 5.

Halide perovskites have a cubic structure (Fig. 5), which leads to ferroelectric polarization within the piezoelectric material. This ferroelectric polarization is achieved by the halide perovskites due to structural phase transition that depends on the twisting of the octahedral lattice from the centrosymmetric crystal. So, to understand the stability and polarization

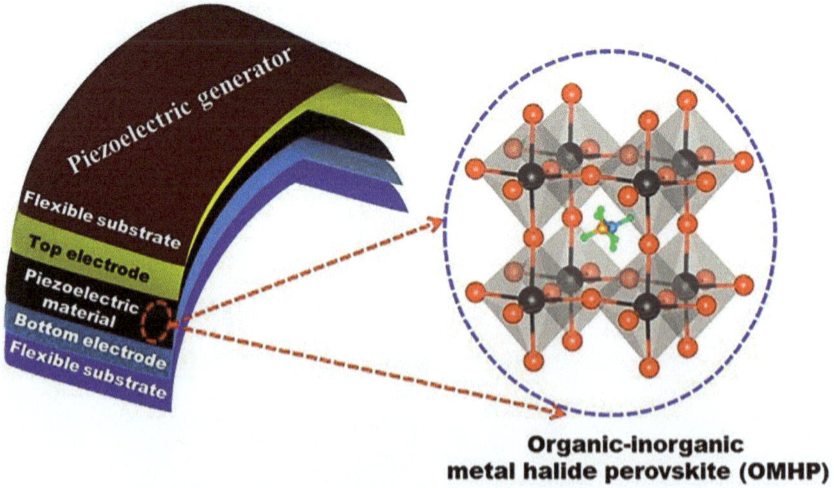

FIG. 5 Device structure of Halide perovskite-based energy harvester and crystal structure of halide perovskite [93]. *From V. Jella, S. Ippili, J.-H. Eom, S.V.N. Pammi, J.-S. Jung, V.-D. Tran, V.H. Nguyen, A. Kirakosyan, S. Yun, D. Kim, M.R. Sihn, J. Choi, Y.-J. Kim, H.-J. Kim, S.-G. Yoon, A comprehensive review of flexible piezoelectric generators based on organic-inorganic metal halide perovskites, Nano Energy 57 (2019) 22, https://doi.org/10.1016/j.nanoen.2018.12.038.*

in different orientations, i.e., in different crystallographic axis, it is necessary to know the phase transition and crystal structure of halide perovskites. The dielectric constants and polarization are proportional to the piezoelectric coefficient and that provides the piezoelectric behavior in perovskite materials. The direct piezoelectric effect for single crystal of $MAPbI_3$ was observed by Dong et al. [94] with piezoelectric coefficient in different crystallographic axis. Liu et al. [95] have investigated the comparative study of the piezoelectric coefficient of $MAPbI_3$ for polar and nonpolar configuration. Further, they observed an enhancement in the piezoelectric response under the light, likely due to a change in molecular configuration from nonpolar under an electric field. $MAPbI_3$-based piezoelectric energy harvesting device on flexible substrate produced high voltage and current density of 2.7 V and 140 nA cm^{-2} under an applied 0.5 Mpa mechanical stress on active area $1 \times 1\,cm^2$ and poled at 80 KV cm^{-1} [96]. Further, a partial replacement of Pb^{2+} with Fe^{2+} in $MAPbI_3$ perovskite films to from $MAPb_{1-x}Fe_xI_3$ ($0.01 \leq x \leq 0.50$) has been investigated. This mixed B-cations-based perovskite structure showed a significant improvement in piezoelectric performance ($x = 0.07$) with an output voltage of piezoelectric generator 7.29 V and output current is 0.88 μA cm^{-2}. [97] Furthermore, inorganic perovskite materials have also shown promising piezoelectric properties due to a change in bond angles ($CsPbBr_3$), which is associated with the structural elastic deformation of the halide perovskite material. The ionic off-centering of BX_6 octahedra in orthorhombic halide perovskite ABX_3 structure along with rotations of dipoles under an electric field possesses the nonlinear behavior of polarization and shows the piezoelectricity [98]. The current theoretical results have been reported to have high polarization value of 62.9 μC cm^{-2} due to the lattice elastic deformation under compressive stress for formamidinium lead iodide ($FAPbI_3$). For the halide perovskite $CsPbBr_3$, the piezoelectric coefficient has been reported $d_{33} \sim 40.3$ pm V^{-1} with an excellent output voltage of ~16.4 V with output current of ~604 nA (~260 nm thin active layer) [99]. These values are sufficient for the elastic deformation of $PbBr_6$ octahedra and responsible for piezoelectricity in all-inorganic perovskite. On the other hand, high electromechanical coupling performance can be used for sensing the specific body motion with an electric signal [100].

Organic-inorganic metal halide perovskites are brittle and nonflexible; hence, to realize the utilization of piezoelectric properties of organic-inorganic metal halide perovskites in flexible and wearable electronic devices, the composites based on flexible polymers and halide perovskite are considered to be a good combination. In this process, polymers having piezoelectric properties are used as matrix, and halide perovskite nanocrystals are embedded in the polymer matrix to form a halide perovskite composite. Piezoelectric polymer-halide perovskite composite can be produced by blending, growing nanocrystals of halide perovskites in the polymer matrix or in situ polymerization in the presence of halide perovskite nanocrystals. Ding et al. fabricated a piezoelectric nanogenerator by using composites of $FAPbBr_3$ halide perovskite and PDMS. Solution of $FAPbBr_3$-PDMS was spin-coated on an ITO-coated PET substrate and then sandwiched with Al foil. The piezoelectric nanogenerator recorded a maximum output voltage of 8.5 V and a current density of 3.8 μA cm^{-2} upon application of 0.5 MPa vertical compression [101]. Dhar et al. fabricated $MAPbI_3$-PDMS composite-based piezoelectric nanogenerator and observed that the defects in $MAPbI_3$ played a vital role in the piezoelectric effect. Using positron annihilation spectroscopy, they found that increasing defects, especially Pb^{2+}, enhanced the piezoelectric property of the material. The piezoelectric nanogenerator achieved a maximum piezovoltage of 108 V and a power density of 0.3 mW cm^{-3} [102]. Due to the high viscosity of PDMS, obtaining a homogeneous mixture with perovskite nanoparticles is not feasible, which reduces the performance of the piezoelectric generator. Alternatively, PVDF, which showed excellent piezoelectric properties, has less viscosity compared to PDMS. Ding et al. fabricated piezoelectric generators using $FAPbBr_3$-PVDF composite. The device showed an excellent output voltage of ~30 V and a current density of 6.2 μA cm^{-1} with a stress of 0.5 MPa owing to the homogeneous dispersion of $FAPbBr_3$ nanoparticles in the PVDF matrix [103]. Interestingly, a study done by Sultana et al. showed that $MAPbI_3$ enhanced the nucleation of the electroactive phase of PVDF. The combined effect of homogeneous dispersion and increased electroactive matrix phase improved the device performance [104]. In a nutshell, halide perovskite-based composite piezoelectric nanogenerator has the potential for various applications owing to the flexibility, low cost of production, feasibility in fabrication.

Recently, the combination of halide-based perovskite as piezoelectric generators has gained strong research interest. Ippili et al. synthesized environmentally safe lead-free methylammonium tin perovskite ($MASnI_3$) using an antisolvent collision technique, which ensures stability at room conditions within 24 h. At ~30°C, a phase transformation from a tetragonal structure to a cubic structure is observed for methylammonium tin perovskite. The films of polycrystalline $MASnI_3$ exhibit an increased dielectric constant of ~65 at 100 kHz, a less-leakage current density 7×10^{-7} A cm^{-2} with 50 kV cm^{-1}, an advanced P-E hysteretic loops, with a strong piezoelectrical coefficient (d_{33}) of 20.8 V^{-1} $MASnI_3$ showed good dielectric properties. Along with the above-mentioned results, $MASnI_3$ piezoelectric nanogenerator produced an output voltage of 3.4 V and an output current density of 0.35 μA cm^{-2}. A composite of PVDF-$MASnI_3$ was synthesized, which showed an output current density of ~4.0 μA cm^{-2} and an output voltage of ~12.0 V, which was further used to power a light-emitting diode (LED) without any need for storage [105]. Further, formamidinium tin iodide ($FASnI_3$) composite with a soft polymer exhibited an output current density of 35.05 mA cm^{-2} and an output voltage of ~23 V. The local piezoelectric coefficient (d_{33}) was measured to be 73 pm V^{-1} for these films. The study suggested that the factor responsible for increased piezoelectric response in $FASnI_3$ is soft polar optic phonons [106]. Song et al. studied that when films of polycrystalline $CH_3NH_3PbI_3$ were interfaced with ferroelectric $PbZr_{0.2}Ti_{0.8}O_3$, their piezoelectric coefficient (d_{33}) increased by more than 10 times. For $MAPbI_3$ deposited on indium tin oxide, Au and $SrTiO_3$, d_{33} values were 0.3–0.4 pm V^{-1} [107].

Researchers are still working for further improvement in energy harvesting through halide perovskite to understand the effect of illumination, pyroelectric effect, photo-induced piezoelectric energy harvesting, ferroelectric response, piezoelectric response, the effect of poling, phase boundaries, and electromechanical coupling of various halide perovskites. So far, it has been observed that piezoelectric energy harvesters are widely employed in various biomedical sensors, ignitors, and ultrasonic wave detection applications [108–111]. Among all the piezoelectric materials so far explored, halide perovskite has been extensively preferred material on the account of its excellent piezoelectric properties compared to that of the traditionally used piezoelectric materials such as lead zirconate titanate, barium titanate, zinc oxide, etc. [112–114], facile low-temperature fabrication, spontaneous polarization, feasible electrochemical conversion, and their biocompatible nature [93,115]. Besides the explicit performance of halide perovskite as a piezoelectric material, some inevitable challenges affect the performance of the piezoelectric property of halide perovskites. First and foremost is the mechanical stability issue faced by the halide perovskite-based piezoelectric materials when they are incorporated in flexible and portable devices for energy harvesting purposes. Further, hydration of perovskite materials in the presence of moisture or humid environment for a longer duration; it has been observed that the water molecules form strong hydrogen bonds with organic cations [116] such as MA^+, which subsequently leads to disturbance in the perovskite crystal structure. Apart from these, higher toxicity in the case of lead-based halide perovskites, temperature-dependent change in the crystal structure, and effect of the electrical poling process. In order to generate high-performance halide perovskite-based piezoelectric materials, it is imperative to focus on these challenges and advance research in this direction. First, it is necessary to explore the material characteristics of the halide perovskite comprehensively; for instance, to design efficient piezoelectric devices, one needs to focus on the effect of polarization, photo-induced phase transition, and morphotropic phase boundary on the piezoelectricity. It has been experimentally demonstrated in various ongoing research that under illumination, there is a structural transition that occurs from a nonpolar one to a polar one, which greatly influences the dielectric and piezoelectric responses of halide perovskite. Further, the structural tunability aspects of the halide perovskite material and its effect on the piezoelectric property could be taken into consideration; for instance, it was reported that the substitution of MA^+ with more polar cations such as $CF_3NH_3^+$, etc., effectively improves the piezoelectric response and enhances the chance of halide perovskite-based piezoelectric devices to get commercialized [95]. It was theoretically investigated that replacing Pb with other cations such as Sn, etc., further enhances the piezoelectric property of halide perovskite and henceforth paved a way for the development of environmentally compatible and sustainable materials, which could consequently use in medical or biomedical applications.

7. Conclusions

In summary, this book chapter reviewed the recent progress in the development of organic-inorganic metal halide perovskite for various applications. Section 2 focused on organic-inorganic halide perovskite's potential application as electrode materials in Li-ion batteries. Halide perovskites possess higher capacities compared to conventional electrode materials like graphite. All three mechanisms for Li-ion insertion, i.e., intercalation, conversion, and alloying, have been proposed in the case of halide perovskite electrodes. However, irreversible conversion and alloying reactions leading to capacity fading over cycles is a challenge to overcome. Interestingly, lower-dimensional halide perovskites show even higher capacity owing to the larger area of interaction with Li-ions. Halide perovskite is also under consideration for Li-S batteries. Besides, halide perovskite materials are used for supercapacitor applications, owing to their high surface area and remarkable ionic conductivity, which has proved quite beneficial for energy-efficient supercapacitors. Recently, photo-batteries or photo-supercapacitors are also implementing the halide perovskite, where a single perovskite material is used for both energy generation and storage purposes.

Section 3 discussed the use of halide perovskite semiconductors for gas sensing. We discussed how the involvement of OIHPs as sensing elements has increased the sensing abilities for the detection of moisture and various other gases. Sensing parameters (such as fast response and recovery time along with high sensitivity and operation at room temperature) are promising by utilizing this material as an active sensor. The important aspect is the interaction between the target gas and the perovskite platform can be easily tuned, which is the reason behind the excellent sensing results of these interesting materials.

Section 4 focused on the progress of *ReRAM* devices using perovskite semiconductors. Metal halide perovskites accommodate nanoscale movements of various defects; this makes the possibility of ultrahigh-speed resistance switching devices. Further, we have discussed various switching mechanisms for observing *ReRAM* characteristics, and based on switching behavior, *ReRAM* devices are classified into three different categories, namely, ECM, VCM, and TCM. We have discussed various quality factors, writing voltage, reading voltage, current ratio, switching speed, retention, and endurance capability of the *ReRAM* devices.

In Section 5, we discussed the emergence and importance of organic halide perovskites for piezoelectric energy harvesters. In this section, the basic understanding of the mechanism of electromechanical coupling, polarization mechanism, and phase transformation of OIHPs with temperature was mentioned. Furthermore, a comprehensive discussion of various electromechanical properties such as piezoelectric coefficient, strain, voltage, and current of piezoelectric devices based on OIHPs has been covered.

Acknowledgment

This work was supported by a grant from Science Engineering Research Board (SERB), India with grant No. SRG/2019/01303.

References

[1] Y. Fu, H. Zhu, J. Chen, M.P. Hautzinger, X.Y. Zhu, S. Jin, Metal halide perovskite nanostructures for optoelectronic applications and the study of physical properties, Nat. Rev. Mater. 4 (3) (2019) 169–188.

[2] C.R. Kalaiselvi, N. Muthukumarasamy, D. Velauthapillai, M. Kang, T.S. Senthil, Importance of halide perovskites for next-generation solar cells—a review, Mater. Lett. 219 (2018) 198–200.

[3] T. Singh, T. Miyasaka, Stabilizing the efficiency beyond 20% with a mixed cation perovskite solar cell fabricated in ambient air under controlled humidity, Adv. Energy Mater. 8 (3) (2018) 1700677.

[4] M. Winter, R.J. Brodd, What are batteries, fuel cells, and supercapacitors? Chem. Rev. 104 (10) (2004) 4245–4269.

[5] F. Sahli, J. Werner, B.A. Kamino, M. Brauninger, R. Monnard, B. Paviet-Salomon, L. Barraud, L. Ding, J.J. Diaz Leon, D. Sacchetto, G. Cattaneo, M. Despeisse, M. Boccard, S. Nicolay, Q. Jeangros, B. Niesen, C. Ballif, Fully textured monolithic perovskite/silicon tandem solar cells with 25.2% power conversion efficiency, Nat. Mater. 17 (9) (2018) 820–826.

[6] C. Eames, J.M. Frost, P.R. Barnes, B.C. O'Regan, A. Walsh, M.S. Islam, Ionic transport in hybrid lead iodide perovskite solar cells, Nat. Commun. 6 (2015) 7497.

[7] J.-W. Lee, S.-G. Kim, J.-M. Yang, Y. Yang, N.-G. Park, Verification and mitigation of ion migration in perovskite solar cells, APL Mater. 7 (4) (2019), 041111.

[8] J. Haruyama, K. Sodeyama, L. Han, Y. Tateyama, First-principles study of ion diffusion in perovskite solar cell sensitizers, J. Am. Chem. Soc. 137 (32) (2015) 10048–10051.

[9] S. Ahmad, P.K. Kanaujia, W. Niu, J.J. Baumberg, G. Vijaya Prakash, In situ intercalation dynamics in inorganic-organic layered perovskite thin films, ACS Appl. Mater. Interfaces 6 (13) (2014) 10238–10247.

[10] H.R. Xia, W.T. Sun, L.M. Peng, Hydrothermal synthesis of organometal halide perovskites for Li-ion batteries, Chem. Commun. (Camb.) 51 (72) (2015) 13787–13790.

[11] J.A. Dawson, A.J. Naylor, C. Eames, M. Roberts, W. Zhang, H.J. Snaith, P.G. Bruce, M.S. Islam, Mechanisms of lithium intercalation and conversion processes in organic–inorganic halide perovskites, ACS Energy Lett. 2 (8) (2017) 1818–1824.

[12] N. Vicente, D. Bresser, S. Passerini, G. Garcia-Belmonte, Probing the 3-step lithium storage mechanism in $CH_3NH_3PbBr_3$ perovskite electrode by operando-XRD analysis, ChemElectroChem 6 (2) (2019) 456–460.

[13] M. Tathavadekar, S. Krishnamurthy, A. Banerjee, S. Nagane, Y. Gawli, A. Suryawanshi, S. Bhat, D. Puthusseri, A.D. Mohite, S. Ogale, Low-dimensional hybrid perovskites as high performance anodes for alkali-ion batteries, J. Mater. Chem. A 5 (35) (2017) 18634–18642.

[14] P. Pandey, N. Sharma, R.A. Panchal, S.W. Gosavi, S. Ogale, Realization of high capacity and cycling stability in Pb-free A2 $CuBr_4$ (A = CH_3NH_3/Cs, 2D/3D) perovskite-based li-ion battery anodes, ChemSusChem 12 (16) (2019) 3742–3746.

[15] L.T. López, D. Ramírez, F. Jaramillo, J.A. Calderón, Novel hybrid organic-inorganic $CH_3NH_3NiCl_3$ active material for high-capacity and sustainable lithium-ion batteries, Electrochim. Acta 357 (2020), 136882.

[16] A. Kostopoulou, D. Vernardou, K. Savva, E. Stratakis, All-inorganic lead halide perovskite nanohexagons for high performance air-stable lithium batteries, Nanoscale 11 (3) (2019) 882–889.

[17] L. Zhang, B. Wu, J. Li, Understanding interactions between lead iodide perovskite surfaces and lithium polysulfide toward new-generation integrated solar-powered lithium battery: an ab initio investigation, J. Phys. Chem. C 123 (1) (2018) 82–90.

[18] A. Walsh, S.D. Strank, Taking control of ion transport in halide perovskite solar cells, ACS Energy Lett. 3 (8) (2018) 1983–1990.
[19] S. Zhou, L. Li, H. Yu, J. Chen, C.-P. Wong, N. Zhao, Thin film electrochemical capacitors based on organolead triiodide perovskite, Adv. Electron. Mater. 2 (7) (2016) 1600114.
[20] W. Zhang, G.E. Eperon, H.J. Snaith, Metal halide perovskites for energy applications, Nat. Energy 1 (6) (2016).
[21] I. Popoola, M. Gondal, L. Oloore, A. Popoola, J. AlGhamdi, Fabrication of organometallic halide perovskite electrochemical supercapacitors utilizing quasi-solid-state electrolytes for energy storage devices, Electrochim. Acta 332 (2020), 135536.
[22] A. Slonopas, H. Ryan, P. Norris, Ultrahigh energy density $CH_3NH_3PbI_3$ perovskite based supercapacitor with fast discharge, Electrochim. Acta 307 (2019) 334–340.
[23] S. Ahmad, C. George, D.J. Beesley, J.J. Baumberg, M. De Volder, Photo-rechargeable organo-halide perovskite batteries, Nano Lett. 18 (3) (2018) 1856–1862.
[24] Y. Wang, L. Feng, X. Miao, Z. Li, J. Wang, X. Deng, Multifunctional energy devices caused by ionic behaviors in perovskite-polymer hybrid films, Synth. Met. 250 (2019) 31–34.
[25] C.H. Ng, H.N. Lim, S. Hayase, Z. Zainal, S. Shafie, H.W. Lee, N.M. Huang, Cesium lead halide inorganic-based perovskite-sensitized solar cell for photo-supercapacitor application under high humidity condition, ACS Appl. Energy Mater. 1 (2) (2018) 692–699.
[26] R. Liu, C. Liu, S. Fan, A photocapacitor based on organometal halide perovskite and PANI/CNT composites integrated using a CNT bridge, J. Mater. Chem. A 5 (44) (2017) 23078–23084.
[27] J. Liang, G. Zhu, C. Wang, P. Zhao, Y. Wang, Y. Hu, L. Ma, Z. Tie, J. Liu, Z. Jin, An all-inorganic perovskite solar capacitor for efficient and stable spontaneous photocharging, Nano Energy 52 (2018) 239–245.
[28] F. Zhou, Z. Ren, Y. Zhao, X. Shen, A. Wang, Y.Y. Li, C. Surya, Y. Chai, Perovskite photovoltachromic supercapacitor with all-transparent electrodes, ACS Nano 10 (6) (2016) 5900–5908.
[29] E. Kymakis, A. Panagiotopoulos, M.M. Stylianakis, K. Petridis, Organometallic Hybrid Perovskites for Humidity and Gas Sensing Applications, Elsevier, 2020, pp. 131–147.
[30] M. Ahmadi, T. Wu, B. Hu, A review on organic-inorganic halide perovskite photodetectors: device engineering and fundamental physics, Adv. Mater. 29 (41) (2017).
[31] S. Priya, J. Halder, D. Mandal, A. Chowdhury, T. Singh, A. Chandra, Hierarchical SnO_2 nanostructures for potential VOC sensor, J. Mater. Sci. 56 (16) (2021) 9883–9893.
[32] A. Maity, A.K. Raychaudhuri, B. Ghosh, High sensitivity NH_3 gas sensor with electrical readout made on paper with perovskite halide as sensor material, Sci. Rep. 9 (1) (2019) 7777.
[33] W. Jiao, J. He, L. Zhang, Synthesis and high ammonia gas sensitivity of $(CH_3NH_3)PbBr_{3-x}I_x$ perovskite thin film at room temperature, Sens. Actuators B 309 (2020), 127786.
[34] S.-H. Kim, A. Kirakosyan, J. Choi, J.H. Kim, Detection of volatile organic compounds (VOCs), aliphatic amines, using highly fluorescent organic-inorganic hybrid perovskite nanoparticles, Dyes Pigments 147 (2017) 1–5.
[35] M. Li, J. Zhou, M.S. Molokeev, X. Jiang, Z. Lin, J. Zhao, Z. Xia, Lead-free hybrid metal halides with a green-emissive $[MnBr_4]$ unit as a selective turn-on fluorescent sensor for acetone, Inorg. Chem. 58 (19) (2019) 13464–13470.
[36] X. Zhang, Y. Sun, Y. Fan, Z. Liu, Z. Zeng, H. Zhao, X. Wang, J. Xu, Effects of organotin halide perovskite and Pt nanoparticles in SnO_2-based sensing materials on the detection of formaldehyde, J. Mater. Sci. Mater. Electron. 30 (23) (2019) 20624–20637.
[37] K.-Y. Lee, J.-C. Hsieh, C.-A. Chen, W.-L. Chen, H.-F. Meng, C.-J. Lu, S.-F. Horng, H.-W. Zan, Ultrasensitive detection of hydrogen sulfide gas based on perovskite vertical channel chemo-sensor, Sens. Actuators B 326 (2021), 128988.

[38] M.-Y. Zhu, L.-X. Zhang, J. Yin, J.-J. Chen, L.-J. Bie, B.D. Fahlman, Physisorption induced p-xylene gas-sensing performance of $(C_4H_9NH_3)_2PbI_4$ layered perovskite, Sens. Actuators B 282 (2019) 659–664.

[39] Y. Chen, X. Zhang, Z. Liu, Z. Zeng, H. Zhao, X. Wang, J. Xu, Light enhanced room temperature resistive NO_2 sensor based on a gold-loaded organic-inorganic hybrid perovskite incorporating tin dioxide, Mikrochim. Acta 186 (1) (2019) 47.

[40] G. Kakavelakis, E. Gagaoudakis, K. Petridis, V. Petromichelaki, V. Binas, G. Kiriakidis, E. Kymakis, Solution processed $CH_3NH_3PbI_{3-x}Cl_x$ perovskite based self-powered ozone sensing element operated at room temperature, ACS Sens. 3 (1) (2018) 135–142.

[41] L. Hu, G. Shao, T. Jiang, D. Li, X. Lv, H. Wang, X. Liu, H. Song, J. Tang, H. Liu, Investigation of the interaction between perovskite films with moisture via in situ electrical resistance measurement, ACS Appl. Mater. Interfaces 7 (45) (2015) 25113–25120.

[42] W. Xu, F. Li, Z. Cai, Y. Wang, F. Luo, X. Chen, An ultrasensitive and reversible fluorescence sensor of humidity using perovskite CH3NH3PbBr3, J. Mater. Chem. C 4 (41) (2016) 9651–9655.

[43] Q. Zhao, A. Hazarika, X. Chen, S.P. Harvey, B.W. Larson, G.R. Teeter, J. Liu, T. Song, C. Xiao, L. Shaw, M. Zhang, G. Li, M.C. Beard, J.M. Luther, High efficiency perovskite quantum dot solar cells with charge separating heterostructure, Nat. Commun. 10 (1) (2019) 2842.

[44] Z. Chen, Y. Zhang, Y. Yu, M. Cao, Y. Che, L. Jin, Y. Li, Q. Li, T. Li, H. Dai, J. Yang, J. Yao, Light assisted multilevel resistive switching memory devices based on all-inorganic perovskite quantum dots, Appl. Phys. Lett. 114 (18) (2019).

[45] C. Gu, J.S. Lee, Flexible hybrid organic-inorganic perovskite memory, ACS Nano 10 (5) (2016) 5413–5418.

[46] Y. Park, M.-K. Kim, J.-S. Lee, Emerging memory devices for artificial synapses, J. Mater. Chem. C 8 (27) (2020) 9163–9183.

[47] J. Choi, S. Park, J. Lee, K. Hong, D.H. Kim, C.W. Moon, G.D. Park, J. Suh, J. Hwang, S.Y. Kim, H.S. Jung, N.G. Park, S. Han, K.T. Nam, H.W. Jang, Organolead halide perovskites for low operating voltage multilevel resistive switching, Adv. Mater. 28 (31) (2016) 6562–6567.

[48] T.S. Sherkar, C. Momblona, L. Gil-Escrig, J. Ávila, M. Sessolo, H.J. Bolink, L.J.A. Koster, Recombination in perovskite solar cells: significance of grain boundaries, interface traps, and defect ions, ACS Energy Lett. 2 (5) (2017) 1214–1222.

[49] S. Prathapani, P. Bhargava, S. Mallick, Electronic band structure and carrier concentration of formamidinium–cesium mixed cation lead mixed halide hybrid perovskites, Appl. Phys. Lett. 112 (9) (2018), 092104, https://doi.org/10.1063/1.501682.

[50] M.A. Green, A. Ho-Baillie, H.J. Snaith, The emergence of perovskite solar cells, Nat. Photonics 8 (7) (2014) 506–514.

[51] T.W. Hickmott, Potential distribution and negative resistance in thin oxide films, J. Appl. Phys. 35 (9) (1964) 2679–2689.

[52] W. Banerjee, Q. Liu, H. Hwang, Engineering of defects in resistive random access memory devices, J. Appl. Phys. 127 (5) (2020).

[53] B. Hwang, J.-S. Lee, A strategy to design high-density nanoscale devices utilizing vapor deposition of metal halide perovskite materials, Adv. Mater. 29 (29) (2017) 1701048.

[54] X. Zhao, Z. Wang, W. Li, S. Sun, H. Xu, P. Zhou, J. Xu, Y. Lin, Y. Liu, Organic–inorganic perovskite memristors: photoassisted electroforming method for reliable low-power organic–inorganic perovskite memristors (Adv. Funct. Mater. 17/2020), Adv. Funct. Mater. 30 (17) (2020) 2070111.

[55] X. Xiao, J. Hu, S. Tang, K. Yan, B. Gao, H. Chen, D. Zou, Recent advances in halide perovskite memristors: materials, structures, mechanisms, and applications, Adv. Mater. Technol. 5 (6) (2020) 1900914.

[56] J.S. Lee, S. Lee, T.W. Noh, Resistive switching phenomena: a review of statistical physics approaches, Appl. Phys. Rev. 2 (3) (2015).
[57] D.B. Strukov, G.S. Snider, D.R. Stewart, R.S. Williams, The missing memristor found, Nature 453 (7191) (2008) 80–83.
[58] J. Suñé, N. Raghavan, K.L. Pey, Dielectric breakdown processes, in: D. Lelmini, R. Waser (Eds.), Resistive Switching, Wiley, 2016, pp. 225–252.
[59] Y. Sun, M. Tai, C. Song, Z. Wang, J. Yin, F. Li, H. Wu, F. Zeng, H. Lin, F. Pan, Competition between metallic and vacancy defect conductive filaments in a CH3NH3PbI3-based memory device, J. Phys. Chem. C 122 (11) (2018) 6431–6436.
[60] E.J. Yoo, M. Lyu, J.-H. Yun, C.J. Kang, Y.J. Choi, L. Wang, Resistive switching behavior in organic–inorganic hybrid $CH_3NH_3PbI_{3-x}Cl_x$ perovskite for resistive random access memory devices, Adv. Mater. 27 (40) (2015) 6170–6175.
[61] M. Patmiou, D. Niraula, V.G. Karpov, The Poole-Frenkel laws and a pathway to multi-valued memory, Appl. Phys. Lett. 115 (8) (2019), 083507.
[62] A. Gehring, S. Selberherr, Modeling of tunneling current and gate dielectric reliability for nonvolatile memory devices, IEEE Trans. Device Mater. Reliab. 4 (3) (2004) 306–319.
[63] W.-J. Yin, T. Shi, Y. Yan, Unusual defect physics in $CH_3NH_3PbI_3$ perovskite solar cell absorber, Appl. Phys. Lett. 104 (6) (2014), 063903.
[64] J.M. Ball, A. Petrozza, Defects in perovskite-halides and their effects in solar cells, Nat. Energy 1 (11) (2016) 16149.
[65] B. Chen, M. Yang, S. Priya, K. Zhu, Origin of J–V hysteresis in perovskite solar cells, J. Phys. Chem. Lett. 7 (5) (2016) 905–917.
[66] A. Beck, J.G. Bednorz, C. Gerber, C. Rossel, D. Widmer, Reproducible switching effect in thin oxide films for memory applications, Appl. Phys. Lett. 77 (1) (2000) 139–141.
[67] X. Wu, H. Yu, J. Cao, Unraveling the origin of resistive switching behavior in organolead halide perovskite based memory devices, AIP Adv. 10 (8) (2020), 085202.
[68] L. Goux, S. Spiga, Unipolar resistive-switching mechanisms, in: D. Lelmini, R. Waser (Eds.), Resistive Switching, Wiley, 2016, pp. 363–394.
[69] S. Menzel, J.-H. Hur, Modeling the VCM- and ECM-type switching kinetics, in: D. Ielmini, R. Waser (Eds.), Resistive Switching, Wiley, 2016, pp. 395–436.
[70] J.Y. Seo, J. Choi, H.S. Kim, J. Kim, J.M. Yang, C. Cuhadar, J.S. Han, S.J. Kim, D. Lee, H.W. Jang, N.G. Park, Wafer-scale reliable switching memory based on 2-dimensional layered organic-inorganic halide perovskite, Nanoscale 9 (40) (2017) 15278–15285.
[71] S. Ge, X. Guan, Y. Wang, C.H. Lin, Y. Cui, Y. Huang, X. Zhang, R. Zhang, X. Yang, T. Wu, Low-dimensional lead-free inorganic perovskites for resistive switching with ultralow bias, Adv. Funct. Mater. 30 (25) (2020).
[72] W. Li, H. Dong, G. Dong, L. Wang, Hystersis mechanism in perovskite photovoltaic devices and its potential application for multi-bit memory devices, Org. Electron. 26 (2015) 208–212.
[73] Z. Xu, Z. Liu, Y. Huang, G. Zheng, Q. Chen, H. Zhou, To probe the performance of perovskite memory devices: defects property and hysteresis, J. Mater. Chem. C 5 (23) (2017) 5810–5817.
[74] J.H. Heo, D.H. Shin, S.H. Moon, M.H. Lee, D.H. Kim, S.H. Oh, W. Jo, S.H. Im, Memory effect behavior with respect to the crystal grain size in the organic-inorganic hybrid perovskite nonvolatile resistive random access memory, Sci. Rep. 7 (1) (2017) 16586.
[75] N. Park, Y. Kwon, J. Choi, H.W. Jang, P.-R. Cha, Plausible carrier transport model in organic-inorganic hybrid perovskite resistive memory devices, AIP Adv. 8 (4) (2018).
[76] X. Zhao, Z. Wang, W. Li, S. Sun, H. Xu, P. Zhou, J. Xu, Y. Lin, Y. Liu, Photoassisted electroforming method for reliable low-power organic–inorganic perovskite memristors, Adv. Funct. Mater. 30 (17) (2020).

[77] X. Guan, Y. Wang, C.-H. Lin, L. Hu, S. Ge, T. Wan, A. Younis, F. Li, Y. Cui, D.-C. Qi, D. Chu, X.D. Chen, T. Wu, A monolithic artificial iconic memory based on highly stable perovskite-metal multilayers, Appl. Phys. Rev. 7 (3) (2020).

[78] F. Matteocci, L. Cinà, E. Lamanna, S. Cacovich, G. Divitini, P.A. Midgley, C. Ducati, A. Di Carlo, Encapsulation for long-term stability enhancement of perovskite solar cells, Nano Energy 30 (2016) 162–172.

[79] Q. Jiang, Y. Zhao, X. Zhang, X. Yang, Y. Chen, Z. Chu, Q. Ye, X. Li, Z. Yin, J. You, Surface passivation of perovskite film for efficient solar cells, Nat. Photonics 13 (7) (2019) 460–466.

[80] B. Hwang, C. Gu, D. Lee, J.S. Lee, Effect of halide-mixing on the switching behaviors of organic-inorganic hybrid perovskite memory, Sci. Rep. 7 (2017) 43794.

[81] F. Xia, Y. Xu, B. Li, W. Hui, S. Zhang, L. Zhu, Y. Xia, Y. Chen, W. Huang, Improved performance of $CH_3NH_3PbI_{3-x}Cl_x$ resistive switching memory by assembling 2D/3D perovskite heterostructures, ACS Appl. Mater. Interfaces 12 (13) (2020) 15439–15445.

[82] X. Yu, T. Shen, C. Zhu, Q. Zeng, A. Yu, S. Liu, R. Yi, Z. Weng, Y. Zhan, X. Hou, J. Qin, Memory devices via unipolar resistive switching in symmetric organic–inorganic perovskite nanoscale heterolayers, ACS Appl. Nano Mater. 3 (12) (2020) 11889–11896.

[83] A. Younis, L. Hu, P. Sharma, C.H. Lin, Y. Mi, X. Guan, D. Zhang, Y. Wang, T. He, X. Liu, B. Shabbir, S. Huang, J. Seidel, T. Wu, Enhancing resistive switching performance and ambient stability of hybrid perovskite single crystals via embedding colloidal quantum dots, Adv. Funct. Mater. 30 (31) (2020).

[84] F. Zhou, Y. Liu, X. Shen, M. Wang, F. Yuan, Y. Chai, Low-voltage, optoelectronic $CH_3NH_3PbI_{3-x}Cl_x$ memory with integrated sensing and logic operations, Adv. Funct. Mater. 28 (15) (2018) 1800080.

[85] S. Ge, Y. Huang, X. Chen, X. Zhang, Z. Xiang, R. Zhang, W. Li, Y. Cui, Silver iodide induced resistive switching in $CsPbI_3$ perovskite-based memory device, Adv. Mater. Interfaces 6 (7) (2019) 1802071.

[86] Y.H. Lee, D.H. Kim, C. Wu, T.W. Kim, Memristive devices with a large memory margin based on nanocrystalline organic-inorganic hybrid CH3NH3PbBr3 perovskite active layer, Org. Electron. 62 (2018) 412–418.

[87] J.-M. Yang, E.-S. Choi, S.-Y. Kim, J.-H. Kim, J.-H. Park, N.-G. Park, Perovskite-related $(CH_3NH_3)_3Sb_2Br_9$ for forming-free memristor and low-energy-consuming neuromorphic computing, Nanoscale 11 (13) (2019) 6453–6461.

[88] X.F. Cheng, W.H. Qian, J. Wang, C. Yu, J.H. He, H. Li, Q.F. Xu, D.Y. Chen, N.J. Li, J.M. Lu, Environmentally robust memristor enabled by lead-free double perovskite for high-performance information storage, Small 15 (49) (2019), e1905731.

[89] P. Cheng, Y. Zhu, J. Shi, H. Wang, Y. Liu, R. Xiong, J. Wei, H. Ma, M. Yin, One-step solution deposited all-inorganic perovskite $CsPbBr_3$ film for flexible resistive switching memories, Appl. Phys. Lett. 115 (22) (2019), 223505.

[90] A. Solanki, A. Guerrero, Q. Zhang, J. Bisquert, T.C. Sum, Interfacial mechanism for efficient resistive switching in Ruddlesden–Popper perovskites for non-volatile memories, J. Phys. Chem. Lett. 11 (2) (2020) 463–470.

[91] H.H. Law, P.L. Rossiter, G.P. Simon, L.L. Koss, Characterization of mechanical vibration damping by piezoelectric materials, J. Sound Vib. 197 (4) (1996) 489–513.

[92] I. Chopra, Review of state of art of smart structures and integrated systems, AIAA J. 40 (11) (2002) 2145–2187.

[93] V. Jella, S. Ippili, J.-H. Eom, S.V.N. Pammi, J.-S. Jung, V.-D. Tran, V.H. Nguyen, A. Kirakosyan, S. Yun, D. Kim, M.R. Sihn, J. Choi, Y.-J. Kim, H.-J. Kim, S.-G. Yoon, A comprehensive review of flexible piezoelectric generators based on organic-inorganic metal halide perovskites, Nano Energy 57 (2019) 74–93.

[94] Q. Dong, J. Song, Y. Fang, Y. Shao, S. Ducharme, J. Huang, Lateral-structure single-crystal hybrid perovskite solar cells via piezoelectric poling, Adv. Mater. 28 (14) (2016) 2816–2821.
[95] S. Liu, F. Zheng, I. Grinberg, A.M. Rappe, Photoferroelectric and photopiezoelectric properties of organometal halide perovskites, J. Phys. Chem. Lett. 7 (8) (2016) 1460–1465.
[96] Y.-J. Kim, T.-V. Dang, H.-J. Choi, B.-J. Park, J.-H. Eom, H.-A. Song, D. Seol, Y. Kim, S.-H. Shin, J. Nah, S.-G. Yoon, Piezoelectric properties of $CH_3NH_3PbI_3$ perovskite thin films and their applications in piezoelectric generators, J. Mater. Chem. A 4 (3) (2016) 756–763.
[97] S. Ippili, V. Jella, J. Kim, S. Hong, S.-G. Yoon, Enhanced piezoelectric output performance via control of dielectrics in Fe2+−incorporated MAPbI3 perovskite thin films: flexible piezoelectric generators, Nano Energy 49 (2018) 247–256.
[98] M. Coll, A. Gomez, E. Mas-Marza, O. Almora, G. Garcia-Belmonte, M. Campoy-Quiles, J. Bisquert, Polarization switching and light-enhanced piezoelectricity in lead halide perovskites, J. Phys. Chem. Lett. 6 (8) (2015) 1408–1413.
[99] D.B. Kim, K.H. Park, Y.S. Cho, Origin of high piezoelectricity of inorganic halide perovskite thin films and their electromechanical energy-harvesting and physiological current-sensing characteristics, Energ. Environ. Sci. 13 (7) (2020) 2077–2086.
[100] S. Shahrokhi, W. Gao, Y. Wang, P.R. Anandan, M.Z. Rahaman, S. Singh, D. Wang, C. Cazorla, G. Yuan, J.-M. Liu, T. Wu, Emergence of ferroelectricity in halide perovskites, Small Methods 4 (8) (2020) 2000149.
[101] R. Ding, H. Liu, X. Zhang, J. Xiao, R. Kishor, H. Sun, B. Zhu, G. Chen, F. Gao, X. Feng, J. Chen, X. Chen, X. Sun, Y. Zheng, Flexible piezoelectric nanocomposite generators based on formamidinium lead halide perovskite nanoparticles, Adv. Funct. Mater. 26 (42) (2016) 7708–7716.
[102] J. Dhar, S. Sil, N.A. Hoque, A. Dey, S. Das, P.P. Ray, D. Sanyal, Lattice-defect-induced piezo response in methylammonium-lead-iodide perovskite based nanogenerator, ChemistrySelect 3 (19) (2018) 5304–5312.
[103] R. Ding, X. Zhang, G. Chen, H. Wang, R. Kishor, J. Xiao, F. Gao, K. Zeng, X. Chen, X.W. Sun, Y. Zheng, High-performance piezoelectric nanogenerators composed of formamidinium lead halide perovskite nanoparticles and poly(vinylidene fluoride), Nano Energy 37 (2017) 126–135.
[104] A. Sultana, P. Sadhukhan, M.M. Alam, S. Das, T.R. Middya, D. Mandal, Organo-Lead halide perovskite induced electroactive β-phase in porous PVDF films: an excellent material for photoactive piezoelectric energy harvester and photodetector, ACS Appl. Mater. Interfaces 10 (4) (2018) 4121–4130.
[105] S. Ippili, V. Jella, J.-H. Eom, J. Kim, S. Hong, J.-S. Choi, V.-D. Tran, N. Van Hieu, Y.-J. Kim, H.-J. Kim, S.-G. Yoon, An eco-friendly flexible piezoelectric energy harvester that delivers high output performance is based on lead-free MASnI3 films and MASnI3-PVDF composite films, Nano Energy 57 (2019) 911–923.
[106] R. Pandey, G. Sb, S. Grover, S.K. Singh, A. Kadam, S. Ogale, U.V. Waghmare, V.R. Rao, D. Kabra, Microscopic origin of piezoelectricity in lead-free halide perovskite: application in nanogenerator design, ACS Energy Lett. 4 (5) (2019) 1004–1011.
[107] J. Song, Z. Xiao, B. Chen, S. Prockish, X. Chen, A. Rajapitamahuni, L. Zhang, J. Huang, X. Hong, Enhanced piezoelectric response in hybrid lead halide perovskite thin films via interfacing with ferroelectric PbZr0.2Ti0.8O3, ACS Appl. Mater. Interfaces 10 (22) (2018) 19218–19225.
[108] C.-H. Hong, H.-S. Han, J.-S. Lee, K. Wang, F.-Z. Yao, J.-F. Li, J.-H. Gwon, N.V. Quyet, J.-K. Jung, W. Jo, Ring-type rotary ultrasonic motor using lead-free ceramics, J. Sens. Sci. Technol. 24 (4) (2015) 228–231.

[109] J.H. Lee, K. Heo, K. Schulz-Schonhagen, J.H. Lee, M.S. Desai, H.E. Jin, S.W. Lee, Diphenylalanine peptide nanotube energy harvesters, ACS Nano 12 (8) (2018) 8138–8144.
[110] H. Park, C. Ha, J.-H. Lee, Advances in piezoelectric halide perovskites for energy harvesting applications, J. Mater. Chem. A 8 (46) (2020) 24353–24367.
[111] L. Persano, C. Dagdeviren, Y. Su, Y. Zhang, S. Girardo, D. Pisignano, Y. Huang, J.A. Rogers, High performance piezoelectric devices based on aligned arrays of nanofibers of poly(vinylidenefluoride-co-trifluoroethylene), Nat. Commun. 4 (2013) 1633.
[112] M. El Achaby, F.Z. Arrakhiz, S. Vaudreuil, E.M. Essassi, A. Qaiss, Piezoelectric β-polymorph formation and properties enhancement in graphene oxide—PVDF nanocomposite films, Appl. Surf. Sci. 258 (19) (2012) 7668–7677.
[113] K.I. Park, J.H. Son, G.T. Hwang, C.K. Jeong, J. Ryu, M. Koo, I. Choi, S.H. Lee, M. Byun, Z.L. Wang, K.J. Lee, Highly-efficient, flexible piezoelectric PZT thin film nanogenerator on plastic substrates, Adv. Mater. 26 (16) (2014) 2514–2520.
[114] K. Verma, D.K. Bharti, S. Badatya, A.K. Srivastava, M.K. Gupta, A high performance flexible two dimensional vertically aligned ZnO nanodisc based piezoelectric nanogenerator via surface passivation, Nanoscale Adv. 2 (5) (2020) 2044–2051.
[115] T. Singh, T. Miyasaka, High performance perovskite solar cell via multi-cycle low temperature processing of lead acetate precursor solutions, Chem. Commun. (Camb.) 52 (26) (2016) 4784–4787.
[116] J.A. Christians, P.A. Miranda Herrera, P.V. Kamat, Transformation of the excited state and photovoltaic efficiency of $CH_3NH_3PbI_3$ perovskite upon controlled exposure to humidified air, J. Am. Chem. Soc. 137 (4) (2015) 1530–1538.

CHAPTER

10

Halide perovskite for photodetector applications

C. Rajkumar[a], P. Vengatesh[b], T.S. Shyju[b], A. Arulraj[c,d], and R.V. Mangalaraja[d]

[a]Center for High Energy Systems and Sciences (CHESS), DRDO, Hyderabad, India, [b]Centre for Nanoscience and Nanotechnology, Sathyabama Institute of Science and Technology (Deemed to be University), Chennai, India, [c]Instituto de Investigaciones Científicas y Tecnológicas (IDICTEC), Universidad de Atacama, Copiapó, Chile, [d]Faculty of Engineering & Sciences, Universidad Adolfo Ibáñez, Peñalolén, Santiago, Chile

1. Introduction

Nanomaterials with tailored structures hold great promise to be utilized as active layers in various applications, including memory devices, photodetectors, and sensors, due to their high surface-to-volume ratio and more surface-active sites. Nanomaterials such as nanoporous materials, nanotubes, nanoparticles, nanocrystals, and so forth have been rendered as active material in a wide range of applications [1–3]. These materials could be susceptible to specific variations in the chemical composition and hence could be actively engaged in sensory behavior. Various nanostructured materials have been widely used to detect harmful pollutants in the past few decades. However, the factors such as operating temperature, low-cost fabrication processes, and detection limit tend to find alternative nanostructured materials that can accomplish the above factors. The response (recovery) time of the sensors is defined as the time interval in which the variation in the desired properties reaches 90% of the total change upon introducing (evacuating) the target gases. Among the

Low-Dimensional Halide Perovskites
https://doi.org/10.1016/B978-0-323-88522-5.00003-X

reported materials, perovskite nanostructured materials are of great concern owing to their potential sensory behavior in addition to their astonishing optoelectronic behavior [3].

Perovskites are a class of semiconducting nanomaterials consisting of noncentrosymmetric unit cells that develop permanent electric polarization by ion displacements. For many years, the inorganic perovskite oxides (ABO_3) were being utilized as active material in gas sensors, where the sensing mechanism depends on the change in the electrical conductivity when it gets contact with the toxic gases in the environment. Moreover, the presence of two cations (A and B) in different sizes allows the materials to be prone to compositional engineering, which helps control the properties of materials such as transport and catalysis [4], since most of the chapters provide detailed information on the properties, synthesis, and applications (energy conversion, energy storage, etc.) of halide perovskites. This chapter focuses directly on the other halide perovskite applications, including memory devices, photodetectors, and various sensors.

2. Photodetectors

Photodetectors are devices that detect and measure the light properties via the photoelectric effect, which is usually expressed as photocurrent [5–7]. Nowadays, they get more interest in a broad range of applications like chemical/biological detection, optical communication, environmental monitoring, and image sensing. During photoexcitation, a semiconductor material is necessary for absorbing incident photons as well as generating electron-hole pairs. An applied electric field or built-in electric field is also essential for separating electrons and holes for producing an electric current. The inorganic semiconductors like InGaAs, Si, and GaN are used to prepare predominant photodetectors, wherein InGaAs, Si, and GaN are employed to detect near-infrared (0.9–1.7 μm), visible (0.45–0.8 μm), and ultraviolet (0.25–0.4 μm) sub-bands, respectively. Although photodetector technologies and processes have become reliable as well as mature, they have mechanical inflexibility and complex as well as expensive manufacturing. In addition, they required high driving voltage, which limited their versatility, spreading as well as compatibility for new technologies and broad applications. In the past few years, low-cost, simple, large-area, and flexible photodetectors have been prepared using solution-processable optoelectronic materials, viz., nanocomposites, nanomaterials, and organic materials. However, they have poor charge carrier mobility that hinders further improvement in their photodetection performance. In recent years, halide perovskites have gotten more interest in developing high-performance photodetectors due to their advantages, such as facile solution processability, effective light absorption, and high charge carrier mobility. As per the latest reports, the usage of perovskites greatly enhances the detectivity, responsivity, linear dynamic range, response speed, and noise equivalent power of photodetectors. Based on their device architectures, photodetectors are classified into phototransistors, photodiodes, and photoconductors, wherein photodiodes and photoconductors are two-terminal devices whereas phototransistors are three-terminal devices [7,8]. Perovskite photodetectors can be categorized into two major types based on the spatial layout of electrodes and photoactive medium: the lateral and vertical types. The photodiode configuration is very similar to that of photovoltaic, i.e., electrons and holes are generated and separated within a photosensitive layer during photoexcitation; then the generated charges are transported toward opposite electrodes through a built-in electric field. This section explains the photoconductor and its performance, which is prepared using versatile halide perovskites (inorganic compositions and organic-inorganic hybrid). In the case of photoconductors, the photosensitivity layer behaves as a channel between two lateral metal electrodes, i.e., charge carriers are created and separated within channels during photoexcitation, and then the generated charge carriers are collected by electrodes at bias voltage.

The photoconductor is very simple to manufacture, and it does not require transparent electrodes for fabrication. Due to this, it has great benefits in choosing materials for flexible optoelectronics. Due to the recycling of photoexcited electrons (holes), the amplification of photocurrent in the photoconductor makes it possible to acquire high responsivity R_λ and external quantum efficiency (EQE). However, the EQE and large gain come at the cost of response time because the lifetime of trapped carriers helps to determine both response time and gain time. Hence, the trade-off between response time and gain time is required to achieve the best device performance. As shown in Fig. 1A, perovskite photoconductors can be both vertical and lateral devices in which lateral devices generally use perovskite film as well as electrode spacing ($>5\,\mu m$) which is wider as compared to photodiodes ($1\,\mu m$). In addition, devices of lateral structure achieve flexibility very easily due to simple configuration. In vertical structure devices, a high gain can be achieved with low driving voltage owing to smaller electrode spacing [12]. Moreover, active layers are perovskite single crystals that provide low dark current due to fewer defects and high detectivity [12,13].

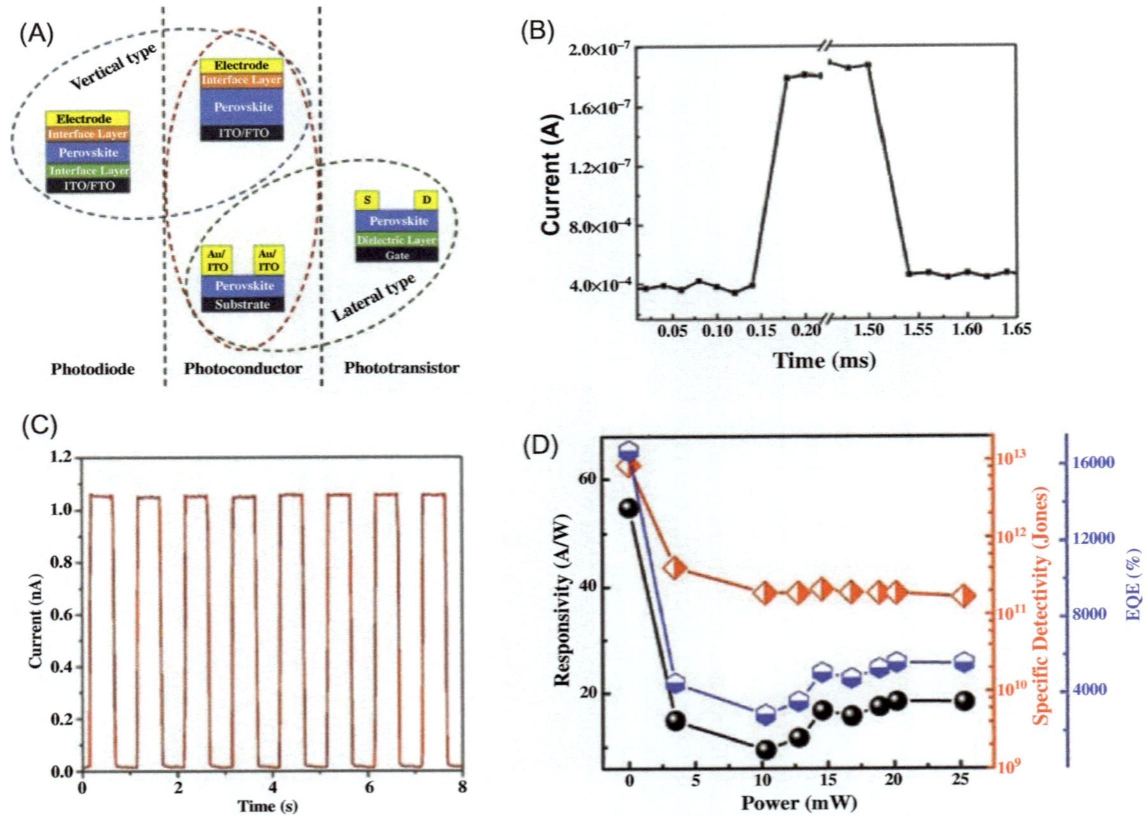

FIG. 1 (A) Categories of configurations of perovskite photodetector. (B) Time response of device under pulsed light irradiation at frequencies of 800 Hz [9]. (C) Time response of photodetector under pulsed laser (450 nm, 1 Hz) with 1 V bias ($P = 13.0\,mW/cm^2$) [10]. (D) Photoresponsivity, specific detectivity, and EQE of photodetectors versus illuminated light irradiance [11].

2.1 Vertical photoconductors

The photocurrent was amplified strongly in reverse bias by introducing a porous TiO_2 blocking layer (BL) in perovskite solar cells, which permits the cell to work as a high gain photodetector at low voltage [7]. Moehl et al. [14] used two types of TiO_2 (BL), one made by spin coating for forming a porous structure and the other by spray pyrolysis with a compact morphology. The device using a sample with a porous BL shows better dark and photocurrent behavior than different types of devices in reverse bias. The hole injection is facilitated into the valence band (VB) of the perovskite under reverse bias because the porous BL is placed with perovskite in direct contact with FTO, similarly, in the case without any BL as well. In contrast, because the VB of perovskite was not accessible from FTO without direct contact, the hole transfer was hindered by a compact sprayed TiO_2 BL under reverse bias. The amplification of the measured photocurrent signal showed six times gain under reverse bias at 0.9V. Such photocurrent amplification indicates that high-gain photodetectors were prepared at low voltage under reverse bias. Chen et al. employed this technique to prepare a planar perovskite device (FTO/TiO_2/$CH_3NH_3PbI_{3-x}Cl_x$/Spiro-MeOTAD/Au) wherein the compact TiO_2 BL played an important role in regulating the performance [15]. Similar to Moehl's work, gas-phase spray pyrolysis was used to prepare high-density CL, whereas spin-coating method was used to prepare less compact CL. The EQE action spectrum of the giant current was found in a planar device with spin-coated CL at a bias voltage of −0.9V along with J_{SC} EQE spectrum at a bias voltage of zero. It is found that the high current EQE value is reached 1700% or more while J_{SC} EQE is reached around 80%. The amplified photocurrent was explained as photoconductive current of $CH_3NH_3PbI_{3-x}Cl_x$ (substantially $CH_3NH_3PbI_3$), whose amplitude was intensely regulated by TiO_2 CL in reverse bias based on the results.

2.2 Lateral photoconductors

Lateral photoconductors are different from vertical photoconductors because lateral photoconductors require uniformity and surface coverage of the active area. By constructing microelectrodes, the lateral-structure devices (metal-semiconductor-metal and MSM architecture) can be used perovskite 2D nanosheets and 1D nanowires. Lateral geometry $CH_3NH_3PbI_3$ film photodetector with a perovskite-filled channel between a pair of conductive contacts was first demonstrated by Hu et al. [16]. This device detects the range of ultraviolet to visible wavelength. At 3V bias voltage, it shows photoresponsivity of 3.49 AW^{-1} at 365nm and 0.0367 AW^{-1} at 780nm. Moreover, after bending several times, it also shows excellent robustness and flexibility with no obvious variation in photocurrent. The authors indicated that the modified Schottky barrier at the ITO/perovskite interface is used to increase the electron injection, increasing the observed photocurrent. Wang et al. reported that a UV photodetector prepared using $CH_3NH_3PbCl_3$ film based on the MSM configuration shows high photoresponsivity of 7.56 A W^{-1} at 360 nm wavelength at 4 V [17]. Guo et al. used fluorous polymer as a light protection layer using spin-coating for preparing air-stable perovskite $CH_3NH_3PbI_{3-x}Cl_x$ photodetector [18]. It shows better photoresponsivity of 14.5 A W^{-1} at white light and 7.85 A W^{-1} at solar-blind UV light ($\lambda = 254$ nm). It indicates that the fluorous polymer coating increased the lifetime of the device up to 100 days. Zhou et al. prepared photodetectors using 2D perovskite materials with different layer numbers: $(C_4H_9NH_3)_2PbI_4$ (one layer), $(C_4H_9NH_3)_2(CH_3NH_3)Pb_2I_7$ (two layers), and $(C_4H_9NH_3)_2(CH_3NH_3)_2Pb_3I_{10}$ (three layers) [19]. These photodetectors showed highly reproducible on-and-off switching properties. The photodetector prepared using three-layered perovskites showed better performance than others, which may be due to more condensed and compact microstructure as well as smaller optical bandgap. These kinds of layer-structured perovskite materials can be new candidates for preparing functional, flexible, and tunable semiconductors in optoelectronics applications including photodetectors [7].

2.3 Factors influencing the performances of photodetectors

The output performance of photodetectors is quantitatively evaluated via numerous figures of merit, which are mentioned below [8].

The photon-electron conversion efficiency is indicated using external quantum efficiency (EQE), which is calculated as follows:

$$\mathrm{EQE} = \frac{\mathrm{R}h\upsilon}{e} \tag{1}$$

where e is the elementary electron charge, and $h\upsilon$ is the energy of the incident photon.

The photoresponsivity (R) of photodetectors is closely associated with the quantum yield of photodetectors and is described as follows:

$$R = \frac{I_{\mathrm{light}} - I_{\mathrm{dark}}}{P_{\mathrm{in}}A} \tag{2}$$

where A, P_{in}, I_{dark}, and I_{light}, are active area, incident light intensity, dark current, and photocurrent, respectively.

The detectivity of photodetectors is described as follows:

$$D^* = \frac{\sqrt{A\Delta f}}{\mathrm{NEP}} \tag{3}$$

where NEP is the noise equivalent power and Δf is the detection bandwidth. This expression can be simplified as below when shot noise from I_{dark} is a key component of overall noise of photodetector:

$$D^* = \frac{\sqrt{A\Delta f}}{\sqrt{2eI_{dark}}} \tag{4}$$

The photosensitivity of photodetector is calculated from I_{light}/I_{dark} (or on/off ratio) at bias voltage, wavelength, and fixed incident power conditions.

The response time of photodetectors toward an incident light is provided by temporal response time and is stated as rise (τ_{rise}) as well as decay (τ_{decay}) times. These parameters are closely associated with recombination processes and charge trapping/detrapping.

The photoconductive gain (G) is used to indicate the number of charge carriers traveling via external circuit per incident photon and can be expressed as follows:

$$G = \frac{\tau_1}{\tau_t} = \frac{\tau_1}{d^2/\mu V} \tag{5}$$

where V is bias voltage, μ is the mobility of charge carrier, d is device thickness, τ_t is charge carrier transit time of electrons, and τ_1 is charge lifetime of holes.

The photosensitivity linearity or linear dynamic range provides the linear relationship between P_{in} and I_{light} in a particular range. Moreover, it is stated in logarithmic scale as follows:

$$\text{linear dynamic range} = 20 \log \frac{J^*_{ph}}{J_d} \tag{6}$$

where J_d is the dark current density and J^*_{ph} is the photocurrent density which is measured at $1\,\mathrm{mW\,cm^{-2}}$ light illumination.

Recently, researchers paid much attention to hybrid (organic/inorganic) perovskites owing to their superb properties. In layered perovskite framework, the metal halide octahedral cages offer distinct advantages such as long charge carrier lifetime, high mobility, ultrafast charge generation, and diffusion length; meanwhile, several useful properties such as structural diversity and capacity of solution processing are offered by organic components [20]. During low-cost solution-based process, only shallow-level defects are created by organometallic halide perovskites intrinsic defects, even with defects inevitably formed. Gap states are not found within the bandgap [21,22]. Therefore, the response speed of devices is increased because inorganic semiconductors persistent photoconductivity could be prevented by organometallic halide perovskites. The photodetector shows high responsivity and fast response speed when hybrid perovskite materials act as a functional layer. Still, the use of electron transport layer (ETL) and hole transport layer (HTL) increases the cost of fabrication and four-time spin-coating processes, increasing the difficulties during fabrication in their work. Fabrication of photoconductive devices is a very simple and efficient way because they need only one cycle of spin-coating, metal electrodes, and perovskite materials [23]. Wang et al. have used a simple and effective method for applying hybrid perovskite materials onto the IDT (interdigitated)-patterned Au electrodes to fabricate photoconductive photodetectors with symmetrical structure. The electrodes with the perovskite layer form ohmic contacts. Moreover, the device current increases quickly when the light is switched on. In addition, the device has a fast response speed which is less than 20 μs (Fig. 1B [9].

3. Inorganic halide perovskite photoconductor

All inorganic halide perovskite $CsPbBr_3$ has high thermal and air stabilities as well as high mobility of $2000\,cm^2\,V^{-1}\,s^{-1}$ as compared to its counterpart $MAPbBr_3$, which raises the researchers' interest in optoelectrical devices with high stability and high performance [24,25]. 2D $CsPbBr_3$ nanosheets (NS) photodetector has excellent photosensitivity and higher on/off ratio than 10^3, excellent electronic transport properties, outstanding flexibility (>10,000 cycles), and high stability with less than 2.6% fluctuation after 12 h irradiation [26]. Furthermore, ultrathin 2D $CsPbBr_3$ NS film photodetector possesses high carrier mobility of $77.9\,cm^2\,V^{-1}\,s^{-1}$ and high absorption cross-section, which enhance the responsivity up to $0.53\,A\,W^{-1}$. In addition, it also has stability for up to 3 months, which increases the performance of the perovskite-based device [27]. Lv et al. also fabricated $CsPbBr_3$ NS film photodetector, which also has a high on/off ratio and fast response time (Fig. 1C) [10]. Li et al. fabricated CsPbBr3 thin film-based photodetectors using the one-step spin coating method. This photodetector has EQE of 16,700%, high R of $55\,A\,W^{-1}$, specific detectivity of up to 0.9×10^{13} Jones, fast response speed of 430/318 μs, and on/off photocurrent ratio of 1.06×10^5. Such performances are better as compared to most of the perovskite photodetectors prepared from $CsPbBr_3$ nanocrystals. Moreover, such performances are better than the highest results reported on organic-inorganic perovskite photodetectors [28]. To meet the need for ultrasensitive photodetection, ultrahigh responsivity is pursued. For this, perovskite/conducting layer heterostructure is constructed, where perovskite layer absorbs light efficiently while one type of photogenerated carrier is transported to conducting layer and high gain is achieved from one carrier which recirculates several times within opposite charges lifetimes. In spite of using transition metal dichalcogenide (TMD) materials such as MoS_2, WS_2, and WSe_2 as conducting layers, high R can be reached only at high applied voltage since low conductivity can be possessed by atomically thin TMD [29–32]. Therefore, wide bandgap hole-conducting materials like NiO and CuSCN or electron transport materials like ZnO with favorable band alignment and high conductivity are standard conducting layers for fabricating perovskite heterostructures [33,34]. Besides, manufacturing of flexible perovskite photodetectors has been the hotspot in research in recent days because of their potential applications in wearable optoelectronics as well as portable consumer electronics.

4. Organic-inorganic hybrid perovskite photodetector

Despite having reliable and mature fabrication processes as well as technologies, their mechanical inflexibility, expensive, and complex manufacture cannot meet the prerequisite of low-cost and large-area application. To overcome the problems of inorganic photodetectors, solution-processable semiconductor materials like organic materials can act as good active layers in flexible and simple photodetectors in recent years. However, their poor charge carrier mobilities hinder their further improvement in photodetection performance. Therefore, to develop high-performance photodetectors, researchers have used organometallic halide perovskites that have advantages of inorganic semiconductors (high charge carrier mobility) and organic semiconductors (facile solution processability), and effective light absorption [35].

Hu et al. proposed the first organic-inorganic hybrid perovskite (OIHP)-based photodetector. The perovskite precursor solution was spin-coated on a flexible substrate of polyethylene terephthalate (PET) patterned with ITO electrodes. When light is switched on at a fixed wavelength (365 nm) and different light intensities, the photocurrent is significantly increased from a closely insulating state at dark conditions and exhibits nonlinear and asymmetrical characteristics caused by the Schottky barrier ITO/perovskite interface. This hybrid also has an excellent on/off ratio of 152 at 1 V, EQE of 1.19×10^3%, R of 3.49 A W^{-1}, and response time lower than 0.2 s [36]. During almost the same period, Dou et al. fabricated photodetectors from glass/FTO/PEDOT:PSS/$CH_3NH_3PbI_{3-x}Cl_x$/PCBM/HBL/Al device architecture wherein conjugated polymers (PFN and BCP) were employed as HTLs. This photodetector with HTLs has a very low dark current at reverse bias as related to that without HTLs. These devices have suppressed dark current characteristics, reduced hysteresis effect, noise current lower than 1 pA $Hz^{-1/2}$, fast response times (τ_{decay} of $\approx$160 ns, τ_{rise} of $\approx$180 μs) originating from precise interface modification.

Several outstanding works have further enhanced the output performances of hybrid perovskite photodetectors because of the encouragement of these two pioneering works. Heterostructures (hybridizing perovskites with other materials) have been used as an effective and alternative approach to enhance the broadening of light detection range and performance of photodetectors. Several organic-inorganic composites or inorganic semiconducting nanomaterials such as graphene, PbS quantum dots, ZnO, and TiO_2 have been used as light harvesters in these hybrid devices. So far, graphene has been widely used in optoelectronics due to its high operating bandwidth and wavelength-independent light absorption [35,37,38]. In the photodetector, graphene usage has been limited due to its drawbacks, such as a fast recombination rate and low absorption cross-section. In the pure perovskite layer, photon absorption generates electron-hole pairs, which would recombine very fast, i.e., within a few picoseconds, which is lesser than the photoexcited electrons lifetime. Thus, the performance of the photodetector is restricted seriously. However, this issue can be effectively solved by integrating perovskite nanostructures and graphene as photoactive materials. Photon absorption produces the empty states in the perovskite valence band that would be filled due to the charge carriers' transfer process at the heterostructure interface, limiting the recombination of electron-hole pairs [39]. In situ synthesis method, the reduced graphene oxide (rGO) has been applied into $CH_3NH_3PbI_3$-based photodetector by He et al. [40]. This hybrid heterostructure photodetector shows a faster response speed and six times higher ON/OFF ratio than a pure $CH_3NH_3PbI_3$ device. Wang et al. introduce nanocrystalline graphite (NCG) into a solution-processed $CH_3NH_3PbI_3$ to form a photodetector (Fig. 2A) [40,43]. The photo-generated electrons are transferred from $CH_3NH_3PbI_3$ to NCG due to the mediation of NCG, which decreases the recombination of photo-induced band-to-band or electron-hole. The rise time and decay time of this photodetector are lesser than 25 ms, which revealed that electrons and holes extraction trapped in $CH_3NH_3PbI_3$ are accelerated by NCG (Fig. 2B) [40]. To extend the detection range of photodetector in the near-infrared (NIR) region, Chen et al. used conjugated polymer (PDPP3T) as photosensitizers because the conjugated polymer is able to absorb NIR light. The composite film ($CH_3NH_3PbI_3$/PDPP3T) also has an absorption range of 550–940 nm. This flexible photodetector has better specific detectivity in the region of UV-visible at low voltage (1 V), and it has at least eightfold higher photoresponsivity than pure $CH_3NH_3PbI_3$ photodetector. Due to its narrow absorption range, Rhodamine B dye has also been used in photodetector [41]. Teng et al. have used perovskite-Rhodamine B bulk-heterojunction architecture for preparing hybrid planar flexible photodetector. This device shows the high I_{light}/I_{dark} ratio of ~286 to visible light at 500 μW cm^{-2} power and high

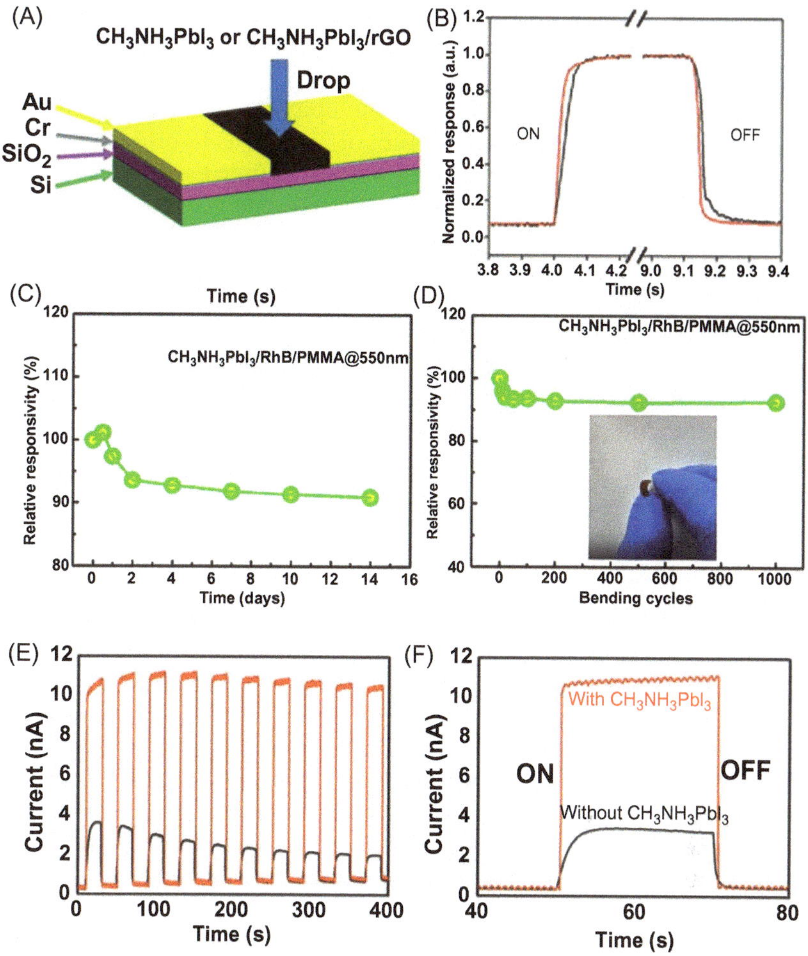

FIG. 2 (A and B) Structure of photodetector with its (B) rise and decay time [40]. (C) Relative responsivity of photodetector in air with 20%–30% relative humidity for 2 weeks [41]. (D) Relative responsivity to 550 nm light of photodetector by bending at radius of 9 mm for 1000 cycles [41]. (E, F) Time response behaviors of photodetectors with (*red line*)/without (*black line*) $CH_3NH_3PbI_3$ [42].

responsivity of 43.6 mA W^{-1} to 550 nm monochromatic light (Fig. 2C and D) [41].

The Rhodamine B electrons can be activated from HOMO to LUMO due to illumination of specific visible light, which are subsequently inserted into perovskites conducting band bottom. In the meantime, the holes move equally from the perovskite valance band top to Rhodamine B HOMO. So, Rhodamine B introduction would favor electrons and holes separation, efficiently reducing different recombination of carriers. Researchers used TiO_2 materials as ultraviolet response semiconducting material due to its wide-bandgap [42,44]. The $CH_3NH_3PbI_3/TiO_2$ hybrid photodetector has been fabricated by spin-coating the solution of $CH_3NH_3PbI_3$ onto mesoscopic thin film of TiO_2. This hybrid photodetector shows a much faster response time and three-time enhancement I_{light}/I_{dark} ratio as compared to pure perovskite photodetector (Fig. 2E and F) [42]. A. Popoola et al. fabricated a bifacial sandwiched heterojunction photoconductor [45]. The fabricated photodetectors have maximum responsivity in either face is 1.94×10^3 and 1.61×10^3 A/W, respectively. They also have maximum detectivity values in the order of 10^{14} Jones and 10^{12} Jones in both faces, respectively. The LDR of fabricated photodetectors is 55 dB as compared to that of some commercial photodetectors. In general, photodetectors with cations or holes collection sites show better performance than anions or electrons collection sites [45].

5. Memory devices

Halide perovskites offer broader scope in optoelectronics such as photovoltaics, photodetectors, etc., due to their fascinating futuristic (tunable bandgap, high absorption coefficient, mobility, and large carrier diffusion length) [46,47]. These intriguing properties are commonly addressed in the optoelectronic applications of halides perovskites, specifically in photovoltaic devices. Besides these intriguing properties, halide perovskites also show contradicting undesirable carrier behaviors (i.e., hysteresis) at the interfaces or terminals. The presence of hysteresis along with the current density-voltage (*J-V*) characteristics significantly influences the reliability and stability assessment of the corresponding optoelectronic devices [48]. However, the undesirable unique hysteresis carrier behavior enables the pathway in employing halide perovskite materials for the next-generation data storage (memory) devices with an outstanding performance [49].

Recently, rapid headway of digital world stimulates the exploration of next-generation memory devices, among various contenders, resistive-switching memories (RSM), also known as resistive-switching random access memory (RSRAM), spotlighted to be one of the outstanding data storage devices owing to its downscaling potential, faster operational speed, low power consumption, and large endurance [50–52]. The architecture of RSRAM is composed of metal-insulator-metal structures, and its data storage operations rely on resistive switching at reversible conditions between two resistances (high and low) states. Based on the resistivity activity between these high (HRS) and low (LRS) states, two kinds of electrical polarity switching, viz., SET and RESET functions. When resistance fluctuates from HRS to LRS, it performs SET function, and in its reverse operation, i.e., from LRS to HRS, it performs RESET [50]. For RSM applications, enormous materials, including binary metal oxides, organic materials, chalcogenides, and halide perovskites, have been tested. Among them, halide perovskites have promoted the development of cost-effective nonvolatile high-density memory devices (RSM) due to their exceptional mechanical and electronic features [53]. Moreover, the smaller exciton band energy and high carrier mobility of the halide perovskites make it a more potential candidate for memory devices than other semiconducting materials [51]. However, there are limited reports and work on the halide perovskite-based RSM devices. The current state-of-the-art with recent advancements in halide perovskite for RSM is elaborated in the following sections.

The aforementioned high carrier mobility of halide perovskite greatly supports storing data; more specifically, the halide perovskites with defects favor in possible reduction of recombination rate with enhancement in carrier mobility [53]. Halide perovskites have various defects, including lead vacancy, iodide ion vacancy, methyl ammonium ion vacancy, etc., attributing to the formation of various energy levels where charges are trapped [54]. Further, the charge carrier traps, ferroelectricity, and/or migration of ions in electronic applications of halide perovskites exhibit distinct hysteresis [55,56]. These distinctive features of halide perovskites meet the essential requirements such as fast switching response, high ON/OFF ratio, nonvolatility, etc., of memory (data storage) devices [54]. Generally, the perovskite was sandwiched between the two electrodes (electrode/perovskite/electrode) in the configuration of RSM device, as depicted in Fig. 3A. Initially, the memory device will be in "OFF" state (HRS), and upon exceeding the threshold voltage, it transits from "OFF" to "ON" state (LRS). The threshold transition functions the writing or SET (HRS to LRS) and erasing or REST (LRS to HRS) in the memory cell. Mostly, halide perovskite-based RSM shows a nonvolatile feature which means the device can store data even in the OFF state. In contrast, RESET or erasing can be recovered at the reverse threshold voltage and this kind of process is classified as resistive random access memory (RRAM). If the state of "RESET/OFF" cannot be recovered, then the stored data will be held permanently and read for many times, which cannot be modified further. This kind of characteristic was classified as write-once read-many times (WORM). The behavior of WORM is similar to that of storing archival standards in CDs and DVDs. Finally, if the threshold voltage (polarity) existing between SET and REST is equivalent, the storage device exhibits the unipolar memory. The schematic illustrations of different memory characteristics are depicted in Fig. 3A [53].

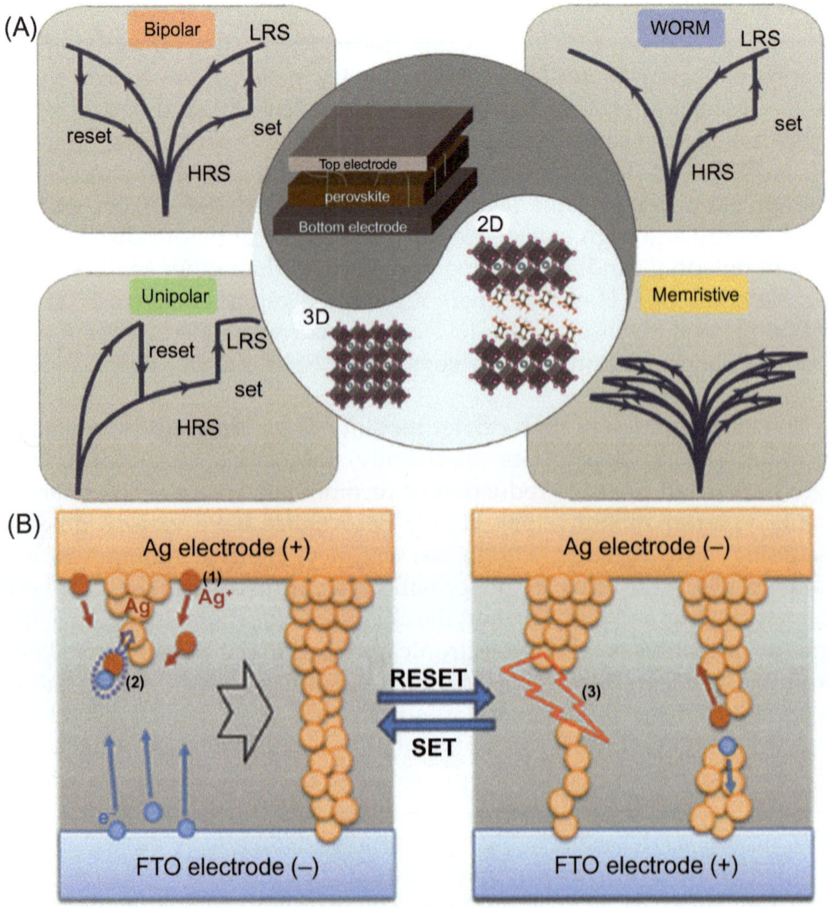

FIG. 3 Schematic illustration of (A) 2D/3D perovskite device and crystal structures with its nonvolatile memory characteristics (*I-V* curve) [53]. (B) Formation and rupturing of Ag conductive filament during electrochemical process [57].

The first research article on halide perovskite ($MAPbI_{3x}Cl_x$)-based resistive switching RAM (RSM) device with a configuration of Au/perovskites/FTO was reported by Yoo and co-workers in 2015. The halide perovskite ($CH_3NH_3PbI_{3-x}Br_x$) sandwiched between two electrodes acts as a dielectric layer and exhibits a typical bipolar resistive switching effect showing 0.8 V as SET voltage and −0.6 V as RESET voltage with longer retention times of 4104 s and a stable endurance of 4100 times. The results satisfy the requirements of nonvolatile RSM with trapped SCLC controlled resistive switching behavior [58]. The report uncovered the correlative investigations of halide perovskites for digital data storage devices (memory). Moreover, the report also suggests that metal electrodes used in the configuration of RSM play an important role in deciding its characteristics behavior. It was further demonstrated by subsequently replacing Au in the configuration of metal electrode/perovskite/FTO with silver (Ag) [57]. During the electrochemical reaction, the metal electrode (Ag) greatly influences in exhibiting resistive switching effect in the fabricated memory devices (Fig. 3B). When a positive voltage is applied to the metal electrode, the metal Ag gets oxidized into Ag ions, thereby migrating into the perovskite layer ($MAPbI_{3x}Cl_x$), forming a bridge between the two contacts resulting in LRS [57]. Upon applying negative bias, the memory device recovered to HRS by rupturing the conducting filament caused due to the Joule heating. As a result, the memory device fabricated using halide perovskite with Ag as metal electrodes exhibits a retention time of 4×10^4 s with an endurance of above 1000 times [57]. Following this report, systematic investigations have been made on a series of metal electrodes (Cu, Ti, Al, and Zn). The obtained results support evidence for using metal electrodes in the configuration of RSM, which benefices its practical applications [59]. Besides the metal electrode usage in the halide perovskite-based memory devices, other factors that influence the performances of the devices are its thickness of the films, flexibility (bending radius), rupturing activity, and so on. Generally, during the electrochemical reaction of the fabricated halide perovskite RSM, there are higher possibilities for the formation of halide anion defect due to the migration of cations and anions in the perovskites. In this concern, the film thickness (perovskite) involved in the RSM induces the possibility of many defects leading to poor performances. Sun et al. comprehensively demonstrated the effect of ions (anion and cation) migration in the perovskite by fabricating a series of perovskites films with different thicknesses (90, 150, 240, and 300nm) with a configuration of Ag/$MAPbI_3$/FTO [60]. The results suggest that films fabricated with larger thickness create a huge gap between the two contacting electrodes, thereby hindering the ion migrations and formation of metal conductive filaments. With reduced thickness (90nm), conductive filament (Ag cation) features the characteristics of RS behavior. Moreover, the film thickness (halide perovskite) in RSM induces an effect in drifting the carrier concentration and iodide vacancies under a gradient electric field [60,61].

Nowadays, flexibility is considered to be one of the essential criteria in device architecture, including memory devices. In this regard, a flexible halide perovskite-based RSM fabricated over the cycloolefin polymer (COP) with high mechanical strength has been demonstrated. The RSM configuration comprises two metal electrodes (Ag and Pt) sandwiching the halide perovskite ($CH_3NH_3PbI_3$) performing the resistive switching equivalent to that of rigid RSM devices [54,62]. The mechanical strength and electrical properties measurements of the fabricated flexible RSM

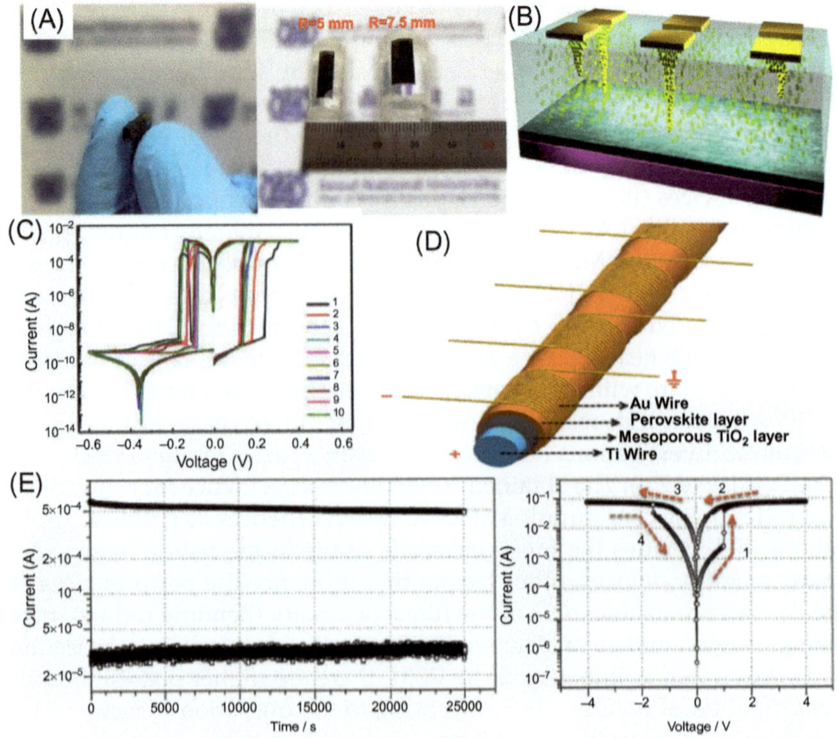

FIG. 4 (A) Photographic image of a fabricated perovskite resistive switching device on cycloolefin polymer substrate with its bent states; (B) its resistive switching mechanism illustration of the RSM; (C) its *I-V* characteristics of flexible device at 5mm bending stress [63]; (D) fiber-shaped perovskite ($Ti/TiO_2/MAPbI_{3x}Cl_x/Au$) memristor schematics with its (E and F) *I-V* and state-retention characteristics [64].

devices were carried out under different tensile states (5 and 7.5mm radii) as shown in Fig. 4A.

The electrical property of the flexible RSM shows a slight variation in the measurement value; hence different SET/REST voltages have been applied to optimize the conditions. When a voltage sweep of 0V→ +0.4V→0V→−0.6V→0V was applied, the flexible RSM (Ag/$CH_3NH_3PbI_3$/Pt) exhibits equivalent *I-V* characteristics of the rigid substrates depicted in Fig. 4C. With this existing feature, the flexible RSM delivers a high ON/OFF ratio of 10^6 with an endurance and operation speed of over 1300cycles and 640s, respectively. The results suggest the resistive switching mechanism without any cause of electro-forming process due to the formation of conducting filament at the condition of external bias (Fig. 4B) [63,65,66].

Similarly, fiber-shaped RSM device with a configuration of Ti/TiO_2/$MAPbI_{3x}Cl_x$/Au was demonstrated as next-generation smart wearable devices (e-textile) illustrated in Fig. 4D [64]. In this configuration, initially, mesoporous structured titanium dioxide (TiO_2) was deposited over the titanium (Ti) wire, followed by which the halide perovskite layer ($CH_3NH_3PbI_{3x}Cl_x$) was deposited in open atmospheric condition. The fabricated flexible RSM exhibits clear bipolar RS characteristics with retention stability of 4104 s, as shown in Fig. 4E and F [64]. Lee et al. fabricated halide perovskite ($MAPbI_3$) over the conducting polyethylene terephthalate (PET) substrates as a flexible RSM device. The fabricated flexible RSM devices exhibit reproducible and reliable bipolar resistive switching characteristics under the tensile strength of 1.5 cm (compressive bending radius) over 100 bending cycles. Moreover, the report also suggests the resistive switching behavior of the fabricated RSM attributes the mechanism of iodide vacancies (native defects) in the perovskite layer [54].

Recently, flexible fibrous cross-point (crossbar) architectured wearable memories were demonstrated with a configuration of Al@$MAPbI_3$/Al (Fig. 5A). The fabrication process involves low-temperature solution processing and dip-coating techniques. By optimizing the halide perovskite condition pin-hole free, uniformly patterned perovskite ($MAPbI_3$) was deposited on the aluminum (Al) fiber. Finally, Al@$MAPbI_3$/Al configured fibrous cross-point RSM devices were assembled at a certain bending angle by perpendicularly crossing the bare fiber (Al). The configured flexible cross-point fibrous perovskite RSM device exhibits a remarkable resistive switching effect with good retention. The possible switching (resistive) mechanism of the fabricated fibrous perovskite RSM was proposed using thermionic emission. In most of the RSM devices, the switching mechanism was explained through either ion migration and/or charge trapping [70,71]. Herein, a clear carrier role in the RSM was processed by plotting *I-V* curves of SET/REST and fitted at a double logarithmic scale, respectively. Therefore, applying positive sweep, the devices are at HRS state without complying with any space charge limitation in *I-V* cure curves that fit in linear order indicating the thermionic emission [72,73]. The thermionic emission can be described using the following relation:

$$J = A^{*}T^{2}\exp\left[-\frac{q}{k_{B}T}\left(\varphi_{B}-\sqrt{\frac{qV}{4\pi\varepsilon d}}\right)\right]$$

where J is the current density, A^* is Richardson's constant, T is the absolute temperature, k_B is the Boltzmann constant, q is the electronic charge, φ_B is the height of the potential barrier, V is the voltage, ε is the dielectric constant, and d is the thickness of dielectric film [74]. At this stage (HRS), huge amount of charge (electrons) accumulated over the surface of the

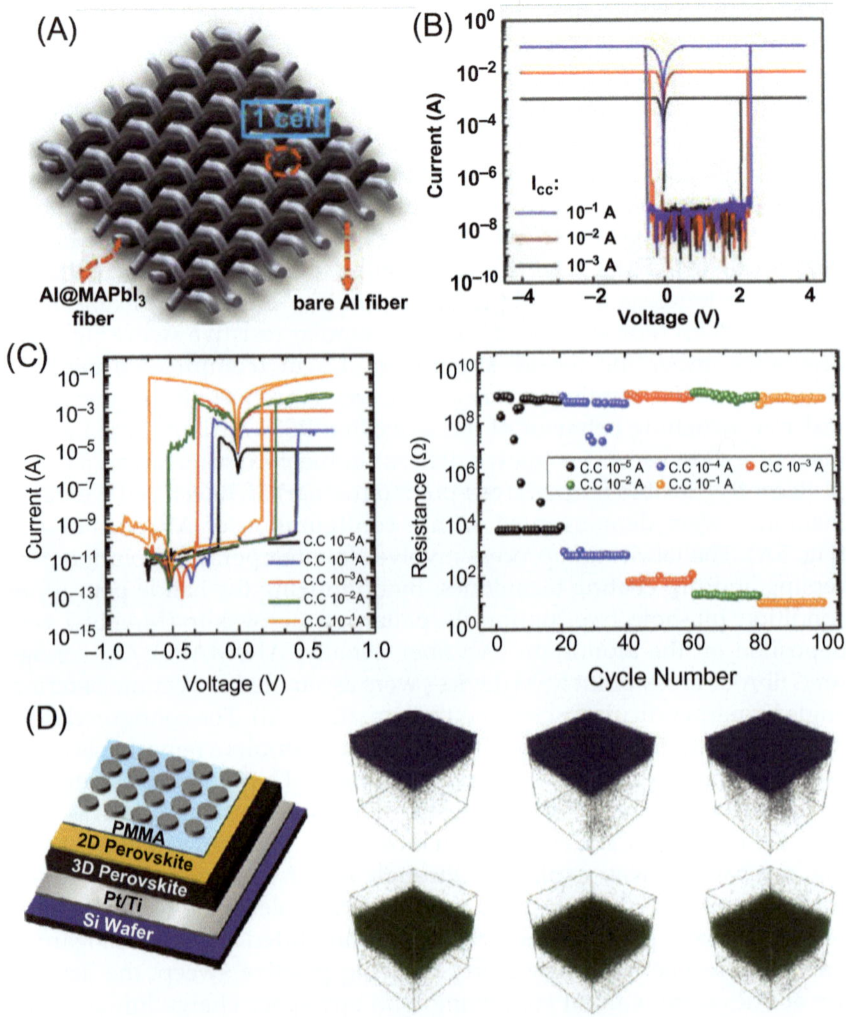

FIG. 5 (A and B) Schematic illustration of fibrous cross-point RSM device with its *I-V* characteristics at different current rate (10^{-1} to 10^{-3} A) [67]. (C) Multilevel *I-V* characteristics of Ag/PMMA/$(BzA)_2CuBr_4$/Pt with its endurance measured at various compliance current [68]. (D) Schematics and ToF images of 2D/3D heterointerface perovskite RSM devices [69].

metal electrodes (Al) owing to the insufficient energy to carry over the electrons between the interface of the perovskite ($MAPbI_3$) and metal electrode (Al). Upon applying an external voltage at this stage (HRS to LRS), more electrons penetrate through the perovskite film providing the suitable condition for Pb^{2+} to reduce into conductive metallic filaments suggesting the Ohmic conduction mechanism. There might be a misperception with the RS behavior of the fabricated RSM devices, whether it is caused due to the metallic electrode (Al) or perovskite layer ($MAPbI_3$). In order to clear the misperception, the active metal electrode Al was replaced with Au via electrochemically forming RSM with a configuration of Au@$MAPbI_3$/Au. The fabricated RSM with a configuration of Au electrodes also exposes similar RS behavior as compared with RSM with Al electrode, which undoubtedly clears that fast resistive switching effect of the fibrous RSM caused mainly due to the perovskite layer. Moreover, the interconnected (cross-point) fibrous network supports in smooth passage of current with its adjacent contacts during the mechanical deformations (stress and strain) [75]. Besides, the proposed architecture can also be used as a high-density energy storage device by implementing analog memristor and complying with multilevel resistive switching [76]. Thus, the fabricated cross-point fibrous perovskite-based RSM exposes a high ON/OFF ratio of $\sim 10^6$ at different current rates (10^{-1} to 10^{-3} A), as shown in Fig. 5B. The fibrous RSM exhibits bipolar resistive switching with lower voltage (SET/RESET) and good retention time, depicting the paves for future wearable high energy-density storage density [67].

Although halide perovskite performs to be a potential candidates in the RSM devices, the switching ratio (ON/OFF) and the retention time of perovskite alone seem to be quite less (mostly ~103) compared with other benchmarking materials in RSM. In order to tackle this, researchers developed double-layered perovskite, two-dimensional/three-dimensional (2D/3D) heterojunction, etc., in the architecture of RSM. Recently, layered perovskite ($(B_zA)_2CuBr_4$) [$B_zA = C_6H_5CH_2NH_3$] has been demonstrated as effective halide perovskite materials for fabrication of RSM with long endurance and high ON/OFF ratio. The configuration of fabricated RSM comprises Ag/PMMA/($(B_zA)_2CuBr_4$)/Pt; upon applying positive sweep to the configured RSM devices, it exposes bipolar resistive switching effect. *I-V* curve measurements under positive sweep compliance different current rates with regular intervals ranging from 10^{-5} A to 10^{-1} A (Fig. 5C). During the measurement, the voltage (RESET) increases gradually from high to low compliance current, which may be due to the expansion in the conductive metal filaments [77,78]. Additionally, multilevel endurance switching was also measured at each compliance current rate in both HRS and LRS, as shown in (Fig. 5C). At the low current rate (10^{-5} A), the conductive filament formed becomes unstable, or there might be the possibility for rupturing of filament, making the HRS level relatively poor [77,79]. A high ON/OFF ratio of 108 with long endurance of over 2000 cycles and retention time of ~1000 s has been achieved, enabling the multilevel storage characteristics. Moreover, the fabricated device retained its bipolar resistive switching effect of about 50% even for 2 weeks under relative humidity depicting the stable memory material [68].

Similarly, a low-temperature solution process demonstrated interesting heterojunction film with 2D/3D perovskite. The report mainly suggests the possibility of improving endurance by controlling the rupture of the conductive filaments. Further, the report also demonstrates the possible mechanism of bipolar resistive switching with support of hopping activation energy (due to thermal assisted ion) and the time of flight (ToF) secondary ion mass spectroscopy. The results obtained from the mass spectroscopy suggest the prevention of rupture in the conductive filament (Ag ion migration) into the 3D perovskite layer. A three-dimensional ToF secondary ion mass spectroscopy image recorded at different endurance rates for the ion distributions (Ag and Pb) in the perovskite layer is shown in Fig. 5D [69]. From the recorded image, a small distribution of Ag^- ions in the perovskite layer has been observed due to the instantaneous electric field caused during ion etching [80]. The ions (Ag-) distribution was also observed at the fresh device, but at the pristine perovskite (3D) layer, very small distribution was observed during the 300 cycles. The small distribution of ions in the perovskite layer was the main cause for the longer endurance rate (~2700 cycles) of the 2D/3D heterointerface perovskite-based RSM devices. Thus, the fabricated 2D/3D interface perovskite operates at a speed of 640 μs with ON/OFF ratio of 10^6 [69].

6. Sensors

Gas sensing technologies are of particular interest in modern sensors owing to the increase in global environmental pollutants. A variety of toxic gases such as ammonia (NH_3), nitrogen di-oxide (NO_2), carbon di-oxide (CO_2), ozone (O_3), and so forth are more harmful to human health [81]. Therefore, it is imperious to develop potential devices that can detect these toxic gases present in the atmosphere and human surroundings. The development of gas sensors will eventually fulfill the need for safety concerns. The gas sensors are categorized by the principle of detection such as optical, electrical, electrochemical, catalytic, semiconducting metal oxides, and infrared (IR) types. Some of the basic requirements for gas sensors are high sensitivity, fast response, room temperature operation, and low-cost fabrication. A suitable material for gas sensors should possesses an appropriate electronic structure along with a good affinity toward the target reactant and thereby could satisfy the other requirements such as manufacturability, hydrothermal stability, poisoning resistance, and adaptability with existing technologies [81]. However, the requirement is still confined to microsize devices owing to the emergence of smart sensor technology. Although oxide semiconductors have demonstrated strong sensing behavior in the recent years, their operating temperature and cost effectiveness limit their effective use, and as a result, researchers have been looking for potential alternative materials to fill the role effectively occupied by halide perovskites [3,82]. Metal halide perovskites of type ABX_3 have been explored into diverse directions such as solar cells, light emitting diodes, transistors, lasers, photodetectors, and sensors. Since perovskites have high charge carrier mobility, more chances are there to obtain an electrical response due to the ambient composition variation. However, certain factors affect the sensory behavior of halide perovskites: crystal structure, stability, environmental affordability, and quantum yield [83]. Hence, one must be keen on the aforementioned factors to fabricate efficient gas sensors based on perovskite nanostructured materials. This section pays more attention to the utilization of various nanostructured organic-inorganic halide perovskites as sensing elements for different target gases.

Organometallic halide perovskites have demonstrated remarkable progress in the photovoltaic industry with power conversion efficiency greater than 25%. Despite its optoelectronic behavior, the fundamental physical and chemical properties induce the material to be utilized in other fields beyond photovoltaics. One such field is the sensors and, in particular, gas sensors, where these nanostructured perovskite materials are being used as active layers [82]. The important aspect is the response or interaction of these materials with the surrounding environment, such as moisture, atmospheric gases, and temperature, which explores the sensory behavior of halide perovskites. It is well known that the halide perovskites, particularly $MAPbI_3$, was found to change their color from brown to yellow when exposed to water, and the reverse will not happen when the water is removed. This phenomenon is nothing but the degradation behavior of the perovskite materials that the researchers used to denote.

However, Zhao et al. studied the optical bleaching of $MAPbI_3$ via phase transformation induced by NH_3, which was the key factor to use this material as an optical based NH_3 sensor at room temperature [86]. They showed that the introduction of 3% NH_3 solution to the film induced the phase transformation, which was indicated by the sudden change of brown to colorless film (<1 s), and the phase transformation was confirmed by XRD and absorbance spectrum from UV-visible analyses. They also observed the reversible reaction when the gas is removed within a few seconds. Moreover, the transient photoresponse (probed with 660 nm illumination using a silicon detector connected to a potentiostat) upon introducing and removing NH_3 source to $MAPbI_3$ film clearly demonstrated the potential use of halide perovskite films as NH_3 sensors.

Later in 2015, Bao et al. reported the resistance change in $MAPbI_3$ films on exposure to NH_3 atmosphere at the same room temperature, which is the driving force to fabricate electrical based NH_3 sensors using halide perovskites [84]. They found that the electrical resistance in NH_3-exposed perovskite film (for 5 and 50 s) was reduced, leading to the formation of a new phase, and the high resistance was returned back within seconds upon the removal of NH_3 atmosphere. They attributed these electrical changes to the adsorption or intercalation of NH_3 molecules onto the crystal lattice of $MAPbI_3$ films. The test system with the perovskite electrode is schematically outlined with its *I-V* characteristics under exposure and removal of NH_3 gas in Fig. 6A. Fig. 6B represented the time-dependent current characteristics with ON and OFF NH_3 switching with its enlarged view. They observed the increased current from 10 to ~110 nA in 3.0 s and recovery of the same in 4.5 s.

Barnali Ghosh group in 2018 reported the fabrication of cheap, disposable rapid room temperature NH_3 sensor based on $MAPbI_3$ films coated on papers obtained by spin coating of PbI2 followed by solution immersion process MAI solution [85]. Thus, the fabricated paper could detect even at low concentrations (10 ppm) of NH_3 gas in open or closed conditions. They also claimed that the sensor is merely sensitive to NH_3, leaving behind other toxic gases such as methane (CH_4), nitrous oxide (N_2O) and CO_2, and also moisture. They also attributed the fast response of this optical sensor (based on $MAPbI_3$ films) to the nanorod morphology attained, which is depicted in Fig. 6C and D. The color change from brown to yellow and then from yellow to brown on exposure and removal of NH_3 and the respective XRD patterns are also shown in Fig. 6E, whereas the sensory behavior (response time and stability) of paper-based gas sensors is depicted in Fig. 6F. They also found that the reversible was not observed when the concentration was about 30 ppm, and hence, those devices could act as a disposable sensor. The same group reported the fabrication of the same disposable sensor in 2019 but with the electrical response studied using *I-V* characteristics [87]. The sensor showed 55% sensitivity when exposed to 1 ppm of NH_3. The same decomposition of $MAPbI_3$ to PbI_2 was responsible for increasing conductivity on exposure of targeted gas molecules.

Sheikh et al. successfully developed $MAPbI_3$ pellet-based NH_3 sensor where the pellet was made out of the perovskite powders synthesized via two-step antisolvent precipitation (TSAP) method [88]. The surface of the pellet served as an active layer for sensing performance at room temperature itself. Upon exposure to 10 ppm of NH_3 gas, the pellets displayed high responsivity (200% in 0.3 s) and high recovery (3.9 s) due to the electrical behavior changes, attributed to electron exchange between perovskite and NH_3 molecules. They also tested the stability and lifetime of the pellet-based sensors and exhibited a lifetime of 60 days. Apart from the methyl ammonium cation in ABX_3 perovskites, ethylenediamine (as A-site cation, bidentate-ligand)-based perovskite was synthesized by Gupta et al. reported the fabrication of NH_3 sensor based on the perovskite thin film [89]. The same optical changes, as well as the electrical changes, were observed upon exposure to this toxic gas. The detection level was

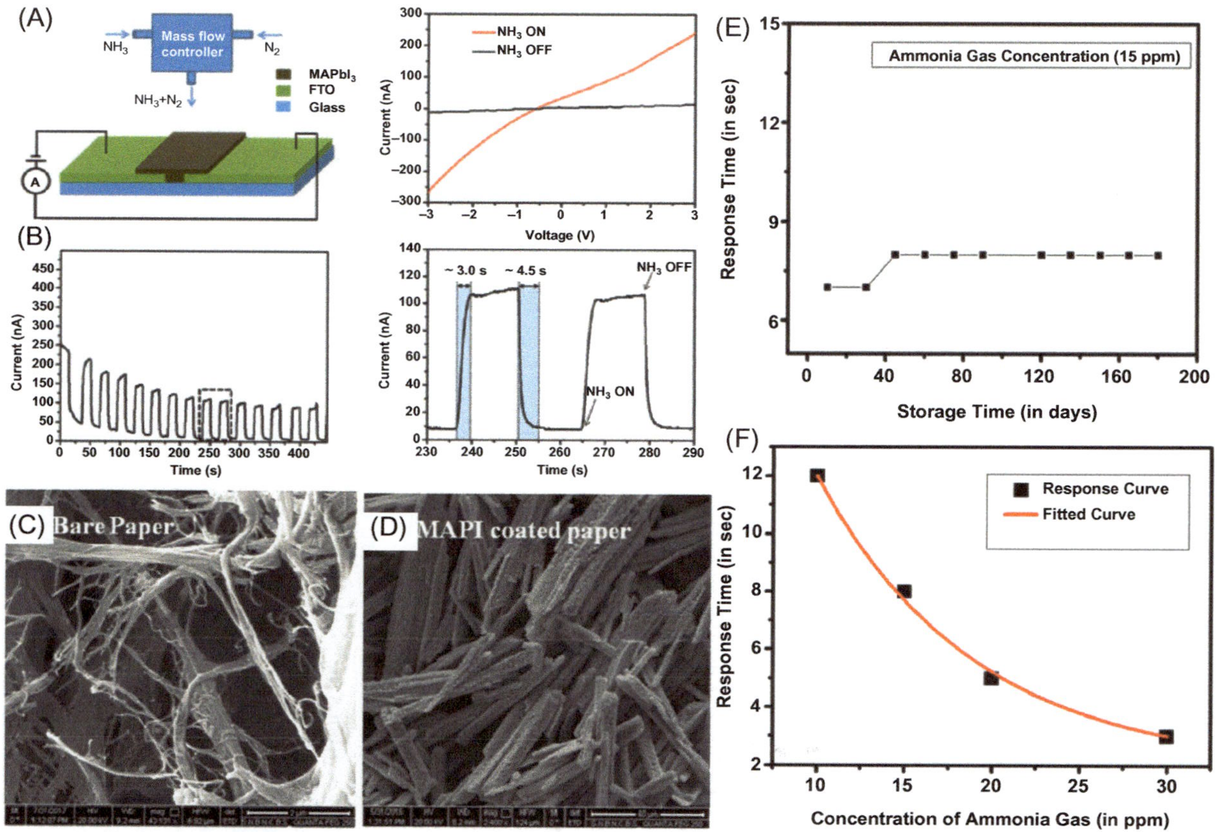

FIG. 6 (A) Schematics of the $MAPbI_3$ electrode-based gas sensor with its *I-V* curves. (B) *I-t* curve ($MAPbI_3$) under the NH_3 switching (ON/OFF) at RT with its enlarged view [84]. (C and D) FESEM micrographs bare and $MAPbI_3$-coated paper. (E and F) Response time and stability of the sensor at different/fixed NH_3 concentrations [85].

checked for this sensor from 50 to 200 ppm and found that the response time was much faster (16.5 s) at higher concentrations, whereas it is slightly higher (39 s) at low concentrations. The NH_3 molecules enter into the voids of the perovskite structure followed by the donation of the free electron and thus leading to increased conductivity. Moreover, they studied the sensing behavior of films at different exposure time levels and found the increased response from 30.18% to 64.57% for 1 to 5 min, respectively.

Tuning the halides in halide perovskites results in nanostructured materials with enhanced gas sensing properties, particularly for NH_3 sensors. In this way, Jiao et al. (2020) obtained $CH_3NH_3PbBr_{3-x}I_x$ perovskite thin film using a one-step spin-coating method from $PbBr_2$ and MAI precursor solutions [90]. The obtained red color of the spin-coated films immediately turned into gray while it was recovered upon the introduction and evacuation of NH_3 gas. This optical change was endorsed to the redox reactions that the authors have proposed where MA^+ is reduced to MA and NH_3 is oxidized to NH_4^+. Therefore, they reported the replacement of A-site MA^+ ions by NH_4^+ ions upon exposure of NH_3 gas and regaining of MA^+ on the removal of the target gas. Furthermore, they also tested the sensing toward ethanol and acetone gases and observed a considerable sensing performance, where the physical reversible electrons transfer to intrinsic bromide vacancies and uncoordinated Pb^{2+} was accountable for the sensing behavior.

Designing and developing oxygen (O_2) sensors using nanostructured materials with high sensitivity and selectivity become indispensable. In this regard, Stoeckel et al. reported for the first time a fully reversible, fast-response sensor device based on hybrid metal halide perovskite to sense the O_2 gas [91]. They studied the electrical response of $MAPbI_3$ films to O_2 gas and found a 3000-fold increase in the conductivity of perovskite sensors in the O_2 atmosphere. They also ascribed this effect to a dynamic trap-healing mechanism where the oxygen molecules get occupied at the iodine vacancies, as shown in Fig. 7A. When CH_3I vapors partially filled these iodine vacancies (shown in Fig. 7A), the response to oxygen decreased. Hence, they demonstrated the above-said mechanism for the sensory behavior of $MAPbI_3$ films to the targeted O_2 gas molecules.

Ozone (O_3) is also considered one of the major pollutants to the environment, and hence, the detection of this toxic gas using potential devices is also becoming essential. In this aspect, Kakavelakis et al. developed a portable, flexible, self-powered, ultrasensitive O_3 sensor based on spin-coated hybrid lead mixed halide perovskites ($CH_3NH_3PbI_{3-x}Cl_x$) (Fig. 7B) [92]. The same electrical resistance changes were observed when the films were exposed to O_3 and after the removal of O_3. The observed initial high current was ascribed to the large grain size and high carrier diffusion length of mixed halide perovskite. They also varied the gas concentrations starting from 2500 to 5 ppb and analyzed the sensing performance such as response time and sensitivity with the electrical changes to the films.

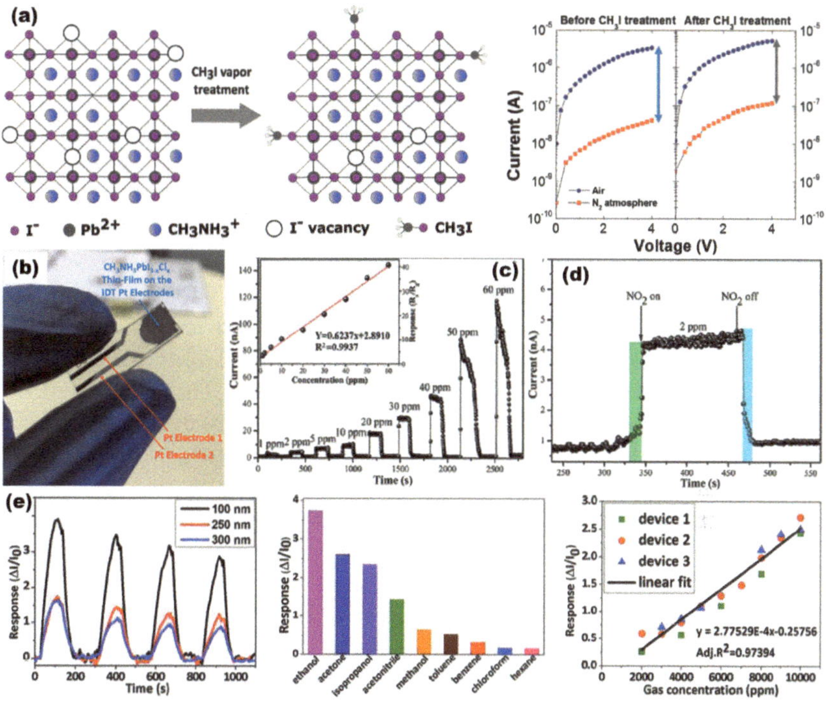

FIG. 7 (A) Mechanism of iodine vacancy filling by CH_3I vapors with its current measured in N_2 atmosphere and in air [91]; (B) photographic image of the prepatterned interdigitated perovskite films on the Pt substrate [92]; (C & D) dynamic response and response recovery curves of the $MAPbI_3$ film sensor [93]; (E) sensor response for different thickness (100, 250, and 300nm) perovskite films with its comparison responses toward 10,000ppm of each gas and sensor response [94].

They also observed substantial enhancement in PL intensity after exposure which was due to the passivation of surface traps in the perovskite films that increased the intensity of radiative recombination. The other toxic gas, NO_2, is also harmful to the environment, and hence, more focus is also turned toward the detection of this gas using nanostructured perovskite materials.

Because of this, Fu et al. fabricated room temperature NO_2 sensor based on $MAPbI_3$ films with quick response (~5s) and recovery (~25s) as well as high selectivity toward NO_2 even at extremely low concentrations (1ppm) and at high pressure [93]. They tested the gas sensing performance at different gas concentrations (Fig. 7C and D) and revealed a linear relationship with that of the sensor response, whereas the average sensitivity was high at 0.62ppm. They also checked the reproducibility by repeated injection and evacuation of NO_2 in the testing chamber and found that the film remained stable for more than 12cycles. They attributed the superior and highly selective sensing behavior to the adsorbed

NO_2 molecules onto the active sites of the porous perovskite films and the electron transfer process leading to the decreased resistance.

The halide engineering concept plays a crucial role in attaining various nanostructured halide perovskites for gas sensing applications. Li and his co-workers developed bromide-based perovskite ($MAPbBr_3$) films as NO_2 sensors, which showed high performance even at room temperature [95]. Also, they had prepared two different films that resulted directly from precursor solutions and also from surface capped $MAPbBr_3$ nanoparticles. The effect of surface capping by the alkyl chain on the NO_2 sensing performance of the respective films was also reported. They concluded that the sensor fabricated from surface-capped nanoparticles showed better sensing performance with the detection limit of 0.1 ppm compared to the films obtained directly from precursor solutions. They proposed the sensing mechanism to the electron transfer process involved in the exposure of the gas. Briefly, physical adsorption of oxidizing NO_2 gas molecules onto the film's surface occurs upon exposure of NO_2 gas to the perovskite film, followed by the extraction of electrons from the conduction band of $MAPbBr_3$ (Pb-s and Br-p orbitals) because of the strong electron-withdrawing nature of NO_2 molecules. Due to this phenomenon, the conductivity of $MAPbBr_3$ perovskite film is reduced. They also attributed the long-term sensing stability to the protected surface of perovskite nanoparticles. In addition to the sensing behavior of $MAPbI_3$ and $MAPbBr_3$ films, the methylammonium tin iodide ($MASnI_3$) films (replacement of B-site cation by Sn) were also shown the promising sensing activity toward toxic gases. One such work was reported recently by Hein et al., where the $MASnI_3$ films were grown by vacuum deposition technique and utilized as a sensor material for the selective detection of the toxic NO_2 gas (detection limit around 25 ppb) [96]. The problem associated with Sn based perovskites is the oxidation of Sn^{2+} to Sn^{4+} upon illumination of light or subjected to other environmental conditions. However, the films deposited by the vacuum process exhibited excellent sensitivity and selectivity toward NO_2 gas. They discussed the mechanism of sensing (chemisorption of NO_2 to the surface of the films) in detail.

Volatile organic compound (VOC) gases also contribute to air pollution due to their high volatility, where these compounds are commonly used in many household products. Hence, it is crucial to make device sensors to detect these kinds of toxic gases. Nur'aini et al. developed VOC gas sensors based on $MAPbI_3$ films obtained through the conventional one-step spin-coating method along with the antisolvent (toluene) dripping [94]. The responses were originated from the conductivity change in the presence and absence of VOC gases. Although several VOC polar (ethanol, acetone, isopropanol, acetonitrile, and methanol) and nonpolar (toluene, benzene, chloroform, and hexane) gases were tested for its sensitivity, ethanol showed the highest response and hexane being the lowest (shown in Fig. 7E). Table 1 outlines the sensing behavior of various halide

TABLE 1 Summarized table of halide perovskite based various sensors.

Halide perovskite	Deposition method	Target gas	Response time (s)	Recovery time (s)	Sensor type	Ref.
$MAPbI_3$	Spin coating	NH_3	<1	1–2	Optical sensor	[71]
$MAPbI_3$	Spin coating	NH_3	3	4.5	Electrical sensor	[72]
$MAPbI_3$	Spin coating followed by solution-immersion process	NH_3	~10	–	Optical sensor	[73]
$MAPbI_3$	Dip coating followed by solution-immersion process	NH_3	~250 (total)	~250 (total)	Electrical sensor	[74]
$MAPbI_3$ (pellet)	Two-step antisolvent precipitation	NH_3	0.3	3.6	Electrical sensor	[75]
$MAPbBr_{3-x}I_x$	One-step spin coating	NH_3	10–20	10–20	Electrical sensor	[67]
$MAPbI_3$	One-step as well as two-step spin coating	O_2	0.4	–	Electrical sensor	[68]
$MAPbI_{3-x}Cl_x$	Spin coating	O_3	188–225	40–60	Electrical sensor	[77]
$MAPbI_3$	Spin coating	NO_2	~5	~25	Electrical sensor	[78]
$MAPbBr_3$	Spin coating/drop-casting	NO_2	–	–	Electrical sensor	[79]
$MASnI_3$	Thermal evaporation	NO_2	200	600	Electrical sensor	[80]
$MAPbI_3$	Spin coating with toluene dripping	VOCs (ethanol)	66	67	Electrical sensor	[69]
TBA-$MAPbBr_3$	Spin coating with toluene dripping	NH_3	61	65	Photoluminescence	[97]
$MAPbI_{3-x}(SCN)_x$	Spin coating	Acetone and NO_2	4.5 and 3.7	4.0 and 6.0	Chemiresistive sensor	[98]

perovskites toward different gases. In general, the perovskites are more sensitive toward polar gases than nonpolar gases. Perovskites adsorb more polar gas molecules onto their defect sites due to their polar ionic nature, resulting in higher responses.

7. Summary

In summary, the rapid advancement of halide perovskite in various applications (such as switching memories, photodetectors, sensors, etc.) apart from photovoltaics has been achieved owing to their unique characteristics. Here, interesting reports on halide perovskite for RSM, photodetectors, and various sensors were reviewed, and their overall view has been presented. Halide perovskites hold different properties, including tunable bandgap, photo/electroluminescence, ion/defect migration, carrier diffusion length, hysteresis, etc. Based on the exciting properties of the halide perovskite, the chapter postulates halides' potential reach in switching memories, photodetectors, and sensors. Besides, the chapter also reveals the new and exciting scope of the perovskite is highlighted in their corresponding applications.

Acknowledgment

The author Rajkumar C acknowledges University Grant Commission (UGC), INDIA, for providing the fellowship. The authors P. Vengatesh and T.S. Shyju would like to thank MHRD, INDIA, for establishing Centre of Excellence for Energy Research in Sathyabama Institute of Science and Technology (No. 5-8/2014-TS.VII). A. Arulraj and R.V. Mangalaraja duly acknowledge ANID-FONDECYT Postdoctoral Fellowship (Project No.: 3200076), Government of Chile, Universidad Adolfo Ibáñez, Chile for financial support. A. Arulraj acknowledge Universidad de Atacama, Copiapó for their support.

References

[1] J.A. La, J. Kang, J.Y. Byun, I.S. Kim, G. Kang, H. Ko, Highly sensitive and fast perovskite photodetector functionalized by plasmonic Au nanoparticles-alkanethiol assembly, Appl. Surf. Sci. 538 (2021), 148007.

[2] A. Arulraj, N. Ilayaraja, V. Rajeshkumar, M. Ramesh, Direct synthesis of cubic shaped Ag2S on Ni mesh as binder-free electrodes for energy storage applications, Sci. Rep. 9 (2019) 10108.

[3] E. Kymakis, A. Panagiotopoulos, M.M. Stylianakis, K. Petridis, Organometallic hybrid perovskites for humidity and gas sensing applications, in: 2D Nanomaterials for Energy Applications, Elsevier, 2020, pp. 131–147.

[4] A. Arulraj, M. Ramesh, Rational strategies for large-area perovskite solar cells, in: Rational Design of Solar Cells for Efficient Solar Energy Conversion, John Wiley & Sons, Inc., Hoboken, NJ, USA, 2018, pp. 307–337.

[5] Y. Zhang, R. Tao, X. Zhao, Z. Sun, Y. Wang, L. Xu, A highly photoconductive composite prepared by incorporating polyoxometalate into perovskite for photodetection application, Chem. Commun. 52 (2016) 3304–3307.
[6] V. Bhatia, K.S. Narayan, Photoconducting devices, in: Wiley Encyclopedia of Biomedical Engineering, John Wiley & Sons, Inc., Hoboken, NJ, USA, 2006, pp. 1–6.
[7] H. Wang, D.H. Kim, Perovskite-based photodetectors: materials and devices, Chem. Soc. Rev. 46 (2017) 5204–5236.
[8] J. Zhou, J. Huang, Photodetectors based on organic-inorganic hybrid lead halide perovskites, Adv. Sci. 5 (2018) 1700256.
[9] F. Wang, J. Mei, Y. Wang, L. Zhang, H. Zhao, D. Zhao, Fast photoconductive responses in organometal halide perovskite photodetectors, ACS Appl. Mater. Interfaces 8 (2016) 2840–2846.
[10] L. Lv, Y. Xu, H. Fang, W. Luo, F. Xu, L. Liu, B. Wang, X. Zhang, D. Yang, W. Hu, A. Dong, Generalized colloidal synthesis of high-quality, two-dimensional cesium lead halide perovskite nanosheets and their applications in photodetectors, Nanoscale 8 (2016) 13589–13596.
[11] Y. Li, Z.-F. Shi, S. Li, L.-Z. Lei, H.-F. Ji, D. Wu, T.-T. Xu, Y.-T. Tian, X.-J. Li, High-performance perovskite photodetectors based on solution-processed all-inorganic CsPbBr 3 thin films, J. Mater. Chem. C 5 (2017) 8355–8360.
[12] Y. Wang, M.-L. Gao, J.-L. Wu, X.-W. Zhang, Metal halide perovskite photodetectors: material features and device engineering, Chin. Phys. B. 28 (2019), 018502.
[13] Z. Sun, L. Aigouy, Z. Chen, Plasmonic-enhanced perovskite–graphene hybrid photodetectors, Nanoscale 8 (2016) 7377–7383.
[14] T. Moehl, J.H. Im, Y.H. Lee, K. Domanski, F. Giordano, S.M. Zakeeruddin, M.I. Dar, L.-P. Heiniger, M.K. Nazeeruddin, N.-G. Park, M. Grätzel, Strong photocurrent amplification in perovskite solar cells with a porous TiO_2 blocking layer under reverse Bias, J. Phys. Chem. Lett. 5 (2014) 3931–3936.
[15] H.-W. Chen, N. Sakai, A.K. Jena, Y. Sanehira, M. Ikegami, K.-C. Ho, T. Miyasaka, A switchable high-sensitivity photodetecting and photovoltaic device with perovskite absorber, J. Phys. Chem. Lett. 6 (2015) 1773–1779.
[16] X. Hu, X. Zhang, L. Liang, J. Bao, S. Li, W. Yang, Y. Xie, High-performance flexible broadban photodetector based on organolead halide perovskite, Adv. Funct. Mater. 24 (2014) 7373–7380.
[17] W. Wang, H. Xu, J. Cai, J. Zhu, C. Ni, F. Hong, Z. Fang, F. Xu, S. Cui, R. Xu, L. Wang, F. Xu, J. Huang, Visible blind ultraviolet photodetector based on CH_3NH_3PbCl_3 thin film, Opt. Express 24 (2016) 8411.
[18] Y. Guo, C. Liu, H. Tanaka, E. Nakamura, Air-stable and solution-processable perovskite photodetectors for solar-blind UV and visible light, J. Phys. Chem. Lett. 6 (2015) 535–539.
[19] J. Zhou, Y. Chu, J. Huang, Photodetectors based on two-dimensional layer-structured hybrid lead iodide perovskite semiconductors, ACS Appl. Mater. Interfaces 8 (2016) 25660–25666.
[20] C.S. Ponseca, T.J. Savenije, M. Abdellah, K. Zheng, A. Yartsev, T. Pascher, T. Harlang, P. Chabera, T. Pullerits, A. Stepanov, J.-P. Wolf, V. Sundström, Organometal halide perovskite solar cell materials rationalized: ultrafast charge generation, high and microsecond-long balanced mobilities, and slow recombination, J. Am. Chem. Soc. 136 (2014) 5189–5192.
[21] J. Kim, S.-H. Lee, J.H. Lee, K.-H. Hong, The role of intrinsic defects in methylammonium lead iodide perovskite, J. Phys. Chem. Lett. 5 (2014) 1312–1317.
[22] W.-J. Yin, T. Shi, Y. Yan, Unusual defect physics in CH3NH3PbI3 perovskite solar cell absorber, Appl. Phys. Lett. 104 (2014), 063903.
[23] L. Dou, Y.M. Yang, J. You, Z. Hong, W. Chang, G. Li, Y. Yang, Solution-processed hybrid perovskite photodetectors with high detectivity, Nat. Commun. 5 (2014) 5404.

[24] J. Song, Q. Cui, J. Li, J. Xu, Y. Wang, L. Xu, J. Xue, Y. Dong, T. Tian, H. Sun, H. Zeng, Ultralarge all-inorganic perovskite bulk single crystal for high-performance visible-infrared dual-modal photodetectors, Adv. Opt. Mater. 5 (2017) 1700157.
[25] H. Cho, Y. Kim, C. Wolf, H.-D. Lee, T. Lee, Improving the stability of metal halide perovskite materials and light-emitting diodes, Adv. Mater. 30 (2018) 1704587.
[26] J. Song, L. Xu, J. Li, J. Xue, Y. Dong, X. Li, H. Zeng, Monolayer and few-layer all-inorganic perovskites as a new family of two-dimensional semiconductors for printable optoelectronic devices, Adv. Mater. 28 (2016) 4861–4869.
[27] Z. Yang, M. Wang, H. Qiu, X. Yao, X. Lao, S. Xu, Z. Lin, L. Sun, J. Shao, Engineering the exciton dissociation in quantum-confined 2D CsPbBr3 nanosheet films, Adv. Funct. Mater. 28 (2018) 1705908.
[28] Y. Li, Z.-F. Shi, S. Li, L.-Z. Lei, H.-F. Ji, D. Wu, T.-T. Xu, Y.-T. Tian, X.-J. Li, High-performance perovskite photodetectors based on solution-processed all-inorganic CsPbBr3 thin films, J. Mater. Chem. C 5 (2017) 8355–8360.
[29] B. Xu, S. Zhu, X. Xie, J. Shen, Q. Zhou, Asymmetric N-H insertion reaction cooperatively catalyzed by rhodium and chiral Spiro phosphoric acids, Angew. Chem. Int. Ed. 50 (2011) 11483–11486.
[30] C. Ma, Y. Shi, W. Hu, M. Chiu, Z. Liu, A. Bera, F. Li, H. Wang, L. Li, T. Wu, Heterostructured WS 2/CH3 NH3 PbI3 photoconductors with suppressed dark current and enhanced photodetectivity, Adv. Mater. 28 (2016) 3683–3689.
[31] D. Kang, S.R. Pae, J. Shim, G. Yoo, J. Jeon, J.W. Leem, J.S. Yu, S. Lee, B. Shin, J. Park, An ultrahigh-performance photodetector based on a perovskite-transition-metal-dichalcogenide hybrid structure, Adv. Mater. 28 (2016) 7799–7806.
[32] H.-C. Cheng, G. Wang, D. Li, Q. He, A. Yin, Y. Liu, H. Wu, M. Ding, Y. Huang, X. Duan, van der Waals heterojunction devices based on organohalide perovskites and two-dimensional materials, Nano Lett. 16 (2016) 367–373.
[33] H. Liu, X. Zhang, L. Zhang, Z. Yin, D. Wang, J. Meng, Q. Jiang, Y. Wang, J. You, A high-performance photodetector based on an inorganic perovskite–ZnO heterostructure, J. Mater. Chem. C 5 (2017) 6115–6122.
[34] Z. Yang, J. Dou, M. Wang, J. Li, J. Huang, J. Shao, Flexible all-inorganic photoconductor detectors based on perovskite/hole-conducting layer heterostructures, J. Mater. Chem. C 6 (2018) 6739–6746.
[35] X. Wang, M. Li, B. Zhang, H. Wang, Y. Zhao, B. Wang, Recent progress in organometal halide perovskite photodetectors, Org. Electron. 52 (2018) 172–183.
[36] X. Hu, X. Zhang, L. Liang, J. Bao, S. Li, W. Yang, Y. Xie, High-performance flexible broadband photodetector based on organolead halide perovskite, Adv. Funct. Mater. 24 (2014) 7373–7380.
[37] T. Mueller, F. Xia, P. Avouris, Graphene photodetectors for high-speed optical communications, Nat. Photonics 4 (2010) 297–301.
[38] R.R. Nair, P. Blake, A.N. Grigorenko, K.S. Novoselov, T.J. Booth, T. Stauber, N.M.R. Peres, A.K. Geim, Fine structure constant defines visual transparency of graphene, Science 320 (2008) 1308.
[39] V.Q. Dang, G.S. Han, T.Q. Trung, L.T. Duy, Y.U. Jin, B.U. Hwang, H.S. Jung, N.E. Lee, Methylammonium lead iodide perovskite-graphene hybrid channels in flexible broadband phototransistors, Carbon 105 (2016) 353–361.
[40] M. He, Y. Chen, H. Liu, J. Wang, X. Fang, Z. Liang, Chemical decoration of CH3NH3PbI3 perovskites with graphene oxides for photodetector applications, Chem. Commun. 51 (2015) 9659–9661.
[41] C. Teng, D. Xie, M. Sun, S. Chen, P. Yang, Y. Sun, Organic dye-sensitized CH3NH3PbI3 hybrid flexible photodetector with bulk heterojunction architectures, ACS Appl. Mater. Interfaces 8 (2016) 31289–31294.

[42] H.-R. Xia, J. Li, W.-T. Sun, L.-M. Peng, Organohalide lead perovskite based photodetectors with much enhanced performance, Chem. Commun. 50 (2014) 13695–13697.
[43] Y. Wang, Z. Xia, S. Du, F. Yuan, Z. Li, Z. Li, Q. Dai, H. Wang, S. Luo, S. Zhang, H. Zhou, Solution-processed photodetectors based on organic–inorganic hybrid perovskite and nanocrystalline graphite, Nanotechnology 27 (2016), 175201.
[44] L. Su, Z.X. Zhao, H.Y. Li, J. Yuan, Z.L. Wang, G.Z. Cao, G. Zhu, High-performance organolead halide perovskite-based self-powered triboelectric photodetector, ACS Nano 9 (2015) 11310–11316.
[45] A. Popoola, M.A. Gondal, I.K. Popoola, L.E. Oloore, O.M. Bakr, Fabrication of bifacial sandwiched heterojunction photoconductor—type and MAI passivated photodiode—type perovskite photodetectors, Org. Electron. 84 (2020), 105730.
[46] N.J. Jeon, J.H. Noh, Y.C. Kim, W.S. Yang, S. Ryu, S. Il Seok, Solvent engineering for high-performance inorganic–organic hybrid perovskite solar cells, Nat. Mater. 13 (2014) 897–903.
[47] Q. Chen, H. Zhou, Y. Fang, A.Z. Stieg, T.-B. Song, H.-H. Wang, X. Xu, Y. Liu, S. Lu, J. You, P. Sun, J. McKay, M.S. Goorsky, Y. Yang, The optoelectronic role of chlorine in CH3NH3PbI3(Cl)-based perovskite solar cells, Nat. Commun. 6 (2015) 7269.
[48] Z. Xiao, Q. Dong, C. Bi, Y. Shao, Y. Yuan, J. Huang, Solvent annealing of perovskite-induced crystal growth for photovoltaic-device efficiency enhancement, Adv. Mater. 26 (2014) 6503–6509.
[49] Z. Xiao, Y. Yuan, Y. Shao, Q. Wang, Q. Dong, C. Bi, P. Sharma, A. Gruverman, J. Huang, Giant switchable photovoltaic effect in organometal trihalide perovskite devices, Nat. Mater. 14 (2015) 193–198.
[50] D. Kumar, R. Aluguri, U. Chand, T.Y. Tseng, Metal oxide resistive switching memory: materials, properties and switching mechanisms, Ceram. Int. 43 (2017) S547–S556.
[51] Y. Liu, F. Li, Z. Chen, T. Guo, C. Wu, T.W. Kim, Resistive switching memory based on organic/inorganic hybrid perovskite materials, Vacuum 130 (2016) 109–112.
[52] B. Hwang, C. Gu, D. Lee, J.-S. Lee, Effect of halide-mixing on the switching behaviors of organic-inorganic hybrid perovskite memory, Sci. Rep. 7 (2017) 43794.
[53] B. Li, W. Hui, X. Ran, Y. Xia, F. Xia, L. Chao, Y. Chen, W. Huang, Metal halide perovskites for resistive switching memory devices and artificial synapses, J. Mater. Chem. C 7 (2019) 7476–7493.
[54] C. Gu, J.-S. Lee, Flexible hybrid organic–inorganic perovskite memory, ACS Nano 10 (2016) 5413–5418.
[55] H.J. Snaith, A. Abate, J.M. Ball, G.E. Eperon, T. Leijtens, N.K. Noel, S.D. Stranks, J.T.-W. Wang, K. Wojciechowski, W. Zhang, Anomalous hysteresis in perovskite solar cells, J. Phys. Chem. Lett. 5 (2014) 1511–1515.
[56] W. Tress, Metal halide perovskites as mixed electronic–ionic conductors: challenges and opportunities—from hysteresis to memristivity, J. Phys. Chem. Lett. 8 (2017) 3106–3114.
[57] E. Yoo, M. Lyu, J.-H. Yun, C. Kang, Y. Choi, L. Wang, Bifunctional resistive switching behavior in an organolead halide perovskite based Ag/CH3NH3PbI3−xClx/FTO structure, J. Mater. Chem. C 4 (2016) 7824–7830.
[58] D.T. Moore, K.W. Tan, H. Sai, K.P. Barteau, U. Wiesner, L.A. Estroff, Direct crystallization route to methylammonium lead iodide perovskite from an ionic liquid, Chem. Mater. 27 (2015) 3197–3199.
[59] K. Yan, M. Peng, X. Yu, X. Cai, S. Chen, H. Hu, B. Chen, X. Gao, B. Dong, D. Zou, High-performance perovskite memristor based on methyl ammonium lead halides, J. Mater. Chem. C 4 (2016) 1375–1381.
[60] Y. Sun, M. Tai, C. Song, Z. Wang, J. Yin, F. Li, H. Wu, F. Zeng, H. Lin, F. Pan, Competition between metallic and vacancy defect conductive filaments in a CH3NH3PbI3-based memory device, J. Phys. Chem. C 122 (2018) 6431–6436.

[61] X. Zhu, J. Lee, W.D. Lu, Iodine vacancy redistribution in organic-inorganic halide perovskite films and resistive switching effects, Adv. Mater. 29 (2017) 1700527.

[62] W. Xu, H. Cho, Y.-H. Kim, Y.-T. Kim, C. Wolf, C.-G. Park, T.-W. Lee, Organometal halide perovskite artificial synapses, Adv. Mater. 28 (2016) 5916–5922.

[63] J. Choi, Q. Van Le, K. Hong, C.W. Moon, J.S. Han, K.C. Kwon, P.-R. Cha, Y. Kwon, S.Y. Kim, H.W. Jang, Enhanced endurance organolead halide perovskite resistive switching memories operable under an extremely low bending radius, ACS Appl. Mater. Interfaces 9 (2017) 30764–30771.

[64] K. Yan, B. Chen, H. Hu, S. Chen, B. Dong, X. Gao, X. Xiao, J. Zhou, D. Zou, First fiber-shaped non-volatile memory device based on hybrid organic-inorganic perovskite, Adv. Electron. Mater. 2 (2016) 1600160.

[65] R. Waser, R. Dittmann, G. Staikov, K. Szot, Redox-based resistive switching memories—nanoionic mechanisms, prospects, and challenges, Adv. Mater. 21 (2009) 2632–2663.

[66] W.H. Xue, W. Xiao, J. Shang, X.X. Chen, X.J. Zhu, L. Pan, H.W. Tan, W.B. Zhang, Z.H. Ji, G. Liu, X.-H. Xu, J. Ding, R.-W. Li, Intrinsic and interfacial effect of electrode metals on the resistive switching behaviors of zinc oxide films, Nanotechnology 25 (2014), 425204.

[67] P. Shu, X. Cao, Y. Du, J. Zhou, J. Zhou, S. Xu, Y. Liu, S. Cao, Resistive switching performance of fibrous cross-point memories based on an organic–inorganic halide perovskite, J. Mater. Chem. C 8 (2020) 12865–12875.

[68] S. Kim, J. Yang, E. Choi, N. Park, Layered (C6H5CH2NH3)2CuBr4 perovskite for multi-level storage resistive switching memory, Adv. Funct. Mater. 30 (2020) 2002653.

[69] S. Lee, H. Kim, D.H. Kim, W. Bin Kim, J.M. Lee, J. Choi, H. Shin, G.S. Han, H.W. Jang, H.-S. Jung, Tailored 2D/3D halide perovskite heterointerface for substantially enhanced endurance in conducting bridge resistive switching memory, ACS Appl. Mater. Interfaces 12 (2020) 17039–17045.

[70] G. Sun, J. Liu, L. Zheng, W. Huang, H. Zhang, Preparation of weavable, all-carbon fibers for non-volatile memory devices, Angew. Chem. Int. Ed. 52 (2013) 13351–13355.

[71] E. Ercan, J.-Y. Chen, P.-C. Tsai, J.-Y. Lam, S.C.-W. Huang, C.-C. Chueh, W.-C. Chen, A redox-based resistive switching memory device consisting of organic-inorganic hybrid perovskite/polymer composite thin film, Adv. Electron. Mater. 3 (2017) 1700344.

[72] S. Lee, J. Choi, J.B. Jeon, B.J. Kim, J.S. Han, T.L. Kim, H.S. Jung, H.W. Jang, Conducting bridge resistive switching behaviors in cubic MAPbI3, orthorhombic RbPbI3, and their mixtures, Adv. Electron. Mater. 5 (2019) 1800586.

[73] Y.H. Lee, D.H. Kim, C. Wu, T.W. Kim, Memristive devices with a large memory margin based on nanocrystalline organic-inorganic hybrid CH3NH3PbBr3 perovskite active layer, Org. Electron. 62 (2018) 412–418.

[74] H. Xu, M. Feng, M. Liu, X. Sun, L. Wang, L. Jiang, X. Zhao, C. Nan, A. Wang, H. Li, Strain-mediated converse magnetoelectric coupling in La 0.7 Sr 0.3 MnO 3/Pb(Mg 1/3 Nb 2/3)O3 –PbTiO3 multiferroic heterostructures, Cryst. Growth Des. 18 (2018) 5934–5939.

[75] A. Jo, Y. Seo, M. Ko, C. Kim, H. Kim, S. Nam, H. Choi, C.S. Hwang, M.J. Lee, Textile resistance switching memory for fabric electronics, Adv. Funct. Mater. 27 (2017) 1605593.

[76] J.S. Han, Q. Van Le, J. Choi, K. Hong, C.W. Moon, T.L. Kim, H. Kim, S.Y. Kim, H.W. Jang, Air-stable cesium lead iodide perovskite for ultra-low operating voltage resistive switching, Adv. Funct. Mater. 28 (2018) 1705783.

[77] J. Choi, S. Park, J. Lee, K. Hong, D.-H. Kim, C.W. Moon, G. Do Park, J. Suh, J. Hwang, S.Y. Kim, H.S. Jung, N.-G. Park, S. Han, K.T. Nam, H.W. Jang, Organolead halide perovskites for low operating voltage multilevel resistive switching, Adv. Mater. 28 (2016) 6562–6567.

[78] D. Ielmini, Resistive switching memories based on metal oxides: mechanisms, reliability and scaling, Semicond. Sci. Technol. 31 (2016), 063002.

[79] E.-S. Choi, J.-M. Yang, S.-G. Kim, C. Cuhadar, S.-Y. Kim, S.H. Kim, D. Lee, N.-G. Park, The effect of compositional engineering of imidazolium lead iodide on the resistive switching properties, Nanoscale 11 (2019) 14455–14464.
[80] J.L.S. Lee, I.S. Gilmore, I.W. Fletcher, M.P. Seah, Topography and field effects in the quantitative analysis of conductive surfaces using ToF-SIMS, Appl. Surf. Sci. 255 (2008) 1560–1563.
[81] M. Enhessari, A. Salehabadi, Perovskites-based nanomaterials for chemical sensors, in: W. Wang (Ed.), Progresses in Chemical Sensor, IntechOpen, 2016.
[82] Z. Zhu, Q. Sun, Z. Zhang, J. Dai, G. Xing, S. Li, X. Huang, W. Huang, Metal halide perovskites: stability and sensing-ability, J. Mater. Chem. C 6 (2018) 10121–10137.
[83] M. Shellaiah, K.W. Sun, Review on sensing applications of perovskite nanomaterials, Chem. Aust. 8 (2020) 55.
[84] C. Bao, J. Yang, W. Zhu, X. Zhou, H. Gao, F. Li, G. Fu, T. Yu, Z. Zou, A resistance change effect in perovskite CH3NH3PbI3 films induced by ammonia, Chem. Commun. 51 (2015) 15426–15429.
[85] A. Maity, B. Ghosh, Fast response paper based visual color change gas sensor for efficient ammonia detection at room temperature, Sci. Rep. 8 (2018) 16851.
[86] Y. Zhao, K. Zhu, Optical bleaching of perovskite (CH3NH3)PbI3 through room-temperature phase transformation induced by ammonia, Chem. Commun. 50 (2014) 1605.
[87] A. Maity, A.K. Raychaudhuri, B. Ghosh, High sensitivity NH3 gas sensor with electrical readout made on paper with perovskite halide as sensor material, Sci. Rep. 9 (2019) 7777.
[88] A.D. Sheikh, V. Vhanalakar, A. Katware, K. Pawar, P.S. Patil, Two-step antisolvent precipitated MAPbI3-pellet-based robust room-temperature ammonia sensor, Adv. Mater. Technol. 4 (2019) 1–10.
[89] N. Gupta, O. Nanda, R. Grover, K. Saxena, A new inorganic-organic hybrid halide perovskite thin film based ammonia sensor, Org. Electron. 58 (2018) 202–206.
[90] W. Jiao, J. He, L. Zhang, Synthesis and high ammonia gas sensitivity of (CH3NH3) PbBr3-xIx perovskite thin film at room temperature, Sens. Actuators B Chem. 309 (2020), 127786.
[91] M.-A. Stoeckel, M. Gobbi, S. Bonacchi, F. Liscio, L. Ferlauto, E. Orgiu, P. Samorì, Reversible, fast, and wide-range oxygen sensor based on nanostructured organometal halide perovskite, Adv. Mater. 29 (2017) 1702469.
[92] G. Kakavelakis, E. Gagaoudakis, K. Petridis, V. Petromichelaki, V. Binas, G. Kiriakidis, E. Kymakis, Solution processed CH3NH3PbI3–xClx perovskite based self-powered ozone sensing element operated at room temperature, ACS Sens. 3 (2018) 135–142.
[93] X. Fu, S. Jiao, N. Dong, G. Lian, T. Zhao, S. Lv, Q. Wang, D. Cui, A CH3NH3PbI3 film for a room-temperature NO2 gas sensor with quick response and high selectivity, RSC Adv. 8 (2018) 390–395.
[94] A. Nur'aini, I. Oh, Volatile organic compound gas sensors based on methylammonium lead iodide perovskite operating at room temperature, RSC Adv. 10 (2020) 12982–12987.
[95] R. Zhu, Y. Zhang, H. Zhong, X. Wang, H. Xiao, Y. Chen, X. Li, High-performance room-temperature NO2 sensors based on CH3NH3PbBr3 semiconducting films: effect of surface capping by alkyl chain on sensor performance, J. Phys. Chem. Solids 129 (2019) 270–276.
[96] V.X. Hien, P.T. Hung, J. Han, S. Lee, J.-H. Lee, Y.-W. Heo, Growth and gas sensing properties of methylammonium tin iodide thin film, Scr. Mater. 178 (2020) 108–113.
[97] G. Li, W. Zhang, C. She, S. Jia, S. Liu, F. Yue, C. Jing, Y. Cheng, J. Chu, Stable fluorescent NH3 sensor based on MAPbBr3 encapsulated by tetrabutylammonium cations, J. Alloys Compd. 835 (2020), 155386.
[98] Y. Zhuang, W. Yuan, L. Qian, S. Chen, G. Shi, High-performance gas sensors based on a thiocyanate ion-doped organometal halide perovskite, Phys. Chem. Chem. Phys. 19 (2017) 12876–12881.

CHAPTER

11

Techno-economic analysis and toxicity of halide perovskites

Pandiyarajan Mariyappan[a] *and*
Towhid H. Chowdhury[b]

[a]Department of Electrical and Electronics Engineering, Sri Venkateswara College of Engineering (SVCE), Chennai, Tamil Nadu, India,
[b]Organic Printed Electronics Laboratory (OPEL), Department of Chemical Engineering, Pohang University of Science & Technology, Pohang, Republic of Korea

1. Introduction

Since the inception of solar cells (SCs) in the early 1950s, material chemists, physicists, engineers, and scientists have tried various approaches to develop the best power conversion efficiency (PCE) SCs with high throughput [1–4]. The development of SCs from their inception to date can be categorized into various generations according to the technology development and respective timeline. The first generation of solar cells was designed with single-layer *p–n* junction diodes [5]. These *p–n* junction diodes were fabricated with the then developed silicon-wafer-based technology. Surprisingly, even to date, these kinds of solar cells govern the photovoltaic (PV) market due to their robust stability and highly efficient photovoltaic response. However, as the source of silicon used in these PVs requires a complex purification process, they remain expensive. With an aim to develop SCs at a reduced cost, the second generation of SC thin-film technology has been developed and adopted. Thin-film technology-based solar cells often contain amorphous silicon, polycrystalline silicon, microcrystalline silicon, cadmium telluride, and copper indium selenide/sulfide [6,7] compounds as the key light absorber material. Compared to the earlier generation of SCs, these thin-film-based SCs could be

Low-Dimensional Halide Perovskites
https://doi.org/10.1016/B978-0-323-88522-5.00011-9

fabricated with much cheaper materials. However, the processing costs of these SCs are still high due to the fabrication condition dependency on high vacuum and temperature conditions. To overcome these challenges, third-generation SCs have been considered, which are lighter in weight, flexible in shape, and can offer versatile custom material synthesis choices with low cost for large-scale production [8–11]. These third-generation solar cells include organic photovoltaics (OPVs), dye-sensitized solar cells (DSSCs), quantum dot sensitized solar cells (QDSSCs), and perovskite solar cells (PSCs).

PSCs, which have seen rapid development of photovoltaic performances and stability on a laboratory scale over the past decade, already have a compatible photovoltaic performance with silicon SCs. However, to upscale this technology and enable its daily use, some challenges must be addressed. To overcome the bottleneck of the existing PV technologies, the upscaling of PSCs for mass production should be cost-effective. In this regard, a levelized cost of electricity (LCOE) can be a viable tool to understand the potential real-life applications of PSCs on a large scale. A precise LCOE could direct potential manufacturers to develop fabrication and marketing for the implementation of PSCs at a feasible price. Another major concern for commercialization is the toxicity of PSCs. The presence of a vast amount of lead (Pb) as the key component in PSCs is still under debate. Although the research and development of PSCs have seen significant advancement in terms of PCE, the key issues of toxicity and economic feasibility have rarely been addressed. In this chapter, we aim to discuss the techno-economic analysis and feasibility of PSCs to understand their potential implementation in real-life applications and compare them with existing technologies.

2. Methods

For any new technology to be adopted for commercialization and daily life usage, it is important to understand the detailed techno-economic assessment. For this reason, key technological analysis tools such as LCOE have emerged as a viable source to understand the potential successful implementation of perovskite solar cells (PSCs). The System Advisor Model (SAM, version 2016.3.14) was created by the National Renewable Energy Laboratory (NREL) to calculate the LCOE of a photovoltaic (PV) system using PSC modules to examine the cost of solar energy. In general, LCOE is the ratio of the entire lifecycle cost of the PV module to the total electricity produced throughout the module's lifespan [12].

$$\mathrm{LCOE} = \frac{\sum_{i=10}^{n} \frac{C_i}{(1+d)^i}}{\sum_{i=10}^{n} \frac{E_i}{(1+d)^i}} \tag{1}$$

where C_i is the total module cost and E_i is the energy produced (kWh) throughout the year, d is the discount rate, and n is the system's lifespan.

The annual energy production (E_t) of the PV power station with the same set of modules' (solar array) capacity is computed using the following equation:

$$E_t = P \times S \times (1 - L) \times (1 - r)^t \tag{2}$$

where P is the power rating of the solar array, S is the annual solar insolation at the installed site, L is the array's constant loss factor owing to shading, soiling, AC conversion, and DC connection, and r denotes the system degradation rate.

The system's annual cost (C_t) can be calculated using the following equation:

$$C_t = I_t \times O_t \times F_t \tag{3}$$

where I_t represents the initial PV module installation cost, O_t represents the operating cost of the system, and F_t represents the maintenance charges for year t.

The initial installation charge (I_0) is the most expensive factor of a PV system. The cost of the PV modules, DC-AC inverters, other balance of system components (such as wiring, switching, and mounting systems), manpower charges, project executing overhead, grid connectivity, land cost, and government taxes are all included in this initial cost. The annual costs of maintaining the system's regular operation, as well as the costs of servicing faulty modules and other electrical components, are included in the system operating charge. Interest on project debt, taxes, insurance, incentives, and internal return on initial investment are all included in the financial cost [13].

3. Techno-economic analysis of different types of perovskite-based PV modules

Among the various structures of PSCs, two types of structures are considered seriously for the commercialization aspects: the single-junction perovskite modules and perovskite/tandem heterojunction modules. In this section, we discuss the progress of techno-economic aspects of various types of PSCs.

3.1 Perovskite single-junction PV module

Cai et al. [14] analyzed the LCOE for a single-junction perovskite module with 3.5–4.9 US cents/kWh with a module lifetime of 15 years. Although this was the first report on the LCOE of PSCs for commercialization, the authors remarked on enhancing photovoltaic performance rather than focusing on the cost. However, in their studies, the material costs, which constitute one of the most critical parameters, have not been considered [12,15]. Chang et al. made more precise calculations based on material cost and analyzed the LCOE of the single-junction PSC material and projected the module cost at US \$107/m^2. However, Song et al. projected an LCOE of 5 US cents/kWh with a module lifetime of 30 years [16]. The detailed analysis of the LCOE cost of PSCs has been summarized in some recent articles [17,18]. These studies encourage further discussion on cost estimation and justification for the commercialization of PSCs.

A single-junction PSC module with the balance of module (BOM) components including frame, interconnection busbars, sealant, lamination film, junction box, and wiring has been identified as a potential structure. To avoid the complexity of the fabrication process, this single-junction PSC structure (Fig. 1) implements methylammonium lead iodide

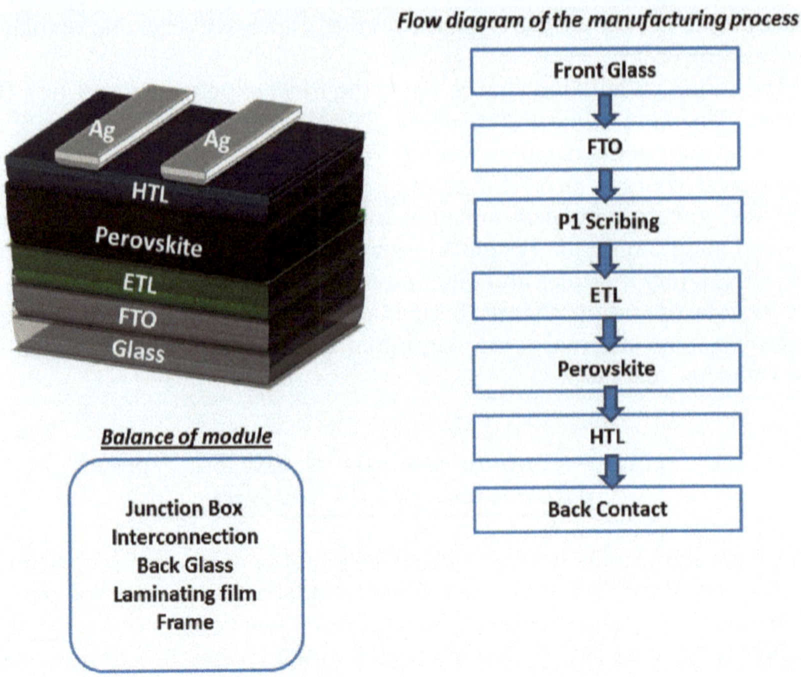

FIG. 1 Device structure and manufacturing process flow of single-junction perovskite solar cell module. *No permission required.*

($MAPbI_3$) perovskite as the main absorber layer, SnO_2 as an electron transport layer (ETL), poly(3-hexylthiophene-2,5-diyl) (P3HT) as a hole transport layer (HTL), silver (Ag) as a back contact, and fluorine-doped tin oxide (FTO) as a front contact. This type of module can be fabricated by low-temperature processing techniques such as screen printing and dip coating [12,14]. When the individual costs of each fabrication step are summed up, the cost per unit area of the perovskite single-junction module can be obtained. For each manufacturing step, five important cost segments need to be considered: (i) raw materials, (ii) utilities, (iii) equipment maintenance, (iv) labor source for the production line, and (v) building and processing for modules. Based on this model, the estimated manufacturing cost of the single-junction perovskite PV module is $\$28.7 \pm 4.9/m^2$ [12]. For front and back contact individually, the cost has been estimated at 10 $\$/m^2$. For the ETL of TiO_2, 9 \$/kg has been considered a reasonable price. In this model, the perovskite compound of CH_3NH_3I and PbI_2 cost a total of 1.046 $\$/m^2$. The cost of the absorber layer can also be cheaper than the other existing absorber layer materials. However, one of the drawbacks of this model is the usage of Spiro-OMeTAD as the HTL, which costs 1621 \$/kg. Considering the large-scale production, this might be a critical issue. As highlighted by many research articles, the synthesis and purification cost of Spiro-OMeTAD cannot be reduced any further due to the necessary purification steps [19–22]. Although numerous alternatives of HTLs to Spiro-OMeTAD [23,24] have been suggested, in terms of real-life implementation, the use of these alternative HTLs remains arguable. In addition to the required materials for PSCs fabrication for upscaling, it is important to have the necessary sealing, lamination of the PSCs, solder wire to interconnect the cells in arrays, and the junction box to make a proper module. These components are readily available and can be transferred from existing technologies. The total cost of the single-junction PV modules can be reduced by using cheaper glass substrates without sacrificing photovoltaic performance. However, considering the present state of the module fabrication, it is estimated that the value of a single-junction perovskite module can be in the range of 21.9 to 22.3 $\$/m^2$ [13,16]. Single-junction PSCs have already been adopted for commercialization trials. As a result of fast development in perovskite PV technology, in-line production efficiency is expected to reach more than 20% in future. The major process flow and a schematic diagram of the device structure are shown in Fig. 1, and the probable costs of the associated components are provided in Table 1. (A few parameters have been adapted from Li et al. [13] with permission.)

The standard single-junction perovskite device fabricated by roll-to-roll (R2R) configuration with a flexible substrate has been highlighted as a potential solution for the commercialization of large-scale modules. R2R fabrication of solar cells has been successfully implemented in organic photovoltaics (OPVs) and polymer solar cells, and many flexible solar cells

TABLE 1 Material costs for single-junction perovskite solar cells module fabrication.

Component	Raw material	Price ($/kg)	Weight (g/m^2)	Material cost ($\$/m^2$)
Front glass	3mm glass	0.95	7500	7.000
FTO	FTO	720 (550–990)	1.79	1.298
SnO_2	SnO_2	100	0.29	0.031
Perovskite component	CH_3NH_3I	450	0.23	0.066
Perovskite component	PbI_2	700	2.30	0.980
Spiro-OMeTAD	Spiro-OMeTAD	1621	0.20	0.324
LiF	LiF	8000	0.003	0.054
Ag (silver)	Ag	521	1.07	0.557
Edge sealing	Butyl rubber	2.2	10.5	0.037
Lamination	EVA	11	138	1.535
Solder wire	Sn	10	1.065	0.011
Back glass	2mm glass	0.6	5000	3.000
Junction box	N/A	N/A	N/A	7.500
Total cost				**22.3**

From Z. Li, Y. Zhao, X. Wang, Y. Sun, Z. Zhao, Y. Li, H. Zhou, Q. Chen, Cost Analysis of perovskite tandem photovoltaics, Joule 2 (8) (2018) 1559–1572. https://doi.org/10.1016/j.joule.2018.05.001.

fabricated by this method are gaining momentum in the PV market [25,26]. Inspired by the research in OPVs, many recent findings highlight the probability of fabricating PSCs with high power conversion efficiency (PCE) [27–29]. One of the most popular structures for R2R production is configured with indium-doped zinc oxide (IZO) coated on polyethylene terephthalate (PET). Flexible substrates are used as the front contact [30]. These flexible substrates will pave the way for lightweight solar cell applications with lower manufacturing costs. The active material in this module is mixed iodide-bromide formamidinium cesium lead perovskite, which is more stable with flexible substrates compared to $MAPbI_3$ [31]. To reduce the material and processing costs, nickel oxide (NiO_x) is used as an HTL and phenyl-C_{61}-butyric acid methyl ester (PCBM) is used as an ETL. For the back contact, aluminum (Al) is suggested. The full device structure and a flow diagram of the manufacturing process are shown in Fig. 2. The front contact IZO and back contact Al can be deposited by using the sputtering technique, followed by laser scribing. Similarly, both ETLs and

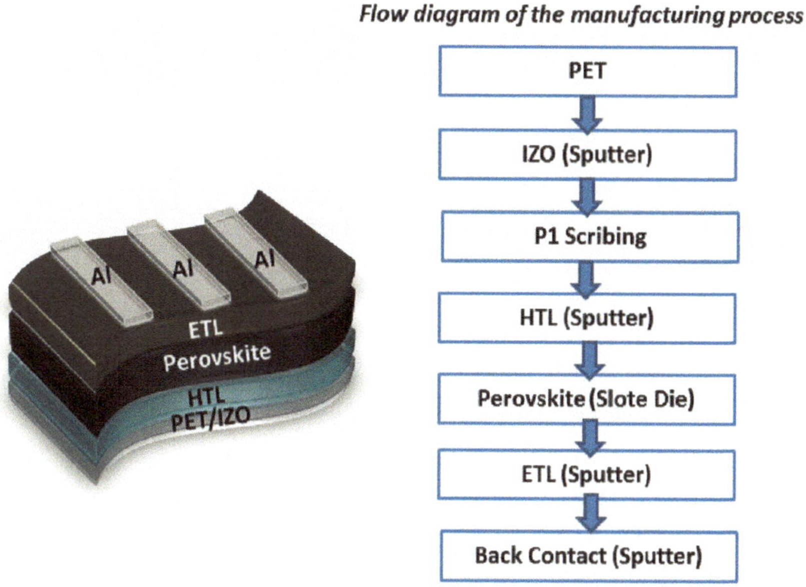

FIG. 2 Device structure and manufacturing process flow of R2R processing for the flexible perovskite solar cell module. *No permission required.*

HTLs can be deposited by physical vapor deposition techniques (e.g., sputtering). For the R2R process, a perovskite layer deposited by the slot-die coating technique can be considered the most suitable method. The total manufacturing cost of the R2R processed PV module can be calculated by adding the cost of each manufacturing step (Table 2). The cost analysis of the R2R process reveals that the barrier film dominates the overall cost. The sputtering process and instruments are also costly in terms of maintenance. However, a very interesting and encouraging thing in this process is that the perovskite material used in this model is very cheap, at 0.50 $\$/m^2$ [16].

However, in real-life applications, the PV module cost is estimated by means of $/W. Hence another method to decrease PV module cost is to increase the PCE. If the PCE of R2R processed perovskite single-junction is improved to the real world PCE of 20%, the cost of the single-junction perovskite module can be reduced. This strategy will enable PSCs to be priced competitively in relation to silicon PV manufacturing costs [29,30].

3.2 Perovskite/silicon tandem PV module

Perovskite/silicon tandem PV modules are generally fabricated by combining PSCs with silicon SCs by forming a heterojunction. The device configuration and the manufacturing process flow are shown in Fig. 3.

TABLE 2 Material cost for R2R processing perovskite solar cell module fabrication.

Component	Raw material	Price (\$/kg)	Weight (g/m^2)	Material cost (\$/m^2)
PET	PET	4	1	2
IZO	IZO (sputter)	–	1.405	1.979
NiO	NiO (sputter)	200	0.667	0.143
Perovskite	CsBr	0.563	0.134	0.26
Perovskite	$PbBr_2$	70 (50–150)	0.341	0.157
Perovskite	PbI_2	700	1.84	0.078
Perovskite	FABr	200	0.385	0.584
PCBM	PCBM	35,000	0.068	2.4
Al	Al	2.2	1.07	1.26
Laser scribing	Laser scribing	–	–	0.8
Edge sealing	Butyl rubber	2.2	10.5	0.037
Lamination	EVA	11	138	1.535
Total cost				**11.23**

This type of PV module showed the highest PCE of 29.52% in 2020 [32], and is one of the most promising PV technologies that is moving toward commercialization. The overall manufacturing cost of this type of PV module increased by 5 \$/m^2 from the single-junction perovskite PV module manufacturing cost due to the multilayer manufacturing process. For the fabrication of perovskite/silicon tandem modules, the *n*-type c-silicon wafers can be double-side polished by an industrial-scale polishing machine. The silicon nitride layer is then deposited on one side by plasma-enhanced chemical vapor deposition (PECVD), followed by texturing of the uncoated side with potassium hydroxide (KOH). Intrinsic and *p*-type a-Si:H films can be fabricated by the conventional PECVD method on the textured side of the wafer, and intrinsic and *n*-type a-Si:H films are then deposited on the polished side [13]. The ITO layer is usually deposited by sputtering on both sides of the device, followed by a heating treatment. After successful completion, a screen-printed NiO_x as the HTL and a screen-printed perovskite layer can be deposited sequentially. C_{60} as ETL can be thermally evaporated on the module, followed by stoichiometric SnO_2 deposition and ITO sputtering on top of the C_{60}. It has been suggested that Cu could be deposited as the bottom electrode.

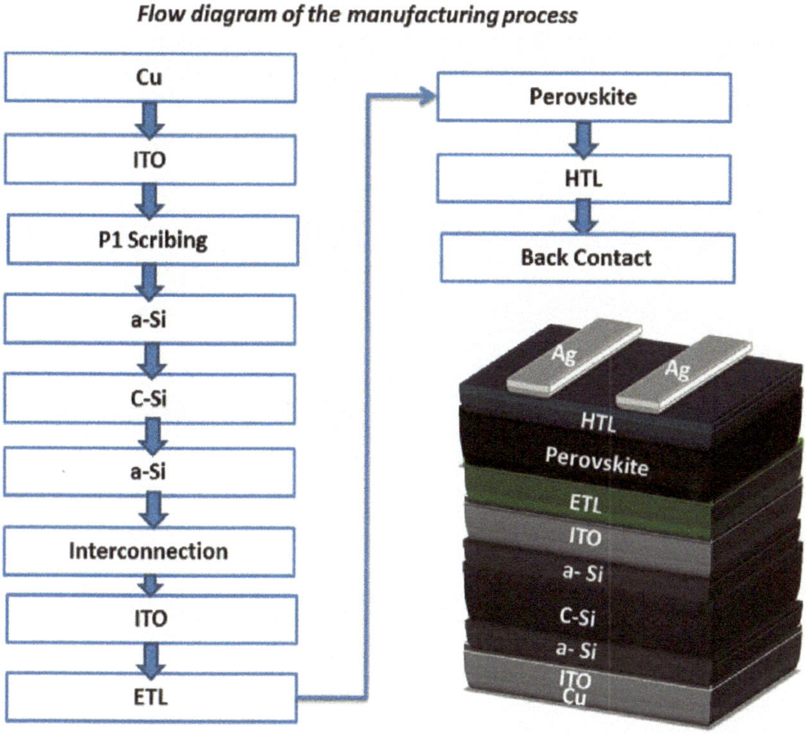

FIG. 3 Device structure and manufacturing process flow of perovskite/silicon solar cell module. *No permission required.*

The total cost for the fabrication of the perovskite/silicon tandem PV module has been estimated at 87.3 $\$/m^2$ [32]. Although the overall cost of this technology is almost three times higher than that of single-junction PSCs, the durability of these tandem devices is expected to be greater than that of single-junction PSCs. Additionally, the bottom cell of Si technology has matured over the last three decades. In these aspects, the perovskite/silicon tandem PV is the front-runner among the existing technologies evaluated for large-scale production. The major process flow is summarized in Fig. 3, and the material costs of key components used in the manufacturing process are listed in Table 3 adopted from [12,13,16,18,29,30].

We compared the overall LCOE of the PSCs based on existing literature. In our analysis, the manufacturing cost describes the cost of raw materials, and the cost of module production includes building, equipment processing, etc. It is apparent that the R2R processed perovskite PVs can have the lowest material costs, which reduces the overall LCOE for single-junction PSC modules. If PSCs can be fabricated by R2R processing, together with the silicon devices, a cost-effective tandem perovskite module can be

TABLE 3 Material costs for perovskite/silicon solar cells module fabrication.

Component	Raw material	Price ($/kg)	Weight (g/m^2)	Material cost ($/m^2)
Front glass	3mm glass	0.95	7500	7.000
HIT bottom cell	Silicon wafer, KOH, *n*-type silicon and *p*-type silicon, Ag back contact	–	–	60
ITO	ITO	700	209	1.50
NiO	NiO	200	0.667	0.143
C_{60}	C_{60}	160,000	0.026	5.467
Perovskite	CH_3NH_3I	450	0.23	0.066
Perovskite	PbI_2	700	2.30	0.980
SnO_2	SnO_2	100	0.035	0.004
LiF	LiF	8000	0.003	0.054
Cu	Cu	9.0	1.11	0.011
Edge sealing	Butyl rubber	2.2	10.5	0.037
Lamination	EVA	11	138	1.535
Solder wire	Sn	10	1.065	0.011
Back glass	2mm glass	0.6	5000	3.000
Junction box	N/A	N/A	N/A	7.500
Total cost				**87.3**

From Z. Li, Y. Zhao, X. Wang, Y. Sun, Z. Zhao, Y. Li, H. Zhou, Q. Chen, Cost analysis of perovskite tandem photovoltaics, Joule 2 (8) (2018) 1559–1572. https://doi.org/10.1016/j.joule.2018.05.001.

fabricated. As larger sizes of modules with higher PCEs and stability are constantly being reported these days, it is very likely that the balance between the material cost and performance will be achieved within a few years.

4. Toxicity of perovskite solar cells

Although PSCs are providing a solid platform for future greener energy harvesting at a reduced cost, the presence of Pb in PSCs is highly concerning. It is obviously important to avoid or prevent Pb leakage into the environment during the deconstruction process of perovskite modules. In comparison to existing technologies, the coal industry emits 80,000 metric tons of Pb into the atmosphere per year. However, the global installed capacity of coal power plants is around 1.75 TW. When compared to Pb-based PSCs, if a 1 TW solar power plant were installed with 20% efficient Pb-based PSCs, it would require approximately $0.8\,g\,m^{-2}$ of Pb, or 20,000 metric tons of Pb. It must be noted that this amount of Pb would be within the SCs and would not have an effect directly on the environment until the end of life [33]. Additionally, the usual concentration of Pb in the soil may be as high as 36 mg/kg. It should be noted that if the Pb is contained in a 1 m^2 panel leakage, this would increase the concentration of Pb by 4 mg/kg in solid form. Therefore, strict measures for preventing Pb leakage must be introduced. The aim should always be to dispose of Pb safely, but leakages may occur under various circumstances. Pb-based PSCs can produce electricity in a more environmentally friendly way than by burning coal. As a rough comparison, to demonstrate the potential risk of Pb, there are approximately 250 million lead-acid batteries in the US alone [34]. However, a recent study shows that the Pb recycled from these batteries can be used in PSCs without sacrificing the PCE [34]. With the rapid growth of lithium batteries using current technologies, the potential usage of Pb acetate batteries is decreasing and the Pb from these batteries can be used to develop perovskite PVs. In the aforementioned study by Chen et al., it was projected that the quantity of Pb recovered from a single car battery would be enough to produce about 700 m^2 of perovskite-based PV panels. Therefore, Pb recycled from existing technologies can be a popular choice in terms of reducing costs and providing environmental benefits. Bearing all this in mind, using only Pb as the metal in a perovskite absorber may not have a significant environmental impact if the necessary steps are taken. The estimated cost of PSCs compared to other existing technologies means that PSCs can provide alternative solutions for the next generation of energy harvesting. If all aspects of PSCs can be considered and their various parts can be recycled, it is highly likely that PSCs will be able to be marketed at a cheaper price.

5. Comparison to other existing PV technologies

To evaluate the total manufacturing cost of an SC module, minimum sustainable price (MSP) cost analyses are often adopted. The MSP of the PV module includes the manufacturing cost of the module and overhead costs such as sales, research and development, taxes, and bank interests [12,14]. Fig. 4 compares the MSP of the perovskite-based PV module with other existing commercial PV technologies. The average MSP of the all perovskite-based PV modules (including single-junction PSCs and

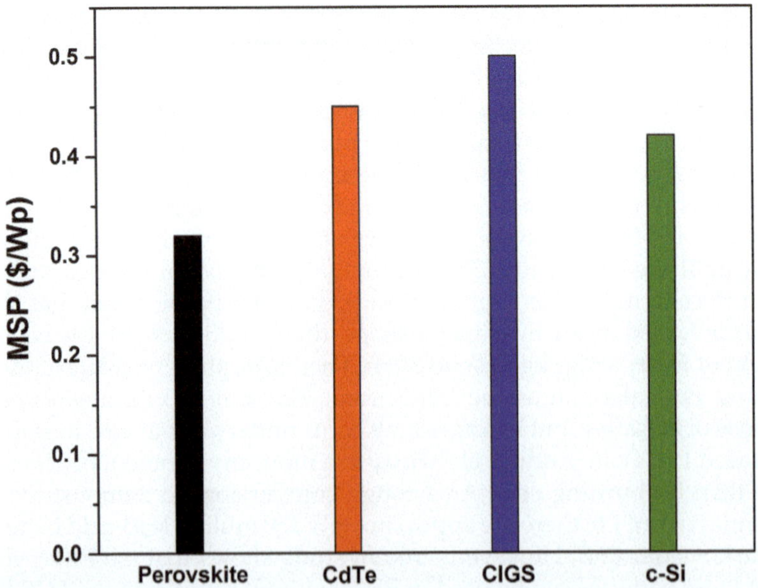

FIG. 4 MSP comparison of different PV technologies. *No permission required.*

perovskite/silicon tandem SCs) is about 0.32 $/Wp with a PCE of 18%, which is lower than for well-established PV technologies such as CdTe (0.45 $/Wp), CIGS (0.5 $/Wp), and c-Si (0.4 $/Wp) [14]. In addition, the environmental impacts of perovskite PV technology, for 1 kWh as a function unit, are higher than for any other PV technology. Although these performances are significantly lower than what has been demonstrated in the laboratories, these perovskite-based PV technologies have the ability to achieve a PCE of 30%.

6. Challenges and future perspective

In accordance with the current sustainable goal for a greener and safer future, the end of life of PSC modules should be considered. As the main component for the fabrication of highly efficient modules is still Pb-based, alternative materials with similar optoelectronic properties are being sought [35–37]. However, due to the more complex issues involved, Pb-free absorbers are not on a par with current Pb-based PSC technology. Considering this, Pb-based PSCs are still the only feasible choice for fabricating modules. A recent study has evaluated the potential risk factors of implementing Pb modules for large-scale production, and highlights that if the potential risk hazards are not taken into account now, the PSCs may impact highly on the environment and daily life [38]. End of life has been considered in some of the LCOE models [39,40]. However, until serious evaluation occurs on a larger scale, this impact cannot be assessed. Only one report highlights two possible end-of-life aspects of the PSCs, and recycling each layer is a probable solution [41]. If each component of the layers can be reused, the end-of-life problem can be solved. However, it is unclear whether the recycled components can yield the same level of performance as the new components. Therefore, multiple thorough studies should focus on this area in future.

The rapid improvement in perovskite PVs compared to any other reported PVs over the course of just a decade is making the commercialization aspects brighter than ever. It has previously been estimated that the market for PSCs has the potential to cross $214 million in a matter of just 4 years [42]. However, in terms of stability and cell lifetime, questions regarding the perovskite PVs' future remain critical. Until now, studies conducted with Pb-based perovskite solar panels have not included any assessment of the recycling of Pb metals and the corresponding environmental impact. As highlighted in this chapter, the toxicity of Pb at the end of life of perovskite panels is potentially threatening to the environment. Even the vast amount of Pb disposal has not been assessed prior to the end of life of PSC panels. As perovskite PVs are also a type of electronic device, they have to be recycled according to European Union regulations in EU member countries. From previous experience gained in the PV community, recycling costs can be high and hazardous at the same time. Some studies show that, unlike other types of SCs, PSCs can be deconstructed easily due to their chemical approach to fabrication [43]. If some of the remaining issues of PSCs, such as long-term stability and the recycling process, are addressed and overcome, it is likely that PSCs will come into daily usage and effectively provide cheaper electricity.

References

[1] R.L. Crabb, F.C. Treble, Thin silicon solar cells for large flexible arrays [11], Nature 213 (5082) (1967) 1223–1224, https://doi.org/10.1038/2131223a0.

[2] S. Günes, H. Neugebauer, N.S. Sariciftci, Conjugated polymer-based organic solar cells, Chem. Rev. 107 (4) (2007) 1324–1338, https://doi.org/10.1021/cr050149z.

[3] H. Hoppe, N.S. Sariciftci, Organic solar cells: an overview, J. Mater. Res. 19 (7) (2004) 1924–1945, https://doi.org/10.1557/JMR.2004.0252.

[4] J. Hou, H.Y. Chen, S. Zhang, G. Li, Y. Yang, Synthesis, characterization, and photovoltaic properties of a low band gap polymer based on silole-containing polythiophenes and 2,1,3-benzothiadiazole, J. Am. Chem. Soc. 130 (48) (2008) 16144–16145, https://doi.org/10.1021/ja806687u.

[5] M.A. Green, Third generation photovoltaics: solar cells for 2020 and beyond, Phys. E: Low-Dimensional Syst. Nanostruct. 14 (1–2) (2002) 65–70, https://doi.org/10.1016/S1386-9477(02)00361-2.

[6] P. Jackson, D. Hariskos, E. Lotter, S. Paetel, R. Wuerz, R. Menner, W. Wischmann, M. Powalla, New world record efficiency for Cu(In,Ga)Se2 thin-film solar cells beyond 20%, Prog. Photovolt. Res. Appl. 19 (7) (2011) 894–897, https://doi.org/10.1002/pip.1078.
[7] B.R. Saunders, M.L. Turner, Nanoparticle-polymer photovoltaic cells, Adv. Colloid Interf. Sci. 138 (1) (2008) 1–23, https://doi.org/10.1016/j.cis.2007.09.001.
[8] G.F. Brown, J. Wu, Third generation photovoltaics, Laser Photonics Rev. 3 (4) (2009) 394–405, https://doi.org/10.1002/lpor.200810039.
[9] G. Conibeer, Third-generation photovoltaics, Mater. Today 10 (11) (2007) 42–50, https://doi.org/10.1016/S1369-7021(07)70278-X.
[10] M.A. Green, Third generation photovoltaics: ultra-high conversion efficiency at low cost, Prog. Photovolt. Res. Appl. 9 (2) (2001) 123–135, https://doi.org/10.1002/pip.360.
[11] R. Szweda, Third generation solar cells, III-Vs Rev. 16 (6) (2003) 53–55, https://doi.org/10.1016/S0961-1290(03)00792-0.
[12] Z. Song, C.L. McElvany, A.B. Phillips, I. Celik, P.W. Krantz, S.C. Watthage, G.K. Liyanage, D. Apul, M.J. Heben, A technoeconomic analysis of perovskite solar module manufacturing with low-cost materials and techniques, Energy Environ. Sci. 10 (6) (2017) 1297–1305, https://doi.org/10.1039/c7ee00757d.
[13] Z. Li, Y. Zhao, X. Wang, Y. Sun, Z. Zhao, Y. Li, H. Zhou, Q. Chen, Cost analysis of perovskite tandem photovoltaics, Joule 2 (8) (2018) 1559–1572, https://doi.org/10.1016/j.joule.2018.05.001.
[14] M. Cai, Y. Wu, H. Chen, X. Yang, Y. Qiang, L. Han, Cost-performance analysis of perovskite solar modules, Adv. Sci. 4 (1) (2017), https://doi.org/10.1002/advs.201600269.
[15] N.L. Chang, A.W.Y. Ho-Baillie, D. Vak, M. Gao, M.A. Green, R.J. Egan, Manufacturing cost and market potential analysis of demonstrated roll-to-roll perovskite photovoltaic cell processes, Sol. Energy Mater. Sol. Cells 174 (2018) 314–324, https://doi.org/10.1016/j.solmat.2017.08.038.
[16] N.L. Chang, A.W. Yi Ho-Baillie, P.A. Basore, T.L. Young, R. Evans, R.J. Egan, A manufacturing cost estimation method with uncertainty analysis and its application to perovskite on glass photovoltaic modules, Prog. Photovolt. Res. Appl. 25 (5) (2017) 390–405, https://doi.org/10.1002/pip.2871.
[17] E. Leccisi, V. Fthenakis, Life-cycle environmental impacts of single-junction and tandem perovskite PVs: a critical review and future perspectives, Prog. Energy 032002 (2020), https://doi.org/10.1088/2516-1083/ab7e84.
[18] A. Urbina, The balance between efficiency, stability and environmental impacts in perovskite solar cells: a review, J. Phys. Energy 2 (2) (2020), https://doi.org/10.1088/2515-7655/ab5eee.
[19] I.M. Abdellah, T.H. Chowdhury, J.J. Lee, A. Islam, M.K. Nazeeruddin, M. Gräetzel, A. El-Shafei, Facile and low-cost synthesis of a novel dopant-free hole transporting material that rivals spiro-OMeTAD for high efficiency perovskite solar cells, Sustain. Energy Fuels 5 (1) (2021) 199–211, https://doi.org/10.1039/d0se01323d.
[20] R. Kaneko, T.H. Chowdhury, K. Sugawa, J.J. Lee, J. Otsuki, A. Islam, Electro-active nanofibers of a tetrathiafulvalene derivative with amide hydrogen bonds as a dopant-free hole transport material for perovskite solar cells, Sol. Energy 194 (2019) 248–253, https://doi.org/10.1016/j.solener.2019.10.078.
[21] K. Narayanaswamy, B. Yadagiri, T.H. Chowdhury, T. Swetha, A. Islam, V. Gupta, S.P. Singh, Impact of A-D-A-structured dithienosilole- and phenoxazine-based small molecular material for bulk heterojunction and dopant-free perovskite solar cells, Chem. Eur. J. 25 (71) (2019) 16320–16327, https://doi.org/10.1002/chem.201903599.
[22] G. Wu, Y. Zhang, R. Kaneko, Y. Kojima, K. Sugawa, T.H. Chowdhury, A. Islam, Q. Shen, M. Akhtaruzzaman, T. Noda, J. Otsuki, Hole-transport materials containing triphenylamine donors with a spiro[fluorene-9,9′-xanthene] core for efficient and stable

large area perovskite solar cells, Solar RRL 1 (9) (2017), https://doi.org/10.1002/solr.201700096.
[23] X. Sun, X. Yu, Z. Li, Recent advances of dopant-free polymer hole-transporting materials for perovskite solar cells, ACS Appl. Energy Mater. 3 (11) (2020) 10282–10302, https://doi.org/10.1021/acsaem.0c01917.
[24] L. Zhang, X. Zhou, C. Liu, X. Wang, B. Xu, A review on solution-processable dopant-free small molecules as hole-transporting materials for efficient perovskite solar cells, Small Methods 4 (9) (2020), https://doi.org/10.1002/smtd.202000254.
[25] L.H. Rossander, H.F. Dam, J.E. Carlé, M. Helgesen, I. Rajkovic, M. Corazza, F.C. Krebs, J.-W. Andreasen, In-line, roll-to-roll morphology analysis of organic solar cell active layers, Energy Environ. Sci. 10 (11) (2017) 2411–2419, https://doi.org/10.1039/c7ee01900a.
[26] R. Søndergaard, M. Hösel, D. Angmo, T.T. Larsen-Olsen, F.C. Krebs, Roll-to-roll fabrication of polymer solar cells, Mater. Today 15 (1–2) (2012) 36–49, https://doi.org/10.1016/S1369-7021(12)70019-6.
[27] Y.J. Heo, J.E. Kim, H. Weerasinghe, D. Angmo, T. Qin, K. Sears, K. Hwang, Y.S. Jung, J. Subbiah, D.J. Jones, M. Gao, D.Y. Kim, D. Vak, Printing-friendly sequential deposition via intra-additive approach for roll-to-roll process of perovskite solar cells, Nano Energy 41 (2017) 443–451, https://doi.org/10.1016/j.nanoen.2017.09.051.
[28] J.L. Hodgkinson, H.M. Yates, A. Walter, D. Sacchetto, S.J. Moon, S. Nicolay, Roll to roll atmospheric pressure plasma enhanced CVD of titania as a step towards the realisation of large area perovskite solar cell technology, J. Mater. Chem. C 6 (8) (2018) 1988–1995, https://doi.org/10.1039/c8tc00110c.
[29] J.E. Kim, S.S. Kim, C. Zuo, M. Gao, D. Vak, D.Y. Kim, Humidity-tolerant roll-to-roll fabrication of perovskite solar cells via polymer-additive-assisted hot slot die deposition, Adv. Funct. Mater. 29 (26) (2019), https://doi.org/10.1002/adfm.201809194.
[30] J. Xiao, Technoeconomic Analysis of Perovskite Photovoltaic Manufacturing for Powering Telecommunications Towers, Massachusetts Institute of Technology, USA, 2018.
[31] T. Leijtens, K. Bush, R. Cheacharoen, R. Beal, A. Bowring, M.D. McGehee, Towards enabling stable lead halide perovskite solar cells; interplay between structural, environmental, and thermal stability, J. Mater. Chem. A 5 (23) (2017) 11483–11500, https://doi.org/10.1039/c7ta00434f.
[32] R. Peleg, Oxford PV Hits New Perovskite-Silicon Tandem World Record, 2020. https://www.perovskite-info.com/oxford-pv-hits-new-perovskite-silicon-tandem-world-record.
[33] S.D. Stranks, H.J. Snaith, Metal-halide perovskites for photovoltaic and light-emitting devices, Nat. Nanotechnol. 10 (5) (2015) 391–402, https://doi.org/10.1038/nnano.2015.90.
[34] P.Y. Chen, J. Qi, M.T. Klug, X. Dang, P.T. Hammond, A.M. Belcher, Environmentally responsible fabrication of efficient perovskite solar cells from recycled car batteries, Energy Environ. Sci. 7 (11) (2014) 3659–3665, https://doi.org/10.1039/c4ee00965g.
[35] M. Abdel-Shakour, T.H. Chowdhury, K. Matsuishi, I. Bedja, Y. Moritomo, A. Islam, High-efficiency tin halide perovskite solar cells: the chemistry of tin (II) compounds and their interaction with Lewis base additives during perovskite film formation, Solar RRL 5 (1) (2021), https://doi.org/10.1002/solr.202000606.
[36] W. Ke, M.G. Kanatzidis, Prospects for low-toxicity lead-free perovskite solar cells, Nat. Commun. 10 (1) (2019), https://doi.org/10.1038/s41467-019-08918-3.
[37] M. Wang, W. Wang, B. Ma, W. Shen, L. Liu, K. Cao, S. Chen, W. Huang, Lead-free perovskite materials for solar cells, Nano-Micro Lett. 13 (1) (2021), https://doi.org/10.1007/s40820-020-00578-z.
[38] J.M. Kadro, A. Hagfeldt, The end-of-life of perovskite PV, Joule 1 (1) (2017) 29–46, https://doi.org/10.1016/j.joule.2017.07.013.

[39] J.A. Alberola-Borràs, R. Vidal, E.J. Juárez-Pérez, E. Mas-Marzá, A. Guerrero, I. Mora-Seró, Relative impacts of methylammonium lead triiodide perovskite solar cells based on life cycle assessment, Sol. Energy Mater. Sol. Cells 179 (2018) 169–177, https://doi.org/10.1016/j.solmat.2017.11.008.

[40] L. Serrano-Lujan, N. Espinosa, T. Larsen-Olsen, J. Abad, A. Urbina, F. Krebs, Tin-and lead-based perovskite solar cells under scrutiny: an environmental perspective, Adv. Energy Mater. 5 (2015).

[41] I. Celik, A.B. Phillips, Z. Song, Y. Yan, R.J. Ellingson, M.J. Heben, D. Apul, Environmental analysis of perovskites and other relevant solar cell technologies in a tandem configuration, Energy Environ. Sci. 10 (9) (2017) 1874–1884, https://doi.org/10.1039/c7ee01650f.

[42] X. He, The Rise of Perovskite Solar Cells, 2015. https://www.offgridenergyindependence.com/articles/8659/the-rise-of-perovskite-solar-cells.

[43] A. Binek, M.L. Petrus, N. Huber, H. Bristow, Y. Hu, T. Bein, P. Docampo, Recycling perovskite solar cells to avoid lead waste, ACS Appl. Mater. Interfaces 8 (20) (2016) 12881–12886, https://doi.org/10.1021/acsami.6b03767.

CHAPTER

12

Recycling of halide perovskites

Md. Farhan Naseh[a], *Choudhary Arjun Sunilbhai*[b], *Mohammad Khalid*[c], *and Jamilur R. Ansari*[a,d]

[a]University School of Basic and Applied Sciences, Guru Gobind Singh Indraprastha University, Dwarka, Delhi, India, [b]Faculty of Physical Sciences, PDM University, Bahadurgarh, Haryana, India, [c]Graphene & Advanced 2D Materials Research Group (GAMRG), School of Engineering and Technology, Sunway University, No. 5, Jalan Universiti, Petaling Jaya, Selangor, Malaysia, [d]Department of Applied Science & Humanities, Dronacharya College of Engineering, Gurugram, Haryana, India

List of abbreviations

5-AVA	5-ammoniumvaleric acid
CIGS	copper indium gallium selenide
CPCB	Central Pollution Control Board
DMAC	N,N-dimethyl-acetamide
DMEU	N,N′-dimethyl-ethylene urea
DMF	N,N-dimethyl-formamide
DMPU	N,N′-dimethyl-propylene urea
DMSO	dimethyl sulfoxide
ELC	electrolysis cell
EoL	end-of-life
EPBT	energy-payback time
ETL	electron-transport layer
EU	European Union
FA	formamidinium
FAI	formamidinium iodide
$FAPbI_3$	formamidinium lead tri-iodide
FESEM	field-emission scanning electron microscopy
FTO	fluorine-doped tin oxide
GBL	gamma-butyrolactone
HPSC	halide perovskite solar cell
HTL	hole-transport layer
HTM	hole-transport material
ICP	inductively coupled plasma

https://doi.org/10.1016/B978-0-323-88522-5.00013-2

I_{sc}	short-circuit current
ITO	indium tin oxide
MA	methylammonium
MABr	methylammonium bromide
MAI	methylammonium iodide
$MAPbI_3$	methylammonium lead tri-iodide
MHP	metal halide perovskite
MMT	million metric tons
MPP	maximum power point
NMP	N-methyl-2-pyrrolidone
NMR	nuclear magnetic resonance
OSC	organic solar cell
P3HT	poly(3-hexylthiphene-2,5-diyl)
PAN	polyacrylonitrile
PCBM	(6,6)-phenyl C61 butyric acid methyl ester
PCE	power conversion efficiency
PEDOT	poly(3,4-ethylenedioxythiophene)
PEG	polyethylene glycol
PESC	photo-electrosynthetic cell
PL	photoluminescence
PLQE	photoluminescence quantum efficiency
PSS	poly(styrenesulfonate)
PV	photovoltaic
PVDF	polyvinylidene difluoride
SEM	Scanning Electron Microscopy
Si-SC	silicon solar cell or silicon-based solar cell
S-Q	Shockley-Queisser
TCG	transparent conducting glass
TENG	triboelectric nanogenerator
THF	tetrahydrofuran
US-EPA	United States Environmental Protection Agency
V_{oc}	open-circuit voltage
WEEE	Waste Electrical and Electronic Equipment
WHO	World Health Organization
XRD	X-ray diffraction

1. Introduction

Solar energy with limitless potential has become the market's cheapest and fastest-growing power source on earth within a period of 25 years, famously called as "the sunrush." The solar industry is the heart of the clean energy revolution transforming world technologies and global economies, with the sun beaming down more energy every hour compared to what the world uses in an entire year. Solar cells based on perovskite materials are in the huge lead compared to other emerging photovoltaic (PV) technologies [1]. A typical metal halide perovskite (MHP) crystal structure has an AMX_3 chemical formula, where A, M, and X represent cations, metal cations, and anions, respectively. For instance, "A" can be methylammonium ($MA= CH_3NH_3^+$) or formamidinium [$FA= CH(NH_2)_2^+$] or cesium (Cs^+), "M" can be Pb^{2+} or Sn^{2+}, and "X" is monovalent halide ion: Cl^-, Br^-, or I^-. Most common examples of organic-inorganic or hybrid MHPs are methylammonium lead tri-iodide ($CH_3NH_3PbI_3$ or $MAPbI_3$) and formamidinium lead tri-iodide [$CH(NH_2)_2PbI_3$ or $FAPbI_3$]. An example of cesium-based MHP is $CsSnI_3$ [2], while perovskites with mixed-halides and/or mixed cations (Table 1) are also possible [24–28]. MHP

TABLE 1 Efficiency and stability of mixed-cation/mixed-halide perovskite based solar cells.

Mixture combination	PCE_{max}	Stability; degradation (%)	Reference
$MA_{0.6}FA_{0.4}PbI_3$ (Cu cathode)	18.3%	20 days; < 5%	Deng et al. [3]
$MA_{0.6}FA_{0.4}PbI_3$ (Au cathode)	14.23%	120 days; 14.4%	Salado et al. [4]
$MA_{0.85}FA_{0.15}PbI_3$ (Au cathode)	14.5%		
$MA_{0.6}FA_{0.4}PbI_3$	11.6%	10 days; < 10%	Tang et al. [5]
$MA_{0.6}FA_{0.4}PbI_{2.975}Br_{0.025}$	16.4%	–	
$MA_{0.6}FA_{0.39}Cs_{0.01}PbI_{2.975}Br_{0.025}$	18.7%	–	
$MA_{0.6}FA_{0.38}Cs_{0.02}PbI_{2.975}Br_{0.025}$	19.3%	30 days; < 10%	
$MA_{0.6}FA_{0.37}Cs_{0.03}PbI_{2.975}Br_{0.025}$	18%	–	
$FAPbI_{2.55}Br_{0.45}$	13.9%	< 7 days; 10%	Tang et al. [6]
$FA_{0.85}MA_{0.15}PbI_{2.55}Br_{0.45}$	15.8%	< 3 days; 10%	
$FA_{0.8}MA_{0.15}Cs_{0.05}PbI_{2.55}Br_{0.45}$	18.2%	< 15 days; 10%	
$FA_{0.8}MA_{0.15}Cs_{0.05}PbI_3$	16.3%	< 7 days; 10%	
$FA_{0.84}MA_{0.16}PbI_{2.52}Br_{0.48}$	18.3%	< 84 days; 17%	Chen et al. [7]
$FA_{0.85}MA_{0.15}PbI_{2.55}Br_{0.45}$	20.4%	60 days; < 7%	Gratia et al. [8]
FAI:PbI_2:MABr:$PbBr_2$ (fluoropolymer-coated)	18.67%	180 days; < 5%	Bella et al. [9]
$Cs_{0.15}FA_{0.85}PbI_{2.49}Br_{0.51}$	20%	–	Saliba et al. [10,11]
$Rb_{0.1}FA_{0.9}PbI_{2.49}Br_{0.51}$	12%	–	
$Rb_{0.05}Cs_{0.1}FA_{0.85}PbI_{2.49}Br_{0.51}$	19.3%	–	
$MA_{0.17}FA_{0.83}PbI_{2.49}Br_{0.51}$	17.42%	<5h; 15%	
$Cs_{0.1}(MA_{0.17}FA_{0.83})_{0.9}PbI_{2.49}Br_{0.51}$	20.96%	–	
$Cs_{0.05}(MA_{0.17}FA_{0.83})_{0.95}PbI_{2.49}Br_{0.51}$	21.17%	10 days; < 15%	
$Rb_{0.05}(MA_{0.17}FA_{0.83})_{0.95}PbI_{2.49}Br_{0.51}$	19.2%	–	
$Rb_{0.05}(Cs_{0.05}MA_{0.16}FA_{0.79})_{0.95}PbI_{2.49}Br_{0.51}$	21.8%	< 21 days; 5%	
$FA_{0.83}MA_{0.17}PbI_{2.49}Br_{0.51}$	15.3%		Dang et al. [12]
$Cs_{0.05}(FA_{0.83}MA_{0.17})_{0.95}PbI_{2.49}Br_{0.51}$	18.6%		
		–	

Continued

TABLE 1 Efficiency and stability of mixed-cation/mixed-halide perovskite based solar cells—cont'd

Mixture combination	PCE_{max}	Stability; degradation (%)	Reference
$Rb_{0.05}(FA_{0.83}MA_{0.17})_{0.95}PbI_{2.49}Br_{0.51}$ $Cs_{0.07}Rb_{0.03}(FA_{0.83}MA_{0.17})_{0.9}PbI_{2.49}Br_{0.51}$	16.5% 20.1%		
$MAPbI_{2.7}Br_{0.3}$ $MAPbI_{3-x}Cl_x$ $MA_{0.8}FA_{0.2}PbI_3$ $MA_{0.75}FA_{0.25}PbI_{3-x}Cl_x$ $MA_{0.8}FA_{0.2}PbI_{3-x}Cl_x$ $MA_{0.85}FA_{0.15}PbI_{3-x}Cl_x$	13.37% 15.95% 13.78% 17.74% 18.14% 17.41%	–	Isikgor et al. [13]
$MAPbI_{3-x}Cl_x$	20.04%	–	Lu et al. [14]
$MAPbI_2Br$ $MAPbI_{2-x}BrCl_x$ (Indoor light harvesting)	^{ip} 33% ^{ip} 36.2%	10 days; 10% 85 days; 5%	Cheng et al. [15]
$Cs_{0.05}MA_{0.05}FA_{0.9}PbI_{2.85}Br_{0.15}$ (a) $MA_{0.3}FA_{0.7}Pb_{0.5}Sn_{0.5}I_3$ (b) $Cs_{0.2}FA_{0.8}PbI_{1.8}Br_{1.2}$ All-perovskite: (a) + (b)	22.2% 21.1% 16.5% 24.8%	– – – < 20 days; 10%	Lin et al. [16]
$GA_{0.05}MA_{0.95}PbI_3$ $GA_{0.1}MA_{0.9}PbI_3$ $GA_{0.14}MA_{0.86}PbI_3$ $GA_{0.17}MA_{0.83}PbI_3$ $GA_{0.2}MA_{0.8}PbI_3$ $GA_{0.25}MA_{0.75}PbI_3$	18.81% 19.09% 20.15% 19.29% 18.43% 18.3%	– – – – – < 13 days; 15%	Jodlowski et al. [17]
$(GA)(MA)_nPb_nI_{3n+1}$	14.7%	240 days; 11.9%	Zhang et al. [18]
$(FAPbI_3)_{60}(BA_2PbI_4)$ $(FAPbI_3)_{60}(BA_2PbCl_4)$ $(FAPbI_3)_{60}(BA_2PbBr_4)$ ITIC passivated-$(FAPbI_3)_{60}(BA_2PbBr_4)$ IDTBR passivated-$(FAPbI_3)_{60}(BA_2PbBr_4)$	18.84% 17.66% 18.76% 20.1% 20.62%	~ 17 days; 10% < 25 days; 10% ~ 19 days; 10% 60 days; 9.9% 60 days; 12.7%	Niu et al. [19]
$(5\text{-}AVA)_xMA_{1-x}PbI_3$ $(5\text{-}AVA)_xMA_{1-x}PbI_3$	12.84% 14.89%	42 days; < 5% 375 days; < 5%	Mei et al. [20,21]
2D/3D $(5\text{-}AVA)_2PbI_4/MAPbI_3$	14.6%	> 400 days; 0%	Grancini et al. [22]
$(5\text{-}AVA)_2MA_{56}Pb_{57}I_{172}$	16.9%	–	Zheng et al. [23]

GA= guanidinium, i.e., $C(NH_2)_3^+$; FA= formamidinium, i.e., $CH(NH_2)_2^+$; MA= methylammonium, i.e., $CH_3NH_3^+$; BA= n-butylammonium, i.e., $CH_3(CH_2)_3NH_3^+$; 5-AVA= 5-ammoniumvaleric acid, i.e., $HOOC(CH_2)_4NH_3^+$; ITIC= 3,9-bis(2-methylene-(3-(1,1-dicyanomethylene)-indanone))-5,5,11,11-tetrakis (4-hexylphenyl)-dithieno[2,3-d:2′,3′-d′]-s-indaceno[1,2-b:5,6-b′]dithiophene, i.e., $C_{94}H_{82}N_4O_2S_4$; IDTBR= rhodanine-benzothiadiazole-coupled indacenodithiophene, i.e., $C_{72}H_{88}N_6O_2S_8$; ^{ip} = indoor PCE.

materials have a good potential of absorbing light in thin-film PV devices [29–31]. Thin films of MHPs can be prepared by either vapor-based or solution-based processes or by combining both these processes [2,32–47]. The different types of fabrication processes lead to MHP films with different optoelectronic properties and, consequently, different output efficiencies [45,48–50]. Halide perovskite solar cells (HPSCs) are typically composed of MHP light-absorbers, counter electrodes, and charge-transport layers. The target price for PV modules set by the United States Department of Energy is 0.5 USD per Wp. Silicon solar cells (Si-SCs) cost between 0.6 and 1.0 USD per Wp, while the target price of HPSCs is lower than 0.2 USD per Wp [51]. Thus, HPSCs have gained enormous attention because of their proven great potential for different configurations of PV devices.

Advantages of HPSCs over other thin-film photovoltaics:

- Perovskite materials are earth-abundant
- Low-temperature fabrication is possible
- Low-cost production
- Better optoelectronic properties
- Better power conversion efficiency (PCE)
- Mechanical flexibility and low weight
- Self-healing or self-recovery is possible

Disadvantages of HPSCs:

- Shorter device lifetime
- Long-term instability due to high temperature and moisture
- Hazardous lead waste (for lead-containing HPSCs)

Lead (Pb), a naturally occurring heavy metal, is very harmful to all living organisms due to open exposure. Lead can interact with proteins and cause variations in proteins' folding, making them inactive [52]. Lead exposure can cause developmental and cardiovascular diseases and damage neurological and reproductive systems. Lead also causes cancers and is very disruptive if it enters the food chain, as it can damage the whole ecosystem [53]. The United States Environmental Protection Agency (US-EPA) has set the maximum permissible limits of lead in air and drinking water as 0.15 ppb and 15 ppb, respectively [54,55]. For safety concerns, the lead content in paints, gasoline, water pipe solder, and ceramic products has been limited to minimize the concentrations of lead which can be released into the environment. During the long-term commercial operation of HPSCs, the potential leakage of lead due to bad weather conditions or other reasons is a huge public health and environmental risk concerning the stability and durability of HPSCs. Therefore, the use of lead metal in HPSCs is very hazardous to humans as well as nature, as it can become a potential new source of environmental contamination [56]. This is one of the big reasons preventing the commercial expansion of the HPSCs on a global scale. A report mentions that for 1 GW/year of solar power (for 20 years lifetime), it requires 30 tons of lead [54]. So, for the global commercial expansion of HPSCs, issues such as cost-effectiveness, economic feasibility in manufacture, long-term stability, eco-friendly disposal, or reuse or recycling of lead waste, and efficacy of lead-free perovskite solar cells need to be considered. Consequently, the development of HPSCs with lead alternatives came into being [2,49,57–64], but the lead-free HPSCs were observed to be less stable and had low photovoltaic performance in general, hence cannot be used at large scales. So, to maintain good performance together without causing lead poisoning, proper device encapsulation is essential to prevent any leakage of lead into the environment. After using lead-based HPSCs, those solar cells could be collected and disposed of or recycled to prevent environmental damage [53]. The utilization of MHPs is not just limited to solar cell devices but extends to other electronic/optoelectronic applications as well such as for lasers, solid-state lighting, next-generation displays, light-emitters (LEDs), photodetectors, radiation detectors, photocatalysts, and transistors [65–71], and also in power devices like nanogenerators [70,72–76]. So, proper disposal or recycling of such devices is also important. Electronic wastes are liabilities whose management responsibility lies on the product's manufacturer. Regulatory framework within countries requires recycling and economic policies for the recycling technology, environmental protection, and waste management, including extending the duties of the manufacturers for disposal or reuse [77]. Effective regulatory management of production, deployment, operation, disposal, reuse, or reprocessing/recycling of Pb-containing technologies can reduce associated environmental and societal risks [78].

Photon recycling is a phenomenon through which GaAs solar cells achieved the highest PCE reached by any single-junction solar cells (~30.5%). This phenomenon is also experienced in HPSCs and, therefore, can help improve the output performance of the HPSCs. Energy-payback time (EPBT) for PV modules is the amount of time a PV module takes to generate energy equivalent to the amount of energy that was used to produce the PV module. HPSCs have an average EPBT of 0.22 years, significantly less than that of other types of solar cells (CdTe: 0.9 years, organic: 0.3–3 years, crystalline silicon: 3–4 years) [79,80]. Through recycling of degraded HPSC modules, shorter EPBT can be realized. The average device lifetime of existing commercial silicon and CdTe solar panels is around 25 years. Shorter EPBT along with longer device lifetime is desirable and investment-wise beneficial but current HPSCs do not have such long device lifetimes. So, the research for improvements in HPSCs not only relates to higher PCE but also upscalability, reproducibility, long-term stability, and modularization. Significant barriers halt the progress of HPSCs, and the development of stable and environmentally benign HPSCs is the major industrial challenge to overcome. A market-based technology needs to exhibit versatility in the context of flexibility, portability, cost, and most importantly, recyclability. More attention is drawn toward recyclable PV technology in order to generate high amounts of energy for more extended periods of time. Economic recycling of valuable materials and preventing the release of hazardous constituents promote long-term sustainability and environmental safety. Recycling also reduces the production cost by reducing the requirement for new modules and simultaneously minimizing the harmful emissions released during the production of such modules. Reusing the recovered material(s) as secondary new supply material(s) to manufacture new devices or selling them into global commodity markets also consistently secures the supply of raw materials. This chapter describes how HPSCs degrade, stability of HPSCs, how to recycle parts or entire perovskite solar cells in the context of the environment and circular economy, and further advancement through photon recycling and different integrated device system with HPSCs.

2. Stability and degradation of halide perovskite solar cells (HPSCs)

The operational stability of any solar cell is crucial for its global commercialization. Nonstandardized stability assessments result in incomparable data needing to adjust the common standards. The stability tests vary from measuring the PCE of unencapsulated devices kept in the dark and under unrestricted ambient conditions to tests done under a controlled environment [81,82]. However, stability in the dark might not always be interpreted as operational stability for devices with or without any encapsulation, even if they were under inert conditions [81,83]. Hence, the operational stability of the HPSCs must be studied under light illumination to mimic the solar cell's natural environment. Since outdoor assessment of commercial solar panels under actual operating conditions is not practical due to long time (> 20 years) requirements, stringent industrial standards like IEC 61215 exist for pass-fail qualification of the solar modules [84]. These include accelerated stress tests to find any material and/or design errors leading to field failures but are not very suitable for determining a novel material system's stability and failure mechanisms [81,85]. For better relevance, the HPSC research community may consider the organic photovoltaics' ISOS testing protocols to compare and report the stability results or develop a similar and separate stability measurement standard solely for HPSCs [81,84–86]. For investigating the stability of HPSCs, it is beneficial to gather statistics by averaging many stability curves, plotting results with standard deviations, and comparing the device performance before and after ageing keeping the recovery period into consideration [81]. As ageing studies of HPSCs demonstrating stability over 40 days to one year of operation are on record [4,7–9,15,18–22,83,87–89], accelerated ageing studies prove to be a second step for assessing the device-relevant operational stability of HPSCs [81]. An acceleration factor is commonly extracted if the process is temperature-activated and follows an Arrhenius-type behavior [86]. If the degradation mechanisms get triggered, accelerated ageing can be brought about by doing the measurements at elevated light intensities. For example, Law et al. reported on the stability of lead-perovskite and other types of solar cells at extreme light intensities of up to 70 suns [90].

This section describes the effects of the different extrinsic and intrinsic factors leading to the degradation of the HPSCs, explaining the need for standardized stability measurement protocols for HPSCs. The extrinsic parameters are the harmful environmental factors such as atmosphere, humidity, temperature, and UV light. These can be blocked by proper encapsulation, so determining the need and costs of the encapsulant is required. Encapsulation is a primarily important means to radically increase the operational stability of the solar cells by improving the mechanical strength against extraneous impact and inhibiting environmental degradation [91]. Intrinsic parameters such as light illumination and electrical load help exclude certain materials from being used for a long-term stable PV device. To observe the sole effect of intrinsic factors, the operational stability measurements are usually conducted under an inert atmosphere to avoid any superposition with the extrinsic parameters. The identification and knowledge of the influences of these parameters could also be exploited for accelerated ageing studies. The effects of such parameters on the operational stability, ageing, and degradation behavior of the HPSCs are described below:

2.1 Effects of light illumination, atmosphere, and humidity

Light illumination acts as a critical intrinsic factor in the ageing of the HPSCs. Comparing the illumination results of UV light-rich unfiltered Xe lamp and UV-free white LED lamp on TiO_2-based triple-cation perovskite solar cell, it was observed that UV-rich Xe illumination accelerated the degradation of the device though the difference was not statistically significant [81]. This may be attributed to devices based on TiO_2 scaffolds being vulnerable to UV light [87,92]. Conversely, under inert conditions, the degrading effect of UV light was not very catastrophic but relatively slow, and the degraded cell performance was recoverable [93]. However, in the presence of H_2O and/or O_2, the degrading effect of UV light on the HPSCs becomes much more severe [9,94,95]. The two apparent sources of H_2O are the surrounding moisture and rainwater.

Illumination of light and O_2 have undeniable degrading effects on the operational stability of the HPSCs. When $MAPbI_3$ ($CH_3NH_3PbI_3$) MHP absorber gets optically excited, a species $MAPbI_3$* (which carries both photo-induced electrons and holes) is formed. The electron transfer from the $MAPbI_3$* species to molecular O_2 creates superoxide (O_2^-). The degradation of the $MAPbI_3$ MHP into methylamine (CH_3NH_2), lead iodide (PbI_2), iodine (I_2), and water is then caused by the highly reactive superoxide (O_2^-) formed due to the interaction of the O_2 molecule and photo-excited electrons of the $MAPbI_3$* species [96–99]. The degradation process is described by the reactions given in Eqs. (1)–(3). It should also be noted that the degradation reaction shown in Eq. (4) would require +1.6 eV enthalpy per O_2 molecule. In other words, despite the presence of O_2, the reaction would be unfavorable in the absence of light illumination. On the contrary, in the presence of both illumination of light and O_2, superoxide would form, and the enthalpy requirement would be negative, i.e., -1.4 eV per O_2 molecule, inferring that the reactants of Eq. (3), $MAPbI_3$* and O_2^-, are unstable with respect to the reaction products, so the degradation reaction would be highly favorable [96].

$$CH_3NH_3PbI_3 \xrightarrow[\text{No moisture}]{h\nu} CH_3NH_3PbI_3^* \quad (1)$$

$$O_2 \xrightarrow[e^-]{CH_3NH_3PbI_3^*} O_2^- \quad (2)$$

$$4\,CH_3NH_3PbI_3^* + O_2^- \rightarrow 4\,PbI_2 + 4\,CH_3NH_2 + 2\,I_2 + 2\,H_2O \quad (3)$$

$$4\,CH_3NH_3PbI_3 + O_2 \rightarrow 4\,PbI_2 + 4\,CH_3NH_2 + 2\,I_2 + 2\,H_2O \quad (4)$$

It has also been shown that molecular O_2 had minimal impact on the MHP stability and that it was the formation of interstitial charged oxygen species (superoxide and peroxide) that accelerated the degradation by cleaving the Pb–I bond [100,101]. This light and O_2-induced degradation can be avoided by using suitable absorbing or barrier interlayers to prevent superoxide formation. Since iodide defects/vacancies in the $MAPbI_3$ perovskite structure were probable sites of superoxide formation, thin-film passivation with iodide salts would help compensate the iodide defects/vacancies, limiting the formation of superoxide and improving the stability [96,97]. Further, investigations on the effect of light illumination, O_2, and H_2O showed that the degradation was slow under an inert N_2 environment but accelerated if atmospheric O_2 was present [81,98,99]. The $MAPbI_3$ MHP gets decomposed when it interacts with water, where the methylammonium iodide (MAI = CH_3NH_3I) gets reverted to its forming constituents, methylamine (CH_3NH_2) and hydro-iodic acid (HI), leaving behind lead iodide (PbI_2) depicted in Eqs. (5) and (6) [54,94,102]. Next, the hydroiodic acid breaks down, and iodine (I_2) is generated when exposed to light illumination and oxygen (Eqs. 7 and 8).

$$CH_3NH_3PbI_3 \overset{H_2O}{\Leftrightarrow} CH_3NH_3I\ (aq) + PbI_2 \tag{5}$$

$$CH_3NH_3I\ (aq) \Leftrightarrow HI\ (aq) + CH_3NH_2(aq) \tag{6}$$

$$2\,HI\ (aq) \overset{h\nu}{\Leftrightarrow} I_2(s) + H_2 \uparrow \tag{7}$$

$$4\,HI + O_2 \Leftrightarrow 2\,I_2(s) + 2\,H_2O \tag{8}$$

Within a few seconds of exposure to water, the black or dark-brown coloured films turn yellow. In the presence of water but in the absence of sunlight, they get hydrated and become yellow though annealing at a high temperature changes them back to the black/dark-brown dehydrated phase. However, in the presence of both water and sunlight, the resultant is PbI_2 which lacks the methylammonium component and is hence nonreversible even after annealing [54]. The presence of PbI_2 in MHP minimizes charge carrier recombination. Still, practically, only a critical amount of residual or excess PbI_2 in MHP improves the performance of the HPSCs beyond which it contrarily reduces the HPSC performance [103]. The continuous illumination of PbI_2 photo-decomposes it into I_2 and Pb, where the metallic Pb becomes a quencher for carriers. Hence, excess PbI_2 in MHP degrades the device's performance [104]. Researchers also demonstrated that the degradation of $MAPbI_3$ MHP films started at the smaller grains and grain boundaries and was initially independent of the presence of excess PbI_2 in the MHP films. Without affecting the conversion dynamics, they varied the content of excess PbI_2 in $MAPbI_3$ by varying the annealing temperature from 105°C to 135°C, avoiding the formation of nonconverted PbI_2 [99,105]. Their research conclusion was that light-illumination and humidity induced subbandgap defects in the MHP, and that resulted in the loss in photo-voltage, i.e., degraded the solar cell performance. The reduced cell performance was caused due to the degradation of the interface between perovskite and spiro-OMeTAD hole-transport layer (HTL). In addition, excess PbI_2 in the MHP films also lowered the device performance as it reduced the photoluminescence (PL) lifetimes such that the reduction was greater if a larger area of the perovskite surface (which was exposed to light illumination and humidity) was excited/illuminated [99]. A previous report mentioned that despite reconverting the PbI_2 to perovskite and renewing the degraded perovskite/spiro-OMeTAD interface, the initial device performance was not recoverable [106].

Fig. 1 displays an example of the effects of different atmospheric conditions and light illumination on the degradation of HPSCs. It can be seen

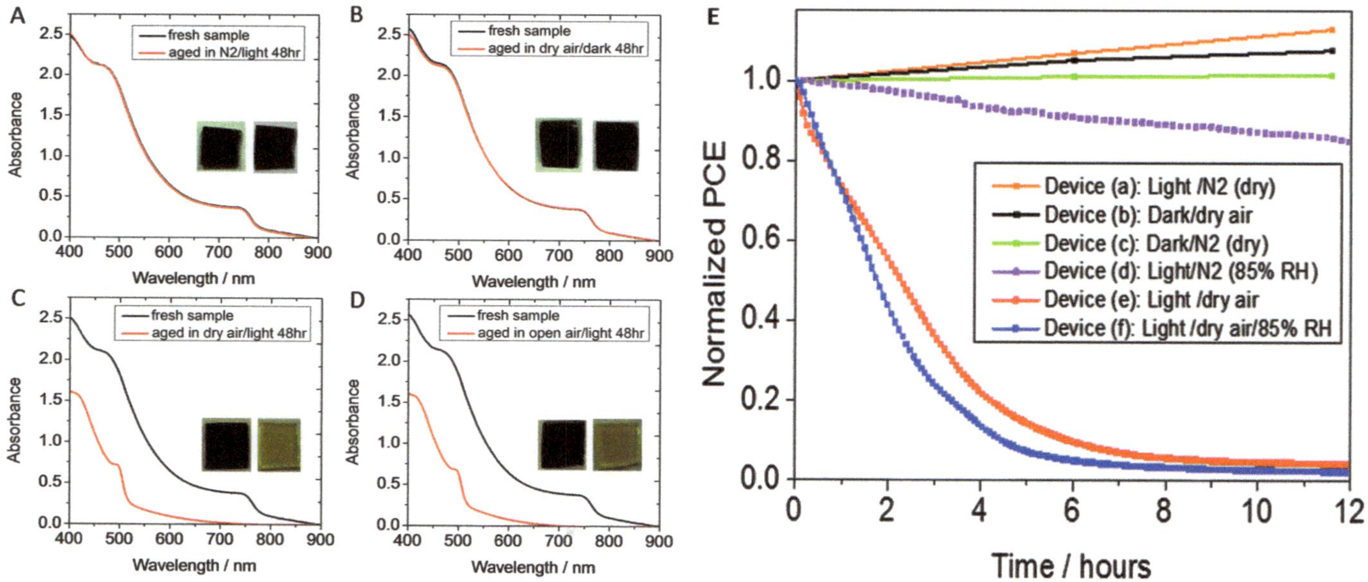

FIG. 1 Degrading effects of atmosphere and light illumination on HPSCs (RH: relative humidity and open air: 48% RH). *Reproduced from D. Bryant, N. Aristidou, S. Pont, I. Sanchez-Molina, T. Chotchunangatchaval, S. Wheeler, J.R. Durrant, S.A. Haque, Light and oxygen induced degradation limits the operational stability of methylammonium lead triiodide perovskite solar cells, Energy Environ. Sci. 9 (2016) 1655–1660. https://doi.org/10.1039/C6EE00409A with permission from the Royal Society of Chemistry.*

that there was negligible degradation if the device was aged in an N_2 environment with or without light (dark), or in clean dry air (i.e., presence of O_2) under dark. However, in the presence of both light and dry air, the degradation was rapid and fatal. Although the effect of high humidity in combination with light and dry air (O_2) was similarly severe, it only produced a minor effect in the presence of N_2 and light. This infers that though humidity contributes to degradation, the primary cause for the HPSC degradation is the presence of both light and oxygen. Other reported results also suggest that storing the devices in dry conditions avoids degradation even if O_2 or small amounts of H_2O were present, again indicating as earlier that O_2 resulted in degradation only after the illumination of light, through photo-oxidation in HPSCs and superoxide related mechanisms [81,97,99]. On the other hand, high humidity quickly degraded the HPSCs irrespective of light illumination [81,107].

Contrarily, the presence of humidity can also be beneficial for the perovskite films/HPSCs. Experiments show that the beneficial or detrimental effect of humidity actually depends on the circumstance. For instance, controlled humidity (of about 30%) during the crystallization phase of the MHP films improved the HPSC performance [108,109], whereas more than 50% ambient humidity after the crystallization deteriorated the HPSC [110]. Preventive measures against humidity include modifying the perovskite synthesis route, use of hydrophobic polymer layers, and device encapsulation [9,111–116]. Niu's team reported that molecule passivation of Ruddlesden-Popper/3D heterostructured film with semiconducting molecules resulted in outstanding long-term ambient air stability (~60 days) of their nonencapsulated $FAPbI_3$-based HPSCs with PCE as high as 20.62%. They also tested the operational stability of their unencapsulated devices under open-circuit conditions and 130 h of continuous light illumination in the air (relative humidity = 50%–60%, temperature = 50°C), concluding that the devices maintained more than 68% of their initial performance after the continuous operation [19].

2.2 Effects of temperature

Temperature is a very detrimental factor causing degradation of the perovskite materials, making thermal instability of HPSCs – a crucial matter of concern. In complex nanostructured hybrid materials, different thermal degradation pathways get activated at different temperatures [117]. Fig. 2 display a research team's results on the degrading effects of temperature on HPSCs. It can be seen that the device parameters degraded severely at temperatures beyond 100°C. It was reported that high temperatures may decompose the perovskite into lead iodide. The resulting PbI_2 gets dispersed in the MHP layer and creates defect states which limits the transfer of charge

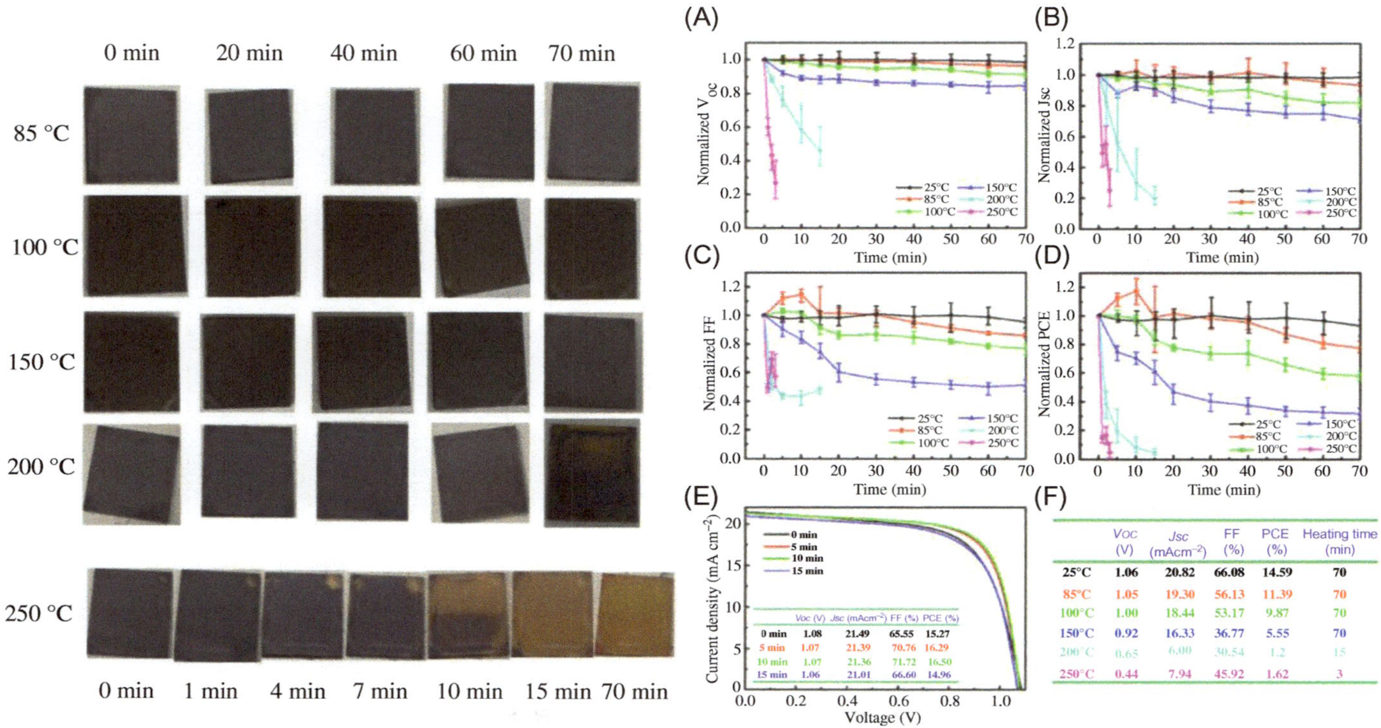

FIG. 2 Effect of temperature on the degradation of HPSCs. *Reproduced with permission from Q. Meng, Y. Chen, Y.Y. Xiao, J. Sun, X. Zhang, C.B. Han, H. Gao, Y. Zhang, H. Yan, Effect of temperature on the performance of perovskite solar cells, J. Mater. Sci. Mater. Electron. 32 (2021) 12784–12792. https://doi.org/10.1007/s10854-020-03029-y.*

carriers, and aggravates electron-hole recombination decaying the device performance rapidly. However, a proper heat treatment time at a certain temperature can passivate the defects and relatively improve the device efficiency by 15% [118]. For example, a 10-min heat treatment time at 85°C passivated the defects and improved the device efficiency to 16.5%, as shown in Fig. 2E. This was because during that 10 min at 85°C, a critical amount of PbI_2 was formed which acted as a passivation center and reduced charge carrier recombination in the MHP film [119].

Niu's group reported that their unencapsulated devices kept in a controlled environment at 80°C had suffered 10% degradation after 24 h of operation [88]. Jena and coworkers reported that the degradation of the $MAPbI_3$ solar cells having spiro-OMeTAD (as HTL) was not due to the deterioration of MHP but because of the changes in the perovskite/spiro-OMeTAD interface at high temperatures [106]. They noticed that the solar performance was further degraded as high-temperature induced severe deformations in the spiro-OMeTAD layer, and despite replacing the degraded spiro-OMeTAD layer with a fresh film, the initial device performance was not recovered as the reconstructed perovskite/spiro-OMeTAD interface worsened. In addition, they hinted that the MA ($CH_3NH_3^+$) ions played a certain role in altering the perovskite/spiro-OMeTAD interface causing faster degradation in the $MAPbI_3$ solar cells containing excess MAI (CH_3NH_3I), because better stability was observed in the MA-free HPSCs. However, degradation at elevated temperatures can be prevented, and HPSCs can remain stable even at temperatures greater than 100°C by the use of alternative contact materials, metal-oxide electrodes, and/or carbon-based materials [112,120–124].

On the other hand, it is interesting that the devices do not degrade in the dark at low temperatures but degrade at high temperatures even without light illumination [81]. This makes elevated temperature a critical degradation factor (often more because of the device stack rather than the perovskite alone) and a thermal stability test in dark can be a very useful criteria for screening and selecting devices for further operational stability assessments. It is important to note here that the thermal stability measured in dark does not always correspond to the operational stability of the solar device and it was simply a feasible practical way for screening-out devices before other operational stability measurements. Reports suggest that grain-boundary passivation improved the solar cell's thermal and environmental stability and also enhanced the device's PCE [88].

2.3 Effects of light cycling

The investigations on the effect of light soaking on HPSCs postulate that the degradation in the optoelectronic performance occurs due to photochemical or field-assisted ion migration [125]. However, light cycling naturally happens in the solar cells over the day-night periods making losses reversible and recovery during the dark [126]. Also, light cycling produces no fatal effect on the stability of HPSCs, and the dark recovery was highly reproducible after each cycle [81]. The performances of two devices were compared, one was light-exposed continuously and the other one was cycled between 6h of light and 6h of dark. It was observed that for the same cumulative operating time of 135h, the continuously aged device operated at about 88%, while the cycled device operated at about 96% of the initial efficiency [81]. Another group of researchers reported that the formation of deep-level meta-stable photo-activated trap states slowly degraded the device's photo-current, but by keeping the devices in the dark for just less than a minute, the devices self-healed entirely, while operating the devices at 0°C completely prevented it from degradation [125].

2.4 Effects of electrical load

A research group measured PCE derived from periodic J-V (current density-voltage) scans for several HPSCs aged at short-circuit, open-circuit, and maximum power point (MPP) conditions [81]. They observed that the device degradation was least while MPP tracking, as reported by other researchers [125]. The degradation was faster if the device was aged at short-circuit and very severe if it was aged at open circuit. However, after keeping the short-circuit aged HPSCs in the dark for 24h, the dark recovered value was similar to the devices aged at MPP, but contrarily recovery for open-circuit aged devices was highly limited. As initial potential reversible losses occur in the initial phase of measurements, the researchers should judiciously select the time duration for performing the ageing experiments [81]. The presumable source for the degradation may be the unextracted photo-generated charge carriers causing high concentrations of radicals in the HPSCs, similar to degradation under O_2 [98,127]. The presence and migration of ionic defects in the perovskite crystal lattice are other possible causes of the degradation of HPSCs [126]. So, ageing at short- or open-circuits provides a conservative estimation while ageing at MPP simulates the solar cell's native working conditions. Thus, the most suitable way for ageing HPSCs is to age the devices at the MPP and determine the PCE tracking the MPP [81].

Finally, it is evident that the afore-mentioned parameters strongly influence the HPSC's stability, albeit for measuring the operational stability of the HPSCs imitating their natural working environment, the most appropriate way of ageing would be under 1 solar light illumination at both room and elevated temperatures preferentially keeping the load at the MPP [81]. As an exception, some conditions could be relaxed without compromising the overall conclusions of the measurements, like white

LEDs could be used for light illumination and keeping the load at open- or short-circuit would give a conservative stability estimation [81]. Applying these conditions to a variety of HPSC architectures with various perovskite compositions could verify the governance of these suppositions. For device protection against H_2O and O_2, the use of an inert environment or proper encapsulation are the two practical solutions. However, continuous flushing of gases might decompose the perovskite material at elevated temperatures [128,129], and on the other hand, stability assessments at ambient conditions by only encapsulating the devices would investigate the quality of encapsulation more than that of the device. So, the safest way for stability assessments would be ageing the encapsulated devices in an inert environment to avoid the undesirable impacts of trace gases [81]. In addition, a strong and well-designed encapsulation of any HPSC should not only effectively shield it from external moisture, rainwater, and oxygen but also limit the escape of lead and volatile compounds if the organic-inorganic MHP layer gets decomposed or damaged [116,130].

3. Impact of environment on HPSCs and impact of HPSCs on environment

Some perovskite materials can get dissociated in the presence of highly polar solvents (e.g., water), i.e., HPSCs may be sensitive to water due to chances of dissociation and draining out of lead into the environment [116,131]. As solar cell panels are commonly kept outdoors, their exposure to water, moisture, or rain is unavoidable. How atmosphere and humidity (i.e., oxygen and water) degrade the perovksite, leaving behind lead in the form of PbI_2, was already detailed in Section 2.1. Though the solar cells are generally encapsulated but unexpected module damage may make them vulnerable to rain and leakage of lead, as shown in Fig. 3. The solubility product (K_{sp}) of PbI_2 is much higher than that of other materials like CdTe/CdS/PbS used for making solar cells [54]. Therefore, accidental leeching out of the damaged panels can potentially be very harmful. Based on the typical PV performance of $MAPbI_3$ containing HPSC devices, device geometries, and crystal structure of $MAPbI_3$, the total Pb content per unit of produced electricity (in the USA) is around 38 μg/kWh. This would be equivalent to 160 tons of Pb content per year of electricity generation in the USA. Even if 1% of HPSC modules broke during deployment, 1.6 tons per year of Pb might be leeched into the environment [78]. The maximum permissible limits set by the United States Environmental Protection Agency (US-EPA) for Pb in air and drinking water are 0.15 ppb and 15 ppb, respectively [54]. The World Health Organization (WHO) safety limit for Pb in drinking water is 10 ppb. The concentration

FIG. 3 Possible ways for lead to leech into the environment. *Reproduced with permission from Y. Jiang, L. Qiu, E.J. Juarez-Perez, L.K. Ono, Z. Hu, Z. Liu, Z. Wu, L. Meng, Q. Wang, Y. Qi, Reduction of lead leakage from damaged lead halide perovskite solar modules using self-healing polymer-based encapsulation, Nat. Energy 4 (2019) 585–593. https://doi.org/10.1038/s41560-019-0406-2.*

limits set by the Indian Central Pollution Control Board (CPCB) for Pb in inland surface water and public sewers are 0.1 ppm and 1 ppm, respectively. The WHO permissible limits for Pb in agricultural soil and wastewater are 0.1 ppm and 10 ppb, while US-EPA recommended limits for Pb in garden soil and wastewater are 200 ppm and 6 ppb, respectively [55]. Considering a typical 300 nm thin $MAPbI_3$ film containing approximately 0.4g of Pb, a 1 m^2 HPSC panel will contain less than 1g of Pb. So, for an upper limit of 1g Pb per m^2, 15% efficient array of HPSC modules will require 30 tons of Pb to produce 20 GW-yr of energy, i.e., 1 GW/year of energy for 20 years lifetime [54]. The natural levels of Pb in the uncontaminated soil are around 10–50 ppm. In urban areas, it maybe around 50–200 ppm or even higher. Lead drainage from the HPSC modules due to rain can increase the Pb concentration in the soil (of 1.5 g/mL density) just below the panels by about 70 ppm. Continuous rain may distribute the lead throughout the depth of the soil lowering the overall Pb concentration. However, for many soil types, due to the strong adsorbing tendency of Pb to the soil particles, about 20 ppm of Pb may remain concentrated in the uppermost layers of the soil [54]. Generally, compared to other lead pollution sources (e.g., coal power stations, leaded-fuels, electronic solders), lead leakage from the HPSCs to the soil is not considered very disastrous, but the seeping of Pb from the soil to groundwater, surface water, or other inland water sources can result in a great hazard [54,78]. Since HPSCs can become a potential new source of environmental lead pollution, it is important to investigate how it might leach into the surroundings to prevent or at least reduce the leaching to a nonharmful level [54,132]. Researchers investigated the impact of rain on $MAPbI_3$ perovskite films by exposing the films to water(s) of different pH strengths, simulating the exposure of HPSCs to rain. Their 5 mL/h simulated rain for 1 h was equivalent to an hour of heavy rainfall or 3 h of moderate rainfall. Typically, more acidic rain dissolves PbI_2 more rapidly, and among the common acid rain constituents, HNO_3 dissolves PbI_2 better than H_2SO_4. They determined the extent of Pb lost to the surroundings in the worst case of encapsulation/device damage, i.e., direct exposure of perovskite to rain [54]. They concluded that rain causes a complete and irreversible degradation of the MHP layer, and if measures (such as making other components of the HPSCs above the MHP layer, like hole conductor and/or contacts, impermeable to water) are not undertaken to restrain the loss of Pb, all the Pb can be leached out of the damaged modules with continuous rainfall.

Lead is present in the solid-state inside the MHP photovoltaics, encapsulated, and for technological viability, it should ideally remain inside only [78]. For complete safety, lead pollution should be considered at all stages of HPSC lifetime, starting from their manufacture to damage during deployment or operation to disposal or recycling. Constructing encapsulation with Pb-absorbing materials, mixing Pb-adsorbing materials like cation-exchange resin with carbon electrode or alkoxy-PTEG-based hole-transport layer (HTL), and on-device Pb sequestration can significantly limit the irreversible drain-out of Pb from the damaged modules into the external environment during extreme weather conditions or at times of other natural calamities [91,132–134]. Employing photo-curable fluoropolymers as hydrophobic barriers or adding large hydrophobic cations to Pb-based MHPs also enhances the environmental stability of the HPSCs made using them [9,135].

On one side, where rain was a potential agent incrementing the Pb pollution; on the other side, it can effectively generate energy alongside solar power and create electricity during the rainy weather and also at night if it rains. So, rain or shine, the same device may harvest kinetic energy from the motion of falling raindrops as well as solar power from the sun and generate more electricity. For instance, Liu and coworkers had integrated a heterojunction Si-SC with a triboelectric nanogenerator, TENG (a power generator that converts mechanical energy into electrical energy), using a film of PEDOT:PSS [Poly(3,4-ethylenedioxythiophene):Poly(styrenesulfonate)] as a mutual electrode [136]. The PEDOT:PSS film reduced the light reflection enhancing the output power, while the TENG protected the solar cell by acting as a waterproof barrier. The TENG produced a power density of 1.74 mW/m^2 from waterdrops dripping at 13 mL/s. Their hybrid system configuration combined the merits of high voltage of TENG with high current level of solar cell, establishing a convenient means for collecting energy from the environment during varied weather conditions. Adding wind for power generation, flexible solar cells coupled with flexible piezoelectric films or

TENG can harness energy from sunlight as well as wind and rainfall [137–139]. Vatansever and group made an inexpensive nontransparent hybrid structure by depositing P3HT:PCBM [Poly(3-hexylthiphene-2,5-diyl): (6,6)-Phenyl C61 butyric acid methyl ester]-based organic solar cell (OSC) on a commercial piezoelectric PVDF (polyvinylidene difluoride) substrate at ambient room conditions, and had used Cu and Al electrodes instead of the costly indium tin oxide (ITO) [138]. Their hybrid device harvested 85 μW peak power from a simulated turbulent wind of 36 km/h, and the structure was able to withstand without causing any damage to its solar harnessing ability. The innovative Si-SC+TENG hybridized power panel made by Zheng's group could simultaneously harness power from sunlight, wind, and rain in the availability of all or any of them [139]. The hybrid device produced a power density of about 86 mW/m^2 from raindrops dripping at a rate of 13.6 mL/s and 8 mW/m^2 from 9.72 km/h moving wind without affecting the device's solar conversion efficiency. For high energy harvesting, different recently reported TENG based on carbon nanomaterials, Al/Ti-embedded silicone, sepiolite-doped polyacrylonitrile (PAN), and PAN/$AgSbS_2$:Nylon 6.6 nanofibers demonstrate great potential through enhancement in the TENG's output voltage [140–143]. Researchers might test the compatibility of some or all of the aforementioned TENG by integrating them with HPSCs (or other types of solar cells) for multiple energy harvesting using the same device. Adopting the hybridized way of generating power could enhance and change the traditional method of solar energy harvesting.

Reports for perovskite-based TENGs, and perovskite-TENG working as solar cells by coupling the photovoltaic and triboelectric effects, are also on records [72–76]. For example, the barium-doped all-inorganic $CsPb_{0.91}Ba_{0.09}Br_3$ perovskite-based TENG produced an ultra-high open-circuit voltage (V_{oc}) of 220 V and maximum power density of 3.07 W/m^2 capable of powering 80 LED lights [75]. The $2 \times 2\,cm^2$ perovskite-TENG of Su's team produced a voltage of 15.3 V and short-circuit current (I_{sc}) of 0.29 μA without any light illumination, and increased voltage and I_{sc} (17 V and 0.38 μA) with light illumination [74]. This research provided a new approach to enhance the TENG's mechanical-to-electrical conversion efficiency by simple sunlight absorption, introducing a new principle in the field of TENG. Wang's group came up with different TENG configurations such as dielectric/perovskite TENG and carbon/perovskite TENG for harnessing mechanical energy. They reported that the 0.3 μA triboelectric current of carbon/perovskite TENG increased to 270 μA upon light illumination. They also demonstrated the working of the carbon/perovskite TENG as a solar cell in a full-contact situation under light illumination, producing a photo-voltage of 487 mV and photo-current of 10.6 μA [76]. In addition, works on piezoelectric nanogenerators, HPSC-piezoelectric hybrid integrations, and Pb-free-based integrated systems are also in progress [63,72].

Furthermore, lead hazard was just one of the many environmental risks of the lead-based MHP technologies. Other factors included impacts on surrounding natural resources due to long-term industrial usage, greenhouse gas emissions, and other toxic materials [78,79]. For example, common organic solvents used for mass fabrication of MHP films, namely, DMF (N,N-dimethyl-formamide), DMAC (N,N-dimethyl-acetamide), NMP (N-methyl-2-pyrrolidone), DMEU (N,N′-dimethyl-ethylene urea), DMPU (N,N′-dimethyl-propylene urea), GBL (gamma-butyrolactone), and THF (tetrahydrofuran), were toxic to human health as well as the environment [45,116,144,145]. DMF, DMAC, and NMP caused harmful effects on the human reproductive system, while DMEU, DMPU, GBL, and THF were also classified as dangerous. DMSO (dimethyl sulfoxide) was a possible alternative solvent for fabricating MHP films as it was the least hazardous among all the solvents mentioned above. Still, since its removal from the wastewater requires more energy, it also has some effects on the environment [145]. Therefore, the compositional development and use of green solvent systems (i.e., nonhazardous solvent mixtures or solvents with additives, e.g., solvent/alcohol/acid system, DMSO+2-methylpyrazine+1-pentanol, polyethylene glycol (PEG) + GBL, 3-methylcyclohexanone) [133,144–148] and development of solvent-free or other alternative strategies for large-area MHP-film fabrication [45,89,149] should be promoted. Recycling of the MHP technologies is another promising sustainable, efficient, and cost-effective strategy which should be adopted.

4. Recycling measures to counter the environmental hazards

The average service life of existing commercial (Si/CdTe/CIGS) solar panels is expected to be of about 25 years. The global demand for PV exhibited a drastic increase of up to 586GW (in 2019) from just 1.3GW (in 2000), with the annual growth rate being 20%, the highest among all other renewable energy industries. The International Technology Roadmap for Photovoltaic predicts that the global PV installations will reach a huge 4500GW capacity by 2050 [150,151]. The rise in demand leads to more solar panel installations which are potentially large amounts of waste after the panel's end-of-life (EoL). The solar panel wastes are generated during their 4 primary life cycle phases: (i) production, (ii) transportation, (iii) installation and operation, and (iv) EoL panel disposal [77]. The waste during panel production is usually collected and assumed to be easily treated by the manufacturers or waste treatment contractors, so not considered a societal waste management issue. The PV modules contain valuable materials such as Si, Ag, Cu, Al and hazardous materials such as Pb and Cd. Recycling valuable materials and preventing the release of hazardous ones promote the long-term sustainability of the supply chain as well as environmental safety. Recycling also reduces the production cost by reducing the requirement for new solar modules and simultaneously minimizing the harmful emissions released during the solar module production. It is expected that the total amount of global PV waste will be 1.35 million metric tons (MMT) by 2030 and 12.3 MMT by 2040, where 2.8 MMT waste would be in China, 2.2 MMT in Germany, and 1.8 MMT in Japan, creating an essential need for recycling strategies. The processes for the recycling industry include decommissioning of PV panels, their collection, dismantling, and recycling. An efficient waste recycling management scheme is formulated by economic, administrative, and legislative rulings, which requires balancing high recovery/recycling targets with a high collection rate and negligible module damage during the recycling process. For Si-SC panels and thin-film based (CdTe and CIGS) solar panels, the recycling protocols have already been established, though the recycling processes are different due to the different module structures of Si-SC panels and thin-film solar panels, respectively [77,150–153]. However, the basic recycling procedure started with the collection and segregation/dismantling of the modules. The next step was stripping off the metals via chemical dissolution, mechanical separation, electro-deposition, precipitation, or other means [152]. And then, the toxic metals were isolated before they were finally decided to be recycled/reused. Countries like Germany, USA, Japan, Italy, and France are researching and developing solar panel recycling with more focus on Si-SC panels to recycle silicon and rare metals. European organization PV Cycle, French company Veolia, and German company Solar World recycled Si-SC panels, while American company First Solar and German Antec Solar recycled CdTe solar panels. Germany's Loser Chemie recycled materials from Si-SC, CdTe, CIGS, and GaAs solar modules [150].

Since large-scale commercial production of the HPSCs (as the fastest growing third-generation photovoltaics) has started, their disposal/recycling due to damage or after their EoL will require significant attention. The equivalent low cost of the HPSCs is due to cheaper material cost and vacuum-free process of fabrication of perovskite light absorbers [154]. However, the use of costly gold or silver electrode, electron/hole transport materials, and transparent conducting glass (TCG) makes it difficult to hold the target price less than 0.2 USD per Wp [51]. For example, a gold electrode costs around 40 USD per gram, and fluorine-doped tin oxide (FTO) conducting glass costs around 100 USD per m^2. Spiro-OMeTAD, a commonly used hole-transport material, costs around 200 USD per gram. These constitute 98% of the total cost associated with a 1 x 1 sq. inch small device. Recovering and recycling device modules is considered an alternative method of reducing the cost. Reusing or recycling is always a better option since the disposal of hazardous materials is also very expensive. The average cost for disposing of hazardous wastes is as high as 1.10 USD per kg [53]. So, a well-formulated strategy for collecting different solar cell layers separately and then reusing it could help solve or at least minimize the waste problem cost-effectively. The lead waste/hazard problem was highlighted in the last section, imposing the need and importance for recycling. The degradation of the $MAPbI_3$ perovskite to CH_3NH_2 and HI due to water, leaving behind lead in the form of PbI_2, was illustrated in Section 2.1. A crystal structure schematic of degradation of the $MAPbI_3$ perovskite to PbI_2 due to water, and recycling strategy to reconvert it back to perovskite is depicted in Fig. 4. Thus, recycling the MHP-based devices is the most sustainable solution not only from an environmental perspective but also from resource and economic market perspectives, to utilize some toxic materials and expensive parts of the MHP devices. Also, the fabrication of reproducible HPSCs should be favored for higher long-term performance [6,9,11,12,22,28,33–35,49,85,105,155].

Jena and coworkers reported that the HPSC performance recovery and thermal stability upon recycling were dependent on the perovskite composition. They experimented on recovering the device efficiency by reusing the perovskite films of the HPSC devices degraded because of high temperature and humidity. After reconstructing the degraded perovskite/spiro-OMeTAD interface and reconverting the PbI_2 to perovskite, the performance of the recycled devices improved but was lower

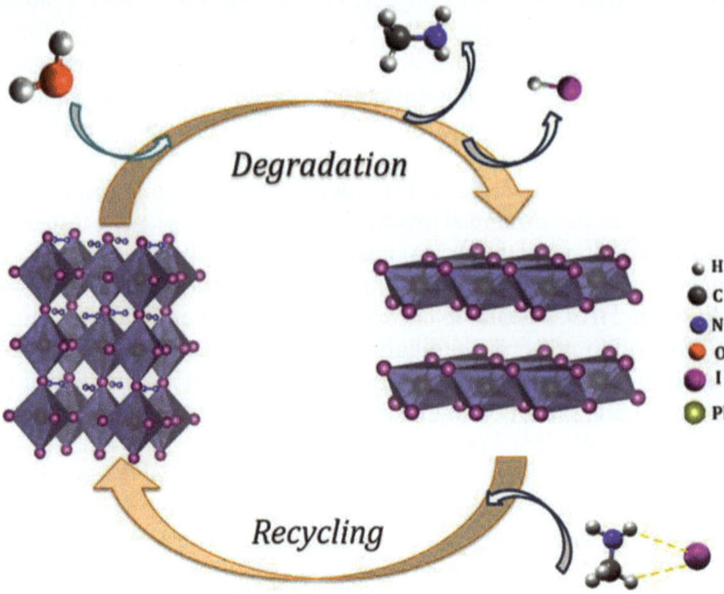

FIG. 4 Crystal structure schematic for degradation and recycling of the $MAPbI_3$ perovskite. *Reproduced with permission from P. Chhillar, B.P. Dhamaniya, V. Dutta, S.K. Pathak, Recycling of perovskite films: route toward cost-efficient and environment-friendly perovskite technology, ACS Omega 4 (2019) 11880–11887. https://doi.org/10.1021/acsomega.9b01053.*

than the initial device performance. The reason for the reduced recovered efficiency was attributed to inefficient carrier transport leading to lower photo-current in the recycled HPSC devices [106]. Binek and group demonstrated that recycling different layers of planar HPSCs was feasible in which the layers could simply be removed one by one and recovered in an isolated manner. They also showed that the toxic lead precursor and the costly conductive glass could also be reused with comparable performance [53]. The possibility to recycle HPSCs by reusing their TCGs and gold electrodes helps in reducing the production cost [156]. Another group of researchers reported recycling the TCG of HPSCs by removing the MHP layer through a selective dissolution process. The perovskite materials included tri-halides like $MAPbI_3$, and mixed halide of $FAPbI_3$ and $MAPbBr_3$. This selective process was capable of dissolving the perovskite layers such that the FTO transparent conducting glass and gold electrodes were separated through the dissolution of the halide layers. They found that even after ten cycles of recycling, the value of PCE remained constant [51]. Chhillar and coworkers reported the cost-reducing reuse of lead, FTO, and recycling degraded perovskite films without compromising the device's optoelectronic performance. They also analyzed the feasibility of recycling perovskite films fabricated through different standard procedures [102]. Such researches pave the way to use lead-based HPSCs commercially mitigating the lead hazard problem. The technical processes for recycling different parts of the HPSCs are elaborated in the subsections.

4.1 Recycling of PbI_2

Perovskite thin-films can be fabricated using the solution-based method via a one-step deposition technique [38,43], or a two-step deposition technique [33,34,44,46]. The two-step deposition technique is more frequently adopted, involving the dissolution of PbI_2 powder in DMF solvent, and spin-casting it on a substrate. After that, the system is annealed and exposed to a methylammonium iodide, MAI (CH_3NH_3I), or formamidinium iodide, FAI [$CH(NH_2)_2I$] solution to form thin films of $MAPbI_3$ or $FAPbI_3$, respectively [44]. During the transformation of PbI_2 to $MAPbI_3$, it is supposed that the perovskite phase is formed through a conversion reaction at the PbI_2–CH_3NH_3I interface. The factors that influence this process are surface morphology/microstructure of PbI_2 film, MAI solution concentration, and annealing temperature [46,47]. For developing HPSCs, Barrit and coworkers synthesized porous thin films of PbI_2 using the solution-based method. They demonstrated that the deposition of PbI_2 proceeded via a sol-gel process, which involved the formation of intermediate solvates, and showed that the air-dried film samples were more porous than the thermally annealed film samples [35]. For fabricating thin films of $MAPbI_3$ and $FAPbI_3$ using a two-step solution process, they emphasized that proper control over the solvation state of the PbI_2 was very important, to obtain higher quality films and higher PCEs with reproducibility [33,34]. More effective perovskite-based PV devices can be fabricated using the solution-based methods, but these methods include some major problems like difficulty in upscaling and use of toxic materials like Pb [45,157]. Among nonprecious metals, the global EoL recycling rate of Pb (the percentage of Pb in discards which is recycled) is already very high, ~greater than 50% [158]. High recycling rates usually correspond to easily recoverable uses of materials which were in large quantities as the recycling of materials in complex products in small amounts is very challenging. For Pb, the Pb in lead-acid batteries is easily recycled [78,158]. In the context of HPSCs, the toxic lead (i.e., PbI_2) can also be recycled/reused to avoid any hazardous Pb waste. It has been shown that the PbI_2 can be recovered in high yields, and its reuse in new HPSCs can produce comparable performance [53]. The recycling process starts with removing the PbI_2 from the substrate via DMF extraction. The recovered PbI_2 amount can be checked quantitatively through inductively coupled plasma (ICP) and UV-vis absorption analyses techniques. Recycling $0.7\,m^2$ of an MHP film yields about 600 mg of PbI_2, which is sufficient to make a 1 mL PbI_2 solution of 1.25 molar concentration and produce a recycled $MAPbI_3$ film of $0.02\,m^2$ area. A large quantity of MHP films is required to prepare the recycled lead iodide solution since a lot of material is lost while preparing the films via the spin-coating method. Application of the slot-die coating (instead of spin-coating) to avoid material loss can make the recycling process more efficient at the industrial level.

The J-V curves of the HPSC devices with fresh, recycled, and recycled + recrystallized PbI_2 are shown in Fig. 5A. Although the devices with recycled PbI_2 exhibited good performance, their >12% PCE was lower than the PCE of the device fabricated using fresh/highly pure PbI_2. The lower PCE relates to lower V_{oc}, the reason for which may be very tiny amounts of impurities (possibly TiO_2 particles and/or MAI) which might have been present in the lead iodide solution [53]. To validate this and remove the contaminations, the recycled PbI_2 was recrystallized from water [159]. As a result, the V_{oc} and the PCE of the device with recycled + recrystallized PbI_2 increased from 0.95 V to 1.02 V and 12.1% to 13.5%, respectively. Figs. 5B and C display similar scanning electron microscopy (SEM) cross-sectional morphology of the devices fabricated using fresh and recycled PbI_2, which is in agreement with the comparable performance explained above. Therefore, this method of PbI_2 recycling effectively demonstrates how to isolate the toxic PbI_2 from the HPSCs and reuse it in new solar cells with equivalent performance.

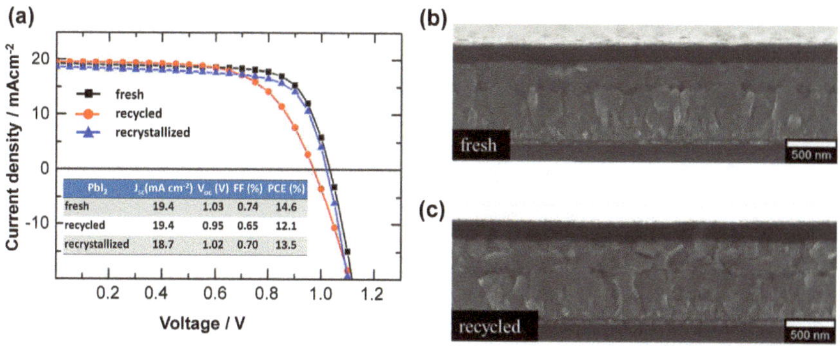

FIG. 5 (A) J-V characteristics of HPSCs fabricated using fresh, recycled, and recycled + recrystallized PbI_2. SEM cross-sections of HPSCs prepared from (B) fresh PbI_2 and (C) recycled PbI_2. *Reproduced with permission from A. Binek, M.L. Petrus, N. Huber, H. Bristow, Y. Hu, T. Bein, P. Docampo, Recycling perovskite solar cells to avoid lead waste, ACS Appl. Mater. Interfaces 8 (2016) 12881–12886. https://doi.org/10.1021/acsami.6b03767.*

4.2 Recycling of FTO conducting glass

The FTO conducting glass constitutes about 20%–60% of the module cost for solar technologies like CdTe, organic, and dye-sensitized solar cells [53]. Likewise, FTO/glass substrate is the most expensive part of the HPSCs. Alternatives for the FTO substrates such as PEDOT:PSS [160] and graphene [161] have shown great potential; however, best performances are still achieved using FTO/glass substrates. Therefore, recycling of the FTO/glass substrate will be economically very beneficial. The titanium dioxide (TiO_2)-coated FTO/glass substrate can be recycled directly after the removal of the lead iodide via DMF [53]. However, HPSC devices fabricated using such recycled substrates displayed a significant decrease in the output performance when they were recycled over and over again. The reason for the reducing performance was that DMF extraction for PbI_2 removal significantly deteriorated the underlying TiO_2 layer. Because TiO_2 is insoluble in DMF [162], there is a possibility that the DMF permeates through the pores of the TiO_2 layer delaminating the TiO_2 underlayer completely, if it remained immersed in the DMF for a longer time [53]. To prevent this, a fresh electron-selective TiO_2 layer was deposited on a clean and recycled FTO/glass substrate. Following which, the modified recycled devices displayed no significant loss in the output performance even after recycling multiple times. Moreover, the device with the three-times recycled substrate resulted in the highest PCE, possibly due to an improved TiO_2 coverage over the FTO substrate in the modified recycled devices [53]. Fig. 6A shows the J-V characteristics of the HPSC devices with one-, two-, and three-times recycled FTO/glass substrate.

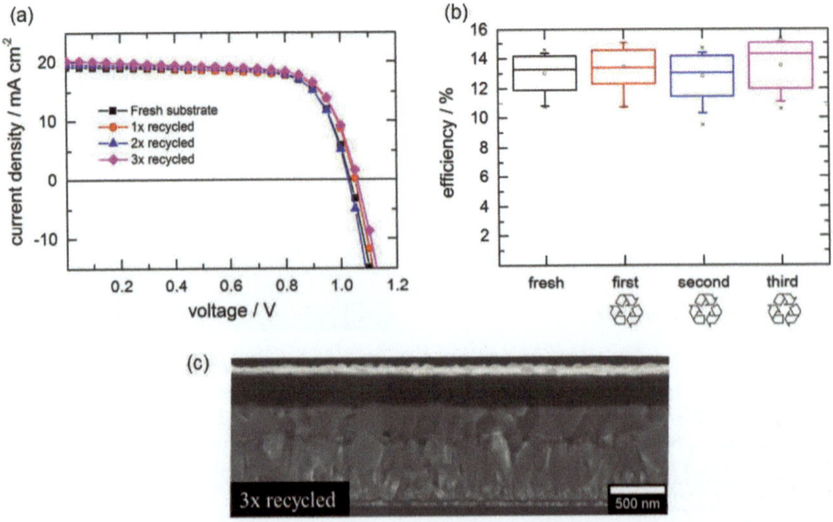

FIG. 6 (A) J-V characteristics and (B) PCE distribution of HPSC devices fabricated on one-, two-, and three-times recycled FTO/glass substrate. (C) SEM cross-section of an HPSC with three-times recycled FTO/glass substrate. *Reproduced with permission from A. Binek, M.L. Petrus, N. Huber, H. Bristow, Y. Hu, T. Bein, P. Docampo, Recycling perovskite solar cells to avoid lead waste, ACS Appl. Mater. Interfaces 8 (2016) 12881–12886. https://doi.org/10.1021/acsami.6b03767.*

Fig. 6B shows a PCE distribution of 24 individual cells for each time the substrate was recycled, where the bar diagram represents the average (circle), median (middle-line), minimum/maximum (crosses), 75% (box), and 95% (small error bar) distribution. Fig. 6C displays the SEM cross-sectional morphology of an HPSC with a three-times recycled FTO/glass substrate. Hence, this clearly illustrates that the expensive FTO/glass substrate of an HPSC can be recycled/reused several times without any performance loss. In addition, low-temperature processable multirecycling of the used FTO/glass substrates (of degraded HPSC devices) paves the way for sustainable commercial production/reproduction of the HPSCs [156,163,164].

4.3 Recycling of $MAPbI_3$ perovskite

In relation to recycling, the HPSC's performance recovery and thermal stability depend on the perovskite composition [106]. The different types of fabrication processes lead to MHP films with different morphologies, different optoelectronic properties, and consequently, different output efficiencies. So, Chhillar and coworkers analyzed the feasibility of recycling the perovskite films fabricated through the 3 different standard

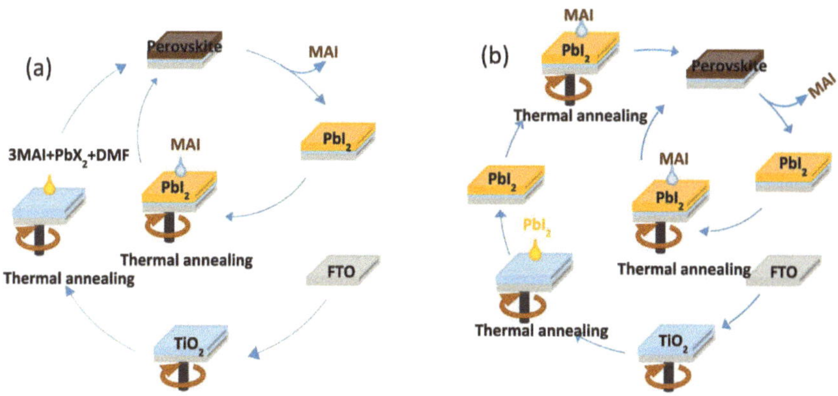

FIG. 7 Schematic of the recycling processes for $MAPbI_3$ film(s) deposited by (A) single-step acetate and chloride routes, and (B) sequential deposition route. *Reproduced with permission from P. Chhillar, B.P. Dhamaniya, V. Dutta, S.K. Pathak, Recycling of perovskite films: route toward cost-efficient and environment-friendly perovskite technology, ACS Omega 4 (2019) 11880–11887. https://doi.org/10.1021/acsomega.9b01053.*

procedures, namely, the single-step acetate route, the single-step chloride route, and the sequential deposition route. They reported cost-effective recycling of the as it is degraded perovskite films without scratching out the MHP films or dissolving anything [102]. Fig. 7 depicts schematic representations of the processes for recycling the degraded $MAPbI_3$ films fabricated via (A) single-step acetate or single-step chloride and (B) sequential deposition routes, respectively.

To observe the structural changes during the degradation of the MHP films fabricated using the three different routes, the films were left for 26 days to degrade under an ambient environment at 25%–35% humidity. The $MAPbI_3$ perovskite slowly degrades to PbI_2 and a complete conversion to PbI_2 happens in about 20 days of 25%–35% humidity exposure [102]. PbI_2 is the remnant of the degraded $MAPbI_3$ perovskite which can be reconverted back to $MAPbI_3$ by reacting it with methylammonium iodide (MAI) [94,102], as shown in Fig. 7. A better understanding of the reconversion of PbI_2 to $MAPbI_3$ (i.e., the extent of recycling the degraded $MAPbI_3$ films) is established through structural characterizations depicted in Fig. 8. Fig. 8A–C shows X-ray diffraction (XRD) patterns of the freshly prepared, degraded, and recycled $MAPbI_3$ films fabricated using the three different deposition routes. Due to the degradation of $MAPbI_3$ to PbI_2 under ambient humid environment for 26 days, the characteristic peak of $MAPbI_3$ gets diminished, while the PbI_2 peak appears and grows at around 12.6° [165]. A total color change from dark brown to yellow signifying complete degradation of the $MAPbI_3$ perovskite to PbI_2 upon exposure to water was already explained in Section 2.1. Fig. 8D displays the color change from dark brown to yellow and back to dark brown upon recycling the degraded films.

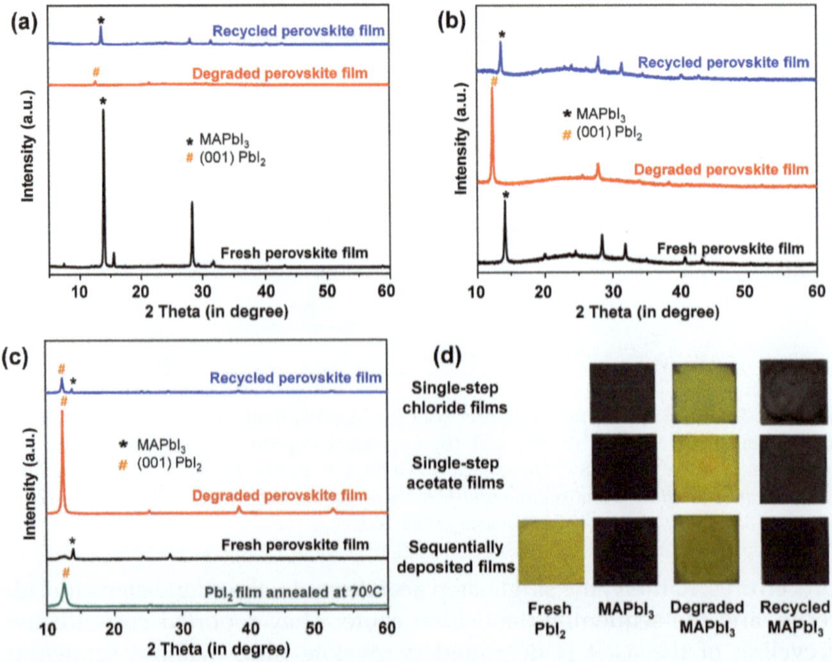

FIG. 8 XRD patterns of freshly deposited $MAPbI_3$ films, degraded $MAPbI_3$ films (left with PbI_2), and recycled $MAPbI_3$ films prepared via (A) single-step chloride route, (B) single-step acetate route, and (C) sequential deposition route; (D) analogous pictures of the freshly deposited, degraded, and recycled films corresponding to A–C. *Reproduced with permission from P. Chhillar, B.P. Dhamaniya, V. Dutta, S.K. Pathak, Recycling of perovskite films: route toward cost-efficient and environment-friendly perovskite technology, ACS Omega 4 (2019) 11880–11887. https://doi.org/10.1021/acsomega.9b01053.*

As evident from Fig. 8A, the high peak intensity of the freshly deposited $MAPbI_3$ film via the single-step chloride route suggests that the MHP film is well oriented and highly crystalline. The reduced peak intensity of the resultant PbI_2 clearly proves that degradation eroded the crystallinity of the chloride route-$MAPbI_3$ film. In contrast, when the acetate route-$MAPbI_3$ film was exposed to humidity, the peak intensity of the resultant PbI_2 and, henceforth, crystallinity was enhanced in the degraded film, depicted in Fig. 8B. For the sequentially deposited $MAPbI_3$ film, the degradation drastically increased the peak intensity and crystallinity of the resultant PbI_2 when exposed to ambient environment. The peak intensity was also much higher than that of a bare lead iodide film annealed at 70°C, shown in Fig. 8C. The low-intensity PbI_2 peaks correspond to randomly oriented PbI_2 film, whereas high-intensity peaks originate from oriented PbI_2 film. All the degraded samples were recycled by reacting each of the degraded film (i.e., PbI_2) with MAI, as shown in Fig. 7. The characteristic $MAPbI_3$ peaks reappeared, and the PbI_2 peaks disappeared after coating with 10 mg/mL MAI solutions. The PbI_2 peaks vanished entirely in the recycled films deposited through single-step chloride/acetate routes, showing complete recovery. However, the coexistence of both $MAPbI_3$ and PbI_2 peaks in the recycled film fabricated via sequential deposition indicated that all the lead iodide was not consumed during the recycling process. It can also be seen through Fig. 8C that a large quantity of unconverted lead iodide remained in the recycled film. Therefore, recycling of the sequentially deposited films may be further optimized for complete recovery of the degraded perovskite. Nevertheless, it was mentioned in Section 2.1 that the presence of a critical amount of residual or slight excess PbI_2 in the MHP improved the HPSC performance, and only beyond that it deteriorated the device performance [99,104]. The critical remnant PbI_2 in MHP acts as a blocking layer between the MHP and TiO_2, and prevents recombination and back-injection of the charge carriers [103,166]. Hence, despite the presence of unconverted PbI_2, i.e., incomplete conversion of PbI_2 to $MAPbI_3$ in sequentially deposited films, the recycled perovskite can produce performance comparable to that of fresh $MAPbI_3$ films [102].

The morphology of the degraded MHP films governs the morphology and optoelectronic properties of the recycled MHP films. The crystal structure and morphology of the degraded chloride route-film deteriorated to the extent that the initial film quality and optoelectronic performance were not retrievable. Moreover, the peak intensities of the degraded film were the least, and the peak intensities after recycling were not sufficiently recovered. So, among the three methods, recycling of the chloride route-film was comparatively not effective; however, complete reconversion of PbI_2 to $MAPbI_3$ was possible. On the other hand, for both sequential deposition and acetate routes, the film morphologies were enhanced. The explicit increase in the crystallinity of the degraded films is the reason behind good photo-performance recovery after the recycling of the degraded films. So, recycling was efficient for the films fabricated via these two routes with an exception that a full recovery of the degraded MHP was possible for the acetate route-films, while it was difficult for the sequentially deposited films. Developing a simple and easy way to completely reconvert the remnant PbI_2 to perovskite in the sequentially deposited films will result in far superior optoelectronic properties in the recycled films.

4.4 Recycling of perovskite solar cells

Poor long-term stability of the HPSCs is a critical issue; however, if the HPSCs are recycled effectively, then this issue might become less critical. Kim and coworkers reported recycling or repair methodology for HPSCs.

Their recycling procedure was capable of dissolving the perovskite layers selectively such that the FTO transparent conducting glass and gold electrodes were separated and collected to be reused [51]. Generally, the electron-transport layer (ETL) is composed of compact and mesoporous TiO_2 layers. The mesoporous TiO_2 layers are coated on the FTO glass which is used as the TCG substrate. Fig. 9 displays the steps for fabrication and recycling of the HPSCs. For HPSC fabrication, first, the ETL is deposited on the FTO glass and annealed. After that the perovskite light absorber layer is coated on the substrate, and the hole-transport layer (HTL) and gold electrode are deposited. During the recycling process, the deposited-metal electrode is peeled-off when the substrate is immersed in a polar aprotic solvent (like DMF, DMSO, or GBL) and shaken, as shown in Fig. 9B. The HTL and the MHP layer dissolve leaving behind a clear ETL-coated TCG substrate and recyclable gold electrode in the resultant liquid, as shown in Fig. 10. Gold of HPSC does not react with the polar aprotic solvent and, hence, can be collected. The collected gold electrodes occasionally contain lead impurities in minute concentrations which can easily be removed through a refining process, and the refined gold can be reused. Costly metals are usually recovered from the industrial electronic and solution wastes, and refined. The refined gold constitutes about 4% of the total gold of the industrial market [167]. To clean and

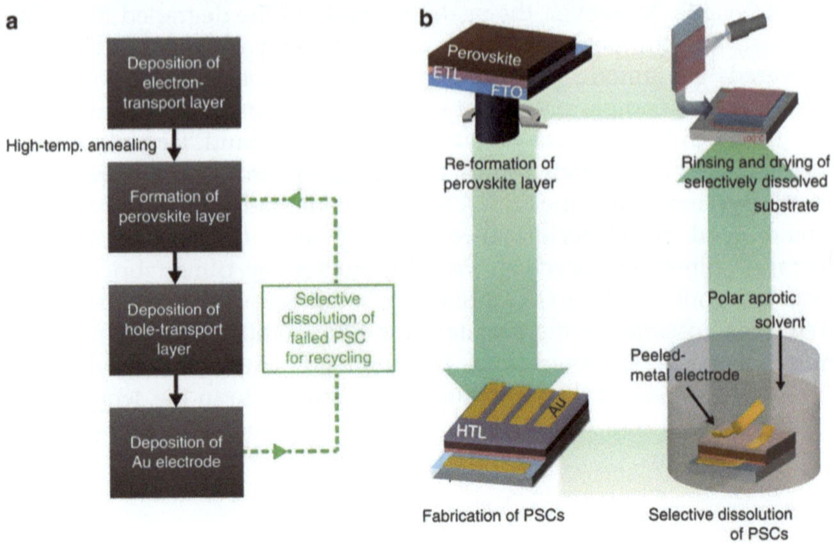

FIG. 9 Recycling of HPSCs. (A) Flow chart of processes for fabrication and recycling of HPSCs. (B) Detailed schematic of the HPSC recycling process done by selective dissolution. *Reproduced with permission from B.J. Kim, D.H. Kim, S.L. Kwon, S.Y. Park, Z. Li, K. Zhu, H.S. Jung, Selective dissolution of halide perovskites as a step towards recycling solar cells, Nat. Commun. 7 (2016) 11735. https://doi.org/10.1038/ncomms11735.*

FIG. 10 Digital images before and after the selective dissolution of HPSCs. The fabricated perovskite layer easily dissolves in the polar aprotic solvent DMF. The expensive gold electrode and the cleaned ETL-coated TCG substrate can be collected and recycled. *Reproduced with permission from B.J. Kim, D.H. Kim, S.L. Kwon, S.Y. Park, Z. Li, K. Zhu, H.S. Jung, Selective dissolution of halide perovskites as a step towards recycling solar cells, Nat. Commun. 7 (2016) 11735. https://doi.org/10.1038/ncomms11735.*

reuse the TCG substrate, it is rinsed with deionized water and then dried. For optimal performance, it may be annealed at 500°C for an hour. Finally, redepositing the MHP layer, HTL, and Au electrode produces a new HPSC on the recycled substrate.

To understand the solvent effect, the recycling scheme was examined using different types of solvents. It was observed that as compared to the nonpolar solvents (e.g., diethyl ether, toluene, chlorobenzene), the polar solvents produced better dissolution results. The nonpolar solvents dissolved only the HTL, and not the perovskite. Contrarily, both protic and aprotic polar solvents dissolved the HTL as well as the perovskite, although with different dissolution kinetics. The polar protic solvents (e.g., ethanol, water) took more time to dissolve the perovskite, while polar aprotic solvents (e.g., DMF) dissolved the perovskite instantly. The impact of polar aprotic solvents causing the removal of the $MAPbI_3$ perovskite materials was analyzed by field-emission scanning electron microscopy (FESEM). Fig. 11A displays a plane-view of the mesoporous TiO_2 layer-coated TCG substrate, and Fig. 11B represents a planar view of the $MAPbI_3$ layer deposited atop the coated TCG substrate. During the recycling process through selective dissolution of the $MAPbI_3$-deposited substrate via DMF, the top $MAPbI_3$ layer dissolves. A cleaned mesoporous TiO_2 layer-coated TCG substrate remains, as shown in Fig 11C. The efficacy of the recycling procedure was determined by repeated recycling of the mesoporous TiO_2 layer-coated TCG substrate.

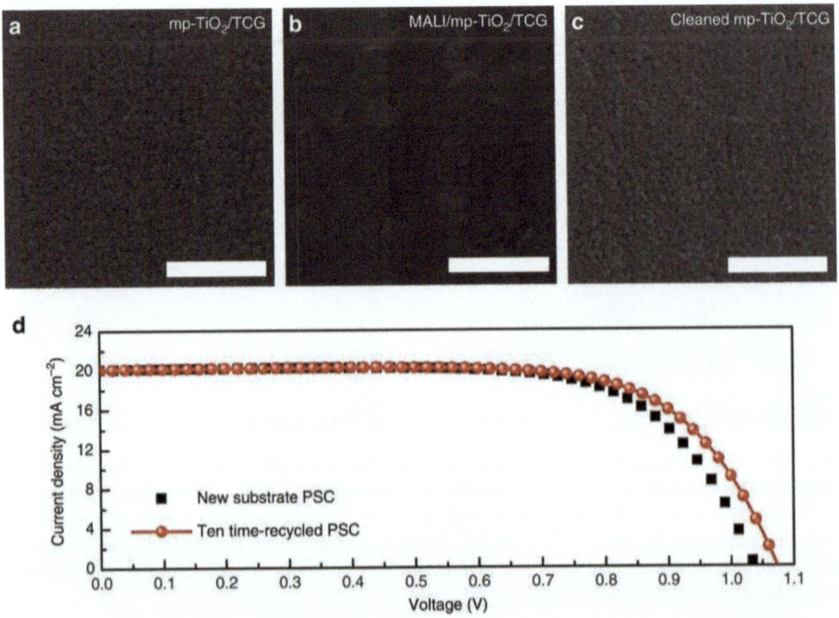

FIG. 11 Planar FESEM morphology of the TCG substrate (A) coated with a fresh mesoporous TiO_2 (mp-TiO_2) layer, (B) $MAPbI_3$ (MALI)-layer deposited on top of the mp-TiO_2 layer, and (C) a cleaned mp-TiO_2 layer-coated TCG substrate after selective dissolution of the $MAPbI_3$-layer using DMF (All scale bars: 1μm). (D) Current density-voltage (J-V) curve comparison of an HPSC fabricated on ten times recycled TCG substrate against an HPSC fabricated on a new substrate. *Reproduced with permission from B.J. Kim, D.H. Kim, S.L. Kwon, S.Y. Park, Z. Li, K. Zhu, H.S. Jung, Selective dissolution of halide perovskites as a step towards recycling solar cells, Nat. Commun. 7 (2016) 11735. https://doi.org/10.1038/ncomms11735.*

It was observed that even after recycling for ten times, the J-V curve of the recycled device was more or less similar to that of a device fabricated using a fresh substrate (Fig. 11D). The recycled device also showed stability comparable to that of a device with a new substrate. Thus, this proves that the HPSCs can be efficiently recycled.

In addition, it was observed that negligible amount of PbI_2 (which was difficult to remove) still remained on the mesoporous TiO_2 layer even after the dissolution of the perovskite layer in DMF. As mentioned in Sections 2.1 and 4.3, residual PbI_2 can improve or degrade the HPSC performance based on a critical amount. However, in this case, the residual PbI_2 did not produce any effect on the performance of the regenerated device, or accumulated with repeated recycling. It was also mentioned in Section 4.2 that DMF extraction method for removing remnant perovskite (i.e., PbI_2) deteriorated the substrate's TiO_2 coating layer which reduced the device performance each time the substrate was recycled. So, if the TiO_2 coating was damaged, a fresh TiO_2 layer needed to be deposited on the recycled substrate before fabricating another HPSC using it, in order to achieve comparable performance after each recycling. Following this recycling procedure, equivalent device performance was achieved for up to 3 times of recycling [53]. However, the selective dissolution method of recycling mentioned here had no effect on the TiO_2 coating layer or the device performance even after recycling the substrate 10 times [51]. It is interesting that despite using the same material (i.e., DMF), the TiO_2 coating gets affected while removing the remnant PbI_2 in one case, while neither residual PbI_2 gets removed completely nor the TiO_2 layer is affected in the other case. A logical interpretation for this difference can be that the stoichiometric amounts of DMF and PbI_2, the device structure, and the recycling method may play some role to determine whether the TiO_2 coating would get damaged or not, and residual PbI_2 would get removed or not.

Further, nuclear magnetic resonance (NMR) measurements were conducted to investigate the role of the bonding between an aprotic solvent and a metal halide facilitating the recycling process. ^{13}C NMR was performed at 500 MHz with DMF, 1M $MAPbI_3$ in DMF, and 1M PbI_2 in DMF. Figs. 12A and B display that in comparison to the peaks of the DMF (marked by blue lines), the main ^{13}C NMR peaks of the lead iodide

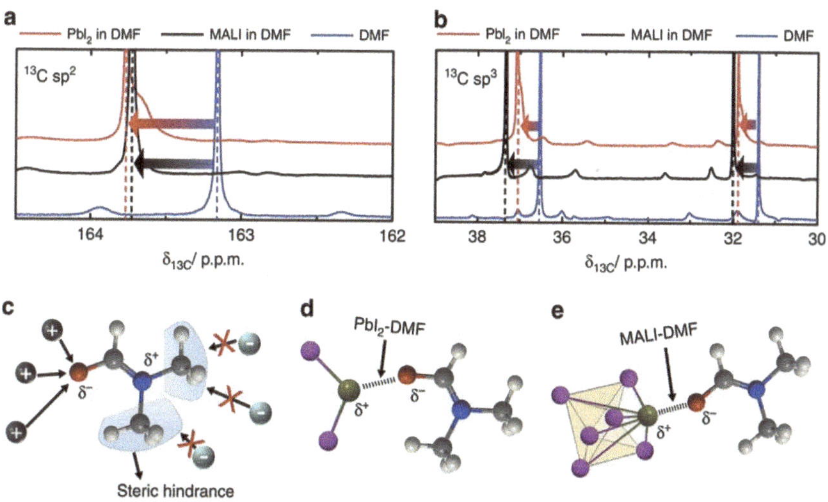

FIG. 12 Dissolution mechanism of $MAPbI_3$ (MALI) perovskite in DMF (a polar aprotic solvent). Chemical shifts of PbI_2−DMF and $MAPbI_3$−DMF signals versus pure DMF signal for NMR (A) ^{13}C sp^2, and (B) ^{13}C sp^3. Schematic illustration of the (C) resonance structure of DMF (with partially negative and positive polarities); chemical bonding structures of (D) PbI_2, and (E) PbI_6 octahedral frames of $MAPbI_3$ with partially negative oxygen of DMF; C: gray spheres, O: red spheres, H: white spheres, N: blue spheres, Pb: yellow spheres, and I: violet spheres. *Reproduced with permission from B.J. Kim, D.H. Kim, S.L. Kwon, S.Y. Park, Z. Li, K. Zhu, H.S. Jung, Selective dissolution of halide perovskites as a step towards recycling solar cells, Nat. Commun. 7 (2016) 11735. https://doi.org/10.1038/ncomms11735.*

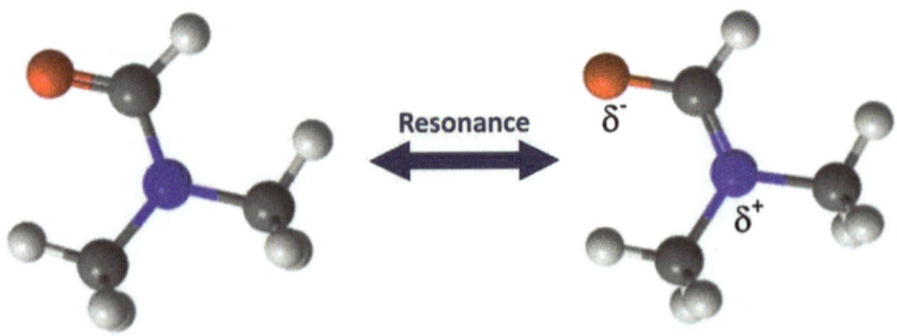

FIG. 13 Chemical resonance structure of DMF (a polar aprotic solvent), generating a partially negative oxygen atom that can bind to the Pb^{2+} of the MHP material; C: gray spheres, O: red spheres, H: white spheres, and N: blue spheres. *Reproduced with permission from B.J. Kim, D.H. Kim, S.L. Kwon, S.Y. Park, Z. Li, K. Zhu, H.S. Jung, Selective dissolution of halide perovskites as a step towards recycling solar cells, Nat. Commun. 7 (2016) 11735. https://doi.org/10.1038/ncomms11735.*

in DMF solution were shifted downward (marked by red lines). These results indicate the formation of the metal halide−DMF bonds. Lead iodide bonds with DMF molecule through a Pb−O bond forming a PbI_2−DMF structure [168–170]. Due to the chemical resonance structure of the DMF (Fig. 13), its partially negative oxygen atom can bond with partially positive atoms (like lead cations, Pb^{2+}), as shown in Fig. 12C. This confirms that aprotic solvents like DMF are capable of forming PbI_2−DMF complexes, as illustrated in Fig. 12D. Also, from Fig. 12A and B, the chemical shifts in the ^{13}C sp^2/sp^3 signals corresponding to the $MAPbI_3$−DMF solution (marked by black lines) are similar to the signals corresponding to the PbI_2−DMF solution (red lines). These results of ^{13}C NMR signify that the Pb^{2+} cations present in the $MAPbI_3$ perovskite layer can facilely form bonds with the partially negative oxygen atoms of the aprotic solvents, as shown in Fig. 12E. This bonding allows the dissociation of the PbI_6 octahedral-frame which is the basis of organic-inorganic MHP materials. This causes the high solubilities of MHP materials in polar aprotic solvents. After the recycling process, a problem that arises is of removing the Pb contents from residual polar aprotic solvent due to environmental concerns. There are a number of methods reported for removing Pb from the Pb-containing solution wastes such as electrodialysis [171], biosorption [172], solvent extraction [173], ion exchange [174], and solvent extraction-ion exchange combination [51]. The use of any of these methods would help in Pb-removal from the residual solvent making the recycling process overall effective. To add, this method of selective dissolution can also be extended to recycle (a) HPSCs with MHPs deposited using different processes or having various perovskite compositions such as $(FAPbI_3)_{0.85}(MAPbBr_3)_{0.15}$, and (b) planar-structured and flexible HPSCs.

Alternatively, an economic recycling strategy has been developed to isolate, collect, and reuse almost all the different layers of the HPSC panels

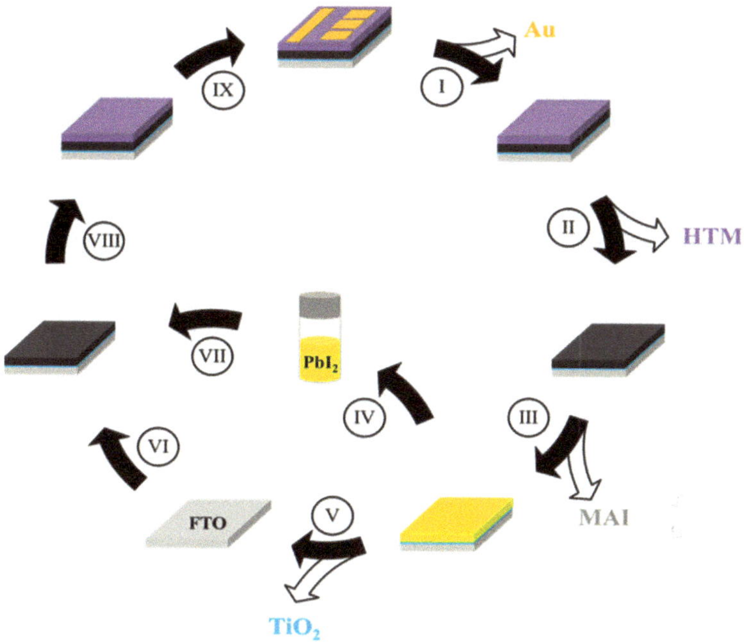

FIG. 14 Layer-by-layer recycling procedure for HPSCs: (I) etching of Au electrode with adhesive tape, (II) removal of HTM by dipping in chlorobenzene, (III) decomposition of MHP into MAI and PbI_2, and extraction of MAI in water, (IV and V) exclusion of PbI_2 and TiO_2 using DMF, (VI) coating a fresh TiO_2 film, (VII) developing perovskite film on the recycled FTO substrate using recycled PbI_2, (VIII) preparing the HTL, and (IX) deposition of Au top-electrode. *Reproduced with permission from A. Binek, M.L. Petrus, N. Huber, H. Bristow, Y. Hu, T. Bein, P. Docampo, Recycling perovskite solar cells to avoid lead waste, ACS Appl. Mater. Interfaces 8 (2016) 12881–12886. https://doi.org/10.1021/acsami.6b03767.*

[53]. The recycling process involves stripping down the device layer by layer, as shown in Fig. 14. This methodology was demonstrated by recycling the $MAPbI_3$-based planar HPSCs which were fabricated following the interdiffusion synthetic technique of Xiao et al. [175]. The first step of recycling begins with removing the HPSC's gold top-contact using an adhesive tape (Fig. 14, step I). The second step involves selective removal of the spiro-OMeTAD ($C_{81}H_{68}N_4O_8$)-based HTL by immersing the HPSC in chlorobenzene (Fig. 14, step II). This step is crucial for removing the hole-transport material (HTM) selectively without removing the perovskite structure. Additionally, steps I and II can be combined by using an eco-friendlier solvent, ethyl acetate, which would dissolve the HTL and delaminate the gold electrode. The gold can then be isolated via filtration, and later recycled. In the third step, the substrate is immersed briefly in ultra-pure water to decompose the MHP layer into PbI_2 and MAI

(CH_3NH_3I), where the MAI component is directly extracted in the aqueous medium (Fig. 14, step III). The short span of immersion in conjugation with the low solubility of PbI_2 in water ultimately results in a low concentration of Pb in the water remaining below the characteristic toxicity level suggested by the US-EPA [176]. The PbI_2 is eliminated from the substrate in the fourth step by immersing the device in DMF for a short period of time (Fig. 14, step IV). This PbI_2 may be collected and recycled after performing a recrystallization process to remove any possible impurities [53]. Further, if washing of the substrate using DMF is continued, the TiO_2 layer also gets eliminated, and a clean FTO substrate is obtained (Fig. 14, step V). The cleaned FTO substrate is dried at 100°C and cooled before reuse. To prepare a new HPSC, a fresh TiO_2 film can be coated on the recycled FTO substrate (Fig. 14, step VI). The perovskite layer can be fabricated using the recycled PbI_2 collected in the fourth step (Fig. 14, step VII). Finally, depositing a new HTL (Fig. 14, step VIII), and top electrode using fresh/recycled gold (Fig. 14, step IX), will produce a new HPSC using recycled components.

A raw cost analysis of planar HPSCs suggests that gold electrode (~25 USD per m^2), spiro-OMeTAD HTM (~40 USD per m^2), and FTO substrate (~100 USD per m^2) are the three expensive components of HPSCs, while perovskite (~1 USD per m^2) and TiO_2 (~4 USD per m^2) are not that costly [53]. Now, since FTO substrate is the most expensive part of HPSCs, its recycling is of highest economic importance. Next comes spiro-OMeTAD which is also bit expensive. So, many other materials have been produced at a much cheaper price to work as efficient HTLs [177–179]. This makes the recycling of spiro-OMeTAD less critical. Gold can be recycled; however, materials like silver paste can be expected to replace gold for making top contacts when HPSCs are scaled up [180,181]. Since TiO_2 and perovskite (consisting of MAI and PbI_2) are not that expensive, their recycling is less interesting from an economical viewpoint. However, from an environmental perspective, isolation/disposal of the toxic PbI_2 is essential, while reusing the collected PbI_2 is much more beneficial as it will not leave any toxic waste behind. Therefore, the possibility of recycling the perovskite, costly gold, and FTO substrate will make HPSCs commercially more reliable. In addition, reproducible HPSCs may become a consumer product and be used as batteries. Industrial start-ups, commercial progress, regulatory frameworks, economical and legislative aspects, and future scope of the recycling of MHP-based technologies will be discussed in the upcoming sections.

5. Circular economy and recycling policies

Many new types of solar cells are in progress, such as printable-solar cells [182,183], quantum-dot solar cells [184–188], organic solar cells [189–193], organic-plastic solar cells [194,195], polymer solar cells [196–199], and solar cells including different materials like 2D-MoS_2 [200,201], various metal oxide-graphene nanocomposites [202] and biomaterials [203]. However, solar cells based on perovskite materials (HPSCs) have taken a considerable lead among the other emerging third-generation PV technologies [1]. Reports suggest that incorporating MXene and graphene-based nanomaterials into the HPSCs [204–206] or using perovskite quantum dots or ordered 1D-nanowires/nanotubes [207,208] further enhances the output performance. Converting a novel new material system-based research into a technology requires scalability assessments under economic funding and policy and regulatory restrictions. Additionally, the final product should ensure a good lifetime based on the stability estimations of the entire product and/or its subcomponents. All solar technologies create their own markets and the third-generation perovskite PV technology demonstrates great potential but HPSCs and other new solar researches remain at status quo, i.e., just as lab-bench novelties because of improper attention of the solar industry and investors toward academia. With fast rising of severe damage to the environment, time is running short for public/private sectors and governments to combine forces and weave research with business and nonpartisan policies, and thus economically tie the researchers to investors and legislators. Solar energy harvesting can bring a clean-energy transition through flexible/hybridized energy systems, innovative and transforming technologies, and constructive financing. Through technological innovation like employing artificial photosynthesis to store the infrequent sunshine as convenient fuels and replacing the existing in-use solar panels with efficient solar coatings as inexpensive as paint and/or printable like newspaper, systemic innovation by adding flexibility to the power grids and the energy systems worldwide connecting them dependably to harness the solar energy, and financial innovation for funding solar projects around the globe extending even to the sunniest far deserts or the poorest villages [1,144]. All these innovative measures will necessarily need visionary public policy for funding researchers working on next-generation solar technologies, restructuring the energy systems, establishing strong economic markets, and laying down a diverse clean energy infrastructure.

The PV technology based on solar energy holds approximately 23% share in the global renewable energy market, with more than 73% of the PV market comprising Si-SC modules. Any improvements in perovskite PV technology independently or in terms of integration with Si-SCs will significantly impact the consumer market [112]. Although China has made Si-SCs so cheap and the cost of the solar panels has reduced over the last ten years resulting in a situation where the cost of installation is just a little greater than 1/2 of the cost of putting in a group of panels, but still improving the efficiency of the solar panels is highly desirable. Methods for efficiency improvements have been discussed in the next section. Continuous improvements in the efficiency would keep reducing the number of panels needed to be installed, indirectly lowering

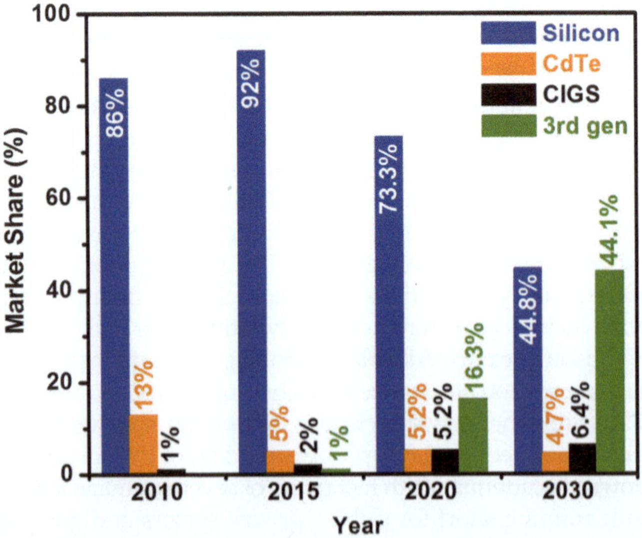

FIG. 15 Solar panel market distribution. *Reproduced using data from Md.S. Chowdhury, K.S. Rahman, T. Chowdhury, N. Nuthammachot, K. Techato, Md. Akhtaruzzaman, S.K. Tiong, K. Sopian, N. Amin, An overview of solar photovoltaic panels' end-of-life material recycling, Energy Strategy Rev. 27 (2020) 100431. https://doi.org/10.1016/j.esr.2019.100431 and D. Sica, O. Malandrino, S. Supino, M. Testa, M.C. Lucchetti, Management of end-of-life photovoltaic panels as a step towards a circular economy, Renew. Sustain. Energy Rev. 82 (2018) 2934–2945. https://doi.org/10.1016/j.rser.2017.10.039.*

the installation costs. As shown in Fig. 15, the Si-SC market share is predicted to decrease to 44.8% by 2030, while third-generation photovoltaics will increase to 44.1% [77,151]. China and other countries are rigorously working toward the large-scale production of HPSCs. The Chinese company, Hangzhou's Microquanta Semiconductor, established the world's first pilot manufacturing facility for MHP photovoltaics with 5GW-capacity. European Oxford PV reached 29.5% PCE for MHP-silicon tandems and currently building a 125MW-capacity manufacturing unit in Berlin to produce MHP-based residential roof-top panels. Hong Kong's GCL New Energy is working on a 100MW-capacity production unit in Kunshan to manufacture 18%-efficient HPSC modules at 70% less production cost than Si-SCs [116]. Another Chinese company, Hubei's Wonder Solar, with the expertise of fully printable-HPSCs, aims to develop a 10GW-capacity production base in Ezhou city. At the same time, the American Cubic PV plans to invest in building a 10GW-capacity production line in India in the upcoming 5 years. Such collaborations of the PV industry with academia must be strengthened to boost the global commercialization of HPSCs.

A market-based technology needs to exhibit versatility in the context of flexibility, portability, cost, and most importantly, recyclability. PV recycling for PV waste management is a big necessity to retain PV technology's position as a "clean energy technology." Though the PV sector is expanding comparatively faster in Asia, there is no Asian country with adequate legislative solar panel waste recycling laws/regulations to manage the enormous amounts of PV waste accumulating with time. Only the European Union (EU) has a clear and strict regulatory framework that classifies the solar panels explicitly under the Waste Electrical and Electronic Equipment (WEEE) directive, adopting ~85% recovery rate and ~80% recycling rate of the PV waste [209]. The countries that follow the EU's directive include UK, Germany, Czech Republic and Italy. As electronic wastes are like liability whose management responsibility lies on the product's manufacturer, the EU commission entails all the EU market's solar PV material supplying manufacturers and importers to finance the costs for the collection and recycling of the EoL PV products in Europe and subjecting hefty fines if not complied [77]. Sadly, most other countries classify solar PV wastes as just industrial or general wastes. However, the USA has upgraded the facilities and companies for recycling the PV wastes, and Australia is also taking measures for PV waste management. From the Asian nations, South Korea is getting serious with PV waste regulations, China's Trina Solar and Yingli Solar companies are studying PV recycling, and Japan has subsidized recycling equipment in order to promote solar panel recycling. For reprocessing all of Japan's PV waste produced by 2020, the Toshiba Environmental Solutions expects the task to take approximately 19 years [209]. Still, due to lack of specific regulations for disposal/recycling, most PV wastes end up in landfills. Only a few global companies are exclusively focused on solar panel recycling, for example, the American First Solar is into end-to-end recycling of thin-film CdTe solar panels, while France's Veolia recycles Si-SC panels. Some of the other companies/organizations involved with direct recycling and/or recycled material trading are Recycle Solar Technologies (UK), Solar World (Germany), Antec Solar (Germany), Rema System (Czech Republic), La Mia Energia (Italy), PV Cycle (Belgium), Recycle PV Solar (USA), PV Industries (Australia), and Eiki Shoji (Japan). The development of PV recycling technology in a country requires efforts of both its private and public sectors, by assessing the patents filed by the private sector and reviewing the investment and Research and Development plans of the public sector [209]. However, strategical improvements in practicality, environmental sustainability, high recovery rate, and economic viability of the PV recycling industry are also indispensable requirements. Since the MHP-based device market is at an early stage, making manufacturers liable for recycling EoL MHP-devices will promote the conservative use of resources and development of readily recyclable and/or easily disposable environment-friendly MHP devices. This will lead to cost-effective and sustainable commercial production of the MHP devices in the long term and reduce the environmental hazard. Apart from pressurizing the manufacturers, mass media and social media support for promoting the importance of recycling are also needed. Enforcing strict policies and regulations at a stage when the MHP-based market and industries are not that widespread will help create logistical and industrial networks. This will support the development and designing of recyclable MHP-devices, avoiding a future situation of massive amounts of unmanaged MHP waste. However, suppose the production of the recyclable MHP devices and the recycling technology are made economically viable creating a positive revenue (i.e., a good business), in that case, the recycling infrastructure will be adopted regardless of any formal regulations.

6. Improving the efficiency of HPSCs

After incorporation of MHP thin films into the single-junction PV devices, the new PCE obtained was 25.5%, which is comparable to conventional single-crystalline Si-SCs and higher than previously reported thin-film PV technologies based on materials like cadmium telluride (CdTe) and copper indium gallium selenide (CIGS) [116,157,210]. For single-junction PV devices, the reported values of PCE are 22.1% for CdTe solar cells, 23.35% for CIGS solar cells, and 21.2%–27.6% for Si-SCs [210]. Si-SCs had seen a massive increase in PCE, which was just 6% when Bell Lab first innovated them in 1954, and it is now more than quadrupled ~27.6% [210,211]. On the other hand, the maximum PCE of HPSCs based on the MHP materials rose rapidly from 3.8% to 25.5% in just 12 years (2009–2021) [210,212]. HPSCs are typically composed of MHP light absorbers, counter electrodes, and charge-transport layers. The efficiency of the HPSC devices is determined by the (a) device's charge extraction and transportation processes and (b) energy loss in interface and bulk of MHP layers. Therefore, conceptual design of charge-transport layers, better crystal quality of MHP films, and minimum or negligible nonradiative recombination at the interface can further improve the efficiency of the HPSC devices [116]. For any solar cell device, it is also crucial that the power conversion efficiency must approach the Shockley-Queisser (S-Q) limit. Single-crystalline Si-SCs, single-crystalline gallium arsenide (GaAs) solar cells, and single-junction solar cells can easily achieve 25% PCE, which is 75% of the S-Q limit [213]. HPSCs rank first in reaching the 25% PCE value using polycrystalline films and have become the third-generation solar cells after silicon- and GaAs-based solar cells to reach that efficiency level. A PCE of 30.5% has already been achieved by GaAs solar cells, the highest PCE reached by single-junction solar cells

[210]. It was due to the photon recycling technology through which GaAs solar cells could reach a PCE as closest to the marginal theoretical value of efficiency [214]. Light absorption in HPSCs only needs a few hundred-nanometer thick film, similar to GaAs solar cells having a direct band gap. In contrast, single-crystalline Si-SCs required around 100μm thick light-absorbing layer due to the indirect band gap of silicon. The further development of HPSCs follows the GaAs solar cells, but there are vital material and structural differences between both of them which are considered in order to progress with the HPSCs. The strategy of multication/multihalide perovskite (illustrated in Table 1) can also improve the efficiency of the HPSCs. The combination of several technologies through integration of different energy devices in a single unit is another adjacent and powerful method to enhance the performance of the hybrid system widening its applicability and simultaneously reducing its size and weight. The concept of hybrid or integrated system of perovskite with triboelectric and piezoelectric nanogenerators for multiple energy harvesting has already been talked about in Section 3. In addition, the high efficiency ability of HPSCs has the potential to produce more efficient operation in an integrated system comprising solar cells [63,215,216]. Such integration of HPSCs can be done with three classes of energy devices:

(a) other solar cells: Si-SCs, CIGS, organic solar cells (OSCs) or another HPSC;
(b) electrochemical devices for energy storage (in solar fuels): photo-electrosynthetic cell (PESC) and electrolysis cell (ELC);
(c) electrical energy storage devices: supercapacitor.

6.1 Photon recycling in halide perovskites

Metal halide perovskites (MHPs) are promising materials owing to attributes such as low trap densities, long carrier lifetimes, high absorption coefficient, strong absorption in the visible region, free charge generation, bipolar transport, large charge diffusion and excitation transport lengths, high mobilities of charge carriers, and tunable optical/electrical properties [154,217–222]. For example, by changing the chemical composition from tri-iodide to tri-bromide, the optical absorption can be shifted from visible to near-IR wavelengths [110]. The semiconducting property of MHPs and recombination of photo-excited charge carriers have a dependence on their transient photoluminescence (PL) [223], and the bimolecular process can provide efficient radiative recombination [224,225]. High radiative efficiencies can be achieved using nanocrystalline materials and laser cooling [226,227]. But, these suggestions conflict in thin films where lower radiative efficiencies are observed due to non-radiative losses around defects and surfaces [228]. This raises questions about the extent to which the external radiative efficiency measurements relate to the internal recombination processes and whether the efficiencies were also influenced by other processes such as light out-coupling [229]. Lead halide perovskites give strong PL and sharp band edges, which are requisites for facilitating photon recycling, i.e., the re-absorption of radiatively recombined charge pairs to regenerate another excitation. Over the years, rapid improvement in the PCE of HPSCs was brought about through various technologies like device structure having thin films, surface engineering, etc., and noting from GaAs solar cells, photon recycling is a technology through which the S-Q limit can be approached [213]. In photon recycling, the concentration of photons in the light-harvesting layer is kept high which helps open-circuit voltages get close to the level of theoretical limit. The applicability of photon recycling in all types of HPSCs is uncertain, but based on the ideologies and reports of photon recycling, there are following three requirements for application of this technology in HPSCs: (i) high yield of radiative emission, (ii) high yield of re-absorption, and (iii) good optical management. Thus, high photoluminescence quantum efficiency (PLQE), small Stokes shift of light-absorbing halide layer, and efficient optical design suitable for the multilayered structure of HPSCs are important factors for expecting photon recycling and getting close to the S-Q limit using HPSCs [213].

A group of researchers observed photon recycling in the thin films of lead tri-iodide perovskite solar cells and reported that the energy transport could occur over large distances through events in the sequence of multiple absorption-diffusion-emissions, and was not prevented by diffusive charge transport. This created high excitation densities inside the perovskite layer and high V_{oc} in the solar cells [230]. In another report, the same group showed that photon recycling strongly affected the externally measured recombination constants, leading to significant differences in the PLQEs measured internally and externally. In their thin films, the PLQEs measured internally exceeded the ones measured externally, but by optimizing photonic structures with light out-coupling, the high internal efficiencies were also recovered in the external measurements, proving efficient radiative recombination inside the MHPs [228]. Kirchartz and coworkers examined the effect and relation of photon recycling with open-circuit voltages in HPSCs. They showed that the voltage boost resulting from photon recycling has a dependence on the Shockley-Read-Hall lifetimes, parasitic absorption, and light out-coupling efficiency [231]. Dursun's group gave theoretical as well as experimental evidence for photon recycling in the cesium lead bromide ($CsPbBr_3$) microwires [232], and Motti's team demonstrated that charge diffusion and heterogeneous photon recycling enhanced the charge transport in quasi 2D Pb-MHP films [135].

6.2 Integrating HPSCs with other power technologies

Multijunction PV devices with GaAs can achieve PCE of about 30%–47% [210]. Multijunction PV devices with HPSCs can be fabricated by using S-Q limit extraction by tuning the band gap of MHP thin films. Multijunction PV devices can be formed in either a two-terminal configuration or a four-terminal configuration [215]. For the fabrication of small- or large-scale perovskite thin films, coevaporation is considered the best method, especially for two-terminal PV devices [31]. Two-terminal PV devices are ideal for solar cells with thin films because they give the least optical losses, require a lesser number of semitransparent electrodes, and the same number of supporting components as single-junction PV devices, while four-terminal PV devices require at least double of those expensive supportive components [215,233]. Also, the two-terminal PV devices must match the two subcell currents according to the recombination layer's electron-hole balance [234,235]. Practical two-terminal PV devices combine the perovskite-based cells (HPSCs) with silicon cells [116,236,237]. On the other hand, entirely perovskite-based PV devices prove to be more advantageous as they are very thin and can be made cost-effective. However, each layer of all-perovskite PV device and/or silicon-perovskite mix PV device requires intricate designs to ensure maximum efficiency [16,48,237–242]. Material's chemical stability, light management in the device, and current-voltage (I-V) characteristics of the subcells are the most critical aspects for the fabrication of any multijunction PV device [241,243]. For the light management through the layers, it is essential to avoid scattering losses due to the reflections of light so that maximum light gets absorbed in each layer [244]. Present all-perovskite-based two-terminal PV devices are made up of 12 layers, each having a different refractive index value [233]. Fine-tuning the thickness of the perovskite thin films is essential for absorbing light in the layers and collecting photo-current in order to improve the PCE of the multijunction PV devices [245].

No existing conventional solar cell gives a PCE of 100% because the entire spectrum of sunlight does not get absorbed and there are optical and electrical losses due to parasitic absorption, heat dissipation, internal resistance, and other reasons. Since the entire light spectrum is not absorbed, tandem solar cells are fabricated with a top cell absorbing light's shorter wavelengths and a bottom cell absorbing the remaining longer wavelengths to maximize sunlight absorption and improve the device's output performance. Wide bandgap HPSCs as top cells promise great benefits in integrated solar cell devices. Thus, HPSCs have been integrated with other power technologies such as Si-SCs, CIGS, OSCs, other HPSC, PESC, ELC, and supercapacitors [63,216]. The reported output efficiencies achieved were 29.5% for HPSC/Si-SC, 24.2% for HPSC/CIGS, 24.8% for HPSC/HPSC, 16.36% for HPSC/OSC, 20% for HPSC/supercapacitor [16,210,216,246], and energy storage efficiency of 6.2% for HPSC/PESC, 12.7% for HPSC/ELC and 10% for HPSC/supercapacitor [246–248]. Theoretical simulations for a perovskite-CIGS tandem predict PCEs up to 42% using a 0.9–1.2 eV bandgap bottom cell and a 1.6–1.9 eV bandgap top cell [216,234]. Predictions for HPSC/Si-SC and HPSC/HPSC hybrids are above 35% [245]. Due to the large bandgap tunability of perovskite from 1.17 to 2.3 eV, multijunction devices with three or more junctions may reach efficiencies of more than 50% in the future [49,216].

7. Conclusion and future scope

Metal halide perovskites are promising materials owing to their remarkable attributes such as abundance of materials, low weight, mechanical flexibility, tunable optoelectronic properties, low trap densities, long carrier lifetimes, high absorption coefficient, strong absorption in visible region, free charge generation, bipolar transport, large charge diffusion and excitation transport lengths, high mobilities of charge carriers imparting novel characteristics such as high power conversion efficiency (PCE), feasibility of low-temperature fabrication, low-cost production, self-healing/self-recovery to the solar cells (HPSCs) made using these materials. Different methods for fabricating the HPSCs included solution-based, vapor-based, or a combination of both these processes. Among them, coevaporation vapor-based technique was considered the best method for fabricating small- or large-scale perovskite thin films, especially with respect to two-terminal PV devices. HPSCs reaching high PCEs are considered good competitors against the other conventional solar cells. For commercialization of the HPSCs, factors such as cost-effectiveness, economic feasibility in manufacture, long-term stability, eco-friendly disposal, or reuse of lead waste, and efficacy of lead-free perovskite solar cells are needed to be assessed. Chinese companies such as Microquanta Semiconductor, GCL New Energy and Wonder Solar, and European company Oxford PV have initiated large-scale production of HPSCs, while an American company named Cubic PV plans to invest in building a 10-GW-capacity production line in India in the upcoming 5 years.

Factors that influenced the degradation behavior of HPSCs gave hints toward making HPSCs more robust and durable. Since the disposal of hazardous wastes was very costly ($\sim$1.10 USD per kg) and the HPSCs with lead alternatives showed comparatively poor solar performance, recycling or reusing the lead-based HPSCs was a better option. For this, different procedures for recycling and reusing the different parts such as gold electrodes, conductive transparent glass (FTO), lead precursor, perovskite film or entire perovskite solar cells were detailed in this chapter. This recycling approach indeed decreases the perovskite solar cells' average production cost, countering lead hazards and solving the long-term instability. Moreover, using photon recycling technology, solar cells could reach higher values of efficiencies, so the improvements in HPSCs due to photon recycling are in constant research. Besides photon recycling, HPSC-advancement via different integrated systems (containing HPSCs) was also elaborated. Through this work, we hope that the societal, environmental, and economic significance of recycling and further development of perovskite-based devices help strategize futuristic policies leading to rapid and sustainable global commercial expansion. This chapter also aims to inspire future researchers to develop cost-effective, eco-friendly, stable, and recyclable technologies based on perovskite materials.

Acknowledgments

J. R. Ansari gratefully acknowledges Dronacharya College of Engineering, Gurugram, India, for the financial support and cooperation during the process of writing of this chapter.

Declaration of competing interest

The authors declare that they have no known competing financial interests or personal relationships that could have appeared to influence the work reported in this chapter.

References

[1] V. Sivaram, Taming the Sun: Innovations to Harness Solar Energy and Power the Planet, The MIT Press, Cambridge, MA, 2018.

[2] P. Zhu, C. Chen, S. Gu, R. Lin, J. Zhu, $CsSnI_3$ solar cells via an evaporation-assisted solution method, Sol. RRL 2 (2018) 1700224, https://doi.org/10.1002/solr.201700224.

[3] Y. Deng, Q. Dong, C. Bi, Y. Yuan, J. Huang, Air-stable, efficient mixed-cation perovskite solar cells with Cu electrode by scalable fabrication of active layer, Adv. Energy Mater. 6 (2016) 1600372, https://doi.org/10.1002/aenm.201600372.

[4] M. Salado, L. Calio, R. Berger, S. Kazim, S. Ahmad, Influence of the mixed organic cation ratio in lead iodide based perovskite on the performance of solar cells, Phys. Chem. Chem. Phys. 18 (2016) 27148–27157, https://doi.org/10.1039/C6CP03851D.

[5] S. Tang, Y. Deng, X. Zheng, Y. Bai, Y. Fang, Q. Dong, H. Wei, J. Huang, Composition engineering in doctor-blading of perovskite solar cells, Adv. Energy Mater. 7 (2017) 1700302, https://doi.org/10.1002/aenm.201700302.

[6] M.-C. Tang, Y. Fan, D. Barrit, X. Chang, H.X. Dang, R. Li, K. Wang, D.-M. Smilgies, S.(. F.). Liu, S. De Wolf, T.D. Anthopoulos, K. Zhao, A. Amassian, Ambient blade coating of mixed cation, mixed halide perovskites without dripping: in situ investigation and highly efficient solar cells, J. Mater. Chem. A 8 (2020) 1095–1104, https://doi.org/10.1039/C9TA12890E.

[7] B.-X. Chen, W.-G. Li, H.-S. Rao, Y.-F. Xu, D.-B. Kuang, C.-Y. Su, Large-grained perovskite films via $FA_xMA_{1-x}Pb(I_xBr_{1-x})_3$ single crystal precursor for efficient solar cells, Nano Energy 34 (2017) 264–270, https://doi.org/10.1016/j.nanoen.2017.02.034.

[8] P. Gratia, G. Grancini, J.-N. Audinot, X. Jeanbourquin, E. Mosconi, I. Zimmermann, D. Dowsett, Y. Lee, M. Grätzel, F. De Angelis, K. Sivula, T. Wirtz, M.K. Nazeeruddin, Intrinsic halide segregation at nanometer scale determines the high efficiency of mixed cation/mixed halide perovskite solar cells, J. Am. Chem. Soc. 138 (2016) 15821–15824, https://doi.org/10.1021/jacs.6b10049.

[9] F. Bella, G. Griffini, J.-P. Correa-Baena, G. Saracco, M. Gratzel, A. Hagfeldt, S. Turri, C. Gerbaldi, Improving efficiency and stability of perovskite solar cells with photocurable fluoropolymers, Science 354 (2016) 203–206, https://doi.org/10.1126/science.aah4046.

[10] M. Saliba, T. Matsui, K. Domanski, J.-Y. Seo, A. Ummadisingu, S.M. Zakeeruddin, J.-P. Correa-Baena, W.R. Tress, A. Abate, A. Hagfeldt, M. Grätzel, Incorporation of rubidium cations into perovskite solar cells improves photovoltaic performance, Science 354 (2016) 206–209, https://doi.org/10.1126/science.aah5557.

[11] M. Saliba, T. Matsui, J.-Y. Seo, K. Domanski, J.-P. Correa-Baena, M.K. Nazeeruddin, S.-M. Zakeeruddin, W. Tress, A. Abate, A. Hagfeldt, M. Grätzel, Cesium-containing triple cation perovskite solar cells: improved stability, reproducibility and high efficiency, Energ. Environ. Sci. 9 (2016) 1989–1997, https://doi.org/10.1039/C5EE03874J.

[12] H.X. Dang, K. Wang, M. Ghasemi, M.-C. Tang, M. De Bastiani, E. Aydin, E. Dauzon, D. Barrit, J. Peng, D.-M. Smilgies, S. De Wolf, A. Amassian, Multi-cation synergy suppresses phase segregation in mixed-halide perovskites, Joule 3 (2019) 1746–1764, https://doi.org/10.1016/j.joule.2019.05.016.

[13] F.H. Isikgor, B. Li, H. Zhu, Q. Xu, J. Ouyang, High performance planar perovskite solar cells with a perovskite of mixed organic cations and mixed halides, $MA_{1-x}FA_xPbI_{3-y}Cl_y$, J. Mater. Chem. A 4 (2016) 12543–12553, https://doi.org/10.1039/C6TA03381D.

[14] C. Lu, I.T. Choi, M.Y. Woo, C.K. Kim, C.H. Kim, H.K. Kim, Well-grown low-defect $MAPbI_{3-x}Cl_x$ films for perovskite solar cells with over 20% efficiency fabricated under controlled ambient humidity conditions, Electrochim. Acta 326 (2019) 134950, https://doi.org/10.1016/j.electacta.2019.134950.

[15] R. Cheng, C. Chung, H. Zhang, F. Liu, W. Wang, Z. Zhou, S. Wang, A.B. Djurišić, S. Feng, Tailoring triple-anion perovskite material for indoor light harvesting with restrained halide segregation and record high efficiency beyond 36%, Adv. Energy Mater. 9 (2019), https://doi.org/10.1002/aenm.201901980.

[16] R. Lin, K. Xiao, Z. Qin, Q. Han, C. Zhang, M. Wei, M.I. Saidaminov, Y. Gao, J. Xu, M. Xiao, A. Li, J. Zhu, E.H. Sargent, H. Tan, Monolithic all-perovskite tandem solar cells with 24.8% efficiency exploiting comproportionation to suppress Sn(ii) oxidation in precursor ink, Nat. Energy 4 (2019) 864–873, https://doi.org/10.1038/s41560-019-0466-3.

[17] A.D. Jodlowski, C. Roldán-Carmona, G. Grancini, M. Salado, M. Ralaiarisoa, S. Ahmad, N. Koch, L. Camacho, G. de Miguel, M.K. Nazeeruddin, Large guanidinium cation mixed with methylammonium in lead iodide perovskites for 19% efficient solar cells, Nat. Energy 2 (2017) 972–979, https://doi.org/10.1038/s41560-017-0054-3.

[18] Y. Zhang, P. Wang, M.-C. Tang, D. Barrit, W. Ke, J. Liu, T. Luo, Y. Liu, T. Niu, D.-M. Smilgies, Z. Yang, Z. Liu, S. Jin, M.G. Kanatzidis, A. Amassian, S.F. Liu, K. Zhao, Dynamical transformation of two-dimensional perovskites with alternating cations in the interlayer space for high-performance photovoltaics, J. Am. Chem. Soc. 141 (2019) 2684–2694, https://doi.org/10.1021/jacs.8b13104.

[19] T. Niu, J. Lu, M.-C. Tang, D. Barrit, D.-M. Smilgies, Z. Yang, J. Li, Y. Fan, T. Luo, I. McCulloch, A. Amassian, S.(.F.). Liu, K. Zhao, High performance ambient-air-stable $FAPbI_3$ perovskite solar cells with molecule-passivated Ruddlesden–Popper/3D heterostructured film, Energ. Environ. Sci. 11 (2018) 3358–3366, https://doi.org/10.1039/C8EE02542H.

[20] A. Mei, X. Li, L. Liu, Z. Ku, T. Liu, Y. Rong, M. Xu, M. Hu, J. Chen, Y. Yang, M. Grätzel, H. Han, A hole-conductor-free, fully printable mesoscopic perovskite solar cell with high stability, Science 345 (2014) 295–298, https://doi.org/10.1126/science.1254763.

[21] A. Mei, Y. Sheng, Y. Ming, Y. Hu, Y. Rong, W. Zhang, S. Luo, G. Na, C. Tian, X. Hou, Y. Xiong, Z. Zhang, S. Liu, S. Uchida, T.-W. Kim, Y. Yuan, L. Zhang, Y. Zhou, H. Han, Stabilizing perovskite solar cells to IEC61215:2016 standards with over 9,000-h operational tracking, Joule 4 (2020) 2646–2660, https://doi.org/10.1016/j.joule.2020.09.010.

[22] G. Grancini, C. Roldán-Carmona, I. Zimmermann, E. Mosconi, X. Lee, D. Martineau, S. Narbey, F. Oswald, F. De Angelis, M. Graetzel, M.K. Nazeeruddin, One-year stable perovskite solar cells by 2D/3D interface engineering, Nat. Commun. 8 (2017) 15684, https://doi.org/10.1038/ncomms15684.

[23] D. Zheng, C. Tong, T. Zhu, Y. Rong, T. Pauporté, Effects of 5-ammonium valeric acid iodide as additive on methyl ammonium lead iodide perovskite solar cells, Nanomaterials 10 (2020) 2512, https://doi.org/10.3390/nano10122512.

[24] D. Barrit, Y. Zhang, T. Yang, M.-C. Tang, R. Li, D.-M. Smilgies, S. (Frank) Liu, T.D. Anthopoulos, A. Amassian, K. Zhao, Sequential formation of tunable-bandgap mixed-halide lead-based perovskites: in situ investigation and photovoltaic devices, Sol. RRL 5 (2021) 2000668, https://doi.org/10.1002/solr.202000668.

[25] D.J. Kubicki, D. Prochowicz, A. Hofstetter, S.M. Zakeeruddin, M. Grätzel, L. Emsley, Phase segregation in Cs-, Rb- and K-doped mixed-cation $(MA)_x(FA)_{1-x}PbI_3$ hybrid perovskites from solid-state NMR, J. Am. Chem. Soc. 139 (2017) 14173–14180, https://doi.org/10.1021/jacs.7b07223.

[26] L.K. Ono, E.J. Juarez-Perez, Y. Qi, Progress on perovskite materials and solar cells with mixed cations and halide anions, ACS Appl. Mater. Interfaces 9 (2017) 30197–30246, https://doi.org/10.1021/acsami.7b06001.

[27] M.-C. Tang, Y. Fan, D. Barrit, R. Li, H.X. Dang, S. Zhang, T.J. Magnanelli, N.V. Nguyen, E.J. Heilweil, C.A. Hacker, D.-M. Smilgies, K. Zhao, A. Amassian, T.D. Anthopoulos, Efficient hybrid mixed-ion perovskite photovoltaics: in situ diagnostics of the roles of cesium and potassium alkali cation addition, Sol. RRL 4 (2020) 2000272, https://doi.org/10.1002/solr.202000272.

[28] K. Wang, M. Tang, H.X. Dang, R. Munir, D. Barrit, M. De Bastiani, E. Aydin, D. Smilgies, S. Wolf, A. Amassian, Kinetic stabilization of the sol–gel state in perovskites enables facile processing of high-efficiency solar cells, Adv. Mater. 31 (2019) 1808357, https://doi.org/10.1002/adma.201808357.

[29] H.-S. Kim, C.-R. Lee, J.-H. Im, K.-B. Lee, T. Moehl, A. Marchioro, S.-J. Moon, R. Humphry-Baker, J.-H. Yum, J.E. Moser, M. Grätzel, N.-G. Park, Lead iodide perovskite sensitized all-solid-state submicron thin film mesoscopic solar cell with efficiency exceeding 9%, Sci. Rep. 2 (2012) 591, https://doi.org/10.1038/srep00591.

[30] M.M. Lee, J. Teuscher, T. Miyasaka, T.N. Murakami, H.J. Snaith, Efficient hybrid solar cells based on meso-superstructured organometal halide perovskites, Science 338 (2012) 643–647, https://doi.org/10.1126/science.1228604.

[31] J.B. Patel, A.D. Wright, K.B. Lohmann, K. Peng, C.Q. Xia, J.M. Ball, N.K. Noel, T.W. Crothers, J. Wong-Leung, H.J. Snaith, L.M. Herz, M.B. Johnston, Light absorption and recycling in hybrid metal halide perovskite photovoltaic devices, Adv. Energy Mater. 10 (2020), https://doi.org/10.1002/aenm.201903653.

[32] J. Ávila, C. Momblona, P.P. Boix, M. Sessolo, H.J. Bolink, Vapor-deposited perovskites: the route to high-performance solar cell production? Joule 1 (2017) 431–442, https://doi.org/10.1016/j.joule.2017.07.014.

[33] D. Barrit, P. Cheng, K. Darabi, M. Tang, D. Smilgies, S.F. Liu, T.D. Anthopoulos, K. Zhao, A. Amassian, Room-temperature partial conversion of α-$FAPbI_3$ perovskite phase via PbI_2 solvation enables high-performance solar cells, Adv. Funct. Mater. 30 (2020) 1907442, https://doi.org/10.1002/adfm.201907442.

[34] D. Barrit, P. Cheng, M. Tang, K. Wang, H. Dang, D. Smilgies, S.F. Liu, T.D. Anthopoulos, K. Zhao, A. Amassian, Impact of the solvation state of lead iodide on its two-step conversion to $MAPbI_3$: an in situ investigation, Adv. Funct. Mater. 29 (2019) 1807544, https://doi.org/10.1002/adfm.201807544.

[35] D. Barrit, A.D. Sheikh, R. Munir, J.M. Barbé, R. Li, D.-M. Smilgies, A. Amassian, Hybrid perovskite solar cells: in situ investigation of solution-processed PbI_2 reveals metastable precursors and a pathway to producing porous thin films, J. Mater. Res. 32 (2017) 1899–1907, https://doi.org/10.1557/jmr.2017.117.

[36] J. Borchert, I. Levchuk, L.C. Snoek, M.U. Rothmann, R. Haver, H.J. Snaith, C.J. Brabec, L.M. Herz, M.B. Johnston, Impurity tracking enables enhanced control and reproducibility of hybrid perovskite vapor deposition, ACS Appl. Mater. Interfaces 11 (2019) 28851–28857, https://doi.org/10.1021/acsami.9b07619.

[37] L. Gil-Escrig, C. Momblona, M.-G. La-Placa, P.P. Boix, M. Sessolo, H.J. Bolink, Vacuum deposited triple-cation mixed-halide perovskite solar cells, Adv. Energy Mater. 8 (2018) 1703506, https://doi.org/10.1002/aenm.201703506.

[38] N.J. Jeon, J.H. Noh, Y.C. Kim, W.S. Yang, S. Ryu, S.I. Seok, Solvent engineering for high-performance inorganic–organic hybrid perovskite solar cells, Nat. Mater. 13 (2014) 897–903, https://doi.org/10.1038/nmat4014.

[39] M. Liu, M.B. Johnston, H.J. Snaith, Efficient planar heterojunction perovskite solar cells by vapour deposition, Nature 501 (2013) 395–398, https://doi.org/10.1038/nature12509.

[40] C. Momblona, L. Gil-Escrig, E. Bandiello, E.M. Hutter, M. Sessolo, K. Lederer, J. Blochwitz-Nimoth, H.J. Bolink, Efficient vacuum deposited p-i-n and n-i-p perovskite solar cells employing doped charge transport layers, Energy Environ. Sci. 9 (2016) 3456–3463, https://doi.org/10.1039/C6EE02100J.

[41] R. Munir, A.D. Sheikh, M. Abdelsamie, H. Hu, L. Yu, K. Zhao, T. Kim, O.E. Tall, R. Li, D.-M. Smilgies, A. Amassian, Hybrid perovskite thin-film photovoltaics: in situ diagnostics and importance of the precursor solvate phases, Adv. Mater. 29 (2017) 1604113, https://doi.org/10.1002/adma.201604113.

[42] J.B. Patel, R.L. Milot, A.D. Wright, L.M. Herz, M.B. Johnston, Formation dynamics of $CH_3NH_3PbI_3$ perovskite following two-step layer deposition, J. Phys. Chem. Lett. 7 (2016) 96–102, https://doi.org/10.1021/acs.jpclett.5b02495.

[43] Y. Rong, Z. Tang, Y. Zhao, X. Zhong, S. Venkatesan, H. Graham, M. Patton, Y. Jing, A.M. Guloy, Y. Yao, Solvent engineering towards controlled grain growth in perovskite planar heterojunction solar cells, Nanoscale 7 (2015) 10595–10599, https://doi.org/10.1039/C5NR02866C.

[44] J. Schlipf, P. Docampo, C.J. Schaffer, V. Körstgens, L. Bießmann, F. Hanusch, N. Giesbrecht, S. Bernstorff, T. Bein, P. Müller-Buschbaum, A closer look into two-step perovskite conversion with X-ray scattering, J. Phys. Chem. Lett. 6 (2015) 1265–1269, https://doi.org/10.1021/acs.jpclett.5b00329.

[45] R. Swartwout, M.T. Hoerantner, V. Bulović, Scalable deposition methods for large-area production of perovskite thin films, ENERGY Environ. Mater. 2 (2019) 119–145, https://doi.org/10.1002/eem2.12043.

[46] H. Zheng, W. Wang, S. Yang, Y. Liu, J. Sun, A facile way to prepare nanoporous PbI_2 films and their application in fast conversion to $CH_3NH_3PbI_3$, RSC Adv. 6 (2016) 1611–1617, https://doi.org/10.1039/C5RA21250B.

[47] Y. Zhou, M. Yang, A.L. Vasiliev, H.F. Garces, Y. Zhao, D. Wang, S. Pang, K. Zhu, N.P. Padture, Growth control of compact $CH_3NH_3PbI_3$ thin films via enhanced solid-state precursor reaction for efficient planar perovskite solar cells, J. Mater. Chem. A 3 (2015) 9249–9256, https://doi.org/10.1039/C4TA07036D.

[48] Y. Bensouda, D. Barrit, Mini-review on all-inorganic lead-based perovskite solar cells: challenges and opportunities for production and upscaling, Emergent Mater. (2022), https://doi.org/10.1007/s42247-022-00364-0.

[49] S. Yang, W. Fu, Z. Zhang, H. Chen, C.-Z. Li, Recent advances in perovskite solar cells: efficiency, stability and lead-free perovskite, J. Mater. Chem. A 5 (2017) 11462–11482, https://doi.org/10.1039/C7TA00366H.

[50] W. Zhang, M. Saliba, D.T. Moore, S.K. Pathak, M.T. Hörantner, T. Stergiopoulos, S.D. Stranks, G.E. Eperon, J.A. Alexander-Webber, A. Abate, A. Sadhanala, S. Yao, Y. Chen, R.H. Friend, L.A. Estroff, U. Wiesner, H.J. Snaith, Ultrasmooth organic–inorganic perovskite thin-film formation and crystallization for efficient planar heterojunction solar cells, Nat. Commun. 6 (2015) 6142, https://doi.org/10.1038/ncomms7142.

[51] B.J. Kim, D.H. Kim, S.L. Kwon, S.Y. Park, Z. Li, K. Zhu, H.S. Jung, Selective dissolution of halide perovskites as a step towards recycling solar cells, Nat. Commun. 7 (2016) 11735, https://doi.org/10.1038/ncomms11735.

[52] L. Patrick, Lead toxicity, a review of the literature. Part 1: Exposure, evaluation, and treatment, Altern. Med. Rev. J. Clin. Ther. 11 (2006) 2–22.

[53] A. Binek, M.L. Petrus, N. Huber, H. Bristow, Y. Hu, T. Bein, P. Docampo, Recycling perovskite solar cells to avoid lead waste, ACS Appl. Mater. Interfaces 8 (2016) 12881–12886, https://doi.org/10.1021/acsami.6b03767.

[54] B. Hailegnaw, S. Kirmayer, E. Edri, G. Hodes, D. Cahen, Rain on methylammonium lead iodide based perovskites: possible environmental effects of perovskite solar cells, J. Phys. Chem. Lett. 6 (2015) 1543–1547, https://doi.org/10.1021/acs.jpclett.5b00504.

[55] G.K. Kinuthia, V. Ngure, D. Beti, R. Lugalia, A. Wangila, L. Kamau, Levels of heavy metals in wastewater and soil samples from open drainage channels in Nairobi, Kenya: community health implication, Sci. Rep. 10 (2020) 8434, https://doi.org/10.1038/s41598-020-65359-5.

[56] I.R. Benmessaoud, A.-L. Mahul-Mellier, E. Horváth, B. Maco, M. Spina, H.A. Lashuel, L. Forró, Health hazards of methylammonium lead iodide based perovskites: cytotoxicity studies, Toxicol. Res. 5 (2016) 407–419, https://doi.org/10.1039/C5TX00303B.

[57] R. Ahmad, G.V. Nutan, D. Singh, G. Gupta, U. Soni, S. Sapra, R. Srivastava, Colloidal lead-free Cs2AgBiBr6 double perovskite nanocrystals: Synthesis, uniform thin-film fabrication, and application in solution-processed solar cells, Nano Res. 14 (2021) 1126–1134, https://doi.org/10.1007/s12274-020-3161-6.

[58] P.P. Boix, S. Agarwala, T.M. Koh, N. Mathews, S.G. Mhaisalkar, Perovskite solar cells: beyond methylammonium lead iodide, J. Phys. Chem. Lett. 6 (2015) 898–907, https://doi.org/10.1021/jz502547f.

[59] F. Hao, C.C. Stoumpos, D.H. Cao, R.P.H. Chang, M.G. Kanatzidis, Lead-free solid-state organic–inorganic halide perovskite solar cells, Nat. Photonics 8 (2014) 489–494, https://doi.org/10.1038/nphoton.2014.82.

[60] F. Igbari, R. Wang, Z.-K. Wang, X.-J. Ma, Q. Wang, K.-L. Wang, Y. Zhang, L.-S. Liao, Y. Yang, Composition stoichiometry of $Cs_2AgBiBr_6$ films for highly efficient lead-free perovskite solar cells, Nano Lett. 19 (2019) 2066–2073, https://doi.org/10.1021/acs.nanolett.9b00238.

[61] N.K. Noel, S.D. Stranks, A. Abate, C. Wehrenfennig, S. Guarnera, A.-A. Haghighirad, A. Sadhanala, G.E. Eperon, S.K. Pathak, M.B. Johnston, A. Petrozza, L.M. Herz, H.J. Snaith, Lead-free organic–inorganic tin halide perovskites for photovoltaic applications, Energy Environ. Sci. 7 (2014) 3061–3068, https://doi.org/10.1039/C4EE01076K.

[62] M.-C. Tang, D. Barrit, R. Munir, R. Li, J.M. Barbé, D.-M. Smilgies, S. Del Gobbo, T.D. Anthopoulos, A. Amassian, Bismuth-based perovskite-inspired solar cells: in situ diagnostics reveal similarities and differences in the film formation of bismuth- and lead-based films, Sol. RRL 3 (2019) 1800305, https://doi.org/10.1002/solr.201800305.

[63] F. Zabihi, M. Tebyetekerwa, Z. Xu, A. Ali, A.K. Kumi, H. Zhang, R. Jose, S. Ramakrishna, S. Yang, Perovskite solar cell-hybrid devices: thermoelectrically,

electrochemically, and piezoelectrically connected power packs, J. Mater. Chem. A 7 (2019) 26661–26692, https://doi.org/10.1039/C9TA08070H.

[64] L. Zdražil, S. Kalytchuk, M. Langer, R. Ahmad, J. Pospíšil, O. Zmeškal, M. Altomare, A. Osvet, R. Zbořil, P. Schmuki, C.J. Brabec, M. Otyepka, Š. Kment, Transparent and low-loss luminescent solar concentrators based on self-trapped exciton emission in lead-free double perovskite nanocrystals, ACS Appl. Energy Mater. 4 (2021) 6445–6453, https://doi.org/10.1021/acsaem.1c00360.

[65] R. Ahmad, L. Zdražil, S. Kalytchuk, A. Naldoni, E. Mohammadi, P. Schmuki, R. Zboril, Š. Kment, Robust dual cationic ligand for stable and efficient warm-white light emission in lead-free double perovskite nanocrystals, Appl. Mater. Today 26 (2022) 101288, https://doi.org/10.1016/j.apmt.2021.101288.

[66] R. Ahmad, L. Zdražil, S. Kalytchuk, A. Naldoni, A.L. Rogach, P. Schmuki, R. Zboril, Š. Kment, Uncovering the role of trioctylphosphine on colloidal and emission stability of Sb-alloyed Cs_2 $NaInCl_6$ double perovskite nanocrystals, ACS Appl. Mater. Interfaces 13 (2021) 47845–47859, https://doi.org/10.1021/acsami.1c10782.

[67] Y.-T. Huang, S.R. Kavanagh, D.O. Scanlon, A. Walsh, R.L.Z. Hoye, Perovskite-inspired materials for photovoltaics and beyond—from design to devices, Nanotechnology 32 (2021) 132004, https://doi.org/10.1088/1361-6528/abcf6d.

[68] F. Li, W. Yu, X. Guan, T. Wu, Emerging transistor applications enabled by halide perovskites, Acc. Mater. Res. 3 (2022) 8–20, https://doi.org/10.1021/accountsmr.1c00103.

[69] L.N. Quan, B.P. Rand, R.H. Friend, S.G. Mhaisalkar, T.-W. Lee, E.H. Sargent, Perovskites for next-generation optical sources, Chem. Rev. 119 (2019) 7444–7477, https://doi.org/10.1021/acs.chemrev.9b00107.

[70] L. Su, Z.X. Zhao, H.Y. Li, J. Yuan, Z.L. Wang, G.Z. Cao, G. Zhu, High-performance organolead halide perovskite-based self-powered triboelectric photodetector, ACS Nano 9 (2015) 11310–11316, https://doi.org/10.1021/acsnano.5b04995.

[71] H. Wang, D.H. Kim, Perovskite-based photodetectors: materials and devices, Chem. Soc. Rev. 46 (2017) 5204–5236, https://doi.org/10.1039/C6CS00896H.

[72] R. Ding, M. Wong, J. Hao, Recent advances in hybrid perovskite nanogenerators, EcoMat 2 (2020), https://doi.org/10.1002/eom2.12057.

[73] M. Sahu, S. Hajra, J. Bijelic, D. Oh, I. Djerdj, H.J. Kim, Triple perovskite-based triboelectric nanogenerator: a facile method of energy harvesting and self-powered information generator, Mater. Today Energy 20 (2021) 100639, https://doi.org/10.1016/j.mtener.2021.100639.

[74] L. Su, Z. Zhao, H. Li, Y. Wang, S. Kuang, G. Cao, Z. Wang, G. Zhu, Photoinduced enhancement of a triboelectric nanogenerator based on an organolead halide perovskite, J. Mater. Chem. C 4 (2016) 10395–10399, https://doi.org/10.1039/C6TC03513B.

[75] Y. Wang, J. Duan, X. Yang, L. Liu, L. Zhao, Q. Tang, The unique dielectricity of inorganic perovskites toward high-performance triboelectric nanogenerators, Nano Energy 69 (2020) 104418, https://doi.org/10.1016/j.nanoen.2019.104418.

[76] Y. Wang, X. Yang, X. Yu, J. Duan, Q. Yang, Y. Duan, Q. Tang, Triboelectric charging behaviors and photoinduced enhancement of alkaline earth ions doped inorganic perovskite triboelectric nanogenerators, Nano Energy 77 (2020) 105280, https://doi.org/10.1016/j.nanoen.2020.105280.

[77] M.S. Chowdhury, K.S. Rahman, T. Chowdhury, N. Nuthammachot, K. Techato, M. Akhtaruzzaman, S.K. Tiong, K. Sopian, N. Amin, An overview of solar photovoltaic panels' end-of-life material recycling, Energ. Strat. Rev. 27 (2020) 100431, https://doi.org/10.1016/j.esr.2019.100431.

[78] D. Fabini, Quantifying the potential for lead pollution from halide perovskite photovoltaics, J. Phys. Chem. Lett. 6 (2015) 3546–3548, https://doi.org/10.1021/acs.jpclett.5b01747.

[79] J. Gong, S.B. Darling, F. You, Perovskite photovoltaics: life-cycle assessment of energy and environmental impacts, Energ. Environ. Sci. 8 (2015) 1953–1968, https://doi.org/10.1039/C5EE00615E.

[80] C.A. Wolden, J. Kurtin, J.B. Baxter, I. Repins, S.E. Shaheen, J.T. Torvik, A.A. Rockett, V.-M. Fthenakis, E.S. Aydil, Photovoltaic manufacturing: present status, future prospects, and research needs, J. Vac. Sci. Technol. Vac. Surf. Films 29 (2011), 030801, https://doi.org/10.1116/1.3569757.

[81] K. Domanski, E.A. Alharbi, A. Hagfeldt, M. Grätzel, W. Tress, Systematic investigation of the impact of operation conditions on the degradation behaviour of perovskite solar cells, Nat. Energy 3 (2018) 61–67, https://doi.org/10.1038/s41560-017-0060-5.

[82] J. You, L. Meng, T.-B. Song, T.-F. Guo, Y.(.M.). Yang, W.-H. Chang, Z. Hong, H. Chen, H. Zhou, Q. Chen, Y. Liu, N. De Marco, Y. Yang, Improved air stability of perovskite solar cells via solution-processed metal oxide transport layers, Nat. Nanotechnol. 11 (2016) 75–81, https://doi.org/10.1038/nnano.2015.230.

[83] W. Chen, Y. Wu, Y. Yue, J. Liu, W. Zhang, X. Yang, H. Chen, E. Bi, I. Ashraful, M. Gratzel, L. Han, Efficient and stable large-area perovskite solar cells with inorganic charge extraction layers, Science 350 (2015) 944–948, https://doi.org/10.1126/science.aad1015.

[84] H.J. Snaith, P. Hacke, Enabling reliability assessments of pre-commercial perovskite photovoltaics with lessons learned from industrial standards, Nat. Energy 3 (2018) 459–465, https://doi.org/10.1038/s41560-018-0174-4.

[85] M.V. Khenkin, E.A. Katz, A. Abate, G. Bardizza, J.J. Berry, C. Brabec, F. Brunetti, V. Bulović, Q. Burlingame, A. Di Carlo, R. Cheacharoen, Y.-B. Cheng, A. Colsmann, S. Cros, K. Domanski, M. Dusza, C.J. Fell, S.R. Forrest, Y. Galagan, D. Di Girolamo, M. Grätzel, A. Hagfeldt, E. von Hauff, H. Hoppe, J. Kettle, H. Köbler, M.S. Leite, S. Liu, Y.-L. Loo, J.M. Luther, C.-Q. Ma, M. Madsen, M. Manceau, M. Matheron, M. McGehee, R. Meitzner, M.K. Nazeeruddin, A.F. Nogueira, Ç. Odabaşı, A. Osherov, N.-G. Park, M.O. Reese, F. De Rossi, M. Saliba, U.S. Schubert, H.J. Snaith, S.D. Stranks, W. Tress, P.A. Troshin, V. Turkovic, S. Veenstra, I. Visoly-Fisher, A. Walsh, T. Watson, H. Xie, R. Yıldırım, S.M. Zakeeruddin, K. Zhu, M. Lira-Cantu, Consensus statement for stability assessment and reporting for perovskite photovoltaics based on ISOS procedures, Nat. Energy 5 (2020) 35–49, https://doi.org/10.1038/s41560-019-0529-5.

[86] Reese, M.O., Gevorgyan, S.A., Jørgensen, M., Bundgaard, E., Kurtz, S.R., Ginley, D.S., Olson, D.C., Lloyd, M.T., Morvillo, P., Katz, E.A., Elschner, A., Haillant, O., Currier, T.-R., Shrotriya, V., Hermenau, M., Riede, M., R. Kirov, K., Trimmel, G., Rath, T., Inganäs, O., Zhang, F., Andersson, M., Tvingstedt, K., Lira-Cantu, M., Laird, D., McGuiness, C., Gowrisanker, S.(.J.)., Pannone, M., Xiao, M., Hauch, J., Steim, R., DeLongchamp, D.M., Rösch, R., Hoppe, H., Espinosa, N., Urbina, A., Yaman-Uzunoglu, G., Bonekamp, J.-B., van Breemen, A.J.J.M., Girotto, C., Voroshazi, E., Krebs, F.C., 2011. Consensus stability testing protocols for organic photovoltaic materials and devices. Sol. Energy Mater. Sol. Cells 95, 1253–1267. doi:https://doi.org/10.1016/j.solmat.2011.01.036.

[87] N. Arora, M.I. Dar, S. Akin, R. Uchida, T. Baumeler, Y. Liu, S.M. Zakeeruddin, M. Grätzel, Low-cost and highly efficient carbon-based perovskite solar cells exhibiting excellent long-term operational and UV stability, Small 15 (2019) 1904746, https://doi.org/10.1002/smll.201904746.

[88] T. Niu, J. Lu, R. Munir, J. Li, D. Barrit, X. Zhang, H. Hu, Z. Yang, A. Amassian, K. Zhao, S.(.F.). Liu, Stable high-performance perovskite solar cells via grain boundary passivation, Adv. Mater. 30 (2018), https://doi.org/10.1002/adma.201706576.

[89] G. Tong, D. Son, L.K. Ono, Y. Liu, Y. Hu, H. Zhang, A. Jamshaid, L. Qiu, Z. Liu, Y. Qi, Scalable fabrication of >90 cm^2 perovskite solar modules with >1000 h operational stability based on the intermediate phase strategy, Adv. Energy Mater. 11 (2021) 2003712, https://doi.org/10.1002/aenm.202003712.

[90] C. Law, L. Miseikis, S. Dimitrov, P. Shakya-Tuladhar, X. Li, P.R.F. Barnes, J. Durrant, B.-C. O'Regan, Performance and stability of lead perovskite/TiO_2, polymer/PCBM, and dye sensitized solar cells at light intensities up to 70 Suns, Adv. Mater. 26 (2014) 6268–6273, https://doi.org/10.1002/adma.201402612.

[91] Y. Jiang, L. Qiu, E.J. Juarez-Perez, L.K. Ono, Z. Hu, Z. Liu, Z. Wu, L. Meng, Q. Wang, Y. Qi, Reduction of lead leakage from damaged lead halide perovskite solar modules using self-healing polymer-based encapsulation, Nat. Energy 4 (2019) 585–593, https://doi.org/10.1038/s41560-019-0406-2.

[92] T. Leijtens, G.E. Eperon, S. Pathak, A. Abate, M.M. Lee, H.J. Snaith, Overcoming ultraviolet light instability of sensitized TiO2 with meso-superstructured organometal trihalide perovskite solar cells, Nat. Commun. 4 (2013) 2885, https://doi.org/10.1038/ncomms3885.

[93] S.-W. Lee, S. Kim, S. Bae, K. Cho, T. Chung, L.E. Mundt, S. Lee, S. Park, H. Park, M.C. Schubert, S.W. Glunz, Y. Ko, Y. Jun, Y. Kang, H.-S. Lee, D. Kim, UV degradation and recovery of perovskite solar cells, Sci. Rep. 6 (2016) 38150, https://doi.org/10.1038/srep38150.

[94] Q.-D. Dao, R. Tsuji, A. Fujii, M. Ozaki, Study on degradation mechanism of perovskite solar cell and their recovering effects by introducing CH_3NH_3I layers, Org. Electron. 43 (2017) 229–234, https://doi.org/10.1016/j.orgel.2017.01.038.

[95] W. Li, W. Zhang, S. Van Reenen, R.J. Sutton, J. Fan, A.A. Haghighirad, M.B. Johnston, L. Wang, H.J. Snaith, Enhanced UV-light stability of planar heterojunction perovskite solar cells with caesium bromide interface modification, Energ. Environ. Sci. 9 (2016) 490–498, https://doi.org/10.1039/C5EE03522H.

[96] N. Aristidou, C. Eames, I. Sanchez-Molina, X. Bu, J. Kosco, M.S. Islam, S.A. Haque, Fast oxygen diffusion and iodide defects mediate oxygen-induced degradation of perovskite solar cells, Nat. Commun. 8 (2017) 15218, https://doi.org/10.1038/ncomms15218.

[97] N. Aristidou, I. Sanchez-Molina, T. Chotchuangchutchaval, M. Brown, L. Martinez, T. Rath, S.A. Haque, The role of oxygen in the degradation of methylammonium lead trihalide perovskite photoactive layers, Angew. Chem. Int. Ed. 54 (2015) 8208–8212, https://doi.org/10.1002/anie.201503153.

[98] D. Bryant, N. Aristidou, S. Pont, I. Sanchez-Molina, T. Chotchunangatchaval, S. Wheeler, J.R. Durrant, S.A. Haque, Light and oxygen induced degradation limits the operational stability of methylammonium lead triiodide perovskite solar cells, Energ. Environ. Sci. 9 (2016) 1655–1660, https://doi.org/10.1039/C6EE00409A.

[99] E. Ugur, E. Alarousu, J.I. Khan, A. Vlk, E. Aydin, M. De Bastiani, A.H. Balawi, S.P. Gonzalez Lopez, M. Ledinský, S. De Wolf, F. Laquai, How humidity and light exposure change the photophysics of metal halide perovskite solar cells, Sol. RRL 4 (2020) 2000382, https://doi.org/10.1002/solr.202000382.

[100] J. He, W.-H. Fang, R. Long, O.V. Prezhdo, Why oxygen increases carrier lifetimes but accelerates degradation of CH3NH3PbI3 under light irradiation: time-domain ab initio analysis, J. Am. Chem. Soc. 142 (2020) 14664–14673, https://doi.org/10.1021/jacs.0c06769.

[101] J. He, W.-H. Fang, R. Long, O.V. Prezhdo, Superoxide/peroxide chemistry extends charge carriers' lifetime but undermines chemical stability of CH3NH3PbI3 exposed to oxygen: time-domain ab initio analysis, J. Am. Chem. Soc. 141 (2019) 5798–5807, https://doi.org/10.1021/jacs.8b13392.

[102] P. Chhillar, B.P. Dhamaniya, V. Dutta, S.K. Pathak, Recycling of perovskite films: route toward cost-efficient and environment-friendly perovskite technology, ACS Omega 4 (2019) 11880–11887, https://doi.org/10.1021/acsomega.9b01053.

[103] J.I. Khan, A.D. Sheikh, M.A. Alamoudi, D. Barrit, E. Ugur, F. Laquai, A. Amassian, Impact of residual lead iodide on photophysical properties of lead triiodide perovskite solar cells, Energ. Technol. 8 (2020), https://doi.org/10.1002/ente.201900627.

[104] G. Tumen-Ulzii, C. Qin, D. Klotz, M.R. Leyden, P. Wang, M. Auffray, T. Fujihara, T. Matsushima, J. Lee, S. Lee, Y. Yang, C. Adachi, Detrimental effect of unreacted PbI_2 on the long-term stability of perovskite solar cells, Adv. Mater. 32 (2020) 1905035, https://doi.org/10.1002/adma.201905035.

[105] E. Ugur, A.D. Sheikh, R. Munir, J.I. Khan, D. Barrit, A. Amassian, F. Laquai, Improved morphology and efficiency of n–i–p planar perovskite solar cells by processing with glycol ether additives, ACS Energy Lett. 2 (2017) 1960–1968, https://doi.org/10.1021/acsenergylett.7b00526.

[106] A.K. Jena, Y. Numata, M. Ikegami, T. Miyasaka, Role of spiro-OMeTAD in performance deterioration of perovskite solar cells at high temperature and reuse of the perovskite films to avoid Pb-waste, J. Mater. Chem. A 6 (2018) 2219–2230, https://doi.org/10.1039/C7TA07674F.

[107] J.A. Christians, P.A. Miranda Herrera, P.V. Kamat, Transformation of the excited state and photovoltaic efficiency of $CH_3NH_3PbI_3$ perovskite upon controlled exposure to humidified air, J. Am. Chem. Soc. 137 (2015) 1530–1538, https://doi.org/10.1021/ja511132a.

[108] G.E. Eperon, S.N. Habisreutinger, T. Leijtens, B.J. Bruijnaers, J.J. van Franeker, D.W. de Quilettes, S. Pathak, R.J. Sutton, G. Grancini, D.S. Ginger, R.A.J. Janssen, A. Petrozza, H.J. Snaith, The importance of moisture in hybrid lead halide perovskite thin film fabrication, ACS Nano 9 (2015) 9380–9393, https://doi.org/10.1021/acsnano.5b03626.

[109] H. Zhou, Q. Chen, G. Li, S. Luo, T. Song, H.-S. Duan, Z. Hong, J. You, Y. Liu, Y. Yang, Interface engineering of highly efficient perovskite solar cells, Science 345 (2014) 542–546, https://doi.org/10.1126/science.1254050.

[110] J.H. Noh, S.H. Im, J.H. Heo, T.N. Mandal, S.I. Seok, Chemical management for colorful, efficient, and stable inorganic–organic hybrid nanostructured solar cells, Nano Lett. 13 (2013) 1764–1769, https://doi.org/10.1021/nl400349b.

[111] Y. Fan, J. Fang, X. Chang, M.-C. Tang, D. Barrit, Z. Xu, Z. Jiang, J. Wen, H. Zhao, T. Niu, D.-M. Smilgies, S. Jin, Z. Liu, E.Q. Li, A. Amassian, S.(.F.). Liu, K. Zhao, Scalable ambient fabrication of high-performance CsPbI2Br solar cells, Joule 3 (2019) 2485–2502, https://doi.org/10.1016/j.joule.2019.07.015.

[112] M.A. Green, T. Bein, Perovskite cells charge forward, Nat. Mater. 14 (2015) 559–561, https://doi.org/10.1038/nmat4301.

[113] I. Hwang, I. Jeong, J. Lee, M.J. Ko, K. Yong, Enhancing stability of perovskite solar cells to moisture by the facile hydrophobic passivation, ACS Appl. Mater. Interfaces 7 (2015) 17330–17336, https://doi.org/10.1021/acsami.5b04490.

[114] M. Long, T. Zhang, Y. Chai, C.-F. Ng, T.C.W. Mak, J. Xu, K. Yan, Nonstoichiometric acid–base reaction as reliable synthetic route to highly stable CH3NH3PbI3 perovskite film, Nat. Commun. 7 (2016) 13503, https://doi.org/10.1038/ncomms13503.

[115] T. Niu, J. Lu, X. Jia, Z. Xu, M.-C. Tang, D. Barrit, N. Yuan, J. Ding, X. Zhang, Y. Fan, T. Luo, Y. Zhang, D.-M. Smilgies, Z. Liu, A. Amassian, S. Jin, K. Zhao, S. Liu, Interfacial engineering at the 2D/3D Heterojunction for high-performance perovskite solar cells, Nano Lett. 19 (2019) 7181–7190, https://doi.org/10.1021/acs.nanolett.9b02781.

[116] T. Wu, Z. Qin, Y. Wang, Y. Wu, W. Chen, S. Zhang, M. Cai, S. Dai, J. Zhang, J. Liu, Z. Zhou, X. Liu, H. Segawa, H. Tan, Q. Tang, J. Fang, Y. Li, L. Ding, Z. Ning, Y. Qi, Y. Zhang, L. Han, The main progress of perovskite solar cells in 2020–2021, Nano-Micro Lett. 13 (2021) 152, https://doi.org/10.1007/s40820-021-00672-w.

[117] G. Divitini, S. Cacovich, F. Matteocci, L. Cinà, A. Di Carlo, C. Ducati, In situ observation of heat-induced degradation of perovskite solar cells, Nat. Energy 1 (2016) 15012, https://doi.org/10.1038/nenergy.2015.12.

[118] Q. Meng, Y. Chen, Y.Y. Xiao, J. Sun, X. Zhang, C.B. Han, H. Gao, Y. Zhang, H. Yan, Effect of temperature on the performance of perovskite solar cells, J. Mater. Sci. Mater. Electron. 32 (2021) 12784–12792, https://doi.org/10.1007/s10854-020-03029-y.

[119] Q. Jiang, L. Zhang, H. Wang, X. Yang, J. Meng, H. Liu, Z. Yin, J. Wu, X. Zhang, J. You, Enhanced electron extraction using SnO2 for high-efficiency planar-structure HC (NH2)2PbI3-based perovskite solar cells, Nat. Energy 2 (2017) 16177, https://doi.org/10.1038/nenergy.2016.177.

[120] K. Aitola, K. Domanski, J.-P. Correa-Baena, K. Sveinbjörnsson, M. Saliba, A. Abate, M. Grätzel, E. Kauppinen, E.M.J. Johansson, W. Tress, A. Hagfeldt, G. Boschloo, High temperature-stable perovskite solar cell based on low-cost carbon nanotube hole contact, Adv. Mater. 29 (2017) 1606398, https://doi.org/10.1002/adma.201606398.

[121] K.A. Bush, C.D. Bailie, Y. Chen, A.R. Bowring, W. Wang, W. Ma, T. Leijtens, F. Moghadam, M.D. McGehee, Thermal and environmental stability of semi-transparent perovskite solar cells for tandems enabled by a solution-processed nanoparticle buffer layer and sputtered ITO electrode, Adv. Mater. 28 (2016) 3937–3943, https://doi.org/10.1002/adma.201505279.

[122] K.A. Bush, A.F. Palmstrom, Z.J. Yu, M. Boccard, R. Cheacharoen, J.P. Mailoa, D.P. McMeekin, R.L.Z. Hoye, C.D. Bailie, T. Leijtens, I.M. Peters, M.C. Minichetti, N. Rolston, R. Prasanna, S. Sofia, D. Harwood, W. Ma, F. Moghadam, H.J. Snaith, T. Buonassisi, Z.C. Holman, S.F. Bent, M.D. McGehee, 23.6%-Efficient monolithic perovskite/silicon tandem solar cells with improved stability, Nat. Energy 2 (2017) 17009, https://doi.org/10.1038/nenergy.2017.9.

[123] S. Gholipour, J.-P. Correa-Baena, K. Domanski, T. Matsui, L. Steier, F. Giordano, F. Tajabadi, W. Tress, M. Saliba, A. Abate, A. Morteza Ali, N. Taghavinia, M. Grätzel, A. Hagfeldt, Highly efficient and stable perovskite solar cells based on a low-cost carbon cloth, Adv. Energy Mater. 6 (2016) 1601116, https://doi.org/10.1002/aenm.201601116.

[124] X. Zhao, H.-S. Kim, J.-Y. Seo, N.-G. Park, Effect of selective contacts on the thermal stability of perovskite solar cells, ACS Appl. Mater. Interfaces 9 (2017) 7148–7153, https://doi.org/10.1021/acsami.6b15673.

[125] W. Nie, J.-C. Blancon, A.J. Neukirch, K. Appavoo, H. Tsai, M. Chhowalla, M.A. Alam, M.Y. Sfeir, C. Katan, J. Even, S. Tretiak, J.J. Crochet, G. Gupta, A.D. Mohite, Light-activated photocurrent degradation and self-healing in perovskite solar cells, Nat. Commun. 7 (2016) 11574, https://doi.org/10.1038/ncomms11574.

[126] K. Domanski, B. Roose, T. Matsui, M. Saliba, S.-H. Turren-Cruz, J.-P. Correa-Baena, C.R. Carmona, G. Richardson, J.M. Foster, F. De Angelis, J.M. Ball, A. Petrozza, N. Mine, M.K. Nazeeruddin, W. Tress, M. Grätzel, U. Steiner, A. Hagfeldt, A. Abate, Migration of cations induces reversible performance losses over day/night cycling in perovskite solar cells, Energ. Environ. Sci. 10 (2017) 604–613, https://doi.org/10.1039/C6EE03352K.

[127] A.J. Pearson, G.E. Eperon, P.E. Hopkinson, S.N. Habisreutinger, J.T. Wang, H.J. Snaith, N.C. Greenham, Oxygen degradation in mesoporous $Al_2O_3/CH_3NH_3PbI_{3-x}Cl_x$ perovskite solar cells: kinetics and mechanisms, Adv. Energy Mater. 6 (2016), https://doi.org/10.1002/aenm.201600014.

[128] A.F. Akbulatov, S.Y. Luchkin, L.A. Frolova, N.N. Dremova, K.L. Gerasimov, I.S. Zhidkov, D.V. Anokhin, E.Z. Kurmaev, K.J. Stevenson, P.A. Troshin, Probing the intrinsic thermal and photochemical stability of hybrid and inorganic lead halide perovskites, J. Phys. Chem. Lett. 8 (2017) 1211–1218, https://doi.org/10.1021/acs.jpclett.6b03026.

[129] A. Dualeh, P. Gao, S.I. Seok, M.K. Nazeeruddin, M. Grätzel, Thermal behavior of methylammonium lead-trihalide perovskite photovoltaic light harvesters, Chem. Mater. 26 (2014) 6160–6164, https://doi.org/10.1021/cm502468k.

[130] L. Shi, M.P. Bucknall, T.L. Young, M. Zhang, L. Hu, J. Bing, D.S. Lee, J. Kim, T. Wu, N. Takamure, D.R. McKenzie, S. Huang, M.A. Green, A.W.Y. Ho-Baillie, Gas chromatography–mass spectrometry analyses of encapsulated stable perovskite solar cells, Science 368 (2020), eaba2412, https://doi.org/10.1126/science.aba2412.

[131] J.M. Frost, K.T. Butler, F. Brivio, C.H. Hendon, M. van Schilfgaarde, A. Walsh, Atomistic origins of high-performance in hybrid halide perovskite solar cells, Nano Lett. 14 (2014) 2584–2590, https://doi.org/10.1021/nl500390f.
[132] S. Chen, Y. Deng, H. Gu, S. Xu, S. Wang, Z. Yu, V. Blum, J. Huang, Trapping lead in perovskite solar modules with abundant and low-cost cation-exchange resins, Nat. Energy 5 (2020) 1003–1011, https://doi.org/10.1038/s41560-020-00716-2.
[133] J. Lee, G. Kim, M. Kim, S.A. Park, T. Park, Nonaromatic green-solvent-processable, dopant-free, and lead-capturable hole transport polymers in perovskite solar cells with high efficiency, Adv. Energy Mater. 10 (2020) 1902662, https://doi.org/10.1002/aenm.201902662.
[134] X. Li, F. Zhang, H. He, J.J. Berry, K. Zhu, T. Xu, On-device lead sequestration for perovskite solar cells, Nature 578 (2020) 555–558, https://doi.org/10.1038/s41586-020-2001-x.
[135] S.G. Motti, T. Crothers, R. Yang, Y. Cao, R. Li, M.B. Johnston, J. Wang, L.M. Herz, Heterogeneous photon recycling and charge diffusion enhance charge transport in quasi-2D lead-halide perovskite films, Nano Lett. 19 (2019) 3953–3960, https://doi.org/10.1021/acs.nanolett.9b01242.
[136] Y. Liu, N. Sun, J. Liu, Z. Wen, X. Sun, S.-T. Lee, B. Sun, Integrating a silicon solar cell with a triboelectric nanogenerator *via* a mutual electrode for harvesting energy from sunlight and raindrops, ACS Nano 12 (2018) 2893–2899, https://doi.org/10.1021/acsnano.8b00416.
[137] R. Kalidoss, A. Bora, V. Natraj, S.V. Jayaraman, Y. Sivalingam, Self-powered, rapid-response, and highly flexible nanosensors, in: Nanosensors for Smart Manufacturing, Elsevier, 2021, pp. 397–415, https://doi.org/10.1016/B978-0-12-823358-0.00020-4.
[138] D. Vatansever, R.L. Hadimani, T. Shah, E. Siores, Hybrid Photovoltaic-Piezoelectric Flexible Device for Energy Harvesting from Nature, in: Presented at the 4th International Conference on Smart Materials, Structures and Systems Symposium A, 2012, pp. 297–301, https://doi.org/10.4028/www.scientific.net/AST.77.297.
[139] L. Zheng, G. Cheng, J. Chen, L. Lin, J. Wang, Y. Liu, H. Li, Z.L. Wang, A hybridized power panel to simultaneously generate electricity from sunlight, raindrops, and wind around the clock, Adv. Energy Mater. 5 (2015) 1501152, https://doi.org/10.1002/aenm.201501152.
[140] Z. Kınas, A. Karabiber, A. Yar, A. Ozen, F. Ozel, M. Ersöz, A. Okbaz, High-performance triboelectric nanogenerator based on carbon nanomaterials functionalized polyacrylonitrile nanofibers, Energy 239 (2022) 122369, https://doi.org/10.1016/j.energy.2021.122369.
[141] A. Okbaz, A. Karabiber, A. Yar, Z. Kınas, A. Sarılmaz, F. Ozel, High-performance triboelectric nanogenerator with optimized Al or Ti-embedded silicone tribomaterial, Energ. Conver. Manage. (2021), https://doi.org/10.1016/j.enconman.2021.115053.
[142] A. Ozen, F. Ozel, Z. Kınas, A. Karabiber, S. Polat, Spring assisted triboelectric nanogenerator based on sepiolite doped polyacrylonitrile nanofibers, Sustain. Energy Technol. Assess. 47 (2021) 101492, https://doi.org/10.1016/j.seta.2021.101492.
[143] A. Yar, Z. Kınas, A. Karabiber, A. Ozen, A. Okbaz, F. Ozel, Enhanced performance of triboelectric nanogenerator based on polyamide-silver antimony sulfide nanofibers for energy harvesting, Renew. Energy 179 (2021) 1781–1792, https://doi.org/10.1016/j.renene.2021.07.118.
[144] X. Chang, Y. Fan, K. Zhao, J. Fang, D. Liu, M.-C. Tang, D. Barrit, D.-M. Smilgies, R. Li, J. Lu, J. Li, T. Yang, A. Amassian, Z. Ding, Y. Chen, S.(F.). Liu, W. Huang, Perovskite solar cells toward eco-friendly printing, Research 2021 (2021), https://doi.org/10.34133/2021/9671892.
[145] R. Vidal, J.-A. Alberola-Borràs, S.N. Habisreutinger, J.-L. Gimeno-Molina, D.T. Moore, T.H. Schloemer, I. Mora-Seró, J.J. Berry, J.M. Luther, Assessing health and environmental impacts of solvents for producing perovskite solar cells, Nat. Sustain. 4 (2021) 277–285, https://doi.org/10.1038/s41893-020-00645-8.

[146] X. Cao, G. Zhang, Y. Cai, L. Jiang, X. He, Q. Zeng, J. Wei, Y. Jia, G. Xing, W. Huang, All green solvents for fabrication of $CsPbBr_3$ films for efficient solar cells guided by the hansen solubility theory, Sol. RRL 4 (2020), https://doi.org/10.1002/solr.202000008.

[147] K.L. Gardner, J.G. Tait, T. Merckx, W. Qiu, U.W. Paetzold, L. Kootstra, M. Jaysankar, R. Gehlhaar, D. Cheyns, P. Heremans, J. Poortmans, Nonhazardous solvent systems for processing perovskite photovoltaics, Adv. Energy Mater. 6 (2016) 1600386, https://doi.org/10.1002/aenm.201600386.

[148] J. Wang, F. Di Giacomo, J. Brüls, H. Gorter, I. Katsouras, P. Groen, R.A.J. Janssen, R. Andriessen, Y. Galagan, Highly efficient perovskite solar cells using non-toxic industry compatible solvent system, Sol. RRL 1 (2017) 1700091, https://doi.org/10.1002/solr.201700091.

[149] H. Chen, F. Ye, W. Tang, J. He, M. Yin, Y. Wang, F. Xie, E. Bi, X. Yang, M. Grätzel, L. Han, A solvent- and vacuum-free route to large-area perovskite films for efficient solar modules, Nature 550 (2017) 92–95, https://doi.org/10.1038/nature23877.

[150] M.M. Lunardi, J.P. Alvarez-Gaitan, J.I. Bilbao, R. Corkish, A review of recycling processes for photovoltaic modules, in: B. Zaidi (Ed.), Solar Panels and Photovoltaic Materials, InTech, 2018, https://doi.org/10.5772/intechopen.74390.

[151] D. Sica, O. Malandrino, S. Supino, M. Testa, M.C. Lucchetti, Management of end-of-life photovoltaic panels as a step towards a circular economy, Renew. Sustain. Energy Rev. 82 (2018) 2934–2945, https://doi.org/10.1016/j.rser.2017.10.039.

[152] V.M. Fthenakis, End-of-life management and recycling of PV modules, Energy Policy 28 (2000) 1051–1058, https://doi.org/10.1016/S0301-4215(00)00091-4.

[153] G.A. Heath, T.J. Silverman, M. Kempe, M. Deceglie, D. Ravikumar, T. Remo, H. Cui, P. Sinha, C. Libby, S. Shaw, K. Komoto, K. Wambach, E. Butler, T. Barnes, A. Wade, Research and development priorities for silicon photovoltaic module recycling to support a circular economy, Nat. Energy 5 (2020) 502–510, https://doi.org/10.1038/s41560-020-0645-2.

[154] M.I. Omer, X. Wang, X. Tang, Low-temperature processed, stable n-i-p perovskite solar cells with indene-C60-bisadduct as electron transport material, J. Mater. Sci. Mater. Electron. 32 (2021) 12872–12880, https://doi.org/10.1007/s10854-020-04167-z.

[155] S. Lee, M.-C. Tang, R. Munir, D. Barrit, Y.-J. Kim, R. Kang, J.-M. Yun, D.-M. Smilgies, A. Amassian, D.-Y. Kim, In situ study of the film formation mechanism of organic–inorganic hybrid perovskite solar cells: controlling the solvate phase using an additive system, J. Mater. Chem. A 8 (2020) 7695–7703, https://doi.org/10.1039/D0TA00048E.

[156] J.M. Kadro, N. Pellet, F. Giordano, A. Ulianov, O. Müntener, J. Maier, M. Grätzel, A. Hagfeldt, Proof-of-concept for facile perovskite solar cell recycling, Energ. Environ. Sci. 9 (2016) 3172–3179, https://doi.org/10.1039/C6EE02013E.

[157] Z. Li, T.R. Klein, D.H. Kim, M. Yang, J.J. Berry, M.F.A.M. van Hest, K. Zhu, Scalable fabrication of perovskite solar cells, Nat. Rev. Mater. 3 (2018) 18017, https://doi.org/10.1038/natrevmats.2018.17.

[158] T.E. Graedel, J. Allwood, J.-P. Birat, M. Buchert, C. Hagelüken, B.K. Reck, S.F. Sibley, G. Sonnemann, What do we know about metal recycling rates? J. Ind. Ecol. 15 (2011) 355–366, https://doi.org/10.1111/j.1530-9290.2011.00342.x.

[159] J.P. Ponpon, M. Amann, Properties of solution grown PbI2 layers embedded in PVA, Cryst. Res. Technol. 42 (2007) 253–259, https://doi.org/10.1002/crat.200610809.

[160] K. Sun, P. Li, Y. Xia, J. Chang, J. Ouyang, Transparent Conductive Oxide-Free Perovskite Solar Cells with PEDOT:PSS as Transparent Electrode, ACS Appl. Mater. Interfaces 7 (2015) 15314–15320, https://doi.org/10.1021/acsami.5b03171.

[161] H. Sung, N. Ahn, M.S. Jang, J.-K. Lee, H. Yoon, N.-G. Park, M. Choi, Transparent conductive oxide-free graphene-based perovskite solar cells with over 17% efficiency, Adv. Energy Mater. 6 (2016), 1501873, https://doi.org/10.1002/aenm.201501873.

[162] J.I. Kroschwitz, A. Seidel (Eds.), Kirk-Othmer Encyclopedia of Chemical Technology, fifth ed., Wiley-Interscience, Hoboken, NJ, 2004.

[163] L. Huang, Z. Hu, J. Xu, X. Sun, Y. Du, J. Ni, H. Cai, J. Li, J. Zhang, Efficient electron-transport layer-free planar perovskite solar cells via recycling the FTO/glass substrates from degraded devices, Sol. Energy Mater. Sol. Cells 152 (2016) 118–124, https://doi.org/10.1016/j.solmat.2016.03.035.

[164] L. Huang, J. Xu, X. Sun, R. Xu, Y. Du, J. Ni, H. Cai, J. Li, Z. Hu, J. Zhang, New films on old substrates: toward green and sustainable energy production via recycling of functional components from degraded perovskite solar cells, ACS Sustain. Chem. Eng. 5 (2017) 3261–3269, https://doi.org/10.1021/acssuschemeng.6b03089.

[165] T. Zhang, M. Yang, Y. Zhao, K. Zhu, Controllable sequential deposition of planar $CH_3NH_3PbI_3$ perovskite films via adjustable volume expansion, Nano Lett. 15 (2015) 3959–3963, https://doi.org/10.1021/acs.nanolett.5b00843.

[166] D.H. Cao, C.C. Stoumpos, C.D. Malliakas, M.J. Katz, O.K. Farha, J.T. Hupp, M.G. Kanatzidis, Remnant PbI_2, an unforeseen necessity in high-efficiency hybrid perovskite-based solar cells? APL Mater. 2 (2014), 091101, https://doi.org/10.1063/1.4895038.

[167] C. Hagelüken, C.W. Corti, Recycling of gold from electronics: cost-effective use through 'Design for Recycling', Gold Bull. 43 (2010) 209–220, https://doi.org/10.1007/BF03214988.

[168] J. Miller, A.J. Parker, Dipolar aprotic solvents in bimolecular aromatic nucleophilic substitution reactions, J. Am. Chem. Soc. 83 (1961) 117–123, https://doi.org/10.1021/ja01462a023.

[169] A.J. Parker, The effects of solvation on the properties of anions in dipolar aprotic solvents, Q. Rev. Chem. Soc. 16 (1962) 163, https://doi.org/10.1039/qr9621600163.

[170] A. Wakamiya, M. Endo, T. Sasamori, N. Tokitoh, Y. Ogomi, S. Hayase, Y. Murata, Reproducible fabrication of efficient perovskite-based solar cells: X-ray crystallographic studies on the formation of $CH_3NH_3PbI_3$ layers, Chem. Lett. 43 (2014) 711–713, https://doi.org/10.1246/cl.140074.

[171] T. Mohammadi, A. Razmi, M. Sadrzadeh, Effect of operating parameters on Pb2+ separation from wastewater using electrodialysis, Desalination 167 (2004) 379–385, https://doi.org/10.1016/j.desal.2004.06.150.

[172] N. Axtell, Lead and nickel removal using Microspora and Lemna minor, Bioresour. Technol. 89 (2003) 41–48, https://doi.org/10.1016/S0960-8524(03)00034-8.

[173] D.B. Shah, A.V. Phadke, W.M. Kocher, Lead removal from foundry waste by solvent extraction, J. Air Waste Manag. Assoc. 45 (1995) 150–155, https://doi.org/10.1080/10473289.1995.10467354.

[174] S. Ahmed, S. Chughtai, M.A. Keane, The removal of cadmium and lead from aqueous solution by ion exchange with Na-Y zeolite, Sep. Purif. Technol. 13 (1998) 57–64, https://doi.org/10.1016/S1383-5866(97)00063-4.

[175] Z. Xiao, C. Bi, Y. Shao, Q. Dong, Q. Wang, Y. Yuan, C. Wang, Y. Gao, J. Huang, Efficient, high yield perovskite photovoltaic devices grown by interdiffusion of solution-processed precursor stacking layers, Energy Environ. Sci. 7 (2014) 2619–2623, https://doi.org/10.1039/C4EE01138D.

[176] Office of the Federal Register, N.A. and R.A, 2002. 40 CFR 261.24—Toxicity characteristic. [WWW Document]. govinfo.gov. URL https://www.govinfo.gov/app/details/CFR-2002-title40-vol22/https%3A%2F%2Fwww.govinfo.gov%2Fapp%2Fdetails%2FCFR-2002-title40-vol22%2FCFR-2002-title40-vol22-sec261-24 (accessed 6.27.21).

[177] J. Liu, S. Pathak, T. Stergiopoulos, T. Leijtens, K. Wojciechowski, S. Schumann, N. Kausch-Busies, H.J. Snaith, Employing PEDOT as the p-type charge collection layer in regular organic–inorganic perovskite solar cells, J. Phys. Chem. Lett. 6 (2015) 1666–1673, https://doi.org/10.1021/acs.jpclett.5b00545.

[178] M.L. Petrus, T. Bein, T.J. Dingemans, P. Docampo, A low cost azomethine-based hole transporting material for perovskite photovoltaics, J. Mater. Chem. A 3 (2015) 12159–12162, https://doi.org/10.1039/C5TA03046C.

[179] M. Saliba, S. Orlandi, T. Matsui, S. Aghazada, M. Cavazzini, J.-P. Correa-Baena, P. Gao, R. Scopelliti, E. Mosconi, K.-H. Dahmen, F. De Angelis, A. Abate, A. Hagfeldt, G. Pozzi, M. Graetzel, M.K. Nazeeruddin, A molecularly engineered hole-transporting material for efficient perovskite solar cells, Nat. Energy 1 (2016) 15017, https://doi.org/10.1038/nenergy.2015.17.

[180] A. Kim, H. Lee, H.-C. Kwon, H.S. Jung, N.-G. Park, S. Jeong, J. Moon, Fully solution-processed transparent electrodes based on silver nanowire composites for perovskite solar cells, Nanoscale 8 (2016) 6308–6316, https://doi.org/10.1039/C5NR04585A.

[181] M. Lee, Y. Ko, B.K. Min, Y. Jun, Silver nanowire top electrodes in flexible perovskite solar cells using titanium metal as substrate, ChemSusChem 9 (2016) 31–35, https://doi.org/10.1002/cssc.201501332.

[182] C.N. Hoth, S.A. Choulis, P. Schilinsky, C.J. Brabec, High photovoltaic performance of inkjet printed polymer:fullerene blends, Adv. Mater. 19 (2007) 3973–3978, https://doi.org/10.1002/adma.200700911.

[183] M.G. Panthani, V. Akhavan, B. Goodfellow, J.P. Schmidtke, L. Dunn, A. Dodabalapur, P.F. Barbara, B.A. Korgel, Synthesis of $CuInS_2$, $CuInSe_2$, and $Cu(In_xGa_{1-x})Se_2$ (CIGS) nanocrystal "Inks" for printable photovoltaics, J. Am. Chem. Soc. 130 (2008) 16770–16777, https://doi.org/10.1021/ja805845q.

[184] R. Ahmad, R. Srivastava, H. Bhardwaj, S. Yadav, V. Nand Singh, S. Chand, N. Singh, S. Sapra, Size-tunable synthesis of colloidal silver sulfide nanocrystals for solution-processed photovoltaic applications, Chem. Select 3 (2018) 5620–5629, https://doi.org/10.1002/slct.201702786.

[185] W. Lee, S.H. Kang, J.-Y. Kim, G.B. Kolekar, Y.-E. Sung, S.-H. Han, TiO_2 nanotubes with a ZnO thin energy barrier for improved current efficiency of CdSe quantum-dot-sensitized solar cells, Nanotechnology 20 (2009) 335706, https://doi.org/10.1088/0957-4484/20/33/335706.

[186] M. Ramar, C.K. Suman, R. Manimozhi, R. Ahamad, R. Srivastava, Study of Schottky contact in binary and ternary hybrid CdSe quantum dot solar cells, RSC Adv. 4 (2014) 32651–32657, https://doi.org/10.1039/C4RA04966G.

[187] P.K. Santra, P.V. Nair, K. George Thomas, P.V. Kamat, $CuInS_2$-Sensitized Quantum Dot Solar Cell. Electrophoretic Deposition, Excited-State Dynamics, and Photovoltaic Performance, J. Phys. Chem. Lett. 4 (2013) 722–729, https://doi.org/10.1021/jz400181m.

[188] T. Sogabe, Q. Shen, K. Yamaguchi, Recent progress on quantum dot solar cells: a review, J. Photonics Energy 6 (2016), 040901, https://doi.org/10.1117/1.JPE.6.040901.

[189] T. Abhijith, M.Y. Ameen, V.S. Reddy, Synthesis of PTCDI-C_8 one dimensional nanostructures for photovoltaic applications, IOP Conf. Ser. Mater. Sci. Eng. 73 (2015), 012052, https://doi.org/10.1088/1757-899X/73/1/012052.

[190] R. Ahmad, V. Arora, R. Srivastava, S. Sapra, M.N. Kamalasanan, Enhanced performance of organic photovoltaic devices by incorporation of tetrapod-shaped CdSe nanocrystals in polymer-fullerene systems: enhanced performance of organic photovoltaic devices, Phys. Status Solidi A 210 (2013) 785–790, https://doi.org/10.1002/pssa.201228347.

[191] A.M. Ali, D.A. Said, M. Khayyat, M. Boustimi, R. Seoudi, Improving the efficiency of the organic solar cell (CuPc/C60) via PEDOT: PSS as a photoconductor layer doped by silver nanoparticles, Results Phys. 16 (2020) 102819, https://doi.org/10.1016/j.rinp.2019.102819.

[192] G.S. Amitha, M.Y. Ameen, V.S. Reddy, S. Vasudevan, Synthesis of peripherally tetra substituted neutral azophenoxy zinc phthalocyanine and its application in bulk hetero junction solar cells, J. Mol. Struct. 1185 (2019) 425–431, https://doi.org/10.1016/j.molstruc.2019.02.086.

[193] M. Yoosuf Ameen, S. Pradhan, M. Remyth Suresh, V.S. Reddy, MoO3 anode buffer layer for efficient and stable small molecular organic solar cells, Opt. Mater. 39 (2015) 134–139, https://doi.org/10.1016/j.optmat.2014.11.012.

[194] S.E. Shaheen, C.J. Brabec, N.S. Sariciftci, F. Padinger, T. Fromherz, J.C. Hummelen, 2.5% Efficient organic plastic solar cells, Appl. Phys. Lett. 78 (2001) 841–843, https://doi.org/10.1063/1.1345834.
[195] S.E. Shaheen, R. Radspinner, N. Peyghambarian, G.E. Jabbour, Fabrication of bulk heterojunction plastic solar cells by screen printing, Appl. Phys. Lett. 79 (2001) 2996–2998, https://doi.org/10.1063/1.1413501.
[196] M.Y. Ameen, P. Shamjid, T. Abhijith, T. Radhakrishnan, V.S. Reddy, Stability enhancement of P3HT:PCBM polymer solar cells using thermally evaporated MoO3 anode buffer layer, Phys. B Condens. Matter 530 (2018) 201–207, https://doi.org/10.1016/j.physb.2017.11.050.
[197] M.Y. Ameen, P. Shamjid, T. Abhijith, V.S. Reddy, Solution processed transition metal oxide anode buffer layers for efficiency and stability enhancement of polymer solar cells, Opt. Mater. 75 (2018) 491–500, https://doi.org/10.1016/j.optmat.2017.11.006.
[198] S.-S. Kim, S.-I. Na, J. Jo, G. Tae, D.-Y. Kim, Efficient polymer solar cells fabricated by simple brush painting, Adv. Mater. 19 (2007) 4410–4415, https://doi.org/10.1002/adma.200702040.
[199] P. Shamjid, S. Anjusree, M. Yoosuf Ameen, V.S. Reddy, Performance enhancement of polymer solar cells by incorporating Ag nanoparticles at an indium tin oxide/MoO $_3$ buffer layer interface, Semicond. Sci. Technol. 32 (2017), 065010, https://doi.org/10.1088/1361-6641/aa6e9e.
[200] R. Ahmad, R. Srivastava, S. Yadav, S. Chand, S. Sapra, Functionalized 2D-MoS_2-incorporated polymer ternary solar cells: role of nanosheet-induced long-range ordering of polymer chains on charge transport, ACS Appl. Mater. Interfaces 9 (2017) 34111–34121, https://doi.org/10.1021/acsami.7b08725.
[201] M.-L. Tsai, S.-H. Su, J.-K. Chang, D.-S. Tsai, C.-H. Chen, C.-I. Wu, L.-J. Li, L.-J. Chen, J.-H. He, Monolayer MoS_2 heterojunction solar cells, ACS Nano 8 (2014) 8317–8322, https://doi.org/10.1021/nn502776h.
[202] S. Bykkam, D. Prasad, M. Maurya, K. Sadasivuni, J.-J. Cabibihan, Comparison study of metal oxides (CeO_2, CuO, SnO_2, CdO, ZnO and TiO_2) decked few layered graphene nanocomposites for dye-sensitized solar cells, Sustainability 13 (2021) 7685, https://doi.org/10.3390/su13147685.
[203] K.K. Sadasivuni, K. Deshmukh, T.N. Ahipa, A. Muzaffar, M.B. Ahamed, S.K.K. Pasha, M.A.-A. Al-Maadeed, Flexible, biodegradable and recyclable solar cells: a review, J. Mater. Sci. Mater. Electron. 30 (2019) 951–974, https://doi.org/10.1007/s10854-018-0397-y.
[204] S. Bykkam, A. Mishra, D.N. Prasad, M.R. Maurya, J.-J. Cabibihan, Z. Ahmad, K.K. Sadasivuni, 2D-MXene as an additive to improve the power conversion efficiency of monolithic perovskite solar cells, Mater. Lett. (2021), https://doi.org/10.1016/j.matlet.2021.131353.
[205] K. Gong, J. Hu, N. Cui, Y. Xue, L. Li, G. Long, S. Lin, The roles of graphene and its derivatives in perovskite solar cells: a review, Mater. Des. 211 (2021) 110170, https://doi.org/10.1016/j.matdes.2021.110170.
[206] J. Yoon, H. Sung, G. Lee, W. Cho, N. Ahn, H.S. Jung, M. Choi, Superflexible, high-efficiency perovskite solar cells utilizing graphene electrodes: towards future foldable power sources, Energ. Environ. Sci. 10 (2017) 337–345, https://doi.org/10.1039/C6EE02650H.
[207] U. Thakur, R. Kisslinger, K. Shankar, One-dimensional electron transport layers for perovskite solar cells, Nanomaterials 7 (2017) 95, https://doi.org/10.3390/nano7050095.
[208] H. Zhao, D. Benetti, X. Tong, H. Zhang, Y. Zhou, G. Liu, D. Ma, S. Sun, Z.M. Wang, Y. Wang, F. Rosei, Efficient and stable tandem luminescent solar concentrators based on carbon dots and perovskite quantum dots, Nano Energy 50 (2018) 756–765, https://doi.org/10.1016/j.nanoen.2018.06.025.

[209] K. Komoto, J.-S. Lee, J. Zhang, D. Ravikumar, P. Sinha, A. Wade, G.A. Heath, End-of-life management of photovoltaic panels: trends in PV module recycling technologies (No. NREL/TP-6A20-73847, 1561523), 2018, https://doi.org/10.2172/1561523.
[210] M.A. Green, E.D. Dunlop, J. Hohl-Ebinger, M. Yoshita, N. Kopidakis, X. Hao, Solar cell efficiency tables (Version 58), Prog. Photovolt. Res. Appl. 29 (2021) 657–667, https://doi.org/10.1002/pip.3444.
[211] D.M. Chapin, C.S. Fuller, G.L. Pearson, A new silicon *p-n* junction photocell for converting solar radiation into electrical power, J. Appl. Phys. 25 (1954) 676–677, https://doi.org/10.1063/1.1721711.
[212] A. Kojima, K. Teshima, Y. Shirai, T. Miyasaka, Organometal halide perovskites as visible-light sensitizers for photovoltaic cells, J. Am. Chem. Soc. 131 (2009) 6050–6051, https://doi.org/10.1021/ja809598r.
[213] S. Lee, K. Choi, C.H. Min, M.Y. Woo, J.H. Noh, Photon recycling in halide perovskite solar cells for higher efficiencies, MRS Bull. 45 (2020) 439–448, https://doi.org/10.1557/mrs.2020.145.
[214] M.A. Green, Radiative efficiency of state-of-the-art photovoltaic cells: radiative efficiency of photovoltaic cells, Prog. Photovolt. Res. Appl. 20 (2012) 472–476, https://doi.org/10.1002/pip.1147.
[215] G.E. Eperon, M.T. Hörantner, H.J. Snaith, Metal halide perovskite tandem and multiple-junction photovoltaics, Nat. Rev. Chem. 1 (2017) 0095, https://doi.org/10.1038/s41570-017-0095.
[216] S. Yun, Y. Qin, A.R. Uhl, N. Vlachopoulos, M. Yin, D. Li, X. Han, A. Hagfeldt, New-generation integrated devices based on dye-sensitized and perovskite solar cells, Energ. Environ. Sci. 11 (2018) 476–526, https://doi.org/10.1039/C7EE03165C.
[217] W. Rehman, D.P. McMeekin, J.B. Patel, R.L. Milot, M.B. Johnston, H.J. Snaith, L.M. Herz, Photovoltaic mixed-cation lead mixed-halide perovskites: links between crystallinity, photo-stability and electronic properties, Energ. Environ. Sci. 10 (2017) 361–369, https://doi.org/10.1039/C6EE03014A.
[218] D. Shi, V. Adinolfi, R. Comin, M. Yuan, E. Alarousu, A. Buin, Y. Chen, S. Hoogland, A. Rothenberger, K. Katsiev, Y. Losovyj, X. Zhang, P.A. Dowben, O.F. Mohammed, E.H. Sargent, O.M. Bakr, Low trap-state density and long carrier diffusion in organolead trihalide perovskite single crystals, Science 347 (2015) 519–522, https://doi.org/10.1126/science.aaa2725.
[219] C.C. Stoumpos, C.D. Malliakas, M.G. Kanatzidis, Semiconducting tin and lead iodide perovskites with organic cations: phase transitions, high mobilities, and near-infrared photoluminescent properties, Inorg. Chem. 52 (2013) 9019–9038, https://doi.org/10.1021/ic401215x.
[220] S.D. Stranks, G.E. Eperon, G. Grancini, C. Menelaou, M.J.P. Alcocer, T. Leijtens, L.M. Herz, A. Petrozza, H.J. Snaith, Electron-hole diffusion lengths exceeding 1 micrometer in an organometal trihalide perovskite absorber, Science 342 (2013) 341–344, https://doi.org/10.1126/science.1243982.
[221] Y. Yamada, T. Yamada, Y. Kanemitsu, Free carrier radiative recombination and photon recycling in lead halide perovskite solar cell materials, Bull. Chem. Soc. Jpn. 90 (2017) 1129–1140, https://doi.org/10.1246/bcsj.20170208.
[222] D. Zhou, T. Zhou, Y. Tian, X. Zhu, Y. Tu, Perovskite-based solar cells: materials, methods, and future perspectives, J. Nanomater. 2018 (2018) 1–15, https://doi.org/10.1155/2018/8148072.
[223] F. Deschler, M. Price, S. Pathak, L.E. Klintberg, D.-D. Jarausch, R. Higler, S. Hüttner, T. Leijtens, S.D. Stranks, H.J. Snaith, M. Atatüre, R.T. Phillips, R.H. Friend, High photoluminescence efficiency and optically pumped lasing in solution-processed mixed halide perovskite semiconductors, J. Phys. Chem. Lett. 5 (2014) 1421–1426, https://doi.org/10.1021/jz5005285.

[224] C.L. Davies, M.R. Filip, J.B. Patel, T.W. Crothers, C. Verdi, A.D. Wright, R.L. Milot, F. Giustino, M.B. Johnston, L.M. Herz, Bimolecular recombination in methylammonium lead triiodide perovskite is an inverse absorption process, Nat. Commun. 9 (2018) 293, https://doi.org/10.1038/s41467-017-02670-2.

[225] W. Rehman, R.L. Milot, G.E. Eperon, C. Wehrenfennig, J.L. Boland, H.J. Snaith, M.B. Johnston, L.M. Herz, Charge-carrier dynamics and mobilities in formamidinium lead mixed-halide perovskites, Adv. Mater. 27 (2015) 7938–7944, https://doi.org/10.1002/adma.201502969.

[226] S.-T. Ha, C. Shen, J. Zhang, Q. Xiong, Laser cooling of organic–inorganic lead halide perovskites, Nat. Photonics 10 (2016) 115–121, https://doi.org/10.1038/nphoton.2015.243.

[227] L. Protesescu, S. Yakunin, M.I. Bodnarchuk, F. Krieg, R. Caputo, C.H. Hendon, R.X. Yang, A. Walsh, M.V. Kovalenko, Nanocrystals of cesium lead halide perovskites ($CsPbX_3$, X = Cl, Br, and I): novel optoelectronic materials showing bright emission with wide color gamut, Nano Lett. 15 (2015) 3692–3696, https://doi.org/10.1021/nl5048779.

[228] J.M. Richter, M. Abdi-Jalebi, A. Sadhanala, M. Tabachnyk, J.P.H. Rivett, L.M. Pazos-Outón, K.C. Gödel, M. Price, F. Deschler, R.H. Friend, Enhancing photoluminescence yields in lead halide perovskites by photon recycling and light out-coupling, Nat. Commun. 7 (2016) 13941, https://doi.org/10.1038/ncomms13941.

[229] Q. Lin, A. Armin, R.C.R. Nagiri, P.L. Burn, P. Meredith, Electro-optics of perovskite solar cells, Nat. Photonics 9 (2015) 106–112, https://doi.org/10.1038/nphoton.2014.284.

[230] L.M. Pazos-Outon, M. Szumilo, R. Lamboll, J.M. Richter, M. Crespo-Quesada, M. Abdi-Jalebi, H.J. Beeson, M. Vrućinić, M. Alsari, H.J. Snaith, B. Ehrler, R.H. Friend, F. Deschler, Photon recycling in lead iodide perovskite solar cells, Science 351 (2016) 1430–1433, https://doi.org/10.1126/science.aaf1168.

[231] T. Kirchartz, F. Staub, U. Rau, Impact of photon recycling on the open-circuit voltage of metal halide perovskite solar cells, ACS Energy Lett. 1 (2016) 731–739, https://doi.org/10.1021/acsenergylett.6b00223.

[232] I. Dursun, Y. Zheng, T. Guo, M. De Bastiani, B. Turedi, L. Sinatra, M.A. Haque, B. Sun, A.A. Zhumekenov, M.I. Saidaminov, F.P. García de Arquer, E.H. Sargent, T. Wu, Y.N. Gartstein, O.M. Bakr, O.F. Mohammed, A.V. Malko, Efficient photon recycling and radiation trapping in cesium lead halide perovskite waveguides, ACS Energy Lett. 3 (2018) 1492–1498, https://doi.org/10.1021/acsenergylett.8b00758.

[233] R. Fu, D.J. Feldman, R.M. Margolis, U.S. Solar Photovoltaic System Cost Benchmark: Q1 2018 (No. NREL/TP-6A20-72399, 1483475), 2018, https://doi.org/10.2172/1483475.

[234] M. Mousa, F.Z. Amer, R.I. Mubarak, A. Saeed, Simulation of optimized high-current tandem solar-cells with efficiency beyond 41%, IEEE Access 9 (2021) 49724–49737, https://doi.org/10.1109/ACCESS.2021.3069281.

[235] A.D. Vos, Detailed balance limit of the efficiency of tandem solar cells, J. Phys. D Appl. Phys. 13 (1980) 839–846, https://doi.org/10.1088/0022-3727/13/5/018.

[236] F. Sahli, J. Werner, B.A. Kamino, M. Bräuninger, R. Monnard, B. Paviet-Salomon, L. Barraud, L. Ding, J.J. Diaz Leon, D. Sacchetto, G. Cattaneo, M. Despeisse, M. Boccard, S. Nicolay, Q. Jeangros, B. Niesen, C. Ballif, Fully textured monolithic perovskite/silicon tandem solar cells with 25.2% power conversion efficiency, Nat. Mater. 17 (2018) 820–826, https://doi.org/10.1038/s41563-018-0115-4.

[237] Q. Xu, Y. Zhao, X. Zhang, Light management in monolithic perovskite/silicon tandem solar cells, Sol. RRL 4 (2020) 1900206, https://doi.org/10.1002/solr.201900206.

[238] W. Ahmad, J. Khan, G. Niu, J. Tang, Inorganic $CsPbI_3$ perovskite-based solar cells: a choice for a tandem device, Sol. RRL 1 (2017) 1700048, https://doi.org/10.1002/solr.201700048.

[239] A. Al-Ashouri, E. Köhnen, B. Li, A. Magomedov, H. Hempel, P. Caprioglio, J.A. Márquez, A.B. Morales Vilches, E. Kasparavicius, J.A. Smith, N. Phung, D. Menzel, M. Grischek, L. Kegelmann, D. Skroblin, C. Gollwitzer, T. Malinauskas, M. Jošt, G. Matič, B. Rech, R. Schlatmann, M. Topič, L. Korte, A. Abate, B. Stannowski, D. Neher, M. Stolterfoht, T. Unold, V. Getautis, S. Albrecht, Monolithic perovskite/silicon tandem solar cell with >29% efficiency by enhanced hole extraction, Science 370 (2020) 1300–1309, https://doi.org/10.1126/science.abd4016.

[240] G.E. Eperon, T. Leijtens, K.A. Bush, R. Prasanna, T. Green, J.T.-W. Wang, D.P. McMeekin, G. Volonakis, R.L. Milot, R. May, A. Palmstrom, D.J. Slotcavage, R.A. Belisle, J.B. Patel, E.S. Parrott, R.J. Sutton, W. Ma, F. Moghadam, B. Conings, A. Babayigit, H.-G. Boyen, S. Bent, F. Giustino, L.M. Herz, M.B. Johnston, M.D. McGehee, H.J. Snaith, Perovskite-perovskite tandem photovoltaics with optimized band gaps, Science 354 (2016) 861–865, https://doi.org/10.1126/science.aaf9717.

[241] J. Hu, Q. Cheng, R. Fan, H. Zhou, Recent development of organic-inorganic perovskite-based tandem solar cells, Sol. RRL 1 (2017), https://doi.org/10.1002/solr.201700045.

[242] H. Uzu, M. Ichikawa, M. Hino, K. Nakano, T. Meguro, J.L. Hernández, H.-S. Kim, N.-G. Park, K. Yamamoto, High efficiency solar cells combining a perovskite and a silicon heterojunction solar cells via an optical splitting system, Appl. Phys. Lett. 106 (2015), 013506, https://doi.org/10.1063/1.4905177.

[243] N.J. Jeon, J.H. Noh, W.S. Yang, Y.C. Kim, S. Ryu, J. Seo, S.I. Seok, Compositional engineering of perovskite materials for high-performance solar cells, Nature 517 (2015) 476–480, https://doi.org/10.1038/nature14133.

[244] J.M. Ball, S.D. Stranks, M.T. Hörantner, S. Hüttner, W. Zhang, E.J.W. Crossland, I. Ramirez, M. Riede, M.B. Johnston, R.H. Friend, H.J. Snaith, Optical properties and limiting photocurrent of thin-film perovskite solar cells, Energ. Environ. Sci. 8 (2015) 602–609, https://doi.org/10.1039/C4EE03224A.

[245] M.T. Hörantner, T. Leijtens, M.E. Ziffer, G.E. Eperon, M.G. Christoforo, M.D. McGehee, H.J. Snaith, The potential of multijunction perovskite solar cells, ACS Energy Lett. 2 (2017) 2506–2513, https://doi.org/10.1021/acsenergylett.7b00647.

[246] X. Xu, S. Li, H. Zhang, Y. Shen, S.M. Zakeeruddin, M. Graetzel, Y.-B. Cheng, M. Wang, A power pack based on organometallic perovskite solar cell and supercapacitor, ACS Nano 9 (2015) 1782–1787, https://doi.org/10.1021/nn506651m.

[247] J. Luo, D.A. Vermaas, D. Bi, A. Hagfeldt, W.A. Smith, M. Grätzel, Bipolar membrane-assisted solar water splitting in optimal pH, Adv. Energy Mater. 6 (2016) 1600100, https://doi.org/10.1002/aenm.201600100.

[248] Y. Qiu, W. Liu, W. Chen, W. Chen, G. Zhou, P.-C. Hsu, R. Zhang, Z. Liang, S. Fan, Y. Zhang, Y. Cui, Efficient solar-driven water splitting by nanocone $BiVO_4$-perovskite tandem cells, Sci. Adv. 2 (2016), e1501764, https://doi.org/10.1126/sciadv.1501764.

CHAPTER

13

Challenges and future prospects

Paulina Carmona-Monroy, Brenda Vargas, and Diego Solis-Ibarra

Instituto de Investigaciones en Materiales, Universidad Nacional Autónoma de México, Mexico

1. Introduction

Low-dimensional halide perovskites offer much promise as potential replacements for three-dimensional perovskites. Among their advantages are the increased stability toward moisture and light, reduced toxicity, and a vastly increased chemical diversity, mainly due to the almost infinite possible organic cations that can be incorporated into their structure. Despite their advantages, many challenges and avenues of research remain unexplored. This chapter will first discuss some of the most pressing issues that low-dimensional perovskites face: environmental, toxicity, stability, and orientation control. Then, we will turn our attention to the frontiers of low-dimensional perovskites development; first, in the implementation of novel, functional organic cations, and then into novel low-dimensional materials such as layered double perovskites and perovskite heterostructures. Finally, we will offer a perspective on future applications for low-dimensional materials, focusing on technologies beyond those discussed in previous chapters.

An important remark for this chapter is the difference between the so-called "morphological" low-dimensional structures and "true" low-dimensional halide perovskites. The first ones stem from different treatments such as exfoliation, epitaxial growth, or surfactant encapsulation of structurally three-dimensional (ABX_3) halide perovskites, resulting in nanosheets (2D), wires (1D), or dots (0D) [1]. On the other hand, structural, or "true," low-dimensional perovskites originate when the canonical 3D ABX_3 structure is reduced at the molecular level by a compositional modification that splits the inorganic metal-halide framework, forming sheets, chains, or dots made out of $[BX_6]^{n-}$ octahedral units, separated by organic

Low-Dimensional Halide Perovskites
https://doi.org/10.1016/B978-0-323-88522-5.00004-1

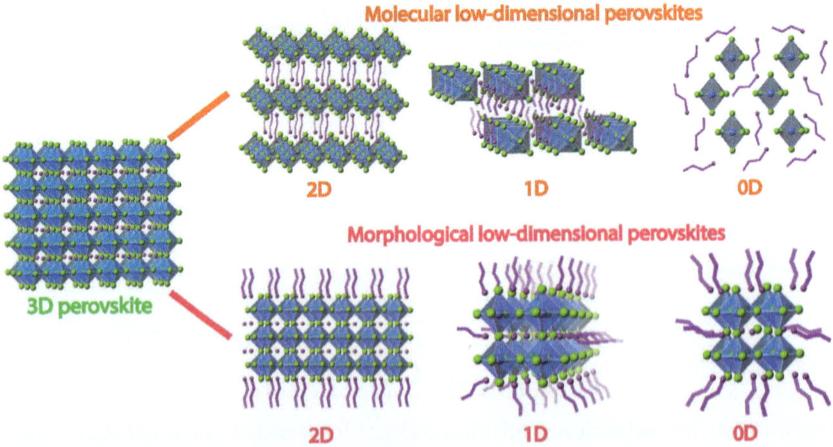

FIG. 1 Schematics representing the difference between molecular and morphological low-dimensional perovskites.

or inorganic cations (Fig. 1). The differences between both types of structures are not only structural but also optoelectronic. While the crystal size control at the nanometric scale bestows the 3D materials with improved optical properties compared to those of the bulk material [2,3], these are completely different to the ones that arise when a molecular low-dimensional halide perovskite is formed [4].

2. Environmental issues and toxicity of halide perovskites

By far, the most studied type of halide perovskites is lead perovskites, which account for roughly 80% of the total halide perovskite reports [5]. However, the toxicity of Pb^{2+} remains a problem, as lead can be leached out of the $APbX_3$ structure due to the decomposition of the materials in water. Environmental and health risks are still a concern for large-scale implementation of technologies based on lead halide perovskites, even when the concentration of lead in a solar panel is just of a few hundred milligrams per square meter [6], mainly because the bioavailability of lead is relatively high [7]. There are several paths for lead intoxication, from water contamination to accumulation in soils and subsequent absorption into the food chain [8], to human exposure during its manufacturing process. Therefore, alternatives have been sought to deal with the toxicity issue, especially when considering that some of the most promising applications of these materials are directed toward clean and efficient energies.

An appropriate encapsulation process of the devices is possible. Still, it requires a more complicated fabrication process, which hampers one of the main advantages of these materials: their simple, low-cost processability. Hence, a possible solution is to substitute the Pb while trying to preserve the halide perovskite structure and its superior optoelectronic properties (see Chapter 6). The obvious potential lead replacements are Sn^{2+} and Ge^{2+}, since they have a similar electronic configuration and are also significantly less toxic than lead, although not entirely innocuous [8]. However, these cations are oxygen-sensitive and are prone to oxidation to the 4+ state, resulting in unstable structures due to self-doping from the M^{4+} metal [9,10]. Nevertheless, Sn^{2+} perovskites can be attained under meticulous conditions where oxygen is excluded and have been implemented in several devices like light-emitting diodes (LEDs) and solar cells [11,12]. Further strategies of Pb^{2+} substitution will be discussed later on in this chapter.

Another strategy to prolongate the long-term stability of the archetypical halide perovskite without replacing lead is to incorporate larger, organic, hydrophobic cations on the A site, generating low-dimensional perovskites. One of the outstanding characteristics of lower dimensional perovskites is the possibility of taking advantage of the individual components' properties when they are combined in one single material. When an aliphatic cation is integrated into the perovskite, it provides moisture stability to the structure, while the inorganic fraction helps it retain its optoelectronic properties. This enhanced stability reduces the risk of lead cations leaching out and effectively deals with its toxicity.

Since lead halide perovskites' unique characteristics have not been met by structures with other metals [13], a discussion has arisen. Given the relatively small quantity of lead contained in the devices (compared to other common lead sources) [7] and the fact that the possible substitutes are not completely harm-free; one must ponder: is eliminating or banning Pb worthwhile? An alternative approach to make these materials truly environmentally benign has been proposed. The efforts are switched from lead replacement to the conscientious implementation of lead halide perovskites. This includes harnessing lead from recycled sources such as spent lead-acid batteries [14], and the subsequent recovery, recycling [15], and remediation once the life cycle of the devices is over.

The straightforward synthesis and film deposition from halide perovskites solutions are convenient, low-cost processes, but solvent toxicity remains a pressing matter. Some of the most used solvents for these purposes, like dimethylformamide (DMF) and dimethyl sulfoxide (DMSO), and some antisolvents for film fabrication like chloroform and chlorobenzene, present considerable health risks and environmental impact [16]. This impact could become unsustainable when considering the large-scale fabrication of these materials and its implicit occupational exposure. However, replacing these solvents is not an easy task, particularly when dealing with the low solubility of lead inorganic salts and the water instability of halide perovskites. Still, several works have reported the use of more environmentally friendly solvents or solvent systems to synthesize both three-dimensional and dimensionally reduced halide perovskites and to deposit films that maintain the quality of those fabricated by traditional methods [17–20]. Finally, another issue when considering larger scale production is the use of inert gases needed to make up for the ambient instability of the materials, which increase fabrication costs and which could be avoided with ambient conditions syntheses.

One example of an environmentally friendly synthesis is a report investigating the effect of water content on the fabrication process of three-dimensional halide perovskites. Under ambient conditions, with a precursor solution with up to 25% water content, Liu et al. synthesized a material with reduced moisture sensitivity, enhanced crystallinity, and optimized performance in photovoltaic devices. The report attributes these properties to the better film quality obtained when water is part of the precursor solution, which prolongs the solvent evaporation time and promotes larger grain sizes, and reduces defects and pinholes on the film [21].

Later, with further exploration of this topic, the total aqueous synthesis of $MAPbBr_3$ was reported. To achieve it, the conversion of Pb(OH)Br into $MAPbBr_3$ was carried out in deionized water by stirring the solution for 12h. The material's characterization showed a successful transformation from the laurionite-type Pb(OH)Br into the $MAPbBr_3$ perovskite. The moisture and thermal stability studies showed that the material was stable after 40 days under a relative humidity of ~60% and thermally stable to up to 150°C. The researchers proposed that the existence of water molecules makes the perovskite crystals less hygroscopic, thus, more stable than the anhydrous ones. Just as Liu and coworkers, they proposed that the larger crystal sizes, a consequence of the prolonged crystallization times, reduced the surface interactions of the material with the environment. The perovskite was further implemented in a resistive memory switching device, which retained its bipolar nonvolatile characteristic for more than 30 days [22].

Both studies prove that it is possible to obtain halide perovskites from greener processes, and while these two are focused on lead halide 3D perovskites, they set a precedent that can be transferred to low-dimensional materials [23]. Such is the case of the two-dimensional Dion-Jacobson (DJ) perovskite $(AMP)PbI_4$ (AMP = 4-(aminomethyl)piperidinium) synthesized in deionized water [24].

In another study, Zhang and coworkers [25] used a solvent system, acetonitrile (ACN) saturated with gaseous MA, to produce $MAPbI_3$ perovskites. While PbI_2 is insoluble in pristine ACN, the presence of MA boosts its solvating power, making it possible to dissolve the lead salt in the greener, more volatile solvent (compared to DMSO and DMF). Upon deposition of the precursor solution (MA^+, Pb^{2+}, I^-) in MA/ACN via spin-coating, smooth films free of pinholes were obtained. These films were implemented in photovoltaic devices that presented efficiencies of 17%, and the method was also tested for larger areas, being able to render a film of ~125 cm^2. They also reported that while annealing is not necessary, given the rapid crystallization of the perovskite at room temperature, a quick thermal treatment further enhances the film quality, yielding more efficient devices.

They also tried the MA system with water and ethanol as solvents, and used other alkylamines, such as propylamine, butylamine, and pentylamine mixed with ACN, but the results were not as successful as those with the MA/ACN system. The water and ethanol systems were unable to dissolve the precursor dispersion even in the presence of MA, and even though the larger amines were able to form a clear precursor solution, the quality of the obtained films was not optimal. Nonetheless, this method opens the path for greener solvents better suited for large-scale fabrication of halide perovskite devices, forming amine/solvent systems of the desired precursor solution [26,27], or systems with larger amines to synthesize low-dimensional perovskites in a more environmentally friendly way.

Another process that completely avoids solvents, at least during the film deposition phase, is the melt processing of 2D hybrid perovskites with several n values reported by Mitzi and colleagues [28]. In this method, the solid perovskite is placed on a substrate heated at a temperature above the melting point of the perovskite, then covered with a Kapton sheet and pressed with another preheated glass substrate. Once the perovskite is melted, it is removed from the heat and left to cool and crystallize at room temperature. This technique is likely to be unique for low-dimensional metal halides, as the organic cation must be judiciously selected to ensure that the melting point is below the decomposition of the material, which is not usually the case with halide perovskites, and particularly for 3D ones. However, the advantage of this method is that solvents are no longer needed to fabricate the thin films with improved crystallinity and higher phase purity than those prepared by spin-coating [29]. Adding to this, potential crystal orientation and one-step self-encapsulation could be achieved in order to implement these high-quality films in optoelectronic devices.

3. Stability and orientation control

The strong anisotropic component of conductivity and mobility in low-dimensional hybrid perovskites is one of the main drawbacks of optoelectronic devices. Their layered (2D), wired (1D), or dotted (0D) structure makes the transport of charges significantly better along the extended

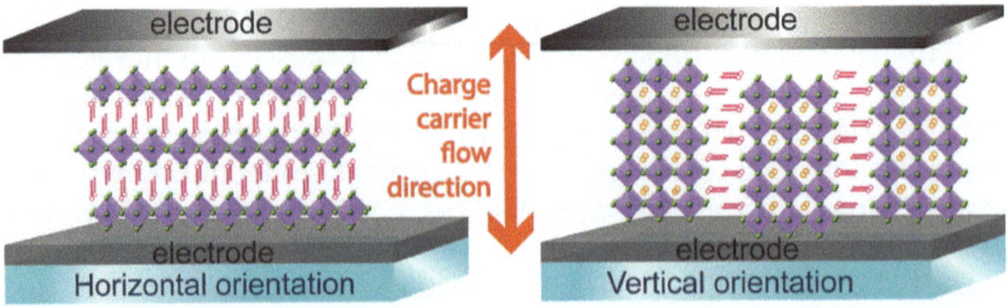

FIG. 2 Horizontal and vertical orientation of low-dimensional perovskite films and charge carrier flow direction within a device.

inorganic structures (layers or chains), while the organic cations usually act as the insulating spacing component. Usually, low-dimensional perovskite films spontaneously crystallize with the extended inorganic fraction (layers or chains) parallel to the substrate, ergo, perpendicular to the charge transport direction of most devices. As some of the key elements for high-performance optoelectronic perovskite devices are linked to optimal conductivity and charge mobility, significant efforts have been made to control the orientation of the deposited films within the devices to achieve better performances, with inorganic slabs parallel to the charge transport direction (Fig. 2) [2].

The primary route to control the orientation of 2D perovskite films has been the compositional control of the materials since it has been observed that the single-layered perovskites align primarily parallel to the substrate. However, for n values greater than 1, the $(BA)_2(MA)_{n-1}Pb_nI_{3n+1}$ ($n=1$–4) perovskite films grew both parallelly and vertically to the substrate, with a growing tendency towards vertical orientation as the n number increased. The photovoltaic devices where the perovskites had n values above 3 had better performances than those with fewer inorganic layers since the tendency to a vertical orientation favored the charge transport in the cells [30]. The same behavior was observed in other two-dimensional perovskites, particularly when the n value was greater than 3 [12,31–34]. However, the distribution within the films was not homogenous, given that a mixture of phases of the 3D material with some 2D perovskite regions took place as the inorganic layer number increases [35].

To better control the vertical orientation of the films, Chen et al. studied the origin of the crystallization in two-dimensional butylammonium halide perovskite films [36]. In this study, they found that the nucleation process and film growth orientation occur at the liquid-air interface where the nuclei first have a vertical orientation and that both the solvent evaporation rate and the concentration of the precursor solutions have a crucial role in the orientation of the film. In a second study [37], they found that the surface nuclei had to follow a continuous crystal growth without more nucleation events to obtain a complete vertical crystallization orientation.

To accomplish the growth rates needed for the material to crystallize without extra nucleation, they modulated the spin-coating time, the solution concentration, and the deposition and annealing temperature. As a result, they obtained films with an orientation degree of over 90% for most experiments, proving the importance crystallization dynamics have in the overall quality of films and, consequently, on devices.

3.1 Molecular engineering

In the same manner, vertical orientation of low-dimensional halide perovskites is pursued certain applications; others require precise control over the thickness of the films and crystals, where only a few nanometers (approximately the thickness of a single quantum well) are desired. For example, once a perovskite microcrystal surpasses a thickness of 100 nm, its properties are those of the bulk material. However, aside from exciting properties associated with quantum confinement, others related to lattice and symmetry distortion, strain, surface, and interfacial effects arise below that scale. For example, at a range from 20 to 60 nm, they have the optimal thickness for nonlinear optics effects, while below 10 nm, single layers are isolated [38].

For certain specific and novel applications, like spintronics, the obtention of molecularly thin 2D perovskites is crucial. For example, richer spin-orbit interactions, such as the Rashba effect—where spin bands present a momentum-dependent splitting—occur due to the confinement and symmetry lowering [39]. Interestingly, this phenomenon does not happen at the bulk or in thick films where the symmetry and level of confinement are not adequate for these interactions [40]. Here is where molecular engineering acquires a more significant role, because chemical modifications can help conduct the microcrystal interactions to obtain molecularly thin halide perovskites. Therefore, to direct the growth of the crystals, one can conduct the surface chemistry of the materials, enhancing inplane lateral growth and suppressing out of plane growth. This can be achieved utilizing bulky, asymmetric, hydrophobic monoammonium organic cations based on tetrathiophene to ensure that the organic fractions lie on the surface of both monocrystals or layers [41]. These organic groups suppress intermolecular self-aggregation and self-crystallization, inducing crystal growth with monolayer unit cell thickness.

3.2 Stability and passivation

Long-term stability is another area where halide perovskites, particularly 3D ones, fall short, as they tend to degrade with moisture and temperature. However, it has been reported that with the incorporation of larger organic cations into the fabrication steps, several processes help improve their stability. The most straightforward, and the one that we have discussed throughout this chapter, is the formation of low-dimensional perovskites, in which the presence of the organic (usually hydrophobic) cation confers the material improved moisture and temperature stability. Nevertheless, heterostructures can be produced depending on the amount of A′ cation introduced into the structure, and the fabrication step in which it is introduced. One possibility is the formation of mixed 2D/3D perovskites, where the incorporation of an interspersed 2D phase within the 3D perovskite grains helps suppress grain boundary defects and traps, where unwanted nonradiative recombination can occur (Fig. 3) [42]. These materials also present improved crystallinity and, in general, behave mainly as the 3D material in terms of their optoelectronic properties but with the improved stability of a 2D perovskite [43]. They are usually prepared in a one-step spin-coating process where the larger organic cation is blended with the 3D precursor solution. To render this type of material, the proportion of the A′/A cations in the precursor solution must be that of a n value of 7 or above [44].

When a 2D perovskite is incorporated into the structure in a second step after the formation of the 3D film, a heterostructure is formed where the lower-dimensional structure acts as a capping layer for the underlying film (Fig. 3). To obtain this kind of structures, a 2D perovskite is spin-coated on top of a 3D perovskite film using a solvent unable to dissolve the 3D perovskite. When incorporated into devices, the 2D perovskite aids in improving the hydrophobicity of the structure, sometimes enhancing

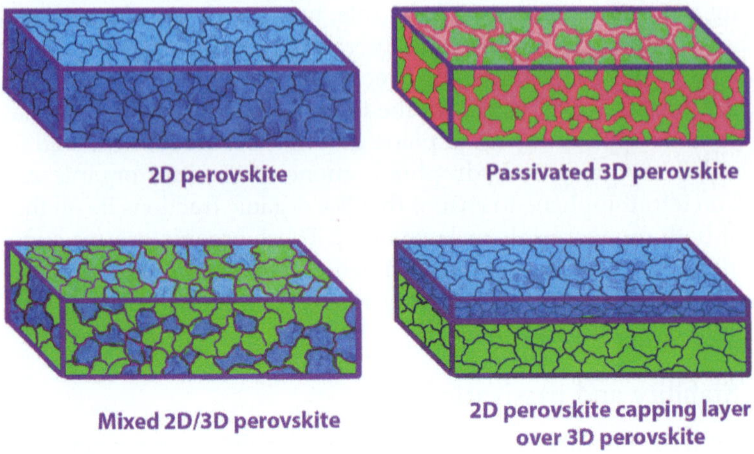

FIG. 3 Different scenarios of the incorporation of large organic cations in the fabrication processes of halide perovskite films.

the contact with the charge transport layers, and in a few cases, acting as a selective charge transport layer itself [45].

Another possible outcome when adding large A′ cations is the passivation of grain boundaries. This scenario generally occurs when just the larger organic cation (without the metal halide) is spin-coated on the surface of another perovskite in a postsynthetic treatment (Fig. 3) [46]. The large molecules tend to bound to the edges of the perovskite crystallites, reducing boundary and edge defects and improving its overall stability to moisture, oxygen, and light [47]. At the molecular level, passivation is associated with the interaction between the organic cations and incomplete PbX_6^- octahedra. Passivated structures present enhanced crystallinity and, therefore, improved optical properties since they have uniform morphology and reduced trap sites, and nonradiative recombination.

4. Novel organic cations

A great virtue of organic-inorganic hybrid materials is that, within a single material, the synergistic effect from the inorganic framework and the organic ligands can be harnessed. As briefly discussed before, in the case of hybrid halide perovskites, the crystallinity, robustness, thermal stability, and unique optoelectronic properties of the inorganic fraction [48] can be unified with tunability, flexibility, hydrophobicity, facile processability, and versatility of the organic components. As previously mentioned, it was believed that due to the aliphatic nature of the organic cations, they had little to no effect on the optical and electrical properties of the materials and rather contributed only to the stability and processability of halide perovskites. Fortunately, recent studies have demonstrated that the electronic interactions between the inorganic framework and the organic cations, and the organic cations themselves, can have a significant impact on the optoelectronic properties and overall performance of the materials [49].

Incorporating organic cations is advantageous for the reasons previously stated, but one of their main drawbacks is that these molecules act as electronic insulators. This may be a desirable property for luminescent applications because quantum wells—in which confinement effects take place—are formed, enhancing properties like exciton confinement and photoluminescence [50]. However, this multiquantum well structure makes conductivity anisotropic; hindering charge transport in the material, where the charge carriers are easily transported within the inorganic layer but not across the organic separators. As will be discussed later, anisotropic conductivity is detrimental for implementing low-dimensional halide perovskites in devices that require an efficient charge transport, such as solar cells or light-emitting diodes (LEDs).

Seminal work has shown that the irradiation of unsaturated organic cations within 2D perovskites results in the topochemical polymerization of said cations while the 2D hybrid perovskite structure is conserved [51]. Derived from this work, the incorporation of ammonium diynes as the organic moiety of 2D halide perovskites was proposed as an alternative to solve the issue of anisotropic conduction. In a study by D. Solis and colleagues [52], the conjugated dialkynyl ammonium salt of deca-3,5-diyn-1-amine ($H_3N^+CH_2CH_2C{\equiv}CC{\equiv}CC_4H_9$) or DDA has incorporated into a 2D halide perovskite with $[DDA]_2PbBr_4$ formula and exposed to a thermal treatment that resulted in the topochemical polymerization of the organic fraction. This rendered a 2D perovskite with highly conjugated polydiacetylenes in the structure, which not only improved the stability and mechanical properties of the material but also resulted in enhanced optical absorption and electronic conductivity.

Similarly, the conjugated 4-vinylbenzylammonium (VBA) cation was incorporated in a 2D/3D perovskite $(VBA)_2PbBr_4/(MAPbBr_3)_{0.15}(F_APbI_3)_{0.85}$ structure, where topochemical cross-linking of the vinyl groups was triggered upon simultaneous thermal treatment (70°C) and photoexcitation (~250nm), to form a cross-linked 2D/3D heterostructure (Fig. 4). This material exhibited improved open-circuit voltages, PCEs exceeding 20%, and better stability in air compared to the 3D-only and the noncross-linked 2D/3D devices, when implemented in a photovoltaic device [53].

These are punctual examples that show the evolution of the organic cation role in these materials. In addition, new reports have appeared on which chiral molecules, organic radical species, organothiolate linkers, and photo-switchable and polymerizable species are incorporated into the

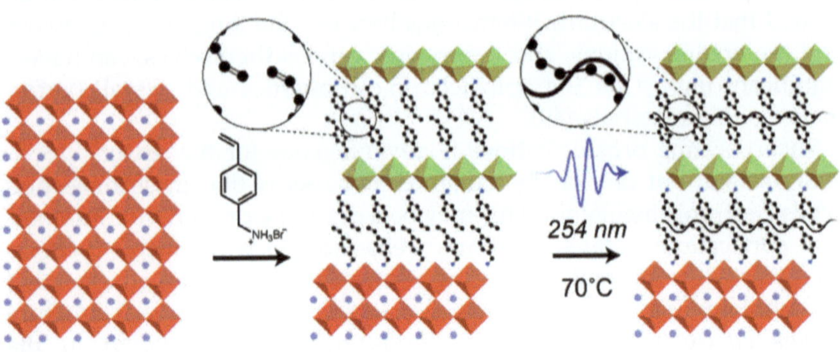

FIG. 4 Schematic illustrating the process by which formation of cross-linked 2D perovskites atop a 3D majority layer is achieved using the ligand 4-vinylbenzylammonium bromide (VBABr). *Reprinted with permission from A.H. Proppe, M. Wei, B. Chen, R. Quintero-Bermudez, S.O. Kelley, E.H. Sargent, Photochemically cross-linked quantum well ligands for 2D/ 3D perovskite photovol- taics with improved photovoltage and stability, J. Am. Chem. Soc. 141 (36) (2019) 14180–14189. Copyright 2019, American Chemical Society.*

inorganic lattice. All these reports add to the historically structure-directing function of the organic molecules [54], switching from the "electronically inert/insulating spacer" role to a more active one that contributes to the optoelectronic properties and performance of the materials and devices based on them. In the *Low Halide Perovskite Heterostructures* section of this chapter, other novel cations that are the backbone of the formation of some of the heterostructures will also be discussed.

5. Halide-layered double perovskites

5.1 Hybrid-layered double perovskites

The first report of a lead halide hybrid organic-inorganic layered perovskite (LP) dates from 1986 when Mayurama and coworkers synthesized $(CH_3(CH_2)_8NH_3)PbI_4$ [55–57]. A few years later, the groundbreaking work of Mitzi and coworkers emphasized the advantages of having materials that combine the properties of inorganic solids with those of the organic molecules in the tuning of optical, magnetic, and mechanical properties [58,59]. Today, the field of layered perovskites has expanded exponentially, and new approaches are being developed to further expand the compositional chemistry and electronic properties of hybrid LPs other than the variation of the organic cations.

One approach is to replace the divalent Pb^{2+} cation with a combination of monovalent B^+ and trivalent B^{3+} cations to generate a mixed-valence hybrid halide layered double perovskite (LDP). Their general formula is $A'_4A_{2n-2}B_nB'_nX_{6n+2}$, where n is the number of octahedral layers forming the inorganic slabs, A is a cation that occupies the cavities between the octahedra within the layers, and A′ is a monocation that separate the inorganic slabs (Fig. 5). Similarly to traditional LPs, LDPs can be classified into three families depending on the direction upon which the inorganic layers are separated, rendering the ⟨100⟩-, ⟨110⟩-, and ⟨111⟩-oriented families, although to date, only the ⟨100⟩- and ⟨111⟩-oriented LDPs have been reported [60]. The realization of such structures was first achieved in 2003 by Guloy and coworkers [61] by using the disproportionation of Au into Au^+ and Au^{3+} to synthesize the $(A)_2[(Au^+I_2)(Au^{3+}I_4)(I_3)_2]$ (A = 1,8-octanediammonium or 1,7-heptanediammonium) LDPs.

This type of structure can also be rationalized as a dimensionally reduced double halide perovskite (DP) (or elpasolite) with general formula $A_2B'BX_6$ or a ⟨001⟩-oriented layered double perovskite of alternating corner-sharing MX_6 and $M'X_6$ octahedra with organic cations in between the layers. Halide elpasolites stole the spotlight of the halide perovskite world back in 2016 when two independent groups synthesized the $Cs_2AgBiBr_6$ double perovskite as a potential alternative lead-free stable solar cell

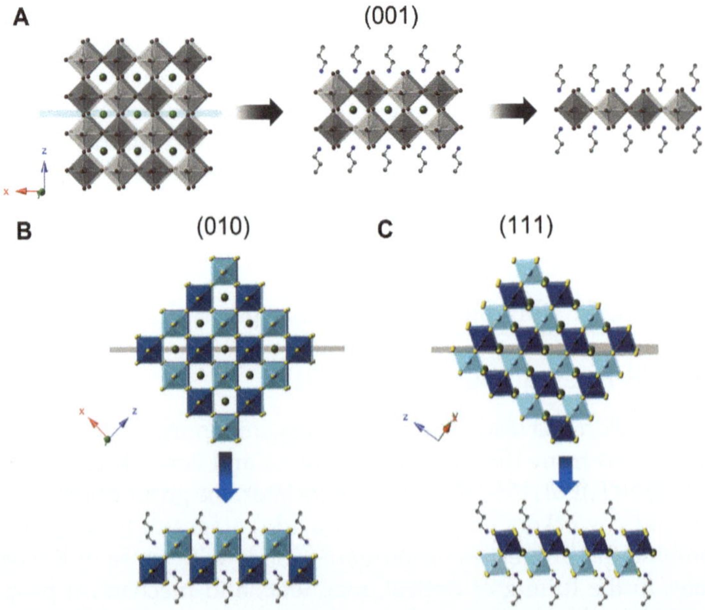

FIG. 5 (A) Insertion of organic cation between an elpasolite inorganic layers to obtain a hybrid layered perovskite $A'_4A_{2n-2}B_nB'_nX_{6n+2}$, where n is the number of octahedral layers forming the inorganic slabs, A is a cation that occupies the cavities between the octahedra within the layers, and A′ is a monocation that separate the inorganic slabs. From right to left: one-octahedron thick slab and two-octahedra slab LDPs. (B) Representations of the hypothetical ⟨110⟩- and ⟨111⟩-oriented hybrid layered perovskites. *Schematic representation reproduced from N.R. Wolf, B.A. Connor, A.H. Slavney, H.I. Karunadasa, Doubling the stakes: the promise of halide double perovskites, Angew. Chem. Int. Ed. 133 (30) (2021) 16400–16414.*

absorber [62,63]. Later in 2018, Karunadasa's group achieved the dimensional reduction of this elpasolite by incorporating butylammonium (BA) between one-octahedron ($n=1$, $(BA)_4AgBiBr_8$) and two-octahedra-thick slabs ($n=2$, $(BA)_2CsAgBiBr_7$) [64]. While new hybrid LDPs have been discovered, they all belong to the $n=1$ and 2 members of the ⟨001⟩-oriented family of LDPs. Assessing whether hybrid LDPs with a larger number of octahedral layers ($n>2$) are stable or achieving dimensional reduction through the incorporation of organic cations for the ⟨010⟩- and ⟨111⟩-oriented LDPs remains a challenge.

The dimensional reduction of a 3D halide DP relieves some geometrical constraints and allows the incorporation of multiple organic spacers. This flexibility enables the tilting and distortion of the octahedral units, which ultimately stabilize combinations of metals that do not usually form 3D DPs (Fig. 6). Such is the case of the parent 3D Cs_2AgBiI_9 elpasolite that has not been synthesized so far, probably because it has an unfavorable tolerance factor value and competes against the favorable formation of the $Cs_3Bi_2I_9$ phase [68]. However, their hybrid layered analogues have been synthesized, for example, $(AE2T)_2AgBiI_8$ [65], $(AMP)_4AgBiI_8$ [69],

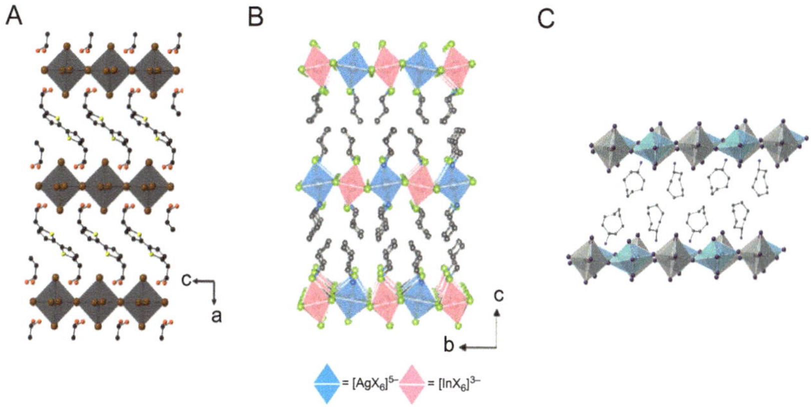

FIG. 6 Some crystal structures of hybrid LDPs that do not have a 3D analog. (A) Schematic of the 2D layered crystal structure of the $[AE2T]_2AgBiI_8$ DP (AE2T = aminomethylpiperidinium) [65], (B) $(PA)_4AgInCl_8$ (PA = propylammonium) [66], and (C) $(CHA)_2CuBiI_8$ (CHA = cycloheptylammonium) [67]. *(A) Copyright © 2019 American Chemical Society. (B) Copyright © 2019 American Chemical Society.*

and $(IPA)_4AgBiI_8$ [70] (where AE2T = aminomethylpiperidinium, CHDA = 1, 4-cyclohexanediammonium, AMP = 4-aminomethylpiperidinium, I-PA = 3-iodopropylammmonium).

Other examples of this phenomenon include $(PA)_4AgInCl_8$ (PA = propylammonium) [66], $(BA)_4AgInCl_8$ [71], whose parent 3D structure, $Cs_2AgInCl_6$, has not been reported to the best of our knowledge. Cu^+-based 3D elpasolites, proposed as very promising solar cell absorbers by Volonakis and coworkers [72], have not been reported in their bulk form as Cu^+ is more energetically favored in a 3- or 4-fold coordination rather than a 6-fold coordination [73]. Their hybrid counterparts incorporate cyclic aliphatic amines (CAA) that stabilize extremely distorted $[Cu^+I_6]^{n-}$ octahedra and enable hybrid LDPs such as $(CHA)_2CuBiI_8$ (CHA = CHA = cycloheptylammonium) [74] or $(PEA)_4CuBiCl_8$ (PEA = phenethylammonium) [75].

2D lead perovskites exhibit a valence band (VB) made up predominantly of halide *p*-orbitals hybridized with some metal *s* orbitals and a conduction band (CB) predominantly made of *p*-orbitals, where the A-site cation rarely contributes to the electronic states near the band edges [4]. Consequently, the optical absorption near the band edge stems mainly from ligand-to-metal-charge-transfer (LMCT) transitions [76]. In HLDPs, both the monovalent M'^+ and trivalent M^{3+} metals contribute to the band edges states, so the bandgap transitions have a metal-to-metal charge transfer (MMCT) contribution [77]. The optoelectronic properties can drastically change depending on the B'/B combinations, but most reported structures are Ag/Bi systems, so there is a clear opportunity to investigate new combinations.

Derived from the B′/B electronic contribution, LDPs exhibit different absorption and PL properties than their Pb-based counterparts. The latter has strong excitonic absorption and photoluminescence (PL) features at room temperature due to their quantum-well-like structure's enhanced exciton binding energies. Most hybrid LDPs exhibit low PL even at low temperatures and an absence of excitonic absorption. Nonetheless, PL emission from self-trapped excitons (STEs) has been induced through pressure in the $(BA)_4AgBiBr_8$ LDP [78], and low-intensity white PL has been detected in $(PA)_4AgSb_{0.5}Bi_{0.5}Br_8$ and $(PA)_4AgIn_{0.5}Bi_{0.5}Br_8$ alloys [66].

Further, as n decreases in Pb-based LPs, the bandgap increases while band dispersion decreases, producing narrower bands [79]. Density functional theory (DFT) calculations have shown that hybrid LDPs typically have flatter bands than Pb-based LPs. In most cases, the energy difference between a direct and indirect transition is in the order of the thermal energy at room temperature (25 meV) or zero due to valley degeneracy [64–66,69,70,80,81]. Yanfa Yan and coworkers proposed that electronic dimensionality [79], defined as the connectivity of the atomic orbital that contributes to the upper VB and lower CB, plays a significant role in the electronic properties of halide perovskites. Based on this definition, Pb-based LPs are electronically and structurally 2D, but hybrid LDPs are electronically 0D and structurally 2D, which would explain the flatter bands.

Several groups have studied the effects of dimensional reduction on the electronic structure of hybrid LDPs. For example, Karunadasa's group calculations showed that $(BA)_4AgBiBr_8$ maintains a direct bandgap until $n = 2$ and then suffers a transition to an indirect bandgap when $n = 1$ [64]. In contrast, the opposite occurs when considering the dimensional reduction of $Cs_2AgTlBr_6$ from $n = 2$ to $n = 1$ [80]. On this matter, Seshadri and coworkers [66] studied the same Ag/Bi system with PA as the organic spacer but did not find that the structure has a true direct bandgap because of the very low band dispersion, and proposed that the structure archetype (RP/DJ) could be more crucial than the dimensional reduction for the direct/indirect transition in this system. More insightful theoretical work ought to be undertaken to generalize some trends in this new field, and determine the role of dimensional reduction and distortions on the optoelectronic properties of hybrid LDPs.

Dimensional reduction is not the only option to tune the electronic structure of hybrid LDPs. Isoelectronic and isovalent alloying, which have been used in DPs to modulate the size [82–84] and symmetry of the bandgap [85], have barely been explored so far in LDPs. Seshadri and coworkers were able to substitute half of the Bi^{3+} sites by In^{3+} or Sb^{3+} in an Ag/Bi system to obtain $(PA)_4AgIn_{0.5}Bi_{0.5}Br_8$, $(PA)_2CsAgIn_{0.5}Bi_{0.5}Br_7$, and $(PA)_4AgSb_{0.5}Bi_{0.5}Br_8$ LDPs [66]. More recently, aliovalent alloying was also explored by Connor in a Cu/In system by replacing half of the octahedral metal sites with Cu^{2+} in the hybrid LDP $(PCA)_4[Cu^{2+}(Cu^+In^{3+})_{0.5}]Cl_8$ (PCA = *p*-chloroanilinium) [75].

Research in hybrid LDPs have been focused mainly on the synthesis and characterization of new structures. However, room temperature ferroelectricity [81,86,87] and photodetection [81,86,88,89] seem common among hybrid LDPs. Ferroelectric performance of LDPs can be improved by incorporating chiral cations [90]. This has been successfully implemented in $A_4M^+M^{3+}(NO_3)_8$ NO_3-LDPs, $(S3HQ)_4KCe\text{-}(NO_3)_8$, and $(R3HQ)_4KCe(NO_3)_8$, where R3HQ and S3HQ stand for R- and S-hydroxylquinuclidinium, respectively [91,92]. Some Ag/Bi layered perovskites have also been proposed as solar cell absorbers with bandgaps of c.a. 2 eV, useful in tandem solar cells, but anisotropic charge transport and low band dispersion need to be addressed to obtain high power conversion efficiencies in actual devices [70].

5.2 All-inorganic layered double perovskites

One strategy to address the toxicity and intrinsic instability of hybrid Pb perovskites is to make an aliovalent substitution of Pb^{2+} with a trivalent cation such as Sb^{3+} or Bi^{3+}. The difference in charge is compensated by the formation of vacancies to form a two-octahedra-thick layered perovskite with the general formula $Cs_3M_2X_9$ (where X = Cl, Br, and I) without the need to incorporate organic cations. These all-inorganic LPs first drew attention in 2015 when $Cs_3Sb_2I_9$ was proposed as a potential lead-free absorber for solar cells [93]. Since then, these LPs have been successfully implemented in solar cells, albeit with low conversion efficiencies [94,95]. Structural dimensionality in these structures can be modulated by incorporating an additional layer of $M^{2+}X_6$ octahedra between the layers of $M'^{III}X_6$ octahedra to form an all-inorganic three-octahedra-thick slab $Cs_4M'^{2+}M_2^{3+}X_{12}$ LDP (Fig. 7). These layered perovskites belong to the $\langle 111 \rangle$-oriented family of LDPs and are also known as quadruple perovskites. Du and colleagues were the firsts to use this term [96], as the structures can be derived from the 3D elpasolite $A_2B'^+B^{3+}X_6$ where B'^+ is replaced by a divalent M'^{2+} metal and a vacancy (V), yielding the general formula $A_4VM'^{2+}M_2^{3+}X_{12}$.

The first demonstration of an all-inorganic fluoride LDP, $K_4Fe_3F_{12}$, was first achieved in 2015 by M. A. Hayward and coworkers [97]. They harnessed the two oxidation states of Fe to synthesize the trigonal $R\overline{3}m$ perovskite $K_4Fe_3F_{12}$ or $K_4Fe^{2+}Fe_2^{3+}F_{12}$, the only layered fluoride analog reported to date. In 2017, our research group reported the synthesis of the first chloride LDP, $Cs_4CuSb_2Cl_{12}$, which showed a nearly direct bandgap of 1.0 eV, similar conductivity values to $MAPbI_3$, and excellent thermal and

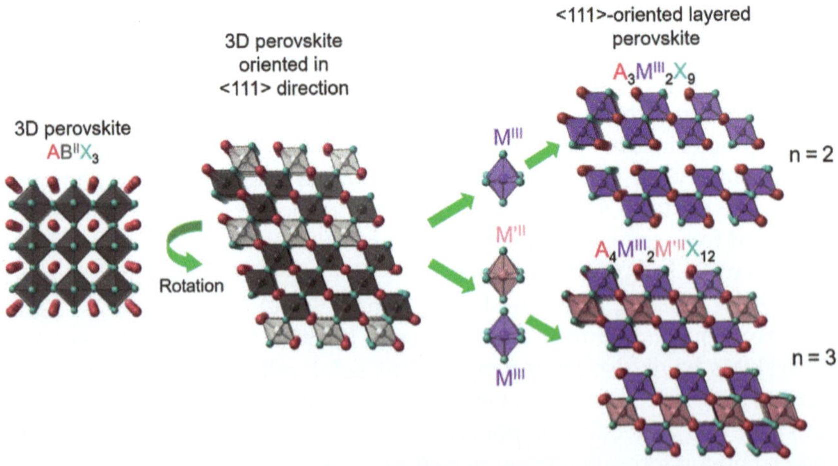

FIG. 7 Schematic representation of a layered perovskite (two-octahedra thick slabs) by substitution of the B^{II} metal in a 3D perovskite by a trivalent metal M^{III}, and layered double perovskite inserting a layer of M'^{II} octahedra between the $A_3M_2X_9$ structure. *Figure from reference B. Vargas, D.T. Reyes-Castillo, E. CoutiÑo-González, C. Sánchez-Aké, C. Ramos, C. Falcony, D. Solis-Ibarra, Enhanced luminescence and mechanistic studies on layered double-perovskite phosphors: $Cs_4Cd_{1-x}Mn_xBi_2Cl_{12}$, Chem. Mater. 32 (21) (2020) 9307–9315. Copyright 2020 American Chemical Society.*

photostability [98,99]. The discovery of this all-inorganic LDP initiated a new avenue to further expand the halide perovskite chemical space by allowing the incorporation of multiple M'^{2+}/M^{3+} cation combinations and extending options of divalent metal to transition metals. In recent years, new members have been discovered by our own and other research groups. To date, the $A_4M'^{2+}M_2^{3+}X_{12}$ perovskites have been synthesized with A = K, Rb, Cs; M′ = Fe, Cu, Mn, Cd; M = Fe, Sb, Bi, In, Yb, Er, Nd; and X = F, Cl, I. Woodward and collaborators have also partially substituted Cl with Br and I in several bulk LDPs [96,99–102].

Interestingly, isoelectronic replacement of the divalent and trivalent metals in low (dopant) and larger (alloys) concentrations can be easily achievedand controlled in these LDPs, which allows the tuning of the optical, electronic, and magnetic properties. For example, the crystal structure, optical absorption, bandgap, and magnetic behavior of $Cs_4MnSb_2Cl_{12}$ were fine-tuned through Cu alloying in the series of $Cs_4Mn_{1-x}Cu_xSb_2Cl_{12}$ ($0 \leq x \leq 1$) solid solutions [99]. The effects of magnetic coupling and temperature over the photoluminescence (PL) properties of the $Cs_4Cd_{1-x}Mn_xBi_2Cl_{12}$ ($0 \leq x \leq 1$) solid solutions were studied by Vargas and collaborators. It was found that dilution of Mn through Cd substitution leads to a maximum PL quantum yield (QY) of 79.5% (Fig. 8) [104,107]. The effects of Bi/Sb substitution in the PL properties have also been studied [108], while Woodward and colleagues explored the stability of mixed-halide LPDs [101].

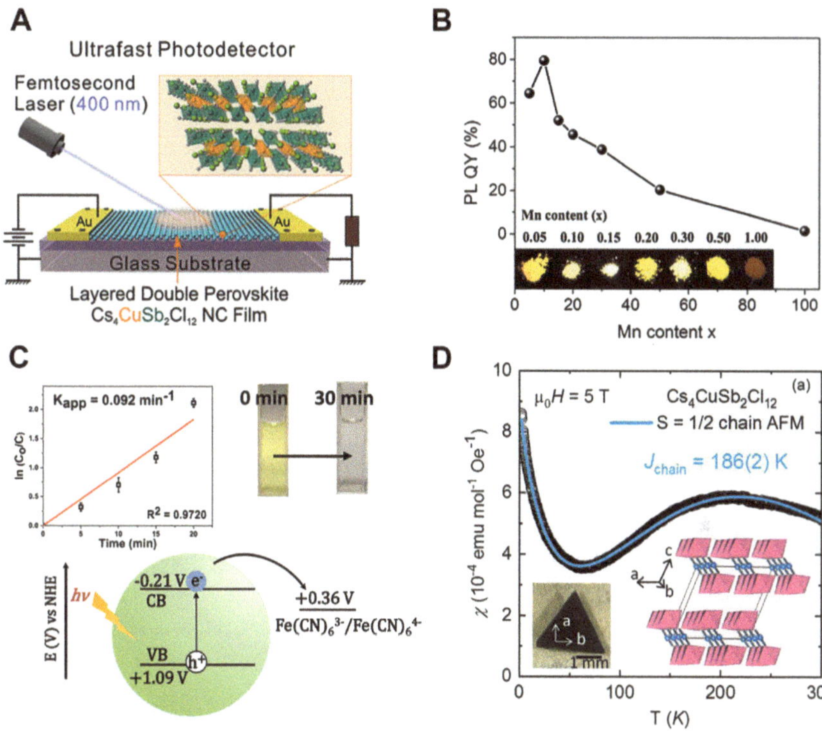

FIG. 8 Some of the applications proposed for all-inorganic LPDs. (A) Proof of concept of an ultrafast photodetector with an active layer made of $Cs_4CuSb_2Cl_{12}$ NCs [103]. (B) Mn dilution with Cd in the series of $Cs_4Cd_{1-x}Mn_xBi_2Cl_{12}$ solid solutions modulates the PL QY [104]. (C) $Cs_4CuSb_2Cl_{12}$ NCs were used for photocatalytic ferricyanide reduction and Congo red dye degradation. The graph shows the first-order kinetic plots for Fe^{3+} reduction. The image shows the visual color change of the solution from yellow to colorless due to Fe^{3+} reduction, and at the bottom, the schematic band alignment and redox potential of the ferricyanide/ferrocyanide redox couple [105]. (D) Susceptibility vs temperature of $Cs_4CuSb_2Cl_{12}$ single crystal [106]. *(A) Copyright © 2020 American Chemical Society. (B) Copyright © 2020 American Chemical Society. (C) Copyright © 2021 American Chemical Society. (D) Copyright ©2020 American Physical Society.*

Isovalent substitution of the M^{3+} sites with lanthanides has recently been reported and could lead to interesting emission properties such as upconversion and downshifting [109]. Li and colleagues manage to incorporate in doping quantities three rare earth trivalent ions, namely Yb, Er and Nd, in the $Cs_4MnBi_2Cl_{12}$ LDP to modulate the photoluminescent properties [110]. Some theoretical calculations predict the stability of perovskites with other M'^{2+} cations, namely, Zn, Fe, Cr, Sn, Ca, Mg, Sr, and Sc [102,111,112]. Finally, aliovalent substitution, such as Tl^+ or In^+, that are known to form hybrid LDPs and DPs could also be investigated [80,113].

Several research groups have probed the effects of downscaling these LDPs to a confined nanometer region to form all-inorganic

LDP nanocrystals (NCs). Kuang and coworkers reported the first $Cs_4CuSb_2Cl_{12}$ NCs by ultrasonic exfoliation and observed a transition from indirect to direct bandgap when passing from the 3D bulk perovskite to its NC form [114]. Then, Chen's group synthesized the same perovskite by a solution-based colloidal synthesis and modulated the crystal structure and electronic band alignment by tuning the composition in the series of $Cs_4Cu_xAg_{2-2x}Sb_2Cl_{12}$ ($0 \leq x \leq 1$) solid solutions [103]. Quantum confinement usually helps to increase the PLQY; however, the $Cs_4Cd_{1-x}Mn_xBi_2Cl_{12}$ NC solid solutions had a maximum PLQY of 4.6% ascribed to fast nonradiative energy relaxation pathways [115]. Han's group was able to improve the PLQY to 47.9% by passivating the nonradiative charge-carrier trapping sites and also studied the effects of aliovalent partial substitution of Bi^{3+} with Na^+ in $Cs_4Mn_xNa_{2-2x}Bi_2Cl_{12}$ NCs [116]. Mandal and colleagues have successfully studied the partial substitution of Cl^- with I^- in $Cs_4CuSb_2Cl_{12}$ NCs, which could be useful in photodetector applications [117]

Computational efforts have attempted to predict new structures and other applications for all-inorganic LDPs. J. Hong found that the photovoltaic potential of all-inorganic LDPs could increase if layers of GeI_6 or SnI_6 were introduced between the $Cs_3Sb_2I_9$ bilayers to form $Cs_{3+n}M_n^{2+}Sb_2X_{9+3n}$ (M=Sn, Ge) [118]. B. Huang proposed that some of these LDPs could work as p-type transparent semiconductors [111,119,120], and more recently, proposed that $Rb_4SnSb_2Br_{12}$ LDP system could be used to design a van der Walls (vdW) tandem solar cell [121]. He also studied the stability of a new type of LDPs with the formula $Cs_4M^{4+}B_2^{2+}X_{12}$ and determined that $Cs_4FePb_2Cl_{12}$ might be a half-metallic ferromagnet useful in spintronic applications [122]. Tran and coworkers noticed the anisotropic magnetic behavior of $Cs_4CuSb_2Cl_{12}$ and determined that its magnetic ordering is characteristic of Heisenberg quasi-1D antiferromagnetic Heisenberg S=1/2 chains that could stabilize a quantum spin liquid (QSL) (Fig. 8) [123]. Other applications of all-inorganic LDPs include photodetectors [103,116,124] and photocatalysis (Fig. 8) [105].

5.3 Other LDPs

In the previous sections, we only discussed LDPs that incorporate two metals with different valences, i.e., B'^{+}/B^{3+} or M'^{2+}/M^{3+} combinations in the B site. One could also consider that two organic species in the A site or two halides in the X site also belong to the LDP category, provided they have a specific ordering (otherwise it would be called a solid solution or alloy). In that sense, Cheetham and colleagues propose that LDPs can be grouped into three categories: type I AA′, type II AA′, BB′, and

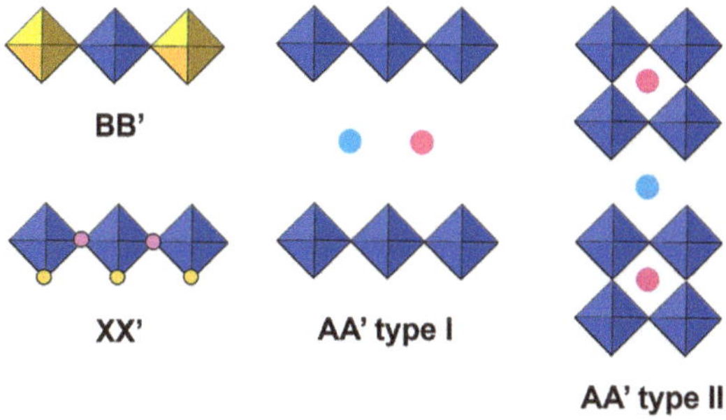

FIG. 9 Schematic representation of layered double perovskites with BB′ ordering, XX′ ordering, and two kinds of AA′: type I, with cations ordering between layers, and type II, with cations ordering within and outside the octahedron cages [125].

XX′ (Fig. 9) [68]. Under this classification, it is worth mentioning two inorganic LDPs with halide ordering, $Cs_2PbI_2Cl_2$ and $Cs_2SnI_2Cl_2$, which show ultrahigh mobilities useful in optoelectronic applications [126,127].

Perovskites with general formula $A_2VB^{4+}X_6$ (where V is a vacancy) are also considered double perovskites by some groups. More generally known as vacancy-ordered perovskites, they can be derived from the $AB^{2+}X_3$ parent structure where the B^{2+} cations have been replaced by B^{4+} cations. The difference in charge is compensated by replacing half of the B sites with vacancies to form a structure where the $B^{4+}X_6$ octahedra alternate with the vacancies [128].

6. Low-dimensional halide perovskite heterostructures

A heterostructure combines two semiconductor materials that present different chemical compositions. It can also be interpreted as a semiconductor structure in which the chemical composition changes with position, where the constituent materials also present different lattice structures and electronic properties, all of this while avoiding the formation of significant chemical disorder that modifies the interfacial electronic properties [129,130].

Low-dimensional halide perovskites are atomic heterostructures themselves, as their structure makes them natural quantum wells. They can be understood as atomically thin inorganic slices separated by large organic cations and derived from the parent three-dimensional structure [131]. However, macroscopic heterojunctions can also be formed between two or more halide perovskites of different dimensionalities, as mentioned in the previous section; or between halide perovskites and other types of semiconductors. These heterostructures can be layer-by-layer stacked structures, lateral epitaxial heterostructures, or heterogenous heterostructures. Moreover, as it will be later discussed, factors like fabrication steps, composition, stacking sequence, and thickness have a primordial role in the properties of the materials.

Heterostructures are the backbone of many modern electronic devices, such as photovoltaics, photodetectors, transistors, and laser diodes, and also a key factor for device miniaturization. One of the reasons they are so useful is because the bandgap mismatch at the junction lets us control and manipulate the generation, recombination, and transport of charge carriers between their components [132].

There are mainly two types of heterostructures: lateral epitaxial heterostructures, and van der Waals (vdW) heterostructures, also known as vertically stacked heterostructures. In the first ones, as their name hints, the change in composition is longitudinal, where the different components show a concentric growth and are held together by chemical bonds, while in the vdW type, they are stacked vertically through weak interactions. In general, for the latter, the components need to have similar crystal symmetry and lattice constant, whereas lateral heterostructures are more tolerant to lattice mismatch and processing limitations.

Creating perovskite heterostructures could be the next big thing in perovskite applications, given that they have less structural and processability restrictions than some of today's most widely used inorganic heterostructures. However, some issues have to be dealt with first, in order to fabricate them. One of them is the soft and ionic nature of halide perovskite crystals, which hinders the formation of perovskite/perovskite heterostructures with an abrupt and stable interface. A clear interface is difficult to maintain in perovskites, especially in 3D ones, because ion diffusion—which causes large defect concentration and poor stability—causes junction interdiffusion and large junction widths [133]. Another issue is that since they are prone to damage after two or more growth steps, the sequential growth of epitaxial halide perovskites structures is complicated. This is due to their poor chemical stability that induces decomposition, especially if the second or third steps are performed under more aggressive conditions, like higher temperatures or using more polar solvents [134]. Lastly, the incompatibility of halide perovskites with current semiconductor fabrication processes, like electron-beam lithography and photolithography, causes material degradation.

Despite these difficulties, several outstanding examples of low-dimensional halide heterostructures have been reported [135,136]. Strategies like lowering the growth temperature, adding antisolvents to the precursor solution, and taking advantage of solubility differences have made it possible to obtain said heterostructures.

6.1 2D van der Waals heterostructures (vertically stacked)

In a broad sense, vertical van der Waals heterostructures are vertically stacked heterostructures bound by weak interactions. Heterostructures are composed of individual layers that have strong bonds within themselves and are stacked on top of each other by van der Waals forces. Due to their two-dimensional morphology and atomically thin thickness, these heterostructures present unique optoelectronic, mechanical, and chemical properties and also unique processability, since they can form highly oriented thin films owing to their 2D structure [129,137].

Since there are no bonding requirements between the different elements of vdW heterostructures, a vast diversity of materials can be integrated within them, from nanodots, nanowires, and nanosheets to bulk three-dimensional materials. Thus, heterostructures can be formed with an almost infinite range of chemical and electronic compositions, as long as the interfaces of the components are flat enough and allow for weak interactions to take place.

This is an advantageous position for halide perovskites, which can form all of the above-stated morphological materials as well as uniform films and atomically thin crystals. They can also have surface molecules that are able to engage in weak interactions and are usually too delicate to be compatible with lithography processes. Since this approach allows the combination of highly disparate materials beyond lattice-matching and processing compatibility limitations, halide perovskites can form vertical heterostructures with other halide perovskites from solution processes, as long as the deposition conditions are meticulously taken care of, i.e., the case of the 2D perovskite capping layer over 3D materials [45], or with other families of materials. In the latter case, multiple processes, like chemical exfoliation or chemical vapor deposition, can be combined to create promising heterostructures with materials such as graphene, oxides, and metal contacts [136,137].

Interesting examples are the 2019 and 2021 reports where vertical heterostructures were formed via exfoliation and stamping between two-dimensional perovskites and other two-dimensional semiconductors: WS_2, a transition metal dichalcogenide (TMDC), and graphene, respectively (Fig. 10) [138,139]. With the formation of the heterostructures, the emission properties of the materials were greatly enhanced (more than an order of magnitude) compared to the individual properties of the components and, in both cases, the stability under ambient conditions was improved.

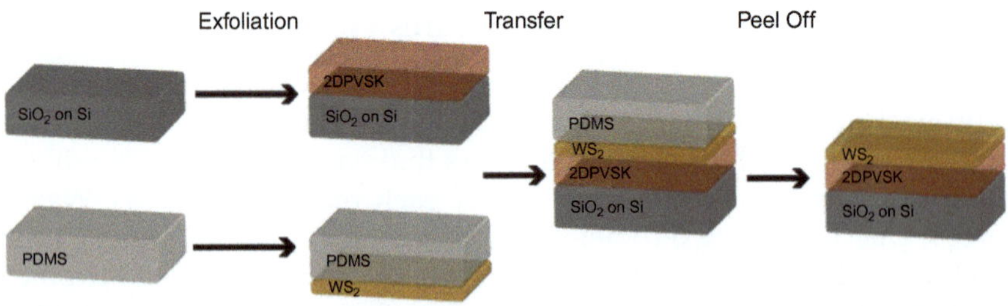

FIG. 10 Schematics of exfoliation and dry transfer process for the $WS_2/BA_2MA_3Pb_4I_{13}$ heterostructure fabrication. The whole process was carried out inside a glovebox under a microscope objective with a remote-controlled mechanical manipulator. *Reprinted with permission from reference A. Yang, J.C. Blancon, W. Jiang, H. Zhang, J. Wong, E. Yan, Y.R. Lin, J. Crochet, M.G. Kanatzidis, D. Jariwala, T. Low, A.D. Mohite, H.A. Atwater, Giant enhancement of photoluminescence emission in WS2-two-dimensional perovskite heterostructures, Nano Lett. 19 (8) (2019) 4852–4860. Copyright 2019 American Chemical Society.*

6.2 Lateral epitaxial heterostructures

In 2020, Shi et al. reported in a ground-breaking study that the formation of atomically sharp two-dimensional halide perovskites lateral heterostructures from a one-pot synthesis. They took advantage of the difference in solubilities to build a concentric heterostructure grown directly over a SiO_2/Si substrate through the control over the solution concentration and growth time and temperature [134]. These heterostructures were fabricated using the combinations between a conjugated ligand based on bithiophenylethylammonium ($2T^+$), and the widely studied butylammonium (BA) cation, different metal cations: Sn and Pb, and different halides: I and Br. Afterward, the researchers were even able to create multicomplex superlattices and multiheterostructures, incorporating other π-conjugated cations like phenylethylammonium (PEA^+), quaterthiophenylethylammonium ($4Tm^+$), and 7-(thiophen-2-yl) benzothiadiazol-4-yl)-[2,2′-bithiophen]-5-yl)ethylammonium (BTm^+) and modulating the different n values of the 2D perovskites (Fig. 11) [132].

The heterostructures were characterized by surface (atomic force microscopy, AFM), optical (transmission electron microscopy, TEM, and scanning electron microscopy, SEM), structural (X-ray diffraction, XRD), and photoluminescence methods. The latter proved to be very useful for these structures since it provided a straightforward way to distinguish the materials that comprised them, which, depending on their intrinsic properties and the nature of the bandgap alignment between them, had different emissions or none. This team was also able to prove that ion interdiffusion between 2D perovskite junctions can be inhibited using appropriate rigid conjugated ligands in the structure, a task that was thought to be complicated given their soft, fragile ionic bonding, and high intrinsic ion mobility of halide ions [132]. Among the surprising discoveries in this report, they found that the soft and weak perovskite

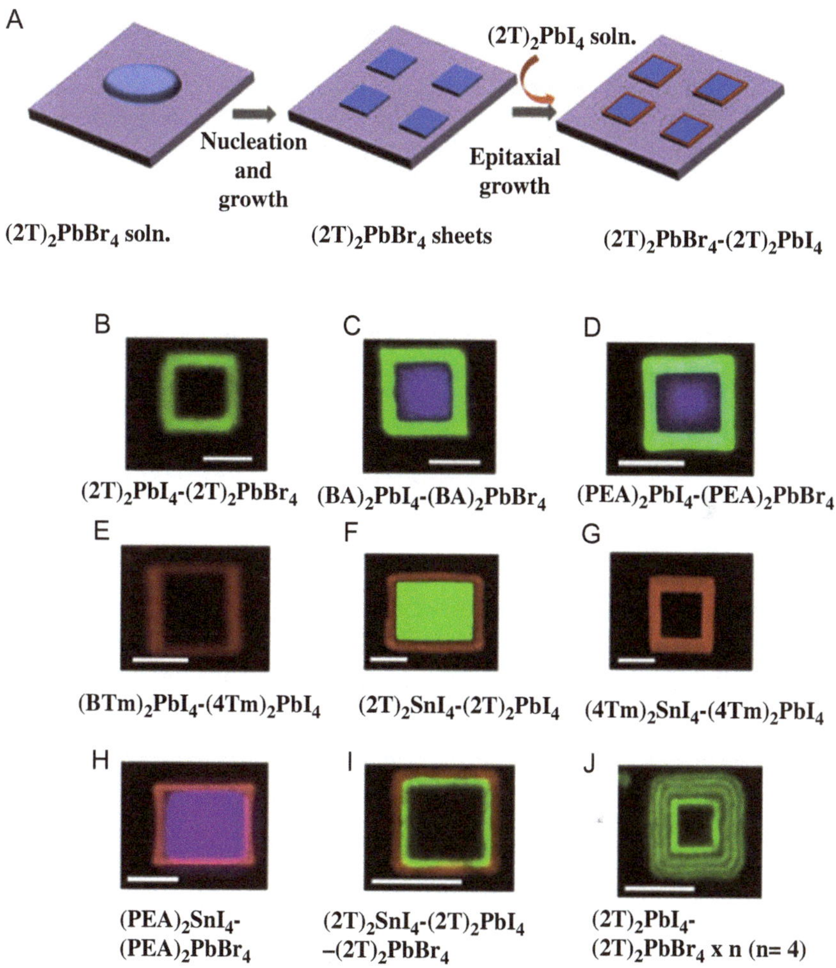

FIG. 11 Synthetic scheme and photoluminescence images of heterostructures. (A) Lateral epitaxial growth technique of 2D perovskite heterostructures. (B) $(2T)_2PbI_4$-$(2T)_2PbBr_4$, (C) $(BA)_2PbI_4$-$(BA)_2PbBr_4$, (D) $(PEA)_2PbI_4$-$(PEA)_2PbBr_4$, (E) $(BTm)_2PbI_4$-$(4Tm)_2PbI_4$, (F) $(2T)_2SnI_4$-$(2T)_2PbI_4$, (G) $(4Tm)_2SnI_4$-$(4Tm)_2PbI_4$, (H) $(PEA)_2SnI_4$-$(PEA)_2PbBr_4$, (I) multiheterostructure $(2T)_2SnI_4$-$(2T)_2PbI_4$-$(2T)_2PbBr_4$, (J) Superlattice, $(2T)_2PbI_4$-$(2T)_2PbBr_4X_n$ ($n = 4$). All the inner 2D perovskites were grown first followed by the outer layer of 2D perovskites. Scale bars for (B), (C), (E), and (I) represent 3 mm; all other scale bars represent 5 mm. *Reprinted with permission from reference T.N. Mandal, A. Jana, Lateral epitaxial heterostructures of halide perovskites for diode application, Matter 3 (3) (2020) 617–619. (Copyright 2020) Elsevier Inc.*

lattice is capable of preserving its coherence at the interface even with the creation of interfacial dislocations and ripples. Also worth mentioning, these materials are so noble that epitaxial growth can be achieved from a simple sequential benchtop synthesis in which the lateral size and vertical thickness of the materials can be controlled [134].

7. Future applications

Many applications of low-dimensional perovskites, such as batteries, gas sensing, memory devices, photodetectors, energy-harvesting devices, and others, were discussed in detail in previous chapters. This section introduces the reader to the state-of-the-art research in low-dimensional halide perovskites and proposes some future applications.

A thriving field in solid-state physics is the field of spintronics, which aims to generate and manipulate the spin polarization of carriers in semiconductors and metallic systems. One of the major achievements of spintronics was the discovery of the giant magnetoresistance effect (GMR), realized by alternating ferromagnetic and nonmagnetic metal layers in layered magnetic thin-film structures [140]. More recently, efforts have been directed to find novel ways to generate and utilize spin-polarized currents [141]. For example, Rashba [142] and Dresselhaus [143] realized that spin-orbit (SO) coupling in noncentrosymmetric compounds leads to an effective magnetic field that can split a crystal's spin-degenerate bands into two spin-polarized bands with opposite spins, an effect known as "Rashba-Dresselhaus splitting" or "Rashba splitting." The electron and hole dispersion relation can then be described by $E_{\pm} = \frac{\hbar^2 k^2}{2m^*} + \alpha_R|k|$ where α_R is the Rashba splitting parameter (Fig. 12). This means that the two

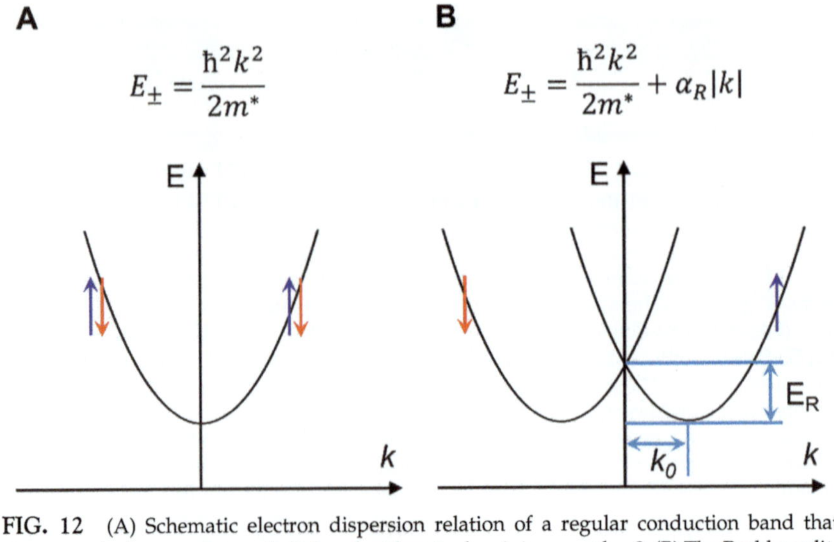

FIG. 12 (A) Schematic electron dispersion relation of a regular conduction band that shows a spin degenerate parabolic band with a single minimum at $k = 0$. (B) The Rashba splitting generates two spin-polarized bands with opposite spins. E_R and k_0 indicate the energy splitting and a momentum offset, respectively.

bands have a momentum offset (k_0) and an energy splitting (E_R) that are related by the Rasha parameter $\alpha_R = 2E_R/k_0$ [144]. Bychkov and Rashba later found that this splitting was also possible in quasi-2D systems. This effect gives an alternative for the electrical manipulation of magnetization and spin directions and can generate novel states of matter such as topological insulators, Dirac materials, Majorana fermions, and more [40].

Materials for spintronic applications require strong SO coupling, i.e., large Rashba splittings, for efficient spin manipulation and long spin-relaxation times for efficient spin transport, properties present in only a handful of materials such as Bi_2Se_3, BiTeI, and ultrathin metallic films [144]. However, theoretical calculations have predicted large Rashba splittings in 2D hybrid lead perovskites as a result of a strong SO coupling, owing to the heavy elements and their quantum-well-type structure that can break the symmetry of the 3D system [145]. Experimentally, Vardeny and collaborators reported a "giant" Rashba energy splitting of 40 meV and a Rashba parameter of 1.6 eV Å in the 2D hybrid perovskite $(C_6H_5C_2H_4NH_3)_2PbI_4$ (PEPI), proving the potential of these 2D perovskites for applications in spintronics [144]. Research in this area is still in its infancy, so reports on the effects of the layer thickness and composition, lattice distortions and dynamics, dimensionality, and excitonic nature of these materials on the Rashba effect are still scarce [146–149].

Another novel application of 2D halide perovskites is in ferroelectrics. Ferroelectric materials present a spontaneous polarization, which can be switched to different orientations under an external electric field. This spontaneous polarization can enhance the separation of photogenerated electron-hole pairs [150], opening the possibility to break the Shockley-Queisser limit through the bulk photovoltaic effect (BPVE). For example, in oxide perovskites, a giant BPVE in thin ferroelectric $BaTiO_3$ films enabled to surpass the power conversion efficiency of single BPVE-solar based cells [151,152]. Combining the semiconducting properties of 2D lead perovskites and their ferroelectric properties is a research avenue that is just starting. For example, Xiong and colleagues [153] designed a narrow bandgap 2D lead iodide ferroelectric perovskite, $(4,4\text{-DFHHA})_2PbI_4$ (4,4-DFHHA = 4,4-difluorohexahydroazepine), using a nonferroelectric 4,4-DFHHA cation. When these cations are confined to the 2D PbI_4^{2-} framework, they order in an asymmetric distribution that eventually leads to spontaneous polarization.

A prerequisite for ferroelectric materials is the absence of centrosymmetry in crystals, which is also a prerequisite for the Rashba splitting. The combination of ferroelectricity and the Rashba effect yields a class of spintronics called Rashba ferroelectrics, a field that is beginning to be explored for 2D halide perovskites. For example, Loh and collaborators [154] propose that the combined effects of both phenomena in the 2D lead perovskite $(AMP)PbI_4$ (AMP=4-(aminomethyl)piperidinium) could lead to enhanced performance in photovoltaic and spintronic applications as the material showed the highest Rashba splitting for 2D lead perovskites of 85 meV and a large saturation polarization Ps = 9.8 μC/cm^2. Further research on this subject could lead to interesting results.

An application that has been recently explored for low-dimensional halide perovskites is thermoelectricity. A temperature gradient is used to move free carriers to generate a thermal voltage in thermoelectric materials. The conversion efficiency of thermoelectric material is expressed by the figure of merit $ZT = \sigma\alpha^2T/\kappa$ where σ, α, and κ are the electrical conductivity, Seebeck coefficient, and thermal conductivity, respectively. To maximize this figure of merit, the material should have a large Seebeck coefficient, a high electrical conductivity—which requires high carrier density and high mobility—and a low thermal conductivity. On the one hand, it has been shown that some 2D hybrid lead perovskites [155] and 0D hybrid perovskite $(MA)_3Bi_2I_9$ [156] possess ultralow thermal conductivities, which makes them attractive for thermoelectric applications (Fig. 13). On the other hand, low charge carrier density and anisotropic carrier transport could hamper the electrical conductivity. As previously mentioned, to enhance electrical conduction on 2D Pb-based perovskites, our research group proposed the incorporation of polydiacetylenes as organic spacers, which resulted in an increase of the electrical conductivity of up to three orders of magnitude [52]. Another approach is to form composites of halide perovskites with high mobility materials [157,158]. Understanding the phonon characteristics, phonon-phonon, and phonon electron coupling, scattering mechanisms, and thermal transport in low-dimensional halide perovskites are of crucial importance to further assess the potential of low-dimensional halide perovskites for thermoelectric applications. This research area is just beginning to be explored [159,160].

As discussed before, the realization of stable epitaxial heterostructures and superlattices with atomically sharp epitaxial interfaces in 2D halide perovskites has the potential to revolutionize yet again the field of halide perovskites. Novel functionalities have emerged from the assembly of complex oxides in superlattices ranging from high-temperature superconductors to colossal magnetoresistance, and synergistic property enhancement can be achieved when layers of different 2D materials, such as transition metal carbides and nitrides (MXenes) and transition metals dichalcogenides and/or graphene-based materials, are assembled in stacked heterostructures [161–163]. In the case

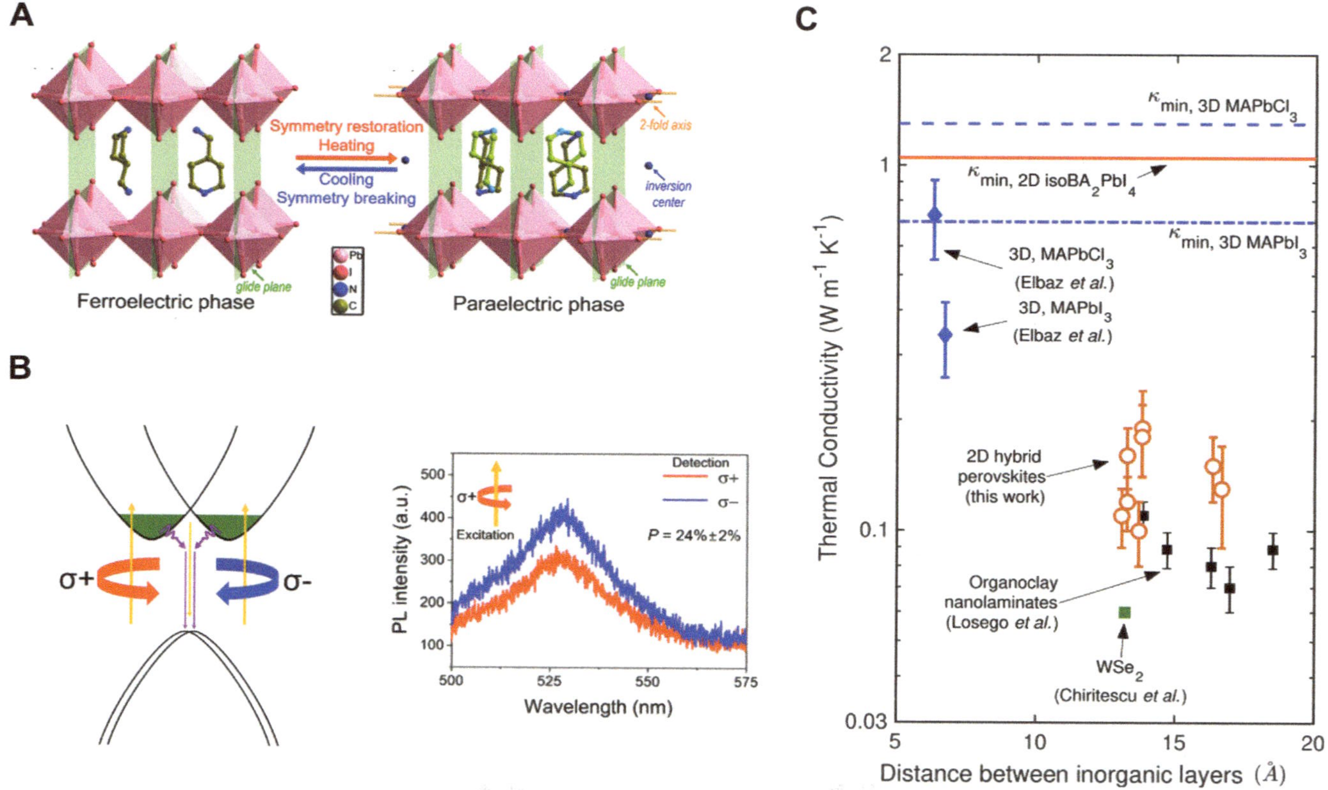

FIG. 13 (A) Transition from the ferroelectric with space group Pc at 298 K to the paraelectric phase with space group $P2_1/c$ at 373 K of $(AMP)PbI_4$. (B) Schematic illustration of the spin splitting valley and left and right circularly polarized PL under left circularly polarized excitation at 473 nm of $(AMP)PbI_4$. (C) Cross-plane thermal conductivities measured via time-domain thermoreflectance (TDTR) for various types of 2D hybrid perovskites as a function of interlayer thickness between the inorganic layers in the film in comparison with some 3D hybrid halide perovskites and WS_2 films. *(B) Adapted from reference I.H. Park, Q. Zhang, K.C. Kwon, Z. Zhu, W. Yu, K. Leng, D. Giovanni, H.S. Choi, I. Abdelwahab, Q.H. Xu, T.C. Sum, K.P. Loh, Ferroelectricity and Rashba effect in a two-dimensional Dion-Jacobson hybrid organic-inorganic perovskite, J. Am. Chem. Soc. 141 (40) (2019) 15972–15976. Copyright (2019) American Chemical Society; (C) Adapted from reference A. Giri, A.Z. Chen, A. Mattoni, K. Aryana, D. Zhang, X. Hu, S.H. Lee, J.J. Choi, P.E. Hopkins, Ultralow thermal conductivity of two-dimensional metal halide perovskites, Nano Lett. 20 (5) (2020) 3331–3337. Copyright (2020) American Chemical Society.*

of the halide perovskite family, a recent report of a TMDC/hybrid 2D perovskite heterostructure found a giant enhancement of PL emission [138].

8. Conclusions

Halide perovskites have come a long way since that first report sparked the scientific community's interest in 2009. In just over a decade, their efficiencies in photovoltaic devices have nearly matched those of technologies that have been in development for decades. Their simple processability, outstanding optoelectronic and photophysical properties, and vast diversity have put them under the spotlight as promising candidates in various applications ranging from energy generators to catalysts. However, with all the upsides that three-dimensional halide perovskites present over traditional semiconductors, some areas present enormous deficiencies that prevent them from achieving their full potential, one of the main ones being their long-term stability.

Low-dimensional halide perovskites were already known for their semiconducting properties and as an alternative with increased stability. Particularly, the incorporation of large hydrophobic organic cations—which reduce the materials dimensionality—has shown to confer additional ambient stability to the materials. This premise has opened the door for a new field with a lot more to offer than just improved stability. With the study of low-dimensional halide perovskites, a virtually infinite number of possibilities in terms of composition, and series of even more attractive optoelectronic properties arose, along with a good deal of challenges. While these materials have proven to be more stable than their three-dimensional counterparts, they have not reached a stage where they can be implemented in commercial devices, owing to their chemical composition, which includes toxic lead in most of the structures, and because their fabrication processing at the laboratory scale has important problems to solve like the film orientation or morphological control, and their scale-up processing is still in its infancy.

Also, despite the monumental advances in this field in recent years, there is still a lack of understanding of some of the photophysical processes that take place in low-dimensional perovskites in general, not because of a lack of effort but because within these materials, that seem to be so simple given the easiness of their synthesis, multiple, and intricate interactions take place.

However, with the research of these families of materials going strong and steady for almost a decade, it is already clear that even with all the challenges that need to be solved, they are promising in a wide range of known applications, like solar cells and light-emitting diodes, where they offer remarkable advantages over current technologies. Moreover, a fair share of other applications will arise along with the deeper study and understanding of these materials and with the combination with other semiconductors to form heterostructures. It is interesting to think, for example, about what could stem from the combination of the spin-polarization effects of 2D halide perovskites and the ferroelectricity of other semiconductors, or the superior conductivity of graphene and the photoluminescence of halide perovskite nanodots. Much work is still required to get low-dimensional halide perovskites to reach their full potential, but for the looks of it so far, the efforts will have been worth it once they get there.

References

[1] P. Chen, Y. Bai, M. Lyu, J.H. Yun, M. Hao, L. Wang, Progress and perspective in low-dimensional metal halide perovskites for optoelectronic applications, Sol. RRL (2018) 1700186.

[2] X. Li, Y. Wu, S. Zhang, B. Cai, Y. Gu, J. Song, H. Zeng, CsPbX3 quantum dots for lighting and displays: room-temperature synthesis, photoluminescence superiorities, underlying origins and white light-emitting diodes, Adv. Funct. Mater. 26 (15) (2016) 2435–2445.

[3] J. Song, J. Li, X. Li, L. Xu, Y. Dong, H. Zeng, Quantum dot light-emitting diodes based on inorganic perovskite cesium lead halides (CsPbX3), Adv. Mater. 27 (44) (2015) 7162–7167.

[4] C. Ortiz-Cervantes, P. Carmona-Monroy, D. Solis-Ibarra, Two-dimensional halide perovskites in solar cells: 2D or not 2D? ChemSusChem 12 (8) (2019) 1560–1575.

[5] E.I. Marchenko, S.A. Fateev, A.A. Petrov, V.V. Korolev, A. Mitrofanov, A.V. Petrov, E.-A. Goodilin, A.B. Tarasov, 2D Perovskites Database, 2022. http://pdb.nmse-lab.ru/index.php. (Accessed 03.06.2022).

[6] A. Babayigit, A. Ethirajan, M. Muller, B. Conings, Toxicity of organometal halide perovskite solar cells, Nat. Mater. 15 (3) (2016) 247–251.

[7] K. Dedecker, G. Grancini, Dealing with lead in hybrid perovskite: a challenge to tackle for a bright future of this technology? Adv. Energy Mater. 10 (31) (2020) 2001471.

[8] B. Pourrut, M. Shahid, C. Dumat, P. Winterton, E. Pinelli, Lead uptake, toxicity, and detoxification in plants, Reviews Env. Contamination 213 (2011) 113–136.

[9] D. Ricciarelli, D. Meggiolaro, F. Ambrosio, F. De Angelis, Instability of tin iodide perovskites: bulk p-doping versus surface tin oxidation, ACS Energy Lett. 5 (9) (2020) 2787–2795.

[10] W. Ke, C.C. Stoumpos, M.G. Kanatzidis, "Unleaded" perovskites: status quo and future prospects of tin-based perovskite solar cells, Adv. Mater. 31 (47) (2019), 1803230.

[11] X. Zhang, C. Wang, Y. Zhang, X. Zhang, S. Wang, M. Lu, H. Cui, S.V. Kershaw, W.W. Yu, A.L. Rogach, Bright orange electroluminescence from lead-free two-dimensional perovskites, ACS Energy Lett. 4 (1) (2019) 242–248.

[12] Y. Liao, H. Liu, W. Zhou, D. Yang, Y. Shang, Z. Shi, B. Li, X. Jiang, L. Zhang, L.N. Quan, R. Quintero-Bermudez, B.R. Sutherland, Q. Mi, E.H. Sargent, Z. Ning, Highly oriented low-dimensional tin halide perovskites with enhanced stability and photovoltaic performance, J. Am. Chem. Soc. 139 (19) (2017) 6693–6699.

[13] M.D. Smith, E.J. Crace, A. Jaffe, H.I. Karunadasa, The diversity of layered halide perovskites, Annu. Rev. Mater. Res. 48 (1) (2018) 111–136.
[14] J. Wang, J. Yang, H. Hou, W. Li, J. Hu, M. Li, W. Yu, Z. Wang, S. Liang, K. Xiao, B. Liu, K. Xi, R.V. Kumar, A green strategy to synthesize two-dimensional lead halide perovskite via direct recovery of spent lead-acid battery, Resour. Conserv. Recycl. 169 (2021) 105463.
[15] C.G. Poll, G.W. Nelson, D.M. Pickup, A.V. Chadwick, D.J. Riley, D.J. Payne, Electrochemical recycling of lead from hybrid organic–inorganic perovskites using deep eutectic solvents, Green Chem. 18 (10) (2016) 2946–2955.
[16] Millipore Sigma, N,N-Dimethylformamide MSDS No. 103034, 2022. https://www.emdmillipore.com/US/en/product/msds/MDA_CHEM-103034?bd=1. (Accessed 02.05.2021).
[17] S.M. Jain, T. Edvinsson, J.R. Durrant, Green fabrication of stable lead-free bismuth based perovskite solar cells using a non-toxic solvent, Commun. Chem. 2 (1) (2019) 91.
[18] J. Wang, F. Di Giacomo, J. Brüls, H. Gorter, I. Katsouras, P. Groen, R.A.J. Janssen, R. Andriessen, Y. Galagan, Highly efficient perovskite solar cells using non-toxic industry compatible solvent system, Sol. RRL 1 (11) (2017) 1–9.
[19] C. Zuo, L. Ding, Lead-free perovskite materials $(NH_4)_3Sb_2I_xBr_{9-x}$, Angew. Chem. Int. Ed. 56 (23) (2017) 6528–6532.
[20] K. Sveinbjörnsson, N.K. Kyi Thein, Z. Saki, S. Svanström, W. Yang, U.B. Cappel, H. Rensmo, G. Boschloo, K. Aitola, E.M.J. Johansson, Preparation of mixed-ion and inorganic perovskite films using water and isopropanol as solvents for solar cell applications, Sustain. Energy Fuels 2 (3) (2018) 606–615.
[21] D. Liu, C.J. Traverse, P. Chen, M. Elinski, C. Yang, L. Wang, M. Young, R.R. Lunt, Aqueous-containing precursor solutions for efficient perovskite solar cells, Adv. Sci. 5 (1) (2018) 1700484.
[22] S. Attique, N. Ali, R. Khatoon, S. Ali, A. Abbas, Y. Yu, J. Hou, B. Cao, H. Wu, S. Yang, Aqueous phase fabrication and conversion of Pb(OH)Br into a $CH_3NH_3PbBr_3$ perovskite and its application in resistive memory switching devices, Green Chem. 22 (11) (2020) 3608–3614.
[23] A. Wang, Y. Guo, Z. Zhou, X. Niu, Y. Wang, F. Muhammad, H. Li, T. Zhang, J. Wang, S. Nie, Z. Deng, Aqueous acid-based synthesis of lead-free tin halide perovskites with near-unity photoluminescence quantum efficiency, Chem. Sci. 10 (17) (2019) 4573–4579.
[24] B. Zhou, L. Liang, J. Ma, J. Li, W. Li, Z. Liu, H. Li, R. Chen, D. Li, Thermally assisted Rashba splitting and circular photogalvanic effect in aqueously synthesized 2D Dion–Jacobson perovskite crystals, Nano Lett. 21 (11) (2021) 4584–4591.
[25] N.K. Noel, S.N. Habisreutinger, B. Wenger, M.T. Klug, M.T. Hörantner, M.B. Johnston, R.J. Nicholas, D.T. Moore, H.J. Snaith, A low viscosity, low boiling point, clean solvent system for the rapid crystallisation of highly specular perovskite films, Energy Environ. Sci. 10 (1) (2017) 145–152.
[26] Q. Liu, Y. Zhao, Y. Ma, X. Sun, W. Ge, Z. Fang, H. Bai, Q. Tian, B. Fan, T. Zhang, A mixed solvent for rapid fabrication of large-area methylammonium lead iodide layers by one-step coating at room temperature, J. Mater. Chem. A 7 (31) (2019) 18275–18284.
[27] D. Burkitt, R. Swartwout, J. McGettrick, P. Greenwood, D. Beynon, R. Brenes, V. Bulović, T. Watson, Acetonitrile based single step slot-die compatible perovskite ink for flexible photovoltaics, RSC Adv. 9 (64) (2019) 37415–37423.
[28] T. Li, A.M. Zeidell, G. Findik, W.A. Dunlap-Shohl, J. Euvrard, K. Gundogdu, O.D. Jurchescu, D.B. Mitzi, Phase-pure hybrid layered lead iodide perovskite films based on a two-step melt-processing approach, Chem. Mater. 31 (11) (2019) 4267–4274.
[29] T. Li, W.A. Dunlap-Shohl, Q. Han, D.B. Mitzi, Melt processing of hybrid organic–inorganic lead iodide layered perovskites, Chem. Mater. 29 (15) (2017) 6200–6204.

[30] D.H. Cao, C.C. Stoumpos, O.K. Farha, J.T. Hupp, M.G. Kanatzidis, 2D homologous perovskites as light-absorbing materials for solar cell applications, J. Am. Chem. Soc. 137 (24) (2015) 7843–7850.

[31] L. Mao, W. Ke, L. Pedesseau, Y. Wu, C. Katan, J. Even, M.R. Wasielewski, C.C. Stoumpos, M.G. Kanatzidis, Hybrid Dion-Jacobson 2D lead iodide perovskites, J. Am. Chem. Soc. 140 (10) (2018) 3775–3783.

[32] H. Tsai, W. Nie, J.C. Blancon, C.C. Stoumpos, R. Asadpour, B. Harutyunyan, A.J. Neukirch, R. Verduzco, J.J. Crochet, S. Tretiak, L. Pedesseau, J. Even, M.A. Alam, G. Gupta, J. Lou, P.M. Ajayan, M.J. Bedzyk, M.G. Kanatzidis, A.D. Mohite, High-efficiency two-dimensional Ruddlesden-Popper perovskite solar cells, Nature 536 (7616) (2016) 312–317.

[33] J. Yan, W. Fu, X. Zhang, J. Chen, W. Yang, W. Qiu, G. Wu, F. Liu, P. Heremans, H. Chen, Highly oriented two-dimensional formamidinium lead iodide perovskites with a small bandgap of 1.51 EV, Mater. Chem. Front. 2 (1) (2018) 121–128.

[34] H. Tsai, R. Asadpour, J.C. Blancon, C.C. Stoumpos, J. Even, P.M. Ajayan, M.G. Kanatzidis, M.A. Alam, A.D. Mohite, W. Nie, Design principles for electronic charge transport in solution-processed vertically stacked 2D perovskite quantum wells, Nat. Commun. 9 (1) (2018) 1–9.

[35] R. Quintero-Bermudez, A. Gold-Parker, A.H. Proppe, R. Munir, Z. Yang, S.O. Kelley, A. Amassian, M.F. Toney, E.H. Sargent, Compositional and orientational control in metal halide perovskites of reduced dimensionality, Nat. Mater. 17 (10) (2018) 900–907.

[36] A.Z. Chen, M. Shiu, J.H. Ma, M.R. Alpert, D. Zhang, B.J. Foley, D.M. Smilgies, S.H. Lee, J.J. Choi, Origin of vertical orientation in two-dimensional metal halide perovskites and its effect on photovoltaic performance, Nat. Commun. 9 (1) (2018) 1–7.

[37] A.Z. Chen, M. Shiu, X. Deng, M. Mahmoud, D. Zhang, B.J. Foley, S.-H. Lee, G. Giri, J.J. Choi, Understanding the formation of vertical orientation in two-dimensional metal halide perovskite thin films, Chem. Mater. 31 (4) (2019) 1336–1343.

[38] G. Grancini, M.K. Nazeeruddin, Dimensional tailoring of hybrid perovskites for photovoltaics, Nat. Rev. Mater. (2019) 4–22.

[39] K. Leng, W. Fu, Y. Liu, M. Chhowalla, K.P. Loh, From bulk to molecularly thin hybrid perovskites, Nat. Rev. Mater. 5 (7) (2020) 482–500.

[40] A. Manchon, H.C. Koo, J. Nitta, S.M. Frolov, R.A. Duine, New perspectives for Rashba spin-orbit coupling, Nat. Mater. 14 (9) (2015) 871–882.

[41] Y. Gao, E. Shi, S. Deng, S.B. Shiring, J.M. Snaider, C. Liang, B. Yuan, R. Song, S.M. Janke, A. Liebman-Peláez, P. Yoo, M. Zeller, B.W. Boudouris, P. Liao, C. Zhu, V. Blum, Y. Yu, B.M. Savoie, L. Huang, L. Dou, Molecular engineering of organic–inorganic hybrid perovskites quantum wells, Nat. Chem. 11 (12) (2019) 1151–1157.

[42] Z. Wang, Q. Lin, F.P. Chmiel, N. Sakai, L.M. Herz, H.J. Snaith, Efficient ambient-air-stable solar cells with 2D-3D heterostructured butylammonium-caesium-formamidinium lead halide perovskites, Nat. Energy 2 (9) (2017) 17135.

[43] C. Ma, C. Leng, Y. Ji, X. Wei, K. Sun, L. Tang, J. Yang, W. Luo, C. Li, Y. Deng, S. Feng, J. Shen, S. Lu, C. Du, H. Shi, 2D/3D perovskite hybrids as moisture-tolerant and efficient light absorbers for solar cells, Nanoscale 8 (43) (2016) 18309–18314.

[44] A.H. Proppe, R. Quintero-Bermudez, H. Tan, O. Voznyy, S.O. Kelley, E.H. Sargent, Synthetic control over quantum well width distribution and carrier migration in low-dimensional perovskite photovoltaics, J. Am. Chem. Soc. 140 (8) (2018) 2890–2896.

[45] K.T. Cho, Y. Zhang, S. Orlandi, M. Cavazzini, I. Zimmermann, A. Lesch, N. Tabet, G. Pozzi, G. Grancini, M.K. Nazeeruddin, Water-repellent low-dimensional fluorous perovskite as interfacial coating for 20% efficient solar cells, Nano Lett. 18 (9) (2018) 5467–5474.

[46] X. Yang, X. Zhang, J. Deng, Z. Chu, Q. Jiang, J. Meng, P. Wang, L. Zhang, Z. Yin, J. You, Efficient green light-emitting diodes based on quasi-two-dimensional composition and phase engineered perovskite with surface passivation, Nat. Commun. 9 (1) (2018) 570.

[47] D.S. Lee, J.S. Yun, J. Kim, A.M. Soufiani, S. Chen, Y. Cho, X. Deng, J. Seidel, S. Lim, S. Huang, A.W.Y. Ho-Baillie, Passivation of grain boundaries by phenethylammonium in formamidinium-methylammonium lead halide perovskite solar cells, ACS Energy Lett. 3 (3) (2018) 647–654.

[48] K. Chondroudis, D.B. Mitzi, Electroluminescence from an organic – inorganic perovskite incorporating a quaterthiophene dye within lead halide perovskite layers, Chem. Mater. 11 (11) (1999) 3028–3030.

[49] J.V. Passarelli, D.J. Fairfield, N.A. Sather, M.P. Hendricks, H. Sai, C.L. Stern, S.I. Stupp, Enhanced out-of-plane conductivity and photovoltaic performance in $n = 1$ layered perovskites through organic cation design, J. Am. Chem. Soc. 140 (23) (2018) 7313–7323.

[50] D. Liang, Y. Peng, Y. Fu, M.J. Shearer, J. Zhang, J. Zhai, Y. Zhang, R.J. Hamers, T.L. Andrew, S. Jin, Color-pure violet-light-emitting diodes based on layered lead halide perovskite nanoplates, ACS Nano 10 (7) (2016) 6897–6904.

[51] B. Tieke, Polymerization of butadiene and butadiyne (diacetylene) derivatives in layer structures, in: Analysis/Reactions/Morphology. Advances in Polymer Science, vol. 71, Springer, Berlin, Heidelberg, 1985, pp. 79–151.

[52] C. Ortiz-Cervantes, P.I. Román-Román, J. Vazquez-Chavez, M. Hernández-Rodríguez, D. Solis-Ibarra, Thousand-fold conductivity increase in 2D perovskites by polydiacetylene incorporation and doping, Angew. Chem. Int. Ed. 57 (2018) 13882–13886.

[53] A.H. Proppe, M. Wei, B. Chen, R. Quintero-Bermudez, S.O. Kelley, E.H. Sargent, Photochemically cross-linked quantum well ligands for 2D/3D perovskite photovoltaics with improved photovoltage and stability, J. Am. Chem. Soc. 141 (36) (2019) 14180–14189.

[54] Y. Gao, Z. Wei, S.-N. Hsu, B.W. Boudouris, L. Dou, Two-dimensional halide perovskites featuring semiconducting organic building blocks, Mater. Chem. Front. 4 (12) (2020) 3400–3418.

[55] M.D. Smith, B.A. Connor, H.I. Karunadasa, Tuning the luminescence of layered halide perovskites, Chem. Rev. 119 (5) (2019) 3104–3139.

[56] L. Mao, K.C. Stoumpos, M.G. Kanatzidis, Two-dimensional hybrid halide perovskites : principles and promises, J. Am. Chem. Soc. 141 (3) (2018) 1171–1190.

[57] Y.I. Dolzhenko, T. Inabe, Y. Maruyama, In situ X-ray observation on the intercalation of weak interaction molecules into perovskite-type layered crystals (C9H19NH3)2PbI4 and (C10H21NH3)2CdCl4, Bull. Chem. Soc. Jap. 59 (2) (1986) 563–567, https://doi.org/10.1246/bcsj.59.563.

[58] D.B. Mitzi, Templating and structural engineering in organic-inorganic perovskites, J. Chem. Soc. Dalton Trans. (1) (2001) 1–12.

[59] D.B. Mitzi, S. Wang, C.A. Feild, C.A. Chess, A.M. Guloy, Conducting layered organic-inorganic halides containing <110>-oriented perovskite sheets, Science 267 (5203) (1995) 1473–1476.

[60] B. Vargas, G. Rodrı´guez-López, D. Solis-Ibarra, The emergence of halide layered double perovskites, ACS Energy Lett. 5 (11) (2020) 3591–3608.

[61] L. Castro-Castro, A. Guloy, Organic-based layered perovskites of mixed-valent gold(I)/gold(III) iodides, Angew. Chem. 115 (24) (2003) 2877–2880.

[62] A.H. Slavney, T. Hu, A.M. Lindenberg, H.I. Karunadasa, A bismuth-halide double perovskite with long carrier recombination lifetime for photovoltaic applications, J. Am. Chem. Soc. 138 (7) (2016) 2138–2141.

[63] E.T. McClure, M.R. Ball, W. Windl, P.M. Woodward, Cs_2AgBiX_6 ($X = Br, Cl$): new visible light absorbing, lead-free halide perovskite semiconductors, Chem. Mater. 28 (5) (2016) 1348–1354.

[64] B.A. Connor, L. Leppert, M.D. Smith, J.B. Neaton, H.I. Karunadasa, Layered halide double perovskites: dimensional reduction of $Cs_2AgBiBr_6$, J. Am. Chem. Soc. 140 (15) (2018) 5235–5240.
[65] M.K. Jana, S.M. Janke, D.J. Dirkes, S. Dovletgeldi, C. Liu, X. Qin, K. Gundogdu, W. You, V. Blum, D.B. Mitzi, Direct-bandgap 2D silver-bismuth iodide double perovskite: the structure-directing influence of an oligothiophene spacer cation, J. Am. Chem. Soc. 141 (19) (2019) 7955–7964.
[66] L. Mao, S.M.L. Teicher, C.C. Stoumpos, R.M. Kennard, R.A. DeCrescent, G. Wu, J.A. Schuller, M.L. Chabinyc, A.K. Cheetham, R. Seshadri, Chemical and structural diversity of hybrid layered double perovskite halides, J. Am. Chem. Soc. 141 (48) (2019) 19099–19109.
[67] L.Y. Bi, T.L. Hu, M.Q. Li, B.K. Ling, M.S. Lassoued, Y.Q. Hu, Z. Wu, G. Zhou, Y.Z. Zheng, Template effects in Cu(I)-Bi(III) iodide double perovskites: a study of crystal structure, film orientation, band gap and photocurrent response, J. Mater. Chem. A 8 (15) (2020) 7288–7296.
[68] H.A. Evans, L. Mao, R. Seshadri, A.K. Cheetham, Layered double perovskites, Annu. Rev. Mater. Res. 51 (1) (2021) 092320–102133.
[69] M.S. Lassoued, L.-Y. Bi, Z. Wu, G. Zhou, Y.-Z. Zheng, Piperidine-induced switching of the direct band gaps of Ag(I)/Bi(III) bimetallic iodide double perovskites, J. Mater. Chem. C 8 (16) (2020) 5349–5354.
[70] Y. Yao, B. Kou, Y. Peng, Z. Wu, L. Li, S. Wang, X. Zhang, X. Liu, J. Luo, $(C_3H_9NI)_4AgBiI_8$: a direct-bandgap layered double perovskite based on a short-chain spacer cation for light absorption, Chem. Commun. 56 (21) (2020) 3206–3209.
[71] E.T. McClure, A.P. McCormick, P.M. Woodward, Four lead-free layered double perovskites with the $n = 1$ Ruddlesden–Popper structure, Inorg. Chem. 59 (9) (2020) 6010–6017.
[72] G. Volonakis, M.R. Filip, A.A. Haghighirad, N. Sakai, B. Wenger, H.J. Snaith, F. Giustino, Lead-free halide double perovskites via heterovalent substitution of noble metals, J. Phys. Chem. Lett. 7 (7) (2016) 1254–1259.
[73] Z. Xiao, K.Z. Du, W. Meng, D.B. Mitzi, Y. Yan, Chemical origin of the stability difference between copper(I)- and silver(I)-based halide double perovskites, Angew. Chem. Int. Ed. 56 (40) (2017) 12107–12111.
[74] Z. Fang, W. Chen, Y. Shi, J. Zhao, S. Chu, J. Zhang, Z. Xiao, Dual passivation of perovskite defects for light-emitting diodes with external quantum efficiency exceeding 20%, Adv. Funct. Mater. 30 (12) (2020) 1909754.
[75] B.A. Connor, R.W. Smaha, J. Li, A. Gold-Parker, A.J. Heyer, M.F. Toney, Y.S. Lee, H.I. Karunadasa, Alloying a single and a double perovskite: a $Cu^{+/2+}$ mixed-valence layered halide perovskite with strong optical absorption, Chem. Sci. 12 (25) (2021) 8689–8697.
[76] J.S. Manser, J.A. Christians, P.V. Kamat, Intriguing optoelectronic properties of metal halide perovskites, Chem. Rev. 116 (21) (2016) 12956–13008.
[77] N.R. Wolf, B.A. Connor, A.H. Slavney, H.I. Karunadasa, Doubling the stakes: the promise of halide double perovskites, Angew. Chem. Int. Ed. 133 (30) (2021) 16400–16414.
[78] Y. Fang, L. Zhang, L. Wu, J. Yan, Y. Lin, K. Wang, W.L. Mao, B. Zou, Pressure-induced emission (PIE) and phase transition of a two-dimensional halide double perovskite $(BA)_4AgBiBr_8$ ($BA = CH_3(CH_2)_3NH_3^+$), Angew. Chem. 131 (43) (2019) 15393–15397.
[79] Z. Xiao, M. Weiwei, J. Wang, D.B. Mitzi, Y. Yan, Searching for promising new perovskite-based photovoltaic absorbers: the importance of electronic dimensionality, Mater. Horiz. 4 (2) (2017) 206–216.
[80] B.A. Connor, R.I. Biega, L. Leppert, H.I. Karunadasa, Dimensional reduction of the small-bandgap double perovskite $Cs_2AgTlBr_6$, Chem. Sci. 11 (29) (2020) 7708–7715.
[81] Y. Li, T. Yang, Z. Xu, X. Liu, X. Huang, S. Han, Y. Liu, M. Li, J. Luo, Z. Sun, Dimensional reduction of $Cs_2AgBiBr_6$: a 2D hybrid double perovskite with strong polarization sensitivity, Angew. Chem. 132 (9) (2020) 3457–3461.

[82] K.Z. Du, W. Meng, X. Wang, Y. Yan, D.B. Mitzi, Bandgap engineering of lead-free double perovskite $Cs_2AgBiBr_6$ through trivalent metal alloying, Angew. Chem. Int. Ed. 56 (28) (2017) 8158–8162.
[83] A.H. Slavney, L. Leppert, D. Bartesaghi, A. Gold-Parker, M.F. Toney, T.J. Savenije, J.B. Neaton, H.I. Karunadasa, Defect-induced band-edge reconstruction of a bismuth-halide double perovskite for visible-light absorption, J. Am. Chem. Soc. 139 (14) (2017) 5015–5018.
[84] Z. Li, S.R. Kavanagh, M. Napari, R.G. Palgrave, M. Abdi-Jalebi, Z. Andaji-Garmaroudi, D.W. Davies, M. Laitinen, J. Julin, M.A. Isaacs, R.H. Friend, D.O. Scanlon, A. Walsh, R.-L.Z. Hoye, Bandgap lowering in mixed alloys of $Cs_2Ag(Sb_xBi_{1-x})Br_6$ double perovskite thin films, J. Mater. Chem. A 8 (41) (2020) 21780–21788.
[85] W. Meng, X. Wang, Z. Xiao, J. Wang, D.B. Mitzi, Y. Yan, Parity-forbidden transitions and their impact on the optical absorption properties of lead-free metal halide perovskites and double perovskites, J. Phys. Chem. Lett. 8 (13) (2017) 2999–3007.
[86] W. Guo, X. Liu, S. Han, Y. Liu, Z. Xu, M. Hong, J. Luo, Z. Sun, Room-temperature ferroelectric material composed of a two-dimensional metal halide double perovskite for X-ray detection, Angew. Chem. 59 (33) (2020) 13983–13988.
[87] W. Zhang, M. Hong, J. Luo, Halide double perovskite ferroelectrics, Angew. Chem. Int. Ed. 59 (24) (2020) 9305–9308.
[88] Z. Xu, X. Liu, Y. Li, X. Liu, T. Yang, C. Ji, S. Han, Y. Xu, J. Luo, Z. Sun, Exploring lead-free hybrid double perovskite crystals of $(BA)_2CsAgBiBr_7$ with large mobility-lifetime product toward X-ray detection, Angew. Chem. Int. Ed. 58 (44) (2019) 15757–15761.
[89] X. Liu, S. Wang, P. Long, L. Li, Y. Peng, Z. Xu, S. Han, Z. Sun, M. Hong, J. Luo, Polarization-driven self-powered photodetection in a single-phase biaxial hybrid perovskite ferroelectric, Angew. Chem. Int. Ed. 58 (41) (2019) 14504–14508.
[90] G. Long, R. Sabatini, M.I. Saidaminov, G. Lakhwani, A. Rasmita, X. Liu, E.H. Sargent, W. Gao, Chiral-perovskite optoelectronics, Nat. Rev. Mater. 5 (2020) 423–439.
[91] C. Shi, L. Ye, Z.-X. Gong, J.-J. Ma, Q.-W. Wang, J.-Y. Jiang, M.-M. Hua, C.-F. Wang, H. Yu, Y. Zhang, H.-Y. Ye, Two-dimensional organic–inorganic hybrid rare-earth double perovskite ferroelectrics, J. Am. Chem. Soc. 142 (1) (2020) 545–551.
[92] C. Shi, J.-J. Ma, J.-Y. Jiang, M.-M. Hua, Q. Xu, H. Yu, Y. Zhang, H.-Y. Ye, Large piezoelectric response in hybrid rare-earth double perovskite relaxor ferroelectrics, J. Am. Chem. Soc. 142 (21) (2020) 9634–9641.
[93] B. Saparov, F. Hong, J.P. Sun, H.S. Duan, W. Meng, S. Cameron, I.G. Hill, Y. Yan, D.B. Mitzi, Thin-film preparation and characterization of $Cs_3Sb_2I_9$: a lead-free layered perovskite semiconductor, Chem. Mater. 27 (16) (2015) 5622–5632.
[94] F. Jiang, D. Yang, Y. Jiang, T. Liu, X. Zhao, Y. Ming, B. Luo, F. Qin, J. Fan, H. Han, L. Zhang, Y. Zhou, Chlorine-incorporation-induced formation of the layered phase for antimony-based lead-free perovskite solar cells, J. Am. Chem. Soc. 140 (3) (2018) 1019–1027.
[95] S. Weber, T. Rath, K. Fellner, R. Fischer, R. Resel, B. Kunert, T. Dimopoulos, A. Steinegger, G. Trimmel, Influence of the iodide to bromide ratio on crystallographic and optoelectronic properties of rubidium antimony halide perovskites, ACS Appl. Energy Mater. 2 (1) (2019) 539–547.
[96] Y.-P. Lin, S. Hu, B. Xia, K.-Q. Fan, L.-K. Gong, J.-T. Kong, X.-Y. Huang, Z. Xiao, K.-Z. Du, Material design and optoelectronic properties of three-dimensional quadruple perovskite halides, J. Phys. Chem. Lett. 10 (17) (2019) 5219–5225.
[97] S.W. Kim, R. Zhang, P.S. Halasyamani, M.A. Hayward, $K_4Fe_3F_{12}$: an Fe^{2+}/Fe^{3+} charge-ordered, ferrimagnetic fluoride with a cation-deficient, layered perovskite structure, Inorg. Chem. 54 (13) (2015) 6647–6652.
[98] B. Vargas, E. Ramos, E. Pérez-Gutiérrez, J.C. Alonso, D. Solis-Ibarra, A direct bandgap copper-antimony halide perovskite, J. Am. Chem. Soc. 139 (27) (2017) 9116–9119.

[99] B. Vargas, R. Torres-Cadena, J. Rodríguez-Hernández, M. Gembicky, H. Xie, J. Jiménez-Mier, Y.S. Liu, E. Menéndez-Proupin, K.R. Dunbar, N. Lopez, P. Olalde-Velasco, D. Solis-Ibarra, Optical, electronic, and magnetic engineering of <111> layered halide perovskites, Chem. Mater. 30 (15) (2018) 5315–5321.

[100] M. Liu, S.K. Matta, H. Ali-Löytty, A. Matuhina, G.K. Grandhi, K. Lahtonen, S.P. Russo, P. Vivo, Moisture-assisted near-UV emission enhancement of lead-free $Cs_4CuIn_2Cl_{12}$ double perovskite nanocrystals, Nano Lett. 22 (1) (2022) 311–318.

[101] M.B. Gray, J.D. Majher, N.P. Holzapfel, P.M. Woodward, Exploring the stability of mixed-halide vacancy-ordered quadruple perovskites, Chem. Mater. 33 (6) (2021) 2165–2172.

[102] B. Vargas, R. Torres-Cadena, D.T. Reyes-Castillo, J. Rodríguez-Hernández, M. Gembicky, E. Menéndez-Proupin, D. Solis-Ibarra, Chemical diversity in lead-free, layered double perovskites: a combined experimental and computational approach, Chem. Mater. 32 (1) (2020) 424–429.

[103] T. Cai, W. Shi, S. Hwang, K. Kobbekaduwa, Y. Nagaoka, H. Yang, K. Hills-Kimball, H. Zhu, J. Wang, Z. Wang, Y. Liu, D. Su, J. Gao, O. Chen, Lead-free $Cs_4CuSb_2Cl_{12}$ layered double perovskite nanocrystals, J. Am. Chem. Soc. 142 (27) (2020) 11927–11936.

[104] B. Vargas, D.T. Reyes-Castillo, E. Coutino-Gonzalez, C. Sánchez-Aké, C. Ramos, C. Falcony, D. Solis-Ibarra, Enhanced luminescence and mechanistic studies on layered double-perovskite phosphors: $Cs_4Cd_{1-x}Mn_xBi_2Cl_{12}$, Chem. Mater. 32 (21) (2020) 9307–9315.

[105] A. Paingott Parambil, M. Joshi, D. Verma, S. Jadhav, A.R. Choudhury, D. Jana, Layered $Cs_4CuSb_2Cl_{12}$ nanocrystals for sunlight-driven photocatalytic degradation of pollutants, ACS Appl. Nano Mater. 4 (2) (2021) 1305–1313.

[106] T.T. Tran, C.A. Pocs, Y. Zhang, M.J. Winiarski, J. Sun, M. Lee, T.M. Mcqueen, O.R. Nat, Spinon excitations in the quasi-1D S = ½ chain $Cs_4CuSb_2Cl_{12}$, Phys. Rev. B 101 (23) (2020) 235107.

[107] N.P. Holzapfel, J.D. Majher, T.A. Strom, C.E. Moore, P.M. Woodward, $Cs_4Cd_{1-x}Mn_xBi_2Cl_{12}$—a vacancy-ordered halide perovskite phosphor with high-efficiency orange-red emission, Chem. Mater. 32 (8) (2020) 3510–3516.

[108] B. Vargas, E. Coutiño-Gonzalez, O. Ovalle-Encinia, C. Sánchez-Aké, D. Solis-Ibarra, Efficient emission in halide layered double perovskites: the role of Sb^{3+} substitution in $Cs_4Cd_{1-x}Mn_xBi_2Cl_{12}$ phosphors, J. Phys. Chem. Lett. 12 (2020) 10362–10367.

[109] W.J. Mir, T. Sheikh, H. Arfin, Z. Xia, A. Nag, Lanthanide doping in metal halide perovskite nanocrystals: spectral shifting, quantum cutting and optoelectronic applications, NPG Asia Mater. 12 (2020) 9.

[110] J. Li, J. Xiao, T. Lin, Z. Yan, X. Han, Lanthanide doping enabled multimodal luminescence in layered lead-free double perovskite $Cs_4MnBi_2Cl_{12}$, J. Mater. Chem. C 10 (19) (2022) 7626–7632.

[111] J. Xu, J.-B. Liu, J. Wang, B.-X. Liu, B. Huang, Prediction of novel p -type transparent conductors in layered double perovskites: a first-principles study, Adv. Funct. Mater. 28 (26) (2018) 1800332.

[112] D. Han, T. Zhang, S. Chen, High-throughput first-principles screening of layered magnetic double perovskites $Cs_4MSb_2X_{12}$ for spintronic applications, J. Phys. Condens. Matter 32 (2019) 225705.

[113] X.G. Zhao, D. Yang, J.C. Ren, Y. Sun, Z. Xiao, L. Zhang, Rational design of halide double perovskites for optoelectronic applications, Joule 2 (9) (2018) 1662–1673.

[114] X.D. Wang, N.H. Miao, J.F. Liao, W.Q. Li, Y. Xie, J. Chen, Z.M. Sun, H.Y. Chen, D.B. Kuang, The top-down synthesis of single-layered $Cs_4CuSb_2Cl_{12}$ halide perovskite nanocrystals for photoelectrochemical application, Nanoscale 11 (12) (2019) 5180–5187.

[115] H. Yang, W. Shi, T. Cai, K. Hills-Kimball, Z. Liu, L. Dube, O. Chen, Synthesis of lead-free $Cs_4(Cd_{1-x}Mn_x)Bi_2Cl_{12}$ ($0 \leq x \leq 1$) layered double perovskite nanocrystals with controlled Mn–Mn coupling interaction, Nanoscale 12 (45) (2020) 23191–23199.

[116] T. Bai, B. Yang, J. Chen, D. Zheng, Z. Tang, X. Wang, Y. Zhao, R. Lu, K. Han, Efficient luminescent halide quadruple-perovskite nanocrystals via trap-engineering for highly sensitive photodetectors, Adv. Mater. (2021), 2007215. 1–9.
[117] A. Mandal, A. Mondal, R. Bhattacharyya, S. Bhattacharyya, $Cs_4CuSb_2Cl_{12-x}I_x$ ($x = 0-10$) nanocrystals for visible light photodetection, Nanotechnology 33 (41) (2022) 1–12.
[118] G. Tang, Z. Xiao, H. Hosono, T. Kamiya, D. Fang, J. Hong, Layered halide double perovskites $Cs_{3+n}M(II)_nSb_2X_{9+3n}$ (M = Sn, Ge) for photovoltaic applications, J. Phys. Chem. Lett. 9 (1) (2018) 43–48.
[119] J. Xu, J.B. Liu, J. Wang, B.X. Liu, B. Huang, Response to comment on "Prediction of novel p-type transparent conductors in layered double perovskites: a first-principles study", Adv. Funct. Mater. 31 (2020) 2003149.
[120] S. Hu, B. Xia, Y.P. Lin, T. Katase, J. Fujioka, T. Kamiya, H. Hosono, K.Z. Du, Z. Xiao, P-type transparent quadruple perovskite halide conductors: fact or fiction? Adv. Funct. Mater. 30 (2020) 1909906.
[121] J. Xu, J.B. Liu, B.X. Liu, B. Huang, Exotic structural and optoelectronic properties of layered halide double perovskite polymorphs, Adv. Funct. Mater. 31 (9) (2021) 1–8.
[122] J. Xu, C. Xu, J.-B. Liu, L. Bellaiche, H. Xiang, B.-X. Liu, B. Huang, Prediction of room-temperature half-metallicity in layered halide double perovskites, npj Comput. Mater. 5 (1) (2019) 114.
[123] T.T. Tran, C.A. Pocs, Y. Zhang, M.J. Winiarski, J. Sun, M. Lee, T.M. McQueen, Spinon excitations in the quasi-one-dimensional $S = 1/2$ chain compound $Cs_4CuSb_2Cl_{12}$, Phys. Rev. B 101 (23) (2020) 235107.
[124] P.M. Jayasankar, A.K. Pathak, S.P. Madhusudanan, S. Murali, S.K. Batabyal, Double perovskite $Cs_4CuSb_2Cl_{12}$ microcrystalline device for cost effective photodetector applications, Mater. Lett. 263 (2020) 127200.
[125] H.A. Evans, L. Mao, R. Seshadri, A.K. Cheetham, Layered double perovskites, in: Annual Review of Materials Research, vol. 51, Annual Reviews, 2021, pp. 351–380.
[126] Z. Xu, M. Chen, S.F. Liu, Layer-dependent ultrahigh-mobility transport properties in all-inorganic two-dimensional $Cs_2PbI_2Cl_2$ and $Cs_2SnI_2Cl_2$ perovskites, J. Phys. Chem. C 123 (2019) 27978–27985.
[127] J. Li, Q. Yu, Y. He, C.C. Stoumpos, G. Niu, G.G. Trimarchi, H. Guo, G. Dong, D. Wang, L. Wang, M.G. Kanatzidis, $Cs_2PbI_2Cl_2$, all-inorganic two-dimensional Ruddlesden-Popper mixed halide perovskite with optoelectronic response, J. Am. Chem. Soc. 140 (35) (2018) 11085–11090.
[128] A.E. Maughan, A.M. Ganose, D.O. Scanlon, J.R. Neilson, Perspectives and design principles of vacancy-ordered double perovskite halide semiconductors, Chem. Mater. 31 (4) (2019) 1184–1195.
[129] P. Wang, C. Jia, Y. Huang, X. Van Der Duan, Waals heterostructures by design: from 1D and 2D to 3D, Matter 4 (2) (2021) 552–581.
[130] W.R. Frensley, Heterostructure and quantum well physics, in: Heterostructures and Quantum Devices, vol. 24, Academic Press, Inc., 1994, pp. 1–24.
[131] B. Saparov, D.B. Mitzi, Organic-inorganic perovskites: structural versatility for functional materials design, Chem. Rev. 116 (7) (2016) 4558–4596.
[132] T.N. Mandal, A. Jana, Lateral epitaxial heterostructures of halide perovskites for diode application, Matter 3 (3) (2020) 617–619.
[133] E. Shi, L. Dou, Halide perovskite epitaxial heterostructures, Acc. Mater. Res. 1 (3) (2020) 213–224.
[134] E. Shi, B. Yuan, S.B. Shiring, Y. Gao, Akriti, Y. Guo, C. Su, M. Lai, P. Yang, J. Kong, B.M. Savoie, Y. Yu, L. Dou, Two-dimensional halide perovskite lateral epitaxial heterostructures, Nature 580 (7805) (2020) 614–620.
[135] Akriti, E. Shi, S.B. Shiring, J. Yang, C.L. Atencio-Martinez, B. Yuan, X. Hu, Y. Gao, B.P. Finkenauer, A.J. Pistone, Y. Yu, P. Liao, B.M. Savoie, L. Dou, Layer-by-layer anionic

diffusion in two-dimensional halide perovskite vertical heterostructures, Nat. Nanotechnol. 16 (5) (2021) 584–591.
[136] D. Pan, Y. Fu, N. Spitha, Y. Zhao, C.R. Roy, D.J. Morrow, D.D. Kohler, J.C. Wright, S. Jin, Deterministic fabrication of arbitrary vertical heterostructures of two-dimensional Ruddlesden–Popper halide perovskites, Nat. Nanotechnol. 16 (2) (2021) 159–165.
[137] Y. Wang, Z. Wan, Q. Qian, Y. Liu, Z. Kang, Z. Fan, P. Wang, Y. Wang, C. Li, C. Jia, Z. Lin, J. Guo, I. Shakir, M. Goorsky, X. Duan, Y. Zhang, Y. Huang, X. Duan, Probing photoelectrical transport in lead halide perovskites with van Der Waals contacts, Nat. Nanotechnol. 15 (9) (2020) 768–775.
[138] A. Yang, J.C. Blancon, W. Jiang, H. Zhang, J. Wong, E. Yan, Y.R. Lin, J. Crochet, M.G. Kanatzidis, D. Jariwala, T. Low, A.D. Mohite, H.A. Atwater, Giant enhancement of photoluminescence emission in WS2-two-dimensional perovskite heterostructures, Nano Lett. 19 (8) (2019) 4852–4860.
[139] W. Li, J. Ma, X. Cheng, D. Li, Giant enhancement of photoluminescence quantum yield in 2D perovskite thin microplates by graphene encapsulation, Nano Res. 14 (6) (2020) 1980–1984.
[140] G.A. Prinz, Magnetoelectronics, Science 282 (1998) 1160–1663.
[141] A. Hoffmann, S.D. Bader, Opportunities at the frontiers of spintronics, Phys. Rev. Appl. 4 (4) (2015), 047001.
[142] G. Dresselhaus, Spin-orbit coupling effects in zinc blende structures, Phys. Rev. 100 (2) (1955) 580–586.
[143] Y. Zhai, S. Baniya, C. Zhang, J. Li, P. Haney, C.-X. Sheng, E. Ehrenfreund, Z.V. Vardeny, Giant Rashba splitting in 2D organic-inorganic halide perovskites measured by transient spectroscopies, Sci. Adv. 3 (7) (2017), e1700704.
[144] J. Wang, C. Zhang, H. Liu, R. McLaughlin, Y. Zhai, S.R. Vardeny, X. Liu, S. McGill, D. Semenov, H. Guo, R. Tsuchikawa, V.V. Deshpande, D. Sun, Z.V. Vardeny, Spin-optoelectronic devices based on hybrid organic-inorganic trihalide perovskites, Nat. Commun. 10 (1) (2019) 129.
[145] M. Kepenekian, R. Robles, C. Katan, D. Sapori, L. Pedesseau, J. Even, Rashba and dresselhaus effects in hybrid organic-inorganic perovskites: from basics to devices, ACS Nano 9 (12) (2015) 11557–11567.
[146] X. Chen, H. Lu, Z. Li, Y. Zhai, P.F. Ndione, J.J. Berry, K. Zhu, Y. Yang, M.C. Beard, Impact of layer thickness on the charge carrier and spin coherence lifetime in two-dimensional layered perovskite single crystals, ACS Energy Lett. 3 (9) (2018) 2273–2279.
[147] J. Yin, P. Maity, L. Xu, A.M. El-Zohry, H. Li, O.M. Bakr, J.L. Brédas, O.F. Mohammed, Layer-dependent Rashba band splitting in 2D hybrid perovskites, Chem. Mater. 30 (23) (2018) 8538–8545.
[148] J.-C. Blancon, J. Even, C.C. Stoumpos, M.G. Kanatzidis, A.D. Mohite, Semiconductor physics of organic–inorganic 2D halide perovskites, Nat. Nanotechnol. 15 (12) (2020) 969–985.
[149] X. Liu, A. Chanana, U. Huynh, F. Xue, P. Haney, S. Blair, X. Jiang, Z.V. Vardeny, Circular photogalvanic spectroscopy of Rashba splitting in 2D hybrid organic–inorganic perovskite multiple quantum wells, Nat. Commun. 11 (1) (2020) 323.
[150] R. Nechache, C. Harnagea, S. Li, L. Cardenas, W. Huang, J. Chakrabartty, F. Rosei, Bandgap tuning of multiferroic oxide solar cells, Nat. Photonics 9 (1) (2014) 61–67.
[151] C. Paillard, X. Bai, I.C. Infante, M. Guennou, G. Geneste, M. Alexe, J. Kreisel, B. Dkhil, Photovoltaics with ferroelectrics: current status and beyond, Adv. Mater. 28 (26) (2016) 5153–5168.
[152] A. Zenkevich, Y. Matveyev, K. Maksimova, R. Gaynutdinov, A. Tolstikhina, V. Fridkin, Giant bulk photovoltaic effect in thin ferroelectric $BaTiO_3$ films, Phys. Rev. B 90 (16) (2014) 161409.

[153] X.-G. Chen, X.-J. Song, Z.-X. Zhang, H.-Y. Zhang, Q. Pan, J. Yao, Y.-M. You, R.-G. Xiong, Confinement-driven ferroelectricity in a two-dimensional hybrid lead iodide perovskite, J. Am. Chem. Soc. 142 (22) (2020) 10212–10218.
[154] I.H. Park, Q. Zhang, K.C. Kwon, Z. Zhu, W. Yu, K. Leng, D. Giovanni, H.S. Choi, I. Abdelwahab, Q.H. Xu, T.C. Sum, K.P. Loh, Ferroelectricity and Rashba effect in a two-dimensional Dion-Jacobson hybrid organic-inorganic perovskite, J. Am. Chem. Soc. 141 (40) (2019) 15972–15976.
[155] A. Giri, A.Z. Chen, A. Mattoni, K. Aryana, D. Zhang, X. Hu, S.H. Lee, J.J. Choi, P.E. Hopkins, Ultralow thermal conductivity of two-dimensional metal halide perovskites, Nano Lett. 20 (5) (2020) 3331–3337.
[156] X. Long, Z. Pan, Z. Zhang, J.J. Urban, H. Wang, Solvent-free synthesis of organometallic halides $CH_3NH_3PbI_3$ and $(CH_3NH_3)_3Bi_2I_9$ and their thermoelectric transport properties, Appl. Phys. Lett. 115 (7) (2019), 072104.
[157] M.A. Haque, S. Kee, D.R. Villalva, W. Ong, D. Baran, Halide perovskites: thermal transport and prospects for thermoelectricity, Adv. Sci. 7 (10) (2020) 1903389.
[158] L.M. Cowen, J. Atoyo, M.J. Carnie, D. Baran, B.C. Schroeder, Review—organic materials for thermoelectric energy generation, ECS J. Solid State Sci. Technol. 6 (3) (2017) N3080–N3088.
[159] B. Martín-García, D. Spirito, G. Biffi, S. Artyukhin, F. Bonaccorso, R. Krahne, Phase transitions in low-dimensional layered double perovskites: the role of the organic moieties, J. Phys. Chem. Lett. 12 (1) (2021) 280–286.
[160] Z. Gan, Y. Cheng, W. Chen, K.P. Loh, B. Jia, X. Wen, Photophysics of 2D organic–inorganic hybrid lead halide perovskites: progress, debates, and challenges, Adv. Sci. 8 (6) (2021) 2001843.
[161] E. Pomerantseva, Y. Gogotsi, Two-dimensional heterostructures for energy storage, Nat. Energy 2 (7) (2017) 17089.
[162] X. Zhou, X. Hu, J. Yu, S. Liu, Z. Shu, Q. Zhang, H. Li, Y. Ma, H. Xu, T. Zhai, 2D layered material-based van Der Waals heterostructures for optoelectronics, Adv. Funct. Mater. 28 (14) (2018) 1706587.
[163] F. Withers, O. Del Pozo-Zamudio, A. Mishchenko, A.P. Rooney, A. Gholinia, K. Watanabe, T. Taniguchi, S.J. Haigh, A.K. Geim, A.I. Tartakovskii, K.S. Novoselov, Light-emitting diodes by band-structure engineering in van Der Waals heterostructures, Nat. Mater. 14 (3) (2015) 301–306.

Index

Note: Page numbers followed by *f* indicate figures and *t* indicate tables.

M

Q

R